ORGANIC PHOSPHORUS COMPOUNDS

ORGANIC PHOSPHORUS COMPOUNDS

Volume 4

G. M. KOSOLAPOFF
Auburn University

and

L. MAIER

Monsanto Research S. A.

WILEY-INTERSCIENCE, a Division of John Wiley & Sons, Inc.

New York • London • Sydney • Toronto

Library of Congress Cataloging in Publication Data

Kosolapoff, Gennady M
Organic phosphorous compounds.

1950 ed. published under title: Organophosphorus compounds.
Includes bibliographies.
1. Organophosphorus compounds. I. Maier, L., joint author. II. Title.

QD412.P1K55 1972 547'.07 72-1359
ISBN 0-471-50443-2 (v. 4)

Printed in the United States of America

10 9 8 7 6 5 4 3 2 1

Contents

ORGANIC PHOSPHORUS COMPOUNDS

Chapter 7. Tertiary Phosphine Sulfides, Selenides, and Tellurides

LUDWIG MAIER

Monsanto Research S.A., Zürich, Switzerland

The subject of tertiary phosphine sulfides[233,326] and selenides[326] was last reviewed in 1963, when tertiary phosphine tellurides were not known. Since that time several new methods have become available to prepare this class of compounds, and interesting properties and new reactions have been uncovered. Emphasis in this chapter is placed on these newer developments. Furthermore, efforts have been made to include in the list of compounds all the compounds that have been reported in the literature and adequately identified. References are given to all the chemical and physical properties of the individual compounds. The literature has been covered through May 1970, including patent literature, insofar as abstracts are available in Chemical Abstracts.

The nomenclature of the compounds used in this chapter is in general that established by the Nomenclature Committee of the American Chemical Society and published in Chemical and Engineering News on October 27, 1952. Abbreviations used in formulas and tables are those given in the introduction to Chapter 1.

A. METHODS OF PREPARATION

I. BY ADDITION OF SULFUR, SELENIUM, OR TELLURIUM TO TERTIARY PHOSPHINES

1. TERTIARY PHOSPHINE SULFIDES. The first tertiary phosphine sulfides were prepared by the direct reaction of sulfur with trimethyl- and triethylphosphine.[59] The

$$R_3P + S \longrightarrow R_3PS$$

procedure was used widely in the past and is still one of the best preparations of this class of compounds. The reaction is ordinarily carried out in solution (ether, benzene, toluene, carbon disulfide, or alcohols) with moderate warming. Because the reaction is usually rather exothermic, the addition of the element should be gradual, especially in large runs. The reaction is usually completed by refluxing.[203,233,326] Several phosphine sulfides in which all three of the organic radicals are different have also been prepared by this procedure (see table of compounds), and one of these unsymmetrical phosphine sulfides was resolved into optically active isomers.[90] More recently optically active tertiary phosphine sulfides have been obtained directly in the reaction of optically active tertiary phosphines with elemental sulfur.[154,155] The reaction proceeds with retention of configuration.[154,155,420]

Whereas the rate of the reaction of aliphatic and mixed aliphatic-aromatic substituted phosphines with sulfur is too fast to be measured (it is complete in less than 30 sec[91]) the kinetic of the reaction of triarylphosphines with octatomic sulfur has been studied.[28,123] The rate of the reaction is strongly increased by anion-solvating solvents (relative rate const.: C_6H_{12} 1.25; CCl_4 20.3; C_6H_6 100; C_6H_5Cl 260) and by electron-releasing substituents in the phenyl groups [relative rate const.: Ph_3P 1; $(4\text{-}MeC_6H_4)_3P$ 15] but is retarded by electron-attracting substituents in the phenyl groups (relative rate const.: $Ph_3P > Ph_2PC_6H_4Cl\text{-}p > P(C_6H_4Cl\text{-}p)_3$), the reaction having a value of $\rho = 2.5$ in the Hammett equation[28] (activation energy 16.5 kcal).

Two other forms of sulfur, the hexatomic form and the irradiation-produced amorphous form, react much faster with Ph_3P in benzene.[28] Thus hexatomic sulfur has 25,000 times the reactivity of octatomic sulfur towards Ph_3P in benzene at 7.35°.[26] In a recent kinetic study of the reaction of diphenyl-o-tolylphosphine $Ph_2(2\text{-}MeC_6H_4)P$ with the various sulfur modifications, the following relative rates were observed at 20° in CS_2:[110]

$$S_6 : S_7 : S_8 : S_{12} = 10,700 : 178 : 1 : 187$$

These results are consistent with a mechanism in which the rate-determining step of the reaction is attack of the phosphine on a sulfur atom of the eight-membered sulfur ring to give a dipolar intermediate in which the phosphorus bears a formal positive charge and the terminal sulfur atom of the S_8 open chain a formal negative charge. This dipolar

$$Ph_3PS + Ph_3\overset{\oplus}{P}S_7^{\ominus} \quad \text{etc.}$$

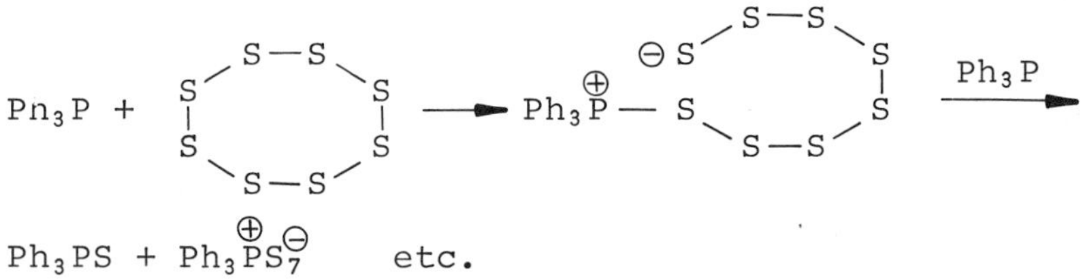

intermediate is then degraded by the attack of additional Ph_3P molecules with successively shorter sulfur chains bonded to phosphorus. This mechanism obtains apparently also with other sulfur modifications.

The formation of optically active tertiary phosphine sulfides with retention of configuration indicates that the dipolar intermediates do not exist in equilibrium with cyclic tautomers.

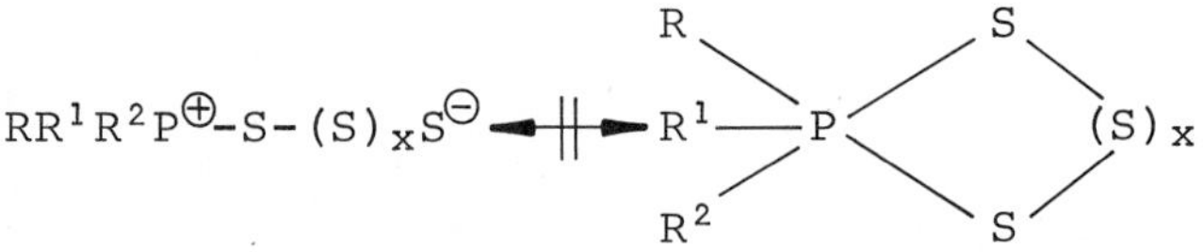

It has been argued that if such cyclic tautomers had been formed, it is highly probably that the phosphine sulfide eventually produced would have been largely or completely racemic.[420]

The rate of the formation of Ph_3PS is extremely sensitive toward impurities in the sulfur,[26] and the different rates recorded by the various groups[26,28,93,123] have been attributed to this observation.[233] Whereas the addition of sulfur to the tertiary phosphines in the absence of a solvent sometimes gave low yields of the phosphine sulfides, the reaction between the mercuric chloride adduct of the tertiary phosphines and ammonium polysulfide in ethanol solution gave the desired phosphine sulfides in good yield and excellent purity.[354] The solid trialkylphosphine-carbon disulfide adducts also yield tertiary phosphine sulfides either on standing or boiling in carbon disulfide solution,[142] or on reaction with carbon tetrachloride.[354]

Only a few tertiary phosphines have been reported not

to react with sulfur. Thus $(CF_3)_3P$,[31] 2,3,5,6,7,8-hexakis-(trifluoromethyl)1,4-diphosphabicyclo-[2.2.2]octa-2,5,7-triene,[211] R_2PCOR,[185] $Ph_2PCOOEt$,[158] and $(cyclo-C_6H_{11})_2$ PCH_2CH_2OH[180] did not add sulfur even under the severest conditions. Their unreactivity toward sulfur is probably due to the strong electronegative groups attached to phosphorus. The cyclic phosphine $P(OCH_2)_3P$ added only one sulfur to give $S=P(OCH_2)_3P$, but the disulfide could not be made.[80]

Tris(pentafluorophenyl) phosphine reacted only with sulfur when heated for six days at 160° (no reaction was observed in benzene).[104] The reaction of dialkylethynyl-phosphines $R_2PC\equiv CH$ (R = Bu, sec-Bu, cyclo-C_6H_{11}) with sulfur or propylene sulfide resulted in the formation of tar; only in the case of R = t-Bu was the desired di-t-butylethynyl-phosphine sulfide obtained in 36% yield.[408] No polymerization problem was encountered in the preparation of $Ph_2P(S)$-$C\equiv CPh$ and $Ph_2P(S)CH=CHPh$, employing sulfurization.[5]

In the preparation of phosphine sulfides, the sulfur can also be donated by sulfur transfer agents; examples include SO_2 (gives oxide also),[114,367] $S_2O_3{}^{2-}$,[27] HSCN,[66,142] $(SCN)_2$,[66] NA_2S_x,[51] $(NH_4)_2S_x$,[354] HgS,[59] $NiS_4C_4R_4$,[87] $PSCl_3$,[1,2,4] episulfides,[50,78,79,92,94,96,352,357,408] and thionocarbonates,[78] mercaptanes,[409] aralkyl hydrodisul-fides,[396] dialkyl and diaryltetrasulfides,[106,272] dialkenyl trisulfides,[271] dialkenyl disulfides,[106,107,271,272] and alkenyl-alkyl disulfides,[106,272] dialkyl disulfides,[88,157,418] dibenzoyldisulfides,[198,348,349] and thiocar-bamoyldisulfides,[348,349,410] or thiofluorenone-S-oxide,[374] epitrithiadioxopiperazines,[325] 5-alkoxy-1,2,3,4,-thiatri-azoles,[195a] and several others.[233] These reactions have been discussed in detail by us previously.[233]

A kinetic study of the reaction of parasubstituted diaryl trisulfides with Ph_3P to give diaryl disulfides and Ph_3PS showed[111] that the rate of the reaction is increased by electron-attracting substituents in the phenyl groups,

$$R\!\!-\!\!\langle\bigcirc\rangle\!\!-\!\!S\text{-}S\text{-}S\!\!-\!\!\langle\bigcirc\rangle\!\!-\!\!R + Ph_3P \longrightarrow R\!\!-\!\!\langle\bigcirc\rangle\!\!-\!\!S\text{-}S\!\!-\!\!\langle\bigcirc\rangle\!\!-\!\!R + Ph_3PS$$

serving to decrease electron density on the sulfur chain and make it thus more susceptible to nucleophilic attack. The rate of reaction is retarded, however, by electron-releasing substituents in the phenyl groups of the trisul-fide (R = NO_2:MeCO : H : Me : NH_2 = 24 : 10.4 : 1 : 0.68 : 0.04), the reaction having a value of ρ= +0.98 in the Hammett equation (E_a = 12 kcal/mole).[111] It should be noted that most of these sulfur transfer reactions have been carried out not for the purpose of preparing tert-phosphine sulfides, but rather to desulfurize organic and inorganic

molecules or to obtain compounds that contain less sulfur
than the starting material. Some of the desulfurization
reactions are of particular value in the synthesis of organic compounds. Thus episulfides[50,92,94] and cyclic
thionocarbonates[78] are converted stereospecifically into
the corresponding olefins, the phosphine being converted to
the phosphine sulfide with retention of configuration.[420]
   Whereas with cis-1,2-vinylenebis(diphenylphosphine)
sulfur gave the cis-disulfide, the sulfur transfer agent
PSCl3 gave only trans-1,2-vinylenebis(diphenylphosphine
sulfide) $Ph_2P(S)CH=CH(S)PPh_2$, and this compound was obtained
starting from either cis-or trans-1,2-vinylenebis(diphenylphosphine).[1,2,4] Since cis-1,2-vinylenebis(phosphine sulfide) was converted to the trans compound by reaction with
PCl3 in refluxing tetrahydrofuran, it can be assumed that
with PSCl3 initial sulfur exchange occurred followed by
isomerization.[2] Other workers also used PSCl3 to prepare
the tetratertiary phosphine sulfide $C[CH_2P(S)Ph_2]_4$,[101]
which could otherwise not be obtained in a pure state. A
kinetic study of the reaction suggested that it is concerted

$$Ph_3PS + Bu_3P \xrightarrow{200°/2 \ hr} Bu_3PS + Ph_3P_{quant}$$

and involves a transition state in which the sulfur is bonded to both phosphorus atoms.[35]
   Most of the diphosphine disulfides have also been prepared by addition of sulfur to the corresponding diphosphines, dissolved in benzene or toluene, and refluxing the
mixture for a short period.[159,173,175,181]
   In several cases the yields of diphosphine disulfides
were rather low, about 15 to 20%,[173,175] especially when
two asymmetric phosphorus atoms were present in the molecule. No explanation for this low yield has been given.

$$RR'P(CH_2)_nPR'R + 2 \ S \longrightarrow RR'\overset{\overset{\displaystyle S}{\|}}{P}(CH_2)_n\overset{\overset{\displaystyle S}{\|}}{P}R'R, \ n = 1-6$$

Although about ten diphosphine disulfides containing two
asymmetric phosphorus atoms have been reported,[173,175] only
one such compound has been resolved into a high-melting
form and a low-melting form one of which must be the racemate (1) and the other the meso form (2).

$$S=\overset{\overset{\displaystyle C_2H_5}{|}}{P}-(CH_2)_6-\overset{\overset{\displaystyle C_6H_5}{|}}{P}=S \quad S=\overset{\overset{\displaystyle C_6H_5}{|}}{P}-(CH_2)_6-\overset{\overset{\displaystyle C_2H_5}{|}}{P} \qquad S=\overset{\overset{\displaystyle C_2H_5}{|}}{P}-(CH_2)_6-\overset{\overset{\displaystyle C_2H_5}{|}}{P}=S$$

$$\underbrace{\qquad\qquad\qquad\qquad\qquad\qquad\qquad\qquad}_{\textstyle Racemate\ (\underline{1})} \qquad\qquad Meso\ (\underline{2})$$

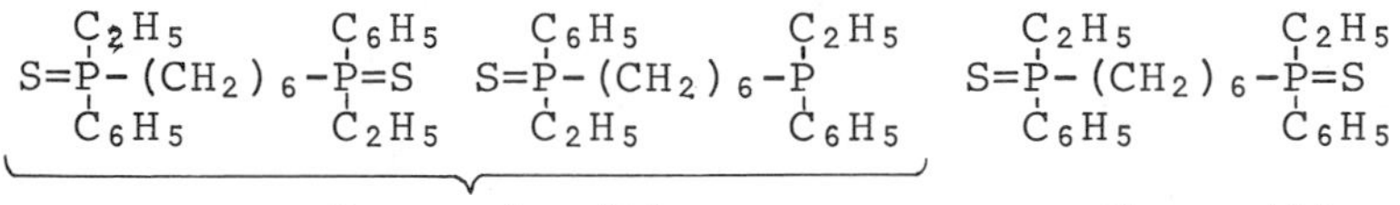

With the exception of three cyclic tertiary phosphine sulfides,[51,222,234] all the other known cyclic compounds were prepared by the addition of sulfur or sodium polysulfide to the cyclic tertiary phosphine in benzene[140,164,170,173,187] or 2-ethoxy-ethanol[51] solution. When the addition of sulfur to the cyclic phosphine was carried out in benzene solution the yield of cyclic tertiary phosphine sulfide (3) was often low,[164,187] whereas with sodium polysulfide and 2-ethoxyethanol as the solvent, a high yield of the cyclic phosphine sulfide was obtained.[51]

$$(CH_2)_n \begin{matrix} CH_2 \\ \\ CH_2 \end{matrix} P-R + S \longrightarrow (CH_2)_n \begin{matrix} CH_2 \\ \\ CH_2 \end{matrix} \overset{\overset{S}{\|}}{P}-R$$

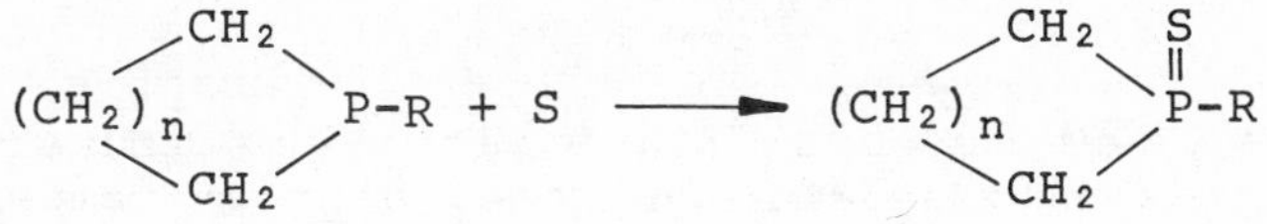

(3)

Some of the cyclic tertiary phosphine sulfides have a remarkable capacity for tenaciously retaining traces of solvent. Thus triethylenediphosphine disulfide (4) prepared from triethylenediphosphine and sulfur in boiling

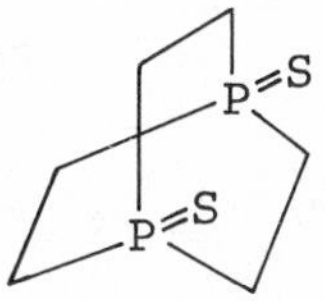

(4)

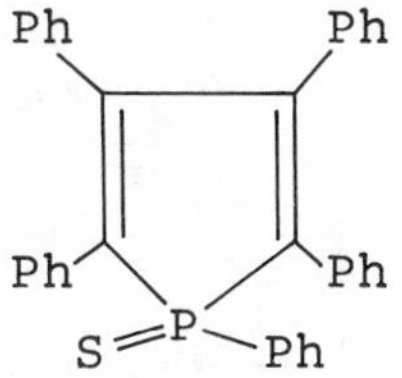

(5)

benzene solution, had the composition $C_6H_{12}P_2S_2 \cdot 1/8\ C_6H_6$, after being heated at 125°/1 mm Hg for 4 hr.[140] Similarly, pentaphenylphosphole sulfide (5), isolated from methylene dichloride solution, crystallized with 1 mole of methylene dichloride which was only lost when the compound was heated to 115 to 125°.[51] Alkyl- and aryl-substituted 1,4-diphosphacyclohexane P-disulfides [(6), R = Ph, cyclo-C_6H_{11}] have been isolated in a high and a low melting form; one must be the cis isomer and the other the trans isomer.[164,187] It has

$$\begin{matrix} S & CH_2-CH_2 & S \\ \| & & \| \\ P & & P \\ | & & | \\ R & CH_2-CH_2 & R \end{matrix}$$

Cis

$$\begin{matrix} R & CH_2-CH_2 & S \\ | & & \| \\ P & & P \\ \| & & | \\ S & CH_2-CH_2 & R \end{matrix}$$

Trans

(6)

been suggested that the high-melting compound might be the trans isomer, although no direct proof was obtained for this assignment. Attempts to convert by heating the low-melting isomer (6) into the high-melting isomer failed.[164,187] In the case of the ethyl derivative [(6), R = Et], only one form was isolated.[164]

2. TERTIARY PHOSPHINE SELENIDES. The addition of selenium to tertiary phosphines is less vigorous than that of sulfur.

$$R_3P + Se \longrightarrow R_3PSe$$

Whereas the lower members of aliphatic-substituted tertiary phosphines add selenium already at room temperature in etheral solution to give the corresponding selenides,[59,322] the higher members of aliphatic-substituted tertiary phosphine selenides and aromatic-substituted phosphine selenides have been produced by fusing the phosphine with elemental selenium in the absence of a solvent[260,266,354,426] or by heating with selenium in a higher boiling solvent such as butanol, carbon disulfide,[90] benzene,[139] or toluene.[131] Much higher yields of tertiary phosphine selenides were obtained in the reaction of tertiary phosphines with potassium selenocyanate in acetonitrile.[283,285] This method was also successful for preparing the tetra-

$$R_3P + KSeCN \xrightarrow{CH_3CN} R_3PSe + KCN$$

selenide $C[CH_2P(Se)Ph_2]_4$, which otherwise could not be obtained in a pure state.[101] In an attempt to dehydrogenate the cyclic phosphine oxide (7) with a mixture of selenium

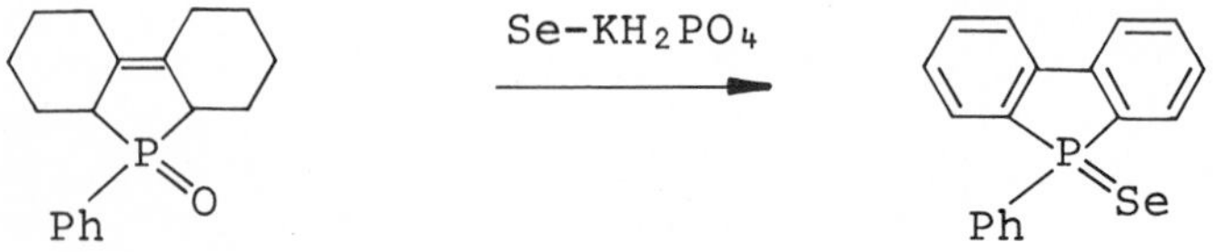

(7) (8)

and potassium dihydrogen phosphate at 270 to 370°, 9-phenyl-9-phosphafluorene-9-selenide (8) was obtained in 25% yield.[60,61] The exchange of selenium for oxygen is surprising and suggests a transitory reduction to the phosphine. Such a transitory reduction seems to be confirmed by the finding that the action of $Se\text{-}KH_2PO_4$ on triphenylphosphine oxide at 330 to 400° produced triphenylphosphine selenide in 16% yield. Other cyclic tertiary phosphine selenides have been prepared by direct addition of selenium to the cyclic phosphine.[51] When selenium dioxide is warmed with triphenylphosphine, reaction occurs, but the product is a mixture of triphenylphosphine oxide and the corresponding selenide.[263]

3. TERTIARY PHOSPHINE TELLURIDES. Only nine tertiary phosphine tellurides are known. They have all been prepared by the direct reaction of elemental tellurium with tertiary phosphines in boiling toluene solution.[424,429] The reaction has to be run in an atmosphere of nitrogen,

$$R_3P + Te \longrightarrow R_3PTe$$

since exposure to air brings about decomposition and separation of tellurium metal. All the compounds prepared possess a pale, golden yellow color and have been obtained in yields ranging from 5 to 45%. The extent to which the reaction proceeds diminishes as the number of phenyl groups on the phosphorus atom increases. Thus the telluride of Ph_3P[429] or of $C[CH_2PPh_2]_4$[101] could not be prepared.

II. BY REACTION OF TERTIARY PHOSPHINE DIHALIDES WITH HYDROGEN SULFIDE

Tertiary phosphine dihalides react readily with dry hydrogen sulfide with the formation of tertiary phosphine sulfides and hydrogen halides.

$$R_3PX_2 + H_2S \longrightarrow R_3P{=}S + 2HCl$$

The addition of tertiary amines as hydrogen halide scavengers seems to offer no advantage. Thus from Ph_3PCl_2 and H_2S in the presence of Et_3N only 53% Ph_3PS was obtained, whereas from Ph_3PBr_2 and H_2S without a tertiary amine 82% $Ph_3P{=}S$ was isolated.[152] Tributylphosphine sulfide was only obtained in 20 to 40% yields from the corresponding dibromide and H_2S.[425] On the other hand, 1-phenyl-3-methyl-1-phospha-3-cyclopentene P-sulfide [also named phospholene sulfide (9)] was produced in 80% yield in the reaction of the corresponding phosphine dichloride with hydrogen sulfide in benzene solution.[222] In this particular case, this

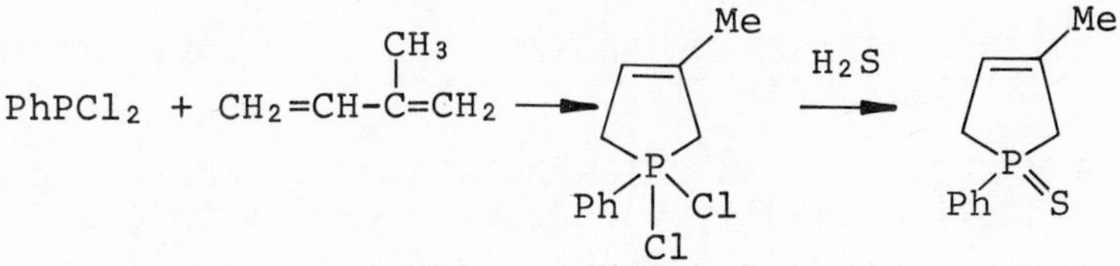

(9)

method of preparing a tertiary phosphine sulfide is of value because the phosphine dichloride is directly obtained in the reaction of phenylphosphonous dichloride with isoprene.[230] Phosphinous halides also react readily with CH_3I,[98] benzyl halides,[144] and activated alkyl halides to yield tertiary phosphine dihalides. By reaction of these dihalides with hydrogen sulfide, several functionally substituted tertiary phosphine sulfides have been prepared,[382-389] e.g.,

$$R_2PCl + ClCH_2SR' \longrightarrow R_2(R'SCH_2)PCl_2 \xrightarrow{H_2S} R_2(R'SCH_2)PS + 2HCl$$

In the reaction of phosphinous halides with 1,2-dichloroethyl butyl ether and H_2S, hydrogen chloride is split off during the reaction and thiophosphinyl-substituted vinyl ethers are obtained.[383] The adducts of α-chloro-substituted aldehydes[381,390] and chloroacetic acid ester[199] with phosphinous halides have also been transformed into tertiary phosphine sulfides by reaction with hydrogen sulfide.

$$R_2PCl + X_2ClCCHO \longrightarrow [R_2\overset{\oplus}{P}(Cl)\text{-}CH(O^{\ominus})CX_2Cl] \xrightarrow{H_2S}$$

$$R_2P(S)CHOHCX_2Cl + HCl$$

Diphosphine disulfides may also be prepared by this procedure.[389] Finally it is worthy of note that tris(tri-

$$2Et_2PCl + BrCH_2CH_2Br \longrightarrow [Et_2P(Cl)(Br)CH_2CH_2P(Cl)(Br)Et_2] \xrightarrow{H_2S}$$

$$Et_2P(=S)CH_2CH_2P(=S)Et_2$$

fluoromethyl) phosphine sulfide, which could not be prepared by the direct addition of sulfur to $(CF_3)_3P$, was obtained from the phosphine dichloride and hydrogen sulfide or silver sulfide.[64]

III. BY REACTION OF ORGANOMETALLIC COMPOUNDS WITH $P(S)Cl_3$, $RP(S)Cl_2$, OR $R_2P(S)Cl$

1. FROM $SPCl_3$ AND A GRIGNARD REAGENT. Whereas phosphorus oxychloride readily yields with excess Grignard reagent tertiary phosphine oxides (see Chapter 6), the

analogous reaction with $P(S)Cl_3$ appears to lead to the
formation of tertiary phosphine sulfides in good yields
only with acetylenic-,[42] vinyl-,[286] styryl-,[43] hindered

$$P(S)Cl_3 + 3RMgX \longrightarrow R_3PS + 3MgXCl$$

alkyl-,[286] benzyl-,[84,375] long-chain aliphatic-,[214] and
aromatic-[84,375] Grignard reagents. The product formed from
lower primary alkyl-magnesium halides and SPX_3 seems to
depend on several factors. The nature of the Grignard
reagent, the phosphorus thiotrihalide, the solvent, and the
reaction temperature appear to be important. Thus with
$SPCl_3$ and alkylmagnesium bromides (RMgBr, R = Me, Et, Pr,
Bu, ratio 1:3.2)[72,190,214,231,286] or α,ω-alkylene
bis(magnesium bromide)[346] in ether or tetrahydrofuran at a
temperature of 5 to 20°, mainly biphosphine disulfides were
formed. On the other hand, the use of $SPBr_3$ instead of
$SPCl_3$ under the same conditions has been said to produce
mainly tertiary phosphine sulfides.[286] Also, the use of
alkylmagnesium chloride RMgCl instead of RMgBr and the use
of higher temperatures (ca.70°) favors formation of tertiary phosphine sulfides.[286] Temperatures below zero and
Grignard reagents prepared from i-Pr, sec-Bu, t-Bu and
cyclohexyl halides seem to produce less substituted products, such as $R_2P(S)Cl$, $R_2P(S)OH$, and $R_2P(S)SH$.[72,174]

Similar results were obtained in reactions involving
phosphonothioic dihalides. The formation of tertiary phosphine sulfides from $PhP(S)Cl_2$ and tolyl-,[273] alkynyl-,[42]
or styrylmagnesium halide,[43] and from $RP(S)Cl_2$ and
cyclohexyl-, or alkynylmagnesium halide[41,44] parallels the
reactions of $P(S)Cl_3$, as does biphosphine disulfide formation from $RP(S)X_2$ and alkylmagnesium bromides.[227] Some
of the restrictions noted in the $P(S)X_3$ reactions, however, no longer apply. Thus tertiary phosphine sulfides
were also obtained from $MeP(S)Br_2$ and α,ω-alkylene di(magnesium bromide)[234] and from α,ω-alkylene bis(phosphonothioic
dichlorides) and ethylmagnesium bromide.[234]

$$MeP(S)Br_2 + BrMg(CH_2)_4MgBr \longrightarrow MeP\overset{\|}{\underset{S}{}}\Big\langle \Big| + 2MgBr_2$$

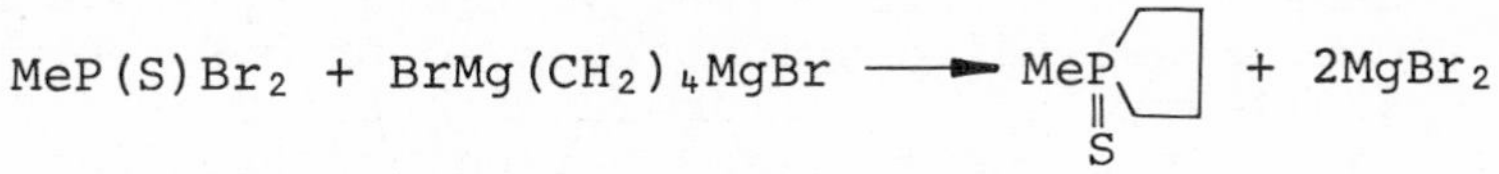

$$Cl_2P(S)(CH_2)_nP(S)Cl_2 + 4EtMgBr \longrightarrow$$

$$Et_2P(S)(CH_2)_nP(S)Et_2 + 4MgClBr$$

On the other hand, mainly biphosphine disulfides were
isolated from the reactions of $MeP(S)Br_2$ with benzyl-,[227]
and surprisingly, phenyl-,[227] or styrylmagnesium bromide,[69]
and also from $PhP(S)Cl_2$ and alkylmagnesium bromides[227,369]
or benzylmagnesium chloride.[84] But in ether styrylphos-
phonothioic dichloride with MeMgBr produced both the

biphosphine disulfide and the tertiary phosphine sulfide.[43] As with $P(S)Cl_3$, the chloro-Grignard reagent MeMgCl gave mainly tertiary phosphine sulfides with $PhP(S)Cl_2$ [82] and t-$BuP(S)Cl_2$.[82] Treatment of the unsaturated 2-methylbutadienylphosphonothioic dichloride with PbMgBr led to complete resinification of the reaction mixture.[253]

Depending on the substituent of the Grignard reagents, phosphinothioic halides containing an aryl group gave both biphosphine disulfides and tertiary phosphine sulfides;[295] but in the same reaction with alkyl-,[82,133,295] aryl-,[133] styryl-,[43] and acetylenic Grignard reagents[44] and

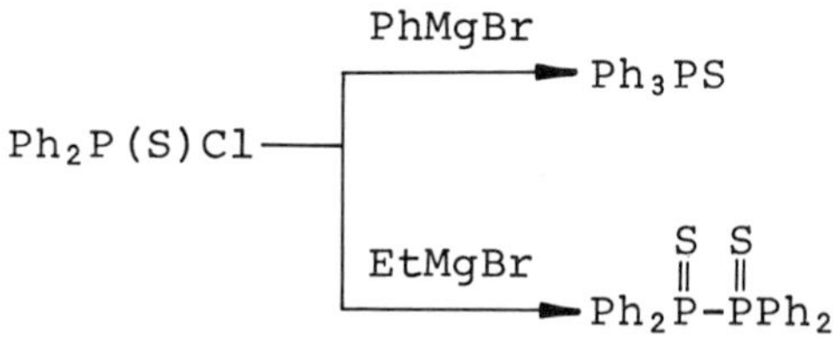

α,ω-alkylene bis(magnesium bromide),[234] dialkylphosphinothioic halides produced tertiary phosphine sulfides only.

$$R_2P(S)X + R'MgX \longrightarrow R_2R'PS + MgX_2$$

$$2R_2P(S)X + BrMg(CH_2)_nMgBr \longrightarrow R_2P(S)(CH_2)_nP(S)R_2 + 2MgXBr$$

A report claiming the formation of biphosphine disulfides from the interaction of $Et_2P(S)Cl$ and EtMgBr and from $Pr_2P(S)Cl$ and PrMgBr [40] could not be confirmed by other workers.[82,295] Methylphenylphosphinothioic chloride, when treated with ethyl- or phenylmagnesium bromide yielded both types of products, the tertiary phosphine sulfide derivative predominating in both cases.[295] (The mechanism of biphosphine disulfide formation is discussed in Chapter 2, Section E.2.)

In contrast to the thiophosphorus halide system, the selenophosphorus halide reaction seems to proceed normally with Grignard reagents. For example with RMgBr (R = Et, Ph) $MeP(Se)Br_2$ gave only the tertiary phosphine selenides,

$$MeP(Se)Br_2 + 2RMgBr \longrightarrow MeR_2PSe + 2MgBr_2$$

$MeEt_2PSe$ and $MePh_2PSe$, respectively.[227]

2. FROM $SPCl_3$ AND ORGANOALKALI COMPOUNDS. In contrast to Grignard reagents, organolithium compounds seem to produce in the interaction with $P(S)Cl_3$ only tertiary phosphine sulfides. Thus the action of methyllithium on $PSCl_3$ (ratio 3.8:1) in ether at -30° and then at room temperature

produced Me_3PS in 49% yield.[354] Similarly, Bu_3PS was isolated in 33% yield[82] from butyllithium and $P(S)Cl_3$ in ether. Diphenylmethylphosphine sulfide was also prepared by the organolithium route from $Ph_2P(S)Cl$ and methyllithium in

$$PSCl_3 + 3RLi \longrightarrow R_3PS + 3LiCl$$

ether.[127] Dialkyl-1-alkynylphosphine sulfides seem to be formed in good yield by the organolithium route.[5]

$$R_2P(S)Cl + LiC{\equiv}CR^1 \longrightarrow R_2P(S)C{\equiv}CR^1 + LiCl$$

In the reaction with diphenylphosphinothioic chloride, alkali derivatives of enolizable compounds (e.g., sodium or potassium diethyl malonate) produce high yields of a functionally substituted tertiary phosphine sulfide.[83,212]

$$Ph_2P(S)Cl + NaCH(CO_2Et)_2 \xrightarrow{H_2O} Ph_2P(S)CH(CO_2Et)_2 + NaCl$$

Finally, the organolithium route has also been used to prepare a cyclic phosphine sulfide. Thus pentaphenylphosphole sulfide was isolated in low yield (3 to 4%) in the reaction of phenylphosphonothioic dichloride with 1,4-dilithio-1,2,3,4-tetraphenylbutadiene in ether.[51] The main product in this reaction was a sulfur-free pentaphenyl-

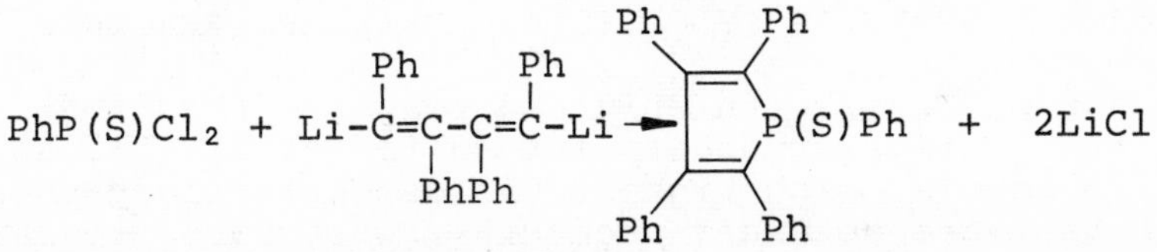
PhP(S)Cl2 + Li-C=C-C=C-Li
Ph Ph
PhPh
Ph Ph
P(S)Ph + 2LiCl
Ph Ph

phosphole. Small amounts of the corresponding pentaphenylphosphole oxide were also isolated. A similar reduction was observed in the reaction of $P(S)Cl_3$ with C_6F_5MgBr, which gave only $(C_6F_5)_3P$ and no sulfide.[113] The mechanism of this reduction reaction is not known.[51,113] The $PhP(S)Cl_2$ was said to react with 1,4-dilithiobenzene with the formation of a polymeric material.[38]

3. FROM $RP(S)Cl_2$ OR $R_2P(S)Cl$ AND ORGANOLEAD COMPOUNDS. Whereas $P(S)Cl_3$ with organolead compounds initially yields phosphonothioic and phosphinothioic chlorides (see Chapter 9), on further treatment with excess organolead compounds, tertiary phosphine sulfides can be prepared by this route.[229,241] It is unattractive as a synthetic method, however, because of the forcing conditions (125 to 180°C), the low yields (6 to 44%), and particularly the long heating times

(66-200 hr) required for the formation of tertiary phosphine sulfides.

4. FROM $P(S)Cl_3$ AND ORGANOALUMINUM COMPOUNDS. Organoaluminum compounds are very attractive starting materials for the alkylation and arylation of phosphorus thio compounds, since many of them are commercially available. Whereas $P(S)Cl_3$ reacts in hexane solution with Et_2AlCl and R_3Al to form mainly dialkylphosphinothioic chlorides (see Chapter 9), it is also possible to prepare tertiary phosphine sulfides under more forcing conditions, such as using a ratio of $P(S)Cl_3:AlR_3$ = 1:1 and also a longer reaction time at elevated temperature.[231,239]

$$P(S)Cl_3 + R_3Al \xrightarrow[140°]{20\ hr} R_3PS + AlCl_3$$

Thus triethylphosphine sulfide was isolated in more than 60% yield when a mixture of $P(S)Cl_3$ and Et_3Al (ratio 1:1) was heated at 140°C for 20 hr. The isolation of Et_3PS was effected by hydrolyzing the reaction mixture with ice water and extracting with benzene.[231,239,270] Similarly, triphenylaluminum and $P(S)Cl_3$ produced Ph_3PS in 67% yield when heated for 30 min. at 125°C.

IV. FROM PHOSPHORUS SULFIDES AND GRIGNARD REAGENTS

Treatment of P_4S_{10} with Grignard reagents resulted in the formation of tertiary phosphine sulfides, as well as many other products. The highest yields of tertiary phosphine sulfides were obtained when P_4S_{10} was caused to react with excess Grignard reagents, first in ether and then after the ether had been distilled at 100 to 120°C for 10 to 12 hr.[247,249] For example, using a ratio of RMgBr : P_4S_{10} = 24.1 gave triethylphosphine sulfide and tris(isobutyl)phosphine sulfide in 76 and 31% yield, respectively. Similarly, triphenylphosphine sulfide was obtained in 68% yield when the ratio used was PhMgBr : P_4S_{10} = 16.1.[249] It should be noted, however, that trimethylphosphine sulfide was obtained in only trace amounts using this procedure.[146,247] The yield was less than 2%.[146] The previously reported melting point of trimethylphosphine sulfide (105°[59,247]) was found to be in error.[146] The correct melting point of $(CH_3)_3PS$ was 155.5 to 155.6°C;[146] this was confirmed by other groups (see table of compounds). The lower phosphorus sulfides gave tertiary phosphine sulfides when treated with Grignard reagents,[248] but only in low yields.

V. FROM $P(S)Cl_3$ OR $R_2P(S)SH$ AND AROMATIC HYDROCARBONS UNDER FRIEDEL-CRAFTS CONDITIONS

Treatment of $P(S)X_3$ with aromatic hydrocarbons in the presence of $AlCl_3$ results in the formation of phosphonothioic and phosphinothioic halides[232,240,260] as well as tertiary phosphine sulfides.[232,260] The yields and the composition of the reaction products depend strongly on the molar ratio of the reactants and also on the reaction time and the reaction temperature.[232] The general reaction may be expressed by the following equation:

$$3P(S)Cl_3 + 6C_6H_6 + xAlCl_3 \longrightarrow PhP(S)Cl_2 + Ph_2P(S)Cl + Ph_3PS + 6HCl + xAlCl_3$$

For the preparation of tertiary phosphine sulfides it is necessary to use a ratio of $P(S)Cl_3$: $AlCl_3$: aromatic component = 1:3:3 and a reaction time of about 8 to 9 hr at ca.80°. Under these conditions tertiary phosphine sulfides are obtained in excellent yields.[232,240] Workup of the reaction mixture is done by pouring it onto ice and extracting the tertiary phosphine sulfide with an organic solvent (usually benzene). Phosphonothioic dichlorides and phosphinothioic chlorides can also be arylated. In this way it is possible to prepare unsymmetrical tertiary phosphine sulfides.

$$RP(S)Cl_2 + 2ArH + AlCl_3 \longrightarrow RAr_2PS + 2HCl + AlCl_3$$

$$RR'P(S)Cl + ArH + AlCl_3 \longrightarrow RR'ArPS + HCl + AlCl_3$$

Thus using this procedure, methyldiphenylphosphine sulfide $MePh_2PS$ and methylethylphenylphosphine sulfide MeEtPhPS have been obtained in 81.3 and 83% yield, respectively.[232] The only active catalysts in this system are the compounds that form stable addition compounds with $P(S)Cl_3$, e.g., $AlCl_3$, $AlBr_3$, and $EtAlCl_2$.[232,233] Conductivity measurements and ^{31}P nuclear magnetic resonance (NMR) studies indicate that the phosphorylating complex has an ionic structure.[238]

$$P(S)Cl_3 + AlCl_3 \longrightarrow [Cl_2(S)P]^{\oplus}[AlCl_4]^{\ominus}$$

Of the 3 moles of $AlCl_3$ necessary to prepare tertiary phosphine sulfides, 1 mole is tied up as $HAlCl_4$. The other 2 moles are necessary to activate $Ph_2P(S)Cl$ (which is an intermediate when starting from $P(S)Cl_3$, see Chapter 9). From conductivity measurements it has been concluded that the complex with 2 moles of $AlCl_3$ $[Ph_2P(S)]^+[Al_2Cl_7]^-$ has about three time the reactivity of that complex with 1 mole of $AlCl_3$ $[Ph_2P(S)]^+[AlCl_4]^-$.[238] The reaction has some limitations. Whereas benzene, flurobenzene, chlorobenzene, toluene, and hexyltoluene gave the normal reaction products no product containing phosphorus could be isolated when trimethoxybenzene, p-methylanisol, thiophene, and furfural were used as the aromatic components.

Antimony pentachloride, which also forms an addition compound with $P(S)Cl_3$, acts as a chlorinating agent and does not produce any organophosphorus compounds.[232] Other Friedel-Crafts catalysts, such as $ZnCl_2$, $FeCl_3$, and $TiCl_3$ are not active in this system.[232] The orientation of the entering group with toluene is mainly para (with little ortho and a trace of meta); with chlorobenzene it is ortho and para in equal amounts. The meta isomer was absent in the latter case.[232] The simple procedure and the high yields make this method very attractive for preparing tertiary phosphine sulfides.

Recently it has been reported that dithio- and thiophosphinic acids also undergo a Friedel-Crafts reaction in the presence of at least 2 moles of $AlCl_3$.[18] Since no yields have been reported, this method cannot be evaluated.

$$R_2P(S)SH[R_2P(S)OH]\cdot AlCl_3 + ArH + AlCl_3 \longrightarrow R_2ArPS$$

VI. FROM SECONDARY PHOSPHINE SULFIDES BY ADDITION TO ALDEHYDES, KETONES, ISOCYANATES, OLEFINS, AND SCHIFF BASES

In direct analogy with secondary phosphine oxides,[34] secondary phosphine sulfides undergo base-catalyzed nucleophilic addition to carbonyl compounds forming α-hydroxy-substituted tertiary phosphine sulfides (10).[8,299] The solid products thus obtained can be purified by

$$R_2P(S)H + O{=}CR'R'' \longrightarrow \underset{(\underline{10})}{R_2P(S)C(OH)R'R''}$$

recrystallization, but the ready reversibility of the reaction rendered complete characterization of the liquid products impossible.[299] By using α-hydroxymethyldialkylamines, stable products have been obtained in every case.[236] α-Hydroxymethyldialkylamines[236] and the more reactive aldehydes such as chloral and formaldehyde required no

$$R_2P(S)H + HOCH_2NR_2(\text{from } CH_2O + HNR_2) \longrightarrow R_2P(S)CH_2NR_2 + H_2O$$

catalyst, nor did secondary phosphine sulfides with electronegative substituents [$R_2P(S)H$, R = Ph, $NCCH_2CH_2$, and $EtO_2CCH_2CH_2$].[299] Also no catalyst was required in the addition of secondary phosphine sulfides to Schiff bases.[245a]

$$R_2P(S)H + ArCH=NR' \longrightarrow R_2(ArCH \cdot NHR')PS$$

Whereas the base-catalyzed addition of secondary phosphine sulfides to isocyanates and activated olefins gave products that could not be satisfactorily characterized,[299] high yields of readily purifiable tertiary phosphine sulfides were obtained when the addition of $R_2P(S)H$ to olefins was radical initiated by AIBN or UV light.[236,242] In the same way, $Et_2P(S)H$ could also be added to triethoxyvinylsilane forming a β-triethoxysilyl-substituted phosphine sulfide.[287] The observed terminal addition, that is,

$$R_2P(S)H + CH_2=CHR \longrightarrow R_2P(S)CH_2CH_2R$$

initial radical attack at the β-carbon atom of the vinyl group, is in accord with all the other radical additions to vinylsilicon compounds investigated so far.[355]

Dibutylphosphine sulfide also reacted with 2,4-dichlorobenzylchloride in methanol in the presence of sodium hydroxide to produce dibutyl-2,4-dichlorobenzylphosphine sulfide (11) in 42% yield.[299] Similarly, when dimethylphosphine sulfide was treated with $EtSCHClCONHCH_3$, it gave the tertiary phosphine sulfide (12).[73]

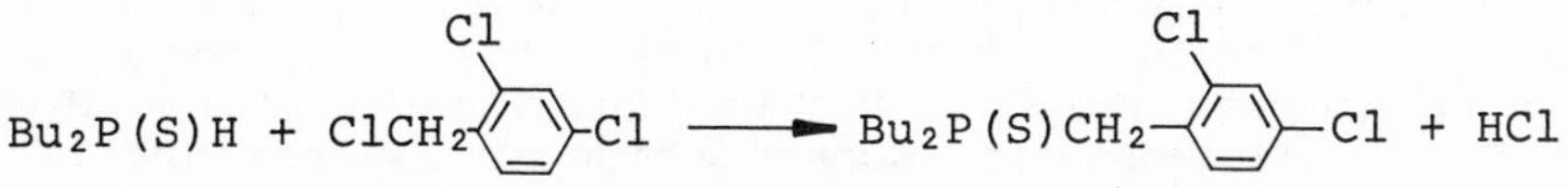

$$Bu_2P(S)H + ClCH_2\text{—}\underset{Cl}{\bigcirc}\text{—}Cl \longrightarrow Bu_2P(S)CH_2\text{—}\underset{Cl}{\bigcirc}\text{—}Cl + HCl$$

$$(\underline{11})$$

$$Me_2P(S)H + ClCH(SEt)CONHMe \longrightarrow Me_2P(S)\underset{SEt}{\underset{|}{C}}HCONHMe + HCl$$

$$(\underline{12})$$

VII. BY ISOMERIZATION OF PHOSPHONODITHIOITES AND PHOSPHINOTHIOITES $RP(SR')_2$ AND R_2PSR'

Phosphinothioites, particularly allyl, benzyl,[12] and phenyl diphenyl-phosphinothioite[223] as well as ethylenebis-(diphenylphosphinothioite)[243] are thermally unstable and suffer self-isomerization to the corresponding tertiary phosphine sulfides and alkylenediphosphine disulfides, respectively, on heating:

$$Ph_2PSR \xrightarrow{\Delta} Ph_2\overset{S}{\overset{\|}{P}}R \qquad R = CH_2=CHCH_2, \; PhCH_2$$

and

$$PhPSCH_2CH_2SPPh_2 \xrightarrow{\Delta} Ph_2\overset{S}{\overset{\|}{P}}CH_2CH_2\overset{S}{\overset{\|}{P}}Ph_2$$

That the rearrangement of allyl diphenylphosphinothioites is concerted was demonstrated by the observation that crotyl diphenylphosphinothioite and α-methylallyl diphenylphosphinothioite readily and completely rearrange to α-methylallyldiphenylphosphine sulfide and crotyldiphenylphosphine sulfide, respectively.[108] Rearrangement of allyl phenylmethylphosphinothioite follows first-order kinetics with

$$Ph_2PSCH_2CH{=}CHR \longrightarrow Ph_2P(S)CHRCH{=}CH_2$$

an activation energy of E_a = 20 kcal/mole.[108] The negative entropy of activation ($\Delta S^{\dagger}$ -11 e.u.) attests to the cyclic character of the transition state in the rearrangement. In this connection it is interesting to note that racemization of (-)-(S)-allylmethylphenylphosphine sulfide has been proposed to involve a [2,3] sigmatropic rearrangement to allyl-(R)-methylphenylphosphinothioite, which is converted to (S)-phosphinothioite by pyramidal inversion at phosphorus. Allylic rearrangement of (S)-phosphinothioite to (R)-phosphine sulfide completes the conversion of (S)-phosphine sulfide into its enantiomer.[108]

It should be pointed out, however, that S-alkyl diphenylphosphinothioites (Ph_2PSR; R = Et, Bu, iso-C_5H_{11}, n-C_6H_{13}, n-C_8H_{17}) distill without isomerization at temperatures around 200° under reduced pressure;[11,306,376] moreover, trifluoromethyl-substituted phosphonodithioites and phosphinothioites cannot be isomerized even in the presence of Arbuzov catalyst.[58]

The Michaelis-Arbuzov reaction has also been used for the preparation of tertiary phosphine sulfides from the corresponding phosphinothioites and alkyl halides.[7,11,324] This reaction involves heating of phosphinothioites with alkyl halides at temperatures around 100°. Although the reactions of phosphinites with alkyl halides are quite

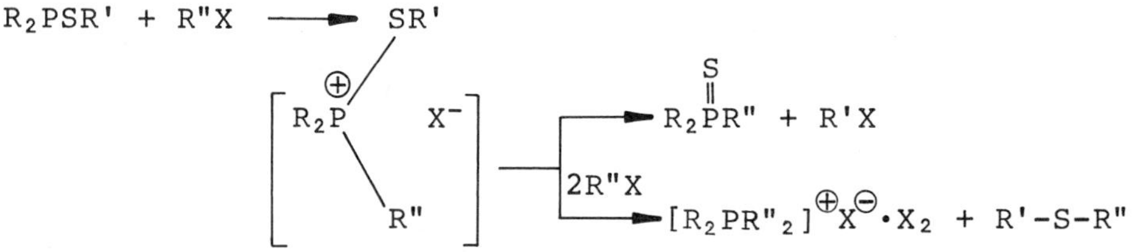

clean-cut, the reactions of phosphinothioites are usually complicated by side reactions in which quaternary phosphonium halides are formed.[11] In some cases, tertiary phosphines and phosphinic acids have also been isolated (see Ref. [203] p. 103).

In a few cases, phosphonodithioites, $RP(SR')_2$ reacted abnormally with alkyl halide and gave tertiary phosphine sulfides, whereas the expected phosphinodithionates were only obtained as by-products or not formed at all. Thus treatment of diethyl ethylphosphonodithioite with ethyl bromide at 140° gave triethylphosphine sulfide and an unidentified liquid product, whereas reaction with benzyl bromide at 140° gave the normal product $Et(PhCH_2)P(S)SEt$ and the abnormal product, $Et(PhCH_2)_2PS$.[13,320]

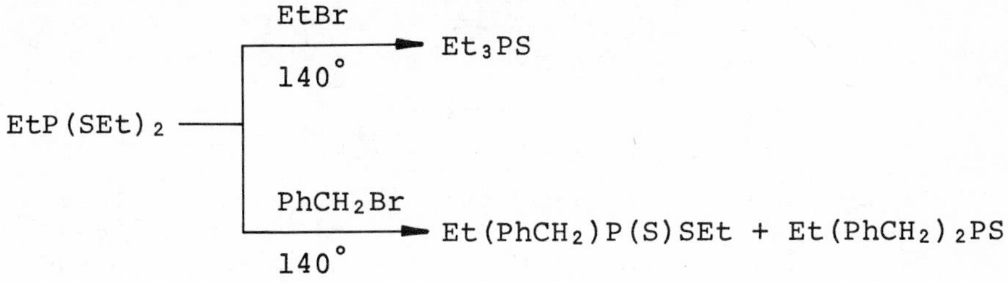

Similarly, treatment of $EtP(SBu)_2$ with MeI [208,210] and of $PhP(SR)_2$ with $PhCH_2Br$ [206] gave in both cases the normal product $RR'P(S)SR$ and the abnormal product RR'_2PS. The yields of RR'_2PS were low. Finally, tertiary phosphine sulfides were also obtained in low yield in the reaction of dialkylphosphinous chlorides with phosphonodithioites.[209] The first step in this reaction is very likely an exchange reaction followed by isomerization.

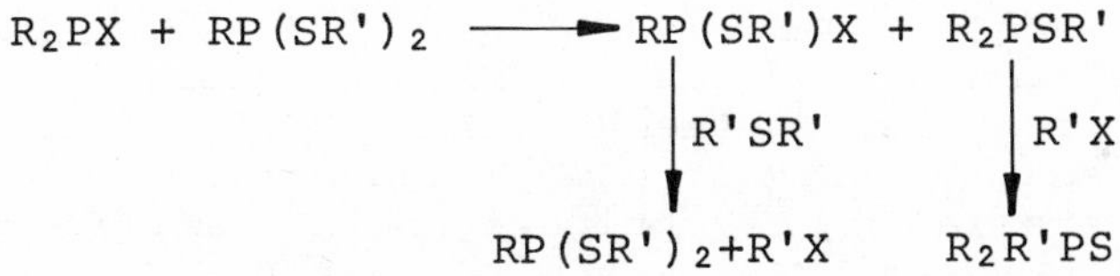

Phenyl diphenylphosphinoselenoite Ph_2PSePh seems to undergo an Arbuzov rearrangement very easily. In an attempt to prepare this compound from Ph_2PCl and NaSePh, only the selenide Ph_3PSe could be isolated.[223] However, the selenoite Ph_2PSePh was obtained from Ph_2PCl and PhSeH in the presence of Et_3N in ether at -10°.[223]

VIII. BY FUNCTIONALIZATION OF TERTIARY PHOSPHINE SULFIDES

Triphenylphosphine sulfide, when treated with methyl- or ethyllithium in an ether-tetrahydrofuran medium,* gave new organolithium reagents, which have been used for further

* In ether alone, no reaction was observed.

synthesis,[358,359] e.g.,

$$Ph_3PS + RCH_2Li \xrightarrow[-PhLi]{Et_2O/THF} Ph_2P(S)CH_2R \xrightarrow[-C_6H_6]{+PhLi}$$

$$Ph_2P(S)CHLiR \xrightarrow{H_2O} Ph_2P(S)CH_2R$$

$$Ph_2P(S)CHLiR \xrightarrow{CO_2} Ph_2P(S)CH(CO_2H)R$$

$$Ph_2P(S)CHLiR \xrightarrow{R_3SnCl} Ph_2P(S)CH(SnR_3)R$$

The mechanism of this reaction was shown to involve a very rapid exchange step followed by a slower metallation reaction.[358] There is no evidence now that allows us to say whether the RCH_2Li-Ph_3PS exchange step proceeds by way of a phosphorus (V) intermediate or an SN_2 displacement at phosphorus. It is clear that tertiary phosphine sulfide containing an α-hydrogen atom will also undergo a metallation reaction when treated with an organolithium reagent, e.g.,

$$Ph_2P(S)CH_3 + RLi \xrightarrow{Et_2O/THF} Ph_2P(S)CH_2Li + RH$$

This was experimentally verified.[300,304,358] Since the metallation of Ph_2MePS proceeds readily in ether, the tetrahydrofuran required in the RCH_2Li-Ph_3PS reaction apparently is necessary only for completion of the exchange reaction.[358] Metallation of optically active MePrPhPS with t-BuLi followed by treatment with carbon dioxide produced optically active $HO_2CCH_2PrPhPS$.[147]

When Ph_3PS was treated with phenyllithium, on the other hand, a reduction was observed, and Ph_3P (56%) and PhSH (41%) were isolated as reaction products.[417]

An interesting variation of the Wittig reaction constitutes the preparation of α,β-unsaturated phosphine sulfides from methylenebis(diphenylphosphine)oxide-sulfide and aldehydes.[121] The carbonyloxygen of the aldehyde always combines with the most positively charged phosphorus atom.[121]

$$Ph_2\overset{S}{\overset{\|}{P}}-\overset{\ominus}{C}H-\overset{O}{\overset{\|}{P}}Ph_2 + RCHO \longrightarrow Ph_2P(S)CH{=}CHR + Ph_2P(O)O^{\ominus}$$

Under normal conditions the hydrolysis of ether-[90,383] or ester groups[83,90,212] containing phosphine sulfides and the condensation[297] and esterification of hydroxy-[90,279] and carboxy groups[304] containing phosphine sulfides leaves the P=S function intact. The oxidation of parahydroxymethylphenyl-diphenylphosphine sulfide with CrO_3/pyridine to paraformyldiphenylphosphine sulfides[331] presents an interesting exception. Normally the corresponding oxides are obtained in this type of reaction.

$$Ph_2P(S)C_6H_4CH_2OH\text{-}p \xrightarrow{CrO_3/\text{pyridine}} Ph_2P(S)C_6H_4CHO\text{-}p$$

When pentafluorophenyldiphenylphosphine sulfide was treated with NaOMe in methanol, it underwent nucleophilic substitution as well as cleavage of the C_6F_5 group (ca.25%).[57] This sulfide also reacted with methylamine in

$$Ph_2P(S)C_6F_5 \xrightarrow{NaOMe} Ph_2P(S)C_6F_4OMe\text{-}4 + Ph_2P(S)C_6F_3\text{-}(OMe)_2\text{-}2,4 + Ph_2P(S)OMe$$

$$Ph_2P(S)C_6F_5 \xrightarrow{MeNH_2/EtOH} Ph_2P(S)C_6F_4NHMe\text{-}4$$

$$Ph_2P(S)C_6F_5 \xrightarrow{MeNH_2/C_6H_6} Ph_2P(S)C_6F_4NHMe\text{-}4 + Ph_2P(S)C_6F_4NHMe\text{-}2$$

benzene to give both ortho- and para- monoreplacement products. When the methylamine reaction was carried out in ethanol, only the para-replacement product was detected.[57]

Normally $LiAlH_4$ reduces tertiary phosphine sulfides to tertiary phosphines (see Section B). Diaryl- and dialkyl-1-alkynylphosphine sulfides are, however, reduced with $LiAlH_4$ under mild conditions (in Et_2O or THF at room temperature) to trans-β-substituted vinylphosphine sulfides. Reduction of $Me_2P(S)C{\equiv}CEt$ with $LiAlD_4$ followed by treatment with water gave β-deuterated $Me_2P(S)CH{=}CDEt$. On the other hand, treatment with $LiAlH_4$ followed by hydrolysis with D_2O led to the α-deuterated sulfide $Me_2P(S)CD{=}CHEt$.[5] These results indicated that the reducing agent adds a hydride ion to the β position in Michael fashion, analogous to the reduction of cinnamate derivatives. Each molecule of $LiAlH_4$ reduces two molecules of alkynylphosphine sulfide.[5] Alkynyl-1-phosphine sulfides (and oxides), but not alkynyl-1-phosphines and alkenyl-1-phosphine sulfides, also undergo a Michael addition with Grignard reagents in the presence of equimolar amounts of cuprous chloride. Reduction of the amount of Cu_2Cl_2 led to greatly reduced yields of product.[6] Since $Ph_2P(S)C{\equiv}CMe$ and PhMgBr

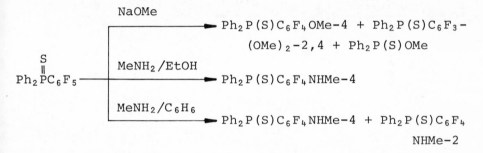

$$2R_2P(S)C\equiv CR^1 + LiAlH_4 \longrightarrow 2[R_2P(S)C=CHR^1]^{\ominus} + Li^{\oplus} + AlH_2^{\oplus}$$

$$\downarrow H_2O$$

$$\underset{H}{\overset{\underset{\displaystyle R_2P}{\overset{\displaystyle S}{\|}}}{\diagdown}} C=C \underset{R^1}{\overset{H}{\diagup}}$$

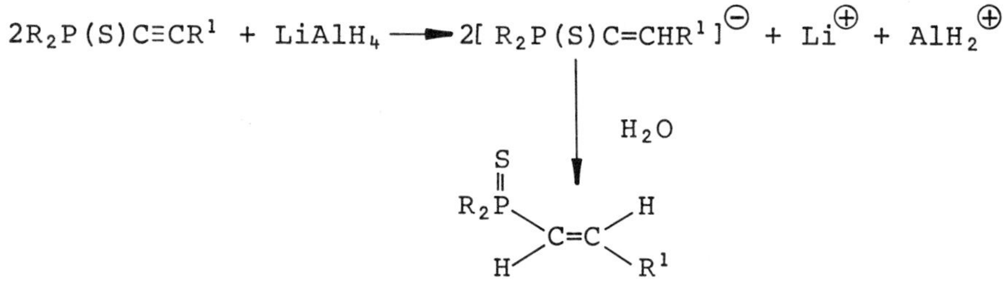

or $Ph_2P(S)C\equiv CPh$ and MeMgI gave identical products, it was concluded that equilibration to the most stable isomer occurs during the reaction. The following reaction path was proposed:[6]

$$\overset{\overset{\displaystyle S}{\|}}{R_2PC\equiv CR^1} \xrightarrow{Cu_2Cl_2} \overset{\overset{\displaystyle S}{\underset{\diagup Cu\diagdown}{\|}}}{R_2P-C\equiv CR^1} \xrightarrow{R^2MgX} \overset{\overset{\displaystyle S}{\underset{\diagup M\diagdown}{\|}}}{R_2P=C=C} \overset{R^2}{\underset{R^1}{\diagup}} \xrightarrow{H^{\oplus}}$$

$$\overset{\overset{\displaystyle S}{\underset{\diagup M}{\|}}}{R_2P-CH=C} \overset{R^2}{\underset{R^1}{\diagup}}$$

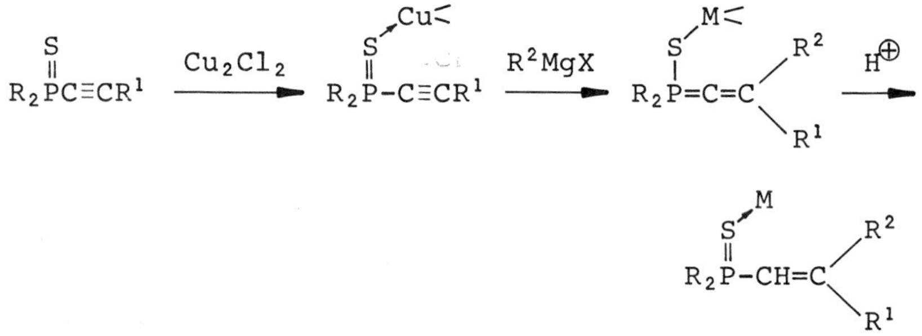

IX.   FROM BIPHOSPHINES AND BIPHOSPHINE DISULFIDES

Heating an equimolar mixture of biphosphines with aliphatic-substituted disulfides for several hours at 130 to 200° gave tertiary phosphine sulfides.[362]

$$R_2P-PR_2 + R'SSR' \longrightarrow [2R_2PSR'] \longrightarrow 2R_2R'PS$$

Intermediates in this process are apparently phosphinothioites, which isomerize under the reaction conditions to tertiary phosphine sulfides.[362,363] Aromatic-substituted disulfides and bisbenzothiazyl disulfide, however, gave phosphinothioites that were isolated. The reaction is initiated by the thioyl radical produced by heating or UV irradiation.[362]

A rather unusual method for preparing diphosphine disulfides consists in the addition of tetraalkylbiphosphine disulfides to ethylene in the presence of catalytic amounts of iodine.[293] The reaction is carried out by

$$\overset{\overset{\displaystyle S}{\|}\ \overset{\displaystyle S}{\|}}{R_2P-PR_2} + CH_2=CH_2 \longrightarrow \overset{\overset{\displaystyle S}{\|}\qquad\ \overset{\displaystyle S}{\|}}{R_2PCH_2CH_2PR_2}$$

heating the biphosphine disulfide and ethylene in the presence of small amounts of iodine in Carius tubes at 275°C for 48 hr.[293] With tetramethylbiphosphine disulfide, the yield of the ditertiary phosphine disulfide was 60%. The reaction has been used successfully for the preparation of cyclic diphosphine disulfides. For example, ethylene bis(cyclotetramethylenephosphine sulfide) has been made in this way.[346]

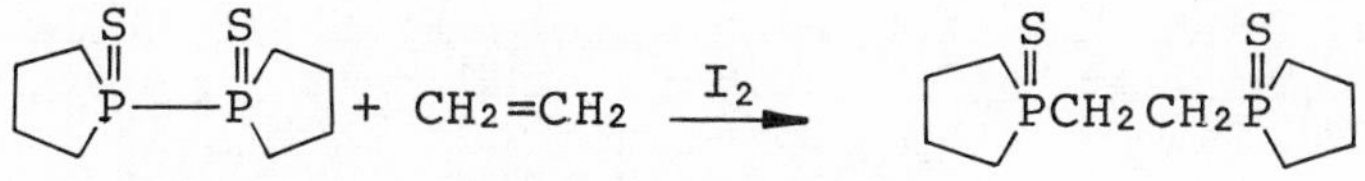

X. BY MISCELLANEOUS REACTIONS

In several cases tertiary phosphine sulfides were obtained in reactions that were performed not for the purpose of preparing these compounds but rather to obtain new organic compounds.[233] Thus triphenylphosphine sulfide was isolated in the interactions of phosphinimines[10,369,372] phosphazenes,[370] and phosphine alkylenes,[350,372,373] with CS_2, RNCS, RNSO, or sulfur.

In a preliminary report it was said that tertiary phosphine oxides can be converted to tertiary phosphine sulfides by reaction with phosphorus pentasulfide[148] or hydrogen sulfide.[128] The latter report contrasts with an earlier paper stating that no reaction was observed when triethylphosphine oxide was caused to react with hydrogen sulfide, potassium sulfide, or ammonium sulfide.[59,142] In an early work[216] it was claimed that heating of tetrabenzylphosphonium sulfate $[(PhCH_2)_4P]^+ SO_4H^-$ produced tribenzylphosphine sulfide. However, the melting point of $(PhCH_2)_3PS$, given in this work as 205 to 206°C does not agree with the more recently reported value (see table of compounds). The melting point given in Ref. 216 is closer to that of the oxide $(PhCH_2)_3PO$ (m. 210-212°; Ref. 203, p. 114) than to that of the sulfide (270°). It thus seems that the oxide and not the sulfide was produced in this decomposition reaction.

A patent[274] describes in a brief way the conversion of diphosphine dioxides into the corresponding disulfides by heating with elemental sulfur at 140° for 1 hr. Several diphosphine disulfides of unusual structure were claimed

$$R_2\overset{O}{\overset{\|}{P}}(CH_2)_6\overset{O}{\overset{\|}{P}}R_2 \xrightarrow[140^\circ C/1\ hr.]{S} R_2\overset{S}{\overset{\|}{P}}(CH_2)_6\overset{S}{\overset{\|}{P}}R_2$$

to have been synthesized, but no physical data were included for them. Another patent reports the conversion of tertiary phosphine oxides into sulfides, by first reducing R_3PO with $SiHCl_3$ to R_3P (See Chapters 1 and 6) and then adding sulfur.[109a,201]

Tertiary phosphine sulfides are also produced in the interaction of phosphinous halides with thioformaldehyde.[268] The scope of this reaction is not known, since only one example has been described.

Adding $P(S)Cl_3$[179] or diphenylphosphinothioyl chloride to triphenylphosphorane-alkylenes gave "ylid"-substituted phosphine sulfides; on treatment with acid these compounds produced phosphine sulfides, substituted by a phosphonium group.[177,179] Similar types of compounds have been obtained by the addition of sulfur to [diphenylphosphino-

$$2Ph_3P=CHR + Ph_2P(S)Cl \longrightarrow Ph_3P=CR-P(S)Ph_2 + [Ph_3\overset{\oplus}{P}-CH_2R]Cl^{\ominus}$$

$$\downarrow HCl$$

$$[Ph_3\overset{\oplus}{P}-CHR-P(S)Ph_2]Cl^{\ominus}$$

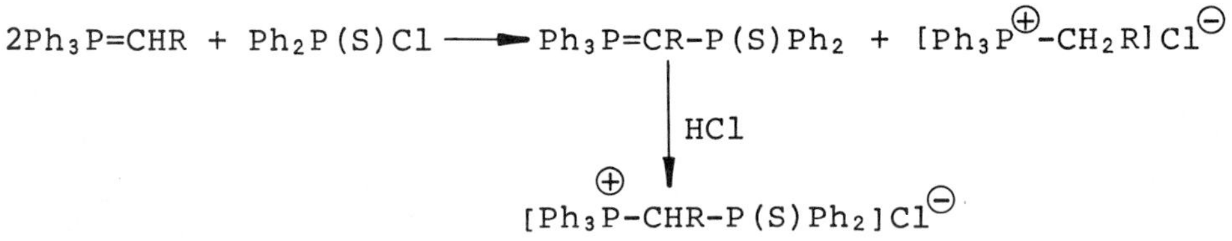

$$6Ph_3P=CH_2 + P(S)Cl_3 \longrightarrow (Ph_3P=CH)_3PS + 3[Ph_3PMe]^{\oplus}Cl^{\ominus}$$

(triphenylphosphoranylidene)methyl]triphenylphosphonium chloride.[36]

$$[(Ph_3P)_2CPPh_2]^{\oplus}Cl^{\ominus} + S \longrightarrow [(Ph_3P)_2C-P(S)Ph_2]^{\oplus}Cl^{\ominus}$$

Unsaturated tertiary phosphine sulfides undergo a 1,3-dipolar addition and produce phosphinothioyl-substituted heterocyclic systems. Thus in the reaction of diphenyl-vinylphosphine sulfide with nitrilimines ($R'C\equiv\overset{+}{N}-NR''$), nitrile N-oxides ($R'C\equiv\overset{+}{N}-O^-$) and nitrones ($R'CH=\overset{+}{N}R''-O^-$), 5-phosphinothioyl-substituted 2-pyrazoline-(13), 2-isoxazoline-(14), and isoxazolidine(15) derivatives, respectively, were obtained.[202]

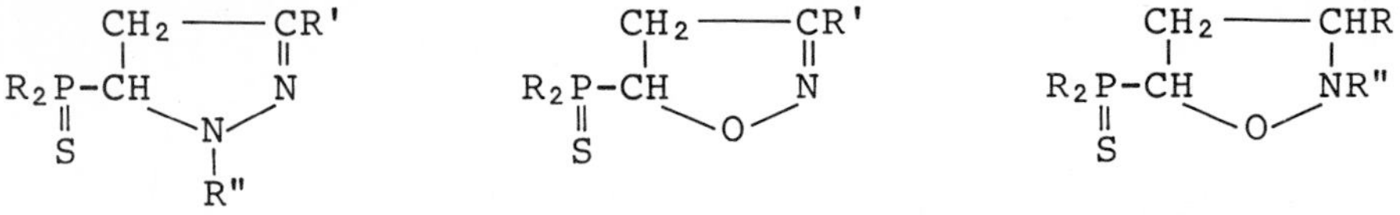

(13)   (14)   (15)

The phosphorinanone sulfide (16) did not yield Δ^3-thiazoline in the simultaneous interaction with sulfur and ammonia.[17] A Δ^3-thiazoline derivative (17) was obtained however, when the sulfide (16) was treated with 3-mercapto-2,2,6,6,-tetramethyl-4-piperidonehydrochloride and ammonia.[17]

Thiophosphinoyldithioformates[85b] were formed in the

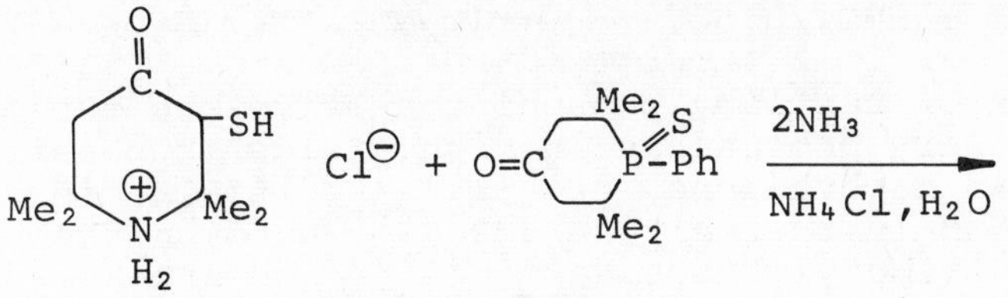

(16)

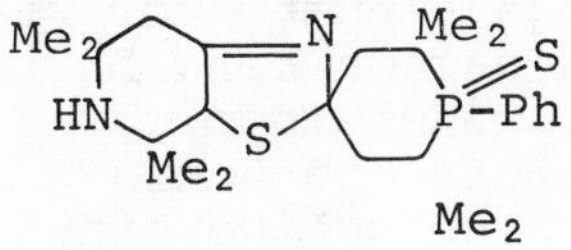

(17)

reaction of the potassium salt of thiophosphinoyldithioformates[85a,205] with alkyl halide.[85b] Aminolysis of these dithioesters afforded the corresponding thiophosphinoylthioformamides.[85c]

$$R_2P(S)CSS^{\ominus}K^{\oplus} + RX \longrightarrow R_2P(S)CSSR + KX$$

$$R_2P(S)CSSR + HNR^1R^2 \longrightarrow R_2P(S)CSNR^1R^2 + RSH$$

B. REACTIONS OF TERTIARY PHOSPHINE SULFIDES, SELENIDES, AND TELLURIDES

Tertiary phosphine sulfides are easily oxidized to the corresponding tertiary phosphine oxides by oxidizing agents such as nitric acid,[142,232,266,273] bromine in alkaline solution,[266] hydrogen proxide,[5] potassium permanganate,[232,273] and thionylchloride.[309]

Excess thionylchloride[309] and antimony pentafluoride[346] convert tertiary phosphine sulfides into dihalophosphoranes R_3PX_2. Tertiary phosphine selenides[423,426] and tellurides[429] are already oxidized by air to the corresponding phosphine oxides. Tertiary phosphine tellurides are rather unstable and deposit gray tellurium on storage.[424]

Oxidation of alkyl side chains on aryl groups of tertiary phosphine sulfides without affecting the thiophosphoryl function has been achieved only once, in the conversion of $Ph_2P(S)C_6H_4CH_2OH$-p to the aldehyde $Ph_2P(S)C_6H_4CHO$-p by treatment with CrO_3/pyridine.[331] In all the other reported examples,[232,273] the corresponding tertiary phosphine oxide was obtained, e.g.,

$$(p\text{-}MeC_6H_4)_3PS \xrightarrow[\text{pyridine}]{KMnO_4} (p\text{-}HOOC_6H_4)_3PO$$

$$Ph_3PS + HNO_3/H_2SO_4 \longrightarrow (m\text{-}NO_2C_6H_4)_3P{=}O$$

Reduction of tertiary phosphine sulfides to the corresponding tertiary phosphines in high yields is possible with sodium in naphthalene,[75,142,148,151] $LiAlH_4$ in dioxane,[136,151,166] NaH in the melt at 300°,[151] Raney nickel in methanol,[151] tributylphosphine,[232a,293] and iron powder.[232a] Reduction of optically active acyclic tertiary phosphine sulfides to optically active tertiary phosphines with hexachlorodisilane proceeds with retention of configuration.[280]

Although a number of reactions are known in which a phosphorus-carbon bond in tertiary phosphines and tertiary phosphine oxides is cleaved (see Chapters 1 and 6), only two such reactions are known with tertiary phosphine sulfides. Bis(diphenylphosphinothioyl) acetylene and tris(p-tolylethynyl)phosphine sulfide were cleaved on boiling with potassium hydroxide solution;[130,131] $(p\text{-}MeC_6H_4C{\equiv}C)_3PS$ was also cleaved with ethanolic $AgNO_3$ and Jlosvay's solution.[131]

The functionalization of tertiary phosphine sulfides and some other reactions have been discussed in Sections A. VIII and A. X, since these reactions lead to new tertiary phosphine sulfides.

Tertiary phosphine sulfides and selenides give stable adducts with halogens, interhalogens, and metal salts (see table of compounds). Tertiary phosphine selenides[343,344] and also sulfides form adducts with alkyl halides,[128,250] dimethylsulfate, and trialkyloxonium salts,[343,344] respectively, of the onium type:

$$R_3PS + RX \longrightarrow [R_3\overset{\oplus}{P}\text{-}SR]\ X^{\ominus} \rightleftharpoons [R_3P{=}\overset{\oplus}{S}\text{-}R]\,X^{\ominus}$$

The same type of salt was obtained in the reaction of tertiary phosphines with sulfenyl chlorides RSCl.[275,306] By metathetical reactions of these adducts with $R_2NH_2SbCl_6$, NH_4PF_6, and NH_4HgI_3, crystalline salts of the type $[R_3PSR']^+X^-$ ($X = SbCl_6^-$, PF_6^-, HgI_3^-) were obtained.[343,344] On being boiled in water, these adducts hydrolyze to give the tertiary phosphine oxide and an alkyl mercaptan in which the alkyl group came from the alkylating agent.

Similar to the corresponding phosphine and phosphine oxide, diphenyl(4-hydroxy-3,5-di-tert-butylphenyl) phosphine sulfide (<u>18</u>) can be dehydrogenated by $K_3Fe(CN)_6$ or PbO_2

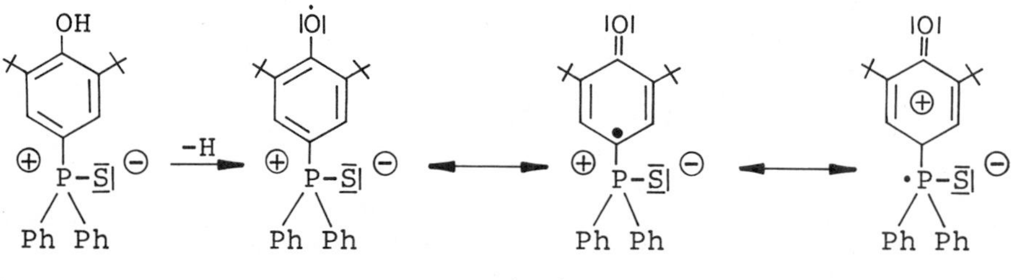

(<u>18</u>)

to give a dark-blue colored radical which is stable in solution for some time and may be formulated as an oxygen or phosphorus radical.[276]

C. APPLICATION OF TERTIARY PHOSPHINE SULFIDES

Tributylphosphine sulfide[141] and in particular tri-n-octylphosphine sulfide[102] are more selective organic extractants for metal ions than the corresponding oxides. More than 20 metals have been separated with $(n\text{-}C_8H_{17})_3P{=}S$.[102] Tributylphosphine sulfide[411] and di-amyl-cyclopentylphosphine sulfide[412] show herbicidal activity, and $Me_2P(S)$-CH(SEt)·CONHMe protects plants from hectoparasites and endoparasites.[220] Tertiary phosphine sulfides,[153] and in particular carbamoyl- and thiocarbamoylphosphine sulfides,[361] are good inhibitors of corrosion for iron and enhance antioxidant and lubricating efficiency of lubricating oils.[361] Triphenylphosphine sulfide and selenide in combination with $EtAlCl_2$ and a transition metal salt give a good catalyst for the polymerization of α-olefins to highly crystalline polymers (e.g., polypropylene, m. 162°).[48]

Tertiary phosphine sulfides[117] such as Ph_3PS [9] and $\{[Ph(CH_2)_4]_2P(S)CH_2\}_2$ [188] have been claimed to be useful as flame retardants for vinyl- and thermoplastic polymers. Adhesion ability toward metals was effectively improved by grafting di-allyl-phenylphosphine sulfide onto polypropylene by irradiation.[218] Several polymeric phosphine sulfides were prepared,[39,69,126,278,287,297] and some of them were said to be useful as hydraulic fluids and lubricants.[69]

Paradimethylaminophenyldimethylphosphine sulfide is effective for the elimination of spark-plug fouling and combustion deposits.[421] Triphenylphosphine sulfide[232] and phospholene sulfides[62] show catalytic activity in the conversion of isocyanates to carbodiimides. And finally, $Me_2(n\text{-}C_{14}H_{29})PS$, $Et_2(n\text{-}C_8H_{17})PS$, and $Et_2(n\text{-}C_{12}H_{25})PS$ show good bacteriostatic and surface active properties[244] and have been suggested as additives to detergents.

D. GENERAL PHYSICAL PROPERTIES OF TERTIARY PHOSPHINE SULFIDES, SELENIDES, AND TELLURIDES

As a class, tertiary phosphine sulfides (selenides and tellurides) are thermally less stable than the tertiary phosphine oxides; for example, Ph_3PS decomposes at 380° and Ph_3PO decomposes only at 454°.[197] This is also indicated in the smaller dissociation energy of the P=S bond [$\bar{D}(R_3P{=}S)$ 91.5 kcal/mole] as compared with that of the P=O bond [$\bar{D}(R_3P{=}O)$ 138 kcal/mole[70]]. The compound (-)-(S)-allylmethylphenylphosphine sulfide racemizes in o-xylene with first-order kinetics ($k_{rac} = 3.53 \times 10^{-6}$, sec^{-1}, at 205°), activation energy $E_a = 33 + 2$ kcal/mole.[108] The nature of bonding between the phosphorus atom and the sulfur (selenium and tellurium bonding being probably similar) has been in dispute for decades and is still attracting wide interest.

In a recent monograph, Hudson[156] has concluded that there is less phosphorus-sulfur than phosphorus-oxygen π bonding. Recent spectroscopic investigations and force constant calculations seem to confirm this (e.g., the bond order of P=S in Me_3PS is 1.33 and that of P=O in Me_3PO is 1.88[124]). Also in accord with this conclusion is the difference between the dissociation energies of the P-X single and double bonds,[70] the different shortening of the bond distance,[401] the electronegativity difference between P and X and, finally, the ^{31}P chemical shifts which are found for phosphine sulfides always at lower field than for the corresponding oxides.[251] On the other hand, basicity measurements of Me_2PhPS and Ph_2MePS suggested that π bonding in the phosphorus-sulfur double bond must be an important factor.[127]

Two full structural analyses of tertiary phosphine sulfides gave 1.86[398] and 1.939 Å[366] for the phosphorus-sulfur double bond distances in Et_3PS and in (cyclo-C_3H_5)$_3$PS, respectively. The phosphorus-sulfur distance in Et_3PS (which, owing to disordered crystals, seems not to be very accurate) would be shorter than the one calculated from the sum of the double bond radii (1.94 Å),[296] whereas that in (cyclo-C_3H_5)$_3$PS agrees with the calculated one. For several other compounds containing a phosphorus-sulfur double bond, the bond distance found is in close agreement with the calculated one; for example, in $[MePS_2]_2$ the P-S distance is 2.1407 Å and the P=S distance is 1.9446 Å.[414] This again suggests that π bonding must be appreciable in the P=S bond.

The large dipole moment of Ph_3PS has been interpreted as indicating a semipolar bond.[195] However, this interpretation is questioned in view of the much higher values of the dipole moment in coordination compounds of the tertiary phosphines with substances like BCl_3 (7 Debyes) or $PtCl_2$ (10.7 Debyes).[308] The dipole moments of parasubstituted phenylphosphine sulfides have been measured recently[122] and compared with calculated values from the dipole moments of the corresponding substituted benzenes.[122] In both cases, electron-attracting groups decrease the observed moment whereas electron-releasing groups increase the dipole moment, in qualitative agreement with the Hammett equation. The dipole moments of tertiary phosphine sulfides containing acetylenic groups correlate quantitatively with the Taft substituent constants according to:[41]

$$\mu = 4.684 + 0.16\ \Sigma\sigma^*$$

The dipole is directed from the acetylenic substituent

toward phosphorus. The UV spectra of tertiary phosphine sulfides[22,41,122,152,327] have been interpreted in terms of the existence of resonance structures such as[122]

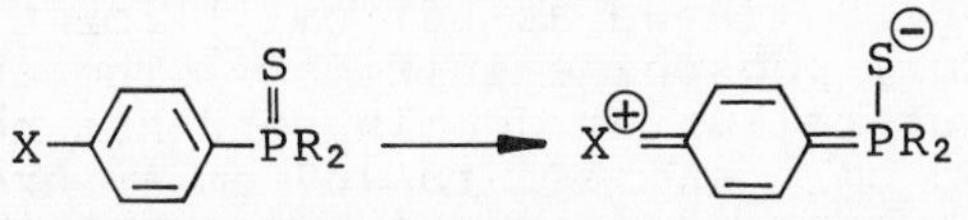

This interpretation has recently been questioned, and it has been said that the P=S group acts as an isolator and does not take part in the chromophor.[327]

A number of reports dealing with IR and Raman spectroscopic investigations of tertiary phosphine sulfides have appeared in the literature.[71,72,116,122,124,125,146,196,213,232,310,408,423] The P=S vibration in tertiary phosphine sulfides is usually found in the region of 600 $\pm$ 20 cm^{-1}. Only in trimethyl and triethylphosphine sulfides has absorption been observed at lower frequency (i.e., in Me_3PS, P=S at 558 cm^{-1} and in Et_3PS, P=S at 535 cm^{-1} [146]). The low frequencies observed for the methyl and ethyl compounds are most probably associated with the very strong intermolecular forces, which tend to enhance the ionic character of the bond and reduce the bond order.

The highest frequencies observed for the phosphorus-sulfur double bond in tertiary phosphine sulfides were those in tricyclopropylphosphine sulfide (727 cm^{-1}),[81] triisopropylphosphine sulfide (701 cm^{-1}),[81] and triphenylphosphine sulfide (637 cm^{-1} [423] or 638 cm^{-1}).[122,213] The high frequency observed for tricyclopropyl- and triphenylphosphine sulfide is very likely due to the electron-withdrawing phosphorus substituents. This leads to an increase in the phosphorus-sulfur double bond order and consequently increases the vibrational frequency. The steric bulk of the isopropyl groups, with consequent opening of the CPC bond angles in $i\text{-}Pr_3PS$, is presumably responsible for the high P=S stretching frequency here.

It has been suggested that the position of the P=S band in $X_3P{=}S$ depends only on the mass of the X atoms and the force constants of the PX bonds, and the hypothesis seems to hold in most cases.[146] More recently an attempt has been made to correlate the fundamental P=S vibration with the steric and polar nature of the phosphorus substituents.[423] A useful correlation was obtained only when the trimethyl and triethylphosphine sulfides were excluded. The electroegativity of the substituents on the phosphorus atom plays a most important role in determining the frequency of the fundamental P=S vibration,[423] which may be expressed in terms of a "Taft equation"

$$\nu_{P-S} = 599 + 14.3\ \Sigma\sigma^*$$

The steric factors appear to be of secondary importance.[423]

The P-Se vibration in tertiary phosphine selenides is normally[423] observed between 496 and 543 cm^{-1}, again excepting Me_3PSe (P=Se at 441 cm^{-1}), Et_3PSe (P=Se at 422 cm^{-1}),[423] and $Ph_3P{=}Se$[196] (P=Se at 561 cm^{-1}) for the same reasons as in tertiary phosphine sulfides. The absorption due to the phosphorus-tellurium double bond has been observed at 400 to 405 cm^{-1} and 445 to 467 cm^{-1}, excepting $(c\text{-}C_6H_{11})_3PTe$, which gave[71a] only one band at 518 cm^{-1}.

Recently it has been shown that the vibrations ν(M=X) are influenced mainly by two factors (viz., the mass of the molecule and the electronegativities of the atoms M and X) and may be expressed[71b] by the equation:

$$\nu(M{=}X) = k\left(\frac{\Sigma M}{m_M m_M}\right)^{1/2} E_M E_X = kZ$$

Nuclear magnetic resonance spectroscopy has proved to be a very valuable tool for the identification of tertiary phosphine sulfides[251] and for determining the basicity[127] and Hammett σ constant of the phosphinothioyl group.[334,334a]

Empirical considerations[269] as well as theoretical calculations[215] lead to the conclusion that the chemical shifts of tetracoordinated phosphorus are mainly determined by bond polarities and by the extent of P_π-d_π bonds. Except for $(C_6F_5)_3PS$, which has ^{31}P of +8.6 ppm,[104] all the ^{31}P chemical shifts of tertiary phosphine sulfides and selenides lie between -30 and -60 ppm (see table of compounds).

The Hammett σ constant for the $Ph_2P(S)$ group is positive, indicating that it is an electron acceptor.[21,334,334a,392,395] Furthermore, the ratio K_{meta}/K_{para} is smaller than one, indicating that the $Ph_2P(S)$ group is meta orienting, as are the Ph_2P- and $Ph_2P(O)$ groups.[392,395]

E. LIST OF COMPOUNDS

The list of compounds is divided into three major types: tertiary phosphine sulfides, tertiary phosphine selenides, and tertiary phosphine tellurides. These primary divisions, which are very unequal in size, are subdivided into 13 structural classes, arranged to permit the bringing together of compounds with similar structure.

Within each structural class, the compounds are listed in order of increasing complexity. Compounds containing functionally modified substituents are listed after the unmodified compound, e.g., Me_3PS, $(HOCH_2)_3PS$, $(MeSCH_2)_3PS$, $(CF_3)_3PS$, Et_3PS, etc.

The entry for each compound indicates the methods by which it has been prepared (Roman numbers I to X), records values of common physical constants, and gives literature references to these and to more complex physical methods. It has been attempted to make the listing as comprehensive as possible.

E.1. Tertiary Phosphine Sulfides

E.1.1. Type: R_3PS

Me_3PS. I.[58a,59,125] III.[82,132,133,229,354] IX.[146,247] IX.[362] Thin plates, m. 153.5-4°[82] m. 155.5-5.7°[58a] m.155-6° (from EtOH),[125,132,133,146,354,362] (the previously reported m. 105°[146,192,247] seems to be in error), crystallizes in the orthorhombic system,[146] IR,[125,146,423] Raman,[116,125] ^{1}H-NMR,[138,288] ^{31}P -30.9 ppm (in $CHCl_3$)[245] [the previously reported shift of -59.1 ppm (in $CHCl_3$), which seems to be in error, might correspond to adduct with Pb-salt![231,269]], forms adducts: $L{\cdot}I_2$, UV,[200] $L_4{\cdot}Co(ClO_4)_2$, $L_2{\cdot}CoCl_2$, $L_2{\cdot}CoBr_2$, $L_2{\cdot}CoI_2$,[53] electronic spectra and magn. moments for these comp-plexes,[53] $L_4{\cdot}Ni(ClO_4)_2$,[54] $L_2{\cdot}ZnI_2$, white crystals, IR,[256] $L{\cdot}CdI_2$, white crystals, IR,[256] $L{\cdot}HgCl_2$, white crystals, IR,[256] $L_2{\cdot}SnCl_4$, solid, m. 196-9°,[380] $L_2{\cdot}SnBr_4$, solid, m. 182-4°.[380]

$(HOCH_2)_3PS$. I. Solid, m. 96.5-8.5°.[323]

$(MeSCH_2)_3PS$. I. Liq., ^{1}H-NMR, mass spect.[301]

$(CF_3)_3PS$. II. Liq., b_{760} 47.4°, log P_{mm} = 7.615-1518/T, m. 6.5-7.2°, alkaline hydrolysis liberates 2 moles of CHF_3, ^{19}F, IR.[64]

Et_3PS. I.[59,85,125,143,146,351,371] III.[72,231,270,295] IV.[249] VII.[7] X.[128] White needles, $b_{1.0}$ 110-19°,[231] m. 93.8-4.0° (from petroleum ether),[59,125,192,231,249,351,371] m. 94.5-5.5°,[72] m. 96-7° (from H_2O),[138] x-ray analysis gave:[398] P=S 1.864 Å, P-C 1.865 Å, C-C 1.38 Å, <S-P-C 112°, <C-P-C 107°, and <P-C-C 114°,[398] UV,[46a] IR,[125,146,423] Raman,[100,125] ^{1}H-NMR,[138] ^{31}P- 51.9 ppm (in acetone).[231,277] With two molecules of iodoform, its methiodide, m. 123°, forms an adduct that melts at 84°.[128] Other adducts: $L_x{\cdot}PtCl_2$ (unknown structure),[59] $L_2{\cdot}SnCl_4$, m. 194-6°,[380] $L_2{\cdot}SnBr_4$ m. 182-4°,[380] $L.2HgCl_2$.[248]

Pr_3PS. I.[426] III.[295] Liq., $b_{1.0}$ 108-10°,[295] $b_{1.1}$ 112°,[426] n_D^{27} 1.5071,[426] $\overline{D}(R_3P{=}S)$ 91.6 kcal/mole,[70] IR,[423] magn. suscept.,[47] μ 4.93 D,[109] adducts: $L_2{\cdot}SnCl_4$, m. 166-6.5°, $L_2{\cdot}SnBr_4$, m. 176-7°.[380]

i-Pr_3PS. I. Solid, m. 35° (by subli.), IR.[81]

(cyclo-C_3H_5)$_3$PS. I. Solid, m. 53° (by subli.), IR.[81]

$(MeC{\equiv}C)_3PS$. III. Solid, m. 198.2-8.8°, IR, ^{1}H-NMR,[42] UV,[41] μ 5.20 D.[41] Space group R3(R$\overline{3}$).[272a]

$(NCCH_2CH_2)_3PS$. I. Solid, m. 141-2°.[319]

$(EtO_2C{\cdot}CH_2CH_2)_3PS$. I. Liq., $b_{1.0}$ 214°, n_D^{25} 1.5005.[319]

Bu_3PS. II.[425] III.[72,82] Liq., $b_{0.01}$ 99°, $b_{0.1}$ 111°,[72]

$b_{0.5}$ 129-31°,[347] $b_{1.1}$ 137-8°,[425] n_D^{20} 1.5045,[72] n_D^{25} 1.4878,[82] 1.4945,[425] d_4^{20} 1.0339, $\overline{D}(Bu_3P{=}S)$ -91 kcal/mole,[7] IR,[72,423] ^{31}P-48 ppm,[227,228,269] magn. suscept.,[47] μ 4.94 D,[109] forms adducts: $L{\cdot}I_2$,[200] $L_2{\cdot}SnCl_4$, solid, m. 93-5°,[380] $L_2{\cdot}SnBr_4$, solid, m. 101-2°,[380] yields with $Me_3O^+SbCl_6^-$ an onium salt $[Bu_3PSMe]^+SbCl_6^-$, m. 119-21°, ^{31}P -63.8 ppm (in CH_2Cl_2).[343]

$(i\text{-}Bu)_3PS$. IV. Solid, m. 59-60°.[247]

$(t\text{-}Bu)_3PS$. I. Solid, m. 220°, ^1H-NMR.[145]

$(Am)_3PS$. I. Liq., $b_{0.9}$ 165-7°, n_D^{29} 1.4902,[426] IR,[423] μ 4.94 D.[109]

$(i\text{-}Am)_3PS$. I. Solid, m. 95.5-6.5°.[369]

$(n\text{-}C_6H_{13})_3PS$. μ 4.98 D,[338] magne. suscept.[53]

$(Me_3C\text{-}C{\equiv}C)_3PS$. I. White solid, m. 112-3°, subl. 305°/760 torr, IR, ^1H-NMR.[321]

$(cyclo\text{-}C_6H_{11})_3PS$, I.[29,161,291,354] III.[190] Solid, m. 172-5°,[291] m. 181-2°,[29,161,190] m. 185° (from EtOH),[354] IR,[423] adduct: $L{\cdot}I_2$, UV.[427]

$(n\text{-}C_8H_{17})_3PS$. I. Liq., distribution coeff. for 13 metal nitrates given,[103] properties as a selective organic extractant described.[102]

$(n\text{-}C_{16}H_{33})_3PS$. III. Solid, m. 64.5°.[214]

$(PhCH_2)_3PS$. I.[217] III.[82,227,375] Needles, m. 268-9° (from acetone),[217,227] m. 274-6° (from $CHCl_3$),[375] m. 282-3.5° (from toluene).[82]

$(Ph{\cdot}MeO{\cdot}CH)_3PS$. I. Solid, m. 121-2°.[105]

$(PhCH{=}CH)_3PS$. I.[109a] II.[217a] III. Solid, m. 225.2-5.7° (from EtOH/benzene),[43] m. 227.5-8°,[109a] IR, μ 4.80 D,[43] UV.[41]

$(PhCCl{=}CH)_3PS$. I. Crystals, m. 188.5-90°.[109a]

$(PhC{\equiv}C)_3PS$. III. Solid, m. 138.4-8.8° (from EtOH), IR, ^1H-NMR,[42] UV,[41] μ 5.32 D.[41] Two modifications, space groups R3 and $P2_1/C$.[272a]

$(4\text{-}MeC_6H_4C{\equiv}C)_3PS$. I.[131] III.[239] Solid, m. 180° (dec.),[131] m. 180-2° (dec.) (from petroleum ether),[239] μ 5.47 D.[41]

$(2,4\text{-}Me_2C_6H_3C{\equiv}C)_3PS$. I. Solid, m. 145° (dec.), μ 5.78 D.[41]

Ph_3PS. I.[26,29,65,66,98,358] II.[152] III.[231,295,375] IV.[249] V.[232,240] VII.[223] X.[369,370,372,373] Monoclinic crystals (from EtOH or acetone/H_2O),[422] m. 158-162,[10,85,92,97,122,149,257,262,264,266,295,354,358,375,428] UV,[122,152] IR,[122,196,213,232,423] mass spect.,[416] ^{31}P -42.6 ppm (in $CHCl_3$),[229,269] dec. temp. ∿380°,[197] μ 4.87 D,[338] 4.73 D,[195,308] 4.89 D.[41,43] Gas chromatography,[33] thin layer chromatography,[341] forms a eutectic with $(PhO)_3PS$. m. 56°, and with Et_3PS, m. 82.5°,[294] and yields with $Me_3O^+SbCl_6^-$ an onium salt, $[Ph_3PSMe]^+SbCl_6^-$, m. 175°, ^{31}P -46.6 ppm (in CH_2Cl_2).[343] Adducts: with potassium,[135] vanillin,[134] $L{\cdot}SO_3$ m. 120-30° (dec.),[29] $L_2{\cdot}3I_2$, blue black, triclinic, m. 140° [428] (x-ray structure[353]), $L{\cdot}ICl$, orange, m. 124°,[428] $L{\cdot}IBr$, red-orange, monoclinic, m. 151.5°,[428] $L{\cdot}(AlCl_3)_{1-3}$, white,[238,397] $L{\cdot}AlBr_3$, cream,

m. 93-5°[397] $L\cdot SnCl_4$, white, m. 163-6° (dec.),[397] $L\cdot SbCl_5$, white or pale yellow, m. 72-3° (dec.)[397] $L\cdot CuCl$, triclinic plates, m. 168°,[86] $L\cdot CuBr$, monoclinic needles, m. 144°,[86] $L\cdot AuCl$, colorless, m. 180°,[200] m. 181-3° (dec.),[312] dec. >130°,[198] $L\cdot AuCl_3$, yellow, m. 160°,[200] m. ca.140°,[198] $L\cdot HgCl_2$, monoclinic crystals, m. 230°,[200] 225° (dec.),[86,397] $L\cdot HgBr_2$, monoclinic crystals, m. 237°,[200] 234° (dec.),[86] $L\cdot HgI_2$, yellow, orthorhombic crystals, m. 173°,[200] m. 166°,[86] $L_4\cdot Hg(ClO_4)_2$,[311] $L_2\cdot TiCl_4$, red, m. 88-90° (?),[397] $L\cdot (TiCl_4)_2$, red, m. 85-8° (dec.),[397] $L\cdot TiBr_4$, brown, m. 240-50° (dec.),[397] $L\cdot NbCl_5$, yellow powder or crystals, m 128-30° (dec.),[54a,77,397] $L\cdot NbBr_5$,[54a] $L\cdot TaCl_5$, pale yellow powder, m. 202-5° (dec.) or violet crystals,[54a,397] $L\cdot TaBr_5$,[54a] $L\cdot (FeCl_3)_{1-1.5}$, red, m. 124-5° (dec.),[397] $L\cdot (FeCl_2)_n$, yellow,[397] $L_2\cdot PdCl_2$, pale orange, m. 235°,[24,200] $L_2\cdot PtCl_2$, pale yellow-orange, m. 211°.[24,200]

$(C_6F_5)_3PS$. I. White solid, m. 170-2°,[104] IR,[104] mass spect. (lost sulfur first),[104,267] UV, ^{19}F-NMR, ^{31}P +8.6 ppm (in $CHCl_3$).[104]

$(4\text{-}MeC_6H_4)_3PS$. I.[22,260] V.[232,240] Crystals, m. 185-6°,[22] m. 181-2° (from petroleum ether),[232,260] $b_{0.4}$ 296-310°,[232] UV,[22,336] IR,[232] σ_p (of $(4\text{-}MeC_6H_4)_2P(S)$ group) + 0.23,[334] ^{31}P -41.1 ppm (in C_6H_6).[232]

$(3\text{-}MeC_6H_4)_3PS$. I. Solid, m. 156°,[336] UV,[336] σ [of 3-$MeC_6H_4)_2P(S)$ group] +0.20,[334a] μ 4.94 D.[338]

$(2\text{-}MeC_6H_4)_3PS$. I. Solid, m. 163°,[336] UV,[336] σ [of $(2\text{-}MeC_6H_4)_2P(S)$ group] +0.21,[334a] μ 4.83 D.[338]

$(4\text{-}ClC_6H_4)_3PS$. I.[331] V.[232,240] Solid, m. 152-3° (from MeOH),[331] m. 146-7° (contained also some ortho; from acetone/H_2O),[232,240] IR,[232] ^{31}P - 40.2 ppm (in C_6H_6).[232]

$(4\text{-}FC_6H_4)_3PS$. I. Solid, m. 139-41° (from C_6H_6/petroleum ether),[339] ^{19}F-NMR, σ_p [of $(4\text{-}FC_6H_4)_2P(S)$ group] +0.540.[339] ^1H-NMR, sign of coupl. const., ^{31}P -37.1 ppm.[222a]

$(4\text{-}MeOC_6H_4)_3PS$. I. Colorless needles, m. 109-10° (from EtOH),[22] UV,[22] σ_p [of $4\text{-}MeOC_6H_4P(S)$+group] +0.65,[330] σ' +0.89.[330] ^1H-NMR, sign of coupl. const., ^{31}P -35.7 ppm.[222a]

$(4\text{-}NC\cdot C_6H_4)_3PS$. Solid, m. 255-60° (from MeOH), IR.[335]

$(3\text{-}NC\cdot C_6H_4)_3PS$. I. Solid, m. 175-6° (from MeOH), IR.[335]

$(4\text{-}HO_2C\cdot C_6H_4)_3PS$. I. Solid, m. 321-3° (from MeOH/H_2O), IR.[335]

$(3\text{-}HO_2C\cdot C_6H_4)_3PS$. I. Solid, m. 294-8° (from MeOH/H_2O), IR.[335]

$(4\text{-}MeO_2C\cdot C_6H_4)_3PS$. I. Solid, m. 178-9° (from MeOH), IR.[335]

$(3\text{-}MeO_2C\cdot C_6H_4)_3PS$. I. Solid, m. 139-40° (from MeOH), IR.[335]

$(4\text{-}Me_2NC_6H_4)_3PS$. I. Blunt, strongly light-reflecting squares, m. 258° (from chlorobenzene), IR,[328] UV,[327] σ' + 0.57.[329]

$(4\text{-Ph}\cdot C_6H_4)_3PS$. I. Plates, M.241-2°,[419] m, 248-53° (from $CHCl_3$/MeOH).[331]
$(2,4\text{-}Me_2C_6H_3)_3PS$. I. Plates, m. 167°.[260]
$(2,5\text{-}Me_2C_6H_3)_3PS$. I. Solid, m. 170°.[260]
$(C_6H_{13}\cdot Me\cdot C_6H_3)_3PS$. V. $b_{0.5}$ 235-320° [contained 30% $R_2P(S)Cl$],[232] ^{31}P -41.0 ppm.[232]
$(2,4,5\text{-}Me_3C_6H_2)_3PS$. I. Plates, m. 192° (from EtOH).[260]
$(1\text{-}C_{10}H_7)_3PS$. I. Solid, m. 255.5-6.5° (from ROH), IR.[379] Adduct: $L\cdot I_2$, light brown, m. 174-6°.[378]
$(4\text{-}MeC_{10}H_6)_3PS$. I. Solid, m. 361-4° (from hexanol),[379] IR,[379] μ 5.39 D.[338] Adduct: $L\cdot I_2$, light brown, m. 233-45°,[378]
$(2\text{-}C_4H_3NH)_3PS$. I. Solid, m. 80°.[162]
$(2\text{-}C_4H_3O)_3PS$. I. Solid, m. 151°, IR, UV.[289] ^{1}H-NMR, sign of coupl. const.[193a]
$(2\text{-}C_4H_3S)_3PS$. I. Solid, m. 138°,[162] ^{1}H-NMR,[193] sign of coupl. const.[193a]
$(3\text{-}C_4H_3S)_3PS$. I. Solid, m. 158-9°, ^{1}H-NMR, Sign of coupl. const.[193a]
$(2\text{-}C_5H_4N)_3PS$. I. Solid, m. 161° (from EtOH),[250] dec. temp. 308°,[197] ^{1}H-NMR,[125b] picrate, m. 156-8°,[250] methiodide m. 156-7°.[250]
Tri-(ferrocenyl)PS. I. Yellow needles, m. 291-3°(dec.).[368]

E.1.2. Type: $R_2R'PS$

Me_2ClCH_2PS. X. ^{31}P - 44.3 ppm.[268]
Me_2CF_3PS. VII. Crystals, m. 49.9-50°, b_{760} 163°(calc.), $\log P_{mm} = 11.9842-3551/T$ (solid), and $\log P_{mm} = 6.8204 + 1.75 \log T - 0.005T-2783/T$ (liq.), ^{1}H-NMR, IR.[58a] Forms no MeI - adduct.[58a]
Me_2EtPS. III.[229,295] VII.[208,210] Solid, m. 103.5-5°,[295] m. 103-6°,[208,210] ^{31}P - 57 ppm,[229] μ 4.66 D.[41,43]
$Me_2(MeNH\cdot CO\cdot CH\cdot SEt)PS$. VI. Solid, m. 146°,[220] useful for protecting plants from hecto- and endoparasites.[220]
Me_2PrPS. III. Solid, m. 89-90° (from cyclohexane or EtOH).[295]
$Me_2(MeC{\equiv}C)PS$. III. Solid, m. 77.5-8.2°, IR, ^{1}H-NMR,[44] UV, μ 4.90 D.[41]
$Me_2t\text{-}BuPS$. III. Solid, m. 203.5° (from petroleum ether).[82]
$Me_2(CH_2{=}CH\text{-}C{\equiv}C)PS$. III. Solid, m. 42-3°, IR, ^{1}H-NMR,[44] μ 4.73 D.[41]
$Me_2(EtC{\equiv}C)PS$. III. Liq., $b_{0.4}$ 80.5-2.0°, IR, ^{1}H-NMR.[5]
$Me_2(trans\text{-}EtCH{=}CH)PS$. VIII. (From above plus $LiAlH_4$). Liq., $b_{0.4}$ 70.5-73.0°, IR, ^{1}H-NMR.[5]
$Me_2(trans\text{-}EtCD{=}CH)PS$. VIII. From $Me_2P(S)C{\equiv}CEt$ + $LiAlD_4$. Liq., $b_{0.45}$ 72-2.5°, ^{1}H-NMR.[5]
$Me_2(trans\text{-}EtCH{=}CD)PS$. VIII. From $Me_2P(S)C{\equiv}CEt$ + $LiAlH_4$ then D_2O. Liq., $b_{0.35}$ 68°, ^{1}H-NMR.[5]
$Me_2(MeEtC{=}CH)PS$. VIII. From $Me_2P(S)C{\equiv}CEt$ + MeMgI. Liq., $b_{0.3}$ 74-5°, IR, ^{1}H-NMR.[6]
$Me_2(Et_2C{=}CH)PS$. VIII. Lig. $b_{0.2}$ 81-3°, IR, ^{1}H-NMR.[6]

$Me_2(n\text{-}C_{14}H_{29})PS$. III.[241] VI[236,242] Solid, $b_{0.2}$ 186-90°, m. 55-6°,[236,239,241] IR,[236] ^{31}P -34.1 ppm (in benzene).[236] Surface tension of 0.01% solution is 35.5 dynes/cm,[244] shows bacteriostatic properties.[244]
Me_2PhCH_2PS. IX. Solid, m. 133.6-4.0°, IR, ^1H-NMR.[362]
$Me_2(PhCH{=}CH)PS$. III.[43] VIII.[5] (From $Me_2P(S)C{\equiv}CPh$ and $LiAlH_4$, yields phosphine sulfide with trans structure.) Crystals, m. 124.8-5.4° (from EtOH/benzene),[43] m. 124-6.5° (from MeOH),[5] μ 4.79 D,[43] IR, ^1H-NMR.[5,43]
$Me_2(PhC{\equiv}C)PS$. III. Crystals, m. 91-3° (from MeOH),[5] m. 92.4-3.2°,[44] IR, ^1H-NMR,[5,44] μ 4.91 D.[41]
$Me_2(PhEtC{=}CH)PS$. VIII. Crystals, m. 104-6°, IR, ^1H-NMR.[6]
Me_2PhPS. I.[127,426] III.[82,127,132,133] Solid, $b_{0.12}$ 87-91°, m. 45.5-7°,[82,132,133] white needles, m. 42°,[127,426] m. 42.5-4.5,[304] IR,[127,423] ^1H-NMR,[194,304,423] ^{31}P -35.2 ppm (in $CHCl_3$),[304] pK_a -4.5.[127]
$Me_2(4\text{-}HOC_6H_4)PS$. I. Solid, m. 104-5°, pK_a 8.44, σ_p[of $Me_2P(S)$ group] + 0.62.[395a]
$Me_2(3\text{-}HOC_6H_4)PS$. I. Solid, m. 69-70°, pK_a 8.87, σ_m[of $Me_2P(S)$ group] + 0.43.[395a]
$Me_2(4\text{-}FC_6H_4)PS$. I. Solid, m. 50-3°, $b_{0.01}$ 119-20°, ^{19}F-NMR.[313]
$Me_2(C_6F_5)PS$. I. Solid, m. 72°, IR, ^{19}F-NMR, ^{31}P -30.5 ppm.[112]
$Me_2(4\text{-}Me_2NC_6H_4)PS$. I. Solid, m. 155° (from EtOH).[265]
$(HOCH_2)_2PrPS$. I. Liq., b_2 135-7°.[305]
$(HOCH_2)_2PhPS$. I.[137,245] Solid, m. 84-6°,[137] m. 80-4°,[245] ^{31}P -41.8 ppm (in MeOH).[245]
$(ClCH_2)_2MePS$. I. Pale yellow oil, ^1H-NMR.[245b]
$(ClCH_2)_2EtPS$. I.[245b] White solid, m. 30.5-1° (from ether/hexane),[245b] ^1H-NMR. ^{31}P -55.2 ppm (in $CHCl_3$).[245b]
$(ClCH_2)_2PhPS$. Crystals, m. 41-1.5° (from Et_2O/hexane), ^1H-NMR ^{31}P -44.2 ppm (in $CHCl_3$).[245b]
$(Et_2NCH_2)_2PhPS$. I. Solid, m. 130-3°.[306]
Et_2MePS. III.[227,229,295] VII.[7] IX[362] Solid, m. 66°,[7,227,229,295,362] IR, ^1H-NMR.[362] Adduct: $L_2{\cdot}3HgBr_2$, m. 93°.[245]
$Et_2(EtO_2C)PS$. I.[158] VII.[207] Oil, decomposes on dist.[158] $b_{0.02}$ 73°, n_D^{20} 1.5099, d_4^{20} 1.0811.[207]
$Et_2[(CH_2{=}CHCH_2)_2NCH_2]PS$. VI. Liq., $b_{0.01}$ 100-5°, IR, ^{31}P -51.4 ppm.[236]
$Et_2[(n\text{-}C_{12}H_{25})_2NCH_2]PS$. VI. Liq., $b_{0.001}$ 150-60°, IR, ^{31}P -52.2 ppm.[236]
$Et_2(KS_2C)PS$. I. Solid, m. 209-10°, IR, ^1H-NMR.[85b]
$Et_2(RS{\cdot}CS)PS$. X. From salt above and RX.[85b]
R = Me. Crystals, m. 29.5-30° (from MeOH), IR, ^1H-NMR.[85b]
R = Et. Oil, IR, ^1H-NMR.[85b]
R = $PhCH_2$. Crystals, m. 50-1° (from MeOH), IR, ^1H-NMR.[85b]
$Et_2(RHN{\cdot}CS)PS$. X. From ester above and RNH_2.
R = H. Yellow crystals, m. 114-6°.[85c]
R = Me. Yellow crystals, m. 67-8.5°, ^1H-NMR.[85c]
R = Et. Yellow crystals, m. 24-4.5°, ^1H-NMR.[85c]

R = Me_2CH. Yellow crystals, m. 22.5-3.5°, ^{1}H-NMR.[85c]

$Et_2(Me_2N{\cdot}CS)PS$. X. Yellow crystals, m. 32-2.5°, ^{1}H-NMR.[85c]

$Et_2[(CH_2)_4N{\cdot}CS]PS$. X. Yellow crystals, m. 78-9°.[85c]
$Et_2(MeOCH_2)PS$. II. Liq., $b_9$120-1°, n_D^{20} 1.5169, d_4^{20} 1.0319.[386]
$Et_2(EtOCH_2)PS$. II. Liq., b_{11} 130-1°, n_D^{20} 1.5072, d_4^{20} 1.0107.[386]
$Et_2(EtSCH_2)PS$. II. Liq., $b_6$150-1°, n_D^{20} 1.5549, d_4^{20} 1.0681.[387]
$Et_2(PrOCH_2)PS$. II. Liq., b_{10} 138-40°, n_D^{20} 1.4999, d_4^{20} 0.9832.[386]
$Et_2(Me_2CHSCH_2)PS$. II. Liq., $b_4$149-0°, n_D^{20} 1.5478, d_4^{20} 1.0444.[387]
$Et_2(BuSCH_2)PS$. II. Liq., $b_{0.5}$ 136-7°, n_D^{20} 1.5403, d_4^{20} 1.0279.[387]
$Et_2(PhCH_2SCH_2)PS$. II. Liq., $b_{3.5}$ 185-7°, n_D^{20} 1.5998, d_4^{20} 1.1231.[387]
$Et_2(BuOMe{\cdot}CH)PS$. II. Liq., b_{10} 154-5°, n_D^{20} 1.4968, d_4^{20} 0.9842.[383]
$Et_2(ClCH_2CH_2OMe{\cdot}CH)PS$. II. Liq., b_9 164-5°, n_D^{20} 1.5127, d_4^{20} 1.1071.[383]
$Et_2(EtSMe{\cdot}CH)PS$. II. Liq., b_8 161-2°, n_D^{20} 1.5553, d_4^{20} 1.0599.[386]
$Et_2(PhSMe{\cdot}CH)PS$. II. Liq., $b_{0.5}$ 166-7°, n_D^{20} 1.6102, d_4^{20} 1.1354.[386]
$Et_2(CH_2{=}C{\cdot}OBu)PS$. II. Liq., b_8 165-6°, n_D^{20} 1.5051, d_4^{20} 1.0270.[383]
$Et_2(MeCO)PS$. VIII. Liq., b_6 149-51°, n_D^{20} 1.5006, d_4^{20} 1.0640.[383]
$Et_2(CCl_3CH{\cdot}OH)PS$. II. Oil, n_D^{20} 1.5335 (yellow liq.).[390]
$Et_2(HOCH_2CH_2)PS$. I. Solid, m. 106-9°.[180]
$Et_2(NaO_2CCH_2)PS$. I. Solid, m. 277-80°.[189]
Et_2PrPS. III. Solid, m. 47-9° (from cyclohexane or EtOH).[295]
$Et_2(Me{\cdot}CO{\cdot}CH_2)PS$. I. Liq., $b_{1.5}$ 123-4°, n_D^{20} 1.5230, d_4^{20} 1.0747,[290] ^{1}H-NMR.[307]
$Et_2(NaO_2CCH_2CH_2CH_2)PS$. I. Solid, m. 160-2°.[189]
$Et_2(n\text{-}C_8H_{17})PS$. III.[239] VI.[236,242] $b_{0.2}$ 126-30°, n_D^{20} 1.5028,[236,239] IR, ^{31}P -52.6 ppm.[236] Surface tension of 0.01% solution is 40.0 dynes/cm,[244] shows bacteriostatic properties.[244]
$Et_2(n\text{-}C_{12}H_{25})PS$. III.[239,241] VI.[236,242] Liq., $b_{0.2}$ 155-65°, n_D^{20} 1.4971,[236,239,241] IR, ^{31}P -51.6 ppm.[236] Surface tension of 0.01% solution is 36.1 dynes/cm.[244] shows bacteriostatic properties.[244]
$Et_2(PhCH_2)PS$. I.[75] II.[385] VII.[7,13] Solid, b. 300-10°,[13] $b_{0.007}$ 140°,[385] m. 51.5°,[45,46,385] m. 94.5°,[75] m. 127-7.5°,[13] (unknown which m.p. is the correct one!).
$Et_2(4\text{-}NO_2C_6H_4CH{\cdot}NHCH_2CH{=}CH_2)PS$. VI. Oil, picrate, m. 131-3°, (from EtOH/ether), ^{31}P -61.1 ppm (in ETOH/acetone).[245a]
$Et_2(3,4\text{-}Cl_2C_6H_3CH{\cdot}NHCH_2CH{=}CH_2)PS$. VI. Oil, picrate, m. 135-6° (from EtOH), ^{31}P -60.5 ppm (in EtOH/acetone).[245a]

$Et_2(1\text{-}C_{10}H_7CH_2)PS$. II. Solid, $b_{0.006}$ 176°, m. 87.5°.[385]
$Et_2(Ph_2CHCH_2)PS$. I. Solid, m. 73-5° (from EtOH).[172]
Et_2(2-pyridyl-CH_2CH_2)PS. II. Liq., b_{12} 203-4°, n_D^{20} 1.5692, d_4^{20} 1.0930, IR, ^{1}H-NMR.[390a]
Et_2(PhCH=C·Me)PS. I. Solid, m. 63-4° (from MeOH), IR.[176]
Et_2(PhCH=C·Ph)PS. I. Solid, m. 62-5° (from EtOH), probably trans structure.[172]
$Et_2(Ph(EtO_2C)C{=}C\cdot Ph)PS$. I. Solid, m. 102-4° (from EtOH)[172] probably trans structure.
Et_2PhPS. I.[259,261,265] III.[229] VII.[324] Liq., $b_{0.8}$ 127.8°, m. ∿25-30°,[229,259,261,265] $b_{0.08}$ 112-3°, n_D^{20} 1.5925, d_4^{20} 1.0869,[324] IR,[423] ^{31}P -52 ppm.[229]
$Et_2(C_6F_5)PS$. I. Solid, m. 40°, IR, ^{19}F-NMR, ^{31}P -33.6 ppm.[112]
$Et_2(4\text{-}Me_2NC_6H_4)PS$. I. Solid, m. 148°.[265]
$(CH_2{=}CHCH_2O_2CCH_2)_2MePS$. I. Liq., $b_{0.03}$ 134°, n_D^{20} 1.5235, d_4^{20} 1.1694.[406a]
$(MeO_2CCH_2)_2EtPS$. I. Liq., $b_{0.5}$ 135-6°, n_D^{20} 1.5205, d_4^{20} 1.2145.[404]
$(CH_2{=}CHCH_2O_2CCH_2)_2EtPS$. I. Liq., $b_{0.02}$ 135°, n_D^{20} 1.5223, d_4^{20} 1.1480.[406a]
$(MeO_2CCH_2)_2BuPS$. I. Liq., $b_{0.5}$ 141°, n_D^{20} 1.5108, d_4^{20} 1.1565.[404]
$(MeO_2CCH_2)_2PhPS$. I. Crystals, m. 61-2°.[404]
$(CH_2{=}CHCH_2O_2CCH_2)$ PhPS. I. Liq., $b_{0.02}$ 162°, n_D^{20} 1.5635, d_4^{20} 1.1946.[406a]
Pr_2MePS. III. Solid, $b_{1.5}$ 106-9°, m. 55-7° (from cyclohexane or EtOH).[295]
Pr_2EtPS. III. Liq., $b_{0.7}$ 113-6°.[295]
Pr_2(2-pyridyl-CH_2CH_2)PS. II. Liq., b_2 174-5°, n_D^{20} 1.5508, d_4^{20} 1.0424, IR, ^{1}H-NMR.[390a]
$(MeC{\equiv}C)_2MePS$. III. Solid, m. 38.2-8.8°, IR, ^{1}H-NMR,[44] UV.[41]
$(MeO_2CCH_2CH_2)_2PhPS$. I. $b_{1.0}$ 200-1°, m. 52-4°.[15,16]
$(NCCH_2CH_2)_2(Me_2C\cdot OH)PS$. VI. Solid, m. 113-5°.[298,299]
$(NCCH_2CH_2)_2PhPS$. I. Solid, m. 74-5°.[297]
$(H_2NCH_2CH_2CH_2)_2BuPS$. I. Liq., $b_{2.5}$ 173°, n_D^{20} 1.5430, d_4^{20} 1.0294.[14,16]
$(H_2NCH_2CH_2CH_2)_2PhPS$. I. Solid, $b_{1.0}$ 199-9.5°, m. 41-2°.[15,16]
$(HOCH_2CH_2CH_2)_2BuPS$. I. Liq., b_2 220-1°, n_D^{20} 1.5330, d_4^{20} 1.0879.[14,16]
$(HOCH_2CH_2CH_2)_2PhPS$. I. Solid, m. 102-3.5°.[15,16]
$(MeCO_2CH_2CH_2CH_2)_2BuPS$. I. Liq., $b_{1.5}$ 188-91°, n_D^{20} 1.5022 d_4^{20} 1.0917.[14,16]
$(MeCO_2CH_2CH_2CH_2)_2PhPS$. I. Liq., $b_{1.0}$ 227°, n_D^{20} 1.5494, d_4^{20} 1.1754.[15,16]
$(CH_2{=}CHCH_2O_2CCH_2CH_2)_2PhP$. I. Liq., $b_{0.016}$ 199-200°, n_D^{20} 1.5581, d_4^{20} 1.1624.[405]
$Bu_2(CH_2{=}CH)PS$. I. Liq., b_1 95-100°, ^{31}P -44 ppm (in $CHCl_3$).[300]
$Bu_2(CCl_3CH\cdot OH)PS$. VI. Liq.[299]

$Bu_2(EtSCH{=}CH)PS$. I. Solid, $b_{0.02}$ 140-5°, m. 36° (from EtOH), n_D^{20} 1.5510 (undercooled melt).[407,408]

$Bu_2(Me{\cdot}CO{\cdot}CH_2)PS$. I. Liq., $b_{1.5}$ 136-7°, n_D^{20} 1.5300, d_4^{20} 1.0410.[290]

$Bu_2(Me_2C{\cdot}OHC{\equiv}C)PS$. I. Liq., $b_{0.005}$ 126°, n_D^{20} 1.5158.[408]

$Bu_2[Me(MeCO)C{\cdot}OH]PS$. VI. Liq., n_D^{20} 1.5020, d_4^{20} 1.0288.[314a]

$Bu_2[Me(Et_2O_3P)C{\cdot}OH]PS$. VI. Liq., n_D^{20} 1.4900, d_4^{20} 1.0818.[314a]

$Bu_2[Ph(EtO_2C)C{\cdot}OH]PS$. VI. Liq., n_D^{20} 1.5280, d_4^{20} 1.0836.[314a]

$Bu_2[Ph(Et_2O_3P)C{\cdot}OH]PS$. VI. Solid, m. 65 .[314a]

$Bu_2[Ph(PhCO)C{\cdot}OH]PS$. VI. Solid, m. 87-8°.[314a]

$Bu_2[4\text{-}MeC_6H_4(Et_2O_3P)C{\cdot}OH]PS$. VI. Liq., n_D^{20} 1.5230, d_4^{20} 1.1029.[314a]

$Bu_2[4\text{-}MeOC_6H_4(Et_2O_3P)C{\cdot}OH]PS$. VI. Liq., n_D^{20} 1.5330, d_4^{20} 1.1445.[314a]

$Bu_2(PhCH{\cdot}OH)PS$. VI. Solid, m. 54-5°, m. 59-62°.[8,298,299]

$Bu_2(2,4\text{-}Cl_2C_6H_3CH_2)PS$. VI. Solid, m. 60-2°.[299]

$Bu_2(4\text{-}NO_2C_6H_4CH{\cdot}NHCH_2CH{=}CH_2)PS$. VI. Oil, picrate, m. 139-40° (from EtOH/ether), ^{31}P -57.4 ppm (in EtOH/acetone).[245a]

$Bu_2(3,4\text{-}Cl_2C_6H_3CH{\cdot}NHCH_2CH{=}CH_2)PS$. VI. Oil, picrate, m. 108-9° (from EtOH), ^{31}P -56.5 ppm (in EtOH/acetone).[245a]

$Bu_2(2\text{-}pyridyl\text{-}CH_2CH_2)PS$. II. Liq., b_2 191-2°, n_D^{20} 1.4503, d_4^{20} 1.0222, IR, ^1H-NMR.[390a]

$Bu_2(PhC{\equiv}C)PS$. I. Solid, m. 39.5-40°.[408]

Bu_2PhPS. I. Solid, m. 47°,[426] m. 50.5-1°,[347] IR.[423]

$Bu_2(4\text{-}ClC_6H_4)$ PS. I. Solid, m. 77.5-8.5° (from i-PrOH).[423]

i-Bu_2 (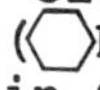NCH_2)PS. VI. Liq., $b_{0.01}$ 140°, IR, ^{31}P -47.4 ppm (in $CHCl_3$).[236]

i-$Bu_2(n\text{-}C_{12}H_{25})PS$. VI.[236,242] Liq., $b_{0.2}$ 150-60°, n_D^{20} 1.4882, IR, ^{31}P -44.2 ppm.[236]

t-$Bu_2(HC{\equiv}C)PS$. I. Solid, m. 130-2°, IR.[408] ^{31}P -67 ppm.

$(CH_2{=}CH{\cdot}C{\equiv}C)_2PhPS$. III. Solid, m. 67.8-8.4°, IR, ^1H-NMR.[42]

$(NCCH_2CH_2CH_2)_2BuPS$. I. Liq., b_1 210-12°, n_D^{20} 1.5320, d_4^{20} 1.0662.[14,16]

$(NCCH_2CH_2CH_2)_2PhPS$. I. Solid, m. 91°.[16]

$(HO_2CCH_2CH_2CH_2)_2BuPS$. X. (By hydrolysis of nitrile.) Solid, m. 135-7°.[14]

$(MeO_2CCH{\cdot}MeCH_2)_2PhPS$. I. Liq., $b_{0.3}$ 184°, n_D^{20} 1.5470, d_4^{20} 1.1703.[15,16]

$(EtO_2CCH{\cdot}MeCH_2)_2PhPS$. I. Liq., $b_{1.5}$ 194-5°, n_D^{20} 1.5403, d_4^{20} 1.1438.[15,16]

$(PrO_2CCH{\cdot}MeCH_2)_2PhPS$. I. Liq., $b_{0.5}$ 195-6°, n_D^{20} 1.5258, d_4^{20} 1.0997.[15,16]

$(CH_2{=}CHCH_2O_2CCH{\cdot}MeCH_2)_2PhPS$. I. Liq., $b_{0.015}$ 176°, n_D^{20} 1.5428, d_4^{20} 1.1252.[405]

$(i\text{-}PrO_2CCH{\cdot}MeCH_2)_2PhPS$. I. Liq., $b_{0.5}$ 188-9°, n^{20} 1.5242, d_4^{20} 1.0997.[15]

$(BuO_2CCH{\cdot}MeCH_2)_2PhPS$. I. Liq., $b_{0.5}$ 208-11°,[16] n_D^{20} 1.5219, d_4^{20} 1.0789.[15,16]

$(i\text{-}BuO_2CCH{\cdot}MeCH_2)_2PhPS$. I. Liq., $b_{1.0}$ 191.5-2°, n_D^{20} 1.5175, d_4^{20} 1.0702.[15,16]

$(n\text{-}C_6H_{13})_2$MePS. I. Liq., $b_{0.02-0.03}$ 135-40°, n_D^{26} 1.4967, ^{1}H-NMR, ^{31}P -42.7 ppm (in $CHCl_3$).[304]
$(n\text{-}C_6H_{13})_2(MeO_2C{\cdot}CH_2)_2$PS. I. VIII.[304] Liq., $b_{0.2-0.25}$ 177-85°, n_D^{26} 1.4943, ^{1}H-NMR, ^{31}P -47 ppm (in $CHCl_3$).[304]
$(c\text{-}C_6H_{11})_2$MePS. III. Solid, m. 99.5-100° (from petrol-ether).[227]
$(c\text{-}C_6H_{11})_2HOCH_2$PS. I. Solid, m. 111-3°.[137]
$(c\text{-}C_6H_{11})_2(EtO_2C)$PS. I. Solid, m. 92-3°.[158]
$(c\text{-}C_6H_{11})_2$(PhNH·CS)PS. I. Solid, m. 120° (from EtOH), IR.[169]
$(c\text{-}C_6H_{11})_2$EtPS. I. Solid, m. 74-5° (from MeOH).[176]
$(c\text{-}C_6H_{11})_2(NaO_2CCH_2)$ PS. I. Solid, m. 306-12° (dec.).[189]
$(c\text{-}C_6H_{11})_2$(PhC≡C)PS. I. Solid, m. 156°.[168]
$(c\text{-}C_6H_{11})_2$PhPS. I. Solid, m. 168° (from EtOH).[354]
$(PhCH_2)_2$MePS. III. Solid, m. 161.5-2°.[227]
$(PhCH_2)_2$PhPS. I.[84] III.[84] VII.[206] Solid, m. 150.5-1.5° (from EtOH),[84] m. 146-7°.[206]
$(PhCH{=}CH)_2$PhPS. I.[109a] III.[43] Solid, m. 164.4-5.4° (from EtOH/benzene),[43] m. 164-5°,[109a] IR.[43]
$(PhCCl{=}CH)_2$PhPS. I. Crystals, m. 181.5-5.5°.[109a]
$(PhCCl{=}CH)_2$(PhCH=CH)PS. I. Crystals, m. 101.5-3°.[109a]
$(PhC{\equiv}C)_2$MePS. III. Solid, m. 91.3-1.6°, IR, ^{1}H-NMR,[44] UV, μ 5.14 D.[41]
$(PhC{\equiv}C)_2$(PhCH=CH)PS. III. Solid, 183-3.4° (from EtOH/benzene), IR,[43] μ 5.12 D.[41,43]
$(PhC{\equiv}C)_2$PhPS. III. Solid, m. 129.6-130.0°, IR, ^{1}H-NMR,[42] UV, μ 5.20 D.[41]
Ph_2MePS. I.[358,426] III.[127,295] V.[232,240] VIII.[358] Crystals, $b_{0.015}$ 145°,[304] $b_{0.1}$ 157-9°,[232] $b_{0.4}$ 162-3°,[358] $b_{1.5}$ 181°,[426] $b_{2.0}$ 183-5°,[127] $b_{1.0}$ 202-4°, m. 66-7°[295] (crystallizes with difficulty), n_D^{20} 1.6533,[232] n_D^{25} 1.6515,[318] n_D^{25} 1.6503,[358] IR,[127,423] ^{1}H-NMR,[127] ^{31}P -32.5 ppm (in $CHCl_3$),[304] pK_a 4.5.[127]
$Ph_2(HOCH_2)$PS. I. Solid, m. 71-2°.[137]
Ph_2(PhNH·CO)PS. I. Solid, m. 32°,[361] m. 139°.[204]
Ph_2(PhCO·NHCO)PS. I. Solid, m. 118°.[361]
$Ph_2(H_2N{\cdot}CO)$PS. I. Solid, m. 143° (from benzene), ^{1}H-NMR.[403]
$Ph_2(MS_2C)$PS. I. M = Li, Na, K, light brown solids, free acid not stable.[85a,205] K-salt, m. 108-9 (dec.), IR.[85b]
Ph_2(RS·CS)PS. X. From $Ph_2(KS_2C)$PS and RX.[85b]
R = Me. Crystals, m. 72-4° (from MeOH), IR, ^{1}H-NMR.[85b]
R = Et. Oil.[85b]
R = $PhCH_2$. Crystals, m. 82.4° (from MeOH) IR, ^{1}H-NMR.[85b]
$Ph_2(H_2N{\cdot}CS)$PS. I.[290a] X. (from thioester and NH_3).[85c] Yellow prism, m. 132-5° (from i-PrOH),[290a] m. 129-30.5°,[85c] mass spect.[290a]
Ph_2(RHN·CS)PS. I. X. (From thioester and RNH_2).
R = Me. I.[169,290a] X.[85c] Yellow crystals, m. 103° (from (EtOH),[169] m. 98-100°[290a] m. 93.5-5.5,[85c] UV,[290a] IR,[169,290a] ^{1}H-NMR.[85c]
R = Et. X. Yellow crystals, m. 87-9°, ^{1}H-NMR.[85c]

R = Ph. I. Yellow crystals, m. 135°,[361] m. 126-8° (from EtOH), UV, IR.[290a]
R = 1-$C_{10}H_7$ VI. Yellow crystals, m. 128-30° (from EtOH), UV, IR.[290a]
$Ph_2(Me_2N{\cdot}CS)PS$. X. Yellow crystals, m. 191-3°, ^{1}H-NMR.[85c]
$Ph_2[(CH_2)_5N{\cdot}CS]PS$. X. Yellow crystals, m. 185-7°.[85c]
$Ph_2(Et_2NCH_2)PS$. VI. Solid, m. 50-2° (from benzene), IR, ^{31}P -34.4 ppm (in benzene).[236]
$Ph_2[(CH_2)_4NCH_2]PS$. VI. Solid, m. 82°, IR, ^{31}P -36.1 ppm (in $CHCl_3$).[236]
$Ph_2[(CH_2)_5NCH_2]PS$. VI. Solid, m. 75-6°, IR, ^{31}P -34.6 ppm (in $CHCl_3$).[236]
$Ph_2(CH_2CH_2OCH_2CH_2NCH_2)PS$. VI. Solid, m. 97-8° (from benzene), IR, ^{31}P -35.3 ppm (in $CHCl_3$).[236]
$Ph_2(MeSCH_2)PS$. I. Solid, m. 82-4° (from EtOH), ^{1}H-NMR, ^{31}P -41.6 ppm (in $CHCl_3$).[301]
$Ph_2(Me_2CHSCH_2)PS$. II. Solid, $b_{0.033}$ 189°, m. 72°.[387]
Ph_2EtPS. I.[174,358,426] VII.[11] VIII.[358] Plates, m. 64-5°,[174] m. 65.5-6°,[174,358] m. 67°,[426] IR.[423]
$Ph_2(CH_2{=}CH)PS$. I. Solid, m. 53-6°,[300] m. 53.3-4.9°,[317] m. 57°,[32] IR,[32] ^{31}P -37.5 ppm (in $CHCl_3$),[300] attempts to polymerize the compound in bulk were unsuccessful.[32] With AIBN at 90° a polymer was, however, obtained, mol. w. 4470, m. range 60-130°.[315]
$Ph_2(HO_2CCH_2)PS$. I.[83,188,189] VIII.[304,358,359] X.[83,212] Solid, m. 188-9°,[188,189] m. 193-5°,[358,359] m. 195-8° (from benzene/acetone),[83,304] m. 201-3°,[212] IR,[358] ^{1}H-NMR,[304] ^{31}P -37.3 ppm (in DMSO),[304] pK_a (in 50% EtOH) 4.76.[176,252]
$Ph_2(EtO_2CCH_2)PS$. I. Liq., b_{5-6} 225-35°.[188]
$Ph_2(HOCH_2CH_2)PS$. I. Solid, m. 53°.[180]
$Ph_2(EtSMe{\cdot}CH)PS$. II. Liq., $b_{0.3}$ 212°, n_D^{20} 1.6451.[386]
$Ph_2(CCl_3CH{\cdot}OH)PS$. II. Yellow liq., n_D^{20} 1.6119.[390]
$Ph_2(CH_2{=}C{\cdot}OBu)PS$. II. Liq., b_4 141-2°, n_D^{20} 1.6318, d_4^{20} 1.1888.[383]
$Ph_2(CH_2{=}CHOMe{\cdot}CH)PS$. II. Liq., b_6 227-8°, n_D^{20} 1.6268.[383]
Ph_2PrPS. VII. Plates, m. 97-8°.[11]
$Ph_2(CH_2{=}CHCH_2)PS$. VII. Solid, b_{1-2} 184-5°, m. 49-50°.[12]
$Ph_2(MeC{\equiv}C)PS$. I. Solid, m. 91.5-4° (from MeOH),[5] m. 88°,[67,68] IR,[5,68] UV,[68] ^{1}H-NMR.[5,68,365]
Ph_2(trans-MeCH=CH)PS. VIII. (From above and $LiAlH_4$.) Crystals, m. 77.5-9.5° (from MeOH), IR, ^{1}H-NMR.[5]
$Ph_2(HOCH_2CH_2CH_2)PS$. VI.[236] [242] Crystals, m. 105-7° (from hexane/acetone), IR, ^{31}P - 42.4 (in acetone).[236]
$Ph_2(Me_2C{\cdot}OH)PS$. VI. Solid, m. 118-20°.[8,298,299]
$Ph_2(NCCH_2CH_2)PS$. I.[314] VI.[236,242] Crystals, m. 119-24° (from petrol-ether),[236] m. 125-7°[314] IR, ^{31}P -34.7 ppm (in acetone).[236]
$Ph_2(HO_2CCH_2CH_2)PS$. I. Solid, m. 128°.[188]
$Ph_2(Me_2NCH_2CH_2CH_2)PS$. I. Solid, m. 78.5°, IR,[415] (compound is toxic[415]).

$Ph_2[(MeO_2C)_2{\cdot}CH]PS$. III.[83,212] White prisms, m. 70°,[212] m. 76.5-5.7°.[83]
Ph_2BuPS. I. Solid, m. 45°,[426] IR.[423]
Ph_2(i-Bu)PS. VII. Rhombic crystals, m. 80-1°.[11]
$Ph_2(Me_2C{=}CH)PS$. VIII. Crystals, m. 112.5-3.5°, IR, ^{1}H-NMR.[6]
$Ph_2(HO_2CCH_2CH_2CH_2)PS$. I. Solid, m. 71°, IR, UV, ^{1}H-NMR.[68,365]
$Ph_2(Me_2C{\cdot}CN)PS$. I. Solid, m. 123-5°.[221]
Ph_2(i-Am)PS. VII. Rhombic crystals, m. 63.5°, b_9 230-40°.[11]
$Ph_2(MeC{\equiv}C{-}C{\equiv}C)PS$. I. Solid, m. 71°, IR, UV,[68] ^{1}H-NMR.[68,365]
$Ph_2(Me_2C{\cdot}OHC{\equiv}C)PS$. I. Solid, m. 127°,[67,68] IR, UV, ^{1}H-NMR.[68]
$Ph_2(n{-}C_6H_{13})PS$. I.[300,376] VI.[236,242] Solid, m. 48-9°,[300] m. 51.2°,[376] m. 53-4° (from petrol. ether),[236] IR,[236] ^{1}H-NMR,[300] ^{31}P -42 ppm (in $CHCl_3$),[300] -41.5 ppm (in benzene).[236]
$Ph_2(c{-}C_6H_{11})PS$. III.[239] VI.[236,242] Solid, m. 180-5° (from CH CN),[236,239] IR, ^{31}P -48.6 ppm (in benzene).[236]
Ph_2(cyclo-2-HOC_6H_{10})PS. I. Solid, m. 178° (dec.) (from EtOH/CS_2).[186]
Ph_2(1-cyclopentenyl-C≡C)PS. I. Solid, m. 93°,[67,68] IR, UV, ^{1}H-NMR.[68]
$Ph_2(Am{\cdot}HO_2C{\cdot}CH)PS$. I. VIII. Solid, m. 130-2° (From benzene/hexane), ^{1}H-NMR, ^{31}P -46 ppm (in $CHCl_3$).[300]
Ph_2(t-$BuCH_2{\cdot}HO_2C{\cdot}CH$)PS. I. Solid, m. 196.5-8.5° (from benzene), ^{31}P -48.3 ppm (in acetone).[300]
$Ph_2(n{-}C_8H_{17})PS$. I. Solid, m. 14-6°, n_D^{20} 1.5862, d_4^{20} 1.0473.[376]
Ph_2(1-cyclohexenyl-C≡C)PS. I. Solid, m. 120°,[67,68] IR, UV, ^{1}H-NMR.[68]
$Ph_2(n{-}C_{12}H_{25})PS$. VI.[236,242] Solid, $b_{0.2}$ 218-20°, m. 42-3°, IR, ^{31}P -41.5 ppm (in benzene).[236]
$Ph_2(PhCH_2)PS$. I.[4] VII.[12] Solid, m. 162-3°,[4] (another reported temp., m. 144-7°,[12] seems to be wrong).
$Ph_2(4{-}HO_2CC_6H_4CH_2)PS$. VI. Crystals, m. 241-2° (from EtOH), pK_a 5.68, σ_I [of 4-$Ph_2P(S)CH_2$ group] + 0.23.[395b]
$Ph_2(3{-}HO_2CC_6H_4CH_2)PS$. VI. Crystals, m. 234.5-5.0° (from MeCOEt), pK_a 5.75, σ_I [of 3-$Ph_2P(S)CH_2$ group] + 0.14.[395b]
$Ph_2(3,4{-}Cl_2C_6H_3CH{\cdot}OH)PS$. VI. Solid, m. 153-4° (from EtOH), ^{31}P -50.3 ppm (in EtOH/acetone).[245a]
$Ph_2(Ph{\cdot}EtNH{\cdot}CH)PS$. VI. Solid, m. 158-9° (from MeOH), ^{31}P -48.8 ppm (in EtOH).[245a]
$Ph_2(4{-}NO_2C_6H_4{\cdot}CH_2{=}CHCH_2NH{\cdot}CH)PS$. VI. Solid, m. 110-1° (from ether/hexane).[245a]
$Ph_2(Ph{\cdot}HO_2C{\cdot}CH)PS$. I. Solid, m. 203-5°.[37]
$Ph_2(Ph{\cdot}Me_2NCO{\cdot}CH)PS$. I. Solid, m. 189-91°.[37]
$Ph_2(PhCH_2CH_2)PS$. I. Solid, m. 90-2° (from EtOH).[172]
Ph_2(2-pyridyl-CH_2CH_2)PS. II. Crystals, m. 97-8°, (from i-PrOH), IR, ^{1}H-NMR.[390a]
$Ph_2(PhCH_2{\cdot}Ph{\cdot}CH)PS$. I. Solid, m. 182-3°.[4]
$Ph_2(Ph_2CHCH_2)PS$. I. Solid, m. 152-7° (from EtOH).[172]
$Ph_2(Ph_2{\cdot}HO{\cdot}CCH_2)PS$. I. Solid, m. 148-50° (from benzene/hexane), ^{1}H-NMR, ^{31}P -34.6 ppm (in $CHCl_3$).[302]

$Ph_2(PhCH=CH)PS$. I.[358] III.[43] VIII.[121] White crystals, m. 110° (from EtOH or MeOH),[121] m. 112.2-2.8° (from EtOH/benzene),[43] UV,[41] IR,[43] μ 4.80 D.[43] Conformation unknown.

$Ph_2(cis\text{-}PhCH=CH)PS$. I. Crystals, m. 110-2° (from EtOH), IR, ^1H-NMR,[5] Is isomerized by $LiAlH_4$ in Et_2O to the trans isomer.[5]

$Ph_2(trans\text{-}PhCH=CH)PS$. I.[3] VIII.[5] Crystals, 105.5-8.0° (from EtOH),[5] m. 106.5-7.5° (from hexane),[3] IR, ^1H-NMR.[3,5]

$Ph_2(PhMeC=CH)PS$. VIII. Crystals, m. 127.5-8.5°, IR, ^1H-NMR.[6]

$Ph_2(PhCH=C{\cdot}Ph)PS$. I. Crystals, m. ∿69° (from EtOH), probably trans structure.[172]

$Ph_2(2\text{-}MeOC_6H_4CH=CH)PS$. I. VIII. Crystals, m. 125-6° (from EtOH or MeOH).[121]

$Ph_2(PhCH=CH\text{-}CH=CH)PS$. I. VIII. Crystals, m. 139-40° (from EtOH or MeOH).[121]

$Ph_2(MeCOCH_2{\cdot}Ph{\cdot}CH)PS$. I. Crystals, m. 150-1° (from EtOH).[171]

$Ph_2(t\text{-}BuCOCH_2{\cdot}Ph{\cdot}CH)PS$. I. Crystals, m. 175-6° (from EtOH).[171]

$Ph_2(PhCOCH_2{\cdot}Ph{\cdot}CH)PS$. I. Crystals, m. 151-2° (from EtOH).[171]

$Ph_2(2\text{-}thienyl\text{-}CH=CH)PS$. I. VIII. Solid, m. 106-7° (from EtOH or MeOH).[121]

$Ph_2(2\text{-}pyridyl\text{-}CH=CH)PS$. I. VIII. Solid, m. 129-30° (from EtOH or MeOH).[121]

$Ph_2(PhC{\equiv}C\text{-}C{\equiv}C)PS$. I. Solid, m. 89°, IR, UV, ^1H-NMR.[68]

$Ph_2(4\text{-}FC_6H_4)PS$. I. Solid, m. 121-2.5° (from EtOH),[339] ^{19}F-NMR,[313,339] σ_p [of $Ph_2P(S)$ group] + 0.579,[339] σ_p + 0.58.[313]

$Ph_2(3\text{-}FC_6H_4)PS$. I. Crystals, m. 94° (from EtOH),[342] ^{19}F-NMR, σ_m [of $Ph_2P(S)$ group] + 0.47.[342]

$Ph_2(4\text{-}ClC_6H_4)PS$. I. Solid, m. 132-3°,[122] m. 129-31° (from MeOH),[331,394] UV, μ 4.81 D.[122] ^{31}P -44.3 ppm.

$Ph_2(3\text{-}ClC_6H_4)PS$. I. Crystals, m. 75.6.5° (from i-PrOH).[219]

$Ph_2(4\text{-}BrC_6H_4)PS$. I. Solid, m. 135°,[122] m. 142-4° (from MeOH),[331] μ 4.71 D,[122] preliminary x-ray structure.[129] ^{31}P -41.8 ppm.

$Ph_2(4\text{-}MeC_6H_4)PS$. I. Solid, m. 139°,[98] σ_p [of $Ph_2P(S)$ group] + 0.05.[334a]

$Ph_2(3\text{-}MeC_6H_4)PS$. σ_m [of $Ph_2P(S)$ group] - 0.32.[334a]

$Ph_2(2\text{-}MeC_6H_4)PS$. σ_o [of $Ph_2P(S)$ group] + 0.22.[334a]

$Ph_2(4\text{-}MeOC_6H_4)PS$. I. Solid, m. 128-9°,[122] IR,[330] UV,[122,327] μ 5.19 D,[122] σ_p [of $Ph_2P(S)$ group] from IR + 0.65, from pK_a + 0.49, σ' + 0.89 (0.73).[330] ^{31}P -41.4 ppm.

$Ph_2(4\text{-}NCC_6H_4)PS$. I. Solid, m. 105-8° (from MeOH).[331]

$Ph_2(3\text{-}NCC_6H_4)PS$. I. Solid, m. 120-2° (from MeOH), IR.[335]

$Ph_2(4\text{-}HO_2CC_6H_4)PS$. I.[331,391] Solid, m. 153-5° (from benzene) and 181-2° (from $MeOH/H_2O$, high temp. modification),[331] m. 177-8.5°,[391] m. 177-8.5,[395] m. 181-2°,[21] pK_a 4.97,[395] pK_a 4.93 (H_2O/EtOH = 1:1),[391] pK_a 6.03

(THF/H_2O = 60:40),[21] σ_p [of $Ph_2P(S)$ group] + 0.54,[391] σ_p + 0.28,[21] σ_p + 0.47,[395] σ^* 0.32.[21]

$Ph_2(3\text{-}HO_2CC_6H_4)PS$. I. Solid, m. 171-2°,[395] m. 170-1° (from MeOH/H_2O),[335] m. 171-2° (from EtOH),[392] IR,[335] pKa 5.24 (H_2O/EtOH = 1:1),[392,395] σ_m [of $Ph_2P(S)$ group] + 0.34,[392] + 0.29.[395]

$Ph_2(4\text{-}HOC_6H_4)PS$. I. Crystals, m. 176.5-7.5° (from benzene).[395] pKa 9.49, σ_p^- 0.63, σ_p (of $Ph_2P(S)$-group) + 0.47.[395]

$Ph_2(4\text{-}MeO_2C\cdot C_6H_4)PS$. I. Solid, m. 84-5° (from MeOH).[331]

$Ph_2(4\text{-}H_2NCO\cdot C_6H_4)PS$. I. Solid, m. 102° and 171°-3° (from benzene),[331] (two modifications).

$Ph_2(3\text{-}H_2NCO\cdot C_6H_4)PS$. Solid, m. 194-6° (from MeOH), IR.[335]

$Ph_2(4\text{-}Me_2NCO\cdot C_6H_4)PS$. I. Solid, m. 149-53° (from benzene/ligroin),[331] IR.[333]

$Ph_2(4\text{-}HOCH_2C_6H_4)PS$. I. Solid, m. 118-22° (from MeOH/H_2O).[331]

$Ph_2(4\text{-}OCHC_6H_4)PS$. VIII. Solid, m. 115-6° (from MeOH), IR, ^1H-NMR.[331]

$Ph_2(4\text{-}Me_2NC_6H_4)PS$. I. Needles, m. 183°,[122,265] m. 188-90° (from benzene)[328] IR,[328] UV,[122] μ 6.42 D,[122] σ_p [$Ph_2P(S)$ group] + 0.76.[329]

$Ph_2(4\text{-}NO_2C_6H_4)PS$. I. Solid, m. 86-8° (from MeOH).[332]

$Ph_2(PhNHC_6H_4)PS$. I. Solid, m. 129-32° and 164-6° (positional isomers, position unknown), IR,[221] two oligomers $[C_{24}H_{20}PSN]_2$, m. 140-7° and $[C_{24}H_{20}PSN]_3$, m. 160-75°, of unknown structure also obtained.[221]

$Ph_2(4\text{-}CH_2{=}CHC_6H_4)PS$. I. Solid, m. 105-7°, UV,[316] was readily polymerized and copolymerized by AIBN, polymers more thermally stable than polystyrene.[316]

$Ph_2(4\text{-}MeCO\cdot C_6H_4)PS$. I. Solid, m. 95-8° (from MeOH).[332]

$Ph_2(2\text{-}CH_2{=}C\cdot Me\cdot CO_2C_6H_4)PS$. I. VIII. Solid, m. 149-52°, IR.[279]

$Ph_2(2,4,6\text{-}Me_3C_6H_2)PS$. I. Solid, m. 166-7°.[337]

$Ph_2(C_6F_5)PS$. I. Solid, m. 131-2° (C_6H_6)[57] m. 132°,[112] IR, ^{19}F-NMR, ^{31}P -26.6 ppm.[112]

$Ph_2(4\text{-}MeOC_6F_4)PS$. VIII. From $Ph_2P(S)C_6F_5$ and NaOMe (2,4-disubstituted methoxy-compound also formed, not isolated pure). Crystals, m. 107-8° (from CCl_4), ^1H-, ^{19}F-NMR.[57]

$Ph_2(4\text{-}MeNHC_6F_4)PS$. VIII. From $Ph_2P(S)C_6F_5$ and H_2NMe in EtOH. Crystals, m. 178-9° (from CCl_4), ^{19}F-NMR.[57]

$Ph_2(2\text{-}MeNHC_6F_4)PS$. VIII. From $Ph_2P(S)C_6F_5$ and H_2NMe in C_6H_6. Gives mixture of ortho and para isomer in the ratio 46:54. Crystals, m. 149-50° (from CH_2Cl_2), ^{19}F-NMR.[57]

$Ph_2(1\text{-}C_{10}H_7)PS$. I. Solid, m. 170-2° (from EtOH), IR,[379] Adducts: $L\cdot I_2$, reddish purple, m. 157-60°;[378] L·IBr, yellow, m. 155-6°.[378]

$Ph_2(4\text{-}MeC_{10}H_6)PS$. I. Solid, m. 174-5° (from EtOH), IR.[379] Adduct: $L\cdot I_2$, reddish brown, m. 161-6°.[378]

$Ph_2(2\text{-}C_5H_4N)PS$. I. Solid, m. 119° (from EtOH).[250]

Ph_2(9-acridyl)PS. I. Solid, m. 212° (from toluene).[163]

Ph_2[5-(3-Ph-2-isoxazolinyl)PS. X. Solid, m. 104-6°, ^1H-NMR.[202]
Ph_2[5-(2,3-Ph_2-isoxazolidinyl)PS. X. Solid, m. 79-80° (from EtOH).[202]
Ph_2[5-(1,3-Ph_2-2-pyrazolinyl)PS. X. Solid, m. 104-5°, IR, ^1H-NMR.[202]
$(4\text{-}FC_6H_4)_2$MePS. I. Crystals, m. 77-8° (from EtOH). ^{19}F-NMR, σ_p [of Me(4-FC_6H_4)P(S) group] + 0.54.[313]
$(3\text{-}FC_6H_4)_2$MePS. I. Crystals, m. 60-2° (from EtOH), ^{19}F-NMR, σ_m [of Me(3-FC_6H_4)P(S) group] + 0.47.[342]
$(4\text{-}MeC_6H_4)_2$PhPS. III. Solid, m. 153-4°,[273] σ_p [of PhP(S)< group] + 0.30.[334]
$(4\text{-}MeC_6H_4)_2(4\text{-}ClC_6H_4)$PS. I. Crystals, m. 149°.[260]
$(3\text{-}MeC_6H_4)_2$PhPS. σ_m [of PhP(S)< group] + 0.09.[334a]
$(2\text{-}MeC_6H_4)_2$PhPS. σ_o [of PhP(S)< group] + 0.47.[334a]
$(4\text{-}MeOC_6H_4)_2$PhPS. IR, σ_p [of PhP(S)< group] + 0.65, σ'+ 0.89 (from IR) + 0.64 (from pK_a).[330]
$(2\text{-}MeOC_6H_4)_2$MePS. I. Solid, m. 87-8°.[199]
$(4\text{-}NCC_6H_4)_2$PhPS. I. Solid, m. 155-7° (from MeOH), IR.[335]
$(3\text{-}NCC_6H_4)_2$PhPS. I. Solid, m. 125-6° (from MeOH), IR.[335]
$(4\text{-}HO_2C{\cdot}C_6H_4)_2$PhPS. I. Solid, m. 189-91° (from MeOH), IR.[335]
$(3\text{-}HO_2C{\cdot}C_6H_4)_2$PhPS. I. Solid, m. 176-8° (from MeOH/H_2O), IR.[335]
$(4\text{-}MeO_2C{\cdot}C_6H_4)_2$PhPS. Solid, m. 86-8° (from MeOH), IR.[335]
$(4\text{-}Me_2NC_6H_4)_2$(PhNHCO)PS. I. Solid, m. 160°.[361]
$(4\text{-}Me_2NC_6H_4)_2$PhPS. I. Blunt crystals, m. 219-20° (from C_6H_6), IR.[328] σ [of PhP(S)< group] + 0.69, UV.[327]
$(C_6F_5)_2$MePS. I. Solid, m. 138°, IR, ^{19}F-NMR, ^{31}P -52.6 ppm.[112]
$(C_6F_5)_2$EtPS. I. Solid, m. 128°, IR, ^{19}F-NMR, ^{31}P -13.1 ppm.[112]
$(C_6F_5)_2$PhPS. I. Solid, m. 110°, IR, ^{19}F-NMR, ^{31}P -15.6 ppm.[112]
$(1\text{-}C_{10}H_7)_2$PhPS. I. Solid, m. 250.5-2.0° (from BuOH), IR.[379] Adducts: L•I_2, dark brown, m. 117-20°, L•IBr, pale yellow, m. 137-40°.[378]
$(4\text{-}MeC_{10}H_6)_2$PhPS. I. Solid, m. 233.5-4.0° (from BuOH), IR.[379] Adduct: L•I_2, brown, m. 177-80°.[378]
$(2\text{-}C_5H_4N)_2$PhPS. I. Solid, m. 141° (from EtOH),[250] dihydrochloride, m. 165-71°, picrate, m. 141.5-2.5°.[250]
(2-Me-5-pyridyl-CH_2CH_2)$_2$PhPS. I. Solid, m. 113°.[16]

E.1.3. Type: RR'R" PS

MeBuPhCH_2PS. I. Liq., isolated in optically active form, not 100% optically pure, $[\alpha]_D^{27}$ + 7.2° and $[\alpha]_D^{27}$ -7.1°.[420]
MeEtPhPS. III.[133,295] V.[232,240] VII.[324] $b_{0.003}$ 87-9°,[324] $b_{0.04}$ 103-5°,[133] $b_{0.2}$ 108-10°,[232] b_6 156-8°,[295] n_D^{20}

1.5996 (undercooled melt),[232] crystallizes with difficulty, m. 33-4° (from EtOH or cyclohexane),[133] m. 41-1.5°,[324] IR, ^{31}P -45.8 ppm.[232]

Me(HO_2CCH_2)PhPS. I. VIII. Solid, m. 120-2° (from benzene/hexane), ^1H-NMR, ^{31}P -37.7 ppm (in acetone).[304]

MePrPhPS. I.[154,430] Liq., isolated in optically active form, not 100% optically pure, $[\alpha]^{20}_{546}$ + 21.5° (in MeOH),[154] and $[\alpha]^{20}_{546}$ -12.8°.[430] UV and rotational dispersion.[23] (+)-(R)-Form, Crystals, m. 58-78°.[430] $[\alpha]_D$ + 12.9° (58% optically pure, abs. rotation is $[\alpha]_D$ 22°.[430] (-)-(S)-form, $[\alpha]_D$ -16.3°, ^1H-NMR.[430] Forms a salt of the type: $[(-)\text{-}(S)\text{-MePrPhPSEt}]^+SbCl_5^-$, m. 107-9°, $[\alpha]_D$ -14.9°, (74% optically pure). On hydrolysis with NaOH, this salt gives (+)-(R)-MePrPhP=O $[\alpha]_D$ + 15°.[430]

Me(CH_2=$CHCH_2$)PhPS. I. Long needles, m. 62-3° (from hexane/benzene).[430] (-)-(S)-Form: $[\alpha]_D$ -18.8°, abs. rotation calculated to be $[\alpha]_D$ 25°, ^1H-NMR.[430] Racemizes in o-xylene with first-order kinetics, k_{rac} = 3.53 x 10^{-6} (sec^{-1}, at 205°), E_a = 33 ± 2 kcal/mole.[108]

$MePhCH_2PhPS$. UV and rotational dispersion.[23]

(HO_2CCH_2)PrPhPS. VIII. Isolated in optically active form, not 100% optically pure $[\alpha]^{20}_{546}$ + 9.02° and $[\alpha]^{20}_{546}$ -7.45°.[147]

Et(CH_2=$CHCH_2$)PhPS. VII. Liq., $b_{0.05}$ 97-8°, n^{20}_D 1.5912, d^{20}_4 1.0800.[324]

$EtPhCH_2PhPS$. VII. Solid, m. 79.5-80°, $b_{0.007}$ 157-9°.[324]

EtPh(4-PhCO·OC_6H_4)PS. I. Crystals, m. 83-4° (from EtOH).[90]

EtPh(4-HOC_6H_4)PS. I. Viscous liq.[90]

EtPh(4-HO_2CCH_2·OC_6H_4)PS. VIII. Crystals, m. 84° (from benzene-cyclohexane).[90] l-Phenylethylamine salt, m.206-7°, d-sec-butylamine salt, m. 189-90° [90] (from water), m. 190-3° (from EtOH/acetone), d-aminocamphor salt, m. 166-8°.[90]

EtPh(2-pyridyl-CH_2CH_2)PS. II. Liq., b_2 218-20°, n^{20}_D 1.6017, d^{20}_4 1.1409, IR, ^1H-NMR.[390a]

$NCCH_2CH_2$($NC(CH_2)_3$) PhPS. I. Crystals, m. 74-5.5°.[406]

($NCCH_2CH_2$)Ph(2-BrC_6H_4)PS. I. Crystals, m. 96.5-7.5° (from benzene).[115]

($MeO_2CCH_2CH_2$)(MeO_2CCH·$MeCH_2$)PhPS. I. Liq., b_1 196-8°, n^{20}_D 1.5551, d^{20}_4 1.1907.[406]

(CH_2=$CHCH_2O_2CCH_2CH_2$)(CH_2=$CHCH_2O_2CCH$·$MeCH_2$)PhPS. I. Liq., $b_{1.0}$ 195-7°, n^{20}_D 1.5515, d^{20}_4 1.1485.[405]

BuPh($PhCO_2C_6H_4$)PS. I. Crystals, m. 66-7° (from EtOH).[90]

BuPh(4-HOC_6H_4)PS. VIII. Crystals, m. 97-8° (from cyclohexane).[90]

BuPh(4-HO_2CCH_2·OC_6H_4)PS. VIII. Solid, insoluble in water.[90] Its phenylethylamine salt, m. 209-10°, was resolved into optical isomers. The free acid has $[M]_D$ + 9.6 and -9.7°.[90]

Ph(4-BrC_6H_4)(4-$Me_2NC_6H_4$)PS. I. Crystals, m. 126° (from MeOH).[90] N-methiodide, m. 158-9°, N-metho-d-camphorsulfonate, m. 224-6°, N-methobromocamphorsulfonate,

m. 198°.[90]
$Ph(4\text{-}MeOC_6H_4)(4\text{-}MeC_6H_4)PS$. I. Crystals, m. 121-4° (from EtOH).[90]
$Ph(4\text{-}BrC_6H_4)(2\text{-}C_5H_4N)PS$. I. Crystals, m. 109° (from EtOH).[90] Methiodide, yellow solid, m. 132-4°.[90]
$Ph(4\text{-}BrC_6H_4)(3\text{-}C_5H_4N)PS$. I. Crystals, m. 115-6° (from EtOH).[90] Does not form a methiodide.

E.2. Diphosphine Disulfides

$Et_2P(S)CH_2(S)PEt_2$. III. Colorless crystals, m. 84.5-5° (from petrol. ether),[234] IR, ^{31}P -49.5 ppm (in EtOH).[234]
$Ph_2P(S)CH_2(S)PPh_2$. I. Colorless plates, m. 178° (from benzene/petrol. ether),[159] m. 183-5°,[302] ^1H-NMR[63,302] ^{31}P -34.5 ppm (in $CHCl_3$).[302]
$Ph_2P(S)CH_2(S)P(CH_2NEt_2)_2$. I. Solid, m. 152-2.5° (from benzene), IR.[235,237]
$Ph_2P(S)CH_2(S)PPh(CH_2NEt_2)$. I. Oil, which did not crystallize, IR.[235,237]
$Me_2P(S)CH_2CH_2(S)PMe_2$. IX. Solid, m. 260-3° (from EtOH).[293]
$Et_2P(S)CH_2CH_2(S)PEt_2$. I.[175] II.[389] Solid, m. 84-5° (from acetone),[175] m. 86.5°.[389]
$(c\text{-}C_6H_{11})_2P(S)CH_2CH_2(S)P(C_6H_{11}\text{-}c)_2$. I. Solid, m. 195-7° (from acetone).[175]
$Ph_2P(S)CH_2CH_2(S)PPh_2$. I.[2,175,181,245] VII.[243,244a] Solid, m. 196-8° (from acetone),[181] 205-7°,[243,244a] m. 224-5°,[2] 226-7° (from acetone/$CHCl_3$)[245] m. 229-31° (from $CHCl_3$/petrol. ether),[340] ^1H-NMR[2,63] ^{31}P -44.1 ppm (in $CHCl_3$)[245] (the previously reported shift of -79.2 ppm seems to be in error).[243] Adducts: $L{\cdot}NbBr_5$)[54a] $L{\cdot}TaBr_5$.[54a]
$Ph(H_2NCH_2CH_2)P(S)CH_2CH_2(S)PPh(CH_2CH_2NH_2)$. I. Solid, m. 107-9° (from ethylacetate).[183]
$Et_2P(S)CH_2OCH_2(S)PEt_2$. I. Solid, $b_{0.3}$ 199-200°, m. 103°.[389]
$Ph_2P(S)C(=NH)S(NH=)C(S)PPh_2$. (from Ph_2PH and NH_4SCN.) Crystals, m. 221-2°, IR, UV, ^1H-NMR.[290a]
$Et_2P(S)(CH_2)_3(S)PEt_2$. II.[389] III.[234] Crystals, m. 81.5°,[389] m. 95-6.5° (from petrol. ether),[234] IR, ^{31}P -53.5 ppm (in EtOH).[234]
$(c\text{-}C_6H_{11})_2P(S)(CH_2)_3(S)PEt_2$. I. Solid, m. 147-8° (from acetone).[181]
$MePhP(S)(CH_2)_3(S)PMePh$. I. Solid, m. 126-9° (from MeOH).[175]
$EtPhP(S)(CH_2)_3(S)PPhEt$. I. Solid, m. 139-40° (from MeOH).[173]
$Ph_2P(S)CH_2CH_2CH{\cdot}C_5H_{11}(S)PPh_2$. I. Solid, m. 114-6° (from AcOEt/hexane),[300] ^1H-NMR, ^{31}P -40.4 and -47.7 ppm (two different P).[300]
$Et_2P(S)(CH_2)_4(S)PEt_2$. III. Colorless crystals, m. 132-2.5° (from benzene/petrol.ether),[234] IR, ^{31}P -52.4 ppm (in $CHCl_3$).[234]
$(c\text{-}C_6H_{11})_2P(S)(CH_2)_4(S)P(C_6H_{11}\text{-}c)_2$. I. Solid, m. 214-5°

(from acetone).[189]

$MePhP(S)(CH_2)_4(S)PPhMe$. I. Solid, m. 138-41° (from MeOH).[175]

$EtPhP(S)(CH_2)_4(S)PPhEt$. I. Solid, m. 179-80° (from acetone).[173]

$(H_2NCH_2CH_2)PhP(S)(CH_2)_4(S)PPh(CH_2CH_2NH_2)$. I. Colorless crystals, m. 164-6° (from EtOH).[183]

$(c\text{-}C_6H_{11})PhP(S)(CH_2)_4(S)PPh(C_6H_{11}\text{-}c)$. I. Solid, m. 248-50° (from benzene).[173]

$(Et_2P(S)(CH_2)_5(S)PEt_2$. I. Solid, m. 111-2° (from acetone).[181]

$(c\text{-}C_6H_{11})_2P(S)(CH_2)_5(S)P(C_6H_{11}\text{-}c)_2$. Solid, m. 162-3° (from acetone).[181]

$EtPhP(S)(CH_2)_5(S)PPhEt$. I. Solid, m. 115-6° (from MeOH).[173]

$(c\text{-}C_6H_{11})PhP(S)(CH_2)_5(S)PPh(C_6H_{11}\text{-}c)$. I. Solid, m. 184-5° (from acetone).[173]

$MePhP(S)(CH_2)_6(S)PPhMe$. I. Solid, m. 130-50° (from MeOH).[175]

$EtPhP(S)(CH_2)_6(S)PPhEt$. I. Solid, m. 88-90° (from MeOH) and m. 163-4° (from acetone), racemate and meso form.[173]

$(c\text{-}C_6H_{11}PhP(S)(CH_2)_6(S)PPh(C_6H_{11}\text{-}c)$. I. Solid, m. 211-4° (from benzene).[173]

$Ph_2P(S)(CH_2)_3S(CH_2)_3S(CH_2)_3(S)PPh_2$. I. Viscous oil, which did not crystallize.[99]

E.3. Diphosphine Disulfides Containing Unsaturated Groups

trans-$Ph_2P(S)CH{=}CH(S)PPh_2$. I. Solid, m. 196-7°, IR, ^1H-NMR.[2]

cis-$Ph_2P(S)CH{=}CH(S)PPh_2$. I. Solid, m. 196-7°, IR, ^1H-NMR.[2]

4-$Ph_2P(S)CH{=}CH\text{-}C_6H_4\text{-}CH{=}CH(S)PPh_2$. I. VIII. Solid, m. 282-3° (dec.) (from EtOH or MeOH).[121]

$Ph_2P(S)C{\equiv}C(S)PPh_2$. I. Solid, m. 186° (from $CHCl_3$/EtOH).[130]

$Ph_2P(S)CH_2C{\equiv}C\ CH_2(S)PPh_2$. I. White crystals, m. 139-41° (from CH_2Cl_2), IR, Raman, mass spect.[200a]

$Ph_2P(S)C{\cdot}Ph{=}N\text{-}N{=}Ph{\cdot}C(S)PPh_2$. I. Solid, m. 198° (from dioxane).[160]

$(c\text{-}C_6H_{11})_2P(S)C(Ph){=}N\text{-}N{=}(Ph)C(S)P(C_6H_{11}\text{-}c)_2$. I. Solid, m. 213-4° (from benzene).[160]

E.4. Cyclic Phosphine Sulfides

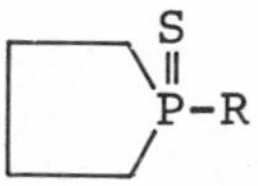

III. R = Me. Colorless liquid, $b_{0.01}$ 73-5°, n_D^{20} 1.5573, IR.[234]

I. R = cyclo-C_6H_{11}. Solid, m. 81° (from petrol. ether).[170]

I. R = Bu. Liq., $b_{0.013}$ 80-2°, n_D^{20} 1.5400, d_{20}^{20} 1.0274.[95]

I. R = Ph. Solid, m. 77° (from petrol. ether),[170] m. 73.5-4° (cyclohexane),[89] IR, mass. spect.[89]

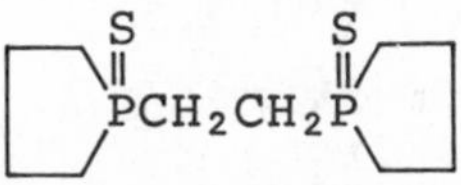

IX. Crystals, m. 174.5° (from EtOH).[346]

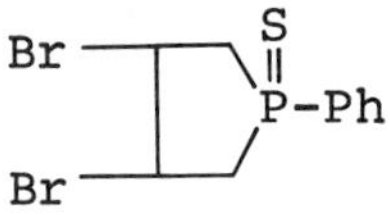

I. Crystals, m. 105-6°.[226]

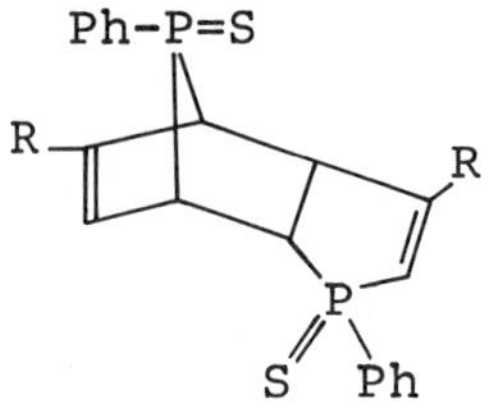

I. R = H. By addition of sulfur to phenylphosphol and by dehydrobromination of above sulfide, monomeric sulfide could not be obtained.[226] Crystals, m. 183-4°.[226]

I. R = Me. From 3-methyl-1-phenylphosphole and sulfur, monomeric sulfide not obtained. Crystals, m. 205-6°.[255]

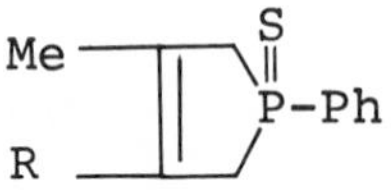

II. R = H. Solid, $b_{1.0}$ 173-5°, m. 69-70° (from $EtOH/H_2O$),[222] m. 69-72°.[347]

I. R = Me. Crystals, m. 77-8° (from hexane).[255]

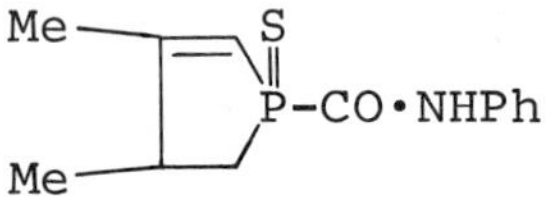

I. Solid, m. 72°.[361]

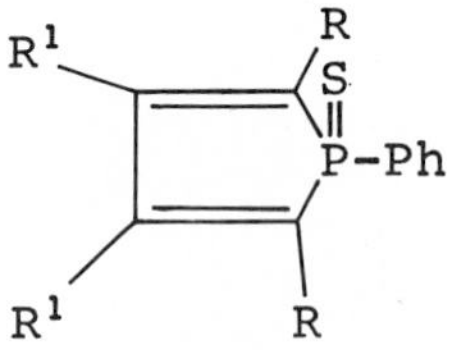

I. R = Me, R¹ = H. Crystals, m. 81-2°. ¹H-NMR.[226] This sulfide is stable and does not dimerize as stated in Chem. Abstr.

I. R = Ph, R¹ = H. Yellow crystals, m. 215-6.5° (from MeOH).[61]

I. R = H, R¹ = Me. Crystals, m. 113-4° (from $EtOH/H_2O$).[255]

I. III. R = R¹ = Ph. Yellow plates, m. 197-8° (from $CH_2Cl_2/EtOH$).[51]

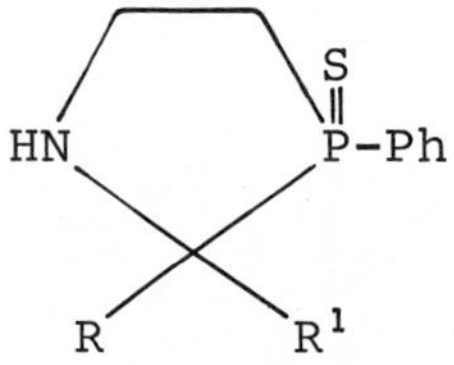

I. R = H, R' = Ph. Solid, m. 158-9° (from EtOH).[182,183]

I. R = R' = Me. Solid, m. 75° (from EtOH),[182,183] N-hydrochloride, m. 224-8° (from EtOH).[182,183]

I. R = Et, R' = Ph. Oil,[182] N-hydrochloride, m. 176-81° (from EtOH).[182,183]

I. R,R' = $-(CH_2)_4-$. Oil,[182] N-hydrochloride, m. 215-20° (from EtOH).[182,183]

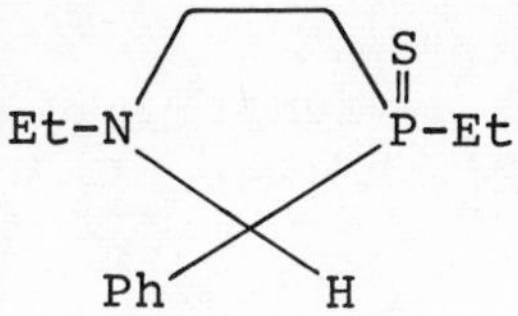

I. Solid, m. 74-6° (from EtOH).[184]

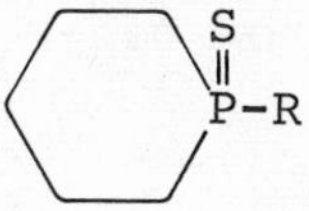

III. R = Me. Crystals, $b_{0.2}$ 113-30°, m. 51-2° (from petrol. ether), IR, ^{31}P -31.3 ppm (in $CHCl_3$).[234]
I. R = Et. Crystals, m. 67° (from petrol. ether).[170]
I. R = cyclo-C_6H_{11}. Crystals, m. 152-3° (from petrol. ether).[170,175]
I. R = Ph. Crystals, m. 86° (from petrol. ether),[170] m. 86°,[89] IR, mass spect.[89]

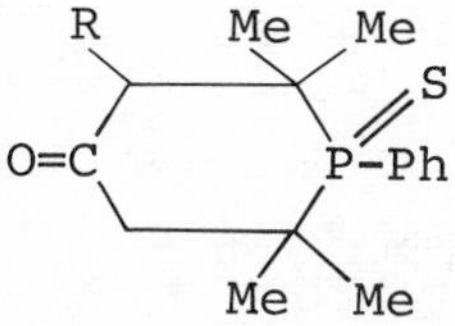

I. R = H. Crystals, m. 138.5-9° (from MeOH),[413] m. 138°,[17] 2,4-dinitrophenylhydrazone derivative, m. 225-7° (orange-red crystals).[17]
R = Br. From phosphine sulfide (R = H) and bromine, m. 155-6°.[17]

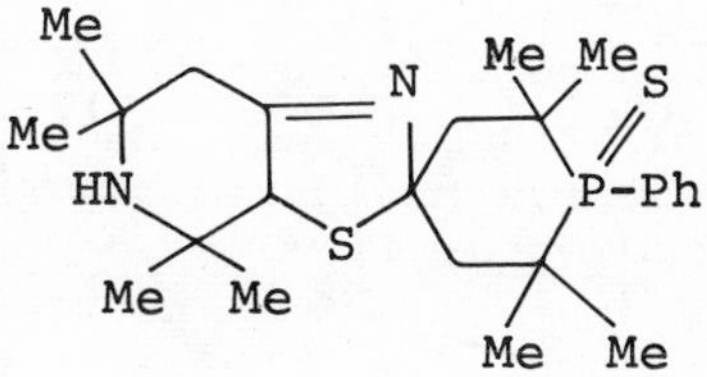

From phosphine sulfide above, NH_3 and 3-mercapto-2,2,6,6-tetramethyl-4-piperidone-hydrochloride. Crystals, m. 155-7°,[17] L•H picrate, dec. 185°.[17]

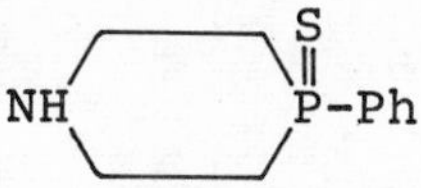

I. Colorless crystals, m. 114-5° (from benzene/petrol. ether).[183]

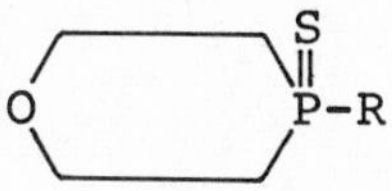

I. R = n-C_8H_{17}. Crystals, m. 35-6° (from pentane);[377]
R = n-$C_{12}H_{25}$. Crystals, m. 31-2° (from MeOH).[377]

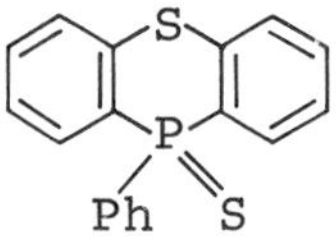

I. Solid, m. 200-1°, useful as fungicide.[52] 5,5'-Dioxide(⁄ SO_2 ⁄), solid, m. 215-6°.[52]

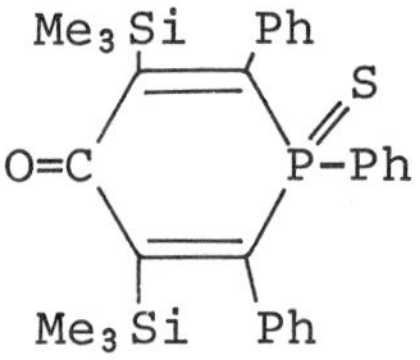

I. Crystals, m. 177-9.5° (from benzene).[225]

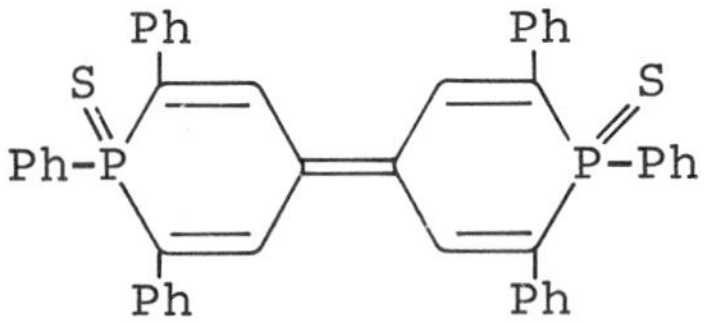

I. Crystals, m. 403-5°,[224] also obtained from tetrachloride

Cl_2C  $P(Ph)Cl_2$ and H_2S or

thioacetic acid.[224]

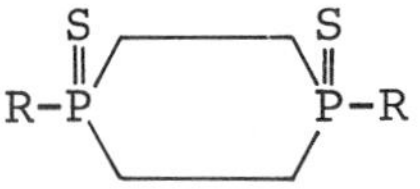

I. R = Et. Crystals, m. 225-35° (from MeOH) only one form isolated.[164]

I. R = cyclo-C_6H_{11}. Crystals, m. 250-5° (acetone/H_2O), and m. 325-6° (from acetone) (one form is cis, the other trans[164]).

I. R = Ph. Crystals, m. 154° (from MeOH/H_2O) and m. 253°,[187] m. 253-5°[191] (from acetone), one form is cis, the other trans.[187]

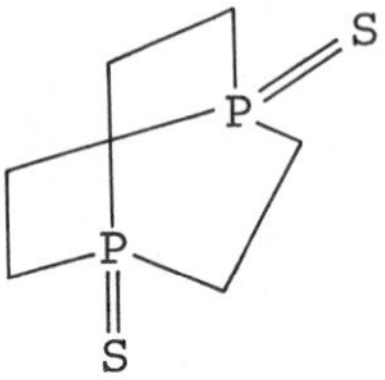

1/8 C_6H_6 I. Solid, dec. 400° without melting.[140]

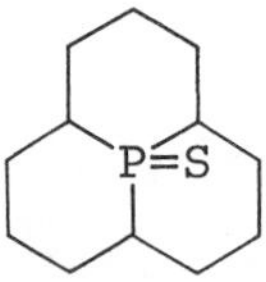

and

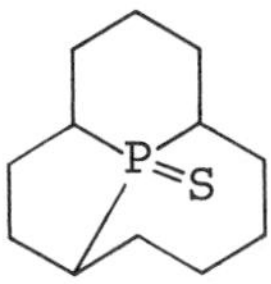

I. Solid, m. 177-8° (isomeric mixture).[254]

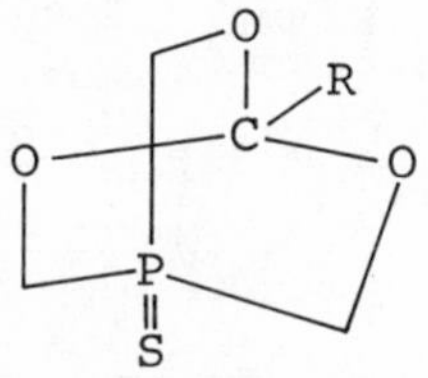

I. R = H. Colorless crystals, subl. at 40° under vacumm, ^{1}H-NMR.[49]
R = Me. Colorless crystals, subl. at 60° under vacuum, IR, ^{1}H-NMR.[49]

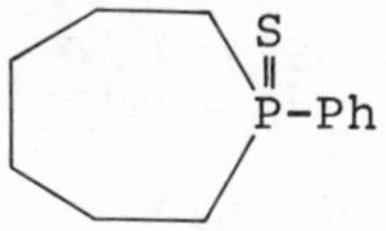

I. Crystals, m. 90.5-1.5° (from CH_2Cl_2),[89] IR, mass spect.[89]

E.5. Phosphine Sulfides Containing other P-Functions, more than two P=S Groups, or Metals

$(c\text{-}C_6H_{11})_2[(c\text{-}C_6H_{11})_2P(O)CH_2]PS$. I. Solid, m. 183° (from toluene).[159]
$Ph_2(Ph_2P(O)CH_2)PS$. I. Needles, m. 209° (from toluene),[159] m. 213-4° (from i-PrOH), IR.[360]
$4\text{-}Ph_2P(S)C_6H_4P(O)(OH)(Ph)$. I. Solid, m. 268-70°,[21] pK_a 3.52, σ_p [of $Ph_2P(S)$ group] +0.28,[21] σ^* +0.32.[21]
$Ph_2P(S)C(=NH)S(HN=)C(O)PPh_2$. (from $Ph_2P(O)H$ and NH_4SCN.) Colorless prism, m. 230-1°, UV, IR, ^{1}H-NMR, mass spect.[290a]
$(4\text{-}MeC_6H_4)_2P(S)C(=NH)S(HN=)C(O)P(C_6H_4Me\text{-}4)_2$. As above. Crystals, m. 226-7°, UV, IR, ^{1}H-NMR, mass spect.[290a]
$4\text{-}Ph_2P(S)C_6H_4P(OH)(O)C_6H_4P(S)Ph_2\text{-}4$ (OH on the central P). I. Solid, >300°.[21]
$Et_2[Ph_4P^{+}\,{}^{-}SC(S)]PS$. I. Dark reddish lilac crystals, m. 166-8°, IR.[85a]
$Ph_2[Ph_4P^{+}\,{}^{-}SC(S)]PS$. I. Red crystals, m. 131-3° (from $EtOH/CS_2$).[85a]
$Ph_2(Cl^-Ph_3P^+Ph\cdot CH)PS$. X. Solid, m. 268-70°,[177] BPh_4^- salt, m. 201-3°,[177] pK_a (in H_2O) 9.01.[177,178]
$Ph_2[Cl^-(Ph_3P^+_{\alpha})_2C]P_{\beta}S$. I. White solid, m. 308.5-10° (from EtOH),[36] IR, ^{1}H-NMR, ^{31}P -24.7 (P_α) and -44.0 (P_β) ppm (in $CDCl_3$).[36]
$(Cl^-Ph_3P^+CH_2)_3PS$. X. Solid, m. 188-91°.[179]
$Ph_2(Ph_3P{=}C\cdot Me)PS$. X. Orange in benzene solution.[177]
$Ph_2(Ph_3P{=}C\cdot Ph)PS$. X. Yellow solid, m. 203-5° (from benzene).[177]
$(Ph_3P{=}CH)_3PS$. X. Brown solid, m. 90-2°.[179]
$(4\text{-}Ph_2P(S)C_6H_4)_2PhPS$. I. Solid, m. 295-8° (from $CHCl_3$/EtOH).[20]
$N_3C_3[P(S)Bu_2]_3$(Triazine). I. Needles, m. 127-9° (from petrol. ether).[139]
$N_3C_3[P(S)Ph_2]_3$(Triazine). I. Yellow solid, m. 223-5° (from benzene).[139]
$C[CH_2P(S)Ph_2]_4$. I. Solid, m. 298-300°, IR.[101]

$Bu_2P(S)CHC_5H_{11}$. I. Oil, ^{31}P -58 ppm (in $CHCl_3$), ^1H-NMR.[300]
$[CH_2-CHP(S)Bu]_x$ with CO_2Me (Degree of polymerization unknown).[300]

$(PhN)_3B_3[CH_2CH_2P(S)Ph_2]_3$ (Borazol). I. Solid, m. 217-9° (from toluene).[356]

$MePh(Me_3SiCH_2)PS$. I. Solid, m. 55-6° (from benzene), ^1H-NMR, mass spect., ^{31}P -36.6 ppm.[303]

$Et_2[(EtO)_3SiCH_2CH_2]PS$. VI. Liq., b_2 137-40°[287] (gives on hydrolysis a phosphine sulfide substituted silicone).[287]

$Ph_2(Me_3SiCH_2)PS$. I. Solid, m. 89.5°.[56]

$Ph_2(Me_2ClSiCH_2CH_2)PS$. I. Solid, m. 95-6° (gives on hydrolysis a disiloxane).[292]

$Ph_2(MeCl_2SiCH_2CH_2)PS$. I. Solid, m. 97-100° (gives on hydrolysis a silicone).[292]

$Ph_2(Me_3SiC{\equiv}C)PS$. I. Colorless needles, m. 161-2° (from acetone), IR.[364]

$Ph_2(Ph_3SiC{\equiv}C)PS$. I. Solid, m. 153° (from acetone), IR.[364]

$(Me_3SiCH_2)_2PhPS$. I. Solid, m. 90-91.5° (from hexane),[303] mass spect., ^1H-NMR, ^{31}P -38.8 ppm.[303]

$(Me_3Si)_2CH(Me_3SiCH_2)PhPS$. I. ^1H-NMR, mass spect.[303]

$Ph_2Si[4-C_6H_4P(S)Ph_2]_2$. I. Solid, m. 305-8°.[20]

$Ph_2(Me_3GeC{\equiv}C)PS$. I. Solid, m. 173° (from acetone), IR.[364]

$Ph_2(Ph_3GeC{\equiv}C)PS$. I. Solid, m. 155° (from acetone), IR.[364]

$Ph_2(Ph_3SnCH_2)PS$. VIII. White crystals, m. 174-6° (from benzene/heptane),[358,359] IR.[358]

$Ph_2(Ph_3SnCH{\cdot}Me)PS$. I. VIII. White crystals, m. 229-30° (from $CHCl_3$), IR.[358]

$Ph_2(Me_3SnC{\equiv}C)PS$. I. Solid, m. 168-70° (from acetone), IR.[364]

$Ph_2(Ph_3SnC{\equiv}C)PS$. I. Solid, m. 133-4° (from acetone), IR.[364]

$[Me_2As(S)CH_2CH_2CH_2]_3PS$. I. Solid, m. 171-1.5° (from dimethylformamide).[30]

$Ph_2(2-Ph_2AsC_6H_4)PS$. III. Solid, m. 190-2° (from BuOH), IR.[284]
Adducts: L·AuCl (coordinates only via arsenic) in all other complexes with "Class b" metals it functions as a bidentate ligand (P=S and As) $L_2{\cdot}Pd(NO_3)_2$, yellow solid; $L{\cdot}PdCl_2$, yellow; $L{\cdot}PdBr_2$, orange; $L{\cdot}Pd(SCN)_2$, orange; $L{\cdot}Pd(SeCH)_2$, red-orange; $L{\cdot}PdI_2$, purple; $L{\cdot}PtCl_2$, yellow; $L{\cdot}PtBr_2$, yellow; $L_2{\cdot}Cu(ClO_4)$, colorless;[284] "Class a" metals form no complexes.

$(c-C_6H_{11})_2[Et_2Sb(S)(CH_2)_4]PS$. I. Solid, m. 125-6° (from benzene).[167]

E.6. Tertiary Phosphine Selenides

E.6.1 Type: R_3PSe

Me_3PSe. I.[59,322] Needles, m. 140.5-1° (from EtOH),[54]

m. 140°,[322] m. 84° (crude),[59] IR.[423] Adducts:[54] L_4CoX_2, X = ClO_4, Cl, Br, I, $L_4Ni(ClO_4)_2$, L_4ZnX_2, X = ClO_4, Cl, Br.[54]

Et_3PSe. I.[59 141a] Prisms, m. 112°[59] m. 108°,[400] IR,[100,423] Raman,[100] ^{31}P -45.8 ppm.[277] x-ray analysis[400] gave: P=Se 1.963 Å, P-C 1.907 Å, C-C 1.40 Å ∠Se-P-C 114°, P-C-C 110°, C-P-C 106°.[399,400]

Pr_3PSe. I. Colorless crystals, m. 32°, $b_{0.95}$ 116°,[426] IR.[423]

(cyclo-$C_3H_5)_3PSe$. I. Solid, m. 35° (purified by subl.) IR.[81]

Bu_3PSe. I. Pale yellow liq., $b_{0.1}$ 140°,[343] $b_{0.8}$ 150-1°, n_D^{27} 1.5150,[426] IR,[423] ^{31}P -37 ppm.[345] Adducts: $L{\cdot}I_2$,[427] $L{\cdot}CdI_2$, colorless solid,[282] $[Bu_3PSeMe]SbCl_6$, m. 117-9°,[343] ^{31}P of adduct -53.2 ppm (in Me_2SO).[345]

Am_3PSe. I. Pale yellow, extremely viscous liq., $b_{0.75}$ 158°, n_D^{25} 1.5055,[426] IR.[423]

$(c\text{-}C_6H_{11})_3PSe$. I. Solid, m. 192°,[354,428] IR.[423] Adducts: L•IBr, gold-brown, m. 176°,[428] $L{\cdot}I_2$.[427]

$(PhCH_2)_3PSe$. I. Needles, m. 256° (from AcOH).[258]

$(PhC{\equiv}C)_3PSe$. Two modifications, $P2_1/C$ and $R3_C$ ($R\bar{3}_C$).[272a]

$(4\text{-}MeC_6H_4C{\equiv}C)_3PSe$. I. Solid, m. 162° (dec.).[131]

Ph_3PSe. I.[263,266,283,285,354,428] VII.[223] Needles, m. 184-5,[263] m. 183-4°,[266] m. 185°,[223] m. 187-8° (from EtOH),[60,283,285,354,428] IR,[196,283,423] thermal stability,[25] P^{31} -35.0 ppm (in CH_2Cl_2),[343] -35.2 ppm (in CH_2Cl_2).[125a] Adducts: $L{\cdot}I_2$, dark maroon. m. 149°,[428] $L{\cdot}SnCl_4$, pale yellow solid, m. 168°,[24] $L_2{\cdot}AgClO_4$, colorless solid,[282] L•AuCl, colorless solid, m. 145°,[200] $L{\cdot}HgCl_2$, colorless solid, m. 231°,[200] x-ray analysis show that this complex is dimeric.[120] $L{\cdot}HgBr_2$, colorless solid, m. 260°,[200] $L{\cdot}HgI_2$, light yellow solid, m. 221°,[200,282] $L{\cdot}NbX_5$ (X=Cl,Bγ),[54a] $L{\cdot}TaX_5$ (X=Cl,Br),[54a] $L_2{\cdot}PdCl_2$, orange brown solid, m. 206°,[24,200] $L_2{\cdot}PtCl_2$, light tan solid, m. 201°,[200,282] gives on treatment with $Me_3O^+SbCl_6^-$ an onium salt $[Ph_3PSeMe]^+SbCl_6^-$, m. 136-7°, ^{31}P -46.6 ppm (in CH_2Cl_2)[343] and forms with SO_3 selenium, SO_2, and $Ph_3PO{\cdot}SO_3$.[29]

$(4\text{-}MeC_6H_4)_3PSe$. I.[260,283] Needles, m. 193° (from EtOH),[260] m. 198-8.5° (from EtOH),[283] IR.[283]

$(3\text{-}MeC_6H_4)_3PSe$. I. Solid, M. 139-40° (from EtOH), IR.[283] Adduct: $L_2{\cdot}PdCl_2$, pale orange.[282]

$(1\text{-}C_{10}H_7)_3PSe$. I. Solid, m. 272-3° (from BuOH or EtOH).[76] Adduct: $L{\cdot}I_2$.[76]

Tri(2-furyl)PSe. I. Solid, m. 174°, IR, UV,[289] ^1H-NMR, sign of coupl const.[193a]

Tri(ferrocenyl)PSe. I. Yellow needles, m. 297-9° (dec.).[368]

Tri(2-thienyl)PSe. I. Crystals, m. 150-1°, ^1H-NMR, sign of coupl. const.[193a]

Tri(3-thienyl)PSe. I. Crystals, m. 171-2°, ^1H-NMR, sign of coupl. const.[193a]

E.6.2. Type: $R_2R'PSe$

Me_2PhPSe. I. Yellow, highly viscous liq., $b_{0.9}$ 129°, n_D^{27} 1.6272,[426] IR.[423]
Et_2MePSe. III. Liq., $b_{0.4}$ 58-64° (crude product, contained Et_2MePO).[227]
Et_2PhPSe. I. Pale yellow, viscous liq., $b_{1.7}$ 149°, n_D^{27} 1.6086,[426] IR.[423]
Bu_2PhPSe. I. Colorless solid, m. 54°,[426] IR.[423]
$(c\text{-}C_6H_{11})_2PhPSe$. I. Colorless solid, m. 166° (from EtOH).[354]
Ph_2MePSe. I. Colorless liq., $b_{1.35}$ 200°, n_D^{26} 1.6780,[426] IR,[423] ^{31}P -22.9 ppm (in CH_2Cl_2).[125a]
Ph_2EtPSe. I. Colorless solid, m. 49°,[426] IR.[423]
$Ph_2(CH_2{=}CHCH_2)PSe$. I. Colorless solid, m. 78-9° (from EtOH).[283]
Ph_2BuPSe. I. Colorless solid, m. 63°,[426] IR.[423]
$Ph_2(Me_3SiCH_2)PSe$. I. Solid, m. 116°.[56]
$Ph_2(Me_2NCH_2CH_2CH_2)PSe$. I. Colorless solid, m. 90-1° (from EtOH), IR.[283]
$Ph_2(1\text{-}C_{10}H_7)PSe$. I. Solid, m. 200-1° (from BuOH or EtOH),[76] gives a charge transfer complex with iodine.[76]
$Ph_2(4\text{-}MeC_{10}H_6)PSe$. I. Solid, m. 198° (from BuOH or EtOH),[76] gives a charge transfer complex with iodine.[76]

E.6.3. Type: RR'R"PSe

$Ph(4\text{-}BrC_6H_4)(4\text{-}Me_2NC_6H_4)PSe$. I. Crystals, m. 135.5-6.5°.[90]

E.6.4. Di-, Tri-, and Tetraphosphine Selenides

$Ph_2P(Se)CH_2(Se)PPh_2$. I. ^1H-NMR.[63]
$Ph_2P(Se)CH_2CH_2(Se)PPh_2$. I. Colorless solid, m. 194-5° (from BuOH), IR,[283] ^1H-NMR.[283]
$[Ph_2P(Se)]_3C_3N_3$. I. Orange cubes, m. 232-4° (from benzene).[139]
$[Ph_2P(Se)CH_2]_4C$. I. Solid, m. 337° (dec.), IR,[101] seems to be intramolecularly associated.[101]

E.6.5. Cyclic Tertiary Phosphine Selenides

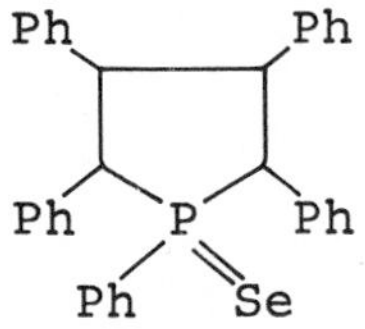

I. Yellow solid, m. 186-8°,[51] crystallizes from CH_2Cl_2 with 1 mole of solvent, which is lost at 100-120°;[51] does not react with maleic anhydride.

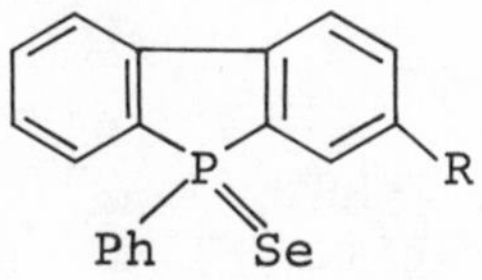

I. R = H. Crystals, m. 160-3°,[60] m. 162-4° (from C_6H_6),[61] gives with MeI Me_3SeI and R_3PI_2, isolated as R_3P=O.[60]
R = Me, Cream needles, m. 162.5-4°.[60,61]

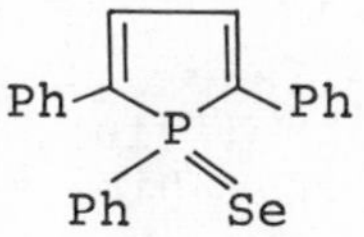

I. Yellow needles, m. 205.5-6.5° (from benzene).[61]

E.7. Tertiary Phosphine Tellurides

E.7.1. Type: R_3PTe

Me_3PTe. I. Yellow crystals (from toluene at -50°).[54] Exposure to air caused rapid dec. Reaction with cobalt or nickel salts in solution gave tellurium deposition.[54]
Et_3PTe. I. Pale, golden yellow crystals, m. 76-8° (from toluene/petrol. ether).[429]
Pr_3PTe. I. Soft, yellow crystals, m. 41-2° (from toluene/petrol. ether),[429] m. 45-6°, IR.[71a]
Bu_3PTe. I. Yellow crystals, m. 35.0-5.5° (from petrol. ether),[424,429] unstable when stored, deposits gray tellurium. IR.[71a]
Am_3PTe. I. IR.[71a]
(cyclo-C_6H_{11})$_3$PTe. I. Yellow crystals, m. 184-7° (from toluene/petrol. ether).[429] IR.[71a]
(n-C_8H_{17})$_3$PTe. I. IR.[71a]

E.7.2. Type: $R_2R'PTe$

Pr_2PhPTe. I. Yellow crystals, m. 67-71° (from toluene/petrol. ether).[429]
Bu_2PhPTe. I. Yellow crystals, m. 51.7-3.5° (from toluene/petrol. ether).[429]

REFERENCES

1. Aguiar, A. M., J. Beisler, and A. Mills, J. Org. Chem., 27, 1001 (1962).
2. Aguiar, A. M., and D. Daigle, J. Am. Chem. Soc., 86, 5354 (1964).
3. Aguiar, A. M., and D. Daigle, J. Org. Chem., 30, 2826 (1965).
4. Aguiar, A. M., J. Giacin, and A. Mills, J. Org. Chem., 27, 674 (1962).

5. Aguiar, A. M., and J. R. S. Irelan, J. Org. Chem., 34, 3349 (1969).
6. Aguiar, A. M., and J. R. S. Irelan, J. Org. Chem., 34, 4030 (1969).
7. Akamsin, V. D., and N. I. Rizpolozhenskii, Dokl. Akad. Nauk, SSSR, 168, 807 (1966); C.A. 65, 8953f (1966).
8. American Cyanamid Co., British Patent 907,497 (1962); C.A., 58, 3459d (1963).
9. American Cyanamid, British Patent 1,028,158 (1966); C.A. 65, 7388e (1966).
10. Appel, R., and A. Hauss, Chem. Ber., 93, 405 (1960).
11. Arbuzov, A., J. Russ. Phys. Chem. Soc., 42, 549 (1910).
12. Arbuzov, A. E., and K. V. Nikonorov, Zh. Obshch. Khim., 18, 2008 (1948); C.A., 43, 3801i (1949).
13. Arbuzov, B. A., N. I. Rizpolozhenskii, and M. A. Zvereva, Izv. Akad. Nauk, SSSR, Otd. Khim. Nauk, 1957, 179; C.A., 51, 11237f (1957).
14. Arbuzov, B. A., and G. M. Vinokurova, Izv. Akad. Nauk, SSSR, Otd. Khim. Nauk, 1963, 502; C.A., 59, 3947d
15. Arbuzov, B. A., G. M. Vinokurova, and I. A. Aleksandrova, Izv. Akad. Nauk, SSSR, Otd. Khim. Nauk, 1962, 290; C.A., 57, 15145c (1962).
16. Arbuzov, B. A., G. M. Vinokurova, I. A. Aleksandrova, and S. G. Fattakhov, Nekotor. Vopr. Organ. Khim. Sb., 1964, 244; C.A. 65, 3900g (1966).
17. Asinger, F., A. Saus, and E. Michel, Monatsh. Chem., 99, 1695 (1968).
18. Bacon, W. E., German Patent 1,211,639 (1966); French Patent 1,414,137 (1965); C.A. 64, 8239e (1966).
19. Baker, J. W., J. P. Chupp, and P. E. Newallis, U.S. Patent 3,052,596 (1963), C.A., 58, 6863e (1963).
20. Baldwin, R. A., and M. T. Cheng, J. Org. Chem., 32, 1572 (1967).
21. Baldwin, R. A., M. T. Cheng and G. D. Homer, J. Org. Chem., 32, 2176 (1967).
22. Baliah, V., and P. Subbarayan, J. Org. Chem., 25, 1833 (1960).
23. Balzer D., Tetrahedron Lett. 1968, 1189.
24. Bannister, E., and F. A. Cotton, J. Chem., Soc., 1960, 1959.
25. Baranauckas, C. F., R. D. Carlson, E. E. Harris and R. J. Lisanke, U.S. Dept. Comm., Office Techn. Serv., AD 263,891 (1961); C.A. 58, 3089e (1963).
26. Bartlett, P. D., E. F. Cox and R. E. Davis, J. Amer. Chem. Soc., 83, 103 (1961).
27. Bartlett, P. D., and R. E. Davis, J. Amer. Chem. Soc., 80, 2513 (1958).
28. Bartlett, P. D., and G. Meguerian, J. Amer. Chem. Soc., 78, 3710 (1956).
29. Becke-Goehring, M., and H. Thielemann, Z. Anorg. Allg. Chem., 308, 33 (1961).

30. Benner, G. S., W. E. Hatfield, and D. W. Meek, Inorg. Chem., 3, 1544 (1964).
31. Bennett, F. W., H. J. Emeléus, and R. N. Haszeldine, J. Chem. Soc., 1953, 1565.
32. Berlin, K. D., Chem. Ind. (London), 1962, 139.
33. Berlin, K. D., T. H. Austin, M. Nagabhushanan, M. Peterson, J. Calvet, L. A. Wilson, and D. Hopper, J. Gas Chromatogr., 3, 256 (1965); C.A. 63, 17155b (1965).
34. Berlin, K. D., and G. B. Butler, Chem. Rev., 60, 243 (1960).
35. Berlin, K. D., and D. G. Hopper, Proc. Okla. Acad. Sci., 46, 85 (1966); C.A. 67, 72941q (1967).
36. Birum, G. H., and C. N. Mathews, J. Amer. Chem. Soc., 88, 4198 (1966).
37. Blicke, F. F., and S. Raines, J. Org. Chem., 29, 204 (1964).
38. Bloomfield, P. R., U.S. Patent 3,044,984 (1962); C.A. 58, 1549c (1963).
39. Bloomfield, P. R., British Patent 983,698 (1965); C.A. 62, 13270f (1965).
40. Bogolyubov, G. M., Zh. Obshch. Khim., 35, 754 (1965); C.A. 63, 4327a (1965).
41. Bogolyubov, G. M., K. S. Mingaleva and A. A. Petrov, Zh. Obshch. Khim., 35, 1566 (1965); C.A. 63, 17860d (1965).
42. Bogolyubov, G. M., and A. A. Petrov, Zh. Obshch. Khim., 35, 704 (1965); C.A. 63, 4330a (1965).
43. Bogolyubov, G. M., and A. A. Petrov, Zh. Obshch. Khim., 36, 724 (1966); C.A. 65, 8954a (1966).
44. Bogolyubov, G. M., and A. A. Petrov, Zh. Obshch. Khim., 35, 988 (1965); C.A. 63, 9981h (1965).
45. Bogolyubov, G. M., and A. A. Petrov, Zh. Obshch. Khim., 36, 1505 (1966); C.A. 66, 10995y (1967).
46. Bogolyubov, G. M., and A. A. Petrov, Dokl. Akad. Nauk, SSSR, 173, 1076 (1967); C.A. 67, 90887e (1967).
46a. Bogolyubov, G. M., and Yu. N. Shlyk, Zh. Obshch. Khim., 39, 1759 (1969).
47. Boigt, D., M. Labarre, and L. Fournes, C. R. Acad. Sci. (Paris), Ser.C, 274, 1613 (1967).
48. Boor, J. Jr., and E. A. Youngman, Belgian Patent 637,449 (1964); C.A. 62, 7889h (1965).
49. Boros, E. J., R. D. Compton, and J. G. Verkade, Inorg. Chem., 7, 165 (1968).
50. Boskin, M. J., and D. B. Denney, Chem. Ind. (London), 1959, 330.
51. Braye, E. H., W. Hübel, and I. Caplier, J. Amer. Chem., Soc., 83, 4406 (1961), Chem. Ind. (London), 1959, 1250.
52. Braye, E. H., U.S. Patent 3,449,426 (1969); C.A. 71, 91,649y (1969).

53. Brodie, A. M., S. H. Hunter, G. A. Rodley, and G. J. Wilkins, J. Chem. Soc. (A), 1968, 2039.
54. Brodie, A. M., G. A. Rodley, and C. J. Wilkins, J. Chem.,Soc. (A), 1969, 2927.
54a. Brown, D., J. Hill, and C. E. F. Rickard, J. Less-Common Metals, 20, 57-65 (1970).
55. Buckler, S. A., and M. Epstein, J. Amer. Chem. Soc., 82, 2076 (1960).
56. Bugerenko, E. F., E. A. Chernyshev, and A. D. Petrov, Izv. Akad. Nauk, SSSR, Ser. Khim., 1965, 286; C.A. 62, 14721h (1965).
57. Burdon, J., I. N. Rozhkov, and G. M. Perry, J. Chem. Soc., (C), 1969, 2615.
58. Burg, A. B., and K. Gosling, J. Amer. Chem. Soc., 87, 2113 (1965).
58a. Burg, A. B., and D. K. Kang, J. Amer. Chem. Soc., 92, 1901 (1970).
59. Cahours, A., and A. W. Hofmann, Ann., 104, 1 (1857).
60. Campbell, I. G. M., R. C. Cookson, and M. B. Hocking, Chem. Ind.(London), 1962, 359.
61. Campbell, I. G. M., R. C. Cookson, M. B. Hocking, and A. N. Hughes, J. Chem. Soc., 1965, 2184.
62. Campbell, T. W., J. J. Monagle, and V. S. Foldi, J. Amer. Chem. Soc., 84, 3673 (1962).
63. Carty, A. J., and R. K. Harris, Chem. Commun., 1967, 234.
64. Cavell, R. G., and H. J. Emeléus, J. Chem. Soc., Suppl. 1, 1964, 5896.
65. Challenger, F., and D. Greenwood, J. Chem. Soc., 1950, 26.
66. Challenger, F., A. L. Smith, and F. J. Paton, J. Chem. Soc., 123, 1046 (1923).
67. Charrier, C., W. Chodkiewicz, and P. Cadiot, Bull. Soc. Chim. (France), 1966, 1002.
68. Charrier, C., M. P. Simonnin, W. Chodkiewicz, and P. Cadiot, C.R. Acad. Sci. (Paris), Ser. C, 258, 1537 (1964); C.A. 60, 10710f (1964).
69. Chemische Fabrik Kalk G.m.b.H., French Patent 1,400,892 (1965); C.A. 64, 2238f (1966).
70. Chernick, C. L., J. B. Pedley, and H. A. Skinner, J. Chem. Soc., 1957, 1851.
71. Chittenden, R. A., and L. C. Thomas, Spectrochim. Acta, 20, 1679 (1964).
71a. Chremos, G. N., and R. A. Zingaro, J. Organomet. Chem., 22, 637 (1970).
71b. Chremos, G. N., and R. A. Zingaro, J. Organomet. Chem., 22, 647 (1970).
72. Christen, P. J., L. M. Van der Linde, and F. N. Hooge, Rec. Trav. Chim. Pays-Bas., 78, 161 (1959).
73. Cölln, R., and G. Schrader, German Patent 1,141,990 (1963).

74. Cölln, R., and G. Schrader, German Patent 1,099,535 (1961).
75. Collie, N., J. Chem. Soc., 1888, 714.
76. Condray, B. R., R. A. Zingaro, and M. V. Kudchadkar, Inorg. Chem. Acta, 2, 309 (1968).
77. Copley, D. B., F. Fairbrother, and A. Thompson, J. Less-Common Metals, 8, 256 (1965); C.A. 62, 15746g (1965).
78. Corey, E. J., and R. A. E. Winter, J. Amer. Chem. Soc., 85, 2678 (1963).
79. Corey, E. J., and E. Block, J. Org. Chem., 34, 1233 (1969).
80. Coskran, K. J., and J. G. Verkade, Inorg. Chem., 4, 1655 (1965).
81. Cowley, A. H., and J. L. Mills, J. Amer. Chem. Soc., 91, 2915 (1969).
82. Crofts, P. C., and I. S. Fox, J. Chem. Soc. (B), 1968, 1416.
83. Crofts, P. C., and K. Gosling, J. Chem. Soc., 1964, 1527.
84. Crofts, P. C., and K. Gosling, J. Chem. Soc., 1964, 2486.
85. Culvenor, C. C. J., W. Davies, and N. S. Heath, J. Chem. Soc., 1949, 282.
85a. Dahl, O., N. C. Gelting, and O. Larsen, Acta Chem. Scand., 23, 3369 (1969).
85b. Dahl, O., and O. Larsen, Acta Chem. Scand., 23, 3613 (1969).
85c. Dahl, O., and O. Larsen, Acta Chem. Scand., 24, 1094 (1969).
86. Dalziel, J. A. W., A. F. C. Holding, and B. E. Watts, J. Chem. Soc. (A), 1967, 358.
87. Davidsohn, A., D. V. Howe, and E. T. Shawl, Inorg. Chem., 6, 458 (1967).
88. Davidson, R. S., J. Chem. Soc. (C), 1967, 2131.
89. Davies, J. H., J. D. Downer, and P. Kirby, J. Chem. Soc. (C), 1966, 245.
90. Davies, W. C., and F. G. Mann, J. Chem. Soc., 1944, 276.
91. Davies, W. C., and W. P. Walters, J. Chem. Soc., 1935, 1786.
92. Davis, R. E., J. Org. Chem., 23, 1767 (1958).
93. Davis, R. E., J. Phys. Chem., 63, 307 (1959).
94. Denney, D. B., and M. J. Boskin, J. Amer. Chem. Soc., 82, 4736 (1960).
95. Derkach, N. Ya., and A. V. Kirsanov, Zh. Obshch. Khim., 38, 331 (1968); C.A. 69, 96832n (1968).
96. Dittmer, D. C., and S. M. Kotin, J. Org. Chem., 32, 2009 (1967).
97. Dodonow, J., and H. Medox, Chem. Ber., 61, 1767 (1928).

98. Dörken, C., Chem. Ber., 21, 1505 (1888).
99. DuBois, T. D., and D. W. Meek, Inorg. Chem. 8, 146 (1969).
100. Durig, J. R., J. S. DiYorio, and D. W. Wertz, J. Mol. Spectrosc., 28, 444 (1968).
101. Ellermann, J., and D. Schirmacher, Chem. Ber., 100, 2220 (1967).
102. Elliott, D. E., and C. V. Banks, Anal. Chim. Acta, 33, 237 (1965).
103. Elliott, D. E., U. S. At. Energy Comm., IS-T-3, 51p. (1965); C.A. 63, 14018e (1965).
104. Emeléus, H. J., and J. M. Miller, J. Inorg. Nucl. Chem., 28, 662 (1966).
105. Ettel, V., and J. Horak, Collect. Czech. Chem. Commun., 26, 1949 (1961); C.A., 56, 5994d (1962).
106. Evans, M. B., G. M. C. Higgins, C. G. Moore, M. Porter, B. Saville, J. F. Smith, B. R. Trego, and A. A. Watson, Chem. Ind.(London), 1960, 897.
107. Evans, M. B., G. M. C. Higgins, B. Saville, and A. A. Watson, J. Chem. Soc., 1962, 5045.
108. Farnham, W. B., A. W. Herriott, and K. Mislow, J. Amer. Chem. Soc., 91, 6878 (1969).
109. Fayet, J. P., P. Mauret, M. C. Labarre, and J. F. Labarre, J. Chim. Phys., 65, 722 (1968).
109a. Fedorova, G. K., L. S. Moskalevskaya, A. V. Kirsanov, Zh. Obshch. Khim., 39, 1227 (1969).
110. Fehér, F., and D. Kurz, Z. Naturforsch., 24b, 1089 (1969).
111. Fehér, F., and D. Kurz, Z. Naturforsch., 23b, 1030 (1968).
112. Fild, M., Z. Anorg. Allg. Chem., 358, 257 (1968).
113. Fild, M., O. Glemser, and G. Christoph, Angew. Chem., 76, 953 (1964).
114. Fluck, E., German Patent 1,271,089 (1968); C.A. 69, 96872j (1968).
115. Gallagher, M. J., E. C. Kirby, and F. G. Mann, J. Chem. Soc., 1963, 4846.
116. Gerding, H., D. H. Zijp, F. N. Hooge, G. Blasse, and P. J. Christen, Rec. Trav. Chim. Pays-Bas, 84, 1274 (1965).
117. Gillham, H. C., and A. E. Sherr, U.S. Patent 3,341,625 (1967); C.A. 67, 100662u (1967).
118. Gillham, H. C., and A. E. Sherr, U.S. Patent 3,284,543 (1966); C.A. 66, 3201d (1967).
119. Gladshtein, B. M., B. L. Zakharov, V. I. Zakharova, and V. M. Zinin, U.S.S.R. Patent 225,186 (1968); C.A. 70, 20215x (1969).
120. Glasser, L. S. D., L. Ingram, M. G. King, and G. P. McQuillan, J. Chem. Soc. (A), 1969, 2501.
121. Gloyna, D., and H. G. Henning, Angew. Chem., 78, 907 (1966).

122. Goetz, H., F. Nerdel, and K. H. Wiechel, Ann., 665, 1 (1963).
123. Goetz, H., G. Nerdel, and E. Busch, Ann., 665, 14 (1963).
124. Goubeau, J., Angew. Chem., 81, 343 (1969).
125. Goubeau, J., and D. Köttgen, Z. Anorg. Allg. Chem., 360, 182 (1968).
125a. Grim, S. O., A. W. Yankowsky, S. A. Bruno, W. J. Bailey, E. F. Davidoff, and T. J. Marks, J. Chem. Eng. Data, 15, 497 (1970).
125b. Griffin, G. E., and W. A. Thomas, J. Chem. Soc. (B), 1970, 477.
126. Gutweiler, K., M. Sander, and C. Schneider, German Patent 1,131,412 (1962); C.A. 57, 12722h (1962).
127. Haake, P., R. D. Cook, and G. H. Hurst, J. Amer. Chem. Soc., 89, 2650 (1967).
128. Hantzsch, A., and H. Hibbert, Chem. Ber., 40, 1508 (1907).
129. Hartley, C., H. J. Kuhn, K. Plieth, and P. Zaeske, Naturwiss., 52, 12 (1965).
130. Hartmann, H., C. Beermann, and H. Czempik, Z. Anorg. Allg. Chem., 287, 261 (1956).
131. Hartmann, H., and A. Meixner, Naturwiss., 50, 403 (1963).
132. Harwood, H. J., and K. A. Pollart, U.S. Patent 3,053,900 (1962); C.A. 58, 1493c (1963).
133. Harwood, H. J., and K. A. Pollart, J. Org. Chem., 28, 3430 (1963).
134. Hecker, H., and F. Hein, Z. Anal. Chem., 174, 354 (1960); C.A. 55, 1297d (1961).
135. Hein, F., and H. Hecker, Z. Naturforsch., 11b, 677 (1956).
136. Hein, F., K. Issleib, and H. Rabold, Z. Anorg. Allg. Chem., 287, 208 (1956).
137. Hellmann, H., J. Bader, H. Birkner, and O. Schumacher, Ann., 659, 49, (1962).
138. Hendrickson, J. B., M. L. Maddox, J. J. Sims, and H. D. Kaesz, Tetrahedron, 20, 449 (1964).
139. Hewerton, W., R. A. Shaw, and B. C. Smith, J. Chem. Soc., 1964, 1020.
140. Hinton, R. C., and F. G. Mann, J. Chem. Soc., 1959, 2835.
141. Hitchcock, R. B., J. A. Dean, and T. H. Handley, Anal. Chem., 35, 254 (1963).
141a. Hofmann, A. W., Phil. Trans., 150, 409 (1860).
142. Hofmann, A. W., Ann.(Suppl.) 1, 1 (1861).
143. Hofmann, A. W., and F. Mahla, Chem. Ber., 25, 2436 (1892).
144. Hoffmann, H., R. Grünewald, and L. Horner, Chem. Ber., 93, 861 (1960).
145. Hoffmann, H., and P. Schellenbeck, Chem. Ber., 100 692 (1967).

146. Hooge, F. N., and P. J. Christen, Rec. Trav. Chim. Pays-Bas, 77, 911 (1958).
147. Horner, L., W. D. Balzer, and D. J. Peterson, Tetrahedron Lett. 1966, 3315.
148. Horner, L., P. Beck, and H. Hoffmann, Chem. Ber., 92, 2088 (1959).
149. Horner, L., and A. Gross, Ann., 591, 117 (1955).
150. Horner, L., and H. Hoffmann, Neuere Methoden der präparativen organischen Chemie, Verlag Chemie, 1960, Vol.II, p. 108.
151. Horner, L., H. Hoffmann, and P. Beck, Chem. Ber., 91, 1583 (1958).
152. Horner, L., and H. Oediger, Ann., 627, 142 (1959).
153. Horner, L., and F. Roettger, Korrosion, 16, 57 (1963).
154. Horner, L., and H. Winkler, Tetrahedron Lett., 1964, 175.
155. Horner, L., H. Winkler, A. Rapp, A. Mentrup, H. Hoffmann, and P. Beck, Tetrahedron Lett. 1961, 161.
156. Hudson, R. F., Structure and Mechanism in Organophosphorus Chemistry, Academic Press, New York, 1965, Chapter 3.
157. Humphrey, R. E., A. L. McCrary, and R. M. Webb, Talanta, 12, 727 (1965); C.A. 63, 6846h (1965).
158. Issleib, K., and H. Anhöck, Z. Naturforsch., 16b, 837 (1961).
159. Issleib, K., and L. Baldauf, Pharm. Zentralhalle, 99, 329 (1960); C.A. 55, 900f (1961).
160. Issleib, K., and A. Balszuweit, Chem. Ber., 99, 1316 (1966).
161. Issleib, K., and A. Brack, Z. Anorg. Allg. Chem., 277, 258 (1954).
162. Issleib, K., and A. Brack, Z. Anorg. Allg. Chem., 292, 245 (1957).
163. Issleib, K., and L. Brüsehaber, Z. Naturforsch., 20b, 181 (1965).
164. Issleib, K., and G. Döll, Chem. Ber., 96, 1544 (1963).
165. Issleib, K., and G. Döll, Z. Anorg. Allg. Chem., 324, 259 (1963).
166. Issleib, K., and G. Grams, Z. Anorg. Allg. Chem., 299, 58 (1959).
167. Issleib, K., and B. Hammann, Z. Anorg. Allg. Chem., 339, 289 (1965).
168. Issleib, K., and G. Harzfeld, Chem. Ber., 95, 268 (1962).
169. Issleib, K., and G. Harzfeld, Z. Anorg. Allg. Chem., 351, 18 (1967).
170. Issleib, K., and S. Häusler, Chem. Ber., 94, 113 (1961).
171. Issleib, K., and K. Jasche, Chem. Ber., 100, 3343 (1967).
172. Issleib, K., and K. Jasche, Chem. Ber., 100, 412 (1967).

173. Issleib, K., and F. Krech, Chem. Ber., 94, 2656 (1961).
174. Issleib, K., and F. Krech, Z. Anorg. Allg. Chem., 328, 21 (1964).
175. Issleib, K., K. Krech, and K. Gruber, Chem. Ber., 96, 2186 (1963).
176. Issleib, K., and R. Lindner, Ann., 699, 40 (1966).
177. Issleib, K., and R. Lindner, Ann., 707, 112 (1967).
178. Issleib, K., and R. Lindner, Ann., 707, 120 (1967).
179. Issleib, K., and M. Lischewski, J. Prakt. Chem., 311, 857 (1969).
180. Issleib, K., and H. M. Möbius, Chem. Ber., 94, 102 (1961).
181. Issleib, K., and D. W. Müller, Chem. Ber., 92, 3175 (1959).
182. Issleib, K., and H. Oehme, Tetrahedron Lett., 1967, 1489.
183. Issleib, K., and H. Oehme, Chem. Ber., 100, 2685 (1967).
184. Issleib, K., H. Oehme, R. Kümmel, and E. Leissring, Chem. Ber., 101, 3619 (1968).
185. Issleib, K., and E. Priebe, Chem. Ber., 92, 3183 (1959).
186. Issleib, K., and R. Roloff, Chem. Ber., 98, 2091 (1965).
187. Issleib, K., and K. Standtke, Chem. Ber., 96, 279 (1963).
188. Issleib, K., and G. Thomas, Chem. Ber., 93, 803 (1960).
189. Issleib, K., and G. Thomas, Chem. Ber., 94, 2244 (1961).
190. Issleib, K., and A. Tzschach, Chem. Ber., 92, 704 (1959).
191. Issleib, K., and K. Weichmann, Chem. Ber., 101, 2197 (1968).
192. Ivin, S. Z., K. V. Karavanov, and V. V. Lysenko, Zh. Obshch. Khim., 34, 852 (1964); C.A. 60, 15902f (1964).
193. Jakobsen, H. J., and J. A. Nielsen, J. Mol. Spectrosc., 31, 230 (1969); C.A. 71, 55294 (1969).
193a. Jakobsen, H. J., and J. A. Nielsen, J. Mol. Spectrosc., 33, 474 (1970).
194. Jenkins, J. M., and B. L. Shaw, J. Chem. Soc. (A), 1966, 770.
195. Jensen, K. A., Z. Anorg. Allg. Chem., 250, 268 (1943).
195a. Jensen, K. A., A. Holm, and E. Huge-Jensen, Acta Chem. Scand., 23, 2919 (1969).
196. Jensen, K. A., and P. H. Nielsen, Acta Chem. Scand., 17, 1875 (1963).
197. Johns, I. B., E. A. McElhill, and J. O. Smith,

J. Chem. Eng. Data, 7, 277 (1962).
198. Keen, I. M., J. Chem. Soc., 1965, 5751.
199. Kennedy, J., E. S. Lane, and J. L. Willians, J. Chem. Soc., 1956 4670.
200. King, M. G., and G. P. McQuillan, J. Chem. Soc. (A), 1967, 898.
200a. King, R. B., and A. Efraty, Inorg. Chem. Acta, 4, 123 (1970).
201. Kirsanov, A. V., A. F. Pavlenko, V. N. Chernyshev, P. P. Akkerman, T. V. Khimchenko, and Ya. N. Ivashchenko, U.S.S.R. Patent 229,506 (1968); C.A. 70, 58003q (1969).
202. Kolokol'tseva, I. G. V. N. Chistokletov, B. I. Ionin, and A. A. Petrov, Zh. Obshch. Khim., 38, 1248 (1968).
203. Kosolapoff, G. M., Organophosphorus Compounds, Wiley, New York, 1950.
204. Kotyanovskii, R. G., V. V. Yakshin, and S. L. Zimont, Izv. Akad. Nauk SSSR, Ser. Khim., 1968, 651; C.A. 69, 14096x (1968).
205. Kramolowsky, R., Angew. Chem., 81, 182 (1969).
206. Krasil'nikova, E. A., O. I. Korol, and A. I. Razumov, Tr. Kazansk. Khim. Tekhnol. Inst. No. 33, 171 (1964); C.A. 66, 10999m (1967).
207. Krasil'nikova, E. A., N. A. Moskva, and A. I. Razumov, Zh. Obshch. Khim., 39, 216 (1969); C.A. 70, 87910p (1969).
208. Krasil'nikova, E. A., A. M. Potapov, and A. I. Razumov, Zh. Obshch. Khim., 38, 609 (1968); C.A. 69, 43993t (1968).
209. Krasil'nikova, E. A., A. M. Potapov, and A. I. Razumov, Zh. Obshch. Khim., 38, 1101 (1968); C.A. 69, 59328r (1968).
210. Krasil'nikova, E. A., A. M. Potapov, and A. I. Razumov, Zh. Obshch. Khim., 38, 1556 (1968); C.A. 69, 87096g (1968).
211. Krespan, C. G., J. Amer. Chem. Soc., 83, 3432 (1961).
212. Kreutzkamp, N., J. Pluhatch, H. Schindler, and H. Kayser, Arch. Pharm., 295, 81 (1962); C.A. 57, 11229e (1962).
213. Kuchen W., and H. Buchwald, Chem. Ber., 91, 2871 (1958).
214. Kuchen, W., H. Buchwald, K. Strolenberg, and J. Metten, Ann., 652, 28 (1962).
215. Letcher, J. H., and J. R. Van Wazer, J. Chem. Phys., 44, 815 (1966).
216. Letts, E. A., and R. F. Blake, Trans. Roy. Soc. Edinburgh, 30, 181 (1883).
217. Letts, E. A., and R. F. Blake, J. Chem. Soc., 58, 766, (1890).
217a. Levin, Ya. A., and V. S. Galeev, U.S.S.R. Patent

253,800 (1969); C.A. 72, 67090j (1970).

218. Lewis, A. F., and L. J. Forrestal, ASTM, Spec. Techn. Publ., No. 360, 59 (1963); C.A. 62, 5389b (1965).

219. Lobanov, D. I., E. N. Tsvetkov, and M. I. Kabachnik, Zh. Obshch. Khim., 39, 841 (1969).

220. Lorenz, W., H. G. Schicke, and G. Schrader, Belgian Patent 616,096 (1962); C.A., 58, 13995a (1963).

221. Low, H., and P. Tavs, Tetrahedron Lett., 1966, 1357.

222. McCormack, W. B., U.S. Patent 2,663,738 (1953); C.A. 49, 7602 (1955).

222a. McFarlane, W., Org. Magn. Resonance, 1, 3 (1969).

223. McLean, R. A. N., Inorg. Nucl. Chem. Lett. 5, 745 (1969).

224. Märkl, G., and H. Olbrich, Angew. Chem., 78, 598 (1966), Int. Ed., 5, 588 (1966).

225. Märkl, G., and H. Olbrich, Tetrahedron Lett., 1968, 3813.

226. Märkl, G., and R. Ruthard, Tetrahedron Lett., 1968, 1755; abstract in C.A. 69, 63980m (1968) is wrong.

227. Maier, L., Chem. Ber., 94, 3043 (1961).

228. Maier, L., J. Inorg. Nucl. Chem., 24, 275 (1962).

229. Maier, L., J. Inorg. Nucl. Chem., 24, 1073 (1962).

230. Maier, L., in Progress in Inorganic Chemistry, F. A. Cotton, Ed., Interscience, New York, 1963, Vol. V, p. 27.

231. Maier, L., Helv. Chim. Acta, 47, 27 (1964).

232. Maier, L., Helv. Chim. Acta, 47, 120 (1964).

232a. Maier, L., Helv. Chim. Acta, 47, 2137 (1964).

233. Maier, L., in Topics in Phosphorus Chemistry, M. Grayson and E. J. Griffith, Eds., Interscience, New York, 1965, Vol. 2, p. 43.

234. Maier, L., Helv. Chim. Acta, 48, 133 (1965).

235. Maier, L., Helv. Chim. Acta, 48, 1034 (1965).

236. Maier, L., Helv. Chim. Acta, 49, 1249 (1966).

237. Maier, L., U.S. Patent 3,253,033 (1966); C.A. 65, 5488f (1966).

238. Maier, L., Z. Anorg. Allg. Chem., 345, 29 (1966).

239. Maier, L., DAS 1,235,911 (1967).

240. Maier, L., DAS 1,238,024 (1967); C.A. 67, 90934t (1967).

241. Maier, L., U.S. Patent 3,321,557 (1967); C.A. 67, 73684u (1967).

242. Maier, L., British Patent 1,101,334 (1968); C.A. 69, 3004g (1968).

243. Maier, L., Helv. Chim. Acta, 51, 405 (1968).

244. Maier, L., Chimia, 23, 323 (1969).

244a. Maier, L., U.S. Patent 3,489,803 (1970); C.A. 72, 79227x (1970).

245. Maier, L., unpublished results.

245a. Maier, L., Phosphorus, 1, 71 (1971).

245b. Maier, L., Helv. Chim. Acta, 54, 1651 (1971).

246. Malatesta, L., Gazz. Chim. Ital., 76, 182 (1946); C.A. 41, 947b (1947).
247. Malatesta, L., Gazz. Chim. Ital., 77, 509 (1947), C.A. 42, 5411i (1948).
248. Malatesta, L., Gazz. Chim. Ital., 77, 518 (1947); C.A. 42, 5413i (1948).
249. Malatesta, L., and R. Pizzotti, Gazz. Chim. Ital. 76, 167 (1946); C.A. 41, 2012a (1947).
250. Mann, F. G., and J. Watson, J. Org. Chem., 13, 502 (1948).
251. Mark, V., C. H. Dungan, M. M. Crutchfield, and J. R. Van Wazer, in Topics in Phosphorus Chemistry, Vol. 5, Chapter 4, 1967.
252. Martin, D. J., and C. E. Griffin, J. Org. Chem., 30, 4034 (1965).
253. Mashlyakovskii, L. N., and V. I. Ionin, Zh. Obshch. Khim., 35, 1577 (1965).
254. Mason, R. F., U.S. Patent 3,435,076 (1969); C.A. 71, 22187j (1969).
255. Mathey, F., C. R. Acad. Sci. (Paris), Ser. C, 269, 1066 (1969).
256. Meek, D. W., and P. Nicpon, J. Amer. Chem. Soc., 87, 4951 (1965).
257. Mel'nikov, N. N., K. D. Shvetsova-Shilovskaya, and T. L. Italinskaya, Zh. Obshch. Khim., 32, 847 (1962); C.A. 58, 1342b (1963).
258. Michaelis, A., Ann., 181, 265 (1876).
259. Michaelis, A., Ann., 181, 355 (1876).
260. Michaelis, A., Ann., 315, 43 (1901).
261. Michaelis, A., and J. Ananoff, Chem. Ber., 8, 493 (1875).
262. Michaelis, A., and L. Gleichmann, Chem. Ber., 15, 801 (1882).
263. Michaelis, A., and H. Köhler, Chem. Ber., 9, 519, 1053 (1876).
264. Michaelis, A., and W. La Coste, Chem. Ber., 18, 2109 (1885).
265. Michaelis, A., and A. Schenk, Ann., 260, 1 (1890).
266. Michaelis, A., and H. v. Soden, Ann., 229, 295 (1885).
267. Miller, J. M., J. Chem. Soc., (A) 1967, 828.
268. Moedritzer, K., U.S. Patent 3,360,556 (1967); C.A. 68, 49755t (1968).
269. Moedritzer, K., L. Maier, and C. D. Groenweghe, J. Chem. Eng. Data, 7, 307 (1962).
270. Monsanto Co., French Patent 1,369,608 (1964); C.A. 62, 16298e (1965).
271. Moore, C. G., and B. R. Trego, Tetrahedron, 19, 1251 (1963).
272. Moore, C. G., and B. R. Trego, Tetrahedron, 18, 205 (1962).

272a. Mootz, D., H. Altenburg, and D. Lücke, Z. Kristallogr., 130, 239 (1969).
273. Morgan, P. W., and B. C. Herr, J. Amer. Chem. Soc., 74, 4526 (1952).
274. Morris, R. C., and J. L. Van Winkle, U.S. Patent 2,642,461 (1953); C.A. 48, 8814a (1954).
275. Morrison, D. C., Abstracts of Papers, 134th ACS Meeting in Chicago, 1958, p. 87P.
276. Müller, E., H. Eggensperger, and K. Scheffler, Ann., 658, 103 (1962); Z. Naturforsch., 16b, 764 (1961).
277. Müller, N., P. C. Lauterbur, and J. Goldenson, J. Amer. Chem. Soc., 78, 3557 (1956).
278. Muslinkin, A. A., and M. S. Lapin, U.S.S.R. Patent 226,852 (1968); C.A. 70, 68893v (1969).
279. Muslinkin, A. A., M. S. Lapin, and A. V. Chernova, Izv. Akad. Nauk, SSSR, Ser. Khim., 1968, 152; C.A. 68, 78365w (1968).
280. Naumann, K., G. Zon, and K. Mislow, J. Amer. Chem. Soc., 91, 2788 (1969).
281. Nesmeyanov, N., V. D. Vil'chevskaya, N. S. Kochetkova, and N. P. Palitsyn, Izv. Akad. Nauk, SSSR, Ser. Khim., 1963, 2051; C.A. 60, 5548g (1964).
282. Nicpon, P., and D. W. Meek, Chem. Commun., 1966, 398.
283. Nicpon, P., and D. W. Meek, Inorg. Chem., 5, 1297 (1966).
284. Nicpon, P., and D. W. Meek, Inorg. Chem., 6, 145 (1967).
285. Nicpon, P., and D. W. Meek, Inorg. Syn., 10, 157 (1967).
286. Niebergall, H., and B. Langenfeld, Chem. Ber., 95, 64 (1962).
287. Niebergall, H., Makromol. Chem., 52, 218 (1962); German Patent 1,118,781 (1959); C.A., 56, 11622b (1962).
288. Nixon, J. F., and R. Schmutzler, Spectrochem. Acta, 22, 565 (1966).
289. Niwa, E., H. Aoki, H. Tanaka, and K. Munakata, Chem. Ber., 99, 712 (1966).
290. Novikova, Z. S., M. V. Proskurnina, L. I. Petrovskaya, I.V. Bogdanova, N. P. Galitskova, and I. F. Lutsenko, Zh. Obshch. Khim., 37, 2080 (1967); C.A. 68, 78392c (1968).
290a. Ojima, I., K. Akiba, and N. Inamoto, Bull. Chem. Soc. Japan, 42, 2975 (1969).
291. Oppegard, A. L., U.S. Patent 2,687,437 (1954); C.A., 49, 11000h (1955).
292. Owen, W. J., and F. C. Saunders, British Patent 1,007,333 (1965); C.A. 63, 18154g (1965).
293. Parshall, G. W., J. Inorg. Nucl. Chem., 14, 291 (1960).
294. Pascal, P., Bull. Soc. Chim., France, 33, 172 (1923).

295. Patel, N. K., and H. J. Harwood, J. Org. Chem., 32, 2999 (1967).
296. Pauling, L., The Nature of the Chemical Bond, Cornell University Press, Ithaca, N.Y., 1960.
297. Pellon, J., and W. G. Carpenter, J. Polym. Sci. (A), 1, 863 (1963); 1, 3561 (1963).
298. Peters, G., J. Amer. Chem. Soc., 82, 4751 (1960).
299. Peters, G., J. Org. Chem., 27, 2198 (1962).
300. Peterson, D. J., J. Org. Chem., 31, 950 (1966).
301. Peterson, D. J., J. Org. Chem., 32, 1717 (1967).
302. Peterson, D. J., J. Organometal. Chem., 8, 199 (1967).
303. Peterson, D. J., and J. H. Collins, J. Org. Chem., 31, 2373 (1966).
304. Peterson, D. J., and H. R. Hays, J. Org. Chem., 30, 1939 (1965).
305. Petrov, K. A., and V. A. Parshina, Zh. Obshch. Khim., 31, 3421 (1961); C.A. 57, 4693c (1962).
306. Petrov, K. A., V. A. Parshina, B. A. Orlov, and G. M. Tsypina, Zh. Obshch. Khim., 32, 4017 (1962); C.A. 59, 657c (1963).
307. Petrovskaya, L. I., M. V. Proskurnina, Z. S. Novikova, and I. F. Lutsenko, Izv. Akad, Nauk SSSR, Ser. Khim., 1968, 1277.
308. Philips, G. M., J. S. Hunter, and L. E. Sutton, J. Chem. Soc., 1945, 146.
309. Pollart, K. A., and H. J. Harwood, J. Org. Chem., 27, 4444 (1962); 28, 3430 (1963).
310. Popov, E. M., M. I. Kabachnik, and L. S. Mayants, Usp. Khim., 30, 846 (1961).
311. Potts, R. A., and A. L. Allred, Inorg. Chem., 5, 1066 (1966).
312. Potts, R. A., and A. L. Allred, J. Inorg. Nucl. Chem., 28, 1479 (1966).
313. Prikoszovich, W., and H. Schindlbauer, Chem. Ber., 102, 2922 (1969).
314. Pudovik, A. N., and E. S. Batyeva, Zh. Obshch. Khim., 39, 334 (1969).
314a. Pudovik, A. N., I. V. Gur'yanova, M. G. Zimin, and A. A. Sobanov, Zh. Obshch. Khim., 39, 2231 (1969).
315. Rabinowitz, R., R. Marcus, and J. Pellon, J. Polym. Sci. (A), 2, 1233 (1964).
316. Rabinowitz, R., R. Marcus, and J. Pellon, J. Polym. Sci. (A), 2, 1241 (1964).
317. Rabinowitz, R., and J. Pellon, J. Org. Chem., 26, 4623 (1961).
318. Rauhut, M. M., and H. A. Currier, J. Org. Chem., 26, 4626 (1961).
319. Rauhut, M. M., I. Hechenbleikner, H. A. Curier, F. C. Schaefer, and V. P. Wystrach, J. Amer. Chem. Soc., 81, 1103 (1959).
320. Razumov, A. I., O. A. Mukhacheva, I. V. Zaikonnikova,

N. N. Godovnikov, and N. I. Rizpolozhenskii, Khim. i. Primenenie Fosforoorgan. Soedin., Akad. Nauk, SSSR, Kazansk. Filial Tr. 1-Konf., 1955, 205; C.A. 52, 293e (1958).
321. Reiff, H. F., and B. C. Pant, J. Organomet. Chem., 17, 165 (1969).
322. Renshaw, R. R., and F. K. Bell, J. Amer. Chem. Soc., 43, 916 (1921).
323. Reuter, M., and F. Jakob, German Patent 1,056,125 (1959); C.A. 55, 10317f (1961).
324. Rizpolozhenskii, N. I., and V. D. Akamsin, Izv. Akad. Nauk, SSSR, Ser. Khim., 1969, 370.
325. Safe, S., and A. Taylor, Chem. Commun., 1969, 1466.
326. Sasse, K., in Houben-Weyl, Methoden der Organischen Chemie, E. Müller, Ed. Vol. XII 1963, "Organische Phosphorverbindungen," part 1.
327. Schiemenz, G. P., Tetrahedron Lett. 1964, 2729.
328. Schiemenz, G. P., Chem. Ber., 98, 65 (1965).
329. Schiemenz, G. P., Angew. Chem., 78, 145 (1966).
330. Schiemenz, G. P., Angew. Chem., 78, 777 (1966).
331. Schiemenz, G. P., Chem. Ber., 99, 504 (1966).
332. Schiemenz, G. P., Chem. Ber., 99, 514 (1966).
333. Schiemenz, G. P., Tetrahedron Lett., 1966, 3023.
334. Schiemenz, G. P., Angew, Chem., 80, 559 (1968).
334a. Schiemenz, G. P., Angew. Chem., 80, 559-560 (1968).
335. Schiemenz, G. P., and H. U. Siebeneick, Chem. Ber., 102, 1883 (1969).
336. Schindlbauer, H., Monatsh. Chem., 96, 1793 (1965).
337. Schindlbauer, H., Monatsh. Chem., 96, 2051 (1965).
338. Schindlbauer, H., Allg. Prakt. Chem., 18, 242 (1967).
339. Schindlbauer, H., Chem. Ber., 100, 3432 (1967).
340. Schindlbauer, H., L. Golser, and V. Hilzensauer, Chem. Ber., 97, 1150 (1964).
341. Schindlbauer, H., and F. Mitterhofer, Fresenius Z. Anal. Chem., 221, 394 (1966).
342. Schindlbauer, H., and W. Prikoszovich, Chem. Ber., 102, 2914 (1969).
343. Schmidpeter, A., and H. Brecht, Z. Naturforsch., 24b, 179 (1969).
344. Schmidpeter, A., and K. Düll, private communication.
345. Schmidpeter, A., private communication.
346. Schmutzler, R., Inorg. Chem., 3, 421 (1964).
347. Schmutzler, R., U.S. Patent 3,246,032 (1966); C.A. 64, 19678f (1966).
348. Schönberg, A., Chem. Ber., 68, 163 (1935).
349. Schönberg, A., and M. Z. Barakat, J. Chem. Soc., 1949, 892.
350. Schönberg, A., E. Frese, and K. H. Brosowski, Chem. Ber., 95, 3077 (1962).
351. Schönberg, A., and H. Knüll, Chem. Ber., 59, 1403 (1926).

352. Schuetz, R. D., and R. L. Jacobs, J. Org. Chem., 23, 1799 (1958).
353. Schweikert, W. W., and E. A. Meyers, J. Phys. Chem., 72, 1561 (1968).
354. Screttas, C., and A. F. Isbell, J. Org. Chem., 27, 2573 (1962).
355. Seyferth, D., in Progress in Inorganic Chemistry, F. A. Cotton, Ed., Interscience New York, 1962, Vol. III, p. 141.
356. Seyferth, D., Y. Sato, and M. Takamizawa, J. Organomet. Chem., 2, 367 (1964); Advan. Chem. Ser. 42, 259 (1964).
357. Seyferth, D., and W. Tronich, J. Amer. Chem. Soc., 91, 2138 (1969).
358. Seyferth, D., and D. E. Welch, J. Organomet. Chem., 2, 1 (1964).
359. Seyferth, D., D. E. Welch, and J. K. Heeren, J. Amer. Chem. Soc., 85, 642 (1963).
360. Seyferth, D., D. E. Welch, and J. K. Heeren, J. Amer. Chem. Soc., 86, 1100 (1964).
361. Shell International Research Maatschappij N. V. Belgian Appl. 654,329 (1965); C.A. 65, 8960h (1966).
362. Shlyk, Yu. N., G. M. Bogolyubov, and A. A. Petrov, Dokl. Akad. Nauk, SSSR, 176, 1327 (1967); C.A. 69 19261g (1968).
363. Shlyk, Yu. N., G. M. Bogolyubov, and A. A. Petrov, Zh. Obshch. Khim., 38, 193 (1968); C.A. 69, 67475b (1968).
364. Siebert, W., W. E. Davidsohn, and M. C. Henry, J. Organomet. Chem., 15, 69 (1968).
365. Simonnin, M. P., J. Organomet. Chem., 5, 155 (1966).
366. Simonsen, S. H., and B. Bowen, in press.
367. Smith, B. C., and G. H. Smith, J. Chem. Soc., 1965, 5516.
368. Sollot, G. P., and W. R. Peterson, J. Organomet. Chem., 4, 491 (1965).
369. Staudinger, H., and E. Hauser, Helv. Chim. Acta, 4, 861 (1921).
370. Staudinger, H., and W. Braunholtz, Helv. Chim. Acta, 4, 897 (1921).
371. Staudinger, H., and J. Meyer, Helv. Chim. Acta, 2, 612 (1919).
372. Staudinger, H., and J. Meyer, Helv. Chim. Acta, 2, 635 (1919).
373. Staudinger, H., G. Rathsam, and F. Kjelsberg, Helv. Chim. Acta, 3, 853 (1920).
374. Strating, J., L. Thijs, and B. Zwanenburg, Rec. Trav. Chim. Pays-Bas, 86, 641 (1967).
375. Strecker, W., and C. Grossmann, Chem. Ber., 49, 63 (1916).
376. Stuebe, C., W. M. LeSuer, and G. R. Norman, J. Amer.

Chem. Soc., 77, 3526 (1955).
377. Tavs, P., Angew. Chem., 81, 742 (1969).
378. Tefteller, W. Jr., and R. A. Zingaro, Inorg. Chem., 5, 2151 (1966).
379. Tefteller, W. Jr., R. A. Zingaro, and A. F. Isbell, J. Chem. Eng. Data, 10, 301 (1965).
380. Teichmann, H. M., Angew. Chem., 77, 809 (1965).
381. Tsivunin, V. S., G. Kh. Kamai, and R. Sh. Khisamutdinova, U.S.S.R. Patent 170,979 (1965); C.A. 63, 9990c (1965).
382. Tsivunin, V. S., G. Kh. Kamai, and R. Sh. Khisamutdinova, U.S.S.R. Patent 173,230 (1965); C.A. 64, 755h (1966).
383. Tsivunin, V. S., G. Kh. Kamai, and R. Sh. Khisamutdinova, Zh. Obshch. Khim., 35, 1815 (1965); C.A. 64, 2122f (1965).
384. Tsivunin, V. S., G. Kh. Kamai, R. Sh. Khisamutdinova, R. R. Shagidullin, and V. V. Kormachev, U.S.S.R. Patent 172,800 (1965); C.A. 64, 756d (1966).
385. Tsivunin, V. S., G. Kh. Kamai, and V. V. Kormachev, Zh. Obshch. Khim., 35, 1819 (1965); C.A. 64, 2122d (1966).
386. Tsivunin, V. S., G. Kh. Kamai, and V. V. Kormachev, Zh. Obshch. Khim., 35, 2190 (1965); C.A. 64, 11243e (1966).
387. Tsivunin, V. S., G. Kh. Kamai, and V. V. Kormachev, Zh. Obshch. Khim., 36, 271 (1966); C.A. 64, 15917h (1966).
388. Tsivunin, V. S., G. Kh. Kamai, V. V. Kormachev, and G. S. Ukader, U.S.S.R. Patent 187,782 (1966); C.A. 67, 54264r (1967).
389. Tsivunin, V. S., G. Kh. Kamai, V. V. Kormachev, and G. S. Ukader, Zh. Obshch. Khim., 36, 1436 (1966); C.A. 66, 11007t (1967).
390. Tsivunin, V. S., G. Kh. Kamai, R. R. Shagidullin, and R. Sh. Khisamutdinova, Zh. Obshch. Khim., 35, 1998 (1965); C.A. 64, 6684c (1966).
390a. Tsivunin, V. S., L. N. Krutskii, T. V. Zykova, and G. Kh. Kamai, Zh. Obshch. Khim., 39, 2666 (1969); C.A. 72, 90570z (1970).
391. Tsvetkov, E. N., D. I. Lobanov, and M. I. Kabachnik, Teor. i Eksp. Khim., Akad. Nauk. Ukr. SSR, 1, 729 (1965); C.A. 64, 12523f (1966).
392. Tsvetkov, E. N., D. I. Lobanov, and M. I. Kabachnik, Teor. i Eksp. Khim., Akad. Nauk. Ukr. SSR, 2, 458 (1966); C.A. 66, 28299r (1967).
394. Tsvetkov, E. N., D. I. Lobanov, and M. I. Kabachnik, Zh. Obshch. Khim., 38, 2285 (1968).
395. Tsvetkov, E. N., D. I. Lobanov, M. M. Makhamatkhanov, and M. I. Kabachnik, Tetrahedron, 25, 5623 (1969).

395a. Tsvetkov, E. N., M. M. Makhamatkhanov, D. I. Lobanov, and M. I. Kabachnik, Izv. Akad. Nauk SSSR, Ser. Khim., 1970, 178; C.A. 72, 110637g (1970).
395b. Tsvetkov, E. N., R. A. Malevannaya, D. I. Lobanov, N. G. Osipenko, and M. I. Kabachnik, Zh. Ohshch. Khim., 39, 2429 (1970).
396. Tsurugi, J., T. Nakabayaski, and T. Jshihava, J. Org. Chem. 30, 2707 (1965); 31, 861 (1966).
397. Van der Veer, W., and F. Jellinek, Rec. Trav. Chim. Pays-Bas, 85, 842 (1966).
398. Van Meerssche, M., and A. Leonhard, Acta Crystallogr., 12, 1053 (1959).
400. Van Meerssche, M., and A. Leonhard, Bull. Soc. Chim. Belg., 69, 45 (1960); C.A. 54, 16982e (1960).
401. Van Wazer, J. R., Phosphorus and its Compounds, Interscience, New York, 1958, Vol. I.
402. Vasil'ev, A. F., Izv. Akad. Nauk, SSSR, Ser. Fiz., 26, 1278 (1962); C.A. 58 7521h (1963).
403. Vaughan, L. G., and R. V. Lindsey, Jr., J. Org. Chem., 33, 3088 (1968).
404. Vinokurova, G. M., Zh. Obshch. Khim., 37, 1652 (1967); C.A. 68, 29798f (1968).
405. Vinokurova, G. M., and S. G. Fattaklov, Zh. Obshch. Khim., 36, 67 (1966); C.A. 64, 14209g (1966).
406. Vinokurova, G. M., and Kh. Nagaeva, Izv. Akad. Nauk, SSSR, Ser. Khim., 1967, 414; C.A. 67, 21977x (1967).
406a. Vinokurova, G. M., and S. G. Fattakhov, Izv. Akad. Nauk, SSSR, Ser. Khim., 1969, 1762.
407. Voskuil, W., and J. F. Arens, Rec. Trav. Chim. Pays-Bas, 81, 993 (1962).
408. Voskuil, W., Doctoral thesis, University of Utrecht, 1963.
409. Walling, C., and M. S. Pearson, J. Amer. Chem. Soc., 86, 2262 (1964).
410. Watson, A. A., J. Chem. Soc., 1964, 2100.
411. Weil, E. D., U.S. Patent 3,158,461 (1964); C.A. 62, 3335c (1965).
412. Weil, E. D., U.S. Patent 3,338,701 (1967); C.A. 67, 107588h (1967).
413. Welcher, R. P., and N. E. Day, J. Org. Chem., 27, 1824 (1962).
414. Wheatley, P. J., J. Chem. Soc., 1962, 300.
415. Wiley, R. A., and H. N. Godwin, J. Pharm. Sci., 54, 1063 (1965); C.A. 63, 13311c (1965).
416. Williams, D. H., R. S. Ward, and R. G. Cooks, J. Amer. Chem. Soc., 90, 966 (1968).
417. Wittig, G., and H. J. Cristau, Bull. Soc. Chim. France, 1969, 1293.
418. Wolfram, L. J., Congr. Int. Rech. Text. Lainiere 3rd, Paris, 2, 505 (1965); C.A. 67, 78941k (1967).
419. Worrall, D. E., J. Amer. Chem. Soc., 52, 2933 (1930).

420. Young, D. P., W. E. McEwen, D. C. Velez, J. W. Johnson, and C.A. VanderWerf, Tetrahedron Lett., 1964, 359.
421. Yust, V. E., and J. L. Bame, U.S. Patent 2,828,195 (1958); C.A. 52, 13244a (1958).
422. Zhdanov, G. S., V. A. Pospelov, M. M. Umanskii, and V. P. Glushkova, Dokl. Akad. Nauk, SSSR, 92, 983 (1953); C.A. 49, 12075a (1955).
423. Zingaro, R. A., Inorg. Chem., 2, 192 (1963).
424. Zingaro, R. A., J. Organomet. Chem., 1, 200 (1963).
425. Zingaro, R. A., and R. E. McGlothlin, J. Org. Chem., 26, 5205 (1961).
426. Zingaro, R. A., and R. E. McGlothlin, J. Chem. Eng. Data, 8, 226 (1963).
427. Zingaro, R. A., R. E. McGlothlin, and E. A. Meyers, J. Phys. Chem., 66, 2579 (1962).
428. Zingaro, R. A., and E. A. Meyers, Inorg. Chem., 1, 771 (1962).
429. Zingaro, R. A., B. H. Steeves, and K. Jrgolic, J. Organomet. Chem., 4, 320 (1965).
430. Zon, G., K. E. DeBruin, K. Naumann, and K. Mislow, J. Amer. Chem. Soc., 91, 7023 (1969).

Chapter 8. Halo- and Pseudohalophosphines

MANFRED FILD AND REINHARD SCHMUTZLER

Lehrstuhl B für Anorganische Chemie der Technischen Universität, Braunschweig, Germany

This chapter deals with phosphonous dihalides RPX_2, with phosphinous halides R_2PX, and with phosphonous dipseudohalides and phosphinous pseudohalides. The pseudohalides covered are cyanides, cyanates, thiocyanates and azides.

In the first part methods of preparation are given; these are followed by a section on the chemistry of these compounds and a general discussion of their physical properties. The compound list contains all the pertinent data up to the end of 1969.

A. SYNTHETIC ROUTES

I. REACTIONS OF PHOSPHORUS (III) HALIDES OR PHOSPHONOUS DIHALIDES WITH ORGANOMETALLIC COMPOUNDS

Organolithium- and Grignard- reagents normally react with phosphorus trihalides or phosphonous dihalides to give tertiary phosphines[155,567] (see also Chapter 1). It was not until recently that the successful synthesis of phosphonous dihalides and phosphinous halides by this method was reported. Using favorable conditions (e.g., low temperatures), highly branched alkyl ligands were inserted in a stepwise fashion. This was shown to occur with tertiary butyl chloride,[374] for example.

$$t\text{-BuMgCl} + PCl_3 \longrightarrow t\text{-BuPCl}_2 + MgCl_2$$

$$2t\text{-BuMgCl} + PCl_3 \longrightarrow t\text{-Bu}_2\text{PCl} + 2MgCl_2$$

1,1,2-Trimethylpropylphosphonous dichloride and other compounds were prepared in this way[141,374,487,559] in yields between 40 and 80%.

With a perfluorophenyl Grignard reagent, C_6F_5MgBr, a similar reaction took place. Depending on the ratio of

Grignard to PCl_3, either phosphonous dihalides or phosphinous halides were formed.[33,91,144,337] It was observed that a mixture of chloro- and bromo derivatives was formed, the bromides being the predominant products.[91,144] In the aromatic series 1-bromonaphthalene as the Grignard component in ether at -30°C gave 1-naphthylphosphonous dichloride.[196]

In another case a long-chain alkyl halide, n-$C_8H_{17}Br$ gave only the phosphinous halide[399] with either PCl_3 or PBr_3. It has been claimed, however, that reaction with BuMgCl, carried out at -70°C, gives the phosphonous dihalide.[428,595] Use of a silicon containing Grignard reagent, $(CH_3)_3SiCH_2MgCl$ and PCl_3 leads to the phosphonous dichloride.[517] Also a complex of a Grignard reagent with an oxygen or nitrogen heterocycle was said to give a phosphonous dichloride; $C_8H_{17}PCl_2$ and p-$C_{18}H_{37}C_6H_4PCl_2$ were cited as examples.[470]

In two cases lithium derivatives have been employed. Trifluoromethylphosphonous dichloride was reacted with n-Bu-lithium to give n-Bu$(CF_3)PCl$,[186] and carboranyl derivatives of phosphorus were synthesized as in the following example:[6,8,528,591]

$$LiCB_{10}H_{10}CLi + PCl_3 \longrightarrow Cl_2PCB_{10}H_{10}CPCl_2$$

Historically the most important synthetic route was by way of organomercury compounds. After some earlier attempts,[115,507] a number of aliphatic and aromatic phosphonous dichlorides have been synthesized by heating PCl_3 with dialkyl- or diarylmercury derivatives in sealed tubes for several hours at temperatures as high as 200-250°C,[35,209,357,382,383]

$$R_2Hg + PCl_3 \longrightarrow RPCl_2 + RHgCl$$

or

$$RHgCl + PCl_3 \longrightarrow RPCl_2 + HgCl_2$$

The reaction gives monochlorophosphines as by-products,[209] which can be somewhat repressed by using PX_3 in substantial excess. In turn, this route can be used for the synthesis of phosphinous chlorides having the same or different substituent.[227,391] Syntheses of a few phosphonous dibromides with organomercury compounds have been reported.[209,387] The use of PBr_3 results in the formation of the corresponding bromo derivatives,[289] but it is difficult to purify the products.[209]

This type of reaction is also useful for obtaining vinyl derivatives such as $CH_2{=}CH{-}PCl_2$[34,35,259] or $CH_2{=}CH{-}PBr_2$[36] from $(CH_2{=}CH)_2Hg$ and PCl_3 or PBr_3.

The alkyl- (or aryl-) mercury halides can be used instead of R_2Hg derivatives.[390,564] In spite of the difficulties associated with obtaining mercury-free preparations and purifying the products, the method remains valuable, especially for the determination of structures of compounds obtained from the Friedel-Crafts reaction.[378,382,383]

Grignard reagents may serve as precursors for organocadmium (R_2Cd) or for organozinc compounds (RZnBr). Fox[155] prepared a series of n-alkylphosphonous dichlorides in 26 to 47% yield by the reaction of PCl_3 and R_2Cd at -20°C.

$$R_2Cd + 2PCl_3 \longrightarrow 2RPCl_2 + CdCl_2$$

Satisfactory results, however, are only obtained with aliphatic compounds.[129,262,264,472,567,588,589] Organozinc derivatives, on the other hand, are useful for both aliphatic and aromatic phosphonous dichlorides and phosphinous chlorides,[106,565-567,582] RZnBr is employed rather than R_2Zn because the latter is much more reactive and gives exclusively tertiary phosphines.[381,566,567]

A more recent method uses aluminum trialkyls in the reaction with PCl_3 or phosphonous dichlorides,[38,212,427,593,594] failing only with $AlMe_3$ (because the methylphosphonous dichloride could not be separated from the reaction products[427]). The alkyl radical is not isomerized in the reaction, and yields can be improved if the aluminum compounds are first complexed with ether or pyridine.[427] Triethyl aluminum, which with excess PCl_3 yields $EtPCl_2$,[427] gives some Et_2PCl as well, but this forms a stable complex with the $AlCl_3$.[427]

For the preparation of aromatic phosphonous dihalides in high yields, arylaluminum halides that were generated from $ArSiCl_3$ and $AlCl_3$ have been employed.[585-587] The aluminum halide is separated by complexing it with $POCl_3$ as in the Friedel-Crafts reaction. The reaction with $LiAlR_4$ to synthesize phosphorus derivatives was also investigated.[38]

Ethyl phosphonous dichloride is conveniently prepared from Et_4Pb and PCl_3.[276-277a,472] The reagents are heated under reflux for 30 hr and the yield is quantitative.

$$Et_4Pb + 2PCl_3 \longrightarrow 2EtPCl_2 + Et_2PbCl_2$$

It was stated, however, that no reaction occurred with Ph_4Pb.[84] In a thorough investigation, the alkylation and arylation of phosphorus-chlorine compounds were examined and described for Me_4Pb, Et_4Pb, Ph_4Pb, and Ph_4Sn.[344] In most cases high yields were obtained. This reaction proceeds in a stepwise fashion and finally leads to the tertiary phosphine.[344] This synthetic route is especially useful for mixed compounds RR'PCl, and there are several examples of these.[36,195,197,266,278,344,350,600]

Syntheses of a few phosphonous dibromides from organolead derivatives have been reported.[36,344,480]

Organotin compounds were used for the synthesis of, for example, $EtPCl_2$.[480] In yet another case, the tin-carbon bond was cleaved to give phosphonous and phosphinous halides,[455,456] e.g.,

$$R_3SnCH_2COOCH_3 + PCl_3 \longrightarrow R_3SnCl + CH_3OCOCH_2PCl_2$$

Some displacement reactions with PCl_3 also occur with triphenylbismuthine and triphenyl-stibine but not with the related arsenic compound.[84]

II. REACTION OF AROMATIC COMPOUNDS WITH PHOSPHORUS (III) HALIDES WITH AND WITHOUT A CATALYST

1. WITH CATALYST. The Friedel-Crafts reaction has been used for the preparation of a vast number of organophosphorus compounds. It is still a versatile route to aryl-substituted derivatives. The basic reaction is represented by the following equation:

$$C_6H_6 + PCl_3 \xrightarrow{AlCl_3} C_6H_5PCl_2 + HCl$$

with $AlCl_3$, $FeCl_3$, or $ZnCl_2$ employed as catalysts. The reaction is conducted by refluxing a mixture of the aromatic compound with PCl_3 in the presence of, for example, anhydrous $AlCl_3$. The latter should be in slight molar excess.[64] It should be noted that it is essential for the $AlCl_3$ to be anhydrous if high yields are to be obtained. The notable exception is the case of alkyl ethers of phenol. To avoid substantial dealkylation, partially hydrated $AlCl_3$ or $FeCl_3$[557,558] is used; otherwise phenyldichlorophosphites are obtained.[316,467] Although PCl_3 has been extensively used, it is also possible to employ PBr_3.[328,329]

The procedure employed by Michaelis[382,383] called for long reaction times, but excellent conversions to dichlorophosphines (70 to 80%) are obtained in periods of 2 to 8 hr. Longer periods may well lead to substantial amounts of phosphinous halides as by-products. Only the polyhalogenated compounds require longer reaction periods.[125,303] Aromatic hydrocarbons containing several alkyl substituents yield diaryl derivatives as the major[383] or only product.[157] In one case even a triarylphosphine was found.[394]

The main difficulty in this method has been in the isolation of the products. Earlier procedures were very inefficient. The original work of Michaelis involved the extraction of the products with a hydrocarbon solvent.[382,383] After subsequent distillation and purification, yields were no higher than 25%.[260,380,382,383,450] A moderate improvement was the vacuum distillation of the entire reaction mixture.[203] The stability of the halophosphine-$AlCl_3$ complex caused the low yields so obtained; -its structure has not been definitely established, however, in contrast to analogous arsenic derivatives that have been studied.[336] The complex $PhPCl_2 \cdot AlCl_3$ was prepared and shown to take up chlorine to form $PhPCl_4 \cdot AlCl_3$, which was then converted to the phosphonic dichloride by hydrolysis.[47]

This problem of low yields was solved by the addition of $POCl_3$ to the mixture at the end of the reaction.[64,125] The $POCl_3$ forms a stronger complex with $AlCl_3$, which can be easily removed by filtration. With this principle yields of higher than 65% were common.[64,65] Only naphthalene failed to give a product because of a too-stable complex. Another variation is the decomposition of the complex with water, adding just the amount required to hydrolyze the complex but not the phosphonous dihalide,[125] otherwise the acid is formed.[300,301]

A base, such as pyridine, has also been employed to break down the complex.[125,167,171] In a more recent patent phosphoric acid was used to destroy the complex and high yields of $C_6H_5PCl_2$ were claimed[162] from the reaction of benzene, $AlCl_3$, and PCl_3. A pilot plant study for production of aromatic chlorophosphines using pyridine as complexing agent was published.[124]

A serious disadvantage of the Friedel-Crafts reaction is that mixtures of isomers are often obtained. Although this reaction has long been known[380,382,383] and improved upon, it was assumed without proof that the orientation of the entering group was para in monosubstituted benzene compounds.[65,170,173,248,253,257,408,449] However, toluene was found to give all three isomers, the orientation being in the ratio 10:27:63 for ortho:meta:para.[302] An extended fractional crystallization procedure of the corresponding acids has been used to establish this isomeric distribution.[302,303] A similar case was found in the reaction of ethylbenzene with PCl_3 and $AlCl_3$[157] to yield isomeric mixtures.

More recent investigations have used physical methods to determine the isomeric distribution. The reaction of ethylbenzene was studied using IR and VPC methods and it was shown that meta- and para isomers were formed in a ratio of 1.0:1.6.[25] The mixtures of arylphosphonous dichlorides were converted into tertiary phosphines with appropriate Grignard reagents, and these derivatives were investigated by VPC; the borane and BBr_3 addition products were prepared and analyzed.[25] The IR spectra of chlorobenzene revealed that the compound yields mostly the para isomer.[169] Other arylphosphonous dichlorides were also studied by this method, pure compounds for comparison being prepared by other routes.[490]

A further method was used to investigate the dichlorophosphination of toluene. The isomeric mixture was converted to fluorophosphoranes, and ^{19}F NMR spectroscopy was used to establish the ratio of the isomers. It was concluded that, for this case, para- and meta isomers were formed; no ortho isomer was detected.[499] Therefore, the literature must be viewed cautiously when a Friedel-Crafts reaction has been employed for substituted aromatic compounds.

Other limitations of this method have been reported. If the reaction is carried out in the presence of an aliphatic hydrocarbon containing a tertiary carbon atom, such as isopentane or methylcyclohexane, the phosphonous dihalide is found to be reduced to a primary phosphine. Thus $C_6H_5PCl_2$ in the presence of methylcyclohexane yields $C_6H_5PH_2$.[496] Furthermore, $AlCl_3$ tends to isomerize aromatic hydrocarbons, as was found in the reaction of amylbenzene with PCl_3. In addition to the expected products, 5% yield of tert-butylphenylphosphonous dichloride was detected.[64] It was also mentioned that alkyl substituents may be split off the aromatic ring.[64]

Failure to take part in the Friedel-Crafts reaction has been reported for stilbene,[384] anthracene,[397] aromatic ketones,[382] ethyl-naphthyl ether,[316] ethyl benzoate,[382] dibenzylaniline and pthalanil,[557] benzonitrile,[382] iodobenzene,[382] and trichlorobenzene.[303]

Woodstock[576] investigated a variation of the method just described, namely, the use of aliphatic hydrocarbons with six or more carbon atoms instead of the aromatic precursors. Other variations include the cyclization of phosphonous dichlorides in the presence of $AlCl_3$,[107,108] e.g.,

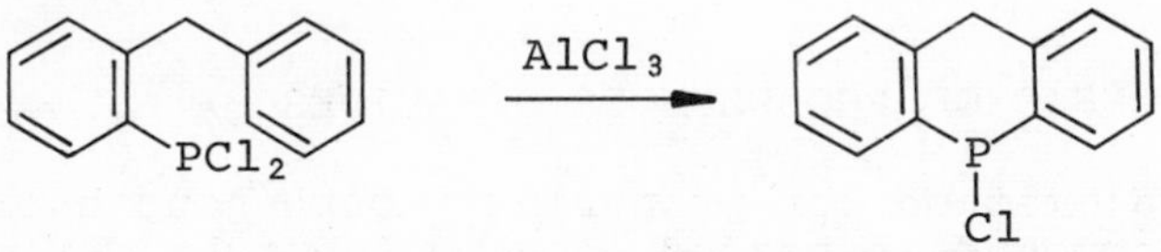

or the synthesis of 2-thienylphosphonous dichloride from thiophene and PCl_3 in the presence of $SnCl_4$. Unlike arylsilanes, which are split at the silicon-aryl bond in the presence of $AlCl_3$,[587] aliphatic-aromatic chlorosilanes, e.g.,

$$R_3Si\text{-}CH_2C_6H_5 + PCl_3 \xrightarrow{AlCl_3} R_3SiCH_2C_6H_4PCl_2$$

condense with PCl_3 and $AlCl_3$ to form silicon-substituted derivatives.[446] In one case polycondensation occurred between PCl_3 and 1,2-diphenylethane.[299]

Ferrocene and R_2NPCl_2 react in n-heptane and produce a variety of products such as $(Fc)_2PCl$ and $FcPCl_2$ (Fc = ferrocenyl-).[530,531] Catalysis and inhibition of this $AlCl_3$-catalyzed reaction were discussed.[531] It is worth mentioning that ligand exchange occurred when $(Fc)_2PNR_2$ was refluxed in the presence of $AlCl_3$ in the absence of ferrocene; the first compound produced was $(Fc)_2PCl$, which then gave $(Fc)_3P$ and $FcPCl_2$.[531] No previous example of such a disproportionation has been reported. Moreover, $(Fc)_2PCl$ does not give this reaction in the absence of $AlCl_3$.

2. WITHOUT CATALYST. Benzene reacts with PCl_3 when a mixture is passed through a hot tube at high temperatures (e.g., 600°C) using a recycling procedure.[375,378] Small amounts of phosphorus and biphenyl are obtained.[291] Numerous improvements on the original procedure of Michaelis have been described.[14,57,125,290,321,330,375,390,393,450] Other compounds employed were thiophene,[482] giving 2-thienylphosphonous dichloride in about 6% yield, and toluene,[382,383,387] giving only a trace of the expected halide, with many cracking products. A very elegant procedure was recently reported; a mixture of benzene and PCl_3 in carbon tetrachloride, refluxed for about 4 hr, was claimed to give 88% yield of $C_6H_5PCl_2$.[288]

Certain anilines can be made to react with PCl_3 in the absence of a catalyst. Dimethyl-aniline, for example, and PCl_3 give 60 to 70% yield of p-dimethylaminophosphonous dichloride.[42,55,471,557] The aniline must be used in excess to bind the hydrogen chloride that is formed. Other dialkylanilines have been used, including two N-alkylcarbazoles.[42,81]

III. PYROLYSIS OF PHOSPHONOUS DIHALIDES

As a special method for preparing phosphinous halides, the dihalide compound is heated in a sealed tube for many hours at about 300°C; it then undergoes ligand exchange to give the phosphinous halide and PCl_3,[14,16,43,59,113,127,176,386] as shown for $C_6H_5PCl_2$,

$$2C_6H_5PCl_2 \longrightarrow (C_6H_5)_2PCl + PCl_3$$

This method has also been used for fluoroalkylphosphonous diiodides. The noncatalytic process gives the halide in about 70% yield by continuous removal of PCl_3 in a stream of nitrogen when arylalkyl-PCl_2 is used.[30,31]

In the presence of catalytic amounts of $AlCl_3$ or $ZnCl_2$ the reaction proceeds at lower temperatures and in better yields.[61,62,234,254,269,282,413] It is necessary to use small amounts of $AlCl_3$, since the chlorophosphine forms a stable complex with $AlCl_3$. To avoid decomposition during the distillation over $AlCl_3$, it is advantageous to dissolve the monochloro compound in a solvent, such as petroleum ether and then to fractionate the solution.[234] It has also been stated that yields drop to 30 to 40% because of further disproportionation to phosphines when aryl-alkyl-PCl_2 is used.[30,31]

This reaction is not generally suitable for substituted derivatives because of isomerization,[384] but it has been employed for tolyl-, halo-, and xylyl-derivatives.[234,254,263,413] Vapor- phase pyrolysis of $C_6H_5PCl_2$ under controlled temperatures (400 to 700°C) and recycling conditions is another variation of this method.[90]

IV. FROM PHOSPHORUS HALIDES AND ALKANES OR CYCLOALKANES

It has been found that at high temperatures (575 to 600°C) and with short contact times, methane or ethane reacts with phosphorus trichloride to form alkylphosphonous dichlorides.[442,443] This reaction is catalyzed by oxygen and inhibited by propylene; a free radical mechanism was proposed.[443] Conversions are moderate (10 to 15%), but the yield, based on consumed trichloride phosphorus is as high as 80%.

Irradiation processes have been investigated for preparing organophosphorus derivatives.[22,222,223] Hydrogen chloride is eliminated and organophosphorus dichlorides are formed in the irradiation with high-energy electrons of mixtures of various hydrocarbons and PCl_3. The yields were often 40 to 50% of the reacted material, whereas conversions were in the order of 20 to 30%. Systems such as cyclohexane/PCl_3, n-heptane/PCl_3, hexene-1/PCl_3, cyclohexene/PCl_3, and ethylbenzene/PCl_3 were investigated.[222]

A free radical mechanism was proposed to explain the reaction course and the various by-products. Isomeric distribution was indicated for the reaction of PCl_3 with ethylbenzene.[222] This reaction is also covered in Section A.II.

In another study $C_6H_{11}{}^{32}PCl_2$ was produced by neutron irradiation of PCl_3 in the presence of cyclohexane, where the ^{32}P activity was induced by the $^{31}P(n, j)$ capture.[224] Further investigations using irradiation procedures and employing phosphorus as starting material[18] are discussed in Section A.XII.

The reaction of PCl_5 with benzene yields $C_6H_5PCl_2$,[527] with ethane $EtPCl_2$,[525,527] and with methane $MePCl_2$.[525]

The hydrocarbon was used in excess (10 to 40 moles) over the PCl_5 by reacting it in a quartz tube at about 500 to 650°C.[525]

V. HEATING PHOSPHORUS HALIDES WITH ORGANOHALIDES

It has been found that when phosphorus trihalides (PCl_3 and PBr_3) are heated to the vicinity of 200°C with alkyl halides (usually iodides), they undergo a reaction that produces mono- and dihalophosphines, as well as a host of other products including tertiary phosphines and the quaternary phosphonium halides. Usually phosphorus iodides give better results than the chlorides or the bromides. The halophosphines have not been isolated as such, but they have been oxidized to the corresponding phosphonic acids after hydrolysis. The dialkyl derivatives appear to be formed to the greatest extent when the molecular ratio of alkyl iodide to phosphorus halide is 3:1. The yields of pure products are very low.[19]

A more recent investigation showed that iodobenzene reacts with both PCl_3 and PBr_3 in the presence of silver powder under photolytic conditions to give $C_6H_5PCl_2$ and $C_6H_5PBr_2$, respectively,[56] but in very low yields.

A related synthetic route to phosphonous dihalides employing phosphorus trihalides, organohalides, and elemental phosphorus is discussed in Section A.XII.

VI. ADDITION OF PHOSPHORUS(III) HALIDES TO OLEFINS AND DIAZOALKANES

Phosphorus trichloride adds to olefins under free radical conditions. The products are β-chloroalkylphosphonous dichlorides. The free radicals can be generated by the

$$RCH{=}CH_2 + PCl_3 \longrightarrow RCHCl{-}CH_2PCl_2$$

decomposition of acetyl peroxide[153,273,274,325] or azobisisobutyronitrile,[598] or by UV irradiation.[153,175,319,325] The first two examples were the preparations of 2-chlorooctyl-PCl_2 from octene in 38% yield[273,274,598] and of 2-chloro-2-(3-cyclohexene-1-yl)ethyl-PCl_2 from 4-vinylcyclohexene in 17% yield.[319] More recent investigations describe the photochemical or thermal initiation of the addition of PCl_3 or PBr_3 to olefins; 1:1 adducts were formed with isobutylene and vinylcyclohexane.[332] Asymmetrical olefins yield both isomers.[153] Excellent separation of the products of the reaction may be achieved by gas chromatography.[521]

β-Chloroalkylphosphonous dichlorides have been

synthesized by γ-irradiation of mixtures of various olefins with PCl_3 at temperatures below 100°C, e.g., for cyclohexene,

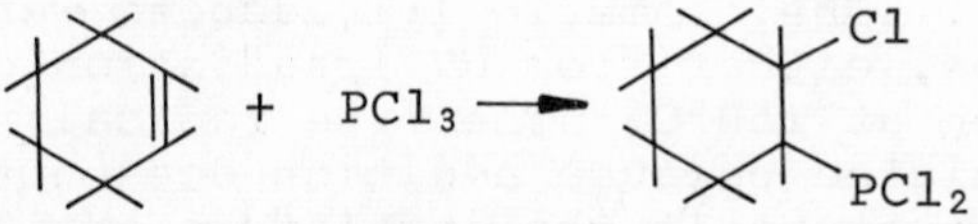

whereas at higher temperatures a different course of reaction is observed

$$2C_6H_{10} + PCl_3 \longrightarrow C_6H_9Cl + C_6H_{11}PCl_2$$

Both reactions are suggested to proceed by way of a radical mechanism.[476]

Several studies of the mechanism of this reaction have been conducted. Kharasch and co-workers suggested a chain-propagating sequence.[274]

$$\cdot PCl_2 + CH_2{=}CH{-}R \longrightarrow PCl_2{-}CH_2\dot{C}HR$$

$$PCl_2CH_2\dot{C}HR + PCl_3 \longrightarrow PCl_2CH_2CHClR + \cdot PCl_2$$

The $C_6H_5PCl_2$, which gives a more stable radical $[C_6H_5\dot{P}Cl]$, undergoes telomerization in the presence of styrene and 1,3-dienes.[364,366] Some doubt has been cast on this sequence by Errede and Pearson[128] as a result of their investigation of the copolymerization of PCl_3, PBr_3, PI_3, $C_6H_5PCl_2$, and $(C_6H_5)_2PCl$ with p-xylylene, which readily occurs to give polymers. In the case of PCl_3 copolymerization was rapid at -78°C, but at room temperature the main product was the corresponding alkylphosphonous dichloride. The investigators postulated that an intermediate phosphoranyl radical was formed which at low

$$R\cdot + PCl_3 \longrightarrow R\dot{P}Cl_3$$

$$R\dot{P}Cl_3 + R'\cdot \longrightarrow RPCl_3R'$$

temperatures coupled to the high polymer but at room temperature decomposed in two different ways.

$$R\dot{P}Cl_3 \longrightarrow RCl + \cdot PCl_2$$

$$R\dot{P}Cl_3 \longrightarrow RPCl_2 + Cl\cdot$$

Therefore they suggested that the reaction proceeds by addition of a hydrocarbon radical to the phosphorus atom.

Little and Hartmann[332] suggested from studies of

UV-induced reactions that both PCl_2 radicals and chlorine atoms are chain carriers. A similar sequence has been put forward recently by Fontal et al.[152] for the addition of PBr_3 to propene. The isomeric 1:1 adducts were of the same composition, either from UV irradiation or from thermal reaction at 150°C. The free radical nature reported is substantiated because addition does not occur at comparable temperature if the initiators are absent.

The addition of PCl_3 to ethylene has also been carried out in the presence of $AlCl_3$. After 30 hr at 35°C, a complex was formed; it was destroyed with potassium chloride to give $ClCH_2CH_2PCl_2$ and small amounts of $(ClCH_2)_2PCl$.[179,544] In the reaction between chloroethylene and PCl_3, substitution occurred to give $CHCl{=}CH\text{-}PCl_2$.[179,544] With PBr_3 and ethylene, in the presence of $AlBr_3$, $BrCH_2CH_2PBr_2$ was formed.[325] The gas-phase reaction of ethylene and PCl_3 has been reported to give $CH_2{=}CH\text{-}PCl_2$.[442]

At low temperature the addition of PCl_3 to diazoalkanes gives α-chloroalkylphosphonous dichlorides in 35 to 50% yield.[164,165,583,584]

$$RCHN_2 + PCl_3 \longrightarrow RCHClPCl_2 + N_2$$

Equally satisfactory results are obtained with PBr_3.[583 584]

VII. HALOGEN EXCHANGE

A very important method for synthesizing organophosphorus halides or pseudohalides is a halogen exchange reaction. Because of the ready availability of the chlorides, this exchange reaction has proved to be a useful procedure.

Phosphonous dibromides and phosphinous bromides are prepared by treating the corresponding chloride with anhydrous hydrogen bromide at reflux temperature.[370] The reaction may also be carried out with phosphorus tribromide serving as a solvent.[289,387,445,516]

$$RPCl_2 + 2HBr \longrightarrow RPBr_2 + 2HCl$$

$$R_2PCl + HBr \longrightarrow R_2PBr + HCl$$

A similar treatment of phosphonous dichlorides with hydrogen iodide at elevated temperatures yields the hydroiodic salts of the phosphonous diiodides.[177,321,378] These dark-colored compounds evolve hydrogen iodide on heating.

Instead of using hydrogen halides, the exchange reaction can be carried out with other covalent halides. From the chlorides and PBr_3 at 100 to 200°C, the corresponding bromides are obtained in high yields,[312,313] as has been shown for $C_6H_5PBr_2$ and $(C_6H_5)_2PBr$. According to Anderson,[9] these reactions occur whenever a more volatile component can be distilled from the reaction mixture. Thus $C_6H_5PCl_2$ reacts with $SiBr_4$ and $GeBr_4$ to give $C_6H_5PBr_2$, and with propyltriiodosilane to give $C_6H_5PI_2$.[9]

$$2C_6H_5PCl_2 + GeBr_4 \longrightarrow 2C_6H_5PBr_2 + GeCl_4$$

$$3C_6H_5PCl_2 + 2C_3H_7SiI_3 \longrightarrow 3C_6H_5PI_2 + 2C_3H_7SiCl_3$$

Phosphonous dichlorides have been converted to the diiodides using alkali iodides.[135,207] A more thorough investigation of this reaction[136] has shown that lithium iodide and aryldichlorophosphines in nonpolar solvents give mixtures of 1,2-diaryldiiododiphosphines and aryldiiodophosphines. When the reaction is conducted in ether solution, only the diphosphine derivatives and polymers of unestablished structure are formed. By reaction with iodine in benzene, the diaryldiiododiphosphines may be converted into the diiodophosphines.

Conversely, chlorides and bromides may be synthesized from the iodine derivatives with AgCl or AgBr, an important route for obtaining perfluorocarbon-substituted halides,[43,72,127] as the following reactions indicate:

$$(CF_3)_2PI + AgBr \longrightarrow (CF_3)_2PBr + AgI$$

$$C_3F_7PI_2 + 2AgCl \longrightarrow C_3F_7PCl_2 + 2AgI$$

The treatment of the halides R_2PX and RPX_2 with silver pseudohalides in inert, polar solvents at reflux temperature (or the use of alkali and ammonium pseudohalides) is the usual route to the cyanide, azide, isocyanate, and isothiocyanate derivatives.[43,127,139,143,219,251,341,345,382,383,431,432,445,541]

It is claimed that exchange occurs between $C_6H_5PCl_2$ and silicon derivatives of the type $R_xSi(NCO)_{4-x}$ and $R_xSi(NCS)_{4-x}$;[9] $C_6H_5P(NCO)_2$ has also been prepared by the action of HCNO on $C_6H_5PCl_2$ in the presence of an amine.[568]

The fluorides are obtained by reaction of antimony or arsenic trifluoride with the appropriate phosphonous dihalides or phosphinous halides.[71,315,339,417,500,501,536,537]

$$3CCl_3PCl_2 + 2SbF_3 \longrightarrow 3CCl_3PF_2 + 2SbCl_3$$

or

$$3(CF_3)_2PI + SbF_3 \longrightarrow 3(CF_3)_2PF + SbI_3$$

This reaction is only applicable to derivatives with electronegative groups on phosphorus (e.g., $ClCH_2$, CF_3), since in other cases a redox reaction occurs to give the fluorophosphorane. This redox reaction may be avoided by carrying out the synthesis in an amine medium such as pyridine. An intermediate, $SbF_3 \cdot 2L$ is formed which decomposes as the chlorine-fluorine exchange proceeds.[117,120]

A preferred method of fluorination is the use of alkali fluorides in an inert solvent.[146,499] Tetramethylenesulfone has been successfully used for the preparation of $C_6H_5PF_2$,[502] whereas CH_3PF_2 and $(CH_3)_2PF$ were prepared by exchange with highly active KF without solvent.[510-512]

VIII. HALOGENATION OF PHOSPHINES

The successful application of the direct halogenation of aliphatic phosphines was first reported by Walling.[561,562] Controlled addition of e.g. chlorine in an inert solvent at low temperatures (usually below 20°C) led to the corresponding chloro compound.

$$RPH_2 + 2X_2 \longrightarrow RPX_2 + 2HX$$

Further halogen addition has to be avoided because tetrachlorophosphoranes form. For example, butylphosphonous dibromide was thus obtained in 44% yield.[561] The halogenation with chlorine or bromine[63,68,148] of phosphines obtained from the addition of PH_3 to olefins is a convenient route to RPX_2 compounds.

Instead of free halogen, phosgene ($COCl_2$) can be used as the source of chlorine, according to the following equations:[385]

$$RPH_2 + 2COCl_2 \longrightarrow RPCl_2 + 2CO + 2HCl$$

$$R_2PH + COCl_2 \longrightarrow R_2PCl + CO + HCl$$

For example, $(C_6H_5)_2PCl$ was prepared in 75% yield by this method.[231] The reaction is generally applicable to alkyl-, aryl-, and cycloalkylphosphines, substituted or unsubstituted.[189,194,220,535] These compounds react preferably in an inert solvent between +8 and -80°C; yields are in the order of 70 to 80%.

A variation of this method is the cleavage reaction of triphenylphosphine with lithium in tetrahydrofuran and the subsequent conversion of the metallophosphine into the phosphinous chloride with phosgene.[238] The cleavage of an organometallic-substituted phosphine with bromine should also be mentioned, although this route is of little synthetic use.[60]

$$(CH_3)_3Ge\text{-}P(C_6H_5)_2 + Br_2 \longrightarrow (CH_3)_3GeBr + (C_6H_5)_2PBr$$

IX. FROM PHOSPHINOUS OR PHOSPHONOUS ACIDS

Phosphonous and phosphinous acids normally exist in the form of the tetracoordinate phosphorus structures, phosphinic acid, (RPH(O)(OH) and phosphine oxide $R_2PH(O)$ (see Chapters 10 and 11). Therefore, the reaction with $SOCl_2$ or PCl_5 gives phosphonic or phosphinic chlorides. Only in one case could a phosphinous acid be obtained.[515]

However, with PCl_3 it was possible to convert phosphonous acids into phosphonous dichlorides in yields of 60 to 80%. The first example was triphenylmethylphosphonous acid,[218] and the method has since been applied to other compounds.[54,156,404,414,415,460,529]

First attempts to use phosphinous acids (secondary phosphine oxides) failed.[156] Later, however, it was reported[460] that these compounds could also be converted with PCl_3. This reaction is applicable to both alkyl and aryl derivatives. The route has considerable synthetic potential, since secondary phosphine oxides are readily available from commercial phosphorus compounds and Grignard reagents.

$$(R'O)_2P(O)H \xrightarrow{RMgX} R_2P(O)H \xrightarrow{PCl_3} R_2PCl$$

For example, dibenzyl- and dioctylphosphinous chlorides were prepared in 76 and 57% yields from the corresponding oxides.[460]

The reaction involves the treatment of the starting material with excess PCl_3 at room temperature for several hours. The product may be obtained directly by distillation or by extracting it with a solvent and then performing distillation.

A tautomeric mechanism was suggested.[156]

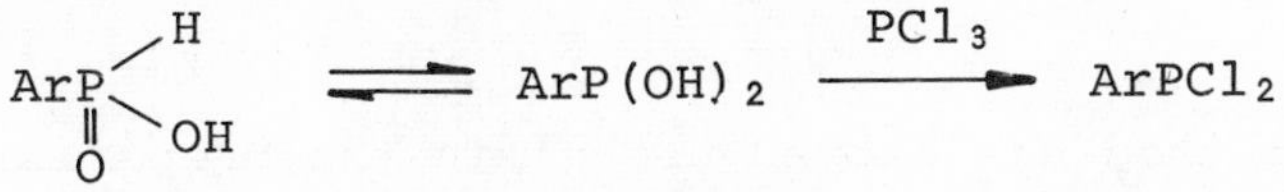

Further aspects of this tautomeric shift and alternative possibilities were discussed,[404] but no experimental evidence for the mechanism is available at present.

X. FROM PHOSPHONOUS DIESTERS AND PHOSPHINOUS ESTERS

The alkyl- or arylesters of phosphonous and phosphinous acids can be converted to the mono- or dihalophosphines by the reaction with PCl_3.[483-485] These reactions take place at low temperatures, for example, at 5°C.

$$R_2POR' + PCl_3 \longrightarrow R_2PCl + R'OPCl_2$$

or

$$RP(OR')_2 + 2PCl_3 \longrightarrow RPCl_2 + 2R'OPCl_2$$

This reaction can give high yields if an excess of PCl_3 is used, since the equilibrium is shifted to the right.

A similar type of cleavage of the phosphorus-oxygen bond was found to occur with hydrogen chloride using a phosphoxane. Thus $(CF_3)_2P\text{-}O\text{-}P(CF_3)_2$ is split into $(CF_3)_2PCl$ and $(CF_3)_2POH$.[199] This method has also been extended to the reaction of thioesters with acylhalides.[3,478]

$$C_2H_5P(SC_2H_5)_2 + 2RCOX \longrightarrow C_2H_5PX_2 + 2RCO(SC_2H_5)$$

XI. FROM PHOSPHINOUS AND PHOSPHONOUS AMIDES

Phosphinous and phosphonous amides are cleaved, in high yields, by anhydrous hydrogen halides in an inert solvent at temperatures of about 0°C.[73,78,80,91,129,144,200,235,242,320,500,536,537,560,590,592]

$$RP(NR_2)_2 + 4HCl \longrightarrow RPCl_2 + 2R_2NH\cdot HCl$$

or

$$R_2PNR_2 + 2HBr \longrightarrow R_2PBr + R_2NH\cdot HBr$$

Excess hydrogen chloride should be avoided because phosphinous chlorides may form adducts. Instead of using hydrogen chloride, it is also possible to cleave the phosphorus-nitrogen bond with PCl_3.[454,553] Mixed fluorochloro compounds have been prepared by this method, e.g., CH_3PFCl and C_6H_5PFCl.[116,121]

$$(CH_3)(F)P\text{-}NR_2 + 2HCl \longrightarrow (CH_3)(F)P\text{-}Cl + R_2NH\cdot HCl$$

The same method has proved useful in the preparation of unsaturated compounds,[52] although care must be taken to avoid an excess of hydrogen chloride, which may add to the double bond, e.g.,

$$C_6H_5\text{-}C{\equiv}C\text{-}P(NR_2)_2 \xrightarrow{4HCl} C_6H_5\text{-}C{\equiv}C\text{-}PCl_2$$

This method is of considerable importance, since the amides are available from the reaction of PCl_3 and amines and subsequent reaction with Grignard or organolithium reagents.

XII. FROM PHOSPHORUS HALIDES AND ORGANOHALIDES

The first synthesis of phosphonous and phosphinous halides from elemental phosphorus was the reaction with trifluoromethyl iodide at 200 to 220°C, under pressure, to yield the perfluoromethyl derivatives.[43,45]

$$P_w + CF_3I \xrightarrow{200\text{-}220°C} CF_3PI_2 + (CF_3)_2PI + (CF_3)_3P + P_xI_y$$

The yields depend on the ratio of the reactants, the temperature, and the reaction time. It is interesting to note that no phosphonium salt was detected. A radical mechanism was suggested.[43] A variation of the method was devised by using the silver salt of CF_3COOH in the presence of iodine to generate the CF_3I in situ.[77]

A further extension of the procedure was investigated by using perfluoropropyl iodide, which gave both iodine derivatives but no phosphine.[87,127] In contrast to this reaction course, a mixture of phosphorus and C_6F_5I under similar conditions gave only the phosphine $(C_6F_5)_3P$.[88]

Two other perfluoroiodophosphines have been synthesized in a similar way. The perfluoroolefin C_2F_4 reacts in the presence of iodine with red phosphorus to give two heterocyclic compounds, the octafluoro-1-iodophospholane and octafluoro-1,4-diiodo-1,4-diphosphorinane; a radical mechanism has been proposed.[307,308]

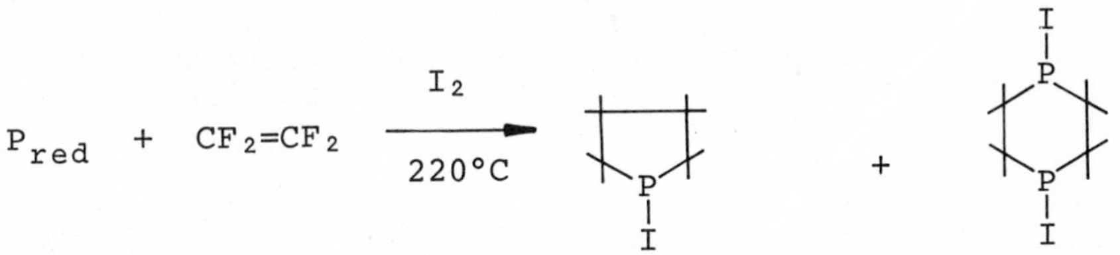

The disadvantage of working under pressure in sealed tubes or metal cylinders was overcome by alkylation of red phosphorus with CF_3I vapor over a copper catalyst.[343,345] The results indicated a different composition (i.e., less phosphine) and reduced yields.

The alkylation discussed previously is not limited to perfluoroalkyl iodides. The alkyl halides were passed through a mixture of red phosphorus and copper between 260 and 400°C to yield phosphonous dihalides and phosphinous halides.[343,345]

$$P_{red} + RX \xrightarrow{350°C} RPX_2 + R_2PX$$

With CH_3Cl, C_2H_5Cl, and CH_3Br, mainly the dihalides were obtained. Because of side reactions, the yields decrease if iodides are employed.

A somewhat better yield has been claimed in the reaction of methyl chloride with phosphorus and hydrogen in the gas phase over an active carbon catalyst.[286] The problem of separating CH_3PCl_2 and $(CH_3)_2PCl$ has been overcome by saturating the mixture with hydrogen chloride; the compounds then separate into two layers.[287]

The mechanism of this catalyzed reaction is thought to be a radical chain process. This hypothesis is supported by the isolation and analysis of various hydrocarbons in the CH_3Cl and $C_6H_5CH_2Cl$ reactions, in addition to the phosphonous dichlorides.[182]

A curious reaction has been found for chloromethylmethylether and red phosphorus: in the presence of CuCl at 360°C CH_3PCl_2 and $(CH_3)_2P(O)Cl$ were formed.[180,181]

The alkylation without catalysts has also been investigated. Reactions in the liquid phase with $C_6H_5CH_2Cl$, C_6H_5Br, 3-$CH_3C_6H_4Br$, and n-$C_8H_{17}Br$ have been reported to yield mainly phosphonous dihalides,[343,345,439,440] but appreciable amounts of phosphinous halides are also formed. Benzyl bromide gave the dibromide and the corresponding phosphonium salt,[543] which was also shown for methyl chloride and phosphorus between 200 and 400°C.[401]

Phosphonous dihalides are also synthesized as the exclusive product from alkyl or aryl halides and white phosphorus in the presence of phosphorus trihalide as solvent.

$$P_4 + 2PX_3 + 6RX \longrightarrow 6RPX_2$$

This reaction is catalyzed by various substances such as I_2, RI, and Br_2.[27-30,49,50,555] Phosphinous chlorides have also been obtained by this route, e.g.,

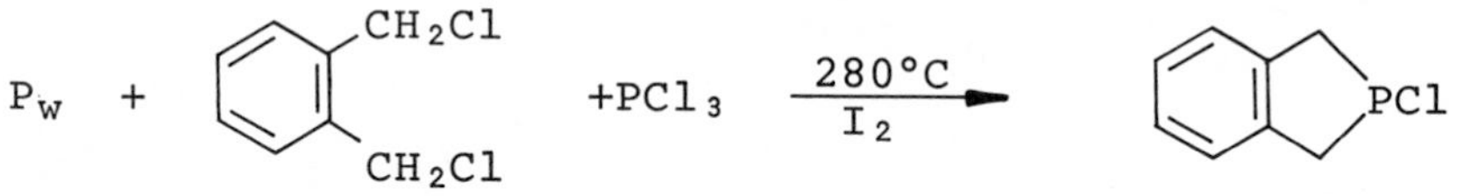

Heating white phosphorus and carbon tetrachloride gave CCl_3PCl_2 as the major product, no other carbon-phosphorus derivative being detected,[436] whereas bromoform at 190°C

$$P_w + CCl_4 \xrightarrow{\;157°C\;} CCl_3PCl_2 + PCl_3 + P_{red}$$

gave phosphonous dibromide and phosphinous bromide.[2] With bromotrichloromethane at 100°C, two diphosphines have been found in yields of about 40%.[1]

$$P_w + CCl_3Br \longrightarrow \underset{Br}{\overset{CCl_3}{>}}P-P\underset{Br}{\overset{CCl_3}{<}} + \underset{Br}{\overset{CCl_3}{>}}P-P\underset{Br}{\overset{Br}{<}}$$

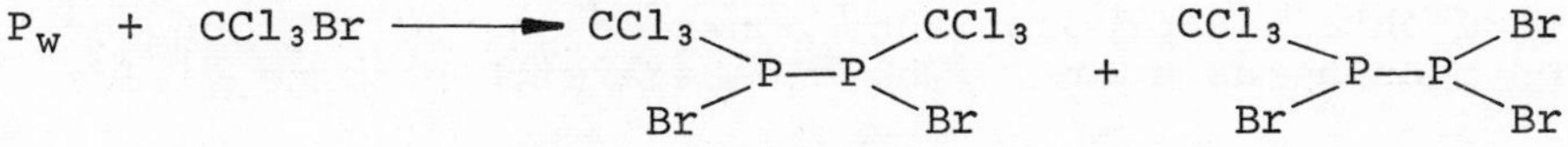

These reactions can also be initiated through γ-irradiation.[1,2]

The reaction of white phosphorus exposed to γ-rays in carbon tetrachloride solutions gives CCl_3PCl_2;[18,436] a radical mechanism is suggested. Similar results have been obtained using visible light.[1,2,18,436] Henglein et al.[224] found that neutron irradiation of the system CCl_4/P_w yields $CCl_3{}^{32}PCl_2$.

XIII. BY CLEAVAGE OF THE PHOSPHORUS-PHOSPHORUS BOND

The phosphorus-phosphorus bond in diphosphines and cyclopolyphosphines can be cleaved with various reagents. This method is related to the one described in the preceding section, since there the phosphorus-phosphorus bond in elemental phosphorus is split. In this section an organic substituent is already attached to phosphorus.

The diphosphines react with an equivalent amount of bromine (or chlorine or iodine) to form the corresponding phosphinous halides, e.g.[72,229,241-243,309,346,556]

$$Ph_2P-PPh_2 + Br_2 \longrightarrow 2Ph_2PBr$$

Excess halogen must be avoided to prevent the formation of phosphoranes. Yields are very high in all cases. This cleavage reaction with bromine was extended to compounds such as $R_2P-P(S)R_2$, where two compounds were obtained: R_2PBr and $R_2P(S)Br$.[346]

The cyclopolyphosphines react in the same way to form phosphonous dihalides,[229,241-243,293,310-312,338,340]

$$(CF_3P)_4 + 4I_2 \longrightarrow 4CF_3PI_2$$

In two cases cleavage reactions have been undertaken with hydrogen chloride rather than halogen, which is probably not a useful synthetic method. The diphosphines $(CH_3)_2PP(CH_3)_2$ and $(CF_3)_2PP(CF_3)_2$ gave the phosphinous chlorides and the corresponding phosphines.[75,76,425] The cleavage of the above-mentioned perfluoro compound with CF_3I and CH_3I has also been investigated. In both cases the phosphinous iodide and the corresponding tertiary phosphine were isolated.[97]

$$(CF_3)_2PP(CF_3)_2 + CF_3I \longrightarrow (CF_3)_2PI + (CF_3)_3P$$

From γ-irradiation of a mixture of white phosphorus and CCl_4 we obtain a product that contains one CCl_3 group per seven phosphorus atoms. The cleavage with bromine yields CCl_3PBr_2.[436]

XIV. THERMAL DECOMPOSITION OF PHOSPHORANES

Dichloro- and trichloroorganophosphoranes of the type R_3PCl_2 and R_2PCl_3 decompose at elevated temperatures, as shown in the following equations:

$$R_3PCl_2 \longrightarrow R_2PCl + RCl$$

$$R_2PCl_3 \longrightarrow RPCl_2 + RCl$$

This decomposition was observed a long time ago[89,396] and has been used for preparative purposes.[216,307,353,445] The heating and subsequent distillation, preferably carried out under an inert atmosphere, give yields of 40 to 80%. The chlorophosphoranes, especially the dichloro compounds, are readily available from chlorination of the corresponding phosphines or phosphine oxides.

Similar observations were made for bromophosphoranes containing perfluoromethyl groups[72] which decompose to trivalent bromophosphines according to the following example:

$$(CF_3)_2PBr_3 \longrightarrow CF_3PBr_2 + CF_3Br$$

Further interaction between $(CF_3)_2PBr_3$ and the phosphonous dibromide yields additional products, so that in this case it is not a useful method of preparation.

The decomposition of a cyclic phosphorane has also been investigated. It was shown that CH_3PCl_2 forms an adduct with norbornadiene which rearranges on heating.[192]

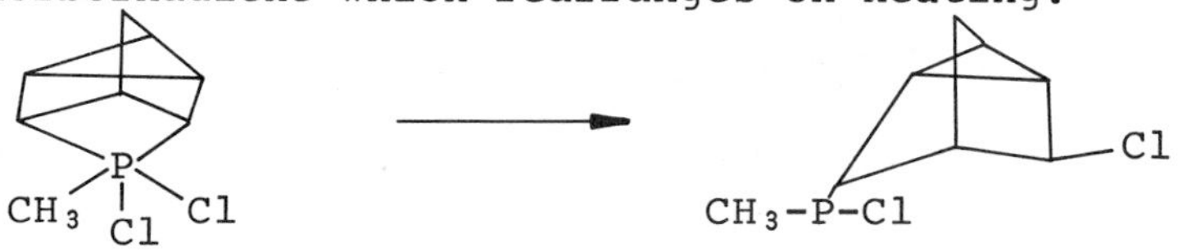

XV. FROM PHOSPHONIC DIHALIDES, PHOSPHINIC HALIDES, OR THEIR THIONO ANALOGS

A very elegant procedure is the abstraction of sulfur from compounds of the types $RP(S)X_2$ and $R_2P(S)X$. This process employs a variety of reducing agents. The thionophosphonic dihalides may be treated with metals such as magnesium, calcium, iron, or mercury to form the corresponding halophosphines.[202] For example, $CH_3P(S)Cl_2$ is heated with mercury in a sealed tube for 66 hr at 140°C to give about 20% CH_3PCl_2. By repeatedly withdrawing the products and continuing the reaction with the remaining reactants, 100% conversion can be obtained.[202]

An even simpler method uses a tertiary phosphine as reducing agent. Thus $CH_3P(S)Cl_2$ is converted to CH_3PCl_2 with tributylphosphine as follows:

$$CH_3P(S)Cl_2 + (C_4H_9)_3P \longrightarrow CH_3PCl_2 + (C_4H_9)_3PS$$

Since the chlorophosphine is of higher volatility, the equilibrium is shifted to the right by subsequent distillation. Yields in these reactions are mostly higher than 60%, and many reach 85%.[201,349,549,550]

The reduction of thiophosphoryl derivatives has also been achieved with elemental phosphorus,[202] $(C_6H_5O)_3P$,[174 351 352] and $C_6H_5PCl_2$.[342,434,435,534,549] The reduction of diphosphine disulfides with $C_6H_5PCl_2$ follows a similar pattern; in this case the phosphorus-phosphorus bond is first cleaved and then, in a second step, the sulfur is removed.[342]

$$R_2P(S){-}P(S)R_2 + C_6H_5PCl_2 \longrightarrow 2R_2P(S)Cl + (C_6H_5P)_n$$

$$2\ R_2P(S)Cl + (C_6H_5P)_n \longrightarrow 2R_2PCl + [C_6H_5P(S)]_n$$

In yet another case a cleavage reaction and subsequent reduction has been used to prepare diethylphosphinous chloride.[304]

$$(C_2H_5)_2P(S)SC_2H_5 + C_2H_5P(SC_2H_5)Cl \longrightarrow (C_2H_5)_2PCl + C_2H_5P(S)(SC_2H_5)_2$$

Furthermore, phosphoryl chlorides were used as starting material, and hydrides such as LiH or CaH_2 [246,297] or metals[247,297] have been employed as reducing agents.

XVI. REDUCTION OF PHOSPHORANES AND COMPLEXES OF PHOSPHORANES

The reduction of pentavalent phosphorus compounds or

their complexes is a very versatile method for the preparation of phosphonous and phosphinous halides.

Alkyl- and aryltrifluorophosphoranes react with potassium fluoride in almost quantitative yields to phosphonous difluorides by abstraction of hydrogen fluoride,[120,245]

$$RPF_3H + KF \longrightarrow RPF_2 + KF \cdot HF$$

A similar removal of hydrogen chloride can be achieved with trialkylamine; the reaction sequence is as follows:[119,120,518]

$$RPCl_2 \xrightarrow{+HF} [RPHFCl_2] \xrightarrow[-HCl]{+HF} [RPHF_2Cl] \xrightarrow{-HCl} RPF_2$$

The conversion of $C_6H_5PF_4$ into the corresponding phosphonous difluoride was observed during the cleavage of a silicon-phosphorus compound.[407]

$$2R_3Si\text{-}PR_2 + C_6H_5PF_4 \longrightarrow C_6H_5PF_2 + 2R_3SiF + (R_2P)_2$$

Chlorophosphoranes have also been reduced with triphenylphosphine,[234] e.g.,

$$R_2PCl_3 + R'_3P \qquad R_2PCl + R'_3PCl_2$$

or with mercury,[339] e.g.,

$$CF_3PCl_4 + Hg \longrightarrow CF_3PCl_2 + HgCl_2$$

Elemental phosphorus has been used extensively as a reducing agent for alkyl- or aryltetrahalophosphoranes.[130,177,178,465,563] This reduction has been achieved either with white phosphorus in carbon disulfide solutions[12,13,178] or with red phosphorus.[130] Catalytic amounts of PI_3 or iodine have been added, but the presence of iodine is not essential.[177] Unsaturated compounds such as $C_6H_5CH{=}CHPCl_2$ were also prepared in this way.[130,154,563,577]

Reduction has also been carried out with various metals[226] such as zinc[334] or antimony.[475] A new versatile reagent is methyldichlorophosphite CH_3OPCl_2, which yields only volatile by-products, e.g.,

$$RPCl_4 + CH_3OPCl_2 \longrightarrow RPCl_2 + POCl_3 + CH_3Cl$$

and was shown to give good yields for various different phosphoranes.[256,257,465]

The products obtained in the Kinnear-Perren synthesis are complexes of tetrachlorophosphoranes with $AlCl_3$[296] (see also Chapter 9).

$$RCl + PCl_3 + AlCl_3 \longrightarrow [RPCl_3][AlCl_4]$$

These complexes may be reduced to the corresponding phosphonous dichlorides with elemental phosphorus,[298,554] sodium,[298] or aluminum.[228,298,493,494] Potassium chloride was used to bind the $AlCl_3$, e.g.,

$$3[RPCl_3][AlCl_4] + 2P + 3KCl \longrightarrow 3RPCl_2 + 2PCl_3 + 3KAlCl_4$$

thus preventing the formation of the chlorophosphine adduct, $RPCl_2 \cdot AlCl_3$.

For the reduction with aluminum, solvents such as CH_3CN have been used.[571] The use of a solvent can be avoided if the liquid complex, $[CH_3PCl_3][Al_2Cl_7]$ is employed. Reduction with aluminum at atmospheric pressure and at 100°C yields $CH_3PCl_2 \cdot AlCl_3$, which is then destroyed with potassium chloride.[495]

An alternative agent for destroying the complex is diethylphthalate, which has been applied to the reduction with powdered antimony, zinc, or magnesium.[132-134,208,331,437] The reaction sequence is as follows:

$$RCl + PCl_3 + AlCl_3 \longrightarrow [RPCl_3][AlCl_4] \xrightarrow{DEP}$$

$$RPCl_4 + DEP \cdot AlCl_3 \xrightarrow{red.} RPCl_2$$

Both phosphonous and phosphinous halides have been prepared by this method.

In one variation, $C_6H_5PCl_2$ in the presence of $POCl_3$ serves as reducing agent for the Kinnear-Perren complexes.[111,244,434] Also, CH_3OPCl_2 has been used to reduce $[CCl_3PCl_3][AlCl_4]$ at 80 to 85°C. An intermediate CH_3OPCl_4 was claimed, which in turn decomposes into CH_3Cl and $POCl_3$; the latter forms the complex $POCl_3 \cdot AlCl_3$.[105] The complex formed by ethyl iodide is destroyed by potassium chloride to give the phosphonous dichloride directly.[205]

$$[C_2H_5PCl_2I][AlCl_4] + KCl \longrightarrow C_2H_5PCl_2 + \tfrac{1}{2}I_2 + KAlCl_4$$

The addition of PCl_5 to olefins, with or without functional groups, and the subsequent reduction has been a very powerful method. The reaction gives β-chloroalkyltetrachlorophosphoranes; when reduced, these compounds lose HCl, giving unsaturated phosphonous dichlorides,[285,563,577]

$$3RCHClCH_2PCl_4 + 2P \longrightarrow 3RCH{=}CH{-}PCl_2 + 2PCl_3 + 3HCl$$

Phosphorus,[433,451] antimony,[268] and CH_3OPCl_2[268,324] have been used as reducing agents. On reaction with PCl_5 and subsequent reduction, styrene yields the phosphinous chloride $(C_6H_5CH{=}CH)_2PCl$.[324] Acetylenes undergo a similar reaction, although HCl is not abstracted.[322a]

$$RC{\equiv}CH \xrightarrow{PCl_5} RCCl{=}CHPCl_4 \xrightarrow{CH_3OPCl_2} RCCl{=}CHPCl_2$$

If excess PCl_5 is used, we find formation of complexes of the form $[RPCl_3][PCl_6]$, which may be reduced with red phosphorus.[130,269,285,519] It is likely that many of the reported reactions with PCl_5 proceed by way of similar complexes.

XVII. FROM ARYLDIAZONIUM COMPLEXES AND PHOSPHORUS (III) HALIDES BY REDUCTION

The preparation of halo- and pseudohalophosphines by reduction from aryldiazonium complexes and phosphorus (III) halides is related to procedures described in the preceding part. It was found that aryldiazonium complexes react with PCl_3 producing a complex that can be reduced to yield arylphosphonous dihalides.[163,287,433,461-464,488]

$$[ArN_2]BF_4 \xrightarrow{PCl_3} [ArPCl_3]BF_4 \xrightarrow{Mg} ArPCl_2$$

The reduction is achieved with magnesium turnings,[461-463] but yields are low (in the order of 30 to 50%). The reduction can also be carried out with aluminum.[464] The addition of the PCl_3 or PBr_3 to the diazonium salt is done in the presence of cuprous halide as a catalyst in an inert solvent. Direct distillation is difficult because of solid residues; solvent extraction and subsequent distillation is better.[462]

This type of preparation serves as a useful route to arylphosphonous dihalides. The extension to the synthesis of diarylphosphinous halides by substituting arylphosphonous dihalides for PCl_3 provides a route to otherwise inaccessible compounds.

The versatility of this method is shown by the following examples. Orthobiphenyldiazonium fluoroborate reacts with PCl_3, and the unisolated intermediate is reduced with powdered aluminum to give a phosphole,

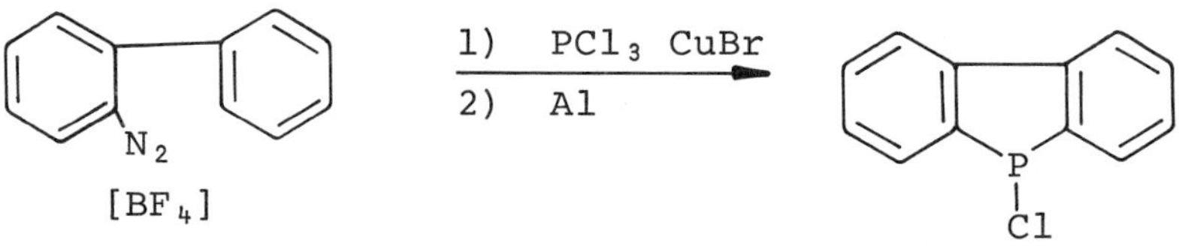

With orthophenoxybenzenediazonium fluoroborate the reaction gives 10-Chlorophenoxaphosphine.[112]

XVIII. MISCELLANEOUS METHODS

The phosphorus analog of HCN, generated in an electric arc, adds hydrogen chloride readily at 110°C[175]

$$HC\equiv P \; + \; 2HCl \longrightarrow CH_3PCl_2$$

Changes in the carbon skeleton have also been used to synthesize phosphonous dichlorides, either by addition of HCl to unsaturated compounds,[93] e.g.,

$$CF_2{=}CF\text{-}PCl_2 + HCl \longrightarrow CF_2ClCFH\text{-}PCl_2$$

or by abstraction of HCl with $BaCl_2$ at elevated temperatures,[256] e.g.,

$$ClCH_2CH_2C_6H_4PCl_2 \xrightarrow[-HCl]{BaCl_2} CH_2{=}CHC_6H_4PCl_2$$

which gave the unsaturated derivative in 98% yield. Removal of HCl is also achieved with triethylamine,[452] e.g.,

$$CH_3COOCHClCH_2PCl_2 \xrightarrow{-HCl} CH_3COOCH{=}CH\text{-}PCl_2$$

In two types of reactions a carbon-phosphorus bond cleavage has been observed. It was found that splitting of tertiary phosphines occurs in the presence of PCl_3 and $AlCl_3$ as a catalyst,[532] e.g.,

$$(C_6H_5)_2PCH_2P(C_6H_5)_2 + 2PCl_3 \xrightarrow{AlCl_3} 2(C_6H_5)_2PCl + Cl_2PCH_2PCl_2$$

Cleavage and subsequent chlorination was found to occur with the phosphinic chloride $(ClCH_2)_2P(O)Cl$, when it was treated with PCl_5.[158]

$$(ClCH_2)_2P(O)Cl \xrightarrow{PCl_5} CCl_3PCl_2$$

In the presence of Lewis acids such as $AlCl_3$ cyclic silanes react with excess PCl_3 at temperatures between 30 and 80°C to give silicon-substituted phosphonous dichlorides[66] (ca. 40-70%).

$$+ \; PCl_3 \xrightarrow{AlCl_3} ClR_2Si(CH_2)_3PCl_2$$

The phosphonium salt, $[(CH_3)_3PP(CH_3)CF_3]I$ was split with HCl to give $CH_3(CF_3)PCl$.[74] Alcoholysis or aminolysis of carboranyl derivatives produced substitution reactions leading to new compounds.[6,8] For example, the partial alcoholysis proceeded as follows:

$$Cl_2P\overline{CB_{10}H_{10}C}PCl_2 \xrightarrow{CH_3OH} Cl_2P[\overline{CB_{10}H_{10}C}P(Cl)]_5OCH_3$$

B. BASIC CHEMISTRY

The hydrolysis of phosphonous and phosphinous halides normally produces the corresponding acids, which are described elsewhere (Chapters 10 and 11). A few phosphonous dihalides, however, have been observed to undergo fission of the carbon-phosphorus bond in conditions under which this bond is usually stable. Thus trichloromethylphosphonous dichloride hydrolyzes in boiling water

$$CCl_3PCl_2 + 3H_2O \longrightarrow CHCl_3 + H_3PO_3 + 2HCl$$

with cleavage of the phosphorus-chlorine and phosphorus-carbon bonds.[178] A similar readiness of hydrolysis was noted in a series of fluoromethyl compounds CF_3PX_2, which yielded fluoroform.[43,44] The strongly electron-withdrawing organic group thus makes the phosphorus atom susceptible to nucleophilic attack and easy carbon-phosphorus cleavage results. In a more recent investigation,[419] the aqueous hydrolysis of CCl_3PCl_2 gave the trimeric acid, $[CCl_3HP(O)(OH)]_3$, probably the intermediate product before complete hydrolysis occurs.

Another study has shown that when perfluoroalkyl derivatives of phosphorus of the type $RCFX\text{-}CF_2\text{-}PZ_2$ are subjected to aqueous alkaline hydrolysis, carbon-phosphorus cleavage occurs.[69] Under neutral conditions the corresponding acids were formed, isolated after oxidation.[69] A difference between $C_3F_7PX_2$ and the trifluoromethyl analogs has been noted in the slower rate of hydrolysis of the former compounds.[127]

Cleavage of the phosphorus-carbon bond is also observed for $(C_6H_5)_3CPCl_2$ during hydrolysis,[218] thus indicating the lability of this bond in compounds with bulky substituents. Splitting can also occur with hydrogen chloride, as has been observed for a dialkylaminophenyl derivative of phosphorus.[395]

$$4\text{-}(CH_3)_2NC_6H_4PCl_2 + 2HCl \longrightarrow 4\text{-}(CH_3)_2NC_6H_5 \cdot HCl + PCl_3$$

The chlorination of CH_3PCl_2 in solution and at temperatures above 50°C is also accompanied by some carbon-phosphorus bond fission.[465] A similar process has been found in the reaction of PCl_5 with $(ClCH_2)_2P(O)Cl$ at 100°C. In a smooth reaction, Frank[158] noted the formation of trichloromethylphosphonous dichloride (CCl_3PCl_2), besides other products, indicating that both fission and chlorination occur.

The esterification of phosphonous dihalides and phosphinous halides with alcohols or mercaptans leads to the corresponding esters or thioesters (Chapters 10 and 11). The aminolysis of halophosphines is another basic reaction to give phosphonous or phosphinous amides. Both reactions are covered in Chapters 10 and 11.

Some interesting variations of these substitution reactions occur with fluorophosphines. It was observed that alcoholysis of CH_3PF_2 yields a fluorophosphorane[117] in an

$$CH_3PF_2 + ROH \longrightarrow CH_3PF_2(OR)H$$

addition reaction. The same type of reaction has been observed with amines,[118] whereas partial aminolysis was

$$CH_3PF_2 + RNH_2 \longrightarrow CH_3PF_2(NHR)H$$

achieved with a fluorochlorophosphine.[116,121]

$$CH_3PFCl + R_2NH \longrightarrow CH_3PF(NR_2)$$

In yet another study, carbon-phosphorus bond splitting was observed on reaction with an amine.[418] In the reaction of $(CH_3)_2NH$ with

$$CCl_3PF_2 + (CH_3)_2NH \longrightarrow HCCl_3 + (CH_3)_2NPF_2$$

CCl_3PF_2, $CHCl_3$, and N,N'-dimethylphosphoramidous difluoride were formed, no substitution of the fluorine atoms being detected.

From the foregoing discussion it may be concluded that chlorides, bromides, and iodides of trivalent phosphorus resemble each other more than the corresponding fluorides. A further contrast exists in the case of oxidation. Alkyl- and arylfluorophosphines decompose readily to the corresponding fluorophosphoranes.[10,146,147,503,512]

$$2RPF_2 \longrightarrow RPF_4 + (RP)_n$$

or

$$3R_2PF \longrightarrow R_2PF_3 + R_2PPR_2$$

This disproportionation depends on the nature of the organic substituent and the temperature. Alkylfluorophosphines decompose faster than aryl-substituted derivatives, whereas perfluoroalkyl- or arylfluorophosphines are stable at room temperature. At higher temperatures and on heating for prolonged periods, the latter also rearrange to fluorophosphoranes. Monofluoro derivatives decompose faster than phosphonous difluorides. The decomposition just discussed has not been studied in great detail, and kinetic and mechanistic investigations are lacking. Under similar conditions the chlorides and bromides are stable.

The dependence of the oxidative behavior on the nature of the organic group has been shown in the reaction of chlorides or bromides of trivalent phosphorus with AsF_3 or SbF_3.[498] In the case of alkyl- and aryl-substituted phosphorus halides, a redox reaction takes place to give fluorophosphoranes.[503]

$$3RPCl_2 + 4SbF_3 \longrightarrow 3RPF_4 + 2SbCl_3 + 2Sb$$

With perfluoro-alkyl and -aryl derivatives of phosphorus, however, only chlorine-fluorine exchange takes place.

The addition of hydrogen fluoride to phosphonous difluorides has been reported for two reactions,[512]

$$(CH_3)_2PF + HF \longrightarrow (CH_3)_2PF_2H$$

$$(CH_3)_2PH + 2(CH_3)_2PF \longrightarrow (CH_3)_2PP(CH_3)_2 + (CH_3)_2PF_2H$$

which again show the difference in behavior between fluorides and other halides

The phosphonous dihalides and phosphinous halides are the precursors for primary and secondary phosphines (Chapter 1), di- and polyphosphines (Chapter 2), and tertiary phosphines (Chapter 1). Halophosphines react with aldehydes to give α-chlorinated ethers, which undergo Arbuzov rearrangement to give phosphinic halides,[255] and with α,β-unsaturated ketones to give cyclic products which, on hydrolysis, lead to γ-keto-phosphonic acids (Chapter 12).

Synthesis of phosphonous anhydrides is achieved from dichlorides and dialkylphosphites or phosphonous monoesters, or from phosphinous chlorides and phosphonous monoesters (Chapter 10). Some other reactions include those with cyclic oxides,[168,172] the oxidative chlorophosphonation,[599,601] and the respective syntheses of phosphoxanes[199] or carboxy derivatives.[438]

Among recent developments has been the application of the Diels-Alder reaction to the synthesis of carbon-phosphorus heterocycles, first observed by McCormack.[365] This method, which utilizes 1,3-dienes and trivalent phosphorus halides, has provided a large number of compounds. On hydrolysis they give

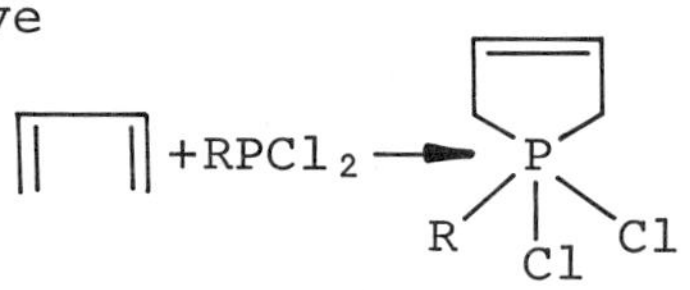

phosphine oxides (Chapter 6). A comprehensive treatment of this subject has been given by Quin.[459]

The addition of oxygen, sulfur, and selenium leads to the corresponding pentavalent phosphorus compounds. These basic reactions are described in detail as synthetic methods for phosphonic and phosphinic halides (Chapter 9). The addition of halogens yields phosphoranes with pentavalent phosphorus, which may be either pentacoordinated covalent phosphoranes or tetracoordinated ionic products [345,508] (Chapter 5B).

A further reaction, leading to an increase in the coordination of phosphorus is the formation of metal complexes, for example, by substitution in metal carbonyls.[345,354,481] These complexes of phosphonous and phosphinous halides are treated elsewhere (Chapter 3).

The pseudohalides of trivalent phosphorus show some differences from the halophosphines. Investigations into their chemistry have been scarce, so only a few reactions are mentioned here. The dicyanides $RP(CN)_2$ are stable toward water but are hydrolyzed by aqueous NaOH to NaCN and the sodium salt of the phosphonous acid.[382] They form adducts with chlorine, but treatment with sulfur dioxide gives no defined product. The diisothiocyanates are decomposed by water; with chlorine organophosphorus tetrachlorides are obtained.[382]

$$RP(NCS)_2 + 3Cl_2 \longrightarrow RPCl_4 + 2S + 2ClCN$$

The diisocyanates form the expected urethan and urea derivatives with alcohols and amines.[219]

The azides are not very stable. Thus only three compounds are listed in the tables: $(CF_3)_2PN_3$,[541] $(C_6H_5)_2PN_3$,[431] and a carboranyl derivative.[528] These compounds decompose at room temperature or above to give polymers of the type $[(C_6H_5)_2PN]_n$,[225,431,541] whereas the carboranyl phosphinous azide decomposes explosively.[528] The attempt to prepare a phosphonous diazide from $C_6H_5PCl_2$ failed, and instead gave a polymer $[C_6H_5PClN]_n$.[225] A complex was formed during the reaction of $(C_6H_5)_2PN_3$ with $(C_6H_5)_3P$ of the composition $[(C_6H_5)_2PN]_3 \cdot P(C_6H_5)_3$.[431]

Other complexes of pseudohalides have also been mentioned. Kirk and Smith[284] have found that $C_6H_5P(CN)_2$ forms a 1:4 and a 1:6 complex with $HCON(CH_3)_2$, as shown by ^1H NMR.

The phosphonous diisocyanate $C_6H_5P(NCO)_2$ polymerizes on standing to $[C_6H_5(NCO)PNC(O)]_3$.[444] A reorganization reaction has also been found. The mixed halide $C_6H_5P(NCO)Cl$ tends to give $C_6H_5PCl_2$ and $C_6H_5P(NCO)_2$.[444]

Since few ^{31}P-NMR data are available for the pseudohalides, there may be some doubt about their structural assignment. In the case of $C_6F_5P(NCS)_2$ it was shown that an oxidation process occured during distillation and a pentafluorophenylthiophosphonic diisothiocyanate $C_6F_5P(S)(NCS)_2$ was formed.[142] Many of the isothiocyanates reported may well be mixtures of trivalent and pentavalent phosphorus or even only pentavalent derivatives.

The reaction of phosphonous dichlorides with phosphine yields a complex that has been assigned a composition $(RPCl)_2P$;[377] $C_6H_5PCl_2$ forms an adduct with $C_6H_5(CH_3)_2P$.[293] No structure has yet been determined for either of these.

The phosphonous dichlorides undergo a redox reaction with antimony pentachloride:

$$RPCl_2 + 2SbCl_5 \longrightarrow RPCl_4 \cdot SbCl_5 + SbCl_3$$

no reaction being observed for $SiCl_4$ or $TiCl_4$.[292] The product mentioned earlier may well be a phosphonium salt of the type $[RPCl_3][SbCl_6]$. A similar reaction has been found to occur with $(C_6H_5)_2PCl$, giving $[(C_6H_5)_2PCl_2][SbCl_6]$.[479] In reactions with antimony pentafluoride, replacement of the chlorine occurs simultaneously with the redox reaction to give the covalent fluorophosphoranes discussed previously.[498]

A few reactions with boron-containing hydrides have been investigated. During the reduction of $C_6H_5PCl_2$ with $LiBH_4$ in a 1:1 ratio, an adduct of the formula $C_6H_5P(Cl)H \cdot BH_3$ (an unstable compound) was observed.[572] Boranes themselves react in a different way. Thus the reaction of $(CF_3)_2PF$ with B_2H_6 to form $[(CF_3)_2PBH_2]_{3,4}$ was reported,[71] whereas $(C_6H_5)_2PCl$ and decaborane gave a complex of the composition $B_{10}H_{12}[(C_6H_5)_2PCl]_2$;[506,596] a similar compound was formed with $(C_6H_5)_2PN_3$. Another type was described as $(C_6H_5)_2P \cdot B_{10}H_{13} \cdot (C_6H_5)_2PCl$.[504]

Further investigations show that $C_6H_5PCl_2$ reacts with S_4N_4 to form saltlike products that can be assigned to the chlorides of imidodiphosphinic acid and of higher imidopolyphosphonic acids.[151] Investigations into phosphorus-nitrogen chemistry are covered elsewhere (Chapter 18).

Two other complexes are worth mentioning. Green[193] has found that CH_3PCl_2 and bicyclo-(2.2.1)heptadiens give a crystalline adduct, and Nesmeyanov et al.[412] discovered that cyclopentadienylmanganese tricarbonyl with PCl_3 in the presence of $AlCl_3$ gives a solid of the following composition:

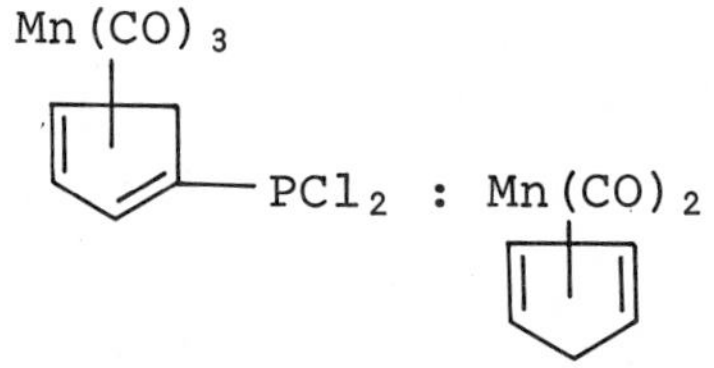

C. GENERAL DISCUSSION OF PHYSICAL PROPERTIES

The halo- and pseudohalophosphines are usually liquids, except for some of the alkyl- and perfluoroalkylfluorophosphines, which are gases. A few compounds containing bulky organic groups or those in which phosphorus is incorporated into a heterocyclic ring are solids.

In several cases vapor pressure data are available, especially for some of the fluorophosphines and low-boiling chloro- and bromophosphines. From these data Trouton constants and heats of evaporation have often been determined. Further characterization includes the recording of refractive indices and densities for the liquids, but often both figures are unreliable - - either the conditions have not been adequately given or impure products (e.g., isomeric mixtures) were measured.

The boiling point and density of various phosphorus compounds, including phosphonous dihalides, have been correlated to classify these compounds into normal or associated liquids.[40] The conclusion reached is supported by the study of the heat of evaporation.[40] However, only a few halides were examined.

A survey and discussion of electron group polarizability and molecular properties of organic phosphorus compounds has been presented,[545] but since only a few halides have been included, it too is of limited value.

In a few cases reactions have been investigated to give thermodynamic parameters. The respective hydrolyses of $C_6H_5PCl_2$ and $C_6H_5PBr_2$ were studied, and the heats of hydrolysis were determined.[148] The oxidation of $C_2H_5PCl_2$ to ethylphosphonic dichloride $C_2H_5P(O)Cl_2$ has been carried out, and the heat of oxidation determined.[411] The complex formation of CH_3PCl_2 with trimethylamine $(CH_3)_3N$, was investigated by phase-diagram studies, and thermodynamic constants were derived.[233] The equilibration of mixtures of $C_6H_5PCl_2$ and $C_6H_5PBr_2$ to yield C_6H_5PClBr was also studied by phase-diagram methods.[148]

A few publications have been concerned with mass spectral analysis. Mass spectra have been recorded, for example, for a series of methylphosphorus halides,[510-512] for trifluoromethyl[82] and pentafluorophenyl derivatives,[398] and for the para- and meta-isomers of monofluorophenyl-phosphonous and -phosphinous chlorides.[106]

Other physical methods employed have been nuclear quadrupole resonance (NQR) of chlorine in CH_3PCl_2 and $C_6H_5PCl_2$,[335] and magnetooptical measurements[317,318] of $C_6H_5PCl_2$, $C_4H_9PCl_2$, $C_5H_{11}PCl_2$, and $C_7H_{15}PCl_2$. UltraViolet absorption data have been recorded for a few compounds.[43,127,489] Solid-state structural data are only available for $CH_3P(CN)_2$. Maier[341] showed that this compound belongs to the space group $P2_1/n$.

Mechanistic and kinetic studies, which have received even less attention, are of speculative character. Flash photolysis of trifluoromethylhalophosphines has been undertaken and has shown that decomposition of these compounds proceeds through the elimination of a CF_2 radical from the CF_3-P unit as well as through dissociation to form CF_3 radicals.

Most of the investigations into the physical properties of organophosphorus compounds have been concerned with IR and Raman studies or with NMR measurements. Infared absorption data have been obtained for a large number of phosphonous and phosphinous halides. However, only a few detailed, high-resolution vibrational studies by IR and Raman spectroscopy have been reported. Most of the published data are limited to the rock salt region; often the compound purity or the accuracy of the published spectra is insufficient. This may to some extent explain the considerable variations that have been reported for some characteristic frequencies. On the other hand, the mass of data is of some value in structural assignment of group frequency characteristics, and it has been helpful in addition to other physical measurements for analyzing the derivatives.

The first major study was undertaken in 1951 by Daasch and Smith.[98] Complete vibrational data for molecules of the type RPX_2 have been given for CH_3PCl_2, CH_3PF_2, and CF_3PCl_2,[198] and most of the fundamental frequencies were assigned. Partial assignments have also been reported for CH_3PX_2 by other authors, as set forth in the compound list.

Infrared absorption studies have also been of value in establishing the structure of the pseudohalides. It has been shown that the cyanide group is linked through carbon to the phosphorus atom, whereas the cyanate and thiocyanate groups are bonded through nitrogen.[43,341,359,444] A survey of IR data in tabular form, listing characteristic frequencies, was given by Corbridge.[86]

The introduction of NMR spectroscopy has been a major advance in the investigation of phosphorus compounds, and halophosphines were among the first phosphorus compounds to be studied.

The study of the ^{31}P nucleus is, of course, applicable to all phosphorus compounds, but investigations of other nuclei, such as ^1H or ^{19}F, are also significant. A comprehensive review of ^{31}P NMR has been given by Crutchfield et al.[95] and a compilation of ^{31}P chemical shift data for phosphorus compounds, including a large number of phosphonous and phosphinous halides, has been presented.[358] Quantum mechanical interpretations of these shifts, including those of trivalent phosphorus halides, have been given by Letcher and Van Wazer.[322,552] More recently, a semiempirical LCAO-MO calculation has been made for the systems R_xPF_{3-x} and R_xPCl_{3-x}

The effective charges of the atoms were found and the average transition energies estimated.[239]

For halophosphines, the recognition of characteristic values of the ^{31}P chemical shift is often an aid in identification of unknown species. For example, phosphonous dichlorides have shifts within the range of -200 to -120 ppm; the ranges for the other types of phosphonous and phosphinous halides are also well to low field of phosphoric acid.

Another use of phosphorus NMR studies is in the investigation of ligand-interchange reactions, and the results of Van Wazer and Maier[553] have been reported for CH_3PCl_2 or CH_3PBr_2 and various phosphorus amides. Dynamic nuclear polarization studies on ^{31}P have been reported; the effects of chemical environment in $C_6H_5PCl_2$, $C_6H_5PBr_2$, and $(C_6H_5)_2PCl$ were investigated.[453]

Much information can also be obtained from the NMR spectra of other nuclei present in halophosphines, particularly ^1H and ^{19}F. The higher sensitivity of these nuclei makes the spectra easier to obtain, and the measurements are generally more accurate. Results from the study of other nuclei in halophosphines, along with other phosphorus compounds, have been reviewed by Mavel.[361]

Determinations of coupling constants to phosphorus, principally from the spectra of other nuclei than phosphorus, are of considerable diagnostic value in the identification of halophosphines.[424] This is because the value of the coupling constants reflects the separation of the two nuclei in the molecule. Fluorine NMR spectra of 4-and 3-fluorophenylhalophosphines were used to examine the inductive and mesomeric effects of substitution at phosphorus in comparison to other derivatives;[454,469] similar studies were also made for pentafluorophenyl derivatives.[232]

D. LIST OF COMPOUNDS

D.1. Phosphonous Dihalides and Phosphonous Dipseudohalides

Phosphonous Difluorides

CCl_3PF_2. VII.[417] b. 73.1 (est.)[417,418] m. 15.2 (est.)[418] 15.8-16.4 (obs.),[417,418] $\log_{10} P_{mm} = 8.543 - 1920/T$ (-9 to 10°)[418] $\log_{10} P_{mm} = 7.777 - 1699/T$ (16 to 40°),[418] Trouton const. 22.4,[418] IR,[418,421] ^{31}P -130.9,[423,542] -131,[417] ^{19}F.[417]

CF_3PF_2. VII.[315] b. -43 to -40,[315] IR,[421] ^{31}P -158.3,[423] -153.3,[542] ^{19}F,[420,423] mass spect.[82]

$ClCH_2PF_2$. VII.[500,501] b. 33.5-34.5,[500,501] ^{31}P -201.8.[423]

$MePF_2$. VII.[117,120,315,511] XVI.[120] b. -28,[117,120,315] -26.9,[511] m. -110,[510,511] $\log P_{mm} = 7.83 - 1220/T$,[511]

Trouton const. 22.6,[511] IR,[120,511] ^{31}P -244.2,[117,120,542] -245,[509] -250.7,[511] ^{1}H,[424,509,511] ^{19}F,[120,509,511] mass spect.[511]

CF_2=CF-PF_2. VII.[93,536,537] b. 2-3,[537] IR,[93] ^{19}F.[92,93]

$CClF_2CHFPF_2$. VII.[93] IR.[93]

$CHF_2CH_2PF_2$. VII.[194] IR.[194]

$EtPF_2$. VII.[117,120,315] XVI.[120] b. 6-7,[117,120,315] ^{31}P -234.1,[117,120,542] -245,[518] ^{19}F.[120,518]

$C_3F_7PF_2$. VII.[420] b. 15.8,[420] $\log P_{mm} = 7.873 - 1443/T$ (-45 to 0°),[420] Trouton const. 22.8,[420] IR,[420] ^{31}P -167.8,[423] ^{19}F.[420,423]

$BuPF_2$. VII.[315] b. 70-71,[315] d_{20} 1.0206.[315]

4-$ClC_6H_4PF_2$. VII.[119,120] XVI.[120] b_5 33-35,[119,120] n_D^{20} 1.4957,[119,120] d^{20} 1.3209,[119,120] ^{31}P -196.3,[542] -196.8,[119,120] ^{19}F.[120]

3-$FC_6H_4PF_2$. VII.[469] b_{22} 47,[469] ^{19}F.[469]

4-$FC_6H_4PF_2$. VII.[469] b_{40} 58,[469] ^{19}F.[469]

$C_6F_5PF_2$. VII.[91,146,398] b. 129-130,[146] vap. press. 2.5 mm/25°,[91] ^{31}P -193.4,[146] ^{19}F,[32,146,232] mass spect.[398]

$PhPF_2$. VII.[119,120,315,502] XVI.[120,245,407] b. 64-70,[315] b_{11} 31,[502] b_{70} 63,[120,245] b_{20} 30-31,[119] n_D^{20} 1.4933,[120,245] 1.4903,[119] d_4^{20} 1.2202,[120,245] 1.2219,[119] IR,[120] ^{31}P -208.3,[502] -206.4,[120,542] -205.3,[119] ^{19}F.[119,120,502]

$PhCH_2PF_2$. VII.[119,120] XVI.[120] b_{10} 43-45,[119,120] n_D^{20} 1.4974,[119,120] ^{31}P -223.8,[119,120,542] ^{19}F.[119,120]

4-$MeC_6H_4PF_2$. VII.[119,120] XVI.[120,245] b_{50} 72,[245] 70-72,[119,120] n_D^{20} 1.5040,[120,245] 1.4990,[119] d_4^{20} 1.2041,[120,245] 1.2054,[119] ^{31}P -205.3,[542] -206.4,[119,120] ^{19}F.[119,120]

Phosphonous Chloride Fluorides

CF_3PClF. XI.[422] ^{19}F.[422]

$MePClF$. XI.[116,121] n_D^{20} 1.4035,[116,121] ^{31}P -240,[518] ^{19}F.[518]

$PhPClF$. XI.[116,121] n_D^{20} 1.5120,[116,121] d_4^{20} 1.2057,[116,121] ^{19}F.[116,121]

Phosphonous Dichlorides

CCl_3PCl_2. XII.[18,436] XVI.[105,111,178,244,269,465,554] XVIII.[158] b_{750} 171-172,[465] b_{26} 73,[18,436] b_{20} 67,[554] 65-67,[111,244] b_{15} 60,[158] b_{23} 69-70,[465] b_7 82-83,[178] b_{13} 51,[105] m. 40,[554] 41,[111,158,244] 40-41,[269] 47[465] 50-52,[178]

$CCl_3{}^{32}PCl_2$. XII.[224] (no data).

CF_3PCl_2. VII.[43,198,339,438] XI.[221] XVI.[339] b. 39.4 (est.),[438] 37,[43] m.-129.6,[438] $\log P_{mm} = 4.2229 + 1.75 \cdot \log T - 0.002743 \cdot T - 1516/T$,[438] Trouton const. 21.75,[438] vap. press. 330.4 mm/17.3°,[198] IR,[43,44,198,422] Raman,[198] ^{19}F,[420] mass spect.[82]

$CHCl_2PCl_2$. XVI.[520] b. 150,[520] b_8 36-37,[520] n_D^{20} 1.5370,[520] d^{20} 1.6302.[520]

$ClCH_2PCl_2$. VI.[164,583,584] XV.[352,534,547-549] b. 128-132,[534,547] 126-133,[547] 127-131,[548] 129,[549] b_{720} 129-135,[352] b_{135} 76,[164] b_{140} 80-81,[583,584] b_{50} 63.5-64.5,[583] n_D^{20} 1.5247,[583,584] 1.5331,[352] n_D^{25} 1.5282,[547,549] 1.5828,[534] 1.5291,[547] 1.511,[164] d_{20}^{20} 1.5289,[583,584] 1.5209,[549] d_{25}^{25} 1.5144,[534,547,549] IR,[259,426] ^{31}P -158.9,[400] ^1H.[327]

$MePCl_2$. I.[344] IV.[327,442,443,525,527] XII.[51,343,345,347] XV.[51,246,297,349,435,550] XVI.[134,204,206,208,228,298,434,437,465,493,495,497,571] XVIII.[175] b. 80-82[204,208,297,527] 81-82,[298,443,465,525] 80.5-81.5,[206,246] 81-84,[51] 78,[345] 81.5[493,495] 74.8 (est.),[233] 80-84,[434] 82,[442] b_{747} 79-80,[550] b_{745} 79-82,[180] b_{729} 80-81,[343] 80.5-82,[349] b_{700} 78-79,[437] m. -67,[233] -80,[510] $\log P_{mm} = -1857/T + 8.20245$ (-44 to 24°),[233] Trouton const. 24.4,[233] n_D^{20} 1.4925,[180,181] 1.4940,[204,297] 1.4948,[206] 1.4959,[246] 1.4960,[208,298,343,349] 1.4967,[527] 1.4988,[51] d_4^{20} 1.3010,[527] 1.3036,[297] 1.3039,[298 343] 1.3040,[206] 1.3043,[204] 1.3052,[208] d_{20} 1.3040,[246] d_4^{25} 1.3039,[349] d_4^{35} 1.280,[442] vap. press. 25.8 mm/0°,[198] IR,[85,123,187,198,343,368,369,466] Raman,[85,123,187,198] ^{31}P -191.0,[434] -191.2,[149,406] -193.0,[138] -192.0,[314,343,344,349,400,550] ^1H,[327 424 509] double resonance,[294] ^{13}C,[362] $J_{^{13}C-H}$[363] ^{35}Cl (NQR) 26.081 MHz.[335]

$CF_2{=}CFPCl_2$. XI.[93,536,537] b. 81.5-82,[536,537] 85.8 (est.),[93] $\log P_{mm} = 7.9236 - 1810.4/T$,[93] n_D^{19} 1.4412,[536,537] d_4^{19} 1.574,[536,537] IR,[93] ^{19}F.[92,93]

$C_2F_5PCl_2$. Signs of J_{P-F}.[355]

$CFClH{-}CF_2{-}PCl_2$. VIII.[68] b_{780} 128-130.[68]

$CF_2ClCHFPCl_2$. XVIII.[93] IR,[93] ^1H,[93] ^{19}F.[93]

$CHCl{=}CHPCl_2$. VI.[179,544] b_{14} 64,[544] b_7 64-65,[179] n_D^{18} 1.5364,[179,544] d_{18} 1.4510.[179,544]

$CHF_2{-}CHFPCl_2$,$CH_2F{-}CF_2PCl_2$ (isomer mixture). VIII.[140] b_{748} 105-107.5.[140]

$CF_2H{-}CF_2{-}PCl_2$. VIII.[68] b_{773} 87-88[68] $b._{773}$ 87-88.

$CH_2{=}CH{-}PCl_2$. I.[34,35,259,373] VI.[442] b. 104,[34,35,259] b_{750} 103,[442] b_{200} 63.5,[35,259] d_4^{35} 1.328.[442]

$CF_2H{-}CH_2{-}PCl_2$. VIII.[68,194] b_{764} 116-117.[68]

$MeCHCl{-}PCl_2$. VI.[583,584] XV.[351] b_{50} 63.5-64.5,[583,584] n_D^{20} 1.5090,[583,584] d_{20}^{20} 1.4232,[583,584] ^{31}P -161.6,[351] ^1H.[351]

$ClCH_2CH_2PCl_2$. VI.[179,544] XV.[351] b. 169-170,[544] b_7 46,[179,544] d_{18} 1.4464,[179,544] ^{31}P -185.0,[351] ^1H.[351]

$EtPCl_2$. I.[104,155,209,262,276-277a,344,381,427,472] IV.[442,443,525,527] X.[3] XI.[537] XII.[347] XV.[297,349] XVI.[204,205,208,298,437,494,495] b. 113-114,[205,525,527] 111-113,[3] 112-114,[208] 111,[104,443] 111.85,[40] 112-113,[472] 112,[155,427,495] 110,115,[537] 94-97,[277,277a] 113-115,[347] 114-117,[209,381] 113-116,[276] b_{752} 113-114,[298] b_{747} 111-112,[297] b_{722} 111-112,[349] b_{150} 64,[437] b_{25}

26,[262] b_{750} 111,[442] n_D^{20} 1.4920,[204,297] 1.4930,[205,262] 1.493,[298] 1.4928,[298] 1.4938,[472] 1.4940,[208,525] 1.4942,[527] 1.4943,[349] 1.4948,[3] 1.4950,[208] n_D^{25} 1.4930,[155,427,472] d^{19} 1.2952,[209,381] d_0^{20} 1.2485,[262] 1.2563,[472] d^{20} 1.2592,[205,298] 1.2526,[3] 1.2586,[527] 1.2594,[208] 1.2950,[297] 1.2589,[298] 1.2600,[427] 1.221,[537] d_4^{20} 1.221,[537] 1.3090,[525] 1.2950,[205] 1.2586,[527] 1.2594,[208] d_4^{25} 1.2627,[349] d_4^{27} 1.2485,[155,472] d_4^{35} 1.250,[442] 1.2485,[40] ΔH_{vap} 8.34 kcal/mole,[40] IR,[85,98,190] Raman,[85] ^{31}P -196.3,[344,349,400] ^{1}H (high-resolution, double resonance),[294] ΔH_{oxide} -40.6 kcal/mole.[411]

$Et^{32}PCl_2$. n_D^{25} 1.4900.[486]

$\overline{HC(B_{10}H_{10})C}$-$PCl_2$ (o-carboranyl-). XI.[592] b_2 100-101.[592]

$C_3F_7PCl_2$. VII.[127] b. 86.4,[127] m. -90,[127] log P_{mm} = 7.744 - 1748/T,[127] Trouton const. 22.3,[127] heat of evap. 8.0,[127] UV,[127] IR,[127] ^{19}F,[420] signs of J_{P-F}.[355]

CH_2=C=CHPCl_2. Relative signs of J_{P-H}.[523]

MeC≡C-PCl_2. XI.[52] b_9 42,[52] n_D^{20} 1.5500,[52] d_4^{20} 1.3024,[52] IR,[52] ^{1}H.[52]

CN-$CH_2CH_2PCl_2$. VIII.[189,220] b_{12} 103-108,[220] $b_{2.5}$ 89.[189]

CH_2=CH-CH_2-PCl_2. XVI.[475] b_{46} 51-52,[475] d_4^{20} 1.2325,[475] n_D^{20} 1.5112.[475]

MeCOCH_2PCl_2. XVI.[154] b_{10} 73-74,[154] n_D^{20} 1.5105,[154] d_{20} 1.3452.[154]

MeOCOCH_2PCl_2. I.[456] VIII.[240] b_{15} 79,[456] b_4 59-61,[240] n_D^{20} 1.4980,[456] d^{20} 1.3860,[456] ^{1}H.[441]

Cl$CH_2CH_2CH_2PCl_2$. VIII.[189] ^{31}P -182.[210]

MeCHClCH_2PCl_2. VI.[179,544] b_7 55,[179,544] n_D^{18} 1.5172,[179,544] d_{18} 1.3250.[179,544]

$Cl_3Si(CH_2)_3PCl_2$. XVIII.[66] b_{23} 125-127,[66] n_D^{20} 1.5156,[66] d_4^{20} 1.4679.[66]

$PrPCl_2$. I.[38,141,155,183,209,262,427] VIII.[189] XI.[80] b. 140-142,[80,209] 133,[427] 128,[38] 134.5,[155] 134.35,[40] b_{15} 43,[262] b_{90} 70,[183] b_{198} 88,[141] d 1.1664,[40] d_0^{20} 1.1664,[262] d_{20} 1.1854,[427] d^{19} 1.1771,[209] d_4^{27} 1.1664,[155] n_D^{20} 1.4842,[262] 1.4860,[427] n_D^{25} 1.4842,[155] ΔH_{vap} 9.36 kcal/mole,[40] ^{31}P -201.[141]

i-$PrPCl_2$. I.[209,427,474] XVI.[437] b. 135-138,[209] b_{745} 130,[427] b_{202} 88,[437] b_{17} 41-42,[474] n_D^{20} 1.4880,[427] n_D^{25} 1.4868,[437] d^{23} 1.2181,[209] d_{20} 1.1922,[427] d_0^{20} 1.2775.[474]

Me$\overline{C(B_{10}H_{10})C}$-$PCl_2$ (methyl-o-carboranyl-). XI.[592] b_3 114-115,[592] m. 46-47.[592]

CH_2=CH-C≡C-PCl_2. XI.[52] b_1 26,[52] n_D^{20} 1.5632,[52] d_4^{20} 1.2573,[52] IR,[52] ^{1}H.[52]

2-Thienyl-PCl_2. II.[46,482] b 218,[482] b_{18} 108,[46] ^{1}H.[67]

CH_2=C=C(Me)PCl_2. Relative signs of J_{P-H}.[523]

MeC(:O)OCH=CHPCl_2. XVIII.[452] $b_{0.5-1}$ 58-59,[452] n_D^{20} 1.5200,[452] d^{20} 1.3744.[452]

MeC(:O)OCHCl-CH_2PCl_2. XVI.[154] b_3 76-77.[154] n_D^{20} 1.5052,[154]

d_{20} 1.4263.[154]

Me_2C=CH-PCl_2. XVI.[268,324,334,563,577] b. 145-148,[334] b_{100} 98-104,[563,577] b_{11} 53-55,[268] 54-55,[324] n_D^{20} 1.5180,[268] 1.5215,[324] n_D^{25} 1.5087,[563 577] d_4^{20} 1.1920,[268] 1.2030,[324] MR_D 39.73.[268]

MeCH=CHCH_2PCl_2. XVI.[331] b_{19} 55-56,[331] b_{20} 58-59,[331] n_D^{20} 1.5126,[331] d_4^{20} 1.1790.[331]

EtOCH=CH-PCl_2. XVI.[12,154,324] b_{15} 84-85.5,[324] b_{13} 78,[154] b_5 67-69,[12] n_D^{20} 1.5230,[324] 1.5228,[154] 1.5234,[12] d^{20} 1.2482,[154] 1.2470.[324]

EtOCOCH_2-PCl_2. I.[456] b_8 73,[456] n_D^{20} 1.4908,[456] d^{20} 1.3103.[456]

MeOC(:O)CH(Me)PCl_2. VIII.[240] b_4 63-64.[240]

$Me_2C(CH_2Cl)PCl_2$. VI.[476] b 204,[476] b_{10} 71.[476]

$Me_2C(Cl)CH_2PCl_2$. VI.[476] b 201,[476] b_{10} 70.[476]

$C_4H_8ClPCl_2$ (chloroisobutyl-). VI.[522 583] n_D^{18} 1.5072.[522]

$BuPCl_2$. I.[114,155,262,264,427,428,472,566,595] X.[483,484] XI.[80] XII.[49,50] b 157-160,[80] 160,[155] 153-155,[427] 156-157,[472] 159.85,[40] 156-159,[49,50] b_{750} 157-160,[114] b_{33} 74,[566] b_{23} 62-64,[264] b_{20} 56,[262] b_{14} 51-52,[484] b_9 39-39.5,[428,595] 38.5-39,[427] n_D^{20} 1.4870,[49,50] 1.4858,[472] 1.4868,[427] 1.4838,[262] 1.4850,[264] 1.4866,[595] 1.488,[566] n_D^{24} 1.4802,[483] n_D^{25} 1.4838,[155] d 1.1341,[40] d_{20} 1.167,[566] 1.1417,[595] 1.1416,[427] d_0^{20} 1.1653,[472] 1.1341,[262] 1.1653,[472] d_4^{27} 1.1341,[155] d_4^{20} 1.1615,[49,50] ΔH_{vap} 10.05 kcal/mole,[40] magnetooptical study.[318]

i-$BuPCl_2$. I.[209,427,472,473,594] VIII.[220] IX.[415] b 155-157,[209] 148-149,[594] 149-151,[220] b_{740} 148-149,[427] b_{100} 97-99,[415] b_{50} 59-60,[427] b_{12} 48.5-49.0,[472-474] b_9 35-36,[427] n_D^{20} 1.4810,[415] 1.4719,[472-474] 1.4818,[427 594] d^{20} 1.1268,[427,594] 1.2010,[415] d_0^{20} 1.1720,[472-474] d^{23} 1.1236.[209]

s-$BuPCl_2$. XVI.[437] b_{40} 70,[437] b_{37} 68,[437] b_{28} 62-62.5,[437] n_D^{25} 1.4909.[437]

t-$BuPCl_2$. I.[374,559] b. 145-150.[559] m. 49,[559] 51.5-52.5.[374]

$MeCl_2Si(CH_2)_3PCl_2$. XVIII.[66] b_8 109-110,[66] n_D^{20} 1.5095,[66] d_4^{20} 1.3409.[66]

$MeOCH_2CH_2CH_2PCl_2$. IX.[414] b_3 53-55,[414] n_D^{20} 1.5100.[414]

$Me_3SiCH_2PCl_2$. I.[517] $b_{1.5}$ 50.[517]

$C_5H_7PCl_2$(cyclopentenyl-). IR.[20]

EtC(:O)OCH=CHPCl_2. XVIII.[452] b_1 65-66,[452] n_D^{20} 1.5120,[452] d^{20} 1.3251.[452]

$C_5H_8ClPCl_2$ (chloro-cyclopentyl-). VI.[476] b 238,[476] b_{18} 114,[476] d^{20} 1.37.[476]

CH_2=C(Me)CHCl-CH_2PCl_2. XVI.[12] b_3 91-91.5,[12] n_D^{20} 1.5412,[12] d_4^{20} 1.3148.[12]

EtOCOCHClCH_2PCl_2. XVI.[451] b_1 86-87,[451] n_D^{20} 1.5015,[451] d^{20} 1.3721.[451]

PrCH=CHPCl_2. XVI.[577] b_{100} 100-102,[577] n_D^{25} 1.5028.[577]

i-PrOCH=CHPCl_2. XVI.[154] b_{12} 84-85,[154] n_D^{20} 1.5153,[154] d_{20} 1.1989.[154]

$MeOCH_2CH_2OCH{=}CHPCl_2$. XVI.[12] b_2 95,[12] n_D^{20} 1.5155,[12] d_4^{20} 1.2737.[12]
$EtOC(:O)CH_2CH_2PCl_2$. VIII.[189]
$Me(CH_2)_2CHCl\text{-}CH_2PCl_2$. VI.[476] b 228,[476] b_{13} 94,[476] d^{20} 1.27.[476]
$MeEtC(Cl)CH_2PCl_2$. VI.[476] b 220,[476] b_{11} 87,[476] d^{20} 1.26.[476]
$Me_2C(Cl)CH(Me)PCl_2$. VI.[476] b 218,[476] b_{21} 87,[476] d^{20} 1.30.[476]
$AmPCl_2$. I.[155] XVI.[134] b. 183.85,[40] 184,[155] n_D^{25} 1.4815,[155] d 1.0997,[40] d_4^{27} 1.0997,[155] $\Delta H_{vap.}$ 10.09 kcal/mole,[40] magnetooptical study.[318]
i-$AmPCl_2$. I.[209,381] XI.[80] b.180-183,[209,381] d^{23} 1.1024.[209,381]
$EtOCH_2CH_2CH_2PCl_2$. VIII.[189]
$ClMe_2Si(CH_2)_3PCl_2$. XVIII.[66] b_9 103-104,[66] n_D^{20} 1.5002,[66] d_4^{20} 1.1941.[66]
$C_6F_5PCl_2$. I.[33,144,337] XI.[33,144] b_{21} 72-74,[144] b_9 81-82,[337] $b_{0.1}$ 39-41,[33] IR,[33,145,337] ^{31}P -137.0,[144] -136.7,[32] ^{19}F[32,232]
2,4,5-$Cl_3C_6H_2PCl_2$. XII.[48] b_1 130-132,[48] n_D^{20} 1.6409,[48] d_4^{20} 1.7110.[48]
3-$BrC_6H_4PCl_2$. XVII.[462] b_6 119-122.[462]
4-$BrC_6H_4PCl_2$. I.[566,586,587] II.[101,253,279,281,382,383] (probably isomeric mixtures). b.271-272[382,383] 270,[586,587] b_{21} 147-148,[382,383] b_{14} 135-136,[253,279] 140,[281] b_{13} 139,[101] b_{12} 136,[281] $b_{0.8}$ 65,[566] m 55-60,[566] n_D^{20} 1.6422,[279] d_{20} 1.7290,[279] 1.6861,[586,587] d^{15} 1.6895.[382,383]
$ClC_6H_4PCl_2$. II[25,54,169,280,281,382,383,416] (probably isomeric mixture). b. 253-255,[382,383] b_{20} 133,[416] b_{15} 121-122,[280,281] b_2 80-81,[54] b_{1-2} 90-92,[169] n_D^{20} 1.6066,[169] 1.6132,[54] d_4^{20} 1.4450,[169] d^{20} 1.447,[54] d^{17} 1.425,[382,383] IR.[54]
2-$ClC_6H_4PCl_2$. XI.[490] XVII.[462] b_9 113-117,[490] $b_{0.53}$ 76-77,[462] n^{20} 1.6118.[490]
3-$ClC_6H_4PCl_2$. I.[581] IX.[287] XI.[490] XVII.[287,333,461,462] b_{35} 142-143,[333] b_{20} 128-129,[287] b_{18} 124-125,[461,462] b_9 113-116,[490] b_5 101-103,[581] n_D^{20} 1.6103,[333] 1.6060,[287] 1.6082,[490] d^{20} 1.4440,[287] d_4^{20} 1.4465.[333]
4-$ClC_6H_4PCl_2$. I.[586,587] IX.[1,54] XI.[490] XII.[28,555] XVII.[461,464,488] b 252-253,[586,587] b_{20} 132-133,[461] b_{12} 141-142,[488] 118-119 (isomeric mixture),[1] $b_{2.5}$ 93-95,[28,555] $b_{1.4-2.8}$ 86-90,[464] b_9 115-118,[490] n_D^{20} 1.6142,[1] 1.6135,[555] 1.6078,[490] d^{20} 1.4203,[586,587] 1.4440,[1] IR.[28]
3-$FC_6H_4PCl_2$. I.[106] XI.[492] b_{37} 114,[492] b_{20} 99-101,[492] $b_{1.3}$ 68-70,[469] b_1 76,[106] n_D^{25} 1.5721,[469] n_D^{20} 1.5733,[492] IR,[106] mass spect.[106] ^{31}P -152,[106] ^1H,[106] ^{19}F.[106]
4-$FC_6H_4PCl_2$. I.[106] XI.[490] b_{11} 85-87,[491] b_{10} 82,[106]

82-83,[490] 90-92,[183] b_3 72,[469] n_D^{25} 1.5704,[469] n_D^{20} 1.5690,[491] 1.5688,[490] IR,[106] mass spect.,[106] Hammett const.,[491,518] ^{31}P -159,[106] ^{1}H,[106] ^{19}F.[106,454,518]

$PhPCl_2$. I.[344,375,376,378,566,567,585-587] II.[14,57,64,65,94,125,148,162,167,171,184,187,260,280,281,288,290,291,321,330,359,375,376,378,380,429,442,450,526,546,574] IV.[385,527] V.[56] IX.[156,404] X.[483,484,595] XI.[80] XII.[49,50] XV.[297] XVI.[130,563] b. 225,[14] 222,[380] 221.85,[40] 221-222,[253] b_{57} 140-142,[389] 140-141,[586,587] b_{56} 134-135,[306] b_{50-52} 128-130,[162] b_{23} 109,[94] b_{20} 107,[184] b_{18} 130,[148] b_{15} 100,[297] b_{14} 102,[453] b_{11} 99.5,[567] 99-101,[566] 99-100,[280,429] b_{10} 89-94,[442] 92.5,[156] 90-92,[65] b_8 80-83,[130] b_6 91-93,[563] b_5 99-100,[49,50] 82-85,[484] b_4 82-85,[483] b_2 73-75,[526] b_1 58-59,[359] 68-70,[64] 59-60,[187] [527] $b_{0.8-1}$ 58.59,[167] $b_{0.7}$ 46-47,[404] $b_{0.5}$ 60-64,[56] m. -48 to -46,[187] n_D^{7} 1.60533,[597] n_D^{18} 1.5940,[130] n_D^{20} 1.5960,[167] 1.5880,[483] 1.600,[567] 1.603,[566] 1.5870,[484] 1.5971,[187] 1.5883,[297] 1.5947,[49,50] 1.5860,[527] 1.5950,[526] n_D^{25} 1.5951,[148] 1.5948,[156] 1.5962,[64] d 1.288,[442] 1.3773,[40] d_4^{20} 1.319,[148,597] 1.3324,[49,50] 1.3186,[527] 1.3181,[526] 1.3199,[297] d_4^{0} 1.341,[14] 1.3428,[597] d_{20}^{20} 1.3180,[587] d_{20} 1.356,[566] 1.3191,[167] ΔH_{hydrol} -42.1 + 0.6 kcal/mole,[148] ΔH_f^{o} (liqu.) -30 kcal/mole,[148] magnetooptical study,[317] UV,[489] IR,[11,85,98,187,188,191] Raman,[85,187,188,191] ^{31}P -161 $\pm$ 1,[148] -166,[106,150,551] -161.5,[344] -161.6,[406] -164,[372] ^{31}P signal enhancement through dynamic polarization,[453] ^{1}H,[270] ^{35}Cl (NQR) 26.557 MHz,[335] ^{13}C.[477]

$C_6H_9PCl_2$ (cyclohexenyl-). IV.[222] b 210-213,[222] n_D^{20} 1.5353,[222] d_{23} 1.236,[222] IR.[20,222]

PrC(:O)OCH=$CHPCl_2$. XVIII.[452] b_1 79-80,[452] n_D^{20} 1.5089,[452] d^{20} 1.2842.[452]

cyclo-$C_6H_{10}ClPCl_2$. IV.[22] VI.[476,544,578,579] b 258,[476] b_{23} 142,[476] b_7 117,[544] $b_{0.6}$ 95-96,[22] n_D^{20} 1.5472,[22] 1.5525,[544] d^{20} 1.35,[476] d_{20} 1.3334,[544] d_4^{20} 1.3502,[22] IR.[20]

n-$PrOCOCHClCH_2PCl_2$. XVI.[451] b_2 106-107,[451] n_D^{20} 1.4970,[451] d^{20} 1.3215.[451]

$C_6H_{11}PCl_2$ (hexenyl-). IV.[222] (isomer mixture). IR.[20]

$C_6H_{11}PCl_2$ (cyclohexyl-). I.[566] IV.[18,22,222] VIII.[189,220] IX.[414,415] XI.[242,513] XIII.[243] b. 219,[222] b_{50} 132-135,[18,222] b_{20} 105-107,[414,415] b_{17} 98-99,[242,243] b_{12} 95-96,[513] 93-94.5,[566] b_{10} 40-41,[220] $b_{0.4}$ 50.5-51,[22] n_D^{20} 1.519,[566] 1.5278,[22] 1.5040,[415] 1.5270,[414] n_D^{25} 1.5262,[222] d_{23} 1.2210,[222] d_{20} 1.173,[566] d_4^{20} 1.2230,[22] d^{20} 1.2065,[415] IR.[20]

$C_6H_{11}{}^{32}PCl_2$ (cyclohexyl-). IV.[224]

EtO-C(Et)=CH-PCl_2. XVI.[154] b_7 85-86,[154] n_D^{20} 1.5140,[154] d_{20} 1.1614.[154]

. XVI.[154] b_9 99-100,[154] n_D^{20} 1.5127,[154] d^{20} 1.1675.[154]

$EtO-CH_2-CH_2-OCH=CH-PCl_2$. XVI.[12] b_2 98,[12] n_D^{20} 1.5140,[12] d_4^{20} 1.2396.[12]

$Et_2C(Cl)CH_2PCl_2$. VI.[476] b. 238,[476] b_{10} 103,[476] d^{20} 1.19.[476]

$Me(CH_2)_3CHClCH_2PCl_2$. VI.[476] b. 240,[476] b_{16} 118,[476] d^{20} 1.20.[476]

n-hexyl-PCl_2. I.[155,427,595] IV.[222] IX.[415] XII.[49,50] b. 207.85,[40] 208,[155] b_{25} 104-107,[49,50] b_{11} 91-92,[415] b_9 83-84,[427] b_2 63,[595] n_D^{25} 1.4800,[155] n_D^{20} 1.4819,[595] 1.4850,[49,50] 1.5010,[415] 1.4820,[427] d 1.0653,[40] 1.1066,[49,50] d_4^{20} 1.1895,[415] d_4^{27} 1.0653,[155] d_{20} 1.0670,[427] 1.0672,[595] ΔH_{vap} 10.83 kcal/mole,[40] IR.[20]

$Me_2CH-C(Me_2)PCl_2$. I.[374] b_{4-5} 50-52,[374] n_D^{25} 1.4978.[374]

$ClMe_2Si(CH_2)_4PCl_2$. XVIII.[66] b_{9-10} 128-133[66] (impure product).

$4-CF_3C_6H_4PCl_2$. I.[582] b_{14} 92-93.[582]

$4-NCC_6H_4PCl_2$. XVII.[433,461,462] $b_{4.5}$ 127-128,[461,462] b_3 119.[433]

$4-CH_2Cl-C_6H_4PCl_2$. XVI.[257] b_1 98-99,[257] n_D^{20} 1.6127,[257] d_4^{20} 1.4111.[257]

$4-ClC_6H_4CH_2PCl_2$. XII.[49,50] b_{15} 143-146,[49,50] n_D^{20} 1.5951,[49,50] d_4^{20} 1.3904.[49,50]

$2-Cl-4-MeC_6H_3PCl_2$. II.[371] b. 265-266,[371] d^{22} 1.373.[371]

$3-Cl-4-MeC_6H_3PCl_2$. XVII.[462] $b_{2.0}$ 117-118.[462]

$Cl_3SiCH_2C_6H_4PCl_2$. II.[446] (isomer mixture). $b_{0.5}$ 128-132,[446] n_D^{20} 1.5860,[446] d_4^{20} 1.4735,[446] IR.[446]

$PhCH_2PCl_2$. I.[250,566] II.[446] XII.[49,50,439,440] b_{15} 113-115,[49,50] b_{12} 110-111,[250] 118-119,[566] 111-113,[440] 110,[439] b_{10} 145-155,[446] n_D^{20} 1.586,[566] 1.5840,[440] 1.5870,[49,50] d_{20} 1.300,[566] d_4^{20} 1.2782,[440] 1.2947.[49,50]

$MeC_6H_4PCl_2$. II.[173,302] (isomer mixture). b_6 110-113,[173] n_D^{20} 1.5836,[173] d_{20} 1.2791.[173]

$2-MeC_6H_4PCl_2$. I.[382,383,393,566] XI.[490] b. 244,[393] b_{12} 127-129,[566] b_{10} 108-109,[490] n_D^{20} 1.5914,[490] 1.598,[566] d_{20} 1.317,[566] d^{18} 1.3067.[393]

$3-MeC_6H_4PCl_2$. I.[382,383] XI.[490] b. 235.[382,383] b_9 102-104,[490] n_D^{20} 1.5883,[490] d^{22} 1.282.[382,383]

$4-MeC_6H_4PCl_2$. I.[393,566,567,587] II.[64,253,265,281,328,380,382,383,390,392,429,468,564] IX.[156] XI.[490] XVI.[130] XVII.[461] b. 245,[393] 243-244,[587] b_{22} 130,[281] b_{18} 117-118,[130] b_{13} 112-113,[265] 120,[283] b_{12} 100,[253] b_{11} 116.5,[467,566] 109-110,[461] 106-109,[490] b_{10} 107-110,[64] 110-113,[156] m. 24,[429,490,566,567] 25,[393] 24-25,[156] 24.5,[587] n_D^{20} 1.591,[566,567] 1.5886,[283] 1.5875,[490] 1.5851,[130] n_D^{25} 1.5884,[156] 1.5865,[64] n_D^{19} 1.5860,[130] d_4^{25} 1.2661,[64] d_0^0 1.2864,[283] d_0^{20} 1.2666.[283]

$2-[Cl_2(Me)Si]C_6H_4PCl_2$. II.[447] $b_{0.5}$ 130,[447] n_D^{20} 1.5809,[447] d^{20} 1.3605,[447] IR.[447]

$4-[Cl_2(Me)Si]C_6H_4PCl_2$. II.[447] $b_{0.5}$ 135-140,[447] m. 41.[447]

$MeOC_6H_4PCl_2$. II.[170] (nothing said about isomers). b_1 104,[170] n_D^{20} 1.6001,[170] d_4^{20} 1.3451.[170]

2-$MeOC_6H_4PCl_2$. XVII.[462] $b_{0.52}$ 88.5-89.[462]

4-$MeOC_6H_4PCl_2$. I.[467] II.[101,183,248,253,261,316,382,383] b. 245-253,[248] b_{21} 153,[261] b_{18} 150,[248] b_{13} 150,[101] b_{11} 140-141,[261] b_{10} 140,[183] d_0^0 1.3604,[261] d_0^{15} 1.3468,[261] d_0^{25} 1.331[248] (the original const.[382,383] were obtained with grossly impure product: b_{12} 130, d^{15} 1.0764).

cyclo-$C_7H_{12}ClPCl_2$. VI.[476] b. 287,[476] b_{0-27} 86,[476] d^{20} 1.32.[476]

BuO-C(Me)=CH-PCl_2. XVI.[154] b_{13} 115-116,[154] n_D^{20} 1.5140,[154] d_{20} 1.1426.[154]

n-$C_7H_{15}PCl_2$. I.[155] IV.[222] b. 228.35,[40] 210-225,[222] 228.5,[155] n_D^{25} 1.4788,[155] n_{20}^D 1.4837,[222] d 1.0636,[40] d_{23} 1.077,[222] d_4^{27} 1.0636,[155] $\Delta H_{vap.}$ 11.45 kcal/mole,[40] magnetooptical study.[318]

PhC≡C-PCl_2. XI.[52] $b_{1.5}$ 94,[52] n_D^{20} 1.6470,[52] d_4^{20} 1.2922,[52] IR,[52] ^1H.[52]

PhCCl=CH-PCl_2. XVI.[323,325] $b_{0.02}$ 112-113,[323] n_D^{20} 1.6396,[323] d_4^{20} 1.4181.[323]

PhCH=CH-PCl_2. XVI.[130,131,285,325,563,577] b_{18} 142-144,[131] b_8 135-138,[563,577] b_7 133-135,[130] $b_{0.03}$ 82-83.[325] n_D^{19} 1.6353,[130] n_D^{20} 1.6355,[325] n_D^{25} 1.6350,[563 577] d^{20} 1.2790.[325]

4-CH_2=CH-C_6H_4-PCl_2. XVIII.[256] b_1 74-75,[256] n_D^{20} 1.6231,[256] d_4^{20} 1.2709.[256]

PhO-CH=CH-PCl_2. XVI.[154,433] b_3 93-94,[154] $b_{0.3}$ 93-94,[433] n_D^{20} 1.5976,[154,433] d_{20} 1.3007,[154] d_4^{20} 1.3006.[433]

4-$ClC_2H_4C_6H_4$-PCl_2. XVI.[256] b_1 105-106,[256] m. 37,[256] n_D^{20} 1.5985,[256] d_4^{20} 1.3550.[256]

$Cl_3SiCH_2CH_2C_6H_4PCl_2$. II.[446,449] (isomer mixture). b_4 159-160,[449] b_2 142,[449] b_1 160,[446] n_D^{20} 1.5748,[446] 1.5740,[449] d_4^{20} 1.4344,[446 449] IR.[446]

$PhC_2H_4PCl_2$. XII.[30] b_3 100-102,[30] n_D^{20} 1.5751,[30] d_4^{20} 1.2524.[30]

$MeC_6H_4CH_2PCl_2$. XII[30] (isomer mixture). b_1 103-105.[30]

2-$EtC_6H_4PCl_2$. XI.[80]

4-$EtC_6H_4PCl_2$. I.[566] II.[25,64,99,157,248,253,382,383,573] (orientation of Et group not clear). IV.[222] b 250-252,[382,383] b_{18} 133,[253] b_{15} 128-129,[99] b_{12} 127,[248] b_{10} 122-125,[64] $b_{0.8}$ 70-82,[157] $b_{0.4}$ 85,[566] n_D^{25} 1.5576,[64] n_D^{20} 1.584,[566] d_{20} 1.239,[566] d^{17} 1.227,[382,383] d_4^{25} 1.225.[248]

2,3-$Me_2C_6H_3PCl_2$. II.[237,382,383,393,569,570] (mixture with 3,4-isomer). b. 278,[382,383,393,569,570] n_D^{20} 1.590,[237] d_4^{20} 1.246.[237]

2,4-$Me_2C_6H_3PCl_2$. I.[569,570] II.[393] (mixture with 3,5-isomer). b. 265,[393] 256-258.[569,570]

2,5-$Me_2C_6H_3PCl_2$. II.[25,249,253,393,569,570,575] b. 253-254,[393,569,570] b_{23} 140,[249] b_{18-20} 134-136,[25] b_{16} 133,[253] b_{12} 129,[249] $b_{0.05}$ 85,[575] n_D^{20}

1.5886 (av.),[575] d_4^{20} 1.2457 (av.),[575] d_{18}^{18} 1.25.[393, 569, 570]

$MeCl_2SiCH_2C_6H_4PCl_2$. II[446,449] (isomer mixture). $b_{0.5}$ 120-125,[446] $b_{0.1}$ 98-99,[449] n_D^{20} 1.5809,[446] 1.5790,[449] d^{20} 1.3605,[446] d_4^{20} 1.3675,[449] IR.[446]

4-$EtOC_6H_4PCl_2$. I.[382,383] II.[382,383] b. 266.[382,383]

$(MeO)_2C_6H_3PCl_2$. II[457,458,558] (orientation of MeO-groups not clear). b_{12-15} 175-180,[558] b_1 132,[458] m. 147,[558] 51.5-52.5.[457,458]

4-$Me_2NC_6H_4PCl_2$. II.[94,185,196,394,395] VIII.[55] b_{120} 250,[394,395] $b_{2.5}$ 165-168,[185] $b_{0.6}$ 158-160,[196] m. 66[394,395] 64-65,[94] 63-65.[185]

$C_8H_{12}ClPCl_2$ [2-chlor-2(3-cyclohexen-1-yl)ethyl-]. VI.[319] b_1 99-102.[319]

$C_8H_{15}PCl_2$(iso-octenyl). XVI.[563,577] b_2 67-70,[563] b_3 70-72,[563] n_D^{25} 1.5035.[563]

PhC($B_{10}H_{10}$)C-PCl_2. XI.[592] m. 79-80.[592]

Me$(CH_2)_5$CHClCH$_2$PCl$_2$. VI.[598] $b_{0.01}$ 75-80.[598]

4-EtO-C_6H_{10}-PCl_2 (cyclohexyl-). VIII.[189]

BuO-C(Et)=CH-PCl_2. XVI.[154] b_7 110-111,[154] n_D^{20} 1.5065,[154] d_{20} 1.1120.[154]

Me$(CH_2)_4$CH(Cl)CH(Me)PCl_2. VI.[476] b 270,[476] $b_{0.8}$ 107,[476] d^{20} 1.17.[476]

Me$(CH_2)_5$CHCl-CH_2-PCl_2. VI.[273,274] $b_{0.5}$ 85-88.[273,274]

n-octyl-PCl_2. I.[155,470] VIII.[189,220] b. 246.85,[40] 247,[155] b_8 106-110,[220] n_D^{25} 1.4778,[155] d 1.0433,[40] d_4^{27} 1.0433,[155] ΔH_{vap} 12.26 kcal/mole,[40] IR.[20]

$C_9H_7PCl_2$ (indenyl-). XVI.[563] b_1 104-105.[563]

PhC(Me)=CHPCl_2. XVI.[324] $b_{0.03}$ 86-87,[324] n_D^{20} 1.6159,[324] d^{20} 1.2605.[324]

$PhCH_2OCH$=CHPCl_2. XVI.[451] b_1 146-148.[451] m. 63-65.[451]

$Cl_3SiCH_2CH_2CH_2C_6H_4PCl_2$. II.[447,449] $b_{0.5}$ 130,[449] $b_{0.3}$ 130,[447] n_D^{20} 1.5655,[447,449] d_4^{20} 1.4054.[447,449]

2,5-$Me_2C_6H_3CH_2PCl_2$. XII.[49,50] b. 138-141,[49,50] n_D^{20} 1.5690,[49,50] d_4^{20} 1.2010.[49,50]

2,4,5-$Me_3C_6H_2PCl_2$. I.[382,383] II.[382,383] b. 280,[382,383] d^{20} 1.2356,[382,383]

2,4,6-$Me_3C_6H_2PCl_2$. I.[58] II.[100,382,383] XI.[80] b 273-275,[382,383] b_{16} 156-157,[100] m. 35-37,[382,383] n_D^{20} 1.581,[58] 1.589,[58] d_4^{20} 1.250,[58] 1.243.[58]

n-$PrC_6H_4PCl_2$. II.[64] b_5 127-131,[64] n_D^{25} 1.5658,[64] d_4^{25} 1.1905.[64]

4-i-$PrC_6H_4PCl_2$. II.[64,253,382,383,405] b. 268-270,[382,383] b_{14} 132-134,[253] b_{10} 125-127,[383,405] 129-132,[64] n_D^{25} 1.5677,[64] d_4^{25} 1.1917.[64]

Cl_2(Me)$SiCH_2CH_2C_6H_4PCl_2$. II[446] (isomer mixture). $b_{0.02}$ 105-109,[446] n_D^{20} 1.5702,[446] d_4^{20} 1.3344,[446] IR.[446]

Cl$(Me)_2SiCH_2C_6H_4PCl_2$. II[446] (isomer mixture). $b_{0.5}$ 120-123,[446] n_D^{20} 1.5740,[446] d_4^{20} 1.2556,[446] IR.[446]

BuO-C(Pr)=CH-PCl_2. XVI.[154] b_7 120-121,[154] n_D^{20} 1.5050,[154] d_{20} 1.0927.[154]

$C_9H_{19}PCl_2$. XII.[49,50] b_{15} 135-138,[49,50] n_D^{20} 1.4846,[49,50] d_4^{20} 1.0364.[49,50]

1-$C_{10}H_7$-PCl_2 (1-naphthyl-). I.[196,271,272,566] II.[329] XI.[80] b. over 360,[271,272] b_{10} 180,[80,329] $b_{0.5}$ 118-120,[196] 135-137,[566] m. 55,[566] 54,[196] 58-59.[80,329]

2-$C_{10}H_7$-PCl_2 (2-naphthyl-). I.[329,566] b_{9-10} 180,[329] $b_{0.2}$ 110,[566] m. 41-55,[566] 50-60.[329]

$(C_5H_5FeC_5H_4)PCl_2$ (ferrocenyl-). II.[530,531] IX.[529] dec. on distillation at 0.5 mm,[529] IR.[529,531]

$C_{10}H_{11}$-PCl_2 (tetrahydronaphthyl-). II.[236] $b_{0.2}$ 99-101,[236] IR,[236] ^{31}P -163.4.[236]

$PhCH_2OC(Me)$=$CHPCl_2$. XVI.[451] $b_{0.5}$ 135-136,[451] n_D^{20} 1.5760,[451] d^{20} 1.3109.[451]

n-$BuC_6H_4PCl_2$. II.[64] b_1 116-119,[64] n_D^{25} 1.5591,[64] d_4^{25} 1.1611.[64]

s-$BuC_6H_4PCl_2$. II.[64] $b_{1.5}$ 116-120,[64] n_D^{25} 1.5644,[64] d_4^{25} 1.1840.[64]

1-Me-4-i-PrC_6H_3-2-(or 3-)PCl_2. II.[382,383] b. 275-278[382,383] (crude).

$Cl(Me)_2SiCH_2CH_2C_6H_4PCl_2$. II[447] (orientation not specified). b_2 148-152,[447] n_D^{20} 1.5648,[447] d_4^{20} 1.2265,[447] IR.[447]

4-$Et_2NC_6H_4PCl_2$. II.[394,395] undistillable oil.[394,395]

$C_{10}H_{15}PCl_2$ (1-adamantyl-). IX.[538] $b_{0.2}$ 100-101,[538] m. 54-55.[538]

$BuCH(Et)CH_2OCH_2$=CH-PCl_2. XVI.[12] b_7 145-147,[12] n_D^{20} 1.5019,[12] d_4^{20} 1.0844.[12]

$Me(CH_2)_7CHClCH_2PCl_2$. VI.[476] b. 315,[476] $b_{0.2}$ 115,[476] d^{20} 1.12.[476]

$C_{10}H_{21}PCl_2$. I.[212]

s-$AmC_6H_4PCl_2$. II.[64] b_1 118-121,[64] n_D^{25} 1.5541,[64] d_4^{25} 1.1437.[64]

$Me(CH_2)_7CH_2$-O-CH=$CHPCl_2$. XVI.[12] b_3 140-141,[12] n_D^{20} 1.4960,[12] d_4^{20} 1.0637.[12]

MeO-(-P(Cl)-$CB_{10}H_{10}C)_5$-PCl_2. XVIII.[6,8] m. 120-135.[6,8]

Dibenzofuran-3-PCl_2. II.[103] b_{25} 245-250.[103]

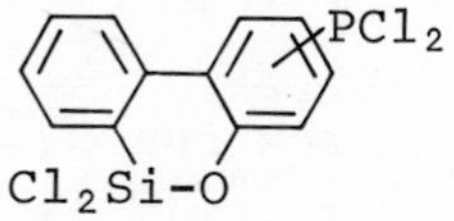

II.[448] $b_{0.1-0.3}$ 155-158,[448] n_D^{20} 1.6050,[448] d_4^{20} 1.4534.[448]

$PhC_6H_4PCl_2$. II[330,384] (orientation of Ph- group not known). b_{10} 200-220,[330,384] m. 5,[330,384] d^{14} 1.3098.[330,384]

4-$PhOC_6H_4PCl_2$. II.[102,253] b_{12} 200,[102,253] d_4^{20} 1.322.[102,253]

n-$C_6H_{13}C_6H_4PCl_2$. II.[64] $b_{1.5}$ 146-149,[64] n_D^{25} 1.5478,[64] d_4^{25} 1.1157.[64]

$C_{12}H_{25}PCl_2$. I.[588,589] $b_{1.5}$ 135-140.[588,589]

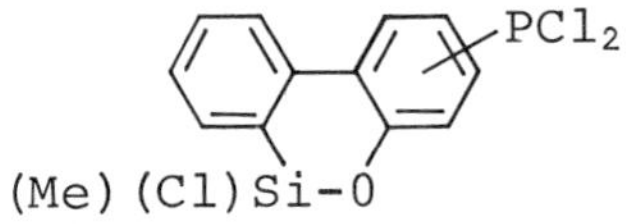

II.[448] $b_{0.1-0.5}$ 140-144,[448] n_D^{20} 1.6120,[448] d_4^{20} 1.3469.[448]

2-$PhCH_2$-C_6H_4-PCl_2. XI.[107] XVII.[112] $b_{0.2}$ 132-137,[112] $b_{0.01}$ 142-143,[107] n_D^{20} 1.6298.[107]
4-$PhCH_2$-C_6H_4-PCl_2. II[384] (probably isomer mixture) b_{20} 221,[384] d^{17} 1.182.[384]
4-$MePhNC_6H_4PCl_2$. II.[394,395] oil dec. 200.[394,395]
4-$PhCH_2CH_2$-$C_6H_4PCl_2$. II[384] (probably isomer mixture) b_{60} 250,[384] m. 2.[384]
4-Me($PhCH_2$)$NC_6H_4PCl_2$. II.[394,395] undistillable oil.[394,395]
4-Et($PhCH_2$)$NC_6H_4PCl_2$. II.[394,395] undistillable oil.[394,395]
Octadecyl-PCl_2. VIII.[189]
Ph_3CPCl_2. IX.[218] dec. 240,[218] m. 138-140.[218]
2-(Ph_2CH)$C_6H_4PCl_2$ (2-benzhydrylphenyl-). XI.[108] $b_{0.001}$ 161,[108] m. 76.5-79.[108]

Bis(phosphonous Dichlorides)

$Cl_2PCH_2PCl_2$. XVIII.[532] b_2 66-67,[532] ^{31}P -187 ± 1.[532]
$Cl_2P(CF_2)_2PCl_2$. VIII.[68] b_{758} 178-180.[68]
$Cl_2P(CH_2)_2PCl_2$. XVIII.[532] b_2 81-82,[532] ^{31}P -191 ± 1.[532]

$Cl_2P\overline{C(B_{10}H_{10})C}PCl_2$. I.[6,8] $b_{0.3}$ 119,[6,8] n_D^{22} 1.6183,[6,8]
$Cl_2P(CH_2)_3PCl_2$. XVIII.[532] b_2 93,[532] ^{31}P -192 ± 1.[532]
$Cl_2P(CH_2)_4PCl_2$. X.[485] XI.[80] XVIII.[532] $b_{2.5}$ 119-123,[485] b_2 104,[532] ^{31}P -192 ± 1.[532]
$Cl_2P(CH_2)_5PCl_2$. XI.[80]
4-Cl_2P-C_6H_4-PCl_2. XI.[80,129] XII.[28,109,555] XVIII.[26] $b_{2.5}$ 145-146,[28] b_2 145-147,[555] b_1 132-133,[109] m. 58-60,[28,80,129] m. (dec.) 58-59,[109] m. 65,[555] IR.[129]
2-Cl_2P-C_6H_{10}-PCl_2 (cyclohexylene-). IV.[22] $b_{0.5}$ 130-135,[22] n_D^{20} 1.5865,[22] d_4^{20} 1.4582,[22] mass spect.[22]
$Cl_2P(CH_2)_6PCl_2$. IV.[22] X.[485] XI.[80] b_1 138-141,[485] $b_{0.5}$ 111-120,[22] n_D^{20} 1.5402,[22] d_4^{20} 1.3205.[22]
4-$Cl_2PCH_2C_6H_4CH_2PCl_2$. XII.[28,555] b_4 85-90,[555] m. 114.5-115.5.[28]

Phosphonous Chloride Bromides

MePClBr. ^{31}P -190.0.[400]
PhPClBr. VII,[149] ^{31}P -158.[148]

Phosphonous Chloride Isocyanate

PhP(NCO)Cl. VII.[444] $b_{0.5}$ 64-65,[444] IR.[444]

Phosphonous Dibromides

CCl_3PBr_2. XIII.[436] b^{14} 98-99.[436]
CF_3PBr_2. VII.[72] XIII.[72] XIV.[72] b. 86.7 (ext.),[72] $\log P_{mm} = 3.8579 - 0.00206 \cdot T + 1.751 \cdot \log T - 1694.74/T$,[72] Trouton const. 21.7,[72] IR,[72] mass spect.,[82] ^{19}F.[420]
$CHBr_2PBr_2$. XII.[2] b.(est.) 269 ± 5,[2] $b_{0.3}$ 65,[2] n_D^{25} 1.622,[2] d^{23} 2.91,[2] IR,[2] ^1H.[2]
$BrCH_2PBr_2$. VI.[126,583,584] b_4 70,[583,584] $b_{3.25}$ 63-65,[126] n_D^{21} 1.6607,[126] d_{20}^{20} 2.6357.[583,584]
$MePBr_2$. I.[344] VII.[313] XI.[80] XII.[343,345,347] b 139-141,[313,345,347] b_{720} 138.5,[343] 139-141,[344] m. -58,[343] n_D^{20} 1.6104,[343] 1.6210,[344] d_4^{20} 2.186,[343] $\log P_{mm} = 7.87364 - 2065.1/T$,[343] IR,[343] ^{31}P -184.0.[343,400]
$CH_2{=}CHPBr_2$. I.[36] XII.[347] b. 163-165,[347] b_{20} 60.[36]
$BrCH_2CH_2PBr_2$. VI.[153,325] $b_{0.5}$ 100.[153] $b_{0.045}$ 61.[325] $b_{0.01}$ 60-62,[153] n_D^{20} 1.6515,[325] d_4^{20} 2.4303.[325]
$EtPBr_2$. I.[344,480] VII.[313] XII.[343,345,347] XIII.[241] b. 161,[241,345,480] 158-165,[347] 161-163,[313] b_{716} 160-162,[345] n_D^{15} 1.5451,[480] d_{20}^{20} 1.8216,[480] ^{31}P -194.0.[343,344,400]
$MeCHBrCH_2PBr_2$. VI[153] [mixture with $MeCH(PBr_2)CH_2Br$] $b_{0.5}$ 84,[153] 79,[153] ^{31}P -183,[153] -189,[153] ^1H.[153]
$BuPBr_2$. VIII.[561] XI.[80] b_{10} 80-95.[80,561]
$C_6F_5PBr_2$. I.[91,144] XI.[144] $b_{1.1}$ 64-65,[91] $b_{0.6}$ 64-65,[144] IR,[145] mass spect.,[398] ^{31}P -113.5.[144]
$PhPBr_2$. I.[289,344,378,387,388] II.[148] V.[56] VII.[9,122,148,289,312,313,378,387,388,410] VIII.[148] XI.[80] XII.[49,50,343,347,439,440] XIII.[310,311] b. 255-257,[80,289,378,387,388] 259-261,[410] b_{14} 132,[310] 132-134,[148] b_{13-14} 132,[148] b_{11} 126-128,[148,440] 124-126,[312] $b_{9.5}$ 122-123,[313] b_5 117-119,[49,50] b_2 95,[439] 93,[310] $b_{0.05}$ 90,[56] $b_{0.005}$ 75-80,[343] n_D^{20} 1.6719,[343] n_D^{25} 1.6534[148,440] d_4^{20} 1.873,[343] 1.872,[148] 1.8732,[440] 1.8772,[49,50] IR,[148] $\Delta H_{hydrol.}$ -41.7 ± 0.5 kcal/mole,[148] ΔH_f^o (liqu.) ∿ -8 kcal/mole,[148] ^{31}P -152.0,[344,400] -152.7,[343] -152 ± 1,[148] ^{31}P signal enhancement through dynamic polarization.[453]
$C_6H_{10}(Br)PBr_2$. VI.[153] b_{10} 114.[153]
cyclo-$C_6H_{11}PBr_2$. XIII.[243] b_{13} 125.[243]
$PhCH_2PBr_2$. XII.[30,543] b_{10} 142-143,[543] b_1 94-96,[30] n_D^{20} 1.6391,[30] d_4^{20} 1.6971.[30]
2-$MeC_6H_4PBr_2$. XII.[49,50] b_1 99-102,[49,50] d_4^{20} 1.7954.[49,50]
3-$MeC_6H_4PBr_2$. XII,[439,440] b_2 110.[439,440]
4-$MeC_6H_4PBr_2$. II.[328] XI.[80] m. 160-161.[328]

n-$C_8H_{17}PBr_2$. XII.[439,440] b_{22} 72.[439,440]
1-$C_{10}H_7PBr_2$. II.[329] XI.[80] m. 65-68.[329]

Bis(phosphonous Dibromides)

4-PBr_2-C_6H_4-PBr_2. XI.[80] no data.

Phosphonous diiodides

CF_3PI_2. XII.[43,77,343] XIII.[338,339,340] b_{413} 133,[43] b_{132} 103,[43] b_{37} 73,[43] b_{29} 69,[43] n_D^{20} 1.6320,[43] UV,[43] IR,[41,43,44] mass spect.,[82] ^{19}F,[420] hydrol.[44]
$MePI_2$. XI.[80] XII.[343,345,347] XVI.[177] b_7 82-85,[343] b_5 73-75,[345,347] b_2 68-70,[177] m. 33-36,[343] n_D^{20} 1.738,[343] d_4^{20} 2.8436,[343] IR,[343] ^{31}P -130.6.[343]
$EtPI_2$. XIII.[241] b_3 70-71.[241]
$C_3F_7PI_2$. XII.[87,127] b. 190 (est.),[127] m. -18,[127] log P_{mm} = 7.350 - 2070/T,[127] Trouton const. 20.5,[127] heat of evap. 9470 cal/mole,[127] UV,[127] IR,[127] ^{19}F.[420]
$C_6F_5PI_2$. XI.[91] $b_{0.7}$ 110-112.[91]
4-$BrC_6H_4PI_2$. VII.[136] $b_{0.05}$ 109-110.[136]
4-$ClC_6H_4PI_2$. VII.[136] $b_{0.06}$ 107-108.[136]
4-$FC_6H_4PI_2$. VII.[136] $b_{0.07}$ 104-105.[136]
$PhPI_2$. VII.[9,135,136,321,378] XI.[80] XIII.[229] $b_{1.3}$ 145-147,[229] b_1 142-143,[136] as HI salt.[321,378]
cyclo-$C_6H_{11}PI_2$. XIII.[137,243] b_{14} 171-172,[243] $b_{0.05}$ 101.[137]
4-$MeC_6H_4PI_2$. VII.[136] b_4 176-177.[136]

Phosphonous Dipseudohalides

$MeP(CN)_2$. VII.[341] m. 72-73,[341] IR,[341] x-ray data,[341] ^{31}P 81.4.[341]
$C_6F_5P(CN)_2$. VII.[143] m. 84,[143] ^{31}P +121.3.[142]
$PhP(CN)_2$. VII,[284,382,383] b_{20} 144-145,[382,383] $b_{0.1}$ 95.[284]
4-$MeC_6H_4P(CN)_2$. VII.[382,383] b_{50} 145.[382]
$MeP(NCO)_2$. VII.[341,345] b_6 33-36,[341,345] IR.[341]
$EtP(NCO)_2$. VII.[139] b_{100} 110-120.[139]
$C_6F_5P(NCO)_2$. VII.[143] $b_{2.1}$ 106,[143] ^{31}P -70.5.[142]
$PhP(NCO)_2$. VII.[9,139,219,359,444,568] b_{10} 140-146,[359] b_3 119-122,[568] 118-122,[219] b_{1-2} 82,[139] $b_{0.5}$ 75-77,[444] IR.[359,444]
$MeP(NCS)_2$. VII.[341] $b_{0.1}$ 55-57,[341] n_D^{20} 1.6741,[341] d_4^{20} 1.3639,[341] IR,[341] ^{31}P -37.9.[341]
$C_6F_5P(NCS)_2$. VII.[143] $b_{2.2}$ 143-145,[143] ^{31}P -62.2.[142]
$PhP(NCS)_2$. VII.[9,382,383] b_{20} 205-207.[382]
4-$MeC_6H_4P(NCS)_2$. VII.[382,383] b_{40} 237-240.[382,383]

D.2. Phosphinous Halides and Phosphinous Pseudohalides

Phosphinous fluorides

$(CF_3)_2PF$. VII.[70,71,315] b. -11 to -8,[315] -11.85 (est.),[71] m. -149.9 to -149.6,[71] log P_{mm} = 6.2919 - 0.007259·T + 1.75·log T - 1501/T,[71] Trouton const. 21.10,[71] IR,[421] mass spect.,[82] ^{31}P -123.9,[423,432] ^{19}F,[420,423,432] relative signs of coupl. const.,[96,355] $J_{^{13}CF}$.[214]

Me_2PF. VII.[510,512] b. 26,[512] m. -109,[510,512] log P_{mm} = 7.46 - 1370/T,[510,512] Trouton const. 21.0,[510,512] IR,[510,512] mass spect.,[510,512] ^{31}P -187,[509,510,512] -185,[518] ^{19}F,[509,510,512,518] ^{1}H.[509,510,512]

Me(Et)PF. ^{31}P -182.[518]

$(CF_2{=}CF)_2PF$. VII.[536,537] b. 63-65.[536,537]

$(C_3F_7)_2PF$. VII.[420] b. 92.2[420] m. -49.2 to -48.8,[420] log P_{mm} = 8.228 - 1954/T (0 to 65°),[420] Trouton const. 24.4,[420] IR,[420] ^{31}P -138.9,[423] ^{19}F.[420,423]

$(C_6F_5)_2PF$. VII.[146] $b_{0.02}$ 104-106,[146] m. 49,[146] ^{31}P -136.0,[146] ^{19}F.[146]

Phosphinous Chlorides

$(CF_3)_2PCl$. VII.[43,432,438] X.[199] XIII.[75] b. 23.2 (ext.),[438] 21,[43] log P_{mm} = 4.1432 + 1.75·log T - 0.002582·T - 1424/T,[438] Trouton const. 22.05,[438] hydrol.,[44] IR,[43,44] mass spect.,[82] ^{31}P -50.0,[432] ^{19}F [420,432] $J_{^{13}CF}$.[214]

$CF_3(CHF_2)PCl$. XI.[217] b. 37.[217]

$Me(CF_3)PCl$. XI.[200] XVIII.[74] b. 51.2,[74] 52-54,[200] log P_{mm} = 6.1707 + 1.75·log T - 0.005962·T - 1865/T,[74] Trouton const. 20.92,[74] IR.[74]

$(ClCH_2)_2PCl$. XV.[174] XVI.[132] b_{22} 76,[132] b_{16-18} 74-77,[174] n_D^{20} 1.5484,[174] d_4^{20} 1.4800.[174]

$Me(Cl_2CH)PCl$. XVI.[132] b_{100} 90.[132]

$Me(ClCH_2)PCl$. VI.[165] XVI.[132,133] b_{130} 63,[165] b_{98} 67,[132] b_9 59,[133] n_D^{25} 1.493,[165] $n_D^{25.5}$ 1.5140.[132]

Me_2PCl. I.[344] IV.[443] XI.[78,80] XII.[180,181,343,345] XIII.[342,425] XV.[187,201,342,349,434,524,550] XVI.[132,133,204,298] b. 72-75,[342] 71,[78,80] 73,[345] 77,[434] 78,[187] 68-74,[524] 77-79,[204] b_{756} 77-79,[298] b_{749} 73-74,[349,550] b_{716} 70-75,[343] b_{697} 72-74,[133] b_{690} 70,[132] m. 0,[342,343,349] -2,[510] -1.4 to -1.0,[78,80] log P_{mm} = 12.1408 - 2887/T (solid),[78,80] log P_{mm} = 7.884 - 1722/T (liqu.),[78,80] Trouton const. 22.9,[78,80] heat of fusion 5.33 kcal/mole,[78,80] n_D^{20} 1.4760,[204] d_4^{20} 1.2281,[343] 1.0707,[204] d_4^{25} 1.2281,[349] IR,[187,510] Raman,[187] mass spect.,[434,510] ^{31}P -93.0,[434] -92,[343,344] -96,[349,509] -96.5,[510] ^{1}H,[424,509,510] J_{H-P}.[215]

$HCF_2CF_2(Me)PCl$. VIII.[63] b_{737} 98.[63]

MeEtPCl. I.[344,600] XI.[80] XIV.[445] XV.[349] XVI.[204,298] b. 157-160,[80,445] 100-102,[600] 100-101,[204] b_{759} 100.5-101.5,[298] b_{720} 100-103,[349] b_{73} 40-41,[298] b_{15} 49-51,[445] n_D^{20} 1.474,[298] 1.4728,[600] 1.4760,[204] d_4^{20} 1.0470,[204] 1.0265,[600] d_{20} 1.0467,[298] ^{31}P -105.2.[344,349,400]

$(CF_2{=}CF)_2PCl$. XI.[93,536,537] b. 94-95,[536,537] n_D^{28} 1.4095,[536] n_D^{23} 1.4095,[537] d_4^{23} 1.550,[537] ^{19}F.[93]

$(CF_2H\text{-}CF_2)_2PCl$. VIII.[68] b_{770} 87.5-89.[68]

$Me(C_3F_7)PCl$. XI.[200] b. 92.5.[200]

$CF_3CH_2CH_2(CF_3)PCl$. XI.[200] b. 100-100.5.[200]

$(CH_2{=}CH)_2PCl$. I.[34,36,373] b. 111-112,[34] b_{20} 38-42.[36]

$(CF_2H\text{-}CH_2)_2PCl$. VIII.[68] b_{180} 120-122.[68]

C_4H_8PCl (1-chlorophospholane). XI.[79,80] $b_{42.5}$ 71.8.[79]

$(ClCH_2CH_2)_2PCl$. VI.[544] b_7 95.[544]

Me(i-Pr)PCl. XVI.[132,133] b_{208} 82-85.[132,133]

Et_2PCl. I.[39,344] XI.[242,348,402] XIV.[89,445] XV.[304,305,314,349,524] XVI.[298] b. 131-132,[39] 133,[242] b_{721} 129-131,[349] b_{714} 128-131,[348] b_{100} 70-73,[304,305] b_{72} 61-62,[298] b_{50} 48-52,[314] b_{15} 60-70,[445] 30-35,[514] n_D^{20} 1.472,[298] 1.4772,[349] d_{20} 1-0233,[298] d_4^{25} 0.9842,[349] vap. press. 4.3 mm/10°,[213] 9.2 mm/23°,[213] UV,[213] ^{31}P -119,[344,349,400] -118.2 ± 0.3.[348]

$Bu(CF_3)PCl$. I.[186] b. 129 (est.),[186] b_{60} 52,[186] log P_{mm} = 7.59 - 1894/T,[186] ΔH_{vap} 8.67 kcal/mole,[186] Trouton const. 21.6,[186] IR,[186] ^{19}F.[186]

$(C_3F_7)_2PCl$. VII.[127] b. 118-119,[127] m. -75,[127] log P_{mm} = 7.883 - 1956/T,[127] Trouton const. 22.9,[127] UV,[127] IR,[127] ^{19}F.[420]

$(CF_3CH_2CH_2)_2PCl$. XI,[320] b. 165,[320] b_{85} 95-96.[320]

Pr_2PCl. XI.[560] XIV.[445] b_{15} 99-101,[445] 65,[560] n_D^{20} 1.4732.[560]

$(i\text{-}Pr)_2PCl$. I.[559] XVI.[132,133] b_{33} 69,[133] b_{10} 46-47,[559] n_D^{20} 1.4752,[559] n_D^{23} 1.4745.[132]

$Ph(CCl_3)PCl$. XVI.[580] b_3 112-114.[580]

Me(Ph)PCl. I.[344,350,403] XIV.[445] b_{15} 150-165,[445] b_2 66-67,[344,350,403] ^{31}P -83.4.[344,400]

$Bu(MeCOCH_2)PCl$. ^{1}H.[441]

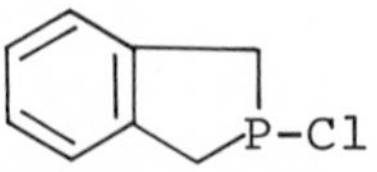

XII.[29] $b_{2.5}$ 124-125,[29] m. 48-49.[29]

$Et(4\text{-}ClC_6H_4)PCl$. I.[278] XI.[278] b_{16} 132-134,[278] b_{12} 127-128,[278] n_D^{20} 1.5869,[278] 1.5834,[278] d^{20} 1.2339,[278] 1.2290.[278]

Et(Ph)PCl. I.[197,266,344,350,403] VIII.[231,535] X.[478] XIV.[445] b. 128-131,[231] b_{15} 180-190,[445] b_{14} 105-106,[535] b_{11} 100-101,[266] b_8 90-91,[478] b_2 73-75,[344,350,403] 76,[197] n_D^{20} 1.5750,[535] 1.5707,[266] 1.5719,[478] d_4^{20} 1.1186,[478] 1.1156,[266] ^{31}P -97.0.[344,400]

$Me(PhCH_2)PCl$. VIII.[535] $b_{0.6}$ 64-67,[535] n_D^{20} 1.5786.[535]

$C_7H_8Cl(Me)PCl$ (exo-3-chloro-endo-5-methylchlorophosphino-tricyclo-[2.2.1.$0^{2,6}$]heptane). XIV.[192] $b_{0.2}$ 100,[192] ν_{max} 3.25, 12.55 μ[192] mass spect.[192] ^{1}H.[192]

$(EtOCOCH_2)_2PCl$. I.[456] b_3 113,[456] n_D^{20} 1.4825,[456] d^{20} 1.2045.[456]

Bu_2PCl. VIII.[220] XI.[80,242,560] XIV.[445] XV.[524] b. 216-217,[242] b_{15} 120-125,[80,445] b_{12} 91-92,[242] 92-93,[560] $b_{1.0}$ 77-81,[220] $b_{0.25}$ 39-47,[524] n_D^{20} 1.4743.[560]

$(i\text{-}Bu)_2PCl$. I.[559] b_{12} 77-77.5,[559] n_D^{20} 1.4670.[559]

$(s\text{-}Bu)_2PCl$. I.[559] b_{10} 80-81,[559] n_D^{20} 1.4819.[559]

$(t\text{-}Bu)_2PCl$. I.[230,487,559] b_{13} 70-72,[230] b_{10} 69-70,[559] b_3 48,[487] m. 2-3,[487] n_D^{20} 1.4830,[559] ^1H.[487]

PhC≡C-PCl(Me). XI.[52] $b_{2.5}$ 105,[52] n_D^{20} 1.6062,[52] d_4^{20} 1.1508,[52] IR,[52] ^1H.[52]

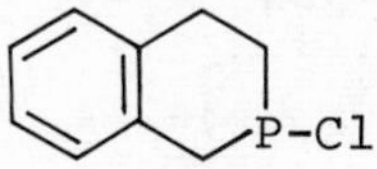

XII.[27] b_2 94-96,[27] n_D^{20} 1.6111.[27]

$Ph(MeOCOCH_2)PCl$. I.[455] b_2 122,[455] n_D^{20} 1.5480,[455] d^{20} 1.1742,[455] ^1H.[441]

$Et(MeC_6H_4)PCl$. II[408] (orientation of Me group not specified). $b_{1.6}$ 82.[408]

$Et(4\text{-}MeC_6H_4)PCl$. I.[267] b_8 107.5-108.5,[267] n_D^{20} 1.5680,[267] d_4^{20} 1.001.[267]

EtOCH=CHPCl(Ph). XVI.[13] b_2 130,[13] n_D^{20} 1.5828,[13] d_4^{20} 1.1419.[13]

Ph(s-Bu)PCl. XVI.[110] $b_{0.35}$ 98-102,[110] n_D^{20} 1.5530.[110]

$(C_5H_{11})_2PCl$. XI.[560] XVI.[132] b_{10} 118-121,[560] b_4 92,[132] n_D^{25} 1.4771,[132] n_D^{20} 1.4750.[560]

$(C_6F_5)_2PCl$. I.[33,337] XI.[33,144] $b_{1.3}$ 111-112,[337] $b_{0.6}$ 98-100,[144] $b_{0.1}$ 80-83,[33] IR,[33,337] ^{31}P -37,[144] ^{19}F.[32,232]

$Ph(C_6F_5)PCl$. XI.[144] $b_{0.6}$ 104,[144] ^{31}P -57.1.[144]

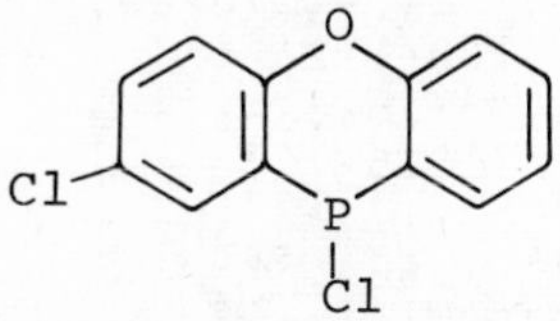

XVII.[326] $b_{ca.\ 5\mu}$ 135,[326] m. 120-122.[326]

5-Chlorodibenzophosphole. XVII.[112] $b_{0.005}$ 146-148,[112] m. 53-56.[112]

$(3\text{-}BrC_6H_4)_2PCl$. XI.[590] b_3 175-176,[590] n_D^{20} 1.6724,[590] d^{20} 1.7242.[590]

$(4\text{-}BrC_6H_4)_2PCl$. III.[263] XI.[590] b_4 205-206,[263] b_3 177-179,[590] m. 70-71.[263]

$(ClC_6H_4)_2PCl$. II[53,234] (isomer mixture). III.[413] $b_{2.5}$ 184-186,[413] $b_{0.3-0.4}$ 144-150,[53] $b_{0.1}$ 189,[234] n_D^{20} 1.5900.[53]

$(2\text{-}ClC_6H_4)_2PCl$. XIV.[445] b_{15} 253-257,[445] m. 37.[445]

$(3\text{-}ClC_6H_4)_2PCl$. XI.[590] b_2 148-150,[590] n_D^{20} 1.6460,[590] d^{20} 1.3722.[590]

$(4\text{-}ClC_6H_4)_2PCl$. III.[282] XI.[590] XIV.[445] XVI.[226] b_{15} 260-265,[445] b_5 178,[282] b_2 134-136,[590] $b_{0.2}$ 147-150,[226] m. 6-8,[445] 50-52,[282] n_D^{20} 1.6455,[590] d^{20} 1.3722.[590]

$3\text{-}ClC_6H_4(4\text{-}ClC_6H_4)PCl$. XVII.[464] $b_{0.4-0.42}$ 140-146.[464]

$(3\text{-}FC_6H_4)_2PCl$. I.[106] XI.[492] b_{22} 174,[492] b_{14} 162,[492] b_1 116.5,[106] n_D^{20} 1.5934,[492] IR,[106] mass spect.,[106] ^{31}P -74.7,[106] ^{19}F,[106] ^1H.[106]

$(4\text{-}FC_6H_4)_2PCl$. I.[106] XI.[454] b_1 110,[106] $b_{0.04}$ 108-110,[454] $b_{0.01}$ 98-105,[454] n_D^{20} 1.5890,[454] Hammett const.,[454] IR,[106] mass spect.,[106] ^{31}P -79,[106] ^{19}F,[106,454] ^1H.[106]

$(4\text{-}O_2NC_6H_4)_2PCl$. XIV.[445] b. over 370,[445] m. 91-93.[445]

10-Chlorophenoxaphosphine. XVII.[112] m. 62-64.[112]

$2\text{-}BrC_6H_4(Ph)PCl$. XVII.[163] $b_{0.5}$ 145-146.[163] n_D^{22} 1.662,[163]

$4\text{-}BrC_6H_4(Ph)PCl$. XVI.[519] XVII.[463,464] b_{11} 203-204,[101] $b_{0.6}$ 136-138,[519] $b_{0.19}$ 127.[463,464]

$2\text{-}ClC_6H_4(Ph)PCl$. XVII.[163] $b_{0.2}$ 132-133,[163] n_D^{21} 1.645.[163]

$3\text{-}ClC_6H_4(Ph)PCl$. XVII.[464] $b_{0.27}$ 123-125.[464]

$4\text{-}ClC_6H_4(Ph)PCl$. IX.[404] XVI.[519] XVII.[463,464] $b_{0.95-1.0}$ 140-145,[404] $b_{0.8}$ 134,[463,464] $b_{0.6}$ 134-135.[519]

5, 10-Dihydro-10-chlorophenophosphazin. IX.[515] yellow oil.[515]

$O_2NC_6H_4(Ph)PCl$. II[408] (orientation of NO_2 group not specified). $b_{0.2}$ 160.[408]

Ph_2PCl. I.[344,356,357,370,379,389,391,565] III.[14,16,59,61,62,90,113,176,211,234,254,386,413,514] VIII.[231,238] IX.[404] X.[483,484] XIII.[556] XIV.[216,234,396,445] XVI.[23,24,187,226,234,539] b. 320,[113,216] 316-320,[176] b_{57} 210-215,[389] b_{26} 193-194,[357] b_{18} 178-180,[356] b_{16} 179-180,[370] b_{15} 178-180,[231] b_{14} 178,[14] b_{13} 175-176,[15] b_{12} 178,[565] b_8 160-162,[238] b_3 150-156,[483] 120,[556] 150-155,[484] b_2 134-136,[413] 133-146,[226] b_1 145-147,[514] 119-120,[254] 90-92,[90] $b_{0.8}$ 117-119,[404] $b_{0.75}$ 70-80,[23,24] $b_{0.6}$ 140,[234] $b_{0.5}$ 112-113,[187] $b_{0.35}$ 108-112,[62] 110-112,[61] $b_{0.3}$ 110-140,[226] 111-112,[539] $b_{0.2-0.3}$ 110-112,[211] $b_{0.2}$ 103-106,[226] $b_{0.036}$ 112.5-116,[306] m. 15-16,[234] 14-16,[254] -18 to -17,[187] n_D^{22} 1.6091,[483] n_D^{20} 1.6360,[254] 1.6091,[484] 1.634,[565] 1.6361,[539] 1.6356,[61] 1.6340,[545] 1.6358,[187] n_D^{25} 1.6340,[556] 1.6341,[90] d_4^{20} 1.1935,[254] 1.1801,[545] d_{25}^{25} 1.2190,[90] d_0^{15} 1.2295,[14] IR,[187,533] Raman,[187] ^{31}P -81.5,[4,106,150,344,400] ^1H,[270] complex $B_{10}H_{12}\cdot 2Ph_2PCl$,[596] ^{31}P signal enhancement through dynamic polarization.[453]

$(C_6H_{11})_2PCl$ (cyclohexyl-). I.[559] IV.[22] X.[258] XI.[242] XIII.[243] b_{17} 173-174,[242] b_{12} 165,[243] b_{10} 164-166,[559] b_8 159-161,[258] b_3 136-138,[242] $b_{0.5}$ 102-108,[22] $b_{0.2}$ 105-113,[21] $b_{0.001}$ 85-87,[559] n_D^{20} 1.5330,[559] 1.5345,[22] 1.5280,[258] d_4^{20} 1.0923,[22] d^{20} 1.0677.[258]

$(C_6H_{13})_2PCl$. IV.[21,22] X.[484] $b_{1.5}$ 98-105,[484] $b_{0.3}$ 80-90,[22] $b_{0.3}$ 105-113,[21] n_D^{20} 1.4720,[484] 1.4862,[22] d_4^{20} 0.959.[22]

$[(Me_2N)_2P\text{-}CB_{10}H_{10}C\text{-}]_2PCl$. XVIII.[6] m. 90-94.[6]
$3\text{-}CF_3C_6H_4(Ph)PCl$. XVII.[463,464] $b_{0.3}$ 95.[463,464]
$4\text{-}CNC_6H_4(Ph)PCl$. XVII.[463,464] $b_{0.2}$ 162.[463,464]

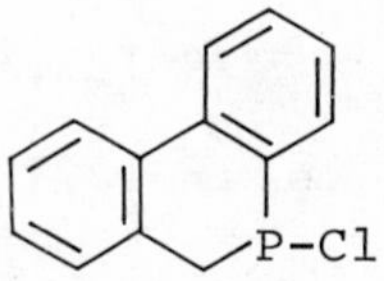

(5-chloro-5,6-dihydrodibenzo [b, d]phosphorin). VIII.[109] $b_{0.001}$ 160,[109] m. 58-62.[109]

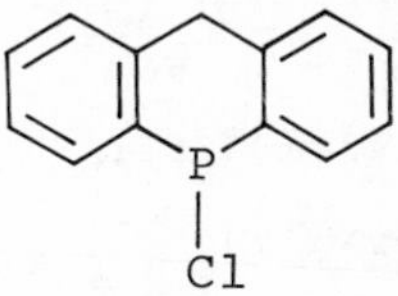

(5-chloro-5,10-dihydrodibenzo [b, e]phosphorin). II.[107,112] XI.[107] m. 78-86,[112] 102-109,[107] mass spect.[107]

$ClC_6H_4(MeC_6H_4)PCl$. II[408] (orientation of Me group not specified). b_2 140.[408]
$4\text{-}ClCH_2C_6H_4(Ph)PCl$. VIII.[409] $b_{0.001}$ 100.[409]
$MeC_6H_4(Ph)PCl$. II[408] (orientation of Me group not specified). $b_{0.2}$ 120.[408]
$4\text{-}MeC_6H_4(Ph)PCl$. I.[384,450,468,564] XIV.[445] XVI.[519] b. 340,[384] b_{100} 230-240,[445] $b_{0.6}$ 128-130.[519]
$PhCH_2(Ph)PCl$. VIII.[535] $b_{0.4}$ 165-170.[535]
$3\text{-}MeOC_6H_4(Ph)PCl$. I.[315]
$4\text{-}MeOC_6H_4(Ph)PCl$. I.[101] XVI.[519] $b_{0.6}$ 143-145,[519] $b_{0.15}$ 149-152,[101] $b_{0.03}$ 137.[101]
$(3\text{-}CF_3C_6H_4)_2PCl$. XI.[590] b_1 109-110,[590] n_D^{20} 1.5192,[590] d^{20} 1.4252.[590]
$(4\text{-}CF_3C_6H_4)_2PCl$. XI.[590] b_1 109-110,[590] n_D^{20} 1.5205,[590] d^{20} 1.4242.[590]
$(MeC_6H_4)_2PCl$. I.[565] $b_{0.5}$ 147-150,[565] $b_{0.1}$ 128-132,[565] n_D^{20} 1.619,[565] d_{20} 1.159.[565]
$(2\text{-}MeC_6H_4)_2PCl$. XIV.[445] m. 4,[445] oil.[445]
$(3\text{-}MeC_6H_4)_2PCl$. XI,[590] b_1 133-135,[590] n_D^{20} 1.6170,[590] d^{20} 1.1382.[590]
$(4\text{-}MeC_6H_4)_2PCl$. I.[384] III.[234,254,413,429] IX.[404] XI.[80,590] XIV.[445] b. 345-350,[384,445] b_3 181-182,[429] $b_{2.5}$ 158-160,[413] b_1 132-133,[254] 165-168,[80] 133-134,[590] $b_{0.21}$ 125-128,[404] $b_{0.15}$ 169,[234] n_D^{20} 1.6190,[254] 1.6160,[590] d^{20} 1.1370,[590] d_4^{20} 1.1470.[254]
$(PhCH_2)_2PCl$. III.[30,31] IX.[460] XII.[440] b_{12} 234-236,[440] b_1 160-170,[30,31] $b_{0.25-0.35}$ 131-135,[460] m. 74.5-75.5,[31] 81.5-82.[30]
$(3\text{-}MeOC_6H_4)_2PCl$. XI.[590] b_2 156-159,[590] n_D^{20} 1.6174,[590] d^{20} 1.2282.[590]
$(4\text{-}MeOC_6H_4)_2PCl$. XI.[590] b_2 190-191,[590] m. 49-53.[590]
$(2,4,5\text{-}Me_3C_6H_2)PhPCl$. I.[379,384,391] b. 356,[379,391] b_{10} 208.[384]

[-$CB_{10}H_9BrC$-P(Cl)-$CB_{10}H_9BrC$-P(OMe)-$]_3$. XVIII.[6] m. 115-120.[6]
1-$C_{10}H_7$(Ph)PCl (1-naphthyl-). XI.[235] $b_{0.1}$ 195.[235]
$(PhCH{=}CH)_2PCl$. XVI.[323] $b_{0.06}$ 229-231.[323]
$(PhCH_2CH_2)_2PCl$. III.[30,31] b_3 180-185.[30,31]
$(MeC_6H_4CH_2)_2PCl$. III[30] (isomer mixture). b_3 179-183.[30]
(2,5-$Me_2C_6H_3)_2PCl$. II.[234,575] $b_{1.5}$ 171.[234]
(4-$Me_2NC_6H_4)_2PCl$. XIV.[445] b_{15} 210-215,[445] m. 16-18.[445]

$(PhC(B_{10}H_{10})C)_2PCl$. I.[591] m. 226-227.[591]
(C_8H_{17})(4-$CH_2{=}CHC_6H_4$)PCl. I.[584]
$(C_8H_{17})_2PCl$. I.[470] IX.[460] X.[484] $b_{1.5}$ 145-150,[484] $b_{0.25-0.35}$ 132-135,[460] n_D^{20} 1.4720.[484]

MeO[$CB_{10}H_9BrC$-P(Cl)-$CB_{10}H_9BrC$-P(OMe)$]_3$OMe. XVIII.[8] m. 115-120.[8]
(2,4,5-$Me_3C_6H_2)_2PCl$. II.[382-384] b. 305.[382-384]
$(Me_3SiC_6H_4)_2PCl$. I.[160,161] b_{43} 72.[160]
5-Chloro-5,10-dihydro-10-phenyldibenzo[b,e]phosphorin. II.[108] $b_{0.001}$ 131,[108] m. 94-101.[108]
(1-$C_{10}H_7)_2PCl$ (1-naphthyl-). XI.[80] XIV.[445] b_{15} 270-280.[80,445]
$(C_5H_5FeC_5H_4)_2PCl$ (ferrocenyl-). II.[530,531] m. 183-184,[530] IR.[531]

Bis(phosphinous Chlorides)

ClP[-$C(B_{10}H_6Cl_4)C$-$]_2PCl$. I.[7,528] m. 426-427.[7,528]

ClP[-$C(B_{10}H_7Br_3)C$-$]_2PCl$. I.[7,528] m. 412-414 [528] 407-409.[7]

ClP[-$C(B_{10}H_8Br_2)C$-$]_2PCl$. I.[528] m. 415.[528]

ClP[-$C(B_{10}H_{10})C$-$]_2PCl$. I.[5,7] m. 240-241.[5,7]

P,P'-Dichlor-P,P'-dimethyl-4-xylylendiphosphine. VIII.[535] m. 70-75.[535]
(cyclo-C_6H_{11})(Cl)PP(Cl)(cyclo-C_6H_{11}). IV.[22] m. 23.[22]

Cl(Ph)$PC(B_{10}H_8Br_2)CP$(Ph)Cl. I.[528] m. 199-200.[528]

Cl(Ph)$PC(B_{10}H_{10})CP$(Ph)Cl. I.[6,8] $b_{0.35}$ 229,[6,8] n_D^{26} 1.6522.[6,8]
P,P'-Dichloro-P,P'-diphenyl-hexamethylendiphosphine. VIII.[535] yellow oil.[535]
P,P'-Dichloro-P,P'-diphenyl-4-xylylendiphosphine. VIII.[535] m. 163-171.[535]

Phosphinous Bromides

$CCl_3(PBr_2)PBr$. XII.[1] $b_{0.4}$ 46,[1] m. 45,[1] d^{23} 2.26,[1] λmax. 290 mμ.[1]

$(CF_3)_2PBr$. VII.[72,432] XIII.[72] XIV.[72] b. 41.8 (est.),[432] 42.2,[72] log P_{mm} = -1510/T + 7.68,[432] log P_{mm} = 5.8940 - 0.00566·T + 1.75·log T - 1766.5/T,[72] Trouton const. 21.0,[72] IR,[72,432] mass spect.,[82] ^{31}P -33.7,[432] ^{19}F.[420,432]

$CCl_3(CHBr_2)PBr$. XII,[1] $b_{0.3}$ 103,[1] IR.[1]

$(CHBr_2)_2PBr$. XII,[2] b. 305 ± 5 (est.),[2] $b_{0.2}$ 95,[2] n_D^{25} 1.682,[2] d^{22} 2.95,[2] IR,[2] ^1H.[2]

Me_2PBr. I.[344] XI.[80] XII.[343,345] XV.[349] b_{720} 100-105,[349] b_{716} 100-105,[343] m. 93-95,[343,349] ^{31}P -87.9,[343,344,400] -90.5 + 0.2.[349]

MeEtPBr. I.[344] VII.[445] XIII.[346] XV.[349,550] b_{740} 129,[346] b_{720} 128-129,[349,550] b_{15} 81-88,[445] ^{31}P -98.5.[344,346,349,400]

Et_2PBr. VII.[445] XI.[80] XII.[345] XIII.[242] b_{745} 153-154,[242] b_{15} 130-135,[80,445] ^{31}P -117,[346] -116.2.[400]

Pr_2PBr. VII.[445] b_{15} 143-147.[445]

MePhPBr. I[344] VII.[445] XIII.[346] b_{15} 200-205,[445] $b_{0.5}$ 65,[346] ^{31}P -77.0.[344,346,400]

EtPhPBr. VII.[445] X.[478] b_{15} 220-225,[445] $b_{0.03}$ 58-59,[478] n_D^{20} 1.6005,[478] d_4^{20} 1.3705.[478]

Bu_2PBr. VII.[445] VIII.[562] XI.[80] XIII.[242] b_{17} 118-119,[242] b_{15} 156-158,[562] b_{10} 126-128.[562]

$(C_6F_5)_2PBr$. I.[144] $b_{3.3}$ 135,[144] mass spect.,[398] ^{31}P -13.0.[144]

$C_6F_5(Ph)PBr$. I.[144] $b_{3.3}$ 145,[143,144] ^{31}P -39.3.[144]

$(2\text{-}ClC_6H_4)_2PBr$. VII.[445] b_{15} 280-290.[445]

$(4\text{-}ClC_6H_4)_2PBr$. VII.[445] b_{15} 300-310.[445] m. 47-48.[445]

$(4\text{-}O_2NC_6H_4)_2PBr$. VII.[445] m. 117-120.[445]

Ph_2PBr. I.[344] VII.[312,313,370,445,516] VIII.[60] XI.[80] XII.[343,439,440] XIII.[309,311,556] b_{18} 183-184,[370,445] b_{15} 183-184,[80] b_{12} 179-180,[440] $b_{2.5}$ 146.5-148,[309] b_2 150,[439] 149-152,[556] $b_{0.35}$ 140-141,[312] $b_{0.3}$ 131,[313] $b_{0.01}$ 105,[60] $b_{0.005}$ 111-115,[343] n_D^{20} 1.6649,[343] 1.6713,[440] d_4^{20} 1.4707,[440] 1.470,[343] ^{31}P -73.2,[343] -70.8.[344,400]

$(C_6H_{11})_2PBr$ (cyclohexyl-). XIII.[242,243] b_{12} 171.[242,243]

$4\text{-}MeC_6H_4(Ph)PBr$. VII.[445] b_{15} 270-280.[445]

$(2\text{-}MeC_6H_4)_2PBr$. VII.[445] b_{15} 210-212.[445]

$(3\text{-}MeC_6H_4)_2PBr$. XII.[439,440] b_2 142,[439] 141-142.[440]

$(4\text{-}MeC_6H_4)_2PBr$. VII.[445] b_{15} 260-270.[445]

$(PhCH_2)_2PBr$. III.[31] b_3 160-165.[31]

$(4\text{-}Me_2NC_6H_4)_2PBr$. VII.[445] m. 102.[445]

$(n\text{-}C_8H_{17})_2PBr$. I.[399] XII.[439,440] b_{11} 140,[439,440] $b_{0.15-0.19}$ 138.5-140,[399] n_D^{25} 1.4856.[399]

$(1\text{-}C_{10}H_7)_2PBr$ (1-naphthyl-). VII.[445] XI.[80] b_{15} 280-300,[80,445] m. 29-30.[80,445]

Bis(phosphinous Bromides)

$CCl_3(Br)PP(Br)CCl_3$. XII.[1] $b_{0.4}$ 77,[1] n_D^{25} 1.628,[1] d^{23}

2.00,[1] IR,[1] UV.[1]
Ph(Br)PP(Br)Ph. VII.[37] XIII.[37] m. 118-121,[37] IR.[37]

Phosphinous Iodides

$(CF_3)_2PI$. XII.[43,77,343] XIII.[97] b 72-73,[43] 71.1 (est.),[438] n_D^{15} 1.403,[43] log Pmm = 6.0085 + 1.75·log T - 0.005407·T - 1964/T,[438] Trouton const. 21.05,[438] IR,[41,43,44] UV,[43] mass spect.,[82] ^{31}P -0.8,[432] ^{19}F,[420,432] hydrol.[44]
$CF_3(Me)PI$. VII.[74] b. 102.3,[74] log Pmm= 5.8116 + 1.75·log T - 0.0047·T - 2129/T,[74] IR.[74]
Me_2PI. XI.[80]
Et_2PI. XIII.[242] b_{17} 70,[242] m. 46-47.[242]
cyclo-C_4F_8PI. XII.[307,308] b. 116-119.[308]
$(C_3F_7)_2PI$. XII.[87,127] b. 135 (est.),[127] m. -108,[127] UV,[127] IR,[127] log Pmm = 8.096 - 2170/T,[127] Trouton const. 23.7,[127] heat of evap. 9920 cal/mole,[127] ^{19}F.[420]
Bu_2PI. XIII.[242] b_{13} 121,[242] b_{2-3} 93,[242] m. 49-51.[242]
Ph_2PI. VII.[135] XIII.[242] b_{23} 215-216,[242] b_{2-3} 167-168,[242] complex with LiCl.[135]
(cyclo-$C_6H_{11})_2PI$. XIII.[242,243] b_{13} 190,[242,243] b_{2-3} 157.[242]

Bis(phosphinous Iodides)

$C_4F_8(PI)_2$ [1,4-diphospha 1.4-diiodo(perfluoro)cyclohexane]. XII.[307,308] m. 73-75.5.[307,308]
(4-$BrC_6H_4PI)_2$. VII.[136] m. 175-177.[136]
(4-$ClC_6H_4PI)_2$. VII.[136] m. 167-168.[136]
(4-$FC_6H_4PI)_2$. VII.[136] m. 138-140.[136]
$(PhPI)_2$. VII.[135,136] XIII.[229] XVIII.[135] m. 170-171,[135] 178-180.[136,229]
(cyclo-$C_6H_{11}PI)_2$. VII.[137] dec. 78.[137]
(4-$MeC_6H_4PI)_2$. VII.[136] m. 154-156.[136]

Phosphinous Pseudohalides

$(CF_3)_2PCN$. VII.[43] b. 48,[43] n_D^{20} 1.3248,[43] IR,[43,44] ^{31}P 40.7,[432] ^{19}F 51.3,[432] hydrol.[44]
MeEtPCN. VII.[445] m. 37-41.[445]
Et_2PCN. VII.[445] m. 42-43.[445]
Pr_2PCN. VII.[445] m. 81-83 (dec.).[445]
Me(Ph)PCN. VII.[445] m. 61.[445]
Et(Ph)PCN. VII.[445] m. 72.3.[445]
Bu_2PCN. VII.[445] m. 89-92.[445]
$(C_6F_5)_2PCN$. VII.[143] m. 62,[143] ^{31}P -100.2.[142]
$C_6F_5(Ph)PCN$. VII.[143] $b_{1.3}$ 139.[143]
(2-$ClC_6H_4)_2PCN$. VII.[445] m. 52.[445]
(4-$ClC_6H_4)_2PCN$. VII.[445] m. 101-102.[445]

Ph_2PCN. VII.[251,252,445] b_{15} 170-175,[445] $b_{13.5}$ 187-188,[251,252] m. 39-40,[445] n_D^{20} 1.6205,[251,252] d_4^{25} 1.1198,[251,252] ^{31}P +33.8.[252]
$4\text{-}MeC_6H_4(Ph)PCN$. VII.[445] m. 37-38.[445]
$(2\text{-}MeC_6H_4)_2PCN$. VII.[445] m. 57-60.[445]
$(4\text{-}MeC_6H_4)_2PCN$. VII.[445] b_{15} 150-153,[445] m. 10-12.[445]
$4\text{-}PhOC_6H_4(Ph)PCN$. VII.[252] $b_{0.05}$ 168-170,[252] n_D^{25} 1.6353,[252] d_4^{25} 1.1725,[252] ^{31}P +35.5.[252]
$(1\text{-}C_{10}H_7)_2PCN$ (1-naphthyl-). VII.[445] m. 129 (dec.).[445]
$(CF_3)_2PNCO$. VII.[432] b. 53.8 (est.),[432] log Pmm = -1775/T + 8.3,[432] IR,[432] ^{31}P -34.5,[432] ^{19}F.[432]
$(C_6F_5)_2PNCO$. VII.[143] m. 69,[143] ^{31}P -17.0.[142]
$C_6F_5(Ph)PNCO$. VII.[143] $b_{1.8}$ 138.[143]
$(Bu_2PNCO)_2$. VII.[295] $b_{0.05}$ 58-61,[295] m. 94-87.[295]
$(CF_3)_2PNCS$. VII.[432] b. 84 (est.),[432] log Pmm = -1715/T + 7.68,[432] IR,[432] ^{19}F.[432]
MeEtPNCS. VII.[445] m. 59-62.[445]
Et_2PNCS. VII.[445] m. 67-69 (dec.).[445]
Pr_2PNCS. VII.[445] m. 71-75.[445]
MePhPNCS. VII.[445] b_{15} 98-100.[445]
EtPhPNCS. VII.[445] $b_{2.5}$ 69-70.[445]
Bu_2PNCS. VII.[445] m. 78-82.[445]
$(C_6F_5)_2PNCS$. VII.[143] $b_{2.6}$ 152-153,[143] ^{31}P -12.4.[142]
$C_6F_5(Ph)PNCS$. VII.[143] $b_{1.5}$ 152-155.[143]
$(2\text{-}ClC_6H_4)_2PNCS$. VII.[445] b_2 110-115.[445]
$(4\text{-}ClC_6H_4)_2PNCS$. VII.[445] m. 120 (dec.).[445]
Ph_2PNCS. VII.[445] b_{15} 270-280.[445]
$4\text{-}MeC_6H_4(Ph)PNCS$. VII.[445] crude crystals, red.[445]
$(2\text{-}MeC_6H_4)_2PNCS$. VII.[445] undistillable oil.[445]
$(4\text{-}MeC_6H_4)_2PNCS$. VII.[445] b_{15} 230-232.[445]
$(1\text{-}C_{10}H_7)_2PNCS$ (1-naphthyl-). VII.[445] m. 68-70.[445]
$(CF_3)_2PN_3$. VII.[540,541] vap. press. 1 mm/-60° 57 mm/0°,[541] dec. at room temp.,[541] IR.[541]
Ph_2PN_3. VII.[430,431] crystals at -78,[430,431] dec. at 13.5-13.8.[431]

$N_3(Ph)P\overline{C(B_{10}H_8Br_2)C}P(Ph)N_3$. VII.[528] m. 125 (explosive dec.).[528]

(received September 17, 1971)

REFERENCES

1. Airey, P. L., Z. Naturforsch., 24b, 1393 (1969).
2. Airey, P. L., H. Drawe, and A. Henglein, Z. Naturforsch., 23b, 916 (1968).
3. Akamsin, V. D. and N. I. Rizpolozhenskii, Izv. Akad. Nauk SSSR, Ser. Khim., 1966, 493; C.A., 65, 5480a (1966).
4. Akitt, J. W., R. H. Cragg, and N. N. Greenwood, Chem. Commun., 1966, 134.

5. Alexander, R. P., and H. Schroeder, Inorg. Chem., 2, 1107 (1963).
6. Alexander, R. P., and H. Schroeder, Inorg. Chem., 5, 493 (1966).
7. Alexander, R. P., and H. Schroeder, U.S. Patent 3,373, 193 (1968); C.A., 68, 105344x (1968).
8. Alexander, R. P., and H. Schroeder, U.S. Patent 3,444,272 (1969); C.A., 71, 39163w (1969).
9. Anderson H. H., J. Amer. Chem. Soc., 75, 1576 (1953).
10. Ang, H. G., and R. Schmutzler, J. Chem. Soc. (A), 1969, 702.
11. Angelelli, J. M., R. T. C. Brownlee, A. R. Katritzky, R. D. Topsom and L. N. Yakhontov, J. Amer. Chem. Soc., 91, 4500 (1969).
12. Anisimov, K. N., and N. E. Kolobova, Izv. Akad. Nauk SSSR, Otd. Khim. Nauk; C.A., 57, 13790f (1962).
13. Anisimov, K. N., and N. E. Kolobova, Izv. Akad. Nauk SSSR, Otd. Khim. Nauk, 1962, 444; C.A., 57, 12529c (1962).
14. Arbuzov, A. E., J. Russ. Phys.-Chem. Soc., 42, 395 (1910); C.A., 5, 1397 (1911).
15. Arbuzov, A. E., and K.V. Nikonorov, Zh. Obshch. Khim., 18, 2008 (1948); C.A., 43, 3801i (1949).
16. Arbuzov, B. A., and N. P. Grechkin, Zh. Obshch. Khim., 20, 107 (1950); C.A., 44, 5832e (1950).
17. Arbuzov, B. A., and L. A. Shapshinskaya, Izv. Akad. Nauk SSSR, Otd. Khim.Nauk, 1962, 65; C.A., 57, 13791b (1962).
18. Asmus, K. D., A. Henglein, G. Meissner, and D. Perner, Z. Naturforsch., 19b, 549 (1964).
19. Auger, V., Compt. Rend., 139, 671 (1904).
20. Babkina, E. I., Zh. Obshch. Khim., 39, 1651 (1969); C.A., 71, 86235f (1969).
21. Babkina, E. I., and I. V. Vereshchinskii, Zh. Obshch. Khim., 37, 513, (1967); C.A., 67, 16710a (1967).
22. Babkina, E. I., and I. V. Vereshchinskii, Zh. Obshch. Khim., 38, 1772 (1968); C.A., 70, 4233w (1969).
23. Bacon, W. E., French Patent 1,451,986 (1966); C.A., 66, PC 65627s (1967).
24. Bacon, W. E., U.S. Patent 3,305,570 (1967); C.A., 67, PC 11584c (1967).
25. Baldwin, R. A., K. A. Smitheman, and R. M. Washburn, J. Org. Chem., 26, 3547 (1961).
26. Baldwin, R. A., C. O. Wilson, and R. I. Wagner, J. Org. Chem., 32, 2172 (1967).
27. Baranov, Yu. I., O. F. Filippov, S. L. Varshavskii, B. S. Glebychev, M. I. Kabachnik, and N. K. Bliznyuk, USSR Patent 213,857 (1968); C.A., 69, 77489g (1968).
28. Baranov, Yu. I., O. F. Filippov, S. L. Varshavskii, and M. I. Kabachnik, Dokl. Akad. Nauk SSSR, 182, 337 (1968); C.A., 70, 11756x (1969).

29. Baranov, Yu. I., O. F. Filippov, S. L. Varshavskii, M. I. Kabachnik, and N. K. Bliznyuk, USSR Patent 210,155 (1968); C.A., 69, 59369e (1968).
30. Baranov, Yu. I., and S. V. Gorelenko, Zh. Obshch. Khim., 39, 836 (1969); C.A., 71, 61494b (1969).
31. Baranov, Yu. I., S. L. Varshavskii, M. I. Kabachnik, and N. K. Bliznyuk, USSR Patent 199,876 (1967); C.A., 68, P59720s (1968).
32. Barlow, M. G., M. Green, R. N. Haszeldine, and H. G. Higson, J. Chem. Soc. (B), 1966, 1025.
33. Barlow, M. G., M. Green, R. N. Haszeldine, and H. G. Higson, J. Chem. Soc. (C), 1966, 1592.
34. Bartocha, B., U.S. Patent 3,100,217 (1963); C.A., 60, 551h (1964).
35. Bartocha, B., F. E. Brinckman, H. D. Kaesz, and F. G. A. Stone, Proc. Chem. Soc., 1958, 116.
36. Bartocha, B., C. M. Douglas, and M. Y. Gray, Z. Naturforsch., 14b, 809 (1959).
37. Baudler, M., O. Gehlen, K. Kipker, and P. Backes, Z. Naturforsch., 22b, 1354, (1967).
38. Becker, S. B., U.S. Patent 3,036,132 (1962); C.A., 57, 11237e (1962).
39. Beeby, M. H., and F. G. Mann, J. Chem. Soc., 1951, 411.
40. Beg, M. A. A., Bull. Chem. Soc. Japan, 40, 15 (1967).
41. Beg, M. A. A., and H. C. Clark, Can. J. Chem., 40, 393 (1962).
42. Benda, L. and W. Schmidt, U.S. Patent 1,607,113 (1926); C.A., 21, 249 (1927).
43. Bennett, F. W., H. J. Emeléus, and R. N. Haszeldine, J. Chem. Soc., 1953, 1565.
44. Bennett, F. W., H. J. Emeléus, and R. N. Haszeldine, J. Chem. Soc., 1954, 3896.
45. Bennett, F. W., H. J. Emeléus, and R. N. Haszeldine, Nature, 166, 225 (1950).
46. Bentov, M., L. David, and E. D. Bergmann, J. Chem. Soc., 1964, 4750.
47. Biddle, P., J. Kennedy, and J. L. Williams, Chem. Ind. (London), 1957, 1481.
48. Bliznyuk, N. K., P. S. Khokhlov, L. I. Markova, Z. N. Kvasha, and S. L. Varshavskii, USSR Patent 237,888(1969); C.A., 71, 50222k (1969).
49. Bliznyuk, N. K., Z. N. Kvasha, and A. F. Kolomiets, USSR Patent 179,316 (1966); C.A., 65, P2297h (1966).
50. Bliznyuk, N. K., Z. N. Kvasha, and A. F. Kolomiets, Zh. Obshch. Khim., 37, 890 (1967); C.A., 68, 13072d (1968).
51. Bliznyuk, N. K., L. D. Protasova, Z. N. Kvasha, and S. L. Varshavskii, USSR Patent 221,697 (1968); C.A., 69, P106877f (1968).
52. Bogolyubov, G. M., and A. A. Petrov, Zh. Obshch. Khim.,

37, 229 (1967); C.A., 66, 105029g (1967).
53. Boisselle, A. P., and N. A. Meinhardt, J. Org. Chem., 27, 1828 (1962).
54. Bokanov, A. I., and V. A. Plakhov, Zh. Obshch. Khim., 35, 350 (1965); C.A., 62, 13174f (1965).
55. Bourneuf, M., Bull. Soc. Chem. France, 33, 1808 (1923).
56. Bowie, R. A., and O. C. Musgrave, J. Chem. Soc. (C), 1966, 566.
57. Bowles, J. A. C., and C. James, J. Amer. Chem. Soc., 51, 1406 (1929).
58. Brazier, J. F., F. Mathis, and R. Wolf, Compt. Rend., 262C, 1393 (1966).
59. Broglie, A., Ber., 10, 628 (1877).
60. Brooks, E. H., F. Glockling, and K. A. Hooton, J. Chem. Soc., 1965, 4283.
61. Brown, M. P., and H. B. Silver, Chem. Ind. (London), 1961, 24.
62. Brown, M. P. and H. B. Silver, British Patent 916,131 (1963); C.A., 59, 3958g (1963).
63. Bruker, A. B., Kh. R. Raver, and L. Z. Soborovskii, Probl. Organ. Sinteza, Akad. Nauk SSSR, Otd. Obshch. i Tekhn. Khim., 1965, 285; C.A., 64, 6681c (1966).
64. Buchner, B., and L. B. Lockhardt, J. Amer. Chem. Soc., 73, 755 (1951).
65. Buchner, B. and L. B. Lockhardt, Org. Syn., 31, 88 (1951).
66. Bugerenko, E. F., E. A. Chernyshev, L. I. Shulgina, and A. S. Petukova, USSR Patent 229,516 (1969); C.A., 70, 58012s (1959).
67. Bulman, M. J., Tetrahedron, 25, 1433 (1969).
68. Burch, G. M., H. Goldwhite, and R. N. Haszeldine, J. Chem. Soc., 1963, 1083.
69. Burch, G. M., H. Goldwhite, and R. N. Haszeldine, J. Chem. Soc., 1964, 572.
70. Burg, A. B., and G. Brendel, U.S. Patent 2,959,620 (1960); C.A., 55, 6375d (1961).
71. Burg, A. B., and G. Brendel, J. Amer. Chem. Soc., 80, 3198 (1958).
72. Burg, A. B., and J. E. Griffiths, J. Amer. Chem. Soc., 82, 3514 (1960).
73. Burg, A. B., and J. Heners, J. Amer. Chem. Soc., 87, 3092 (1965).
74. Burg, A. B., K. K. Joshi, and J. F. Nixon, J. Amer. Chem. Soc., 88, 31 (1966).
75. Burg, A. B., and W. Mahler, U.S. Patent 2,866,824 (1958); C.A., 53, 10037e (1959).
76. Burg, A. B., and W. Mahler, J. Amer. Chem. Soc., 79, 4242 (1957).
77. Burg, A. B., W. Mahler, A. J. Bilbo, C. P. Haber, and D. L. Herring, J. Amer. Chem. Soc., 79, 247 (1957).
78. Burg, A. B., and P. L. Slota, J. Amer. Chem. Soc.,

80, 1107 (1958).
79. Burg, A. B., and P. J. Slota, J. Amer. Chem. Soc., 82, 2148 (1960).
80. Burg, A. B., and R. I. Wagner, U.S. Patent 2,934,564 (1960); C.A., 54, 18437 (1960).
81. Cassella Co., British Patent 258,744 (1925); C.A., 21, 3105 (1927).
82. Cavell, R. G., and R. C. Dobbie, Inorg. Chem., 7, 101 (1968).
83. Cavell, R. G., R. C. Dobbie, and W. J. R. Tyerman, Can. J. Chem., 45, 2849 (1967).
84. Challenger, F., and F. Pritchard, J. Chem. Soc., 125, 864 (1924).
85. Christol, C., and H. Christol, J. Chim. Phys., 62, 246 (1965).
86. Corbridge, D. E. C., Topics Phosphorus Chem., 6, 235 (1968).
87. Codell, M., J. Chem. Eng. Data, 8, 460 (1963).
88. Cohen, S. C., M. L. N. Reddy, and A. G. Massey, Chem. Commun., 1967, 451.
89. Collie, N., and F. Reynolds, J. Chem. Soc., 107, 367 (1915).
90. Cooper, R. S., U.S. Patent 3,094,559 (1963); C.A., 59, 11566e (1963).
91. Cowley, A. H., and R. P. Pinnell, J. Amer. Chem. Soc., 88, 4533 (1966).
92. Cowley, A. H., and M. W. Taylor, J. Amer. Chem. Soc., 91, 1026 (1969).
93. Cowley, A. H., and M. W. Taylor, J. Amer. Chem. Soc., 91, 1929 (1969).
94. Coyne, D. M., W. E. McEwen, and C. D. Vanderwerf, J. Amer. Chem. Soc., 78, 3061 (1956).
95. Crutchfield, M. M., C. H. Dungan, and J. R. Van Wazer, Topics Phosphorus Chem., 5, 1 (1967).
96. Cunliffe, A. V., E. G. Finer, R. K. Harris, and W. McFarlane, Mol. Phys., 12, 497 (1967).
97. Cullen, W. R., Can. J. Chem., 38, 439 (1960).
98. Daasch, L. W., and D. C. Smith, Anal. Chem., 23, 853 (1951).
99. Davankov, A. B., M. I. Kabachnik, V. V. Korshak, Yu. A. Leikin, R. F. Okhovetsker, and E. N. Tsvetkov, Zh. Obshch. Khim., 37, 1605 (1967); C.A., 67, 90884b (1967).
100. Davies, W. C., J. Chem. Soc., 1935, 462.
101. Davies, W. C., and F. G. Mann, J. Chem. Soc., 1944, 276.
102. Davies, W. C., and C. J. O. R. Morris, J. Chem. Soc., 1932, 2880.
103. Davies, W. C., and C. W. Othen, J. Chem. Soc., 1936, 1236.
104. Davis, M., and F. G. Mann, J. Chem. Soc., 1964, 3770.

105. Davydova, V. P., and M. G. Voronkov, USSR Patent 135,485 (1961); C.A., 55, 15350c (1961).
106. De Ketelaere, R., E. Muylle, W. Vanermen, E. Claeys, and G. P. van der Kelen, Bull. Soc. Chim. Belg., 78, 219 (1969).
107. De Koe, P., and F. Bickelhaupt, Angew. Chem., 79, 533 (1967); Angew. Chem. Int. Ed. Engl., 6, 567 (1967).
108. De Koe, P., and F. Bickelhaupt, Angew. Chem., 80, 912 (1968); Angew. Chem. Int. Ed. Engl., 7, 889 (1968).
109. De Koe, P., R. Van Veen, and F. Bickelhaupt, Angew. Chem., 80, 486 (1968).
110. Dietsche, W. H., Tetrahedron, 23, 3049 (1967).
111. Dimitrieva, L. E., K. V. Karavanov, and S. Z. Ivin, Metody Poluch. Khim. Reaktiv. Prep. No. 15, 148 (1967); C.A., 68, 78366x (1968).
112. Doak, G. O., L. D. Freedman, and J. B. Levy, J. Org. Chem., 29, 2382 (1964).
113. Doerken, C., Ber. 21, 1505 (1888).
114. Drake, L. R., and C. S. Marvel, J. Org. Chem., 2, 387 (1937).
115. Dreher, H., and F. Otto, Ann. Chim. Pharm., 154, 130 (1873).
116. Drozd, G. I., Ivin, and V. V. Sheluchenko, Zh. Vses. Khim. Obshch., 12, 472 (1967); C.A., 67, 116917s (1967).
117. Drozd, G. I., S. Z. Ivin, and V. V. Sheluchenko, Zh. Vses. Khim. Obshch., 12, 474 (1967); C.A., 67, 108705f (1967).
118. Drozd, G. I., S. Z. Ivin, V. V. Sheluchenko, and B. I. Tetel'baum, Zh. Obshch. Khim., 37, 957 (1967); C.A., 67, 108712f (1967).
119. Drozd, G. I., S. Z. Ivin, V. V. Sheluchenko, and B. I. Tetel'baum, Zh. Obshch. Khim., 37, 958 (1967); C.A., 68, 39735x (1968).
120. Drozd, G. I., S. Z. Ivin, V. V. Sheluchenko, B. I. Tetel'baum, G. M. Luganskii and A. D. Varshavskii, Zh. Obshch. Khim., 37, 1343 (1967); C.A., 67, 108707h (1967).
121. Drozd, G. I. S. Z. Ivin, V. V. Sheluchenko, B. I. Tetel'baum, G. M. Luganskii, and A. D. Varshavskii, Zh. Obshch. Khim., 37, 1631 (1967); C.A., 68, 78358w (1968).
122. Druce, P. M., M. F. Lappert and P. N. K. Riley, Chem. Commun., 1967, 486.
123. Durig, J. R., F. Block, and I. W. Levin, Spectrochim. Acta, 21, 1105 (1965).
124. Dybovskii, R. K., and V. A. Rogozkin, Zh. Prikl. Khim., 40, 228 (1967); C.A., 67, 21965s (1967).
125. Dye, W. T., J. Amer. Chem. Soc., 70, 2595 (1948).

126. Edmundson, R. S., and E. W. Mitchell, J. Chem. Soc. (C), 1966, 1096.
127. Emeléus, H. J., and J. D. Smith, J. Chem. Soc., 1959, 375.
128. Errede, L. A., and W. A. Pearson, J. Amer. Chem. Soc., 83, 954 (1961).
129. Evleth, E. M., L. D. Freeman, and R. I. Wagner, J. Org. Chem., 27, 2192 (1962).
130. Fedorova, G. K., and A. V. Kirsanov, Zh. Obshch. Khim., 30, 4044 (1960); C.A., 55, 23402c (1961).
131. Fedorova, G. K., and A. V. Kirsanov, Zh. Obshch. Khim., 33, 1011 (1963); C.A., 59, 8785a (1963).
132. Ferron, J. L., Nature, 189, 916 (1961).
133. Ferron, J. L., Can. J. Chem., 39, 842 (1961).
134. Ferron, J. L., B. J. Perry, and J. B. Reesor, Nature, 188, 227 (1960).
135. Feshchenko, N. G., and A. V. Kirsanov, Zh. Obshch. Khim., 31, 1399 (1961); C.A., 55, 27169 (1961).
136. Feshchenko, N. G., T. V. Kovaleva, and A. V. Kirsanov, Zh. Obshch. Khim., 39, 2184 (1969); C.A., 72, 43796t (1970).
137. Feshchenko, N. G., E. A. Mel'nichuk, and A. V. Kirsanov, Zh. Obshch. Khim., 39, 2139 (1969); C.A., 72, 21743w (1970).
138. Fiat, D., M. Halmann, L. Kugel, and J. Reuben, J. Chem. Soc., 1962, 3837.
139. Fielding, H. C., British Patent 907,029 (1962); C.A., 58, 280d (1963).
140. Fields, R., H. Goldwhite, R. N. Haszeldine, and J. Kirman, J. Chem. Soc. (C), 1966, 2075.
141. Fields, R., R. N. Haszeldine, and N. F. Wood, J. Chem. Soc. (C), 1970, 1370.
142. Fild, M., Z. Naturforsch., 23b, 604 (1968).
143. Fild, M., O. Glemser, and I. Hollenberg, Naturwiss., 53, 130 (1966).
144. Fild, M., O. Glemser, and I. Hollenberg, Z. Naturforsch., 21b, 920 (1966).
145. Fild, M., I. Hollenberg, and O. Glemser, Z. Naturforsch., 22b, 248 (1967).
146. Fild, M., and R. Schmutzler, J. Chem. Soc. (A), 1969, 840.
147. Finch, A., P. J. Gardner, A. Hameed, and K. K. Sen Gupta, Chem. Commun., 1969, 854.
148. Finch, A., P. J. Gardner, and K. K. Sen Gupta, J. Chem. Soc. (B), 1966, 1162.
149. Finegold, H., Ann. N.Y. Acad. Sci., 70, 875 (1958).
150. Fluck, E., and H. Binder, Z. Anorg. Allg. Chem., 354, 139 (1967).
151. Fluck, E., and R. M. Reinisch, Z. Anorg. Allg. Chem., 328, 165 (1964).
152. Fontal, B., and H. Goldwhite, Chem. Commun., 1965, 111.

153. Fontal, B., and H. Goldwhite, J. Org. Chem., 31, 3804 (1966).
154. Foss, V. L., V. V. Kudinova, G. B. Postnikova, and I. F. Lutsenko, Dokl. Akad. Nauk SSSR, 146, 1106 (1962); C.A., 58, 7968 (1962).
155. Fox, R. B., J. Amer. Chem. Soc., 72, 4147 (1950).
156. Frank, A. W., J. Org. Chem., 26, 850 (1961).
157. Frank, A. W., J. Org. Chem., 24, 966 (1959).
158. Frank, A. W., Can. J. Chem., 46, 3573 (1968).
159. Frank, A. W., and I. Gordon, Can. J. Chem., 44, 2593 (1966).
160. Frisch, K. C., and H. Lyons, J. Amer. Chem. Soc., 75, 4078 (1953).
161. Frisch, K. C., and H. Lyons, U.S. Patent 2,673,210 (1954); C.A., 49, 4018h (1955).
162. Funatsukuri, G., J. Usui, H. Yoshioka, and M. Ueda, Japanese Patent 17,485 (1960); C.A., 55, 19862a (1961).
163. Gallagher, M. J., E. C. Kirby, and F. G. Mann, J. Chem. Soc., 1963, 4846.
164. Garner, A. Y., U.S. Patent 3,161,607 (1964); C.A., 62, P5405e (1965).
165. Garner, A. Y., U.S. Patent 3,161,687 (1964); C.A., 62, 7957h (1965).
166. Garner, A. Y., and A. A. Tedeschi, J. Amer. Chem. Soc., 84, 4734 (1962).
167. Gefter, E. L., Zh. Obschch. Khim., 28, 1338 (1958); C.A., 52, 19999 (1958).
168. Gefter, E. L., Zh. Obshch. Khim., 31, 949 (1961); C.A., 55, 23399b (1961).
169. Gefter, E. L., Zh. Obshch. Khim., 32, 3401 (1962); C.A., 58, 7966a (1963).
170. Gefter, E. L., Zh. Obshch. Khim., 33, 3548 (1963); C.A., 60, 8056h (1964).
171. Gefter, E. L., USSR Patent 107,266 (1957); C.A., 52, 2065g (1958).
172. Gefter, E. L. and I. A. Rogacheva, Zh. Obshch. Khim., 32, 3962 (1962); C.A., 59, 658a (1963).
173. Gefter, E. L. and I. A. Rogacheva, Zh. Obshch. Khim., 34, 88 (1964); C.A., 60, 10709f (1964).
174. Genkina, G. K. and V. A. Gilyarov, Izv. Akad. Nauk SSSR, Ser. Khim., 1969, 185; C.A., 70, 115265c (1969).
175. Gier, T. E., U.S. Patent 3,051,756 (1962); C.A., 58, 9140d (1963).
176. Gilman, H., and G. E. Brown, J. Amer. Chem. Soc., 67, 824 (1945).
177. Ginsburg, V. A., and N. F. Privezentseva, Zh. Obshch. Khim., 28, 736 (1958); C.A., 52, 17092i (1958).
178. Ginsburg, V. A., and A. Ya. Yakubovich, Zh. Obshch. Khim., 28, 728 (1958); C.A., 52, 17092d (1958).
179. Gitel, P. O., A. I. Titov, and M. V. Sizova, USSR

Patent 172,322 (1965); C.A., 63, P16384d (1965).
180. Gladshtein, B. M., and L. N. Shitov, USSR Patent 207,902 (1967); C.A., 69, P36254z (1968).
181. Gladshtein, B. M., and L. N. Shitov, Zh. Obshch. Khim., 37, 2586 (1967); C.A., 69, 27494a (1968).
182. Gladshtein, B. M., L. N. Shitov, B. G. Kovalev, and L. Z. Soborovskii, Zh. Obshch. Khim., 35, 1570 (1965); C.A., 63, 18141g (1965).
183. Goetz, H. and S. Domin, Justus Liebigs Ann. Chem., 704, 1 (1967).
184. Goetz, H., F. Nerdel, and K. H. Wiechel, Justus Liebigs Ann. Chem, 665, 1 (1963).
185. Goetz, H., and D. Probst, Justus Liebigs Ann. Chem., 715, 1 (1968).
186. Gosling, K., D. J. Holman, J. D. Smith, and B. N. Ghose, J. Chem. Soc. (A), 1968, 1909.
187. Goubeau, J., R. Baumgärtner, W. Koch, and U. Müller, Z. Anorg. Allg. Chem., 337, 174 (1965).
188. Goubeau, J., and D. Langhardt, Z. Anorg. Allg. Chem., 338, 163 (1965).
189. Grayson, M., U.S. Patent 3,074,994 (1963); C.A., 59, 1682d (1963).
190. Green, J. H. S., Spectrochim. Acta, 24A, 137 (1968).
191. Green, J. H. S., and W. Kynaston, Spectrochim. Acta, 25A, 1677 (1969).
192. Green, M., J. Chem. Soc., 1965, 541.
193. Green, M., Proc. Chem. Soc., 1963, 177.
194. Green, M., R. N. Haszeldine, B. R. Iles, and D. G. Rowsell, J. Chem. Soc., 1965, 6879.
195. Green, M., and R. F. Hudson, Proc. Chem. Soc., 1961, 145.
196. Green, M., and R. F. Hudson, J. Chem. Soc., 1958, 3129.
197. Green, M., and R. F. Hudson, J. Chem. Soc., 1963, 540.
198. Griffiths, J. E., Spectrochim. Acta, 21, 1135 (1965).
199. Griffiths, J. E., and A. B. Burg, J. Amer. Chem. Soc., 84, 3442 (1962).
200. Grinblat, M. P., A. L. Klebanskii and V. N. Prons, Zh. Obshch. Khim., 39, 172 (1969); C.A., 70, 106612m (1969).
201. Groenweghe, L. C. D., U.S. Patent 3,071,616 (1963); C.A., 58, 11403b (1963).
202. Groenweghe, L. C. D., U.S. Patent 3,024,278 (1962); C.A., 57, 7311c (1962).
203. Gruettner, G., and M. Wiernik, Ber., 48, 1473 (1915).
204. Gruzdev, V. G., S. Z. Ivin, and K. V. Karavanov, Zh. Obshch. Khim., 35, 1027 (1965); C.A., 63, 9979f (1965).
205. Gruzdev, V. G., S. Z. Ivin, and K. V. Karavanov, Metody Poluch. Khim. Reaktiv. Prep. No. 15, 7 (1967); C.A., 68, 114699u (1968).

206. Gruzdev, V. G., K. V. Karavanov, and S. Z. Ivin, USSR Patent 173, 764 (1965);C.A., 64, PC1701h (1966).
207. Gruzdev, V. G., K. V. Karavanov, and S. Z. Ivin, USSR Patent 186, 467 (1966); C.A., 66, P76153k (1967).
208. Gruzdev, V. G., K. V. Karavanov, S. Z. Ivin, I. S. Mazel, and V. V. Tarasov, Zh. Obshch, Khim., 37, 450 (1967); C.A., 67, 43870j (1967).
209. Guichard, F., Ber., 32, 1572 (1899).
210. Gutowsky, H. S., and D. W. McCall, J. Chem. Phys., 22, 162 (1954).
211. Haber, C. P., D. L. Herring, and E. A. Lawton, J. Amer. Chem. Soc., 80, 2116 (1958).
212. Hall, R. E., A. Kessler, and A. R. McLain, U.S. Patent 3,459, 808 (1969); C.A., 71, 81524q (1969).
213. Halmann, P., J. Chem. Soc., 1963, 2853.
214. Harris, R. K., J. Phys. Chem., 66, 768 (1962).
215. Harris, R. K., and R. G. Hayter, Can. J. Chem., 42, 2282 (1964).
216. Hartmann, H., C. Beermann, and H. Czempik, Z. Anorg. Allg. Chem., 287, 261 (1956).
217. Haszeldine, R. N., H. Goldwhite, and D. G. Rowsell, British Patent 1,072,241 (1967); C.A., 67, P64528u (1967).
218. Hatt, H. H., J. Chem. Soc., 1933, 776.
219. Haven, A.C., J. Amer. Chem. Soc., 78, 842 (1956).
220. Henderson, W. A., S. A. Buckler, N. E. Day, and M. Grayson, J. Org. Chem., 26, 4770 (1961).
221. Heners, J., and A. B. Burg, J. Amer. Chem. Soc., 88, 1677 (1966).
222. Henglein, A., Intern. J. Appl. Radiat. Isotopes, 8, 156 (1960).
223. Henglein, A., "Large Radiation Sources in Ind.," Proc. Conf. Warsaw, 2, 139 (1959).
224. Henglein, A., H. Drawe, and D. Perner, Radiochim. Acta, 2, 19 (1963).
225. Herring, D. L., Chem. Ind. (London), 1960, 717.
226. Higgins, W. A., G. R. Norman, and W. G. Craig, U.S. Patent 2,779,787 (1957); C.A., 51, 8135a (1957).
227. Hinton, B., F. G. Mann, and D. Todd, J. Chem. Soc., 1961, 5454.
228. Hoffmann, F. W., D. H. Wadsworth, and H. D. Weiss, J. Amer. Chem. Soc., 80, 3945 (1958).
229. Hoffmann, H., and R. Grünewald, Chem. Ber., 94, 186 (1961).
230. Hoffmann, H. and P. Schellenbeck, Chem. Ber., 100, 692 (1967).
231. Hofmann, E., British Patent 904,086 (1962); C.A., 57, 16661a (1962).
232. Hogben, M. G., and W. A. G. Graham, J. Amer. Chem. Soc., 91, 283 (1969).
233. Holmes, R. R., and R. P. Wagner, Inorg, Chem., 2,

384 (1963).

234. Horner, L., P. Beck, and V. G. Toscano, Chem. Ber., 94, 2122 (1961).
235. Horner, L., F. Schedlbauer, and P. Beck, Tetrahedron Lett. 1964, 1421.
236. Houalla, D., R. Miquel, and R. Wolf, Bull. Soc. Chem., 1963, 1152.
237. Houalla, D., and R. Wolf, Compt. Rend., 259, 180 (1964).
238. Imperial Chemical Industries Ltd., British Patent 888,398 (1962); C.A., 57, 2256d (1962).
239. Ionov, S. P., and G. V. Ionova, Zh. Fiz. Khim., 43, 825 (1969); C.A., 71, 34903c (1969).
240. Issleib, K., and R. Kuemmel, Chem. Ber., 100, 3331 (1967).
241. Issleib, K., and B. Mitscherling, Z. Naturforsch., 15b, 267 (1960).
242. Issleib, K., and W. Seidel, Chem. Ber., 92, 2681 (1959).
243. Issleib, K., and W. Seidel, Z. Anorg. Allg. Chem., 303, 155 (1960).
244. Ivin, S. Z., L. E. Dmitrieva, and K. V. Karavanov, Zh. Obshch. Khim., 36, 950 (1966); C.A., 65, 10614a (1966).
245. Ivin, S. Z., and G. I. Drozd, USSR Patent 181,111 (1966); C.A., 65, P12239h (1966).
246. Ivin, S. Z., K. V. Karavanov, and V. G. Gruzdev, USSR Patent 159,527 (1963); C.A., 60, 14541a (1964).
247. Ivin, S. Z., K. V. Karavanov, V. G. Gruzdev, and I. P. Komkov, USSR Patent 160,184; C.A., 61, 5962c (1964).
248. Jackson, I. K., W. C. Davies, and W. J. Jones, J. Chem. Soc., 1930, 2298.
249. Jackson, I. K., and W. J. Jones, J. Chem. Soc., 1931, 575.
250. Jerchel, D., Ber., 76, 600 (1943).
251. Johns, I. B., and H. R. DiPietro, J. Org. Chem., 29, 1970 (1964).
252. Johns, I. B., H. R. DiPietro, R. H. Nealey, and J. V. Pustinger, J. Phys. Chem., 70, 924 (1966).
253. Jones, W. J., W. C. Davies, S. T. Bowden, C. Edwards, V. E. Davis, and L. H. Thomas, J. Chem. Soc., 1947, 1446.
254. Kabachnik, M. I., T. Ya. Medved, Yu. M. Polykarpov, and K. S. Yudina, Izv. Akad. Nauk SSSR, Otd. Khim. Nauk, 1961, 2029; C.A., 56, 11609e (1962).
255. Kabachnik, M. I., and E. S. Schepeleva, Izv. Akad. Nauk. SSSR, Otd. Khim. Nauk, 1953, 862; C.A., 49, 843f (1955).
256. Kabachnik, M. I., and E. N. Tsvetkov, Izv. Akad. Nauk SSSR, Otd. Khim. Nauk. 1961, 1896; C.A., 56,

8739a (1962).
257. Kabachnik, M. I., and E. N. Tsvetkov, Zh. Obshch. Khim., 31, 684 (1961); C.A., 55, 23398e (1961).
258. Kabachnik, M. I., L. P. Zhuravleva, M. G. Suleimanova, Yu. M. Polykarpov, and T. Ya. Medved, Izv. Akad. Nauk USSR, Ser. Khim., 1967, 949; C.A., 67, 100200s (1967).
259. Kaesz, H. D. and F. G. A. Stone, J. Org. Chem., 24, 635 (1959).
260. Kamai, G., Zh. Obshch. Khim., 2, 524 (1932); C.A., 27, 966 (1933).
261. Kamai, G., Zh. Obshch. Khim., 4, 192 (1934); C.A., 29, 464 (1935).
262. Kamai, G., and E. A. Gerasimova, Tr. Kazan, 23, 138 (1957); C.A., 52, 9946a (1958).
263. Kamai, G., F. M. Kharrasova, G. I. Rakhimova, and R. B. Sultanova, Zh. Obshch. Khim., 39, 625 (1969); C.A., 71, 50104y (1969).
264. Kamai, G., and L. A. Khismatullina, Zh. Obshch. Khim., 26, 3426 (1956); C.A., 51, 9512h (1957).
265. Kamai, G., and L. A. Khismatullina, Izv. Kazan, Filiala Akad. Nauk SSSR, Ser. Khim. Nauk, 1957, 79; C.A., 54, 6600i (1960).
266. Kamai, G., and G. M. Rusetskaya, Zh. Obshch. Khim., 32, 2848 (1962); C.A., 58, 7965d (1963).
267. Kamai, G., and G. M. Rusetskaya, Zh Obshch. Khim., 32, 2854 (1962); C.A., 58, 7965g (1963).
268. Kamai, G., V. S. Tsivunin, and S. Kh. Nurtdinov, Zh. Obshch. Khim. 35, 1817 (1965); C.A., 64, 3587e (1966).
269. Karavanov, K. V., S. Z. Ivin, and F. I. Ponomarenko, USSR Patent 231,549 (1968); C.A., 70, P68503s (1969).
270. Keat, R., Chem. Ind. (London), 1968, 1362.
271. Kelbe, W., Ber., 9, 1051 (1876).
272. Kelbe, W., Ber., 11, 1499 (1878).
273. Kharash, M. S., U.S. Patent 2,489,091 (1949); C.A., 44, 2009 (1950).
274. Kharash, M. S., E. V. Jensen, and W. H. Urry, J. Amer. Chem. Soc., 67, 1864 (1945).
275. Kharash, M. S., E. V. Jensen, and W. H. Urry, J. Amer. Chem. Soc., 68, 154 (1946).
276. Kharash, M. S., E. V. Jensen, and S. Weinhouse, J. Org. Chem., 14, 429 (1949).
277. Kharash, M. S., and S. Weinhouse, U.S. Patent 2,636,893 (1953); C.A., 47, 7700f (1953).
277a. Kharash, M. S., and S. Weinhouse, U.S. Patent 2,615,043 (1952); C.A., 47, 9346 (1953).
278. Kharrasova, F. M., and G. Kamai, Zh. Obshch. Khim., 38, 617 (1968); C.A., 69, 52215t (1968).
279. Kharrasova, F. M., G. Kamai, and R. R. Shagidullin, Zh. Obshch. Khim., 35, 1993 (1965); C.A. 64, 6680b

(1966).
280. Kharrasova, F. M., G. Kh. Kamai, and R. B. Sultanova, USSR Patent 185,908 (1966); C.A., 67, 3140q (1967).
281. Kharrasova, F. M., G. Kamai, R. B. Sultanova, and G. I. Matveeva, Zh. Obshch. Khim., 37, 902 (1967); 68, 39727w (1968).
282. Kharrasova, F. M., G. Kh. Kamai, R. B. Sultanova, and R. R. Shagidullin, Zh. Obshch. Khim., 37, 687 (1967); C.A., 67 53437n (1967).
283. Khisamova, Z. L., and G. Kamai, Zh. Obshch. Khim., 20, 1162 (1950); C.A., 45, 1531d (1951).
284. Kirk, P. G., and T. D. Smith, J. Chem. Soc. (A), 1969, 2190.
285. Kirsanov, A.V., and G. K. Fedorova, Dopov. Akad. Nauk Ukr. RSR, 1960, 801; C.A., 55, 430i (1961).
286. Knapsack, A.G., French Patent 1,547,575 (1968); C.A., 71, 61546v (1969).
287. Knapsack, A.G., French Patent 1,561,018 (1969); C.A., 72, 79226w (1970).
288. Kodama, Y., M. Nakabayashi, and T. Uehara, Japanese Patent 69,003,354 (1969); C.A., 70, P87956h (1969).
289. Koehler, H., Dissertation, Tübingen, 1877.
290. Koehler, H., Ber., 13, 463 (1880).
291. Koehler, H., Ber., 13, 1623 (1880).
292. Koehler, H., Ber., 13, 1626 (1880).
293. Koehler, H., and A. Michaelis, Ber., 10, 807 (1877).
294. Koehler, H., Z. Phys. Chem., 226, 283 (1964).
295. Kolotilo, M. V., A. G. Matyusha, and G. I. Derkach, Dopov. Akad. Nauk Ukr. SSR, Ser. B, 31, 632 (1969); C.A., 71, 108528h (1969).
296. Komkov, I. P., S. Z. Ivin, and K. V. Karavanov, Zh. Obshch. Khim., 28, 2960 (1958); C.A., 53, 9035e (1959).
297. Komkov, I. P., V. G. Gruzdev, S. Z. Ivin, and K. V. Karavanov, Probl. Organ. Sinteza, Akad. Nauk SSSR, Otd. Obshch. i Tekhn. Khim., 1965, 308; C.A., 64, 12717f (1966).
298. Komkov, I. P., K. V. Karavanov, and S. Z. Ivin, Zh. Obshch. Khim., 28, 2963 (1958); C.A., 53, 9035h (1959).
299. Korshak, V. V., G. S. Kolesnikov, and B. A. Zhubanov, Izv. Akad. Nauk SSSR, Otd. Khim. Nauk, 1958, 618; C.A., 52, 20038a (1958).
300. Kosolapoff, G. M., U.S. Patent 2,632,018 (1953); C.A., 48, 2097i (1954).
301. Kosolapoff, G. M., U.S. Patent 2,594,454 (1952); C.A., 47, 1179g (1953).
302. Kosolapoff, G. M., J. Amer. Chem. Soc., 74, 4119 (1952).
303. Kosolapoff, G. M., and W. F. Huber, J. Amer. Chem. Soc., 69, 2020 (1947).

304. Krasil'nikova, E. A., A. M. Potapov, and A. I. Razumov, Zh. Obshch. Khim., 37, 2365 (1967); C.A., 68, 87375b (1968).
305. Krasil'nikova, E.A., A.M. Potapov, and A. I. Razumov, Zh. Obshch. Khim., 38, 1098 (1968), C.A.; 69, 67471x (1968).
306. Kratzer, R. H., and K. L. Paciorek, Inorg. Chem., 4, 1767 (1965).
307. Krespan, C. G., U.S. Patent 2,931,803 (1960); C.A., 55, 12436 (1967).
308. Krespan, C.G., and C.M. Langkammerer, J. Org. Chem., 27, 3584 (1962).
309. Kuchen, W., and H. Buchwald, Chem. Ber., 91, 2871 (1958).
310. Kuchen, W., and H. Buchwald, Chem. Ber., 91, 2296 (1958).
311. Kuchen, W., and H. Buchwald, Angew. Chem., 68, 791 (1956).
312. Kuchen, W., and W. Gruenewald, Angew. Chem., 75, 576 (1963).
313. Kuchen, W., and W. Gruenewald, Chem. Ber., 98, 480 (1965).
314. Kuchen, W., and B. Knop, Chem. Ber., 99, 1663 (1966).
315. Kulakova, V. N., Yu. M. Zinovev, and L. Z. Soborovskii, Zh. Obshch. Khim., 29, 3957 (1959); C.A., 54, 20846f (1960).
316. Kunz, P., Ber., 27, 2559 (1894).
317. Labarre, J. F., F. Crasnier, and J. P. Faucher, J. Chim. Phys., 63, 1088 (1966).
318. Labarre, M. C., D. Voigt, and F. Gallais, Bull. Soc. Chim. France, 1967, 3328.
319. Ladd, E. C. and J. R. Little, U.S. Patent 2,510,699 (1950); C.A., 44, 7348e (1950).
320. Larionova, M. A., A. L. Klebanskii, and V. A. Bartashev, Zh. Obshch. Khim., 33, 265 (1963); C.A., 59, 656h (1963).
321. Lecoq, H., Bull. Soc. Chim. Belg., 42, 199 (1933).
322. Letcher, J. H., and J. R. Van Wazer, Topics Phosphorus Chem., 5, 75 (1967),
322a. Levin, Ya. A., and V. S. Galeev, USSR Patent 195,452 (1967); C.A., 68, 69125r (1968).
323. Levin, Ya. A., V. S. Galeev, and N. V. Evdokimova, USSR Patent 232,973 (1968); C.A., 70, P106644y (1969).
324. Levin, Ya. A., V. S. Galeev, and E. K. Trutneva, Zh. Obshch. Khim., 37, 1872 (1967); C.A., 68, 29801b (1968).
325. Levin, Ya. A., and R. I. Pyrkin, USSR Patent 213,024 (1968); C.A., 69, P67533u (1968).
326. Levy, J. B., L. D. Freedman, and G. O. Doak, J. Org. Chem., 33, 474 (1968).
327. Lindner, G., P. O. Granbom, and K. Bergquist, FOA

(Foersvarets Forskningsanst.) Rept. 4(6), 1970; C.A., 73, 131085m (1970).
328. Lindner, J., O. Brugger, A. Jenkner, and L. Tschemernigg, Monatsh. Chem., 53/4, 263 (1929).
329. Lindner, J., and M. Strecker, Monatsh. Chem., 53/4, 274 (1929).
330. Lindner, J., W. Wirth, and B. Zaunbauer, Monatsh. Chem., 70, 1 (1937).
331. Liorber, B. G., Z. M. Khammatova, A. I. Razumov, T. V. Zykova, and T. B. Borisova, Zh. Obshch. Khim., 38, 878 (1968); C.A., 69, 67497k (1968).
332. Little, J. R., and P. F. Hartmann, J. Amer. Chem. Soc., 88, 96 (1966).
333. Lobanov, D. I., E. N. Tsvetkov, and M. I. Kabachnik, Zh. Obshch. Khim., 39, 841 (1969); C.A., 71, 61495c (1969),
334. Loper, B. H., and F. S. Seichter, U.S. Patent 3,069,246 (1962); C.A., 58, 7776c (1963).
335. Lucken, E. A. C., and M. A. Whitehead, J. Chem. Soc., 1961, 2459.
336. Lyon, D. R., and F. G. Mann, J. Chem., Soc., 1942, 666.
337. Magnelli, D. D., G. Tesi, J. U. Lowe, and W. E. McQuiston, Inorg. Chem., 5, 457 (1966).
338. Mahler, W., J. Amer. Chem. Soc., 86, 2306 (1964).
339. Mahler, W., and A. B. Burg, J. Amer. Chem. Soc., 80, 6161 (1958).
340. Mahler, W., and A. B. Burg, J. Amer. Chem. Soc., 79, 251 (1957).
341. Maier, L., Helv. Chim. Acta, 46, 2667 (1963).
342. Maier, L., Chem. Ber., 94, 3051 (1961).
343. Maier, L., Helv. Chim. Acta, 46, 2026 (1963).
344. Maier, L., J. Inorg. Nucl. Chem., 24, 1073 (1962).
345. Maier, L., Angew. Chem., 71, 574 (1959).
346. Maier, L., J. Inorg. Nucl. Chem., 24, 275 (1962).
347. Maier, L., German Patent 1,122,522 (1962); C.A., 57, 16660e (1962).
348. Maier, L., Helv. Chim. Acta, 47, 2129 (1964).
349. Maier, L., Helv. Chim. Acta, 47, 2137 (1964).
350. Maier, L., U.S. Patent 3,321,557 (1967); C.A., P73684u (1967).
351. Maier, L., Helv. Chim. Acta, 52, 1337 (1969).
352. Maier, L., and R. Gredig, Helv. Chim. Acta, 52, 827 (1969).
353. Malatesta, L., Gazz. Chim. Ital., 77, 509 (1947).
354. Malatesta, L., and A. Sacco, Ann. Chim. (Rome), 44, 134 (1954); C.A., 48, 13516 (1954).
355. Manatt, S. L., D. D. Elleman, A. H. Cowley, and A. B. Burg, J. Amer. Chem. Soc., 89, 4544 (1967).
356. Mann, F. G., and I. T. Millar, J. Chem. Soc., 1952, 4453.

357. Mann, F. G., and J. Watson, J. Org. Chem., 13, 502 (1948).
358. Mark, V., C. H. Dungan, M. M. Crutchfield, and J. R. Van Wazer, Topics Phosphorus Chem., 5, 227 (1967).
359. Masaaki, Y., E. Akagi, and K. Minami, Kogyo Kagaku Zasshi, 68, 460 (1965).
360. Maslov, P. G., J. Phys. Chem., 72, 1414 (1968).
361. Mavel, G., Progr. NMR Spectrosc., 1, 250 (1966).
362. Mavel, G., and M. J. Green, Chem. Commun., 1968, 742.
363. Mavel, G., and P. Martin, Compt. Rend., 257, 1703 (1963).
364. McCormack, W. B., U.S. Patent 2,671,079 (1954); C.A., 48, 6738c (1954).
365. McCormack, W. B., U.S. Patent 2,663,737 (1953); C.A., 49, 7601a (1955).
366. McCormack, W. B., U.S. Patent 2,671,077 (1954); C.A., 48, 6737h (1954).
367. McCormack, W. B., U.S. Patent 2,671,078 (1954); C.A., 48, 6738b (1954).
368. McIvor, R. A., C. E. Hubley, G. A. Grant, and A. A. Gray, Canad. J. Chem., 36, 820 (1956).
369. McIvor, R. A., and C. E. Hubley, Canad. J. Chem., 37, 869 (1959).
370. Meisenheimer, J., J. Casper, M. Hoering, W. Lauter, L. Lichtenstadt, and W. Samuel, Justus Liebigs Ann. Chem., 449, 213 (1926).
371. Melchiker, P., Ber., 31, 2915 (1898).
372. Meriwether, L. S., and J. R. Leto, J. Amer. Chem. Soc., 83, 3192 (1961).
373. Metal and Thermit Corp., British Patent 869528 (Appl. 1958); C.A., 56, 6004C (1962).
374. Metzger, S. H., O. H. Basedow, and A. F. Isbell, J. Org. Chem., 29, 627 (1964).
375. Michaelis, A., Ber., 6, 601 (1873).
376. Michaelis, A., Ber., 6, 816 (1873).
377. Michaelis, A., Ber., 8, 499 (1875).
378. Michaelis, A., Justus Liebigs Ann. Chem., 181, 265 (1876).
379. Michaelis, A., Ber., 10, 627 (1877).
380. Michaelis, A., Ber., 12, 1009 (1879).
381. Michaelis, A., Ber., 13, 2174 (1880).
382. Michaelis, A., Justus Liebigs Ann. Chem., 293, 193 (1896).
383. Michaelis, A., Justus Liebigs Ann. Chem., 294, 1 (1896).
384. Michaelis, A., Justus Liebigs Ann. Chem., 315, 43 (1901).
385. Michaelis, A., and F. Dittler, Ber., 12, 338 (1879).
386. Michaelis, A., and L. Gleichmann, Ber., 15, 801 (1882).

387. Michaelis, A., and H. Koehler, Ber., 9, 519 (1876).
388. Michaelis, A., and H. Koehler, Ber., 9, 1053 (1876).
389. Michaelis, A., and W. LaCoste, Ber., 18, 2109 (1885).
390. Michaelis, A., and H. Lange, Ber., 8, 1313 (1875).
391. Michaelis, A., and A. Link, Justus Liebigs Ann. Chem., 207, 193 (1881).
392. Michaelis, A., and C. Panek, Ber., 13, 653 (1880).
393. Michaelis, A., and C. Panek, Justus Liebigs Ann. Chem., 212, 203 (1882).
394. Michaelis, A., and A. Schenk, Ber., 21, 1497 (1888).
395. Michaelis, A., and A. Schenk, Justus Liebigs Ann. Chem., 260, 1 (1890).
396. Michaelis, A., and H. Soden, Justus Liebigs Ann. Chem., 229, 295 (1885).
397. Mikhailov, B. M., and N. F. Kutcherova, J. Gen. Chem. USSR 22, 855 (1952).
398. Miller, J. M., J. Chem. Soc. (A), 1967, 828.
399. Miller, R. C., J. Org. Chem., 24, 2013 (1959).
400. Moedritzer, K., L. Maier, and L. C. D. Groenweghe, J. Chem. Eng. Data, 7, 307 (1962).
401. Monsanto Co., French Patent 1,482,337 (1967); C.A., 68, P39814x (1968).
402. Monsanto Co., British Patent 1.068,364 (1967); C.A., 67, 54260m (1967).
403. Monsanto Co., French Patent 1,347,066 (1963); C.A., 60, 12055d (1964).
404. Montgomery, R. E., and L. D. Quin, J. Org. Chem., 30, 2393 (1965).
405. Müller, G., Dissertation, Stuttgart, 1944.
406. Muller, N., P. C. Lauterbur, and J. Goldenson, J. Amer. Chem. Soc., 78, 3557 (1956).
407. Murray, M., and R. Schmutzler, Chem. Ind. (London), 1968, 1730.
408. Nagy, G., and D. Balde, French Patent 1,450,681 (1966); C.A., 67, P3142s (1967).
409. Nagy, G., and D. Balde, French Patent 1,451,377 (1966); C.A., 66, P66015c (1967).
410. Nanelli, P., G. R. Feistel, and T. Moeller, Inorg. Syn., 9, 73 (1967).
411. Neale, E., and L. T. D. Williams, J. Chem. Soc., 1955, 2485.
412. Nesmeyanov, A. N., K. N. Anisimov, and Z. P. Valueva, Izv. Akad. Nauk SSSR, 1964, 763; C.A., 61, 3132f (1964).
413. Niebergall, H., U.S. Patent 3,078,304 (1963); C.A., 59, 5198f (1963).
414. Nifant'ev, E. E., and M. P. Koroteev, USSR Patent 187,018 (1961); C.A., 67, P73680q (1967).
415. Nifant'ev, E. E., and M. P. Koroteev, Zh. Obshch. Khim., 37, 1366 (1967); C.A., 68, 39739b (1968).
416. Nijk, D. R., Rec. Trav. Chim. Pays-Bas, 41, 461 (1922).

417. Nixon, J. F. Chem. Ind. (London), 1963, 1555.
418. Nixon, J. F., J. Chem. Soc., 1964, 2469.
419. Nixon, J. F., J. Chem. Soc., 1964, 2471.
420. Nixon, J. F., J. Chem. Soc., 1965, 777.
421. Nixon, J. F., J. Chem. Soc. (A), 1967, 1136.
422. Nixon, J. F., and R. G. Cavell, J. Chem. Soc., Suppl., 1964, 5983.
423. Nixon, J. F., and R. Schmutzler, Spectrochim. Acta, 20, 1835 (1964).
424. Nixon, J. F., and R. Schmutzler, Spectrochim. Acta, 22, 565 (1966).
425. Noeth, H., Z. Naturforsch., 15b, 327 (1960).
426. Nyquist, R. A., Appl. Spectros. 22 (Pt. 1), 425 (1968).
427. Okhlobystin, O. Yu., and L. I. Zakharkin, Izv. Akad. Nauk SSSR, 1958, 1006; C.A., 53, 1123a (1959).
428. Okhlobystin, O. Yu., L. I. Zakharkin, and B. N. Strunin, USSR Patent 144,484 (1963); C.A., 60, 4183h (1964).
429. Okon, K., J. Sobczynski, J. Sowinski, and K. Niewielski, Biul. Wojsk. Akad. Tech., 13 (7/143), 109 (1964); C.A., 62, 4050g (1965).
430. Paciorek, K. J. L., and R. H. Kratzer, U.S. Patent 3,211,753 (1965); C.A., 63, P18156g (1965).
431. Paciorek, K. J. L., and R. H. Kratzer, U.S. Patent 3,277,170 (1966); C.A., 66, P19023u (1967).
432. Packer, K. J., J. Chem. Soc., 1963, 960.
433. Paidak, P. B., E. F. Grechkin, and A. V. Kalabina, USSR Patent 229,504 (1968); C.A., 70, 58004r (1969).
434. Parshall, G. W., J. Inorg. Nucl. Chem., 12, 372 (1960).
435. Pelchowicz, S., and H. Leader, J. Chem. Soc., 1963, 3320.
436. Perner, D., and A. Henglein, Z. Naturforsch., 17b, 703 (1962).
437. Perry, B. J., J. B. Reesor, and J. L. Ferron, Can. J. Chem., 41, 2299 (1963).
438. Peterson, L. K., and A. B. Burg, J. Amer. Chem. Soc., 86, 2587 (1964).
439. Petrov, K. A., V. V. Smirnov, A. K. Tsareva, and V. I. Emelyanov, USSR Patent 130,512 (1960); C.A., 55, 7356h (1961).
440. Petrov, K. A., V. V. Smirnov, and V. I. Emelyanov, Zh. Obshch. Khim., 31, 3027 (1961); C.A., 56, 12934d (1962).
441. Petrovskaya, L. I., M. V. Proskurnina, Z. S. Novikova, and I. F. Lutsenko, Izv. Akad. Nauk SSSR, Ser. Khim., 1968, 1277; C.A., 69, 86088u (1968).
442. Pianfetti, J. A., U.S. Patent 3,210,418 (1965); C.A., 64, P2128c (1966).
443. Pianfetti, J. A., and L. D. Quin, J. Amer. Chem.

Soc., 84, 851 (1962).
444. Pitts, J. J., M. A. Robinson, and S. I. Trotz, Inorg. Nucl. Chem. Lett., 4, 483 (1968).
445. Plets, V. M., Dissertation, Kazan, 1938.
446. Ponomarev. V. V., S. A. Golubtsov, K. A. Andrianov, and G. N. Kondrashova, Izv. Akad. Nauk SSSR, Ser. Khim. 1969, 1545; C.A., 71, 113061e (1969).
447. Ponomarev, V. V., S. A. Golubtsov, K. A. Andrianov, and G. N. Kondrashova, Izv. Akad. Nauk SSSR, Ser. Khim., 1969, 1743; C.A., 72, 3536b (1970).
448. Ponomarev, V. V., S. A. Golubtsov, E. A. Chernychev, and T. L. Krasnova, USSR Patent 230,153 (1968); C.A., 71, 39164x (1969).
449. Ponomarev, V. V., A. S. Shapatin, and S. A. Golubtsov, USSR Patent 184,856 (1966); C.A., 66, P85855t (1967).
450. Pope, W. J. and C. S. Gibson, J. Chem. Soc., 101, 735 (1912).
451. Postnikova, G. B., A. S. Kostyuk, and I. F. Lutsenko, Zh. Obshch. Khim., 35, 2204 (1965); C.A., 64, 11243g (1966).
452. Postnikova, G. B., A. S. Kostyuk, and I. F. Lutsenko, Zh. Obshch. Khim., 36, 1129 (1966); C.A., 65, 10610f (1966).
453. Potenza, J. A., E. H. Poindexter, P. J. Caplan, and R. A. Dwek, J. Amer. Chem. Soc., 91, 4356 (1969).
454. Prikoszovich, W., and H. Schindlbauer, Chem. Ber., 102, 2922 (1969).
455. Proskurnina, M. V., I. F. Lutsenko, Z. S. Novikova, and N. P. Voronov, Khim. Org. Soedin, Fosfora, Akad. Nauk SSSR, Otd. Obshch. Tekh. Khim., 1967, 8; C.A., 69, 52217v (1968).
456. Proskurnina, M. V., Z. S. Novikova, and I. F. Lutsenko, Dokl. Akad. Nauk SSSR, 159, 619 (1964); C.A., 62, 6508h (1965).
457. Protopopov, I. S., and M. Ya. Kraft, Med. Prom. SSSR, 13; No. 12, 5 (1959); C.A., 54, 10914c (1960).
458. Protopopov, I. S., and M. Ya. Kraft, Zh. Obshch. Khim., 34, 1446 (1964); C.A., 61, 5685f (1964).
459. Quin, L. D., in 1,4-Cycloaddition Reactions, J. Hamer, Ed., Academic Press, New York, 1967, p. 47.
460. Quin. L. D., and H. G. Anderson, J. Org. Chem, 31, 1206 (1966).
461. Quin, L. D., and J. S. Humphrey, J. Amer. Chem. Soc., 82, 3795 (1960).
462. Quin, L. D. and J. S. Humphrey, J. Amer. Chem. Soc., 83, 4124 (1961).
463. Quin, L. D., and R. E. Montgomery, J. Org. Chem., 27, 4120 (1962).
464. Quin, L. D., and R. E. Montgomery, J. Org. Chem., 28, 3315 (1963).

465. Quin, L. D., and C. H. Rolston, J. Org. Chem., 23, 1693 (1958).
466. Quinchon, J., M. LeSech, and E. Gryszkiewicz-Trochimovski, Bull. Soc. Chim. (France), 1961, 735.
467. Rabinerson, J., Ber., 23, 2342 (1890).
468. Radcliffe, L. G., and W. H. Brindley, Chem. Ind. (London), 42, 64 (1923).
469. Rakshys, J. W., R. W. Taft, and W. A. Sheppard, J. Amer. Chem. Soc., 90, 5236 (1968).
470. Ramsden, H. E., U.S. Patent 2,912,465 (1959); C.A., 54, 2170g (1960).
471. Raudnitz, H., Ber., 60, 743 (1927).
472. Razumov, A. I., O. A. Mukhacheva, and Sim-Do-Khen, Tr. Kazan, Khim.-Tekh. Inst., 1953, 151; C.A., 50, 7050e (1956).
473. Razumov, A. I., O. A. Mukhacheva, and Sim-Do-Khen, Izv. Akad. Nauk SSSR, 1952, 894; C.A., 47, 10466c (1953).
474. Razumov, A. I., and Sim-Do-Khen, Tr. Kazan Khim.-Tekh. Inst., 1954-1955, 167; C.A., 51, 6504e (1957).
475. Razumov, A. I., B. G. Liorber, M. B. Gazizov, and Zh. M. Khammatova, Zh. Obshch. Khim., 34, 1851 (1964); C.A., 61, 8334g (1964).
476. Renz, M., K. Wunder, and H. Drawe, Z. Naturforsch., 22b, 486 (1967).
477. Retcofsky, H. L., and C. E. Griffin, Tetrahedron Lett., 1966, 1975.
478. Rizpolozhenskii, N. I. and V. D. Akamsin, Izv. Akad. Nauk SSSR, Ser. Khim., 1969, 370; C.A., 70, 115234s (1969).
479. Ruff, J. K., Inorg. Chem., 2, 813 (1963).
480. Sacco, A., Atti. Accad. Nazl. Lincei, Rend., Classe Sci. Fis. Mat. Nat., 11, 101 (1951); C.A., 49, 158e (1955).
481. Sacco, A., Ann. Chim. (Rome), 43, 495 (1953).
482. Sachs, H., Ber., 25, 1514 (1892).
483. Sander, M., German Patent 1,139,491 (1962); C.A., 13996e (1963).
484. Sander, M., Chem. Ber., 93, 1220 (1960).
485. Sander, M., Chem. Ber., 95, 473 (1962).
486. Saunders, B. C., and T. S. Worthy, J. Chem. Soc., 1953, 2115.
487. Scherer, O. J., and G. Schieder, Chem. Ber., 101, 4184 (1968).
488. Schiemenz, G. P., Chem. Ber., 99, 504 (1966).
489. Schindlbauer, H., Monatsh. Chem., 94, 99 (1963).
490. Schindlbauer, H., Monatsh. Chem., 96, 1936 (1965).
491. Schindlbauer, H., Chem. Ber., 100, 3432 (1967).
492. Schindlbauer, H., and W. Prikoszovich, Chem. Ber., 102, 2914 (1969).
493. Schliebs, R., German Patent 1,119,861 (1961); C.A.,

58, 3125f (1963).
494. Schliebs, R., German Patent 1.165,596 (1964); C.A., 60, 15913f (1964).
495. Schliebs, R., and H. Kaiser, German Patent 1,119,860 (1961); C.A., 58, 6863 (1963).
496. Schmerling, L., U.S. Patent 2902517 (1959); C.A., 54, 2254 (1960).
497. Schmerling, L., U.S. Patent 2,986,579 (1961); C.A., 55, 20960i (1961).
498. Schmutzler, R., Inorg. Chem., 3, 410 (1964).
499. Schmutzler, R., J. Inorg. Nucl. Chem., 25, 3351 (1961).
500. Schmutzler, R., Chem. Ind. (London), 1962, 1868.
501. Schmutzler, R., Advan. Chem. Ser., 1963, No. 37, 150.
502. Schmutzler, R., Chem. Ber., 98, 552 (1965).
503. Schmutzler, R., in "Halogen Chemistry" V. Gutmann, Ed.; Academic Press, New York, 1967, Vol. 2 p. 31.
504. Schroeder, H., Inorg. Chem., 2, 390 (1963).
505. Schroeder, H., J. R. Reiner, and T. A. Knowles, Inorg. Chem., 2, 393 (1963).
506. Schroeder, H., J. R. Reiner, and T. L. Heying, Inorg. Chem., 1, 618 (1962).
507. Schwartze, F., J. Prakt. Chem., 10, 222 (1874).
508. Seel, F., K. Ballreich, and R. Schmutzler, Chem. Ber., 94, 1173 (1961).
509. Seel, F., W. Gombler, and K. H. Rudolph, Z. Naturforsch., 23b, 387 (1968).
510. Seel, F., and K. H. Rudolph, Z. Anorg. Allg. Chem., 363, 233 (1968).
511. Seel, F., K. H. Rudolph, and R. Budenz, Z. Anorg. Allg. Chem., 341, 196 (1965).
512. Seel, F., K. H. Rudolph, and W. Gombler, Angew. Chem., 79, 686 (1967); Angew. Chem. Int. Ed. Engl., 6, 708 (1967).
513. Seidel, W., Z. Anorg. Allg. Chem., 330, 141 (1964).
514. Senear, A. E., W. Valient, and J. Wirth, J. Org. Chem., 25, 2001 (1960).
515. Sergeev, P. G., and D. G. Kudryashov, Zh. Obshch. Khim., 8, 266 (1938); C.A., 32, 5403 (1938).
516. Seyferth, D., and K. A. Braendle, J. Amer. Chem. Soc., 83, 2055 (1961).
517. Seyferth, D., and W. Freyer, J. Org. Chem., 26, 2604 (1961).
518. Sheluchenko, V. V., S. S. Dubov, G. I. Drozd, and S. Z. Ivin, Zh. Strukt. Khim., 9, 909 (1968); C.A., 70, 24524v (1969).
519. Shevchenko, V. I., A. A. Pinchuk, and N. Ya. Kozlova, Zh. Obshch. Khim., 34, 3955 (1964); C.A., 62, 9170b (1965).

520. Shokol, V. A., V. F. Gamaleya, and G. I. Derkach, Zh. Obshch. Khim., 39, 856 (1969); C.A., 71, 61497e (1969).
521. Shostenko, A. G., P. A. Zagorets, and A. M. Dodonov, Tr. Mosk. Khim. Tekhnol. Inst., 1968, No. 58, 242; C.A., 71, 27285z (1969).
522. Shostenko, A. G., P. A. Zagorets, A. M. Dodonov, and A. A. Greish, Khim. Vys. Enrg., 4, 357 (1970); C.A., 73, 87975a (1970).
523. Simonnin, M. P., and C. Charrier, Compt. Rend. 267C, 550 (1968).
524. Sisler, H. H., and S. E. Frazier, Inorg. Chem., 4, 1204 (1965).
525. Smirnov, E. A., Yu. A. Kondrat'ev, V. A. Petrunin, and Yu. M. Zinov'ev, USSR Patent 196,818 (1967); C.A., 68, 69126s (1968).
526. Smirnov, E. A., Yu. M. Zinov'ev, and V. A. Petrunin, Zh. Obshch. Khim., 38, 1197 (1968); C.A., 69, 87110g (1968).
527. Smirnov, E. A., Yu. M. Zinov'ev, and V. A. Petrunin, Zh. Obshch. Khim., 38, 1551 (1968); C.A., 69, 96827y (1968).
528. Smith, H. D., T. A. Knowles, and H. Schroeder, Inorg. Chem., 4, 107 (1965).
529. Sollott, G. P., and B. Howard, J. Org. Chem., 29, 2451 (1964).
530. Sollott, G. P., and W. R. Peterson, J. Organometal. Chem., 4, 491 (1965).
531. Sollott, G. P., and W. R. Peterson, J. Organometal. Chem., 19, 143 (1969).
532. Sommer, K., Z. Anorg. Allg. Chem., 376, 37 (1970).
533. Sosnovsky, G., and D. J. Rawlinson, J. Org. Chem., 33, 2325 (1968).
534. Stauffer Chemical Co., British Patent 934,090 (1963); C.A., 60, 559c (1964).
535. Steininger, E., Chem. Ber., 96, 3184 (1963).
536. Sterlin, R. N., and I. L. Knunyants, Khim. Nauka Prom., 4, 810 (1959); C.A., 54, 10838f (1960).
537. Sterlin, R. N., R. D. Yatsenko, L. N. Pinkina, and I. L. Knunyants, Izv. Akad. Nauk SSSR, 1960, 1991; C.A., 55, 13297h (1961).
538. Stetter, H., and W. D. Last, Chem. Ber., 102, 3364 (1969).
539. Stuebe, C., W. M. LeSuer, and G. R. Norman, J. Amer. Chem. Soc., 77, 3526 (1955).
540. Tesi, G., C. M. Douglas, and C. P. Haber, U.S. Patent 3,087,937 (1963); C.A., 59, 10124c (1963).
541. Tesi, G., C. P. Haber, and C. M. Douglas, Proc. Chem. Soc., 1960, 219.
542. Tetel'baum, B. I., V. V. Sheluchenko, S. S. Dubov, G. I. Drozd, and S. Z. Ivin, Zh. Vses. Khim. Obshch.

12, 351 (1967); C.A., 68, 73847v (1968).
543. Titov, A. I., and P. O. Gitel, Dokl. Akad. Nauk SSSR, 158, 1380 (1964); C.A., 62, 2791a (1965).
544. Titov, A. I., M. V. Sizova, and P. O. Gitel, Dokl. Akad. Nauk SSSR, 159, 385 (1964); C.A., 62, 6509h (1965).
545. Tolkmith, H., Ann. N.Y. Acad. Sci., 79, 189 (1959).
546. Toy, A. D. F., and R. S. Cooper, German Patent 1,095,279 (1960); C.A., 57, 868e (1962).
547. Toy, A. D. F. and K. H. Rattenbury, U.S. Patent 3,244,745 (1966); C.A., 64, 17638f (1966).
548. Uhing. E. H., U.S. Patent 3,314,900 (1967); C.A., 67, P12086d (1967).
549. Uhing, E. H., K. Rattenbury, and A. D. F. Toy, J. Amer. Chem. Soc., 83, 2299 (1961).
550. Ulmer, H. E., L. C. D. Groenweghe, and L. Maier, J. Inorg. Nucl. Chem., 20, 82 (1961).
551. Van Wazer, J. R., C. F. Callis, J. N. Shoolery, and R. C. Jones, J. Amer. Chem. Soc., 78, 5715 (1956).
552. Van Wazer, J. R., and J. H. Letcher, Topics Phosphorus Chem., 5, 169 (1967).
553. Van Wazer, J. R., and L. Maier, J. Amer. Chem. Soc., 86, 811 (1964).
554. Van Winkle, J. L., C. S. Bell, and R. C. Morris, U.S. Patent 2,875,224 (1959); C.A., 53, 13054h (1959).
555. Varshavskii, S. L., Yu. I. Baranov, O. F. Filippov, M. I. Kabachnik, and N. K. Bliznyuk, USSR Patent 209,455 (1968); C.A., 69, P77469s (1968).
556. Vetter, H., and H. Noeth, Ber., 96, 1816 (1963).
557. Viout, M. P., J. Rech. CNRS. Lab. Bellevue, No. 28, 15 (1954); C.A., 50, 7077 (1956).
558. Viout, M. P., and P. Rumpf, Bull. Soc. Chim. (France), 1957, 768.
559. Voskuil, W., and J. F. Arens, Rec. Trav. Chim. Pays-Bas, 82, 302 (1963).
560. Voskuil, W., and J. F. Arens, Rec. Trav. Chim. Pays-Bas, 81, 993 (1962).
561. Walling, C., U.S. Patent 2,437,796 (1948); C.A., 42, 4199 (1948).
562. Walling, C., U.S. Patent 2,437,798 (1948); C.A., 42, (1948).
563. Walsh, E. N., T. M. Beck, and W. H. Woodstock, J. Amer. Chem. Soc., 77, 929 (1955).
564. Wedekind, E., Ber., 45, 2933 (1912).
565. Weil, T., Helv. Chim. Acta, 37, 654 (1954).
566. Weil, T., B. Prijs, and H. Erlenmeyer, Helv. Chim. Acta, 36, 1314 (1953).
567. Weil, T., B. Prijs, and H. Erlenmeyer, Helv. Chim. Acta, 35, 1412 (1952).
568. Weisse, G. K., and R. M. Thomas, German Patent 1,161,559 (1964); C.A., 60, 10719a (1964).

569. Weller, J., Ber., 20, 1718 (1887).
570. Weller, J., Ber., 21, 1492 (1888).
571. White, E. R., PB Report 137075 (1950), Shell Development Company.
572. Wiberg, E., and H. Noeth, Z. Naturforsch., 12b, 125 (1957).
573. Wiley, R. H., and C. H. Jarboe, J. Amer. Chem. Soc., 73, 4996 (1951).
574. Wilson, H. F., and C. E. Glassick, French Patent 1,366,248 (1964); C.A., 61, 16096h (1964).
575. Wolf, R., J. R. Miquel, and F. Mathis, Bull. Soc. Chim. (France), 1963, 825.
576. Woodstock, W. H., U.S. Patent 2,137,792 (1938); C.A., 33, 1763 (1939).
577. Woodstock, W. H., and E. N. Walsh, U.S. Patent 2,685,602 (1953); C.A.,49, 10358 (1955).
578. Wunder, K., U.S. At. Energy Comm. Accession No. 40947, Rept. BMwF-FBK-66-14 (1966); C.A., 66, 70830g (1967).
579. Wunder, K., H. Drawe, and A. Henglein, Z. Naturforsch., 19b, 999 (1964).
580. Yagupolskii, L. M., and P. A. Jufa, Zh. Obshch. Khim., 30, 1294 (1960); C.A., 55, 432a (1961).
581. Yagupolskii, L. M., and P. A. Jufa, Zh. Obshch. Khim., 28, 2853 (1958); C.A., 53, 9109 (1959).
582. Yagupolskii, L. M., and Z. M. Ivanova, Zh. Obshch. Khim., 30, 4026 (1960); C.A., 55, 221962e (1961).
583. Yakubovich, A. Y., and V. A. Ginsburg, Zh. Obshch. Khim., 22, 1534 (1952); C.A., 47, 9254g (1953).
584. Yakubovich, A. Ya., V. A. Ginsburg, and S. P. Makarov, Dokl. Akad. Nauk SSSR, 71, 303 (1950); C.A., 44, 8321b (1950).
585. Yakubovich, A. Y., and G. V. Motsarev, Dokl. Akad. Nauk SSSR, 88, 87 (1953); C.A., 48, 143g (1954).
586 Yakubovich, A. Y., and G. V. Motsarev, Zh. Obshch. Khim., 23, 771 (1953); C.A., 48, 4463a (1954).
587. Yakubovich, A. Ya., and G. V. Motsarev, Zh. Obshch. Khim., 23, 1547 (1953); C.A., 48, 10643c (1954).
588. Yoke, J. T., and R. G. Laughlin, U.S. Patent 3,304,263 (1967); C.A., 68, P60778y (1968).
589. Yoke, J. T., and R. G. Laughlin, U.S. Patent 3,304,330 (1967); C.A., 67, 34100f (1967).
590. Yudina, K. S., T. Ya. Medved, and M. I. Kabachnik, Izk. Akad. Nauk SSSR, Ser. Khim., 1966, 1954; C.A., 66, 76096u (1967).
591. Zakharkin, L. I., V. I. Bregadze, and O. Yu. Okhlobystin, J. Organometal. Chem., 4, 211 (1965).
592. Zakharkin, L. I., A. V. Kazantsev, and M. N. Zhubekova, Izv. Akad. Nauk SSSR, Ser. Khim., 1969, 2056; C.A., 72, 12829b (1970).
593. Zakharkin, L. I., and O. Yu. Okhlobystin, USSR

Patent 110,920 (1958); C.A., 52, 14652h (1958).
594. Zakharkin, L. I., and O. Yu. Okhlobystin, Dokl. Akad. Nauk SSSR, 116, 236 (1957); C.A., 52, 6167c (1958).
595. Zakharkin, L. I., O. Yu. Okhlobystin, and B. N. Strunin, Izv. Akad. Nauk SSSR, 1962, 2002; C.A., 58, 9132b (1963).
596. Zakharkin, L. I., and V. I. Stanko, Izv. Akad. Nauk SSSR, Otd. Khim. Nauk, 1961, 2078; C.A., 57, 8610a
597. Zecchini, F., Gazz. Chim. Ital., 24, I, 34 (1894).
598. Ziegler, K., German Patent 864,866 (1953); C.A., 51, 4416e (1957).
599. Zinovev, Yu. M., and L. Z. Soborovskii, Zh. Obshch. Khim., 26, 3030 (1956); C.A., 51, 8662a (1957).
600. Zinovev, Yu. M., and L. Z. Soborovskii, Zh. Obshch. Khim., 34, 929 (1964); C.A., 60, 15904d (1964).
601. Zinovev, Yu. M., and L. Z. Soborovskii, Zh. Obshch. Khim., 30, 1571 (1960); C.A., 55, 1415b (1961).

Chapter 9. Phosphonyl- (Thiono-, Seleno-) and Phosphinyl- (Thiono-, Seleno-) Halides and Pseudohalides

MANFRED FILD and REINHARD SCHMUTZLER

Lehrstuhl B für Anorganische Chemie der Technischen Universität, Braunschweig, Germany

and STEPHEN C. PEAKE

Department of Inorganic Chemistry, University of Oxford, Oxford, England

This chapter deals with phosphonyl dihalides and dipseudohalides $RP(O)X_2$ and with phosphinyl halides and pseudohalides $R_2P(O)X$. The corresponding thiono- and seleno analogs are also presented. The pseudohalides covered are cyanides, cyanates, thiocyanates, and azides.

Methods of preparation are given in the first part; then follows a section on the chemistry and one on the physical properties of these compounds. The literature concerning this subject is covered through the end of 1969 including patents, so far as abstracts are available in Chemical Abstracts.

A. METHODS OF PREPARATION

I. FROM PHOSPHONIC AND PHOSPHINIC ACIDS

We know of several convenient methods of preparation of phosphonic and phosphinic acids that do not proceed by way of the corresponding halides. From these acids the halides are readily synthesized by warming them with PCl_5 or $SOCl_2$ as shown in the following equations:

$$RP(O)(OH)_2 + 2PCl_5 \longrightarrow RP(O)Cl_2 + 2POCl_3 + 2HCl$$

$$R_2P(O)OH + SOCl_2 \longrightarrow R_2P(O)Cl + SO_2 + HCl$$

The reaction may be conducted with or without an inert solvent, at room temperature or slightly above; the products are isolated by distillation or crystallization.[42b, 74,133,164,185,190,251,252,271,276,288,293,394,491,500,519] Naturally, functional carbon groups at the phosphorus atom will be affected by this reaction. For example, carbonyl groups react to dichloromethylene groups,[496] carboxylic derivatives are converted to acid chlorides,[494,495,529, 723,724] and hydroxy groups will give chloroalkyl compounds.[164,294,754] The last reaction can also give rise to unsaturated phosphonic dichlorides by dehydrohalogenation, e.g.,[251,252,418]

$$(CH_3)_2C(OH)P(O)(OH)_2 \xrightarrow{PCl_5} CH_2{=}C(CH_3)P(O)Cl_2$$

Phosphonic and phosphinic acids may also be converted to the corresponding halides by reaction with substituted chlorophosphoranes, as indicated in the following equation:[424,507]

$$C_6H_5P(O)(OH)_2 + 2C_6H_5PCl_4 \longrightarrow 3C_6H_5P(O)Cl_2 + 2HCl$$

Alkoxyphosphoranes such as $(C_6H_5O)_2PCl_3$ have also been applied.[266]

Limitations in the use of this method have been found in acids that contain electronegative groups. These acids can only be converted to the monochloro derivatives as shown for $CCl_3P(O)(OH)_2$ (which thus yields $CCl_3P(O)(OH)$-Cl[733]). Triarylmethylphosphonic acids give first the monohalide, which at higher temperature is cleaved at the phosphorus-carbon bond to give triarylmethanes.[260,261] It has also been shown that, in reactions employing PCl_5, the by-product $POCl_3$ may take part in the reaction thus:

$$HOCH_2P(O)(OH)_2 + 2PCl_5 \longrightarrow Cl_2P(O)OCH_2P(O)Cl_2 + POCl_3 + 3HCl$$

In this case, $SOCl_2$ in pyridine is the preferred reagent.[42b]

Electronegative groups may also be split from the phosphorus atom as was found in the reaction of $(HOCH_2)_2P(O)(OH)$ with PCl_5,[294] which gave $ClCH_2P(O)Cl_2$ as a by-product. The first step is probably chlorination to $(ClCH_2)_2P(O)Cl$, which is followed by the carbon-phosphorus cleavage; the second stage has been demonstrated independently.[165]

It is also necessary to mention that in many cases isomeric mixtures were obtained in this reaction. Various acids have been synthesized by way of a Friedel-Crafts reaction of halophosphines (Chapter 8); for monosubstituted aromatic rings, these compounds give rise to isomeric distribution in the products.

The synthesis of bromides is achieved by employing PBr_5 in a similar reaction.[1,143] Sulfuryl chloride has also been used to convert acids into chlorides.[568] For the preparation of fluorides, a versatile reagent has been found in sulfur tetrafluoride SF_4 but has only been applied a few times.[658,659]

$$C_6H_5P(O)(OH)_2 + 2SF_4 \longrightarrow C_6H_5P(O)F_2 + 2SOF_2 + 2HF$$

The substitution of hydroxyl groups can also be used for the synthesis of mixed phosphonic halides, as shown below.[274,299]

$$CH_3-P(=O)(F)(OH) + SOCl_2 \longrightarrow CH_3-P(=O)(F)(Cl) + SO_2 + HCl$$

In the case of monosubstituted phosphinic acids, this reaction leads to phosphonic dihalides,[498] e.g.,

$$RP(O)(OH)H + 2PCl_5 \longrightarrow RP(O)Cl_2 + PCl_3 + POCl_3 + 2HCl$$

Thionophosphonic and thionophosphinic acids react with equivalent amounts of PCl_5 in general to form the corresponding thionophosphonyl and -phosphinyl chlorides;[474] the same results have been obtained with $SOCl_2$.[407]

$$R_2P(S)OH + PCl_5 \longrightarrow R_2P(S)Cl + POCl_3 + HCl$$

Compounds of the type $R_2P(S)SH$ react with PCl_5 [274,474] or with HCl[109,274] to give the thionophosphinyl chloride, whereas HF with $CH_3P(S)(SH)F$ forms the corresponding fluoride.[600]

In reactions of thionyl chloride with thioacids, care must be taken to avoid the exchange of sulfur by oxygen; this reaction has been shown, for example, with $C_6H_5(C_2H_5)$-$P(S)SH$.[557]

II. FROM ESTERS AND ANHYDRIDES OF PHOSPHONIC AND PHOSPHINIC ACIDS

The method of preparation from esters and anhydrides of phosphonic and phosphinic acids is related to the preceding method; the only difference is that an organohalide is formed instead of hydrogen chloride. Esters of these acids, which are frequently more readily obtained in direct reactions than the free acids, can be converted at elevated temperatures with PCl_5 to the corresponding halides[601] according to the equation

$$CH_3P(O)(OCH_3)_2 + 2PCl_5 \longrightarrow CH_3P(O)Cl_2 + 2POCl_3 + 2CH_3Cl$$

This reaction usually requires temperatures in excess of 100°C;[327,328] at lower temperatures it may stop at the conversion of one ester group.[327,328] It is probable that the reaction course is more complex and may involve the formation of intermediates with polyphosphonate structures.

Phosphorus pentachloride and $SOCl_2$ have been used extensively, and examples given are plentiful.[116,117,299,331,388,393,468,509,587,590] Sulfuryl chloride[334] and phosgene have been employed[209,284,721] with similar success.

The method is very useful in reactions with unsaturated compounds, since there is no evolution of hydrogen chloride, which could attack a multiple bond. Thus olefinic and acetylenic derivatives have been successfully prepared by this method.[468,579] Another versatile application, the synthesis of mixed halides or pseudohalides of phosphonic acids, has been used extensively;[233,299,642] an example is

$$RP(O)(NCO)OR' + PCl_5 \longrightarrow RP(O)(NCO)Cl + POCl_3 + R'Cl$$

The conversion of diphosphonates to the diphosphonic tetrachlorides was less successful;[166,388] the products were difficult to isolate. For this synthesis it is better to use an equimolar mixture of the diphosphonate and the free diphosphonic acid.[451,595a]

Further variations were found in the use of the salt $C_2H_5P(O)(OC_2H_5)ONa$[578] and the monester $CH_3P(O)(OCH_3)OH$;[512] both compounds reacted with PCl_5 to give the expected products. Benzalchloride, $C_6H_5CHCl_2$ has also been used in this method.[255]

$$CH_3(C_6H_5)P(O)OCH_3 + C_6H_5CHCl_2 \longrightarrow$$

$$CH_3(C_6H_5)P(O)Cl + C_6H_5CHO + CH_3Cl$$

Similar behavior was found in the reaction of anhydrides of phosphonic and phosphinic acids with PCl_5.[230,509] This reaction has also been successfully applied to the synthesis of thiophosphonic dichlorides by reacting the anhydrides with either elemental chlorine, SO_2Cl_2, or PCl_5.[410]

III. REACTION OF ORGANOMETALLIC COMPOUNDS WITH PHOSPHORYL AND THIOPHOSPHORYL CHLORIDE

The reaction between a Grignard reagent and $POCl_3$ has been employed in several procedures for the synthesis of phosphonic and phosphinic acids. During the course of the reaction the halide forms an adduct with the magnesium salt; on hydrolysis, the acid is produced. In a few cases these halides have been separated from the magnesium halide but often in low yields.[312,390] Normally this reaction leads to tertiary phosphine oxides (Chapter 6).

In more recent investigations, other organometallic reagents have been studied. It was found that organolead and organoaluminum derivatives can successfully be employed in the synthesis of phosphoryl and thiophosphoryl compounds,[446,449,454] as indicated in two representative equations

$$2POCl_3 + Pb(C_6H_5)_4 \longrightarrow 2C_6H_5P(O)Cl_2 + (C_6H_5)_2PbCl_2$$

$$2C_6H_5P(S)Cl_2 + Pb(CH_3)_4 \longrightarrow 2C_6H_5(CH_3)P(S)Cl + (CH_3)_2PbCl_2$$

The second reaction shows the versatility of the method for the preparation of asymmetric compounds.

Terminally reactive polymers have also been treated with $RP(O)Cl_2$ and $RP(S)Cl_2$ compounds by using metals to split off chlorine.[279]

IV. ARYLATION OF THIOPHOSPHORYL CHLORIDE WITH CATALYSTS

Aromatic hydrocarbons react with $PSCl_3$ in the presence of $AlCl_3$ to form both thionophosphonic and thionophosphinic chlorides.

$$PSCl_3 \xrightarrow[-HCl]{+RH,\ AlCl_3} RPSCl_2 \xrightarrow[-HCl]{+RH,\ AlCl_3} R_2PSCl$$

This method has been known for a long time,[496] but the actual reaction conditions have only recently been investigated.[450,455] It was found that yields and composition of the products depend on the molar ratio of the reactants, the temperature, and the reaction time. Difficulty is encountered especially in preparing the dichlorides, whereas mono-chlorides may be obtained in near-quantitative yields. The arylation can also lead to tertiary phosphine sulfides (see Chapter 7).

The reaction has also been applied to $PSBr_3$ and benzene in the presence of $AlBr_3$. In this case the dihalide is the major product.

Activation has been achieved mainly with $AlCl_3$. It was shown that the only efficient catalysts are those which form stable adducts with $PSCl_3$. Compounds such as $ZnCl_2$, $FeCl_3$, or $TiCl_4$ are ineffective; $SbCl_5$ forms a complex but acts as a chlorinating agent.[450,455]

Hydrocarbons used with success have been chlorobenzene, benzene, toluene, hexyltoluene, and fluorobenzene. An indication of versatility is that $RPSCl_2$ derivatives can be used instead of $PSCl_3$, and therefore asymmetric compounds, RR'PSCl are easily obtained.

Investigation of orientation of the products from monosubstituted benzenes has shown that isomeric mixtures are formed; similar problems are encountered as in the preparation of phosphonous dihalides (Chapter 8).

V. CLEAVAGE OF PHOSPHORUS-PHOSPHORUS BONDS

A useful synthetic route has been found in the cleavage reaction of diphosphine disulfides with elemental chlorine or bromine to yield thionophosphinic halides,[102,258,400,401,441,442,444,445,557] as follows:

$$R_2P(S)P(S)R_2 + Cl_2 \longrightarrow 2R_2P(S)Cl$$

This reaction gives excellent yields and is often conducted in inert solvents. Excess halogen has to be avoided because the phosphorus-sulfur bond may be attacked. The cleavage has also been achieved with SO_2Cl_2 in a controlled reaction with equimolar amounts,[103,442,610] e.g.,

$$R_2P(S)P(S)R_2 + SO_2Cl_2 \longrightarrow 2R_2P(S)Cl + SO_2$$

Thionyl chloride,[557] SCl_2,[442] and Hg_2Cl_2[442] have also served in this reaction.

With excess SO_2Cl_2 and $SOCl_2$ and at elevated temperatures, the sulfur is replaced by oxygen and phosphinic halides are obtained,[103,443,557] e.g.,

$$[R_2P(S)]_2 + 2SOCl_2 \longrightarrow 2R_2P(O)Cl + 2S + S_2Cl_2$$

All compounds prepared in this way have been listed under this method rather than under Section A.VII to indicate that the starting material is the diphosphine disulfide. The method is important because the diphosphine disulfides are conveniently prepared from Grignard reactions with $PSCl_3$ or $RP(S)Cl_2$, which may give symmetric or asymmetric derivatives (Chapter 2).

VI. CONVERSION OF PHOSPHONIC AND PHOSPHINIC HALIDES INTO THIOPHOSPHONIC AND THIOPHOSPHINIC HALIDES

Aliphatic and aromatic phosphonic dihalides can be converted to the corresponding thio derivatives with P_4S_{10},[104,322,326,430,451,473,585,638] e.g.,

$$10\ R\text{-}P(O)Cl_2 + P_4S_{10} \longrightarrow 10\ R\text{-}P(S)Cl_2 + P_4O_{10}$$

This is a very general route. The reaction is conducted in an inert gas atmosphere at temperatures between 100 and 160°C. Since the thionophosphonic dichlorides are hydrolytically stable, they are easily separated by washing the reaction mixtures with water. This method also applies to phosphinic halides.[474a,607]

Instead of P_4S_{10}, $PSCl_3$ may be used,[585] e.g.,

$$PSCl_3 + ClCH_2\text{-}P(O)Cl_2 \rightleftharpoons ClCH_2\text{-}P(S)Cl_2 + POCl_3$$

This exchange (which is in fact an equilibrium reaction) has been followed by NMR in mixtures of $RPSCl_2$ and $POCl_3$ in sealed tubes at 200°C.[225] If $POCl_3$ is distilled off in the course of the reaction, yields higher than 70% may be obtained. Catalytic amounts of $AlCl_3$ have been used to speed up this reaction. In the case of phosphonic difluorides and P_4S_{10}, only small yields of the thio products are obtained.

VII. CONVERSION OF THIOPHOSPHONIC AND THIOPHOSPHINIC HALIDES INTO PHOSPHONIC AND PHOSPHINIC HALIDES

Thiophosphinic halides react with SO_2Cl_2 to give the corresponding phosphinic halides,[103] e.g.,

$$(CH_3)_2P(S)\text{-}Cl + SO_2Cl_2 \longrightarrow (CH_3)_2P(O)\text{-}Cl + S + SOCl_2$$

This method gives excellent yields, mostly higher than 90%. The reaction is often combined with that of method V, the cleavage of diphosphine disulfides. Thionyl chloride has also been used with similar results.[256,557] The reaction has not been applied to thiophosphonic dichlorides, but the results should be the same.

Both SO_2Cl_2 and Cl_2 have also been used to convert $R_2P(S)SH$ into $R_2P(O)Cl$[543] or to convert $R_2P(S)OH$ into

$R_2P(O)Cl$ via $R_2P(S)Cl$.[107] This reaction is discussed under I.

VIII. FROM PHOSPHONOUS AND PHOSPHINOUS ACIDS AND THEIR DERIVATIVES

Phosphonous acids $RP(OH)_2$ normally exist in the form of monosubstituted phosphinic acids RP(O)(OH)H. Their conversion into phosphonic dihalides was therefore discussed in method I.

Phosphinous acids react with PCl_5 or $SOCl_2$ to give phosphinyl halides,[285,399,403] e.g.,

$$(C_4H_9)_2P(O)H + PCl_5 \longrightarrow (C_4H_9)_2P(O)Cl + HCl + PCl_3$$

The thio analogs may be chlorinated with carbon tetrachloride in the presence of a tertiary base,[428] as follows:

$$(C_2H_5)_2P(S)H + CCl_4 \xrightarrow{R_3N} (C_2H_5)_2P(S)Cl + HCCl_3$$

Chlorine has also been used in this reaction, but yields were poor;[543] SO_2Cl_2 caused replacement of sulfur to give the phosphinyl chloride, $R_2P(O)Cl$.[543]

Esters of phosphinous acids containing three-coordinate phosphorus could be converted to phosphinyl chloride by SO_2Cl_2.[334] The anhydride of bis(trifluoromethyl)phosphinous acid and its thio analog could be cleaved by halogens;[198,215] among the products are $(CF_3)_2P(O)X$ and $(CF_3)_2P(S)X$, respectively.

IX. HALOGEN EXCHANGE

Phosphonic and phosphinic fluorides as well as their thio analogs are conveniently prepared by a halogen exchange reaction. This conversion can be conducted with alkali hydrogen fluoride[4,376,623,624] or by heating the chlorides with NaF and ZnF_2, with or without an inert solvent such as benzene, acetonitrile, or sulfolane;[153,287,469,621,623,626,700,729,731] examples are plentiful. The exchange has also been achieved in high yields with HF.[78,118,512,623]

The most important route employs antimony trifluoride as fluorinating agent for synthesizing phosphonic or phosphinic fluorides or their thio analogs.[66,299,618,669,699] This reagent has also been used in mixtures with NaF.[628] Other fluorinating agents include Na_2SiF_6,[239] benzoyl fluoride,[615,630] and potassium fluorosulfinate.[615,630]

Mixed chlorofluoro derivatives such as RP(O)FCl have been synthesized either by selective fluorination with potassium fluoride in toluene, e.g.,[385]

$$4\text{-}ClC_6H_4P(S)Cl_2 \xrightarrow[\text{toluene}]{KF} 4\text{-}ClC_6H_4P(S)FCl$$

or by equilibration of difluorides with dichlorides at elevated temperatures,[78,139,742] e.g.,

$$CH_3P(O)F_2 + CH_3P(O)Cl_2 \longrightarrow 2CH_3P(O)FCl$$

or by mixing a fluoride with BCl_3,[240] e.g.,

$$3C_6H_5P(O)F_2 + BCl_3 \longrightarrow 3C_6H_5P(O)FCl + BF_3$$

which may proceed further to the dichloride.

The pseudohalides of pentavalent phosphorus have been nearly exclusively prepared by exchange reactions. Reagents used have been alkali pseudohalides (mainly in acetonitrile or benzene as solvents,[126,127,300,316,318,621] in acetone in the case of thiocyanates,[128,300] or with ammonium salts[613] as presented in the following equations:

$$RP(O)Cl_2 + 2NaOCN \longrightarrow RP(O)(NCO)_2 + 2NaCl$$

$$R_2P(S)Cl + KSCN \longrightarrow R_2P(S)NCS + KCl$$

The reaction of azides, especially NaN_3, has been conducted in pyridine with good results.[40,42,42a,43] It has also been carried out successfully in acetone.[622] In the case of LiN_3 the reaction product, $(C_6H_5)_2P(O)N_3$ could only be isolated as an adduct with LiCl.[532]

Other methods include the reaction with silver, lead, or mercury pseudohalides.[36,85,262,263,679,738] Attempts to prepare isocyanates using isocyanic acid failed because of side reactions.[672] In one case a silicon derivative, $(CH_3)_3SiN_3$ was employed in the exchange reaction with $(C_6H_5)_2P(O)Cl$.[533]

$$(C_6H_5)_2P(O)Cl + (CH_3)_3SiN_3 \longrightarrow (C_6H_5)_2P(O)N_3 + (CH_3)_3SiCl$$

The preparation of mixed halopseudohalides of phosphonic acids have been studied mainly in the class of the isocyanates. Either mixed chlorofluoro derivatives are reacted with alkali cyanates,[643,644] e.g.,

$$ClCH_2P(O)FCl + NaOCN \longrightarrow ClCH_2P(O)F(NCO) + NaCl$$

or the diisocyanates are reacted with antimony trifluoride, which leads to the monofluoro product.[300,640]

$$3ClCH_2P(O)(NCO)_2 + SbF_3 \longrightarrow 3ClCH_2P(O)F(NCO) + Sb(NCO)_3$$

This may be followed by the cleavage of the second NCO group to give the corresponding difluoride.[640] Monofluoro compounds, $RP(O)(NCO)F$ can also be obtained from the corresponding chloro derivative and SbF_3.[641] A mixed chloride-azide, $C_6H_5P(O)ClN_3$ has also been reported and was detected by reaction with a phosphine.[40]

X. ADDITION OF OXYGEN TO HALOPHOSPHINES

Several dichlorophosphines have been oxidized to the corresponding phosphonyl dichlorides by the action of air or oxygen.[260,488,490,556,677] The reaction requires rather drastic conditions.[491] The monohalophosphines, on the other hand, are oxidized fairly readily by air or oxygen on ordinary exposure, and the corresponding phosphinyl chlorides are usually found as by-products in the synthesis of the halophosphines. For preparative purposes, air blown through a solution of the halophosphine with moderate warming has been successfully applied.[556]

The passage of oxygen into a mixture of PCl_3 or $RPCl_2$ and organolead compounds has been used for synthetic purposes.[750,751] This reaction has some similarity to the reaction of organometallic compounds and $POCl_3$ discussed as method III.

A more recent preparative route employs nitrogen oxides with or without a solvent; this method has been used, for example, in the preparation of $CF_3P(O)Cl_2$[214] or $C_6H_5P(O)Br_2$ in CCl_4.[554]

Oxidation was also achieved by SO_2Cl_2 at about room temperature,[115,335,426,549,697] e.g.,

$$C_2H_5PCl_2 + SO_2Cl_2 \longrightarrow C_2H_5P(O)Cl_2 + SOCl_2$$

Other sulfur-oxygen compounds have been employed, including SO_2,[89,159,160] SO_3,[614] and dimethylsulfoxide,[7] especially for the conversion of $C_6H_5PCl_2$ and $(C_6H_5)_2PCl$.

Oxidation of methylphosphonous dibromide to $CH_3P(O)Br_2$ with NO_2 is accompanied by a partial phosphorus-carbon bond cleavage. CH_3PBr_2 was oxidized, however, readily with ozone.[448]

A special procedure involves the reaction of an alkyl halide with white phosphorus in the presence of $AlCl_3$ and HCl. The complex formed from a trivalent halophosphine is destroyed by either nitric acid or hydrogen peroxide in aqueous solution.[8,673]

XI. ADDITION OF SULFUR AND SELENIUM TO HALOPHOSPHINES

The addition of sulfur to halophosphines, normally carried out at temperatures above 60°C, may well be spontaneous at temperatures above 100°C. The reaction may be conducted in sealed tubes or inert solvents. Phosphonousdihalides require higher temperatures, whereas phosphinous halides react well in solution in low-boiling solvents. Products are often isolated by vacuum distillation.[234,373,447,448,556,586,716]

$$RPX_2 + S \longrightarrow R\text{-}P(=S)X_2$$

The yields are in general high and may reach 80 to 90%. The reaction proceeds exothermally at lower temperature if catalytic amounts of $AlCl_3$, $FeCl_3$, or $ZnCl_2$ are used.

Thiophosphoryl chloride, $PSCl_3$ may be used as a sulfur donor. This reaction begins at 100°C and is catalyzed by $AlCl_3$ or P_4S_{10}.[56,62,199]

$$RPCl_2 + PSCl_3 \longrightarrow RPSCl_2 + PCl_3$$

This reaction is advantageous since it is difficult to remove excess elemental sulfur from the first reaction. The addition of sulfur to monohalogen derivatives is faster at room temperature or slightly above. Aromatic or perfluoroalkyl phosphinous chlorides are less reactive, therefore catalytic amounts of $AlCl_3$, $ZnCl_2$, etc., are appropriate;[198,209] $(C_6H_5)_3N$ has also been used as a catalyst.[609] In one case CH_3PSCl_2 served as the donor.[58] The reaction with $PSCl_3$ has also proved useful in the preparation of pseudohalides of pentavalent phosphorus,[317,318] e.g.,

$$(C_6H_5)_2PCN + PSCl_3 \longrightarrow (C_6H_5)_2P(S)CN + PCl_3$$

A different course of reaction is encountered using S_2Cl_2 as a sulfur donor. Dihalophosphines, on moderate warming, give a thiophosphonic dichloride and a chlorophosphorane as by-product, according to the following equation:

$$3RPCl_2 + S_2Cl_2 \longrightarrow 2RP(S)Cl_2 + RPCl_4$$

Separation of the products is effected by chilling the mixture to precipitate solid $RPCl_4$. Often satisfactory results cannot be obtained unless selective solvents are used.[370-372]

The complexes of phosphonous dihalides with $AlCl_3$ will also react with sulfur to give complexes of thionophosphonic dihalides; these may be destroyed by addition of alkali halide.[608] Similar complexes of aromatic phosphonous dihalides are obtained directly from the Friedel-Crafts reaction (Chapter 8) and may be allowed to react with sulfur without purification.[313,726] Unfortunately, the isomeric composition of the products from substituted benzenes is generally not specified.

Phosphinous chlorides (and also CH_3PBr_2[448]) react with elemental selenium under an inert gas atmosphere to the corresponding selenium compound,[464] e.g.,

$$(C_2H_5)_2PCl + Se \longrightarrow (C_2H_5)_2P(=Se)Cl$$

In the case of the dihalophosphines, the addition of selenium is fast enough only in the presence of catalytic amounts of $AlCl_3$.[231,311,448] Furthermore, the selenium derivative tends to split at the phosphorus-selenium bond to give elemental selenium again.

XII. OXIDATIVE REACTION OF HYDROCARBONS, ETHERS, AND ORGANOHALIDES WITH PHOSPHORUS HALIDES

The reaction of hydrocarbons with PCl_3 and oxygen has been extensively used in the preparation of phosphonic and phosphinic halides. Alkanes, alkyl chlorides, olefins, and haloolefins all undergo this reaction with PCl_3 and oxygen.[168,218,219,270,404,413,662,709] Basically, the reaction is represented by

$$2PCl_3 + RH + O_2 \longrightarrow RP(O)Cl_2 + POCl_3 + HCl$$

$$2PCl_3 + RCH{=}CH_2 + O_2 \longrightarrow RCHClCH_2P(O)Cl_2 + POCl_3$$

When RH[184] and RCl[413] compounds are used as starting material, a mixture of all possible isomers of the phosphonic dihalides is formed. It has been shown by determining quantities of each isomer in the products of several reactions that the phosphorus-carbon bond is most readily formed at a tertiary carbon atom, followed by a secondary and then by a primary carbon atom.[661] In certain cases the rates and yields are enhanced by increasing the oxygen pressure.[413] Substitution of oxygen by air leads to reduced yields. The PCl_3 may be replaced by a phosphonous dihalide[752] to give a phosphinyl halide.

The reaction is conducted either in excess hydrocarbon

or in PCl_3 as solvent at room temperature or slightly above.[99,201] With PBr_3 no reaction was found to occur. Catalysts such as $AlCl_3$, I_2, Fe, or BF_3 have a negative effect.[99,290]

Benzene and methane did not react with PCl_3 under the conditions just described.[201,555] Sometimes diphosphonyl derivatives are formed by further attack on the initially formed phosphonic dihalides.[168] Alkylbenzenes react in the side chain, and even diphenylmethane reacts at the aliphatic carbon.

It has been proposed that the reaction is initiated by oxidation of PCl_3,[662] and a reaction mechanism has been formulated for the reaction with alkanes.[201] With the exception of the work of Mayo et al.,[479] little attention has been paid to the exact mechanism of the chlorophosphonation or that of the competing reaction, the oxidation of PCl_3. In a more recent investigation, the mechanism was studied by employing cyclohexane where only one isomer is possible.[162] The chlorophosphonation was found to be a competing side reaction in the PCl_3 oxidation. The mechanism is in agreement with that postulated by Mayo et al.[479] The spontaneous nature of the chlorophosphonation apparently arises from the presence of trace amounts of an initiator, probably hydroperoxides, in the hydrocarbon.[162]

If irradiation is employed, $POCl_3$ (or $RP(O)Cl_2$) may be used instead of PCl_3. Both electron beams[267] and electromagnetic irradiation have been applied.[516] No reaction was found between $POCl_3$ and methane in the gas-phase,[555] but with oxygen and ethylene, PCl_3 gave $ClCH_2CH_2P(O)Cl_2$ (and $CH_3CHClP(O)Cl_2$[456]) in reasonable yields.[597] Ethylene also reacted with $PSCl_3$ in the presence of $AlCl_3$[419] to give $ClCH_2CH_2PSCl_2$.

Methylphosphonyl dichloride is prepared commercially from the reaction of Me_2O with PCl_3 at high temperatures and pressures.[404,718] Small yields of Me_2POCl are also formed in this reaction.[404] A catalyst consisting of a mixture of a metal iodide, phosphorus iodide, a lower alkyl iodide, and a complex of the form $M(PX_3)_n$ (M = Ni, Co, or Zn) is used to increase yields.[718] Under similar conditions, dimethylphosphinyl chloride could be obtained in higher yields from the reaction of $ClCH_2OCH_3$ with phosphorus.[193]

Thioethers react with PCl_3 to give principally the thionophosphinyl chlorides;[37,223,518] CH_3SSCH_3 could also be used. With phosphorus halides, the cyclic thioether $(CH_2S)_3$ gives the halomethylthionophosphonyl halides.[510]

XIII. REACTION OF TRIARYL-CARBINOLS WITH PHOSPHORUS(III) HALIDES

A special method for the preparation of phosphonic dichlorides is represented by the following equation:

$$Ar_3COH + PCl_3 \longrightarrow Ar_3CPOCl_2 + HCl$$

Instead of the expected dichlorophosphite, $ROPCl_2$, the product is a phosphonyl dichloride, which forms immediately in the course of the reaction. The same product results from potassium derivatives of the carbinols.[260] It is evident that the reaction proceeds either by way of formation of the dichlorophosphite, which rearranges to the final product, or by way of coordination to the trivalent phosphorus.[72,250]

$$ROH + PCl_3 \longrightarrow ROH{\cdot}PCl_3 \longrightarrow RP(OH)Cl_3 \longrightarrow RP(O)Cl_2 + HCl$$

The latter hypothesis appears to have some support in that a series of triarylcarbinols arranged in ascending order of "basicity" yields, on immediate hydrolysis with water, progressively increasing amounts of the corresponding phosphonic acids.[73]

The reaction is conducted by addition of the carbinols to the phosphorus halide, preferably in benzene, followed by heating until hydrogen chloride evolution is complete. The product is obtained in good yield by crystallization.[20,21,71-73,253,260,261]

XIV. REACTION OF ALDEHYDES AND KETONES WITH PHOSPHORUS(III) HALIDES

The addition of phosphorus halides to aldehydes in the presence of suitable reagents such as acetic or benzoic acid results in the formation of hydroxyphosphonic acids after hydrolysis. However, under pressure and at higher temperatures aldehydes and PCl_3 react to α-chloroalkanephosphonyl dichlorides.[329-331a,629a,706] The reaction sequence is as follows:

$$RCHO \xrightarrow{PCl_3} \left[(\underset{\displaystyle Cl}{\underset{|}{RCH}}\text{-O-})_3P \longrightarrow \underset{\displaystyle Cl}{\underset{|}{RCH}}\text{-P(O)(O-}\underset{\displaystyle Cl}{\underset{|}{CHR}})_2 \right] \xrightarrow{PCl_3} \underset{\displaystyle Cl}{\underset{|}{RCHP}}(O)Cl_2$$

Aliphatic aldehydes give low yields except for formaldehyde; $ClCH_2P(O)Cl_2$ can be obtained from formaldehyde and PCl_3 in 65% yield.[330,629a] Chloral does not enter a reaction. Phosphorus tribromide has been employed, but no product could be isolated.[112b]

Aromatic aldehydes, however, give good yields of α-chloroarylmethanephosphonyl dichlorides.[329,331a] No halide product was obtained from the reaction of PCl_3 with 4-$(CH_3)_2NC_6H_4CHO$ and 2-HOC_6H_4CHO.[329]

Phosphinyl halides may also be synthesized by this method; thus formaldehyde and phosphonous dihalides give the expected products.[332,448,509] The reaction of aldehydes with phosphorus halides may also lead to halogen-free products; trioxan reacts in the presence of iodine or iron turnings to yield $CH_3P(O)Cl_2$.[719a]

Ketones have also been employed in the reaction with PCl_3. In the presence of hydrogen halides, α-haloalkylphosphonyl dihalides were formed in good yields.[706] It has been shown that acetone and PCl_3 reacted in the presence of $AlCl_3$ to give an ester.[19,135] However, diphenylketone reacts with $C_6H_5PCl_2$ in the presence of $AlCl_3$ to yield $C_6H_5[(C_6H_5)_3C]P(O)Cl$ after hydrolysis.[76]

XV. REACTION OF CARBOXYLIC ACIDS AND THEIR DERIVATIVES WITH PHOSPHORUS(III) HALIDES

A synthetic route of potential application has been found in the reaction of halophosphines and α,β-unsaturated acids. This method leads to alkyl- or aryl-(β-chloroformylalkyl)phosphinyl chlorides as given by

$$RPCl_2 + CH_2{=}CHCOOH \longrightarrow R(ClCOCH_2CH_2)P(O)Cl$$

According to the mechanism postulated for this reaction, the first step is an addition of the halophosphine to form a dipolar ion. This ion rearranges to give a phosphorane, which finally decomposes, leaving the product.

In all cases the chlorides are obtained in high yields,[351,358,360,361,574] but they decompose at higher temperature or during distillation to give phosgene and phosphorus-containing polymers. Instead of the free acids, amides can be employed to yield alkyl- (β-cyanoalkyl)phosphinyl chlorides;[355,359,363] acetylenic acids such as propiolic acid give rise to unsaturated derivatives,[361a] e.g.,

$$C_2H_5PCl_2 + HC{\equiv}CCOOH \longrightarrow C_2H_5(ClCOCH{=}CH)P(O)Cl$$

XVI. FROM HALOPHOSPHORANES AND THEIR COMPLEXES BY SOLVOLYSIS

When halophosphoranes are exposed to moist air, hydrolysis occurs. Thus partial hydrolysis can be used to convert the phosphoranes into the corresponding phosphonyl and phosphinyl halides as represented by the following equations:[490,492,493,496,556]

$$RPCl_4 + H_2O \longrightarrow RP(O)Cl_2 + 2\ HCl$$

$$R_2PCl_3 + H_2O \longrightarrow R_2P(O)Cl + 2\ HCl$$

Further hydrolysis is avoided by using cold water or aqueous HCl. Although the hydrolysis of fluorophosphoranes gives similar results,[213,610] it has not been used for preparative purposes.

Complexes of chlorophosphoranes have also been employed. For example, the addition of PCl_5 to olefins yields a chlorophosphorane that complexes with excess PCl_5 to give $[RPCl_3][PCl_6]$. The hydrolysis of these products has been used extensively for the synthesis of phosphonyl halides.[147,558,693] Alkyl chlorides and PCl_3 react in the presence of equimolar amounts of $AlCl_3$ to form complexes of the type $[RPCl_3][AlCl_4]$. These compounds are also useful precursors if phosphonyl dihalides are desired after hydrolysis.[254,321,366,378,541,592,738] Halophosphines have also been complexed with $AlCl_3$, then treated with chlorine, and finally hydrolyzed.[282]

Alkyl halides may be replaced by dialkylethers;[252a,541] dimethylether, for example, thus yields $CH_3P(O)Cl_2$. Phosphorus trichloride can be replaced by phosphorus tribromide, which reacts with alkyl bromide and aluminum tribromide to give phosphonyl dibromides.[84] In the reaction with PCl_3 and $AlCl_3$ and subsequent hydrolysis, 2-chloro-2,3-dimethylbutane yields $(CH_3)_3CP(O)Cl_2$, rather than the expected product.[488]

A more effective method of solvolysis is the reaction of a halophosphorane with a carboxylic acid. In a smooth reaction the phosphonyl dihalide and the acyl halide are formed,[490,491,501,690] e.g.,

$$C_6H_5PCl_4 + CH_3COOH \longrightarrow C_6H_5P(O)Cl_2 + CH_3COCl + HCl$$

This route has also been used in the preparation of a pseudohalo derivative.[195]

$$(CH_3)_2PCl_2CN + CH_3COOH \longrightarrow (CH_3)_2P(O)CN + CH_3COCl + HCl$$

High yields of phosphonyl and phosphinyl halides are also obtained from halophosphoranes by solvolysis with their corresponding acids,[271,424] e.g.,

$$R_2PCl_3 + R_2P(O)OH \longrightarrow 2\ R_2P(O)Cl + HCl$$

The synthesis of thionophosphonyl and thionophosphinyl halides is achieved by solvolysis with hydrogen sulfide. The reaction is conducted in an inert solvent at slightly elevated temperatures by bubbling hydrogen sulfide through the solution.[13,14,170,577,740] The use of Kinnear-Perren type complexes, which gives somewhat reduced yields,[366,558] has not been investigated to any great extent. Hydrogen selenide (H_2Se) was employed in one case.[571]

XVII. FROM HALOPHOSPHORANES AND THEIR COMPLEXES WITH SO_2 AND P_4O_{10}

The reaction with sulfur dioxide is probably the most convenient method for the conversion of halophosphoranes into the corresponding phosphonyl and phosphinyl halides. The general course of the reaction is represented by the following equations:

$$RPX_4 + SO_2 \longrightarrow RP(O)X_2 + SOX_2$$

$$R_2PX_3 + SO_2 \longrightarrow R_2P(O)X + SOX_2$$

The phosphorane is dissolved or suspended in an inert solvent, and sulfur dioxide is bubbled through the solution. The reaction is rapid, and the products are easily recovered.[26,148,234,490,491,493,495,501,558] Chlorophosphoranes have been used almost exclusively, but bromophosphoranes react in the same way;[143,448] the latter may be contaminated by the corresponding thiono derivative.[448] Tetrafluorophosphoranes, however, do not react with SO_2.[729]

This method is of utmost importance because chlorophosphoranes are easily available. The addition of PCl_5 to olefins and acetylenes, for example, yields reaction products that form complexes with excess PCl_5 of the type $[RPCl_3][PCl_6]$. These complexes are treated with sulfur dioxide to give phosphonyl dichlorides in high yields.[9,14,17,337,368,433,485,559,593,690,696,727,728] In the course of the reaction, the complexes may lose hydrogen chloride so that α,β-unsaturated derivatives are obtained. The following examples are illustrative of this useful method:

$$CH_2{=}CHCH{=}CH_2 + PCl_5 \longrightarrow CH_2{=}CHCH(Cl)CH_2PCl_4 \xrightarrow[-SOCl_2]{SO_2}$$

$$CH_2{=}CHCH(Cl)CH_2P(O)Cl_2$$

$$C_6H_5CH{=}CH_2 + 2\ PCl_5 \longrightarrow [C_6H_5CH(Cl)CH_2PCl_3][PCl_6] \xrightarrow{-HCl}$$

$$[C_6H_5CH{=}CHPCl_3][PCl_6] \xrightarrow[-\ 2SOCl_2]{+\ 2SO_2} C_6H_5CH{=}CHP(O)Cl_2 + POCl_3$$

$$C_6H_5C{\equiv}CH + 2\ PCl_5 \longrightarrow [C_6H_5CCl{=}CHPCl_3][PCl_6] \xrightarrow[-\ 2\ SOCl_2]{+\ 2\ SO_2}$$

$$C_6H_5CCl{=}CHP(O)Cl_2 + POCl_3$$

Complexes of chlorophosphoranes with $AlCl_3$ have also been employed in this reaction;[478] as a rule these complexes are mostly converted by hydrolysis, as discussed in method XVI.

Chlorination of a slurry of P_4O_{10} in a chlorophosphine has been reported to give good yields of phosphonyl and phosphinyl chlorides.[115,211,271,607,680,681,727,728] The overall reaction for a phosphonous dichloride is given in the following equation:

$$6RPCl_2 + 6Cl_2 + P_4O_{10} \longrightarrow 6RP(O)Cl_2 + 4POCl_3$$

The reaction is best conducted with a slight excess of the phosphorus oxide at elevated temperatures; for substances with aliphatic groups, the temperature has to be controlled to avoid further chlorination, which can occur at higher temperatures. The halophosphoranes obtained by other methods are converted directly with P_4O_{10} to the corresponding halides; the formation of α,β-unsaturated derivatives is also observed, as discussed previously for the reaction of PCl_5 with certain olefins.

XVIII. FROM HALOPHOSPHORANES AND THEIR COMPLEXES WITH SULFUR COMPOUNDS

The preparation of phosphonothioic and phosphinothioic halides from halophosphoranes and various sulfur reagents has been investigated. Complexes of chlorophosphoranes with $AlCl_3$ have been reacted with sulfur and antimony sulfide[303,380,496,727] or with olefin sulfides.[341,343] To obtain high yields, the reaction is conducted under heating in the presence of potassium chloride; the addition of the alkali halide is required for binding the $AlCl_3$.

Chlorophosphoranes obtained from the reaction of PCl_5 with olefins have also been treated with sulfides (such as P_4S_{10}) to give high yields of thionophosphonyl dichlorides;[714,716,727,728] this reaction is similar to that discussed in the preceding method. Iodine has been found to act as a good catalyst.[716]

Tetrafluorophosphoranes reacted with P_4S_{10} and various metal sulfides such as PbS, ZnS, K_2S at elevated temperatures to yield the corresponding thionophosphonyl difluorides.[381] Elemental sulfur has been used to convert compounds of the type RPF_3H into $RP(S)F_2$ with evolution of hydrogen fluoride.[296] Other sulfur-donating reagents include KSCN[379] and C_2H_5SH.[341]

In one case the reaction of a chlorophosphorane with selenium has been reported; thus $C_2H_5P(Se)Cl_2$ was obtained in good yields.[311]

Treatment of $(CH_3)_2PBr_3$ with sulfur gave $(CH_3)_2P(S)$-Br.[447]

XIX. REACTION OF HALOPHOSPHORANES WITH CARBON-OXYGEN AND SILICON-OXYGEN COMPOUNDS

Halides of phospholenes have been prepared by reacting the corresponding chloro- or bromophosphoranes with acetic anhydride;[24,28,31] isomerization was observed in this reaction.[31] Fluorophosphoranes have also been treated with anhydrides to give phosphonyl difluorides.[438,610] In one case benzaldehyde has been used to convert $(C_6H_5)_2PF_3$ into $(C_6H_5)_2P(O)F$.[610] In an unusual reaction of PCl_5 with the carboxylic ester $CH_2{=}C(CH_3)OCOCH_3$, the new compound, $CH_3(Cl)C{=}C(COCH_3)P(O)Cl_2$ is formed.[163]

From halophosphoranes in their complexed form with $AlCl_3$, phosphonyl dihalides have been prepared using alkylene oxides in the presence of potassium chloride, as shown here for one example.[304,343]

$$[CH_3PCl_3][AlCl_4] + \underset{\diagdown O \diagup}{CH_2{-}CH_2} \longrightarrow [CH_3\underset{\displaystyle |\atop OCH_2CH_2Cl}{P}Cl_2][AlCl_4] \xrightarrow{KCl}$$

$$\longrightarrow CH_3P(O)Cl_2 + ClCH_2CH_2Cl + K[AlCl_4]$$

The same reaction has been observed with carboxylic acid

amides.[305]

The reaction of fluorophosphoranes with silicon-oxygen compounds such as disiloxanes or silyl ethers proceeds by the cleavage of the Si-O-Si or Si-O-C bond. The following examples are illustrative.

$$RPF_4 + R_3'SiOSiR_3' \longrightarrow RP(O)F_2 + 2\ R_3'SiF$$

$$R_2PF_3 + R_3'SiOR' \longrightarrow R_2P(O)F + R_3'SiF + RF$$

This reaction proceeds in particular ease in the case of tetrafluorophosphoranes. It is a convenient route to phosphonic and phosphinic fluorides.[136,155,610,618] Nothing is known about the intermediates in the course of the reaction.

XX.   FROM HALOESTERS OF TRIVALENT PHOSPHORUS

Certain trivalent phosphorus compounds of the type ROPCl2 and RSPCl2 may undergo a Michaelis-Arbuzov rearrangement to RP(O)Cl2 and RP(S)Cl2. Usually elevated temperatures or halide catalysts are required,[39,224] although propargyl esters rearrange readily as shown in the following equation:[286,287,387]

$$R''C \equiv CC\text{-}O\text{-}PCl_2 \longrightarrow \underset{R}{\overset{R'}{>}}C=C=C\text{-}P(O)Cl_2$$

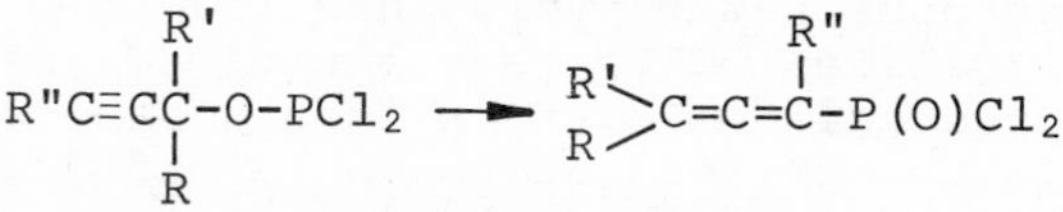

In a similar manner, N-hydroxymethyl carboxylic amides, upon treatment with phosphorus trichloride, form a normal dichlorophosphite, ROPCl2, which isomerizes to the phosphonyl dichloride on prolonged standing. This reaction is affected somewhat by warming and by acetic anhydride. No specific phosphonyl dichlorides have been isolated in these cases; the products were hydrolyzed directly to the phosphonic acids. Hydroxymethylenecamphor reacts analogously.[500]

In the presence of an alkyl halide a similar reaction is found for ester halides. They react at elevated temperatures according to the following equation:[158,404,405,15]

$$R'OPCl_2 + RCl \longrightarrow \left[\underset{R'O}{\overset{R}{>}}\overset{\oplus}{P}Cl_2 \right] Cl^{\ominus} \longrightarrow RP(O)Cl_2 + R'Cl$$

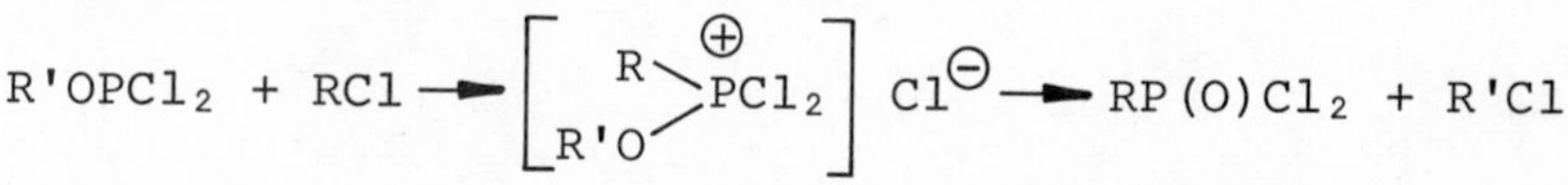

his reaction is performed in the presence of FeCl3 or

NiI_2 as catalysts and sometimes under increased pressure.[404,405]

The reaction has also been applied to phosphonofluoridite esters[402] as shown.

$$CH_3P(F)(OBu) + CH_3I \longrightarrow (CH_3)_2P(O)F + C_4H_9I$$

It is also remarkable that functional alkyl halides could be used in the presence of $FeCl_3$ as a catalyst,[639] e.g.,

$$ROPCl_2 + ClCH_2NCO \longrightarrow OCNCH_2P(O)Cl_2 + RCl$$

A variation is the use of a halogen instead of an alkyl halide,[137] e.g.,

$$CH_3P(F)(OR) + Br_2 \longrightarrow CH_3P(O)FBr + RBr$$

XXI. DECOMPOSITION OF HALOPHOSPHORANES

The application of the Diels-Alder reaction to the synthesis of phosphorus heterocycles has led to a number of new phosphinyl halides. The reaction of dienes and trivalent phosphorus halides gives a phosphorane which, in the case of dihalo-phosphites, is unstable and decomposes to the phosphinyl halide.[24,29,31,54]

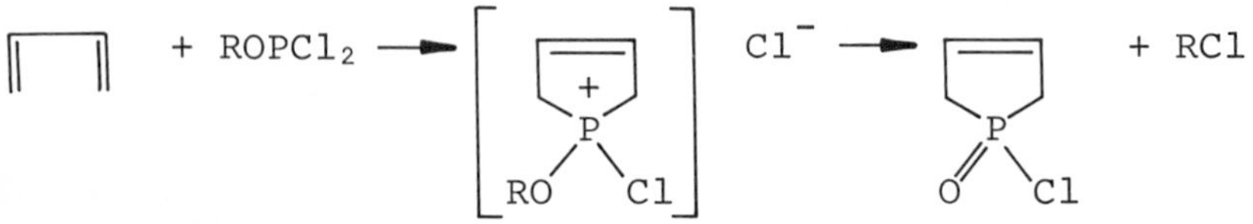

From this equation we see the resemblance to the Michaelis-Arbuzov reaction discussed under method XX. A detailed discussion on scope and mechanism has been presented by Quin.[580]

The reaction is not limited to chlorides and monoesters, as has been shown in the following reaction:[145,544]

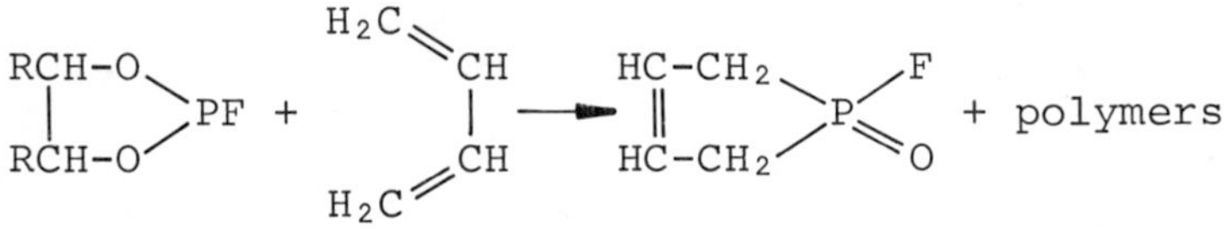

In this manner a number of fluorine compounds have been prepared.

The efficiency of the reactions of dienic hydrocarbons with alkyl phosphorodichloridates can be raised considerably by using PCl_3 as a solvent.[54]

Thioesters have also served to synthesize the corresponding thionophosphinyl halides[374] in the presence of PCl_3. The halophosphorane obtained from the addition of $C_6H_5PCl_4$ to $C_4H_9OCH=CH_2$ also decomposes to give a phosphinyl chloride,[692,695] $C_6H_5(CHCl=CH)P(O)Cl$.

XXII. CHANGE IN THE CARBON SKELETON

A number of halides containing carbo-functional groups may react at the carbon group without affecting the phosphorus grouping, thus giving new halides.

Dehydrochlorination of $4\text{-}ClC_2H_4C_6H_4P(O)Cl_2$ thus yields the unsaturated analog by splitting off HCl,[335]

$$4\text{-}ClC_2H_4C_6H_4P(O)Cl_2 \xrightarrow{-HCl} CH_2{=}CHC_6H_4P(O)Cl_2$$

at elevated temperatures over $BaCl_2$.[325,339,483,559,688] Hydrogen chloride removal may also be achieved by tertiary amines[181,323,468,627,741] as in the following example:[676]

$$CH_3CHClCH_2P(O)Cl_2 \xrightarrow{(C_2H_5)_3N} CH_3CH{=}CHP(O)Cl_2$$

or with copper acetylacetonate[598] at about 270°C. Olefinic derivatives have also been synthesized at 200°C in the presence of 0.1% $(C_6H_5)_3P$.[368]

$$ClCH_2CH_2P(O)Cl_2 \longrightarrow CH_2{=}CHP(O)Cl_2 + HCl$$

The chlorination of the carbon substituent is another useful process for obtaining new compounds, as shown here.[368,593]

$$CH_3CH_2P(O)Cl_2 + Cl_2 \longrightarrow CH_2ClCHClP(O)Cl_2$$

This reaction gives rise to an isomer mixture; $CHCl_2CH_2P(O)Cl_2$ is also obtained.

The chlorination of $4\text{-}C_2H_5C_6H_4P(O)Cl_2$ under UV irradiation[335] leads to chlorinated derivatives. It was claimed that certain conditions yield pure products instead of isomeric mixtures. In the presence of azodiisobutyronitrile, the following reaction was observed:[115]

$$C_2H_5C_6H_4P(O)Cl_2 + Cl_2 \xrightarrow{ABN} CH_3CHClC_6H_4P(O)Cl_2 + HCl$$

Partially chlorinated derivatives [e.g., $ClCH_2P(O)Cl_2$] may be fully chlorinated with chlorine in the presence of PCl_5 or under irradiation.[686]
    Unsaturated compounds such as pholospholenes[54] and vinyl derivatives[338,483,688] can add bromine or chlorine to the double bond, e.g.,

$$C_2H_5(CH_2=CH)P(O)Cl + Br_2 \longrightarrow C_2H_5(CH_2BrCHBr)P(O)Cl$$

The α,β-dihaloethyl derivatives are not very stable and tend to lose hydrogen halide.[338,483]
    Hydrides have also been added to the double bond,[79,91,142] e.g.,

$$(C_2H_5)_3GeH + CH_2=CHCH_2P(O)Cl_2 \longrightarrow (C_2H_5)_3Ge(CH_2)_3P(O)Cl_2$$

For the addition of silicon hydrides, H_2PtCl_6 was used as a catalyst.[79] Further addition reactions have been reported for sulfenyl chlorides[308,309] and diphenyldiazomethane.[570] In the latter case a cyclopropane derivative is obtained.

$$CH_2=CHP(O)Cl_2 + (C_6H_5)_2CN_2 \longrightarrow \begin{matrix} CH_2-CH-P(O)Cl_2 \\ C \\ C_6H_5 \quad C_6H_5 \end{matrix} + N_2$$

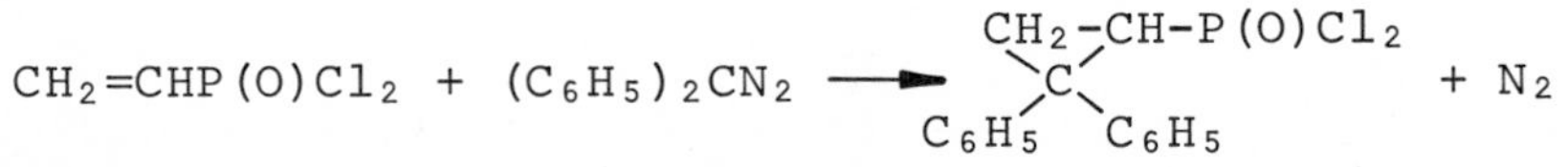

Substitution reactions in the side chain are possible for chlorocarbonyl derivatives, for example, with mercaptans or alcohols in the presence of a base,[528] e.g.,

$$C_2H_5SH + ClC(O)CH_2CH_2P(O)Cl_2 \longrightarrow C_2H_5SC(O)CH_2CH_2P(O)Cl_2 + HCl$$

A more complicated reaction is that of PCl_5 with $NCCH_2P(O)(OC_2H_5)_2$, which gave $Cl_3P=NCCl_2CCl_2P(O)Cl_2$ presumably by way of the intermediate $NCCH_2P(O)Cl_2$.[60] Degradation may also be synthetically useful, as in the following example:

$$CH_3C(O)OC(CH_3)ClCH_2P(O)Cl_2 \longrightarrow CH_3C(O)CH_2P(O)Cl_2 + CH_3C(O)Cl$$

XXIII. MISCELLANEOUS METHODS

A phosphinic amide, $(CF_3)_2P(O)N(CH_3)_2$ could be cleaved with excess hydrogen chloride under pressure to obtain high yields of $(CF_3)_2P(O)Cl$.[81] The corresponding bromide,

$(CF_3)_2P(O)Br$ was obtained from the tertiary phosphine oxide, $(CF_3)_2P(O)C(CH_3)_3$ by carbon-phosphorus bond fission with bromine.[215] A similar cleavage reaction was found to occur with $(ClCH_2)_2P(O)Cl$ and PCl_5, giving the phosphonic dichloride, $ClCH_2P(O)Cl_2$.[165]

From imido derivatives, $R'OC(O)N{=}PRCl_2$, the pseudohalides RP(O)(NCO)Cl are formed along with R'Cl on heating.[637,640,644,645] Some $RPOCl_2$ also occurs as a by-product.

Chlorophosphines, R_2PCl react with chloroamine derivatives such as $(R'O)_2NCl$ to give a phosphorane; subsequent decomposition yields $R_2P(O)NCO$.[125] The pyrolysis of $RP(O)(NHCO_2R')_2$ has also been used to prepare isocyanates, $RP(O)(NCO)_2$.[128]

An alternative synthetic method for isocyanates is the reaction of amide derivatives with $COCl_2$,[678] e.g.,

$$(C_6H_5)_2P(O)NHNa + COCl_2 \longrightarrow (C_6H_5)_2P(O)NCO + HCl + NaCl$$

In a patent the conversion of phosphines into thionophosphonyl or -phosphinyl chlorides by S_2Cl_2 or SCl_2 has been described;[377] it is not clear whether the first step is sulfurization or halogenation.

Selenophosphinic acids undergo an unusual reaction with PCl_5: a chloroseleno compound is produced which, on warming, loses selenium to give a phosphinyl chloride.[463]

$$(C_2H_5)_2P(Se)OH \xrightarrow{PCl_5} (C_2H_5)_2P(O)SeCl \xrightarrow{-Se} (C_2H_5)_2P(O)Cl$$

Diphenylamine reacts with thiophosphoryl chloride at about 200°C to form a phenophosphazine derivative;[479a] no catalyst is necessary.

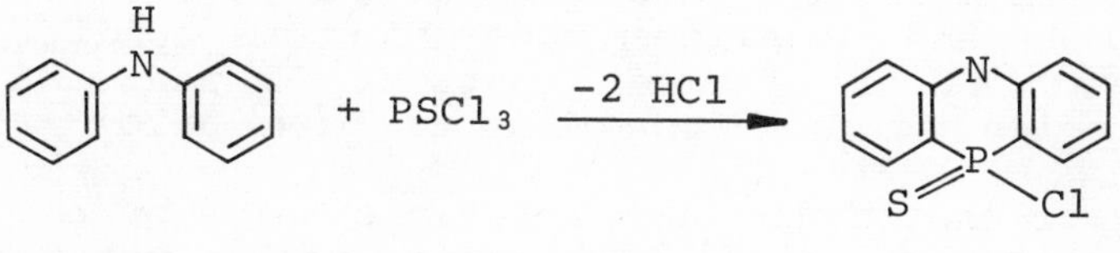

B. BASIC CHEMISTRY

The hydrolysis of phosphonyl and phosphinyl halides leads to the corresponding acids (Chapters 12 and 13). The reaction is slow, partly because of the insolubility of the halides in water. Rates of solvolysis have been presented for many compounds (see Section C). In alkaline solutions the hydrolysis is faster. It is interesting to note that the triarylmethanephosphonic dichlorides hydrolyze slowly on boiling with water but are cleaved at the carbon-phosphorus bond to give carbinols and phosphorous acid.[20,21] To synthesize the phosphonic acids, it is necessary to use alcoholic alkali, followed by the hydrolysis of the resulting monoesters with acetic acid or hydroiodic acid.[20,21,260]

The hydrolysis is also anomalous in the case of CCl_3POCl_2, which is only hydrolyzed by aqueous alkali to the monochloro derivative, $CCl_3P(O)(OH)Cl$, which decomposes on heating to $CHCl_3$ and phosphoric acid.[733,735]

Thionophosphonyl and -phosphinyl halides are stable to water but may be hydrolyzed readily in alkaline solutions to the salts of the corresponding acids (Chapters 12 and 13). Prolonged heating under these conditions or oxidative hydrolysis with, for example, aqueous HNO_3 usually leads to the loss of sulfur and to the formation of phosphonic and phosphinic acids.[494,495,502,503] With alcoholic potassium hydrosulfide solutions, thionophosphinyl chlorides yield dithiophosphinates, R_2PS_2K;[459] with NaSH, monothiophosphinic acids are formed. With hydrogen sulfide, thionophosphonyl dihalides yield dithiophosphonic anhydrides.

The esterification of phosphonyl and phosphinyl halides as well as of their thiono analogs with alcohols or mercaptans or the salts of these gives esters and thioesters (Chapters 12 and 13). Under certain conditions this substitution proceeds stepwise to yield first the monoesters and then, in the presence of a tertiary base, the diesters. Triarylmethane derivatives, however, split at the phosphorus-carbon bond[20] and only sodium alkoxides may be used to form the chloro esters or diesters.[260,261] Hydroquinones form linear polymers.[683]

Robinson[596] used $C_6H_5POCl_2$ to replace -OH groups by chlorine atoms in nitrogen heterocycles.

The aminolysis leads to the corresponding amides of phosphonic or phosphinic acids; the thiono derivatives behave similarly (Chapters 12 and 13). With primary amines and dihalides under certain conditions, imido derivatives are formed (Chapters 12 and 13). Monosubstitution is also possible to give halo amides.

Heating of phosphonic and phosphinic acids or their esters with the corresponding chlorides yields anhydrides (Chapters 12 and 13); thionophosphinyl halides can be directly converted into their anhydrides with water.

The reduction of phosphonyl and phosphinyl halides with hydrides gives primary or secondary phosphines (Chapter 1). Reduction can also be achieved at the phosphorus-oxygen or phosphorus-sulfur bond[221] without conversion of the halide group into hydrogen. In this case, halophosphines are obtained as described in detail in Chapter 8.

Phosphonic dihalides may be converted to phosphinic halides by organometallic reagents, especially Grignard compounds,[118] but further reactions with these give phosphine oxides (Chapter 6). The reaction with Grignard reagents may even lead to reduction of the phosphonic dihalides[389] and not to a phosphinic halide, as follows:

$$RP(O)Cl_2 \xrightarrow{\text{t-BuMgCl}} (CH_3)_3C(R)P(O)H$$

In the case of thionophosphonyl dihalides this reaction takes a different course. Alkyl- or aryl-substituted compounds normally react with alkyl Grignard reagents to give diphosphine disulfides. In certain cases, however, tertiary phosphine sulfides are also obtained.[441] These reactions are comprehensively covered in Chapters 2 and 7.

Various methods of halogen exchange are discussed in Section A. Several reactions are known which give rise to changes in the carbon skeleton; these are also presented as synthetic methods and are discussed under that heading. Conversions of phosphonyl and phosphinyl halides into their thio analogs, and vice versa, have also been treated here as a synthetic method.

Change in coordination is obtained by reacting thionophosphonyl dichlorides with SbF_3 to yield fluorophosphoranes.[342] The reaction of PCl_5 or Cl_2 and phosphinodithioic acids or halides[271] gave chlorophosphoranes, e.g.,

$$(C_6H_5)_2P(S)SH + 3Cl_2 \longrightarrow (C_6H_5)_2PCl_3 + HCl + S_2Cl_2$$

In other cases chlorination of the carbon substituent was found to occur simultaneously.

Since the phosphorus-chlorine bond energy is about 77 kcal, light of less than 220 mμ can split this bond. In this case pentavalent phosphonyl and phosphinyl chlorides react with cyclohexane to give phosphine oxides and phosphinyl halides.[516]

Cleavage of carbon-phosphorus bonds is observed in the reactions of $ClCH_2POCl_2$ and $(ClCH_2)_2POCl$ with PCl_5. In the first case CCl_4 is formed, and in the second reaction CCl_3PCl_2 is obtained, showing that both fission and chlorination occur.[165]

Investigations into the chemistry of pseudohalo derivatives of pentavalent phosphorus are scarce. Reactions have been found to occur at the NCO and NCS groups with alcohols and amines. In a compound such as $CH_3P(O)Cl(NCO)$,[642] this attack takes place before the halogen is substituted.

$$CH_3P(O)Cl(NCO) \xrightarrow{RNH_2} CH_3P(O)Cl(NH\text{-}CONHR)$$

Further hydrolysis with sodium hydroxide leads to $CH_3P(O)(OH)(NHCONHR)$ and finally to methylphosphonic acid.[642] Other studies include the reaction of azide derivatives with tertiary phosphines[40,41,42a] to give imido compounds, e.g.,

$$(C_6H_5)_2P(O)N_3 + (C_6H_5)_3P \longrightarrow (C_6H_5)_2P(O)N{=}P(C_6H_5)_3 + N_2$$

Complexes of $C_6H_5POCl_2$ and $(C_6H_5)_2POCl$ with $TiCl_4$ have been prepared and have been found to contain a donor-acceptor bond between titanium and oxygen.[45] Similar compounds were prepared from $C_6H_5POF_2$ and $C_6H_5POCl_2$ with BF_3, BCl_3, $SbCl_5$, and $SnCl_4$[244] where the oxygen always acts as the donor.

With thionophosphonyl dichlorides, complexes with $AlCl_3$ such as $C_6H_5PSCl_2{\cdot}AlCl_3$ have been obtained. The same reaction occurs with thionophosphinyl halides as well as their oxygen analogs. They all form 1:1 or 1:2 complexes (e.g., $C_2H_5PSCl_2{\cdot}AlCl_3$ and $C_2H_5PSCl_2{\cdot}2AlCl_3$). It was shown from conductivity and NMR measurements that they exist in the ionic form.[453] The existence of a complex between PF_5 and $(C_6H_5)_2POCl$ was demonstrated by conductometric titration, but the complex is unstable.[540]

The compound $C_6H_5POCl_2$ has been extensively investigated as a solvent for chlorides. It has been used for titration of $FeCl_3$ with negative chlorine ion acceptors such as PCl_5, $SbCl_5$, and $HgCl_2$.[34]

No complex formation was observed in the system $C_2H_5POCl_2$ and PCl_5. However, a reaction takes place and a chlorophosphorane complex is formed with excess PCl_5 accompanied by chlorination; the product is $[CH_3CCl_2PCl_3]$-$[PCl_6]$.[367]

C. GENERAL DISCUSSION OF PHYSICAL PROPERTIES

The halo- and pseudohalo derivatives of phosphonic and phosphinic acids are usually liquids. The thio analogs are mostly of higher boiling point; heterocyclic or condensed aromatic compounds are solids.

In some cases vapor pressure data are available, especially for the trifluoromethyl derivatives.[198,213-215] From these data other thermodynamic constants have been derived. For the liquids, further characterizations include the recording of refractive indices and densities. Both figures must be viewed cautiously, since the conditions have not always been given adequately and sometimes isomeric mixtures were measured. For some compounds data have been acquired on such characteristics as viscosity,[237,239] dielectric constants,[237,239] or conductivity;[237,239,538,540] this is especially the case for compounds that have been used as solvents.[33,237] A few dipole moments have also been determined.[61,337]

Mechanistic and kinetic studies of the reactions of phosphonyl and phosphinyl halides have been mainly concerned with their solvolysis. Rate constants for the hydrolysis of many compounds under varied conditions have been determined.[3,134,249,519,521-523] The data have been correlated with substituent constants,[535] and effects of steric hindrance[519,522] and hyperconjugation[522] have been discussed.

The rate of hydrolysis for some phosphonyl chlorides was investigated under conditions close to those of the interfacial polycondensation of polyacrylates.[292] Another special contribution covers the hydrolysis of phosphoric monoesters; this process is catalyzed by enzymes and is inhibited competitively by organophosphorus compounds, including halides.[427]

Some calculations have been carried out using a simple molecular orbital method for P=O compounds, including phosphoryl fluorides, to explain details of the mechanism of these compounds with nucleophilic and electrophilic reagents and to make quantitative comparisons.[406]

Systematic investigations were carried out on donor and acceptor properties of chlorides in solutions of $C_6H_5POCl_2$. Differences between $POCl_3$ and $C_6H_5POCl_2$ as solvents were found to be small.[236] Various physical methods such as potentiometry and conductometry have been used to follow the effects.[33] In a few cases complexes of chlorides with the solvent were obtained and the heat of reaction determined.[244] A quantitative relation was given between the equilibrium constants of extracting agents and the electronegativity of the groups present in the compounds. This was shown for the distribution of uranium(VI) and plutonium(VI), and of uranium(VI) and HNO_3.[602]

Other physical methods employed have been mass spectral analysis,[8] electron-diffraction studies,[710,711] UV absorption measurements,[565,722] and NQR of chlorine in various compounds.[429,523,632,698]

The major contribution to the evaluation of physical properties has come from IR and Raman studies or from NMR measurements. For most of the compounds, IR absorption data are available. The published data are often limited to the rock-salt region, and spectra are often not accurate. However, the amount of data is of some value in the structural assignment of group frequency characteristics and has been helpful in analyzing reaction products. Corbridge has given a survey of IR data in tabular form, listing characteristic frequencies.[106]

Since most of the compounds are of low symmetry and often contain large ligands, vibrational analysis is difficult. Therefore only a few detailed, high-resolution studies by both IR and Raman spectroscopy have been made.

As examples, the trifluoromethyl derivatives of phosphonyl and phosphinyl halides have been investigated,[213-215] as well as the methyl analogs such as CH_3POF_2, CH_3POCl_2,[138,139,581] $(CH_3)_2POCl$,[140] $(CH_3)_2PSCl$,[140,200] and $(CH_3)_2PSBr$.[140] Several studies are concerned with the P=O and P=S vibration. These have received much attention, since the stretching vibration seems to be reasonably independent of the rest of the molecule. Changes, which have been attributed to the electronegativity of the groups at the central atom, have been discussed in some detail.[88,106]

Infrared spectra have also been used to study rotational isomers (e.g., of $ClCH_2POCl_2$ and $ClCH_2PSCl_2$) both in the liquid and the solid phases.[294,530,670] Spectral data have also proved useful in evaluating donor-acceptor interactions between $C_6H_5POF_2$ or $C_6H_5POCl_2$ and various chlorides.[244,245] Band shapes and intensity characteristics have been used in addition to other methods to determine the position of the double bond in phospholenes.[30,772]

The application of NMR spectroscopy to organophosphorus compounds has been a major advance and an aid in analysis and structural assignment. Investigations have been concerned with the ^{31}P, ^{1}H, ^{19}F, and ^{13}C nuclei.

The ^{31}P chemical shifts are characteristic of the structure, and fairly small ranges may be given for the various classes of compounds considered in this chapter. In general the phosphonyl and phosphinyl halides have chemical shifts between -80 and -10 ppm, whereas the thio analogs have slightly lower shifts, in the range -110 to -60 ppm. A comprehensive review of ^{31}P NMR has been given by Crutchfield, et al.,[113] and a compilation of phosphorus chemical shifts including a large number of pentavalent organophosphorus compounds has been presented.[462] Quantum mechanical interpretations of these shifts have been given by Letcher and Van Wazer.[414,705]

Other applications of phosphorus NMR include a study of the halogen exchange reaction in the system $C_6H_5POF_2/BCl_3$.[240] Dynamic nuclear polarization studies on ^{31}P have also been employed to investigate the effects of chemical environment in $C_6H_5POCl_2$, $(C_6H_5)_2POCl$, and $(C_6H_5)_2PSCl$.[566]

The determination of coupling constants can be of considerable value in the identification of compounds,[527,652] since the value of the constants reflects the separation of the two nuclei in the molecule. They are generally determined from the spectra of nuclei other than phosphorus because these may be obtained easily and with greater accuracy. In a typical example, spin decoupling techniques were employed to obtain the proton-phosphorus coupling constants in phospholenes which allowed the isomers to be differentiated.[24,722]

Many data have been obtained from ^{1}H and ^{19}F spectra, and the results have been reviewed by Mavel;[475] correlations of these data with other structural parameters are also summarized. However, ^{13}C NMR measurements have been rare,[595] mostly because of lack of sensitivity.

Fluorine NMR spectra of pentavalent 4- and 3-fluorophenyl-substituted phosphorus halides were used to evaluate the inductive and mesomeric effects of substitution at the central atom in comparison to other derivatives.[567,607] A similar investigation was also made for pentafluorophenyl derivatives.[277] Changes in chemical shifts of ^{19}F and ^{1}H nuclei in $4\text{-}FC_6H_4POCl_2$ were used to determine the acceptor strengths of $SbCl_5$, $TiCl_4$, and $SnCl_4$.[245] The halogen exchange reaction between $C_6H_5POF_2$ and BCl_3 was also studied by ^{19}F NMR measurements.[240]

D. LIST OF COMPOUNDS

D.1. Phosphonyl- (Thio-, Seleno-) Dihalides and Dipseudohalides

Phosphonyl Difluorides

$CCl_3P(O)F_2$. IX.[641] b. 104-105,[641] m. 62-64.[641]

$CF_3P(O)F_2$. XVI.[213] b. -20.1,[213] m. -79.1 to -79.0,[213] log P_{mm} = 10.181 - 1731/T (solid),[213] log P_{mm} = 7.6409 - 0.00968 T + 1.75 log T - 1649/T (liq.),[213] Trouton const. 22.1,[213] $\Delta H_{vap.}$ 5587 ca./mole,[213] IR,[213] ^{19}F.[213]

$CHCl_2P(O)F_2$. IX.[299] b. 126-128,[299] n_D^{20} 1.3987,[299] d_4^{20} 1.6574.[299]

$ClCH_2P(O)F_2$. IX.[615,618,669,699] XIX.[618] b. 128,[669] 112-113,[699] m. -17 to -16,[669] IR,[167,669] Raman,[669] ^{31}P -11.5,[511] -12.0.[526,618]

$MeP(O)F_2$. II.[59] IX.[78,512] XVII.[100,101] XII.[536a] XIX.[438] b. 98,[438] b_{150} 54,[139] b_{14} 48.5,[512] n_D^{20} 1.3165,[438] n_D^{25} 1.3813,[512] d_4^{20} 1.3320,[438] IR,[139,269,675] Raman,[139] ^{1}H,[465] ^{19}F,[82] $J_{^{13}CH}$,[477] thermodynamic properties.[171]

$CH_2{=}CH\text{-}P(O)F_2$. IX.[469] b. 93-95,[469] n_D^{20} 1.3560,[469] d^{20} 1.2741,[469] IR,[469] ^{19}F.[469]

$ClCH_2CH_2P(O)F_2$. IX.[299] b. 124-126,[299] b_{50} 52-54,[299] n_D^{20} 1.3776,[299] d_4^{20} 1.4138.[299]

$EtP(O)F_2$. XIX.[136,618] b. 110-111,[618] n_D^{20} 1.3376,[618] ^{31}P -29.2,[526,618] -29,[136] ^{19}F.[136]

$CH_2{=}C{=}CHP(O)F_2$. IX.[287] b_{14} 36-36.5,[287] n_D^{20} 1.4053,[287] d_4^{20} 1.2989,[287] IR,[287] ^{1}H (anal.).[289]

$MeC{\equiv}C\text{-}P(O)F_2$. IX.[469] b_{15} 34,[469] n_D^{20} 1.3790,[469] d^{20} 1.2442,[469] IR,[469] ^{19}F.[469]

i-PrP(O)F_2. IX.[299] b. 114-115.[299]

$CH_2{=}CHCH{=}CH\text{-}P(O)F_2$. IX.[469] b_8 39-40,[469] n_D^{20} 1.4312,[469] d^{20} 1.2116,[469] IR,[469] ^{19}F.[469]

$EtOCH{=}CHP(O)F_2$. IX.[640] b_9 60-63,[640] n_D^{20} 1.3993,[640] d_4^{20}

1.2441.[640]

$BuP(O)F_2$. I.[658] IX.[618] b_{56} 68-70,[658] ^{31}P -29.3.[526,618]

t-$BuP(O)F_2$. IX.[626] b_{85} 63,[626] m. 40,[626] IR,[620] ^{31}P -31.3,[620] ^{1}H,[620] ^{19}F.[620]

$MeCH{=}CHCH{=}CH{-}P(O)F_2$. IX.[469] b_7 60-62,[469] n_D^{20} 1.4532,[469] d^{20} 1.1456,[469] IR,[469] ^{19}F.[469]

$CH_2{=}CHC(Me){=}CH{-}P(O)F_2$. IX.[469] b_2 29,[469] n_D^{20} 1.4445,[469] d^{20} 1.1631,[469] IR,[469] ^{19}F.[469]

$Me_2C{=}C{=}CH{-}P(O)F_2$. IX.[286,287] $b_{1.5}$ 26,[286,287] n_D^{20} 1.4215,[286,287] d_4^{20} 1.1562,[286,287] IR,[287] ^{19}F.[286]

i-$AmP(O)F_2$. IX.[618] ^{31}P -27.2.[526,618]

$C_6F_5P(O)F_2$. XIX.[155] b_{14} 73,[155] ^{31}P +4.6,[155] ^{19}F.[155,277]

4-$ClC_6H_4P(O)F_2$. IX.[297,729] b. 210-213,[297] 211-213,[729] d_{20}^{20} 1.3381.[729]

$PhP(O)F_2$. I.[658] IX.[239,618,659,729] XIX.[610] b. 186-187,[729] 172-173,[658] b_{13} 73,[610] n_D^{25} 1.4640,[610] d_{20}^{20} 1.2982,[729] d_{20} 1.3052,[239] sp. cond. 204 × 10^{-7} $ohm^{-1}cm^{-1}$,[239] η_{20} 1.77 cP,[239] dielect. const. at 20° 28.6,[239] ^{31}P -11.4,[526,618] complexes: $PhP(O)F_2 \cdot BF_3$ (m. 41-43, IR);[244] $PhP(O)F_2 \cdot BCl_3$ (liq., IR);[244] $PhP(O)F_2 \cdot SbCl_5$ (m. 46-48, IR);[244] $PhP(O)F_2 \cdot TiCl_4$ [m. 68-70 (dec.), IR];[244] $(PhP(O)F_2)_2 \cdot TiCl_4$ [m. 52-54 (dec.), IR];[244] $(PhP(O)F_2)_2 \cdot SnCl_4$ (m. 54-56, IR).[244]

$C_6H_{13}P(O)F_2$. IX.[153] b. 176-177,[153] b_7 58-60,[153] d_4^{20} 1.0700.[153]

4-$CF_3C_6H_4P(O)F_2$. IX.[731] b. 186-187.[731]

4-$MeC_6H_4P(O)F_2$. IX.[729] b. 207-209,[729] d_{20}^{20} 1.2781.[729]

$C_7H_{15}P(O)F_2$. IX.[153] b. 196-197,[153] b_6 74-76,[153] d_4^{20} 1.0500.[153]

$PhCH{=}CHP(O)F_2$. ^{1}H.[676]

$C_8H_{17}P(O)F_2$. IX.[153] b. 210-211,[153] b_6 88-90,[153] d_4^{20} 1.0182.[153]

$PhC(Me){=}CHP(O)F_2$. IX.[676] b_{10} 126-127,[676] n_D^{20} 1.5227,[676] d_4^{20} 1.2515,[676] ^{1}H.[676]

$C_9H_{19}P(O)F_2$. IX.[153] b. 226-227,[153] b_6 96-97,[153] d_4^{20} 1.0202.[153]

$C_{10}H_{21}P(O)F_2$. IX.[153] b. 235-236,[153] b_6 113-115,[153] d_4^{20} 1.0041.[153]

Phosphonyl Fluoride Chlorides

$Cl_2CHP(O)FCl$. II.[299] b_{20} 67-68,[299] n_D^{20} 1.4558,[299] d_4^{20} 1.6875.[299]

$ClCH_2P(O)FCl$. II.[299] b_{22} 64,[299] n_D^{20} 1.4360,[299] d_4^{20} 1.6015,[299] ^{31}P -32.0.[511]

$MeP(O)FCl$. I.[274] II.[299] IX.[139] b. 125-127,[299] b_{50} 56,[139] b_{14} 34-35,[274] IR,[139] Raman,[139] Cl (NQR) 25.84 MHz,[632] thermodynamic properties.[171]

$ClCH_2CH_2P(O)FCl$. II.[299] b_{20} 85-87,[299] n_D^{20} 1.4490,[299] d_4^{20} 1.5178.[299]

$EtP(O)FCl$. II.[299] b_{25} 47-48,[299] n_D^{20} 1.4030,[299] d_4^{20}

1.2968.[299]
i-PrP(O)FCl. I.[299] II.[299] b_8 36-37,[299] n_D^{20} 1.4100,[299] d_4^{20} 1.2250,[299] kinetics of hydrolysis.[521]
PhP(O)FCl. II.[298] IX.[240] b_{15} 95-98,[298] n_D^{18} 1.5080,[298] d_{20}^{18} 1.3560,[298] ^{31}P,[240] ^{19}F.[240]

Phosphonyl Fluoride Bromide

MeP(O)FBr. XX.[137] b_{25} 27-39,[137] n_D^{20} 1.4960,[137] d_4^{20} 1.9699,[137] ^{31}P -34,[137] ^{19}F.[137]

Phosphonyl Fluoride Pseudohalides

$CCl_3P(O)F(NCO)$. IX.[641] b_{756} 178-180,[641] b_{15} 69-71,[641] n_D^{20} 1.4720,[641] d_4^{20} 1.7232,[641] IR.[641]
$CHCl_2P(O)F(NCO)$. IX.[640] b_{10} 80-82,[641] n_D^{20} 1.4620,[640] d^{20} 1.6671.[640]
$ClCH_2P(O)F(NCO)$. IX.[300,644] b_{20} 91-93,[644] n_D^{20} 1.4439,[644] d^{20} 1.5848.[644]
MeP(O)F(NCO). IX.[643] b_{45} 80-82,[643] b_{18} 65-66,[643] n_D^{20} 1.4113,[643] 1.4091,[643] d_4^{20} 1.3615,[643] IR.[643]
$ClCH_2CH_2P(O)F(NCO)$. IX.[299] b_7 80-81,[299] n_D^{20} 1.4550,[299] d_4^{20} 1.4919.[299]
EtP(O)F(NCO). IX.[299] b_{20} 66-67,[299] n_D^{20} 1.4125,[299] d_4^{20} 1.2830.[299]
i-PrP(O)F(NCO). IX.[299,640] b_8 56-57,[299] n_D^{20} 1.4152,[299] d_4^{20} 1.2281.
EtOCH=CHP(O)F(NCO). IX.[300] b_3 54-56,[300] n_D^{20} 1.4447,[300] d_4^{20} 1.2679.[300]
MeP(O)F(NCS). IX.[299] $b_{0.05}$ 40-43,[299] n_D^{20} 1.5037,[299] d_4^{20} 1.3647.[299]

Phosphonyl Dichlorides

$CCl_3P(O)Cl_2$. XVI.[347,366,541,558] XVII.[346,558,733,734] XXII.[686] b. 199-200,[686] b_1 95,[346,347,558] m. 154-156,[686] 155,[733,734] 156.[346,347,366,541,558]
$CF_3P(O)Cl_2$. X.[214] b. 70.2,[214] m. -25.3 ± 0.3,[214] $\log P_{mm} = 4.6775 - 0.00268 \cdot T + 1.75 \log T - 1824.2/T$,[214] Trouton const. 23.5,[214] ΔH_{vap} 5590 cal/mole,[214] IR,[214] Raman,[214] ^{19}F.[214]
$CHCl_2P(O)Cl_2$. XVI.[366,541] b_9 79,[366,541] ^{31}P -34.8,[450a] ^1H.[450a]
$BrCH_2P(O)Cl_2$. XIV.[329] XVI.[84] b_6 123-124,[329] b_7 118-120,[84] n_D^{20} 1.6100,[329] 1.512.[84]
$ClCH_2P(O)Cl_2$. I.[42b,294] II.[509] XIV.[324,329,330,706] XVI.[366,541,735] b_{30} 103,[735] b_{17} 92-94,[698] b_{12} 93,[324] b_{10} 78-79,[294] 77-78,[42b,329,330] $b_{0.5}$ 50,[366,541] b. 40-41,[706] n_D^{20} 1.4990,[706] 1.4978,[329,330] 1.4930,[698] 1.4945,[735] 1.4983,[324] d_4^{20} 1.6429,[698] 1.6361,[329,330] d_{20}^{20} 1.6444,[735] IR,[167,294,530,670] Raman,[294,530,670]

Cl (NQR),[429,698] ^{31}P -38.0.[509]

$Cl_3GeCH_2P(O)Cl_2$. XII.[144] b_1 96,[144] n_D^{20} 1.5305.[144]

$FCH_2P(O)Cl_2$. I.[229] b_{15} 54-55,[229] n_D^{20} 1.4500.[229]

$MeP(O)Cl_2$. I.[185,230,276] II.[51,75,77,116,117,230,509,512,721] III.[312] VIII.[685] X.[194,655] XII.[404,718,719] XIX.[343] b. 163,[276] 162,[253,366] 102,[185] b_{100} 100-110,[685] b_{50} 81-82,[655] b_{22} 64,[721] b_{14} 57.5-58,[230] b_{11} 51-52,[230] m. 33,[253,366] 32.99,[138] IR,[95,138,185,188,269,480-482,581,675] Raman,[95,138,185,188] ^{31}P -44.5,[517] -43.5,[509,511] ^1H,[465,466] $J_{^{13}CH}$,[477] Cl (NQR),[429,522,523,632,698] kinetics of hydrolysis,[521] thermodynamic properties.[171]

$Cl_3P{=}NCCl_2CCl_2P(O)Cl_2$. XXII.[60] $b_{0.05}$ 78,[60] n_D^{20} 1.5610,[60] d^{20} 1.8470.[60]

$CD_2{=}CDP(O)Cl_2$. II.[383] b_{20} 68-71,[383] n_D^{24} 1.4528,[383] d_4^{24} 1.4022,[383] IR.[383]

$CHFCl{-}CFCl{-}P(O)Cl_2$. I.[288] b_8 67-68,[288] n_D^{20} 1.4695,[288] d_4^{20} 1.761.[288]

$CHCl_2{-}CF_2{-}P(O)Cl_2$. I.[288] b_8 67-68,[288] n_D^{20} 1.4690,[288] d_4^{20} 1.751.[288]

$CHCl_2{-}CFCl{-}P(O)Cl_2$. I.[288] b_8 87-88,[288] n_D^{20} 1.5025,[288] d_4^{20} 1.786.[288]

$CF_2H{-}CF_2{-}P(O)Cl_2$. I.[74] b_{300} 95,[74] n_D^{25} 1.3862,[74] IR.[74]

$CH_2{=}CBrP(O)Cl_2$. XXII.[483] b. 194-197,[483] m. 36-39,[483] n_D^{20} 1.5273,[483] d_{20} 1.8393.[483]

$BrCH{=}CHP(O)Cl_2$. XII.[664] b_3 81-82,[664] n_D^{20} 1.4889,[664] d_{20} 1.7415.[664]

$CH_2{=}CClP(O)Cl_2$. XXII.[483] b_8 80-87.[483]

$ClCH{=}CHP(O)Cl_2$. X.[677] b_{10} 99-102,[677] n_D^{18} 1.5202,[677] d^{18} 1.6420.[677]

$C_2H_2FCl_2P(O)Cl_2$. XII[664] (isomer mixture). b_3 66-70.[664]

$Cl_2CHCHClP(O)Cl_2$. XII.[664] b. 88-92,[664] n_D^{20} 1.4990,[664] d_{20} 1.7172,[664] slowly attacked by H_2O.[664]

$Cl_3CCH_2P(O)Cl_2$. XII.[747] XVI.[366] b_5 95,[747] b_1 65-66,[366] d_{20} 1.7398.[747]

$OCNCH_2P(O)Cl_2$. XX.[639] $b_{0.15}$ 59-60,[639] n_D^{20} 1.4980,[639] d_4^{20} 1.5744.[639]

$CH_2{=}CHP(O)Cl_2$. II.[169,605a] XXII.[325,368,598,627,741] b. 165-170,[598] b_{36} 81,[169] b_{21} 67-69,[325] b_{20} 65.5-67.5,[698] b_{18} 63-66,[741] b_{11} 59-61,[605a] b_1 42,[627] m. -58,[169] n_D^{20} 1.4810,[741] 1.4795,[605a] 1.4808,[325] d_4^{20} 1.4098,[698] 1.4096,[741] d_{20} 1.4092,[325] d^{40} 1.4016,[169] IR,[654] Raman,[565] UV,[565] Cl (NQR).[698]

$BrCH_2CHClP(O)Cl_2$. XII.[664] b_3 99-100,[664] n_D^{20} 1.5378,[664] d_{20} 1.9720.[664]

$BrCH_2CHBrP(O)Cl_2$. XXII.[483] b_5 120-122,[483] m. 33-34,[483] n_D^{20} 1.5698,[483] d_{20} 2.2155.[483]

$C_2H_3FClP(O)Cl_2$. XII[664,747] (isomer mixture). b_{30} 109,[664] 98-100.[747]

$FSO_2C_2H_3ClP(O)Cl_2$. XII.[664] b_3 95-97.[664]

$C_2H_3Cl_2P(O)Cl_2$. XII[663,745,747] (isomer mixture). b. 208-210,[663] $b_{2.5}$ 74-75,[663] b_2 78-80.[747]

$ClCH_2CHClP(O)Cl_2$. XII.[661] XXII.[368,483,593] b_{15} 121-124,[368,593] b_3 84-85,[483] b_2 70-72.[661]

$MeCCl_2P(O)Cl_2$. XVII.[367]

$Cl_2HCCH_2P(O)Cl_2$. XXII[368,593] (mixture with $ClCH_2CHClP(O)Cl_2$). b_{15} 121-124.[368,593]

$Cl_3SiC_2H_3ClP(O)Cl_2$. XII.[90] b_4 135-140.[90]

$BrCH_2CH_2P(O)Cl_2$. II.[601] b_{18} 119-120,[601] n_D^{20} 1.5210,[601] d_0^{20} 1.8262,[601] d_4^{20} 1.8242.[601]

$C_2H_4ClP(O)Cl_2$. XII[661] (isomeric mixture). b_2 65-72.[661]

$MeCHClP(O)Cl_2$. XIV.[329,706] XVI.[366,735] b_8 70,[735] b_6 71-72,[329] b_3 57,[366] n_D^{20} 1.4911,[329] 1.4930,[735] d_4^{20} 1.5134,[329] d_{20}^{20} 1.5424,[735] ^{31}P -45.5,[456] ^1H.[456]

$ClCH_2CH_2P(O)Cl_2$. I.[178] II.[327,328] X.[677] XII.[270,395,597,662,709,752] XVI.[366,541,669] XVII.[478] b. 213-217,[327] b_{150} 150,[634] b_{15} 80-84,[366] b_7 85-87,[677] b_6 77,[669] $b_{5.5}$ 82-84,[270,327,709] b_5 84,[597] b_2 68,[327] 86-87,[662] m. < -80,[669] n_D^{16} 1.4977,[327,328] d_0^{16} 1.5443,[327] d_4^{16} 1.5430,[327,328] d_4^{20} 1.5446,[662] IR,[669] Raman,[669] ^{31}P -41.9,[511] -42.9,[456] ^1H.[456]

$Cl_3GeC_2H_4P(O)Cl_2$. XII.[141] $b_{1.5}$ 96.[141]

$Cl_3SiCH_2CH_2P(O)Cl_2$. XII.[92] b_8 130,[92] m. 44.[92]

$EtP(O)Cl_2$. I.[185,266] II.[578] II.[283,312,446] VIII.[578] X.[159,750,751] XVI.[97,98,282,366,378,493,499,541,669,738] XVII.[234] XIX.[343] b. 174.5,[97] 175,[493,499] 179.5,[378] 175-177,[738] b_{50} 75-78,[493,499] b_{31} 91,[185] b_{23} 77.3,[578] $b_{13.5}$ 69.5,[578] b_{11} 156-158,[266] b_4 53-60,[282] b_3 34,[159,366,541,669] m. < -80,[669] n_D^{20} 1.4653,[578] 1.4662,[578] n_D^{25} 1.4639,[185] n_D^{26} 1.4642,[282] d_{20} 1.3707,[578] 1.3675,[578] d^{20} 1.1883,[493,499] d_4^{20} 1.367,[378] d_4^{25} 1.3612,[185] ΔH_{hydrol} at 25° 10.8 ± 0.08 kcal/mole,[186] IR,[95,185,188,480,482,669] Raman,[95,185,188,669] ^{31}P -52.5,[446] ^1H,[467] coupling const.,[652] Cl (NQR),[439,522,698] kinetics of hydrolysis,[521] complexes: $EtP(O)Cl_2 \cdot AlCl_3$ ^{31}P -75.8;[453] $EtP(O)Cl_2 \cdot 2AlCl_3$ ^{31}P -75.8.[453]

$MeOCH_2P(O)Cl_2$. XII.[405]

$CH_2{=}C{=}CDP(O)Cl_2$. XXIII.[384] $b_{1.5}$ 63-65,[384] n_D^{20} 1.5196,[384] d_4^{20} 1.4101,[384] IR.[384]

$NCCH{=}CHP(O)Cl_2$. II.[579] $b_{1-1.5}$ 76-77,[579] n_D^{20} 1.5130,[579] d_4^{20} 1.4521.[579]

$CH_2{=}C{=}CHP(O)Cl_2$. XXIII.[287] b_3 73-74,[306] $b_{2.5}$ 63.5-64,[287] n_D^{20} 1.5226,[287] 1.5232,[306] d_4^{20} 1.4103,[287] d^{20} 1.4105,[306] ^1H (anal.),[289] relative signs of J_{PH},[87] IR.[287]

$MeC{\equiv}CP(O)Cl_2$. II.[468] $b_{1.5}$ 56-57,[468] n_D^{25} 1.4912,[468] d^{20} 1.3953,[468] IR.[674]

$CH_2{=}CHCH_2P(O)Cl_2$. XVI.[79,227,366,541,665] XX.[226] XXII.[559] b_5 77,[79] b_3 55,[226,227,366,541] 55-56,[665] n_D^{20} 1.4830,[226,227] 1.4871,[665] 1.4870,[79] d^{20} 1.3352,[226,227] 1.35875,[665] d_{20} 1.3783,[79] IR,[95,227] Raman.[95]

$MeCH{=}CHP(O)Cl_2$. XVII.[676] XXII.[598,676] b_7 74-75,[676] n_D^{20} 1.4915,[676] d_4^{20} 1.3272,[676] ^1H.[676]

$CH_2{=}C(Me)P(O)Cl_2$. I.[27,251,418] II.[64] b_{35} 83-85,[418] b_{32}

82.5-86,[251] b_{30} 83,[251] b_{14} 71.5-73,[27] b_{13} 64-66,[64] n_D^{20} 1.4818,[64] 1.4823,[27] 1.4820,[418] d_4^{20} 1.3372,[64] 1.3619,[418] d_0^{20} 1.3379.[27]

$EtCCl_2P(O)Cl_2$. XVII.[367]

$ClSCH_2CHClCH_2P(O)Cl_2$. XXII.[308] n_D^{20} 1.5620,[308] d^{20} 1.6527.[308]

$MeOCH{=}CHP(O)Cl_2$. XVII.[17] $b_{1.5}$ 76,[17] n_D^{20} 1.5052,[17] d_{20} 1.4186.[17]

$MeCOCH_2P(O)Cl_2$. XVII.[431] XXII.[433] m. 39-40.[433]

$MeSCH{=}CHP(O)Cl_2$. XVII.[593] $b_{0.2}$ 99,[593] m. 45.[593]

$MeCH(P(O)Cl_2)CH_2Cl$. XVII.[727] b. 190-218.[727]

$Br(CH_2)_3P(O)Cl_2$. I.[391] b_{23} 153-156.[391]

$C_3H_6ClP(O)Cl_2$. XII[661,662,752] (isomeric mixture). b_2 85-87,[662] 80-95,[661] 65-100,[661] n_D^{20} 1.4930,[662] d_4^{20} 1.4615.[662]

$MeCHClCH_2P(O)Cl_2$. X.[677] b_{10} 92-95,[677] n_D^{18} 1.4986,[677] d^{18} 1.404.[677]

$Cl(CH_2)_3P(O)Cl_2$. I.[5,712] II.[265] $b_{0.04}$ 63-64,[5,712] $b_{0.01}$ 65-67,[265] n_D^{20} 1.4982,[5,712] d_4^{20} 1.4534.[5,712]

$Cl_3GeC_3H_6P(O)Cl_2$. XII.[141] b_1 123-125.[141]

$Cl_3Si(CH_2)_3P(O)Cl_2$. XII.[90] XXII.[79] b_6 140-142.5,[79] b_4 134-142,[90] n_D^{20} 1.4985,[79] 1.5003,[90] d_{20} 1.5160,[79] 1.5202.[90]

$C_3H_6FP(O)Cl_2$. XII[747] (isomeric mixture). b_{19} 80-83.[747]

$C_3H_7P(O)Cl_2$. XII[201,752] (isomeric mixture). b_9 62-64.[201]

$C_3H_7P(O)Cl_2$. I.[185] XII.[201] XVII.[234] b_{50} 88-90,[234] b_{15} 83-85,[698] b_{10} 70,[185] b_9 62-64,[201] n_D^{20} 1.4648,[698] n_D^{25} 1.4633,[185] d_4^{20} 1.3039,[698] d_4^{25} 1.2864,[185] d^{20} 1.3088,[234] IR,[185] Raman,[185] ^1H,[467,476] Cl (NQR),[698] kinetics of hydrolysis.[521]

i-$PrP(O)Cl_2$. XVI.[97,98,185,366,378,541,738] XVII.[234] b_{746} 189,[378] b_{50} 104,[378] 82-84,[234,738] b_{23} 76,[97] b_{20} 80.5,[185] b_{15} 78,[378] $b_{1.5}$ 34,[97] n_D^{25} 1.4646,[185] d^{20} 1.3018,[234] d_4^{20} 1.2997,[378] d_4^{25} 1.2899,[185] IR,[95,185,188] Raman,[95,185,188] ^1H,[476] coupling const.,[652] Cl (NQR)[522,698] kinetics of hydrolysis.[521]

$EtSCH_2P(O)Cl_2$. I.[104] b_1 85.[104]

$H(CCl_2{-}CF_2)_2P(O)Cl_2$. I.[288] b_8 120-125,[288] n_D^{20} 1.4812,[288] d_4^{20} 1.778.[288]

$CF_2H(CF_2)_3P(O)Cl_2$. I.[74] b_{30} 75,[74] n_D^{25} 1.3670,[74] IR.[74]

$CH_2{=}CHC{\equiv}CP(O)Cl_2$. II.[468] $b_{2.5}$ 61-65,[468] n_D^{20} 1.5150,[468] d^{20} 1.3784.[468]

$C_4H_3SP(O)Cl_2$ (2-thiophenephosphonyl-). XVII.[605] b. 258-260.[605]

$CH_2{=}CClCH{=}CHP(O)Cl_2$. II.[687] XVII.[690] XXI.[687] b_{11} 135,[690] b_4 114-115,[687] n_D^{20} 1.5485,[687,690] d_4^{20} 1.4850,[687] d^{20} 1.4830.[690]

$CH_2{=}CHCH{=}CHP(O)Cl_2$. XVII.[468] XIX.[468] b_1 65-66,[468] n_D^{20} 1.5413,[468] d^{20} 1.3107.[468]

$CHCl_2CHAcP(O)Cl_2$. XVII.[430,432,433] $b_{2.5}$ 98-99,[432] b_2 94-95,[430,433] n_D^{20} 1.5100,[430,433] 1.5098,[432] d_{20} 1.5464,[430],

[433] 1.5460.[432]

$EtOCH{=}CBrP(O)Cl_2$. XVII.[649] m. 34-35.[649]

$CH_2{=}CHCHClCH_2P(O)Cl_2$. XVI.[646] XVII.[12] b_2 103-105,[12] n_D^{20} 1.5200,[12] d_{20} 1.4452.[12]

$ClCH_2CH{=}CHCH_2P(O)Cl_2$. XVII.[576] b_{12} 142-143,[576] n_D^{20} 1.5210,[576] d_{20} 1.4440.[576]

$ClCH(Me)OCH{=}CHP(O)Cl_2$. XVI.[690] b_3 112-113.5,[690] n_D^{20} 1.5160,[690] d^{20} 1.4510.[690]

$ClCH_2CH_2OCH{=}CHP(O)Cl_2$. XVII.[689] b_3 131-132,[689] n_D^{20} 1.5244,[689] d_{20} 1.483.[689]

$ClCH_2CH(COOMe)P(O)Cl_2$. XII.[736] b_8 124,[736] n_D^{20} 1.4745,[736] d_{20} 1.5238.[736]

$AcOCHClCH_2P(O)Cl_2$. XVII.[431,433] $b_{1.5}$ 99-100,[431,433] n_D^{20} 1.4855,[431,433] d_{20} 1.5035.[431,433]

$Me_2C{=}CHP(O)Cl_2$. XVII.[411,727] b_{17} 99-101,[411,727] m. < -70,[411,727] d 1.302.[411,727]

$MeCH{=}C(Me)P(O)Cl_2$. I.[418] II.[64] b_{15} 94-95,[418] b_{13} 93-95,[64] n_D^{20} 1.4923,[64] 1.4943,[418] d_4^{20} 1.2994,[64] 1.2957.[418]

$PrCCl_2P(O)Cl_2$. XVII.[367]

$ClCH_2SCH_2CHClCH_2P(O)Cl_2$. XXII.[309] n_D^{20} 1.5560,[309] d_4^{20} 1.5661.[309]

$MeSi(Cl_2)CH_2CH{=}CHP(O)Cl_2$. XVII.[559] b_3 105,[559] n_D^{20} 1.4747,[559] d_4^{20} 1.2935,[559] IR.[559]

$EtOCH{=}CHP(O)Cl_2$. XVII.[17,551,553] b_{16} 130-131,[417] b_2 84,[17] b_1 82,[551,553] n_D^{20} 1.4969,[17] 1.4956,[551,552] 1.4960,[417] d_{20} 1.3221,[17] d_4^{20} 1.3320,[551,553] d^{20} 1.3225,[417] ^1H.[412]

$MeOCOCH_2CH_2P(O)Cl_2$. XXII.[528] $b_{0.4}$ 85-86,[528] n_D^{20} 1.4790,[528] d^{20} 1.4259.[528]

$(MeO)_2C{=}CHP(O)Cl_2$. XVII.[515] b_{13} 115,[515] n_D^{20} 1.5010,[515] d_4^{20} 1.4259.[515]

$EtSCH{=}CHP(O)Cl_2$. XVII.[593] $b_{0.05}$ 96,[593] n_D^{20} 1.5669.[593]

$EtCH(P(O)Cl_2)CH_2Cl$. XVII.[727] b_{18} 116-123.[727]

$C_4H_8BrP(O)Cl_2$. XII[747] (isomeric mixture). b_4 109-132.[747]

$C_4H_8ClP(O)Cl_2$. XII[661,746,747] (isomeric mixture). b_5 85-87,[661] b_4 85-120,[661,746] 80-100,[661] $b_{2.5}$ 85-93,[746,747] b_2 85-86,[746] 73-74.[746]

$PrCHClP(O)Cl_2$. XIV.[329] XVI.[735] b_{25} 84,[735] b_{13} 107,[329] n_D^{20} 1.4885,[329] 1.5010,[735] d_4^{20} 1.3598,[329] d_{20}^{20} 1.3236.[735]

$ClCH_2(CH_2)_3P(O)Cl_2$. II.[265] $b_{0.05}$ 90-92.[265]

$MeCHClCHMeP(O)Cl_2$. XII.[746] b_2 70-74,[746] n_D^{20} 1.4820,[746] d_{20} 1.3881.[746]

$MeSCH_2CHClCH_2P(O)Cl_2$. XXII.[309] b_1 118-120,[309] n_D^{20} 1.5405,[309] d_4^{20} 1.4321.[309]

$MeSi(Cl_2)CH_2CHClCH_2P(O)Cl_2$. XVII.[559] $b_{1.5}$ 123,[559] m. 26,[559] IR.[559]

$C_4H_9P(O)Cl_2$. XII[201,219,662,752] (isomeric mixture). b_{20} 95.5-97,[219] $b_{3.5}$ 68-70,[201] b_2 55-57,[662] n_D^{20} 1.4660,[662] 1.4684,[219] d_4^{20} 1.2639,[662] d_{20} 1.2478.[219]

$BuP(O)Cl_2$. I.[185] II.[545] XII.[217,218] b_{23} 105-107,[698] b_{15} 96-97,[545] b_{10} 88,[185] 79-83,[218] b_7 81-82,[217] n_D^{20}

1.4615,[545] 1.4651,[698] 1.4680,[218] 1.4640,[185] 1.4649,[217] d^{20} 1.2403,[217] 1.2415,[545] d_4^{20} 1.2464,[698] 1.2491,[218] d_4^{25} 1.2341,[185] IR,[185] Raman,[185] ^1H,[467] Cl (NQR).[698]

i-BuP(O)Cl_2. XVI.[366] XVII.[234] b_{50} 104-108,[234] b_3 59,[366] d^{20} 1.2333.[234]

MeEtCHP(O)Cl_2. I.[185] XII.[217] XVI.[63,592] b_{20} 92,[592] 90-92,[698] b_{11} 82,[185] b_8 78,[217] b_5 60-70,[63] n_D^{20} 1.4651,[698] 1.4701,[217] n_D^{25} 1.4678,[592] 1.4685,[185] d^{20} 1.2445,[217] d_4^{20} 1.2468,[698] d^{26} 1.2516,[592] d_4^{25} 1.2461,[185] IR,[185] Raman,[185] Cl (NQR),[698] kinetics of hydrolysis.[521]

t-BuP(O)Cl_2. X.[488] XII.[97,98,366,541,661] XVI.[378,488] b. 190,[378] b_{15} 83,[378] m. 120,[378] 123,[97,98] 110,[366,541] 117.5-119.5,[488] 116-117,[488] subl. 110/25 mm,[97] IR,[620] ^{31}P -65.6,[620] ^1H,[620] coupl. const.[652]

$EtSiCl_2CH_2CH_2P(O)Cl_2$. XII.[92] b_8 142,[92] n_D^{20} 1.4960,[92] d_{20} 1.4483.[92]

$Cl_2Si(Me)C_3H_6P(O)Cl_2$. XII.[90] XXII.[79] b_6 137-139,[79] b_4 137-145,[90] n_D^{20} 1.4978,[90] 1.4955,[79] d_{20} 1.4118,[90] 1.4040.[79]

$EtOC_2H_4P(O)Cl_2$. XII.[661] b_2 78-81,[661] n_D^{20} 1.4660,[661] d_4^{20} 1.3073.[661]

PrOCHClCH=CHP(O)Cl_2. XVII.[693] $b_{0.035}$ 112,[693] n_D^{20} 1.5121,[693] d_4^{20} 1.3440.[693]

$Me_3SiCH_2P(O)Cl_2$. XII.[191] XVI.[254] b_{51} 102.8-103.5,[191] $b_{1.4}$ 64,[254] n_D^{25} 1.4710.[254]

┌$(CF_2)_3$┐
Cl-C═══C-P(O)Cl_2. I.[166] II.[166] $b_{1.3}$ 52-55,[166] $b_{0.6}$ 49-50,[166] n_D^{24} 1.4299,[166] n_D^{26} 1.4281,[166] IR.[166]

$C_5Cl_4NP(O)Cl_2$ (tetrachloro-3-pyridyl-). II.[301] $b_{0.15}$ 150,[301] m. 107.[301]

CH_2=C(CH_2Cl)CCl=CHP(O)Cl_2. XVII.[16] b_3 121,[16] n_D^{20} 1.5405,[16] d_{20} 1.5400.[16]

MeCCl=C(Ac)P(O)Cl_2. XIX.[163] XVII.[430,432,433] $b_{2.5}$ 118-120,[432] b_2 113-114,[163] $b_{1.5}$ 112-113,[430,433] n_D^{20} 1.5235,[163] 1.5233,[430] 1.5240,[432,433] d_{20} 1.4416,[163] 1.4413.[430,433]

MeOCH=C(P(O)Cl_2)CCl=CH_2. XVII.[647] $b_{0.002}$ 48-48.5,[235,647] n_D^{20} 1.5826,[235,647] d_{20} 1.4562,[235,647] IR,[235] ^1H.[708]

MeSCH=C(P(O)Cl_2)CCl=CH_2. XVII.[648] $b_{0.02}$ 67-67.5,[235,648] d_{20} 1.4950,[235,648] IR,[235] ^1H.[708]

Me_2C=C=CHP(O)Cl_2. XX.[286] XXIII.[287] b_{15} 79,[286,287] n_D^{20} 1.5140,[286,287] d_4^{20} 1.2553,[286,287] IR,[287] ^1H.[87,286]

MeCH=CHCH=CHP(O)Cl_2. XVII.[468] XIX.[468] $b_{3-3.5}$ 97-100,[468] m. 46-48.[468]

CH_2=CHC(Me)=CHP(O)Cl_2. XVII.[468] b_2 77.5-78,[468] n_D^{20} 1.5400,[468] d^{20} 1.2887.[468]

$CHCl_2$CH(COEt)P(O)Cl_2. XVII.[433,434] $b_{2.5}$ 120-121,[433,434] n_D^{20} 1.5090,[433,434] d_{20} 1.4791.[433,434]

i-PrOCH=CBrP(O)Cl_2. XVII.[649] m. 45-46.[649]

CH_2=CHCMeClCH_2P(O)Cl_2. XVII.[576] b_{12} 146-148,[576] n_D^{20} 1.5240,[576] d_{20} 1.3910.[576]

$CH_2=CMeCHClCH_2P(O)Cl_2$. XVII.[12] b_2 107-108,[12] n_D^{20} 1.5230,[12] d_{20} 1.3918,[12] IR.[654]

$EtOCHClCH=CHP(O)Cl_2$. XVII.[693,694] b_4 134-135,[693,694] n_D^{20} 1.5182,[693,694] d^{20} 1.4041,[694] d_4^{20} 1.4040.[693]

$AcOCMeClCH_2P(O)Cl_2$. XVII.[431,433] $b_{1.5}$ 89-90,[431] m. 39-40,[431] 45.5-46.5.[433]

$EtCO_2CHClCH_2P(O)Cl_2$. XVII.[433,435] $b_{1.5}$ 106-107,[433,435] n_D^{20} 1.4870,[433,435] d_{20} 1.4524.[433,435]

$C_5H_9P(O)Cl_2$ (cyclopentyl-). XII.[290,542] b_{12} 114-116,[542] $b_{6.2}$ 98,[290] n_D^{20} 1.4973,[290] d_{20} 1.3171,[290] IR,[542] Raman.[542]

$Me_2CHCH=CHP(O)Cl_2$. XXII.[470] b_1 50-51,[470] n_D^{20} 1.4964.[470]

$MeEtC=CHP(O)Cl_2$. II.[470] $b_{2.5}$ 68-70,[470] n_D^{20} 1.4922,[470] d^{20} 1.2480.[470]

$ClCH_2CH_2SCH_2CHClCH_2P(O)Cl_2$. XXII.[309] n_D^{20} 1.5565,[309] d_4^{20} 1.5157.[309]

$MeCHClSCH_2CHClCH_2P(O)Cl_2$. XXII.[309] n_D^{20} 1.5519,[309] d_4^{20} 1.5088.[309]

$ClCH_2CH_2SSCH_2CHClCH_2P(O)Cl_2$. XXII.[307,308] n_D^{20} 1.5797,[307,308] d_4^{20} 1.5179.[307,308]

$PrOCH=CHP(O)Cl_2$. XVII.[17,551] b_3 101.[17] $b_{1.5}$ 86.5-87.5,[551] n_D^{20} 1.4945,[17] 1.4922,[551] d_{20} 1.2823,[17] d_4^{20} 1.2355.[551]

$i\text{-}PrOCH=CHP(O)Cl_2$. XVII.[17] b_3 92,[17] m. 52-53.[17]

$EtSCOCH_2CH_2P(O)Cl_2$. XXII.[528] $b_{0.4}$ 103,[528] n_D^{20} 1.5245,[528] d^{20} 1.3750.[528]

$EtOCOCH_2CH_2P(O)Cl_2$. XXII.[528] $b_{0.4}$ 97-99,[528] n_D^{20} 1.4712,[528] d^{20} 1.3482.[528]

$MeOCH_2CH_2OCH=CHP(O)Cl_2$. XVII.[14] b_2 115,[14] n_D^{20} 1.4991,[14] d_{20} 1.3200.[14]

$PrSCH=CHP(O)Cl_2$. XVII.[368,593] $b_{0.15}$ 102,[368,593] n_D^{20} 1.5578.[368,593]

$ClCH_2CH_2CH_2CHMeP(O)Cl_2$. XVI.[366] b_1 112-123.[366]

$PrCH(P(O)Cl_2)CH_2Cl$. XVII.[727] b_{20} 130-132,[727] m. 39-42,[727] d_{25} 1.319.[727]

$i\text{-}PrCHClCH_2P(O)Cl_2$. XVII.[470]

$i\text{-}BuCHClP(O)Cl_2$. I.[164] b_{12} 106-109.[164]

$EtSCH_2CHClCH_2P(O)Cl_2$. XXII.[309] n_D^{20} 1.5424,[309] d_4^{20} 1.3960.[309]

$Me_2SiClCH_2CH=CHP(O)Cl_2$. XVII.[559] b_3 102,[559] n_D^{20} 1.4993,[559] d_4^{20} 1.2806,[559] IR.[559]

$C_5H_{11}P(O)Cl_2$. I.[185] XII[219,662] (isomer mixture). b_8 88.2-99.1,[220] $b_{7.5}$ 87-98,[219] b_2 70,[185] 67-69,[662] 64-65,[662] n_D^{20} 1.4700,[219] 1.4694,[662] 1.4708,[662] n_D^{25} 1.4670,[185] d_{20} 1.2109,[219] d_4^{20} 1.2180,[662] 1.2246,[662] d_4^{25} 1.2116,[185] IR,[185] Raman.[185]

$AmP(O)Cl_2$. I.[185] XII.[218,219] b_{13} 101-108,[218] b_8 96.5-96.6,[219,220] b_5 76,[185] n_D^{20} 1.4665,[219] 1.4703,[218] n_D^{25} 1.4637,[185] d_{20} 1.1942,[219] d_4^{20} 1.2248,[218] d_4^{25} 1.1919.[185]

$i\text{-}AmP(O)Cl_2$. XII.[218] XVII.[234] b_{55} 122-125,[234] b_{14} 92-99,[218] n_D^{20} 1.4713,[218] d^{20} 1.1883,[234] d_4^{20} 1.2294.[218]

$t\text{-}BuCH_2P(O)Cl_2$. XVI.[254] b_5 76.5,[254] n_D^{25} 1.4750.[254]

$Me_2(Et)CP(O)Cl_2$. XVI.[63,651] b_5 60-70,[63] $b_{0.2-0.3}$ 55,[651] m. 88,[63] ^1H.[651]
$Et_2CHP(O)Cl_2$. I.[185] XII.[219] XVI.[63] b_{21} 107,[185] b_8 88.1-88.2,[219] m. 118,[63] n_D^{20} 1.4722,[219] n_D^{25} 1.4691,[185] d_{20} 1.2150,[219] d_4^{25} 1.2194,[185] IR,[185] Raman.[185]
$PrMeCHP(O)Cl_2$. XII.[219] b_8 88.9-89.6,[219] n_D^{20} 1.4710,[219] d_{20} 1.2080.[219]
$Me_2SiClCH_2CHClCH_2P(O)Cl_2$. XVII.[559] $b_{1.5}$ 118,[559] m. 32,[559] IR.[559]
$Cl_2Si(Et)C_3H_6P(O)Cl_2$. XXII.[79] b_3 125-126,[79] n_D^{20} 1.4963,[79] d_{20} 1.3580.[79]
$CF_2H(CF_2)_5P(O)Cl_2$. I.[74] b_{25} 95,[74] b_{20} 102,[74] n_D^{25} 1.3560,[74] IR.[74]
2,5-$Br_2C_6H_3P(O)Cl_2$. I.[121] m. 107-114.[121]
3-$BrC_6H_4P(O)Cl_2$. I.[121] $b_{0.24}$ 94-100.[121]
4-$BrC_6H_4P(O)Cl_2$. I.[472] XVII.[494,495] b. 290-291,[494,495] b_{10} 157-159,[472] m. 46-48.[472]
$ClC_6H_4P(O)Cl_2$. XVII[681] (orientation of Cl not specified). b_{4-5} 104-105.[681]
3-$ClC_6H_4P(O)Cl_2$. I.[121] X.[426] b_{27} 127,[426] $b_{0.4}$ 86-89,[121] n_D^{20} 1.5742,[426] d^{20} 1.5098.[426]
4-$ClC_6H_4P(O)Cl_2$. X.[697] XVII.[494,495,680] b. 284-285,[494,495] b_{15} 152,[697] b_{11} 143-144,[365] b_3 121-123,[680] m. 29.5-30.5,[365] n_D^{20} 1.5775,[697] n_D^{25} 1.5743,[680] d^{20} 1.4892,[494,495] d_4^{20} 1.5070,[697] d_4^{25} 1.302.[680]
4-Cl_3Ge-$C_6H_4P(O)Cl_2$. XII.[144] b_5 197,[144] n_D^{20} 1.5488.[144]
3-$FC_6H_4P(O)Cl_2$. XVII.[607] b_{18} 126-130,[607] n_D^{20} 1.5384,[607] Hammett const.,[607] ^{19}F.[607]
4-$FC_6H_4P(O)Cl_2$. X.[89] XVII.[606] b_{17} 124-125,[606] b_{11} 144-154,[89] n_D^{20} 1.5386,[606] Hammett const.,[567,606] IR,[245] ^1H,[245] ^{19}F,[245,567,606] complexes: 4-$FC_6H_4P(O)Cl_2 \cdot SbCl_5$ (IR, ^1H, ^{19}F);[245] 4-$FC_6H_4P(O)Cl_2 \cdot TiCl_4$ (IR, ^1H, ^{19}F);[245] 4-$FC_6H_4P(O)Cl_2 \cdot SnCl_4$ (IR, ^1H, ^{19}F).[245]
3-$NO_2C_6H_4P(O)Cl_2$. I.[743] b_3 161-163,[743] m. 88-90.[743]
4-$NO_2C_6H_4P(O)Cl_2$. I.[133,743] $b_{0.1}$ 121-134,[133] m. 98-99.[133,743]
$PhP(O)Cl_2$. I.[491] II.[180,211,509] III.[446] IX.[240] X.[7,159,160,490,491,549,614,656] XVI.[490] XVII.[410,490,491,497,501,680,681,738] b. 258,[237,497] b_{17} 138-139,[698] b_{15} 137-138,[211,681] $b_{12-12.5}$ 131-133,[738] b_{12} 126-127,[365] $b_{4.5}$ 91,[7] b_4 104,[680] b_3 104-105,[180] b_2 96-98,[656] b_1 84,[159] m. 3,[237] n_D^{20} 1.5580,[180] 1.5585,[698] 1.5595,[365] n_D^{25} 1.5584,[656] 1.5581,[680] d_4^{20} 1.3959,[698] 1.3987,[365] d^{20} 1.375,[490] d_4^{25} 1.197,[680] 1.3965,[656] d^{25} 1.197,[237] viscosity at 25° 4.44 cP,[237] dielect. const. at 25° 26.0,[237] sp. cond. at 25° 9×10^{-8} $ohm^{-1}cm^{-1}$,[237] IR,[114,369,631] ^{31}P,[240] -34.0,[161,509] -33.7,[446] -44.5,[517,704] ^{31}P signal enhancement through dynamic polarization,[566] ^1H,[349] ^{13}C,[595] electron diffraction study,[710,711] Cl (NQR),[429,523,632,698] complexes: $PhP(O)Cl_2 \cdot BCl_3$ (m. 78-80, IR);[244] $PhP(O)Cl_2 \cdot BF_3$ (liq., IR);[244]

$PhPOCl_2 \cdot SbCl_5$ (m. 88-90, IR);[244] $PhP(O)Cl_2 \cdot TiCl_4$ [m. 115-118 (dec.), IR];[244] $[PhP(O)Cl_2]_2 \cdot TiCl_4$ [m. 109-111 (dec.), IR],[244] (m. 125, IR, ^{31}P -49.2);[46] $(PhP(O)Cl_2)_2 \cdot SnCl_4$ (m. 147-149, IR).[244]

$HC{\equiv}CCH_2CH{=}CClCH_2P(O)Cl_2$. XVII.[485] $b_{0.2}$ 78-81,[485] $n_D^{23.5}$ 1.5292,[485] IR,[485] ^1H.[485]

$EtOCH{=}C(P(O)Cl_2)CCl{=}CH_2$. XVII.[647] $b_{0.012}$ 59-60.5,[235,647] n_D^{20} 1.5730,[235,647] d_{20} 1.3682,[235,647] IR.[235]

$MeCCl{=}C(COEt)P(O)Cl_2$. XVII.[433,434] $b_{1.5}$ 123-124,[433,434] n_D^{20} 1.5220,[433,434] d_{20} 1.3902.[433,434]

$EtSCH{=}C(P(O)Cl_2)CCl{=}CH_2$. XVII.[648,707] $b_{0.02}$ 77-77.5,[235,648] d_{20} 1.4156,[235,648] IR.[235]

$C_6H_9P(O)Cl_2$ (1-cyclohexenyl-). I.[418] b_{10} 115-117,[418] n_D^{20} 1.5230,[418] d_4^{20} 1.3240.[418]

$CHCl_2CH(COPr)P(O)Cl_2$. XVII.[433,434] $b_{2.5}$ 127-128,[433] 126-128,[434] n_D^{20} 1.5045,[433,434] d_{20} 1.4267.[433,434]

$BuOCH{=}CBrP(O)Cl_2$. XVII.[649] m. 31-32.[649]

$PrCH{=}CClCH_2P(O)Cl_2$. XVII.[485] $b_{0.5}$ 82-83,[485] n_D^{22} 1.5020,[485] IR,[485] ^1H.[485]

$2\text{-}ClC_6H_{10}P(O)Cl_2$ (cyclohexyl-). X.[677] b_3 132-134,[677] n_D^{18} 1.5300,[677] d^{18} 1.402.[677]

$PrOCHClCH{=}CHP(O)Cl_2$. XVII.[694] $b_{0.035}$ 112-114,[694] n_D^{20} 1.5121,[694] d^{20} 1.3440.[694]

$PrCO_2CHClCH_2P(O)Cl_2$. XVII.[433,434] b_{2-3} 122-123,[433,434] m. 22-23,[433,434] n_D^{20} 1.4826[433,434] (supercooled liq.), d_{20} 1.3854,[434] 1.4524[433] (supercooled liq.).

$Cl_3GeC_6H_{10}P(O)Cl_2$ (cyclohexyl-). XII[141] (probably isomer mixture). b_5 196-198,[141] n_D^{20} 1.5488,[141] d^{20} 1.6891.[141]

$MeC_5H_8P(O)Cl_2$ (methylcyclopentyl-). XII[290] (isomer mixture). b_5 99.[290]

$C_6H_{11}P(O)Cl_2$ (cyclohexyl-). XII.[99,201,290,315,479,662] XVI.[366,541] b. 260,[201] b_{15-16} 127-128,[99] b_{15} 140-150,[201] b_2 93-94,[662] b_1 90,[366,541] $b_{0.3}$ 85,[479] m. 41,[201] 39.5-40.5,[479] 39-40,[662] 37-37.5,[99] IR,[95] Raman.[95]

$PrOCH{=}CMeP(O)Cl_2$. XVII.[553] b_1 86.5-87,[553] n_D^{20} 1.4906,[553] d_{20} 1.2355.[553]

$BuOCH{=}CHP(O)Cl_2$. XVII.[17,551] b_2 126,[17] b_1 105-106,[551] n_D^{20} 1.4890,[17] 1.4938,[551] d_{20} 1.2083,[17] d_4^{20} 1.1795.[551]

$EtOCH_2CH_2OCH{=}CHP(O)Cl_2$. XVII.[14] b_2 123,[14] n_D^{20} 1.4920,[14] d_{20} 1.2881.[14]

$(EtO)_2C{=}CHP(O)Cl_2$. XVII.[515] b_{12} 121-122,[515] n_D^{20} 1.4925,[515] d_4^{20} 1.3430.[515]

$BuCHClCH_2P(O)Cl_2$. XVII.[148] b_1 99-100,[148] n_D^{20} 1.4863,[148] d_4^{20} 1.2734.[148]

$C_6H_{13}P(O)Cl_2$. I.[153,185] XII.[218,219,662] b_{16} 113-124,[218] b_{15} 124-126,[153] $b_{8.5}$ 107.2-109.5,[219,220] b_3 82-84,[662] b_2 82,[185] n_D^{20} 1.4731,[218] 1.4670,[153,219] n_D^{25} 1.4645,[185] d_4^{20} 1.1910,[218] 1.1627,[153] d_{20} 1.1621,[219] d_4^{25} 1.1605,[185] IR,[185] Raman.[185]

$i\text{-}C_6H_{13}P(O)Cl_2$. XII.[218] b_{13} 103-113,[218] n_D^{20} 1.4739,[218] d_4^{20} 1.1951.[218]

$2\text{-}C_6H_{13}P(O)Cl_2$. I.[185] XII.[219] $b_{12.5}$ 108-108.5,[219] b_2 70,[185] n_D^{20} 1.4711,[219] n_D^{25} 1.4652,[185] d_{20} 1.1742,[219] d_4^{25} 1.1831,[185] IR,[185] Raman.[185]

$3\text{-}C_6H_{13}P(O)Cl_2$. I.[185] XII.[219] b_{11} 103.5-104.6,[219] b_2 73,[185] n_D^{20} 1.4723,[219] n_D^{25} 1.4657,[185] d_{20} 1.1802,[219] d_4^{25} 1.1925,[185] IR,[185] Raman.[185]

$Et_2(Me)CP(O)Cl_2$. XVI.[651] $b_{0.2-0.3}$ 70,[651] ^1H.[651]

$Me_2CHCH(Me)CH_2P(O)Cl_2$. II.[489] XII.[662] b_{12} 107-109,[489] b_2 85-87,[661] 83-84,[489] 75-76,[662] n_D^{20} 1.4688,[489] 1.4715,[662] 1.4720,[661] d_4^{20} 1.1748,[489] 1.1733,[662] 1.1796.[661]

$Me_2CHC(Me)_2P(O)Cl_2$. X.[488,661] b_2 77,[488] 81.5-84,[661] n_D^{20} 1.4855,[488] 1.4728,[661] d_4^{20} 1.2138,[488] 1.1882.[661]

$3\text{-}MeC_5H_{10}P(O)Cl_2$. XII.[99] b_{16-18} 110-120.[99]

$i\text{-}BuOCH_2CH_2P(O)Cl_2$. II.[525] b_2 85-89,[525] n_D^{25} 1.4551.[525]

$Me_3SiCH_2CH{=}CHP(O)Cl_2$. XVII.[559] XXII.[559] b_2 85,[559] n_D^{20} 1.4900,[559] d_4^{20} 1.1472,[559] IR.[559]

$Me_3SiCH_2CHClCH_2P(O)Cl_2$. XVII.[559] b_2 108-110,[559] m. 36,[559] IR.[559]

$Cl(Me)_2Si(CH_2)_4P(O)Cl_2$. XVII.[80] b_9 165-170,[80] n_D^{20} 1.4895,[80] d_4^{20} 1.2519.[80]

$Et_2SiClCH_2CH_2P(O)Cl_2$. XII.[92] b_8 143-148,[92] n_D^{20} 1.4190,[92] d_{20} 1.2499.[92]

$4\text{-}CCl_3C_6H_4P(O)Cl_2$. XVII.[731] $b_{0.05}$ 129-131,[731] n_D^{21} 1.5830,[731] d_{20}^{21} 1.5907.[731]

$2,4,5\text{-}Cl_3C_6H_2OCH_2P(O)Cl_2$. I.[439] $b_{0.4}$ 150-151.[439]

$4\text{-}CF_3C_6H_4P(O)Cl_2$. XVII.[731] n_D^{23} 1.4940,[731] d_{20}^{23} 1.5286.[731]

$3\text{-}NO_2C_6H_4CHClP(O)Cl_2$. XIV.[329] b_1 116,[329] m. 62-64.[329]

$4\text{-}ClC_6H_4CHClP(O)Cl_2$. XIV.[329] $b_{1.5}$ 144-145.[329] m. 58-60.[329]

$PhCCl_2P(O)Cl_2$. XVI.[366] b_2 114-120.[366]

$2,4\text{-}Cl_2C_6H_3OCH_2P(O)Cl_2$. I.[439] $b_{0.3}$ 138.[439]

$PhCHClP(O)Cl_2$. XIV.[329] XVI.[366] b_2 124-126,[329] 132,[366] m. 60-61,[329] 48,[366] n_D^{40} 1.5666,[329] d_5^{40} 1.4534.[329]

$2\text{-}Cl\text{-}4\text{-}MeC_6H_3P(O)Cl_2$. XVII.[486] b. 290-291,[486] m. 36.[486]

$2\text{-}ClC_6H_4OCH_2P(O)Cl_2$. I.[439] $b_{0.5}$ 113.[439]

$4\text{-}ClC_6H_4OCH_2P(O)Cl_2$. I.[439] $b_{0.6}$ 132.[439]

$MeC_6H_4P(O)Cl_2$. II.[182] XVII[681] (orientation of Me group not specified). b_{11} 140-142,[681] b_2 103-105,[182] n_D^{20} 1.5530.[182]

$2\text{-}MeC_6H_4P(O)Cl_2$. XVII.[494,495] b. 273,[494,495] $d^{18.5}$ 1.3877.[494,495]

$3\text{-}MeC_6H_4P(O)Cl_2$. XVII.[494,495] b. 275,[494,495] d^{18} 1.3533.[494,495]

$4\text{-}MeC_6H_4P(O)Cl_2$. XVII.[497,506,680] b. 284-285,[497,506] b_{15} 153-155,[698] b_{14} 149.5-150.5,[365] b_{11} 140-142,[680] m. 26.5-27,[365] n_D^{20} 1.5580,[698] 1.5577,[365] n_D^{25} 1.5542,[680] d_4^{20} 1.3416,[698] 1.3423,[365] d_4^{25} 1.154,[680] Cl (NQR).[698]

$PhCH_2P(O)Cl_2$. X.[554] XVI.[366,541] b_2 130,[366,541,554] m. 57.5.[366,541]

$PhOCH_2P(O)Cl_2$. I.[715] b_2 120-130,[715] n_D^{25} 1.5470.[715]

$4-MeOC_6H_4P(O)Cl_2$. XVII.[410,494,495] b. 300,[494,495] b_{12-15} 173,[494,495] b_{6-7} 149.[410]
MeCCl=C(COPr)P(O)Cl_2. XVII.[432,433,434] $b_{4.5}$ 145-148,[434] $b_{1.5}$ 132-133,[432,433] n_D^{20} 1.5170,[432] 1.5180,[433,434] d_{20} 1.3398,[432] 1.3474.[433,434]
PrOCH=C(P(O)Cl_2)CCl=CH_2. XVII.[647] $b_{0.012}$ 69-69.5,[235,647] n_D^{20} 1.5635,[235,647] d_{20} 1.3152,[235,647] IR,[235] Raman.[44]
PrSCH=C(P(O)Cl_2)CCl=CH_2. XVII.[648] $b_{0.02}$ 83-84,[235,648] d_{20} 1.3634,[235] IR.[235]
$C_7H_{11}OP(O)Cl_2$ (2-EtO-1-cyclopentenyl-). XVII.[26] b_2 139-140.[26]
EtO$(CH_2)_3$CCl=CHP(O)Cl_2. XVII.[15] b_3 139-142,[15] n_D^{20} 1.5112.[15]
$PrCO_2CMeClCH_2P(O)Cl_2$. XVII.[433,435] m. 20-21.[433,435]
$MeC_6H_{10}P(O)Cl_2$. XII[290,638] (isomer mixture). b_7 121.5,[290] b_3 99-101,[638] n_D^{20} 1.5032,[638] d_{20} 1.2534.[638]
$3-MeC_6H_{10}P(O)Cl_2$. XVI[6] (probable isomer). b_2 100-101,[6] n_D^{20} 1.5084,[6] d^{20} 1.2734,[6] Raman.[6]
$Me_2CHCH_2CH_2OCH$=CHP(O)Cl_2. XVII.[17] b_1 112,[17] n_D^{20} 1.4876,[17] d_{20} 1.2058.[17]
$C_5H_{11}CHClCH_2P(O)Cl_2$. XVII.[148] b_3 120-121,[148] n_D^{20} 1.4858,[148] d_4^{20} 1.2423.[148]
$C_7H_{15}P(O)Cl_2$. XII[99,314,315,662] (isomer mixture). b_{15} 166-167,[314] b_{10} 106-108,[99] b_2 96-98,[662] n_D^{20} 1.4830,[662] d_4^{20} 1.1852.[662]
$C_7H_{15}P(O)Cl_2$ (n-heptyl-). I.[153,185] XII.[219,657] $b_{11.5}$ 117-124,[219] b_5 103-108,[657] b_1 87-88,[153] $b_{0.1}$ 82,[185] n_D^{20} 1.4715,[219] 1.4675,[153] n_D^{25} 1.4642,[185] d_{20} 1.1477,[219] d_4^{20} 1.1435,[153] d_4^{25} 1.1276,[185] IR,[185] Raman.[185]
$Pr_2CHP(O)Cl_2$. XII.[219,220] b_{12} 118.2-118.7,[219,220] n_D^{20} 1.4726,[219] d_{20} 1.1512.[219]
BuEtCHP(O)Cl_2. XII.[219] b_{12} 119-123.2,[210] n_D^{20} 1.4719,[219] d_{20} 1.1465.[219]
C_5H_{11}MeCHP(O)Cl_2. XII.[219] b_{12} 119-123.2,[219] n_D^{20} 1.4710,[219] d_{20} 1.1427.[219]
$CF_2H(CF_2)_7P(O)Cl_2$. I.[74] b_{20} 120,[74] b_{12} 112,[74] m. 51.5-55.5,[74] IR.[74]
PhC≡CP(O)Cl_2. II.[468] $b_{0.5}$ 121-124,[468] n_D^{20} 1.5938,[468] d^{20} 1.3650.[468]
cis-$4-BrC_6H_4OCH$=CHP(O)Cl_2. XVII.[337] b_3 164-165,[337] n_D^{20} 1.6200,[337] d^{20} 1.6821,[337] dipole moment 4.34.[337]
trans-$4-BrC_6H_4OCH$=CHP(O)Cl_2. XVII.[337] b_3 162.5-163,[337] n_D^{20} 1.5998,[337] d^{20} 1.6883,[337] dipole moment 3.83.[337]
PhCCl=CHP(O)Cl_2. XVII.[9] b_1 142-143,[9] n_D^{20} 1.6175,[9] d_{20} 1.4675,[9] slowly attacked by H_2O.
$4-ClC_6H_4SCH$=CHP(O)Cl_2. XVII.[368,593] $b_{0.15}$ 155,[368,593] m. 71.[368,593]
$CH_2=CHC_6H_4P(O)Cl_2$. XXII[115] (orientation of CH_2=CH group not specified). $b_{0.9}$ 103-105,[115] n_D^{20} 1.5905,[115] d^{20} 1.3468.[115]
$4-CH_2=CHC_6H_4P(O)Cl_2$. XXII.[335] $b_{1.5}$ 102-103,[335] n_D^{20}

1.5877,[335] d_{20} 1.3412.[335]
CH_2=CPhP(O)Cl_2. I.[415,418] XX.[415] b_3 123-124,[415] $b_{0.04}$ 98-100,[418] m. 37-38,[415] n_D^{20} 1.5738,[418] d_4^{20} 1.4646.[418]
PhCH=CHP(O)Cl_2. XVII.[146,151,617,727] b_{18} 182-184,[727] b_2 129-132,[151] $b_{0.2}$ 107-110,[617] m. 71-72,[617] 70,[727] ^{1}H.[676]
PhOCH=CHP(O)Cl_2. XVII.[17,336] b_2 131,[17] n_D^{20} 1.5708,[17] d_{20} 1.3710.[17]
MeCHClC_6H_4P(O)Cl_2. XXII[115] (orientation of MeCHCl group not specified). b_1 125-127,[115] n_D^{20} 1.5683,[115] d^{20} 1.4164,[115] ^{1}H.[115]
4-Cl$CH_2CH_2C_6H_4$P(O)Cl_2. XXII.[335] b_1 123-124,[335] n_D^{20} 1.5663,[335] d_{20} 1.4067.[335]
4-MeC_6H_4CHClP(O)Cl_2. XIV.[329] $b_{0.5}$ 129-130,[329] m. 52-54.[329]
$Me_2C_6H_3$P(O)Cl_2. XVII[494,495,506] (isomer mixture). m. 145.[494,495,506]
2,5-$Me_3C_6H_3$P(O)Cl_2. XVII.[724] b. 280-281,[724] d^{18} 1.31.[724]
4-Me$C_6H_4CH_2$P(O)Cl_2. XII.[582] b_2 130,[582] n_D^{20} 1.5430,[582] d_{20}^{20} 1.2579.[582]
EtC_6H_4P(O)Cl_2. X.[115] XVII[115] (isomer mixture). b_9 135-137,[115] n_D^{20} 1.5525.[115]
4-EtC_6H_4P(O)Cl_2. X.[335] XVII.[494,495] b. 294,[494,495] $b_{1.5}$ 100-100.5,[335] n_D^{20} 1.5521,[335] d_{20} 1.2949,[335] d^{16} 1.29.[494,495]
4-EtOC_6H_4P(O)Cl_2. XVII.[422] Oil.[422]
2,4-$(MeO)_2C_6H_3$P(O)Cl_2. I.[568] XVII.[713] $b_{1.5}$ 168-170,[568] m. 85-87.[568]
BuOCH=C(P(O)Cl_2)CCl=CH_2. XVII.[647] $b_{0.02}$ 82-82.5,[235,647] n_D^{20} 1.5620,[235,647] d_{20} 1.3078,[235,647] IR,[235] Raman.[44]
BuSCH=C(P(O)Cl_2)CCl=CH_2. XVII.[648] $b_{0.002}$ 81-82,[235,648] d_{20} 1.3222,[235,648] IR.[235]
C_8H_{13}OP(O)Cl_2 (2-EtO-1-cyclohexenyl-). XVII.[25] b_2 136,[25] n_D^{20} 1.5275,[25] d_0^{20} 1.3009.[25]
C_5H_{11}CH=CClCH_2P(O)Cl_2. XVII.[485] $b_{0.25}$ 95,[485] n_D^{26} 1.4939,[485] IR,[485] ^{1}H.[485]
Me_3CCH_2CMe=CHP(O)Cl_2. XVII.[727] b_{13} 128-129,[727] m. -70,[727] d_{25} 1.129.[727]
BuOCH=CEtP(O)Cl_2. XVII.[553] $b_{1.5}$ 123,[553] n_D^{20} 1.4875,[553] d_{20} 1.1795.[553]
i-BuOCH=CEtP(O)Cl_2. XVI.[604] b_2 96-98,[604] n_D^{20} 1.4850.[604]
$BuOCH_2CH_2$OCH=CHP(O)Cl_2. XVII.[14] b_2 142,[14] n_D^{20} 1.4869,[14] d_{20} 1.2133.[14]
BuOCOCH_2CHMeP(O)Cl_2. II.[547] b_2 118-124,[547] n_D^{19} 1.4661.[547]
C_8H_{16}ClP(O)Cl_2. XII[661] (isomer mixture). b_4 140-170.[661]
C_8H_{17}P(O)Cl_2 (n-octyl-). I.[153,185] XII.[219] XVI.[452] b_{14} 136-137,[219] b_{12} 143-145,[153] $b_{2.7-3}$ 95-99,[452] $b_{0.2}$ 78,[185] $b_{0.01}$ 76-78,[452] n_D^{20} 1.4705,[219] 1.4680,[153] 1.4720,[452] n_D^{25} 1.4651,[185] d_{20} 1.1260,[219] d_4^{20} 1.1184,[153] d_4^{25} 1.1052,[185] IR,[185,452] Raman,[185] ^{31}P -28.7.[452]
n-Bu(Me)(Et)CP(O)Cl_2. XVI.[651] $b_{0.2-0.3}$ 90.[651]
(+)-n-C_6H_{13}(Me)CHP(O)Cl_2. I.[110] $b_{0.3}$ 83.[110]

$BuOC_4H_8P(O)Cl_2$. XII.[661] b_2 88-114,[661] d_4^{20} 1.2841.[661]
$Et_3SiCH_2CHClP(O)Cl_2$. XII.[90] b_5 130-131,[90] n_D^{20} 1.4948,[90] d^{20} 1.1844.[90]
$Et_3GeCH_2CH_2P(O)Cl_2$. XII.[141,144] XXII.[142] b_{4-5} 136,[142] $b_{1.5}$ 118-119,[142] b_1 118,[144] n_D^{20} 1.4962,[142] 1.5012,[144] d_4^{20} 1.2947.[142]
$Et_3SiCH_2CH_2P(O)Cl_2$. XII.[92] b_8 140-146,[92] n_D^{20} 1.4895,[92] d_{20} 1.1552.[92]
$C_6H_4CH_2C(P(O)Cl_2)=CH$ (2-indenyl-). XVII.[9] b_1 133,[9] m. 73-74.[9]
$PhCO_2CHClCH_2P(O)Cl_2$. XVII.[433] m. 50-51.[433]
$PhCMe=CHP(O)Cl_2$. XVII.[676] $b_{0.5}$ 115-116,[676] n_D^{20} 1.5928,[676] d_4^{20} 1.3117,[676] ^1H.[676]
$4-CH_2=CHC_6H_4CH_2P(O)Cl_2$. I.[1] II.[1]
$CHCl_2CHBzP(O)Cl_2$. XVII.[434] b_1 159-160,[434] n_D^{20} 1.5560,[434] d_{20} 1.5016.[434]
$MeSiCl_2C_6H_4CH=CHP(O)Cl_2$. XVII[559] (orientation of $MeSiCl_2$ group not specified). XVII.[559] $b_{0.3}$ 145,[559] m. 38,[559] IR.[559]
$4-(MeSiCl_2)C_6H_4CH=CHP(O)Cl_2$. XVII.[560] $b_{0.03}$ 146.[560]
$MeOCH=CPhP(O)Cl_2$. XVII.[603,635] b_2 155-157,[635] m. 76-77.[635]
$2-MeC_6H_4OCH=CHP(O)Cl_2$. XVII.[336] b_{11} 172,[336] n_D^{20} 1.5610,[336] d_{20} 1.3253.[336]
$3-MeC_6H_4OCH=CHP(O)Cl_2$. XVII.[336] b_{33} 154,[336] n_D^{20} 1.5650,[336] d_{20} 1.3252.[336]
$BzOCHClCH_2P(O)Cl_2$. XVII.[435] m. 50-51.[435]
$2,4,5-Me_3C_6H_2P(O)Cl_2$. XVII.[494,495] b. 307-308,[494,495] m. 63.[494,495]
$4-i-PrC_6H_4P(O)Cl_2$. XVII.[494,495] b. 295-300,[494,495] b_{35} 183.[494,495]
$3-Me_3SiC_6H_4P(O)Cl_2$. I.[70] II.[68] $b_{2.5}$ 131-132,[68,70] n_D^{25} 1.5352.[68,70]
$4-Me_3SiC_6H_4P(O)Cl_2$. I.[70] II.[68] $b_{1.12}$ 124-128,[68] $b_{1.1}$ 126-128,[70] m. 44-45.[68,70]
$Cl_3Ge(CH_2)_9P(O)Cl_2$. XII.[141,144] b_1 168-170,[144] 170,[141] n_D^{20} 1.5133.[144]
$C_9H_{19}P(O)Cl_2$. I.[153,185] b_1 115-116,[153] $b_{0.1}$ 98,[185] n_D^{20} 1.4680,[153] n_D^{25} 1.4650,[185] d_4^{20} 1.0895,[153] d_4^{25} 1.0857,[185] IR,[185] Raman.[185]
$Et_3Ge(CH_2)_3P(O)Cl_2$. XXII.[142] b_{20} 77,[142] n_D^{20} 1.4610,[142] d_4^{20} 1.1952.[142]
$CF_2H(CF_2)_9P(O)Cl_2$. I.[74] b_{20} 144,[74] b_{12} 132,[74] m. 90-91.5,[74] IR.[74]
$1-C_{10}H_7P(O)Cl_2$ (1-naphthyl-). I.[424] XVII.[424] b_{20} 280,[424] m. 60.[424]
$2-C_{10}H_7P(O)Cl_2$ (2-naphthyl-). I.[411] II.[211] b_3 173,[211] $b_{0.048}$ 142-143.5,[411] m. 55,[411] ∿46.[211]
$CHCl_2CH(COBz)P(O)Cl_2$. XVII.[433] b_1 159-160,[433] n_D^{20} 1.5560,[433] d_{20} 1.5016.[433]
$PhCO_2CMeClCH_2P(O)Cl_2$. XVII.[433] m. 47-48.[433]

MeC(OBz)=CHP(O)Cl_2. XVII.[434] $b_{1.5}$ 155,[434] n_D^{20} 1.5590,[434] d_{20} 1.3997.[434]
PhC(OEt)=CHP(O)Cl_2. XVII.[27] $b_{2.5}$ 164-164.5,[27] n_D^{20} 1.5719,[27] d_0^{20} 1.3022.[27]
BzOCMeClCH_2P(O)Cl_2. XVII.[435] m. 47-48.[435]
Me_2SiClC_6H_4CH=CHP(O)Cl_2. XVII[559] (orientation of Me_2SiCl group not specified). $b_{0.5}$ 170,[559] m. 36,[559] IR.[559]
4-(Me_2SiCl)C_6H_4CH=CHP(O)Cl_2. XVII.[560] $b_{0.6}$ 170-175.[560]
1-$C_{10}H_{15}$P(O)Cl_2 (1-adamantyl-). XVI.[671] b_2 135-136,[671] subl. 70-80/1 mm,[671] m. 102-103.[671]
3-(Me_3SiCH_2)C_6H_4P(O)Cl_2. I.[70] II.[69] $b_{0.7}$ 119-122.[69,70]
4-(Me_3SiCH_2)C_6H_4P(O)Cl_2. I.[70] II.[69] $b_{0.5}$ 136-138.[69,70]
Me(CH_2)$_6$CH=CClCH_2P(O)Cl_2. XVII.[485] $b_{1.5}$ 147-149,[485] n_D^{20} 1.4923,[485] IR,[485] ^{1}H.[485]
BuEtCHCH_2OCH=CHP(O)Cl_2. XVII.[18] b_3 151-152,[18] n_D^{20} 1.4863,[18] d_{20} 1.087.[18]
PrCCl(OCHPrOEt)P(O)Cl_2. XVII.[192] b_1 150-151 (dec.),[192] n_D^{20} 1.4665,[192] d_{20}^{20} 1.256.[192]
$C_{10}H_{21}$P(O)Cl_2. I.[153,185] b_8 155-157,[153] b_1 115,[185] n_D^{20} 1.4680,[153] n_D^{25} 1.4656,[185] d_4^{20} 1.0750,[153] d_4^{25} 1.0715,[185] IR,[185] Raman.[185]
2-$C_{10}H_{21}$P(O)Cl_2. I.[185] $b_{0.02}$ 47,[185] n_D^{25} 1.4649,[185] IR,[185] Raman.[185]
4-$C_{10}H_{21}$P(O)Cl_2. I.[185] $b_{0.02}$ 50,[185] n_D^{25} 1.4650,[185] IR,[185] Raman.[185]
5-$C_{10}H_{21}$P(O)Cl_2. I.[185] $b_{0.3}$ 70,[185] n_D^{25} 1.4642.[185]
PhO(CH_2)$_3$CCl=CHP(O)Cl_2. XVII.[15] m. 64-66.[15]
$C_{11}H_{13}O_2$P(O)Cl_2 (4,5-methylenedioxy-2-propylbenzene-). XX.[157,158] $b_{4.4\mu}$ 120-122.[157,158]
$Me_3SiC_6H_4$CH=CHP(O)Cl_2. XVII[559] (orientation of Me_3Si group not specified). $b_{0.2}$ 138,[559] m. 46,[559] IR.[559]
4-(Me_3Si)C_6H_4CH=CHP(O)Cl_2. XVII.[560] $b_{0.2}$ 138-140.[560]
PrCCl(OCHPrOPr)P(O)Cl_2. II.[192] b_1 155-160,[192] n_D^{20} 1.4680,[192] d_{20}^{20} 1.163.[192]
CF_2H(CF_2)$_{11}$P(O)Cl_2. I.[74] b_{11} 147,[74] m. 111-115,[74] IR.[74]
4-FC_6H_4S-4-C_6H_4P(O)Cl_2. III.[204]
PhC_6H_4P(O)Cl_2. XVII[425] (isomer mixture). b_{10} 220,[425] m. 90.[425]
$C_{12}H_{25}$P(O)Cl_2. XII[314] (isomer mixture).
$C_{12}H_{25}$P(O)Cl_2 (n-dodecyl-). XII.[219,542] $b_{11.5}$ 211,[542] b_5 161-168,[219] m. 51-52,[542] n_D^{20} 1.4709,[219] d_{20} 1.0529.[219]
4-(PhCCl_2)C_6H_4P(O)Cl_2. XVII.[496] b_{15} 258,[496] m. 64.[496]
4-Ph$CH_2C_6H_4$P(O)Cl_2. XVII.[496] b_{20} 261,[496] d^{20} 1.207.[496]
2-Ph$C_6H_4CH_2$P(O)Cl_2. I.[436] $b_{0.2}$ 150.[436]
CF_2H(CF_2)$_{13}$P(O)Cl_2. I.[74] b_{19} 176,[74] b_{11} 160,[74] m. 116-127,[74] IR.[74]
PhOCH=CPhP(O)Cl_2. XVII.[603]
4-(PhCH_2CH_2)C_6H_4P(O)Cl_2. XVII.[496] m. 75.[496]
$C_{14}H_{29}$P(O)Cl_2. I.[737] XII[314] (isomer mixture).

$Ph_2C(CH_2)CHP(O)Cl_2$. XXII.[570] $b_{0.005}$ 146-148,[570] n_D^{20} 1.6051,[570] IR.[570]
$2\text{-}MeC_6H_4OCH{=}CPhP(O)Cl_2$. XVII.[603]
$3\text{-}MeC_6H_4OCH{=}CPhP(O)Cl_2$. XVII.[603]
$4\text{-}MeC_6H_4OCH{=}CPhP(O)Cl_2$. XVII.[603]
$(PhCH_2)_2CHP(O)Cl_2$. XVII.[500] b_{20} 228,[500] d^{15} 1.036.[500]
$CF_2H(CF_2)_{15}P(O)Cl_2$. I.[74] b_{12} 180,[74] m. 143-147,[74] IR.[74]
$C_{16}H_{33}P(O)Cl_2$. I.[153] b_2 178-180.[153]
$2\text{-}C_{16}H_{33}P(O)Cl_2$. XVII.[366,541] b_1 165.[366,541]
$CF_2H(CF_2)_{17}P(O)Cl_2$. I.[74] b_{11} 230,[74] m. 158-162,[74] IR.[74]
$C_{18}H_{37}P(O)Cl_2$. XII[314] (isomer mixture).
$C_{19}H_{13}P(O)Cl_2$ (9-phenyl-9-fluorenyl-). XIII.[253] m. 153.[253]
$(4\text{-}BrC_6H_4)Ph_2CP(O)Cl_2$. XIII.[73] m. 164.[73]
$(4\text{-}ClC_6H_4)Ph_2CP(O)Cl_2$. XIII.[73] m. 161.5.[73]
$(4\text{-}NO_2C_6H_4)Ph_2CP(O)Cl_2$. XIII.[73] m. 188.5.[73]
$Ph_3CP(O)Cl_2$. X.[260] XIII.[20,71-73,250,260,261] m. 188-189,[250] 188-190,[260] 189-190,[71] 189.5-190,[20] IR.[250]
$(2\text{-}MeC_6H_4)Ph_2CP(O)Cl_2$. XIII.[253] m. 124-125.[253]
$(4\text{-}MeC_6H_4)Ph_2CP(O)Cl_2$. XIII.[73] m. 193.[73]
$(2\text{-}MeOC_6H_4)Ph_2CP(O)Cl_2$. XIII.[253] m. 171.5-173.5.[253]
$(3\text{-}MeOC_6H_4)Ph_2CP(O)Cl_2$. XIII.[73] m. 122-124.[73]
$(4\text{-}MeOC_6H_4)Ph_2CP(O)Cl_2$. XIII.[73] m. 180.[73]
$(4\text{-}MeC_6H_4)_2PhCP(O)Cl_2$. XIII.[253] m. 171.5.[253]
$(4\text{-}MeC_6H_4)_3CP(O)Cl_2$. XIII.[212,253] m. 166-169,[212] 177.5,[253] ^1H.[212]
$(1\text{-}C_{10}H_7)Ph_2CP(O)Cl_2$ (1-naphthyl-). XIII.[73] m. 171-172.[73]
$(2\text{-}C_{10}H_7)Ph_2CP(O)Cl_2$ (2-naphthyl-). XIII.[73] m. 194.[73]
$(4\text{-}PhC_6H_4)Ph_2CP(O)Cl_2$. XIII.[21] m. 139-140.[21]
$(4\text{-}PhC_6H_4)_2PhCP(O)Cl_2$. XIII.[21] m. 100-102.[21]
$(4\text{-}PhC_6H_4)_3CP(O)Cl_2$. XIII.[21] m. 220-222.[21]

Bis(phosphonyl Dichlorides)

$Cl_2P(O)Cl_2CP(O)Cl_2$. XVII.[548] b_7 139-140,[548] m. 75-76.[548]
$Cl_2P(O)CH_2P(O)Cl_2$. II.[264,388,451,595a] m. 104-105,[388] 98-100,[451] 100-102,[264] 101-102,[595a] IR,[595a] ^{31}P -24.2,[451] -24.4.[511]
$Cl_2P(O)(CH_2)_2P(O)Cl_2$. II.[388,451] m. 167-170,[388] 164-165.[451]
$Cl_2P(O)C_3H_6P(O)Cl_2$. XII[168] (isomer mixture).
$Cl_2P(O)CMe_2P(O)Cl_2$. I.[264] m. > 100.[264]
$Cl_2P(O)CHMeCH_2P(O)Cl_2$. II.[546]
$Cl_2P(O)(CH_2)_3P(O)Cl_2$. II.[388,451] $b_{0.005}$ 100-110,[388] m. 59.5-61.5,[388] 59.5-60.5,[388] ^{31}P -46.8,[388] IR.[451]
$Cl_2P(O)C_4H_8P(O)Cl_2$. XII[168] (isomer mixture).
$Cl_2P(O)(CH_2)_4P(O)Cl_2$. II.[388,451] m. 110-112,[388,451] ^{31}P -45.9.[451]
$Cl_2P(O)\overset{\ulcorner(CF_2)_3\urcorner}{C{=\!=}C}P(O)Cl_2$. I.[166] $b_{0.45}$ 89-90,[166] m. 78.5-81.5,[166] IR.[166]
$Cl_2P(O)(CH_2)_5P(O)Cl_2$. II.[388] $b_{0.3}$ 165,[388] m. 60-62.[388]

$Cl_2P(O)C_6H_{10}P(O)Cl_2$. XII[168] (isomer mixture).
$Cl_2P(O)(CH_2)_6P(O)Cl_2$. II.[388] $b_{0.2}$ 166,[388] m. 60-62.[388]
$Cl_2P(O)C_7H_{14}P(O)Cl_2$. XII[168] (isomer mixture).
$Cl_2P(O)(CH_2)_7P(O)Cl_2$. II.[388] $b_{0.2}$ 177-178,[388] m. 32-33.5.[388]
$Cl_2P(O)$-4-$CH_2C_6H_4CH_2P(O)Cl_2$. II.[86] $b_{0.4}$ 190-200,[86] m. 171-173.[86]
$Cl_2P(O)(CH_2)_8P(O)Cl_2$. II.[388] $b_{0.25}$ 183-184,[388] m. 63-65.[388]
$Cl_2P(O)(CH_2)_9P(O)Cl_2$. II.[388] $b_{0.2}$ 191,[388] m. 38.5-40.[388]
$Cl_2P(O)(CH_2)_{10}P(O)Cl_2$. II.[388] $b_{0.25}$ 195-196,[388] m. 70-72.[388]
$Cl_2P(O)CH(C_{11}H_{21})P(O)Cl_2$. I.[264] $b_{0.02}$ 180-210.[264]

Phosphonyl Chloride Bromides

4-$MeC_6H_4P(O)ClBr$. XVII.[423] b_{20} 160-170.[423]

Phosphonyl Chloride Pseudohalides

$CCl_3P(O)Cl(NCO)$. XXIII.[645] b_{12} 88-90,[645] b_7 79-80,[645] n_D^{20} 1.5180,[645] 1.5185.[645]
$CHCl_2P(O)Cl(NCO)$. XXIII.[640] b_{14} 98-100,[640] n_D^{20} 1.5109,[640] d^{20} 1.7086.[640]
$ClCH_2P(O)Cl(NCO)$. XXIII.[644] b_{10} 94-95,[644] n_D^{20} 1.4975,[644] d^{20} 1.6008.[644]
$MeP(O)Cl(NCO)$. II.[642] b. 173-175,[642] b_{17} 79-80,[642] b_8 64-66,[642] IR.[124,642]
$EtP(O)Cl(NCO)$. II.[233] IX.[262] b. 74-76,[233] $b_{0.75}$ 45-46,[262] n_D^{20} 1.4580,[233] d^{20} 1.3078,[233] IR.[124]
$PrP(O)Cl(NCO)$. IR.[124]
i-$PrP(O)Cl(NCO)$. IR.[124]
$BuP(O)Cl(NCO)$. IR.[124]
$PhP(O)Cl(NCO)$. XXIII.[637] b_3 110-115,[637] n_D^{20} 1.5525, d_{20} 1.3761.[637]
$PhP(O)Cl(N_3)$. IX.[40,700] $b_{0.75}$ $\sim$ 65-67,[40] ^{31}P -27.5.[700]

Phosphonyl Dibromides

$BrCH_2P(O)Br_2$. XIV.[706] XVI.[735] XVII.[143] b_2 112,[735] $b_{0.3}$ 103,[143] n_D^{20} 1.6120,[735] n_D^{22} 1.6108,[143] d_{20}^{20} 2.7030,[735] IR.[530]
$MeP(O)Br_2$. X.[448] b_{728} 191-195,[448] n_D^{20} 1.5829,[448] d_4^{20} 1.4278,[448] IR,[448] ^{31}P -8.5.[448]
$PhP(O)Br_2$. X.[554] b_{12} 156-158,[554] n_D^{20} 1.6177,[554] d_{20} 1.9452.[554]
4-$MeC_6H_4P(O)Br_2$. XVII.[423] b_{20} 190,[423] m. 48-50.[423]

Phosphonyl Dipseudohalides

$PhP(O)(CN)_2$. IX.[85]
$CCl_3P(O)(NCO)_2$. IX.[126] m. 82-84.[126]

$CHCl_2P(O)(NCO)_2$. IX.[126] b_{15} 126-127,[126] n_D^{20} 1.5006,[126] d^{20} 1.6206.[126]
$ClCH_2P(O)(NCO)_2$. IX.[126,262,263] b_{15} 128-129,[126] $b_{0.9}$ 80-82,[262] $b_{0.6}$ 76-82,[263] n_D^{20} 1.4955,[126] d^{20} 1.5633.[126]
$MeP(O)(NCO)_2$. IX.[36,127] XXIII.[128] b_7 85-87,[128] 84-86,[127] b_2 71,[36] n_D^{20} 1.4680,[36] d_4^{20} 1.3300.[36]
$CH_2{=}CHP(O)(NCO)_2$. IX.[300] b_6 71-72,[300] n_D^{20} 1.4838,[300] d_4^{20} 1.3651.[300]
$ClCH_2CH_2P(O)(NCO)_2$. IX.[300] $b_{0.1}$ 62-67,[300] n_D^{20} 1.4873,[300] d_4^{20} 1.4364.[300]
$EtP(O)(NCO)_2$. IX.[127,262,263,738] XXIII.[128] $b_{19.20}$ 107-113,[738] b_7 80-81,[127,128] $b_{0.65-1.0}$ 61-65,[263] $b_{0.7}$ 58-59.[262]
$CH_2{=}C{=}CHP(O)(NCO)_2$. IX.[300] b_3 76-77,[300] n_D^{20} 1.3526,[300] d_4^{20} 1.5180.[300]
i-$PrP(O)(NCO)_2$. IX.[262,263,300,738] b_4 60-61,[300] $b_{1-1.75}$ 61-67,[263] b_1 60-61,[262,738] n_D^{20} 1.4628,[300] d_4^{20} 1.2550.[300]
$EtOCH{=}CHP(O)(NCO)_2$. IX.[300] b_5 117-118,[300] n_D^{20} 1.4900,[300] d_4^{20} 1.2926.[300]
4-$ClC_6H_4P(O)(NCO)_2$. IX.[679] $b_{0.4}$ 128,[679] IR.[679]
4-$NO_2C_6H_4P(O)(NCO)_2$. IX.[679] m. 85-88,[679] IR.[679]
$PhP(O)(NCO)_2$. IX.[36,262,263,563,702,738] XXIII.[123] b_3 124-127,[738] b_2 123,[36] $b_{0.75}$ 113-114,[563] $b_{0.1}$ 95-100,[702] m. -10 to 0,[702] n_D^{20} 1.5480,[36] d_4^{20} 1.3530,[36] polymerized above 200.[262,263]
$PhCH_2P(O)(NCO)_2$. IX.[262,263] b_{1-2} 145.[262,263]
4-$MeC_6H_4P(O)(NCO)_2$. IX.[679] b_1 128-130,[679] IR.[679]
4-$MeOC_6H_4P(O)(NCO)_2$. IX.[679] $b_{0.01}$ 113-115,[679] IR.[679]
$Me(CH_2)_{15}P(O)(NCO)_2$. IX.[262,263] Oil.[262,263]
$ClCH_2P(O)(NCS)_2$. IR.[122]
$MeP(O)(NCS)_2$. IX.[128] $b_{0.05}$ 90-102,[128] IR.[122]
$ClCH_2CH_2P(O)(NCS)_2$. IX.[300] $b_{0.02}$ 112-113,[300] n_D^{20} 1.6180,[300] d_4^{20} 1.5294.[300]
$EtOCH{=}CHP(O)(NCS)_2$. IX.[300] $b_{0.03}$ 111-113,[300] n_D^{20} 1.6068,[300] d_4^{20} 1.3130.[300]
$PhP(O)(NCS)_2$. IX.[205,679] $b_{0.3}$ 146-148,[205] $b_{0.2}$ 145-147,[679] IR.[205,679]
$PhP(O)(N_3)_2$. IX.[40,700] $b_{0.1}$ 72-74,[40] n_D^{20} 1.5960,[40] IR,[40] ^{31}P -21.0.[700]
$(N_3)_2P(O)$-4-$C_6H_4P(O)(N_3)_2$. IX.[717]

Thionophosphonyl Difluorides

$CCl_3P(S)F_2$. XVIII.[381] b_{759} 66-67.[381]
$CHCl_2P(S)F_2$. IX.[299] b. 108-109,[299] n_D^{20} 1.4650,[299] d_4^{20} 1.5782.[299]
$ClCH_2P(S)F_2$. IX.[615,669] b. 93.5-94,[615,669] m. -84 to -83,[669] IR,[669] Raman.[669] n_D^{22} 1.4433.[615]
$MeP(S)F_2$. I.[600] IX.[4,66,615,628] XVIII.[381] b. 61-63,[66] 59-61,[4,615] 61-62,[600] 84[628] (in error, see Ref. 4), b_{758} 61-62,[381] n_D^{20} 1.4012,[381] 1.4003,[4] 1.4045,[600] n_D^{25} 1.3930,[66]

1.3940,[66] d_4^{20} 1.2813,[4] d^{20} 1.304,[600] d_{20} 1.2805,[381] IR,[600,675] ^{1}H,[600] ^{19}F.[600]

$ClCH_2CH_2P(S)F_2$. IX.[299] b. 127-130.[299] n_D^{20} 1.4565.[299] d_4^{20} 1.4343.[299]

EtP(S)F_2. I.[600] IX.[376,615] XVIII.[381] b. 89-90,[600,615] b_{755} 89,[376] b_{750} 90-91,[381] n_D^{20} 1.4175,[376] n_D 1.4076,[600] d_4^{20} 1.2123,[376] d^{20} 1.211,[600] 1.2106,[381] IR,[600] ^{1}H,[600] ^{19}F.[600]

$ClC_6H_4P(S)F_2$. IX[730] (orientation of Cl not specified). b_{20} 103-105,[730] n_D^{18} 1.5470,[730] d_{20}^{18} 1.4324.[730]

4-$ClC_6H_4P(S)F_2$. IX.[385] XVIII.[296] b_{40} 114-115,[296] $b_{0.03}$ 72-75,[385] n_D^{24} 1.5390.[296]

PhP(S)F_2. IX.[699,730] b. 186-188,[730] b_{20} 82.[730]

$MeC_6H_4P(S)F_2$. IX[730] (orientation of Me group not specified). b. 208-209,[730] n_D^{18} 1.5180,[730] d_{20}^{18} 1.2543.[730]

Thionophosphonyl Fluoride Chlorides

4-$ClC_6H_4P(S)FCl$. IX.[385] $b_{0.2}$ 62-68.[385]

Thionophosphonyl Dichlorides

$CCl_3P(S)Cl_2$. XI.[165] XVI.[345] XVIII.[130] b_{10} 95,[130,345,558] m. 120,[130,345,558] 166-169,[165] IR.[165]

$CF_3P(S)Cl_2$. XI.[198] b. 85.7,[198] m. -28.7 to -28.5,[198] log $P_{mm} = 5.5578 + 1.75 \cdot \log T - 0.00550 \cdot T - 2026/T$,[198] Trouton const. 20.6,[198] IR.[198]

$ClCH_2P(S)Cl_2$. VI.[322,457,585,619] XI.[53] XII.[510] b_{30} 88,[585] 87-88.5,[457] b_{10} 64,[53] 64-65,[619] b_8 58-59,[322] n_D^{20} 1.5778,[53] 1.5765,[457] 1.5770,[322] n_D^{25} 1.5730-1.5741,[619] d_4^{20} 1.5957,[53] d_{20} 1.5981,[322] IR,[167,530,670] Raman,[670] ^{31}P -73.0,[511] -74.2,[619] -75.7,[510] -76.5.[510]

MeP(S)Cl_2. III.[446,454,514] VI.[322] XI.[53,66,200,275] XII.[37,38,223,487] XVIII.[341,343,380] XX.[39,224] XXIII.[228,377] b. 142-143,[380] b_{50} 60-70,[38,223,224] 70-71,[341] b_{20} 49,[53] b_{19} 49-51,[66] b_{14} 46-47,[228] b_{10} 46,[200] $b_{9.5}$ 46-48,[446] b_9 44-45,[275] b_8 34-35,[322] n_D^{20} 1.5509,[446] 1.5520,[322] 1.5482,[53] 1.548,[61] 1.5450,[228] 1.5460,[341] n_D^{25} 1.5469,[66] 1.5485,[275] n_D^{16} 1.5510,[380] d_{20} 1.4302,[322] d_4^{20} 1.4366,[53] 1.423,[61] 1.4226,[341] d_{25} 1.4178,[275] d_{16} 1.4185,[380] dipole moment 1.00,[61] IR,[95,100,138,268,269,480-482,581,675] Raman,[95,138,200] ^{31}P -79.8,[446,517] Cl (NQR),[523,632] ^{1}H,[465,466] $J_{^{13}CH}$.[477]

$CH_2=CHP(S)Cl_2$. III.[454] VI.[326] XXII.[627] b_{12} 54-55.[326,454] b_1 37,[627] n_D^{20} 1.5623,[326,454] d_{20} 1.3954.[326]

$ClCH_2CH_2P(S)Cl_2$. IV.[419] VI.[326] b_{10} 92,[419] b_9 85-86,[326] n_D^{20} 1.5670,[419] 1.5665,[326] d_4^{20} 1.5120,[419] d_{20} 1.5117,[326] ^{1}H,[67,456] ^{31}P -78.8.[456]

MeCHClP(S)Cl_2. VI.[456] b_{13} 84,[456] ^{1}H, ^{31}P -86.9.[456]

EtP(S)Cl_2. III.[446,454] VI.[322] X.[159] XI.[22,234,275,282,322,517,588,589,608,609] XVIII.[341,343,379] b. 177-

178,[341] b_{740} 177-181,[446,454] b_{50} 80-82,[22,234,588] $b_{45.5}$ 82.5,[589] b_{20} 68-69,[379] b_{14} 62,[608,609] b_8 59-61,[322] b_4 40-50,[282] 55,[159] n_D^{20} 1.5450,[322] 1.5413,[589] 1.5435,[379] 1.5428,[341] n_D^{28} 1.5406,[282] d^{20} 1.3606,[22,234,588] 1.3532,[379] d_{20} 1.3643,[322] d_0^{20} 1.3550,[589] d_4^{20} 1.3534,[341] IR,[95,278,481] Raman,[95] ^{31}P -94.0,[446,454] -94.3,[517] ^1H.[467]

$CH_2{=}CHCH_2P(S)Cl_2$. XI.[586] XVIII.[227] b_{20} 80,[227] b_9 65-66,[586] n_D^{20} 1.5381,[227] 1.5553,[586] d^{20} 1.3056,[227] d_{20} 1.3341,[586] IR,[95,227] Raman.[95]

PrP(S)Cl_2. VI.[322] XI.[234] b_{50} 95-98,[234] b_{10} 66-67,[322] n_D^{20} 1.5360,[322] d^{20} 1.3005,[322] IR.[481,564]

i-PrP(S)Cl_2. XI.[531,609] b_{12} 70,[609] b_{10} 63-64,[531] n_D^{20} 1.5322,[531] d_{20} 1.2956,[531] IR,[95,481] Raman,[95] ^1H,[476] rel. signs of coupling const.[460]

$EtSCH_2P(S)Cl_2$. VI.[104] b_1 85-88.[104]

$ClCH_2CH{=}CHCH_2P(S)Cl_2$. XVI.[577] b_{11} 147,[577] n_D^{20} 1.5566,[577] d_{20} 1.4351.[577]

EtOCH=CHP(S)Cl_2. XVI.[13,552,553] b_2 85,[552] 84,[13] b_1 82,[553] n_D^{20} 1.5223,[552] 1.5275,[553] 1.5422,[13] d_{20} 1.334,[13] d_4^{20} 1.3330.[552]

BuP(S)Cl_2. VI.[218] b_{12-14} 84-87,[218] n_D^{20} 1.5311,[218] d_4^{20} 1.2525,[218] IR.[481]

i-BuP(S)Cl_2. XI.[234] b_{50} 110-113,[234] d^{20} 1.2512.[234]

s-BuP(S)Cl_2. VI.[216] b_8 81.5-82,[216] n_D^{20} 1.5326,[216] d_4^{20} 1.2548.[216]

t-BuP(S)Cl_2. VI.[112] m. 172.5-175.[112]

$CH_2{=}CHCMe{=}CHP(S)Cl_2$. XXII.[468] $b_{1.5-2}$ 86-88,[468] n_D^{20} 1.5842,[468] d^{20} 1.2922.[468]

EtOCHClCH=CHP(S)Cl_2. XVI.[693,694] $b_{1.5}$ 122-123,[693,694] n_D^{20} 1.5557,[693,694] d_4^{20} 1.3974.[693,694]

$C_5H_9P(S)Cl_2$ (cyclopentyl-). IR,[95] Raman.[95]

$ClCH_2CH{=}CHMeCH_2P(S)Cl_2$. XVI.[468] $b_{2.5}$ 118-118.5,[468] n_D^{20} 1.5627,[469] d^{20} 1.3789.[468]

i-PrOCH=CHP(S)Cl_2. XVI.[13] b_2 92,[13] n_D^{20} 1.5224,[13] d_{20} 1.2684.[13]

$MeOCH_2CH_2OCH{=}CHP(S)Cl_2$. XVI.[14] b_2 113,[14] n_D^{20} 1.5413,[14] d_{20} 1.3364.[14]

AmP(S)Cl_2. VI.[216,218] b_{12-13} 100-103,[218] b_7 92-93,[216] n_D^{20} 1.5213,[216] 1.5270,[218] d_4^{20} 1.1961,[216] 1.2211.[218]

i-AmP(S)Cl_2. VI.[218] XI.[234] b_{50} 130-132,[234] b_8 92-93,[218] n_D^{20} 1.5296,[218] d^{20} 1.1771,[234] d_4^{20} 1.2217.[218]

$Et_2CHP(S)Cl_2$. VI.[216] b_7 88.5-89,[216] n_D^{20} 1.5301,[216] d_4^{20} 1.2226.[216]

1,2-$Cl_2C_6H_3P(S)Cl_2$. XI.[726] $b_{0.5}$ 124.[726]

$BrC_6H_4P(S)Cl_2$. XI[726] (orientation of Br not specified). $b_{1.0}$ 115-126.[726]

$ClC_6H_4P(S)Cl_2$. IV.[455] XI[726] (orientation of Cl not specified). $b_{0.6}$ 106-108,[726] $b_{0.02}$ 86-90.[455]

4-$ClC_6H_4P(S)Cl_2$. IV[450] (mixture with 2-isomer). XI.[697,730] b_{14} 155,[730] b_5 135,[697] $b_{0.5}$ 101-102,[196] $b_{0.02}$

86-90,[450] n_D^{20} 1.6360,[450] 1.6348,[697] n_D^{18} 1.6365,[730] d_{18} 1.506,[730] d_4^{20} 1.5166,[697] IR,[196] ^{31}P -68.7.[450]

$FC_6H_4P(S)Cl_2$. IV[455] (orientation of F not specified). $b_{0.04}$ 56-58.[455]

3-$FC_6H_4P(S)Cl_2$. XI.[607] b_{14} 123-126,[607] n_D^{20} 1.5980,[607] Hammett const.,[607] ^{19}F.[607]

4-$FC_6H_4P(S)Cl_2$. IV.[450] XI.[606] b_{10} 121.5-123,[606] $b_{0.05}$ 54-56,[450] n_D^{20} 1.5999,[606] 1.5972,[450] Hammett const.,[567,606] ^{31}P -73.1,[450] ^{19}F.[567]

$PhP(S)Cl_2$. II.[409,410] III.[446,454,514] IV.[450,455] VI.[322] X.[160] XI.[53,199,291,313,370-372,502,537] XII.[510,684] XVI.[170] b. 270,[502] b_{105} 205,[370-372,502] b_{26} 105,[199] b_5 115,[170] 124-125,[53] b_4 110,[537] b_{2-3} 95-100,[313] b_1 100,[446,454] $b_{0.6}$ 76-77,[291] $b_{0.05}$ 72-75,[450] n_D^{20} 1.6227,[450] 1.6292,[53] 1.6250,[322] n_D^{22} 1.6176,[170] n_D^{25} 1.6141,[684] d_4^{20} 1.4071,[53] d^{13} 1.376,[502] d_{20} 1.4042,[322] d_{22} 1.375,[170] IR,[114,197] ^{31}P -74.8,[161,446,450,517,704] -74.5,[454] -81.8,[510] ^{13}C,[595] Cl (NQR),[523,632] complexes: $PhP(S)Cl_2 \cdot AlCl_3$ (^{31}P -109.2, conductivity);[453] $PhP(S)Cl_2 \cdot 2AlCl_3$ (^{31}P -109.2, conductivity).[453]

$C_6H_9P(S)Cl_2$ (2-cyclohexen-1-yl-). II.[409] b_1 94-96.[409]

$C_6H_{11}P(S)Cl_2$ (cyclohexyl-). VI.[216,322] b_8 122,[216] b_2 76-77,[322] n_D^{20} 1.5622,[322] 1.5610,[216] d_{20} 1.3037,[322] d_4^{20} 1.2952,[216] IR,[95] Raman.[95]

PrOCH=CMeP(S)Cl_2. XVI.[552,553] b_3 107,[552] $b_{2.5}$ 107,[553] n_D^{20} 1.5286,[552,553] d_4^{20} 1.2460.[552,553]

BuOCH=CHP(S)Cl_2. XVI.[13] b_2 105,[13] n_D^{20} 1.5234,[13] d_{20} 1.2471.[13]

$EtOCH_2CH_2OCH$=CHP(S)Cl_2. XVI.[14] b_2 120,[14] n_D^{20} 1.5330,[14] d_{20} 1.2887.[14]

$C_6H_{13}P(S)Cl_2$. VI.[218] b_{10} 108-111,[218] n_D^{20} 1.5249,[218] d_4^{20} 1.1885.[218]

i-$C_6H_{13}P(S)Cl_2$. VI.[218] b_{17} 114-121,[218] n_D^{20} 1.5266,[218] d_4^{20} 1.1899.[218]

EtPrCHP(S)Cl_2. VI.[216] b_8 103.5,[216] n_D^{20} 1.5250,[216] d_4^{20} 1.1887.[216]

4-$CF_3C_6H_4P(S)Cl_2$. XI.[731] b_3 89-90,[731] n_D^{21} 1.5445,[731] d_{20}^{21} 1.5254.[731]

2,4-$Cl_2C_6H_3CH_2P(S)Cl_2$. XI.[55] m. 85-88.[55]

2-$ClC_6H_4CH_2P(S)Cl_2$. XI.[55] b_2 130-133,[55] m. 53-55.[55]

4-$ClC_6H_4CH_2P(S)Cl_2$. XI.[56,58] b_2 140-143,[58] 132-135,[56] n_D^{20} 1.6104,[58] 1.6180,[56] d_4^{20} 1.4497,[58] 1.4659.[56]

$PhCH_2P(S)Cl_2$. XI.[56,58] b_2 124-127,[58] b_1 110-112,[56] n_D^{20} 1.6121,[58] 1.6128,[56] d^{20} 1.3620,[58] d_4^{20} 1.3599,[56] ^{31}P -85.3.[511]

$MeC_6H_4P(S)Cl_2$. IV.[455] XI[313] (orientation of Me group not specified). b_2 118-121,[313] $b_{0.05}$ 85-87.[455]

4-$MeC_6H_4P(S)Cl_2$. IV[450] (mixture with 2-isomer). XVIII.[496] b_{22-25} 167,[496] $b_{0.05}$ 85-87,[450] n_D^{20} 1.6130,[450] ^{31}P -74.1.[450]

4-$MeOC_6H_4P(S)Cl_2$. II.[410] b_1 124-128.[410]

$(C_6H_{10})MeP(S)Cl_2$. VI[322,638] (orientation of Me group not specified). b_2 89-90,[322] 109-110,[638] n_D^{20} 1.5570,[322] 1.5558,[638] d_{20} 1.1713,[322] d_{20} 1.2554.[638]
PhC≡CP(S)Cl_2. XI.[62] b_2 125,[62] n_D^{20} 1.6510,[62] d^{20} 1.3236.[62]
PhCH=CHP(S)Cl_2. XVI.[13,714,716,740] XVIII.[727] b_8 162-165,[727] b_1 130,[716] $b_{0.1}$ 110,[740] m. -20,[727] n_D^{20} 1.6439,[13] n_D^{25} 1.6563,[716] n_D^{33} 1.6452,[740] d_{20} 1.3533,[13] d_{26} 1.345.[727]
PhOCH=CHP(S)Cl_2. XVI.[13] b_1 140,[13] n_D^{20} 1.6086,[13] d_{20} 1.1670.[13]
$Me_2C_6H_3P(S)Cl_2$. XI[313] (orientation of Me groups not specified). b_2 121-122.[313]
2,5-$Me_2C_6H_3P(S)Cl_2$. XI.[56] $b_{1.5}$ 124-125,[56] m. 57-59.[56]
$EtC_6H_4P(S)Cl_2$. XI[313] (orientation of Et group not specified). b_1 117-119.[313]
i-$C_8H_{15}P(S)Cl_2$(i-octenyl). II.[409] b_2 78-96.[409]
i-C_6H_{13}CH=CHP(S)Cl_2. XI.[714,716] b_1 86-95,[714,716] n_D^{25} 1.5323.[714,716]
C_6H_{13}OCH=CHP(S)Cl_2. XVI.[13] b_2 128,[13] n_D^{20} 1.5255,[13] d_{20} 1.1841.[13]
BuOCH=CEtP(S)Cl_2. XVI.[552,553] b_3 131,[552] $b_{2.5}$ 131,[553] n_D^{20} 1.1830,[552,553] d_4^{20} 1.1830.[552,553]
$BuOCH_2CH_2$OCH=CHP(S)Cl_2. XVI.[14] b_2 137,[14] n_D^{20} 1.5100,[14] d_{20} 1.2210.[14]
cis-2-MeC_6H_4OCH=CHP(S)Cl_2. XVII.[337] b_3 142,[337] n_D^{20} 1.5990,[337] d^{20} 1.3177.[337]
trans-2-MeC_6H_4OCH=CHP(S)Cl_2. XVII.[337] b_3 146,[337] n_D^{20} 1.6061,[337] d^{20} 1.3156.[337]
i-$PrC_6H_4P(S)Cl_2$. XI[313] (orientation of Pr group not specified). b_1 115-117.[313]
2-$C_{10}H_7P(S)Cl_2$ (2-naphthyl-). II.[409,410] b_4 173-174.[410]

Bis(thionophosphonyl Dichlorides)

$Cl_2P(S)CH_2P(S)Cl_2$. VI.[451] $b_{0.06}$ 96-100,[451] m. 30.5,[451] n_D^{20} 1.6358,[451] IR,[451] ^{31}P -52.3.[451]
$Cl_2P(S)(CH_2)_2P(S)Cl_2$. VI.[451] $b_{0.2}$ 150,[451] m. 99.5-100.5,[451] IR,[451] ^{31}P -79.3.[451]
$Cl_2P(S)(CH_2)_3P(S)Cl_2$. VI.[451] $b_{0.1}$ 148-152,[451] n_D^{20} 1.6158,[451] IR.[451]
4-$Cl_2P(S)C_6H_4P(S)Cl_2$. XI.[57] m. 131-132.[57]

Thionophosphonyl Chloride Bromides

MeP(S)ClBr. ^{31}P -51.0.[511]

Thionophosphonyl Dibromides

$BrCH_2P(S)Br_2$. XII.[510] ^{31}P -16 to -19,[510] -20.1.[511]
MeP(S)Br_2. XI.[440,448,609] b. 203-204,[440] 188,[609] b_{720} 203-204,[448] m. 31-32,[440,448] n_D^{20} 1.6527,[448] d_4^{20}

2.1481,[448] IR,[448] ^{31}P -20.5.[448]

PhP(S)Br$_2$. IV.[450,455] $b_{0.001}$ 88.5-90,[450,455] n_D^{20} 1.6968,[450] ^{31}P -19.8,[450] -20.2.[511]

Thionophosphonyl Dipseudohalides

C_6F_5P(S)(NCS)$_2$. XI.[154a] ^{31}P -6.8.[154a]

PhP(S)(NCS)$_2$. IX.[205] $b_{0.08}$ 128-130,[205] IR.[205]

PhP(S)(N$_3$)$_2$. IX.[40] $b_{0.1}$ $\sim$ 80.[40]

Selenophosphonyl Dihalides

MeP(Se)Cl$_2$. XI.[231,311] b_{25} 74-75,[311] $b_{0.1}$ 20-21,[231] n_D^{18} 1.5970,[311] n_D^{20} 1.5896,[231] d_{18} 1.7540,[311] d_{20} 1.7872,[231] IR,[232,581] ^{1}H.[466]

EtP(Se)Cl$_2$. XI.[2,231,311] XVIII.[311] b_{25} 92-94,[311] b_{20} 95,[311] b_{20} 70,[2] $b_{1.5}$ 38-39,[231] n_D^{20} 1.5894,[231] 1.5850,[311] n_D^{24} 1.5850,[311] d_{20} 1.6987,[231] 1.6800,[311] d_{24} 1.6750,[311] IR.[232]

ClCH$_2$CH=CHCH$_2$P(Se)Cl$_2$. XVI.[571] b_2 108-110,[571] n_D^{20} 1.5702,[571] d^{20} 1.6123.[571]

MeP(Se)Br$_2$. XI.[440,448] b. 179-184,[440] b_{730} 174-176,[448] b_8 63-65,[448] n_D^{20} 1.6349,[448] d_4^{20} 2.2602,[448] IR,[448] ^{31}P +16,[448] coupl. const.[448]

D.2. Phosphinyl- (Thio-, Seleno-) Halides and Pseudohalides

Phosphinyl Fluorides

Me$_2$P(O)F. IX.[615] XX.[402] b_{16} 69-71,[402] b_{20} 69-70,[615] n_D^{20} 1.3920,[402] $n_D^{23.8}$ 1.3940,[615] d^{20} 1.1810,[402] ^{31}P -67.3,[526] -66.7,[636] -66.3,[618] ^{19}F,[636] kinetics of hydrolysis.[249]

C_4H_5ClP(O)F (1-oxo-1-fluoro-3-chloro-phospholene-3). XXI.[145] $b_{0.5}$ 86,[145] m. 32,[145] n_D^{20} 1.4990,[145] d_4^{20} 1.4808,[145] ^{1}H.[145]

C_4H_6P(O)F (1-oxo-1-fluoro-phospholene-3). XXI.[145,544,591] b_2 58,[145,544,591] m. 8,[145] n_D^{20} 1.4608,[145,544] 1.4606,[591] d_4^{20} 1.2860,[145,544] 1.2856,[591] ^{1}H.[145,544]

Et$_2$P(O)F. ^{31}P -71.8,[636] ^{19}F,[636] kinetics of hydrolysis.[3]

C_5H_8P(O)F (1-oxo-1-fluoro-2-methyl-phospholene-3). XXI.[145] $b_{1.0}$ 59,[145] n_D^{20} 1.4630,[145] d_4^{20} 1.2031,[145] ^{1}H.[145]

C_5H_8P(O)F (1-oxo-1-fluoro-3-methyl-phospholene-3). XXI.[145,544] $b_{2.0}$ 60,[145,544] n_D^{20} 1.4662,[145,544] d_4^{20} 1.2540,[145,544] ^{1}H.[145,544]

Me(Bu)P(O)F. ^{31}P -65.2,[636] ^{19}F.[636]

Me(i-Bu)P(O)F. IX.[118] $b_{3.5}$ 71-72.[118]

C_6H_{10}P(O)F (1-oxo-1-fluoro-3,4-dimethyl-phospholene-3). XXI.[145] $b_{1.0}$ 62,[145] m. 39,[145] n_D^{20} 1.4740,[145] d_4^{20} 1.1957,[145] ^{1}H.[145]

(i-Pr)$_2$P(O)F. I.[93] IX.[408] $b_{0.2}$ 50,[93] n_D^{20} 1.4678,[93] n_D^{25} 1.3800,[408] d_{25} 1.055,[408] ^{31}P -78.2,[636] ^{19}F.[636]

Me(Am)P(O)F. IX.[118] b_2 74-75.[118]

$Me(C_6H_{11})P(O)F$. IX.[118] $b_{1-1.5}$ 77.[118]

Et(Ph)P(O)F. ^{31}P -53.4,[636] ^{19}F.[636]

$(i\text{-}Bu)_2P(O)F$. ^{31}P -71.3.[636]

$(C_6F_5)_2P(O)F$. XIX.[155] $b_{0.02}$ 100-110,[155] m. 80,[155] ^{31}P -9.8,[155] ^{19}F.[277]

$Ph_2P(O)F$. XIX.[618] $b_{0.15}$ 132,[618] n_D^{20} 1.5780,[618] ^{31}P -40.5,[526,636] -40.0.[612,618]

Phosphinyl Chlorides

$(CCl_3)_2P(O)Cl$. I.[396] XVI.[396] subl. 70/5 mm,[396] m. 48.5-49.[396]

$(CF_3)_2P(O)Cl$. VIII.[215] XXIII.[81] b. 52.6 (est.),[215] $\log P_{mm} = 6.8443 + 1.75 \cdot \log T - 0.00700 \cdot T - 1980/T$, Trouton const. 20.9, vap. press. 3.5 mm/24°,[81] IR.[81,215]

$Me(CCl_3)P(O)Cl$. XVI.[132] m. 84.[132]

$(ClCH_2)_2P(O)Cl$. I.[165,167,294,295,457] II.[457,509] XIV.[509] b_{11} 131-133,[167] b_{10} 123-124,[294] b_6 120-122,[165] b_2 104-106,[167] b_1 92-98,[457] 95-98,[357] 98-101,[509] $b_{0.02}$ 76-78,[47] n_D^{20} 1.5203,[47,294] 1.5200,[457] n_D^{24} 1.5196,[165] n_D^{25} 1.5196,[167] d^{20} 1.6120,[47,294] IR,[167,294] Raman,[294] ^{31}P -49.3,[457,511] -50.6,[509] ^{1}H.[457]

┌─CH_2─┐
$S\text{-}CH_2\text{-}P(O)Cl$ [3-oxo-3-chloro-1,3-thiaphosphetane(III)]. I.[302] m. 101-103.[302]

$Me(ClCH_2)P(O)Cl$. II.[509] XIV.[509] $b_{0.8}$ 70,[509] ^{31}P -57.0.[509,511]

$Me_2P(O)Cl$. I.[93,276,281,392,594] II.[331,393,509] III.[312] V.[103,257,443,557] VII.[103,256,557] X.[194] XII.[193,404,719] b. 202-204,[393] 207-209,[93] 204-205,[594] 202-206,[392] 204,[276] 203-206,[193] b_{52} 115-118,[256] b_{35} 110-113,[257,557] 107-111,[557] $b_{16.5}$ 128,[281] b_{10} 82,[509] m. 70-72,[557] 68-72,[257] 68-72,[257] 68,[103,443] 68-69,[134] 66.8-68.4,[331] 66.5-68.4,[557] 66,[276] 65-66,[193] 64-66,[594] subl. 150,[134] IR,[52,140] Raman,[140] ^{31}P -62.8,[511] -64.3,[509] Cl(NQR),[522 523 632] kinetics of hydrolysis.[134,521]

$Me(CH{=}CHCl)P(O)Cl$. XII[749] (possibly isomer mixture). b_4 70-75,[749] n_D^{20} 1.4950,[749] d_{20} 1.3657.[749]

$Et(CCl_3)P(O)Cl$. XVI.[132] m. 36.[132]

$Me(C_2H_3Cl_2)P(O)Cl$. XII[749] (isomer mixture). b_4 105-107,[749] n_D^{20} 1.4960,[749] d_{20} 1.4820.[749]

$Et(ClCH_2)P(O)Cl$. III.[446] ^{31}P -43.8.[446]

Me(Et)P(O)Cl. I.[519] V.[443] X.[556,751] b_{20} 105,[519] b_{15} 72-75,[556] 99-102,[443] b_1 55,[751] m. 37-39,[556] n_D^{20} 1.4688,[519] 1.4621,[751] 1.4702,[443] d_4^{20} 1.1854,[519] d_{20} 1.2250,[751] ^{31}P -67.9,[511] Cl (NQR),[522,523,632] kinetics of hydrolysis.[519,521,533]

$Me(F(CF_3)C{=}CF)P(O)Cl$. II.[310] $b_{1.5}$ 40-42,[310] n_D^{20} 1.3940,[310] d_4^{20} 1.5166.[310]

$C_4H_5ClP(O)Cl$ (1-oxo-1,3-dichloro-phospholene-2). IR,[722] UV,[722] ^1H.[722]

$C_4H_5ClP(O)Cl$ (1-oxo-1,3-dichloro-phospholene-3). II.[189] b_{13} 140-143,[189] n_D^{20} 1.5405,[189] d^{20} 1.4700,[189] IR,[722] UV,[722] ^1H.[722]

$(CH_2{=}CH)_2P(O)Cl$. XXII.[323] b_3 68.[323]

$C_4H_6P(O)Cl$ (1-oxo-1-chloro-phospholene-2). II.[32,284] $b_{0.25}$ 83-85,[284] $b_{0.04}$ 84-86,[30] n_D^{20} 1.5242,[284] 1.5252,[30] d_4^{20} 1.3305,[30] IR,[30] UV.[284]

$C_4H_6P(O)Cl$ (1-oxo-1-chloro-phospholene-3). II.[32,284] XIX.[24] XXI.[54] b_{11} 112-113,[24] b_3 100-103,[54] $b_{0.15}$ 78-81,[284] $b_{0.04}$ 72-74,[30] m. 53,[284] 53-54,[30] 52-53,[24] n_D^{20} 1.5198,[54] 1.5203,[30] d_4^{20} 1.3388,[54] 1.3328,[30] IR,[30] UV.[284]

$ClCH_2(NCCH_2CH_2)P(O)Cl$. XV.[363] $b_{0.05}$ 138,[363] n_D^{20} 1.5180,[363] d^{20} 1.4493.[363]

$C_4H_6Cl_2P(O)Cl$ (1-oxo-1,3,4-trichloro-phospholane). XXII.[54] $b_{1.5}$ 140-142,[54] n_D^{20} 1.5390,[54] d_4^{20} 1.5429.[54]

$ClCH_2(ClCOCH_2CH_2)P(O)Cl$. XV.[360] $b_{0.006}$ 128-129,[360] n_D^{20} 1.5195,[360] d_4^{20} 1.5338,[360] IR.[360]

$Et(CHBr{=}CH)P(O)Cl$. XXII.[688] b_{10} 108-110,[688] n_D^{20} 1.5265,[688] d_{20} 1.6470.[688]

$Et(CH_2{=}CBr)P(O)Cl$. XXII.[338] b_{12} 108-112,[338] n_D^{20} 1.5270.[338]

$Et(CHCl{=}CH)P(O)Cl$. XVII.[696] XXI.[695] XXII.[688] b_{18} 119-120,[696] b_{13-14} 113-118,[695] b_{10} 92-94,[688] n_D^{20} 1.5070,[695,696] 1.4935,[688] d_{20} 1.3080,[688] 1.3210.[695]

$CH_2{=}CH(CH_2ClCH_2)P(O)Cl$. II.[323] b_{14} 93-95,[323] n_D^{20} 1.5046,[323] d_{20} 1.3329.[323]

$Me(CH_2CH_2COCl)P(O)Cl$. XV.[361] $b_{0.01}$ 114-115,[361] n_D^{20} 1.5005,[361] d^{20} 1.4010.[361]

$Et(CHCl_2CHCl)P(O)Cl$. XXII.[688] b_{10} 135,[688] n_D^{20} 1.5220,[688] d_{20} 1.5190.[688]

$Me(NCCH_2CH_2)P(O)Cl$. XV.[355] $b_{0.06}$ 136,[355] n_D^{20} 1.4908,[355] d^{20} 1.2866,[355] IR.[355]

$Et(CH_2{=}CH)P(O)Cl$. XXII.[339] b_{10} 90,[339] n_D^{20} 1.4805,[339] d_{20} 1.1695.[339]

$C_4H_8P(O)Cl$ (1-oxo-1-chloro-phospholane). II.[284] b_{12} 137-141,[284] n_D^{20} 1.5001.[284]

$Et(CH_2BrCHBr)P(O)Cl$. XXII.[338] b_{12} 145-157,[338] n_D^{20} 1.5670,[338] d_{20} 2.0511.[338]

$Et(CH_2ClCHCl)P(O)Cl$. XXII.[688] b_{10} 132-133,[688] n_D^{20} 1.5100,[688] d_{20} 1.4290.[688]

$(MeCHCl)_2P(O)Cl$. I.[754] $b_{0.02}$ 68-69,[754] n_D^{20} 1.5030,[754] d^{20} 1.4306.[754]

$Et(ClCH_2CH_2)P(O)Cl$. I.[339] b_{10} 127,[339] n_D^{20} 1.4900,[339] d_{20} 1.2985.[339]

$Me(C_3H_6Cl)P(O)Cl$. XII.[752] b_2 80-90,[752] n_D^{20} 1.4870,[752] d^{20} 1.3063.[752]

$Et_2P(O)Cl$. I.[93,392,519] II.[394,587,588,590] II.[312,446] V.[103,257,557] XVI.[556] XVII.[691] XXIII.[463] b_{52} 134-135,[392] b_{20} 120-122,[519] b_{16} 108-109.5,[93,587,588,590]

b_{15} 102-104,[394] 79-81,[556] $b_{0.7}$ 62.5-64.5,[557] $b_{0.6}$ 45-46,[463] $b_{0.5}$ 64,[103] $b_{0.4-0.55}$ 60.5-63.0,[257] n_D^{20} 1.4688,[463] 1.4647,[587,588,590] 1.4672,[93] 1.4680,[519] n_D^{25} 1.4650,[557] d_4^{20} 1.1373,[519] d_0^{20} 1.1394,[587,588] d_{20} 1.1394,[590] IR,[52] ^{31}P -76.7,[446,511] Cl(NQR),[522,523,632] kinetics of hydrolysis,[249,519,521,535] kinetics of solvolysis.[523]

Me(Pr)P(O)Cl. I.[519] V.[443] XII.[660] b_{12} 113-116,[443] b_{10} 88,[519] b_4 78-81,[660] n_D^{20} 1.4686,[443] 1.4635,[519] 1.4628,[660] d_4^{20} 1.1302,[519] d_{20} 1.1307,[660] Cl(NQR),[522] kinetics of hydrolysis.[519,521,535]

Me(i-Pr)P(O)Cl. I.[519] b_4 71.[519] n_D^{20} 1.4670,[519] d_4^{20} 1.1568,[519] Cl(NQR),[522] kinetics of hydrolysis.[519,521,535]

Et(ClCOCH=CH)P(O)Cl. XV.[361a]

$C_5H_8P(O)Cl$ (1-oxo-1-chloro-3-methyl-phospholene-2). XIX.[24] XXI.[24,29] $b_{0.02}$ 91-92,[24] $b_{0.01}$ 88-90,[29] n_D^{20} 1.5232,[24] 1.5236,[29] d_4^{20} 1.2604,[24] 1.2593,[29] IR,[29] ^1H.[24]

$C_5H_8P(O)Cl$ (1-oxo-1-chloro-2-methyl-phospholene-3). XXI.[375] b_2 96-98,[375] n_D^{20} 1.5159,[375] d_4^{20} 1.2575.[375]

$C_5H_8P(O)Cl$ (1-oxo-1-chloro-3-methyl-phospholene-3). XXI.[29,54] b_2 118-120,[54] $b_{0.01}$ 61-62,[29] n_D^{20} 1.5200,[54] 1.5118,[29] d_4^{20} 1.2858,[54] 1.2541,[29] IR.[29]

$C_5H_8Cl_2P(O)Cl$ (1-oxo-1,3,4-chloro-3-methyl-phospholane). XXII.[54] b_4 160-163,[54] n_D^{20} 1.5480,[54] d_4^{20} 1.5700.[54]

Et($ClCOCH_2CH_2$)P(O)Cl. XV.[350,351] $b_{0.22}$ 118-120,[350,351] n_D^{20} 1.4935,[350,351] d^{20} 1.3412.[350,351]

Me($ClCOCHMeCH_2$)P(O)Cl. XV.[356] $b_{0.01}$ 102,[356] n_D^{20} 1.4915,[356] d_4^{20} 1.3334,[356] IR.[356]

Me($ClCOCH_2CHMe$)P(O)Cl. XV.[354] $b_{0.06}$ 120,[354] n_D^{20} 1.4960,[354] d^{20} 1.3610,[354] IR.[354]

Me($CH_2CHMeCN$)P(O)Cl. XV.[357] $b_{0.01}$ 110,[357] n_D^{20} 1.4845,[357] d^{20} 1.2181,[357] IR.[357]

Et($NCCH_2CH_2$)P(O)Cl. XV.[573] $b_{0.025}$ 138,[573] n_D^{20} 1.4902,[573] d^{20} 1.2362.[573]

Et($Cl_2C_3H_5$)P(O)Cl. XII[660] (isomer mixture). b_2 115-120,[660] n_D^{20} 1.4980,[660] d_{20} 1.3492.[660]

Me(Bu)P(O)Cl. XXI[752] (isomer mixture). b_2 70-73,[752] n_D^{20} 1.4579,[752] d^{20} 1.0996.[752]

Me(Bu)P(O)Cl. I.[519] V.[443] b_{13} 151-157,[443] b_{10} 116,[519] n_D^{20} 1.4630,[519] 1.4668,[443] d_4^{20} 1.0901,[519] Cl (NQR),[522] kinetics of hydrolysis.[519,535]

Me(i-Bu)P(O)Cl. I.[118,519] b_{1-2} 61-62,[118] b_1 68,[519] n_D^{20} 1.4614,[519] d_4^{20} 1.0836,[519] kinetics of hydrolysis.[519,535]

Et(Pr)P(O)Cl. I.[519] b_4 87-88,[519] n_D^{20} 1.4670,[519] d_4^{20} 1.1052,[519] kinetics of hydrolysis,[519,521,535] Cl (NQR).[522]

$(ClCOCH_2CH_2)_2P(O)Cl$. VIII.[403] m. 55-60.[403]

$(CH_2=CHCH_2)_2P(O)Cl$. XXII.[91]

$C_6H_{10}P(O)Cl$ (1-oxo-1-chloro-3,4-dimethyl-phospholene-2).

XIX.[31] $b_{0.05}$ 94-96,[31] n_D^{20} 1.5134,[31] d_4^{20} 1.2074,[31] IR,[722] UV,[722] ^1H.[31,722]
$C_6H_{10}P(O)Cl$ (1-oxo-1-chloro-3,4-dimethyl-phospholene-3). II.[284] XXI.[31] b_{10} 132,[31] $b_{0.02}$ 79,[284] m. 83-86,[31] 74,[284] UV.[284]
Et(ClCH$_2$CH$_2$OCH=CH)P(O)Cl. XVII.[696] b_4 164,[696] n_D^{20} 1.5200,[696] d_{20} 1.3570.[696]
Et(MeClCHOCH=CH)P(O)Cl. XVII.[690] b_2 114-116.5,[690] n_D^{20} 1.5188,[690] d_{20} 1.4380.[690]
Et(ClCOCHMeCH$_2$)P(O)Cl. XV.[350,351] $b_{0.2}$ 116-117,[350,351] n_D^{20} 1.4875,[350,351] d^{20} 1.2817.[350,351]
Et(ClCOCH$_2$CHMe)P(O)Cl. XV.[358] $b_{0.003}$ 116,[358] n_D^{20} 1.4930,[358] d^{20} 1.2954.[358]
ClCH$_2$(EtOCOCH$_2$CH$_2$)P(O)Cl. II.[360] $b_{0.006}$ 120-121,[360] n_D^{20} 1.4950,[360] d_4^{20} 1.3639,[360] IR.[360]
Et(NCCHMeCH$_2$)P(O)Cl. XV.[359] $b_{0.006}$ 131-132,[359] n_D^{20} 1.4830,[359] d^{20} 1.1837.[359]
Me(C_5H_9)P(O)Cl. II.[333] b_1 80-80.5,[333] n_D^{20} 1.4989,[333] d_{20} 1.1800.[333]
Et(CH$_2$=CHCHMe)P(O)Cl. II.[569] b_1 68-69,[569] n_D^{20} 1.4828,[569] d^{20} 1.1045.[569]
Et(MeCH=CHCH$_2$)P(O)Cl. II.[569] b_1 79,[569] n_D^{20} 1.4858,[569] d^{20} 1.1056.[569]
(Cl(CH$_2$)$_3$)$_2$P(O)Cl. I.[5,712] $b_{0.04}$ 134-135,[5,712] n_D^{20} 1.5142,[5,712] d_4^{20} 1.3374.[5,712]
Pr$_2$P(O)Cl. I.[93,519] II.[393,473,587,590] V.[103] XVI.[556] b_{17} 126-128,[393] b_{15} 112-114,[556] b_{11} 118.5-120,[587,590] b_8 110.5-112,[473] b_4 84-88,[519] $b_{0.5}$ 83,[103] $b_{0.2}$ 60,[93] $b_{0.11}$ 56-57,[93] n_D^{20} 1.4681,[519] 1.4638,[587,590] 1.4632,[93] 1.4643,[93] 1.4662,[473] d_4^{20} 1.0820,[519] d_{20} 1.0692,[587,590] 1.0689,[473] IR,[52] Cl (NQR),[522,523,632] kinetics of hydrolysis.[519,521,535]
(i-Pr)$_2$P(O)Cl. I.[519,587,590] b_{12} 109-110.5,[587-590] b_3 77,[519] n_D^{20} 1.4667,[587,590] 1.4669,[519] d_4^{20} 1.0766,[519] d_{20} 1.0806,[587,590] IR,[52] Cl (NQR),[522] kinetics of hydrolysis.[249,519,521,535]
Me(Am)P(O)Cl. I.[118] b_3 95-105.[118]
Ph(CCl$_3$)P(O)Cl. II.[732] XVI.[50,732] m. 87.[50,732]
Ph(ClCH$_2$)P(O)Cl. II.[509] III.[446] XIV.[332,509] b_2 116-117,[332] $b_{0.5}$ 125,[509] m. 47-49,[332] ^{31}P -44.6,[509] -44.4.[446,511]
Me(4-ClC$_6$H$_4$)P(O)Cl. I.[519,524] b_5 155-156,[519,524] n_D^{20} 1.5727,[519,524] d_4^{20} 1.3934.[519,524]
MePhP(O)Cl. I.[96,190,519,556] II.[255,509] III.[446] V.[443,629] b_{22} 167,[190] b_{20} 163,[96] b_{11} 155,[190] b_{10} 152,[509] $b_{1.5}$ 115,[519] b_1 102-107,[255] 98,[629] $b_{0.6}$ 124-128,[443] m. 36-38,[443] n_D^{20} 1.5602,[443] 1.5550,[519] d_4^{20} 1.2683,[519] ^{31}P -51.2,[509] -52.0,[446,511] coupl. const.,[652] kinetics of hydrolysis.[519]
MePhP(^{18}O)Cl. II.[208]
Me(C_6H_{11})P(O)Cl. I.[118] XII.[660] b_3 101-102,[660] b_1 93-

95,[118] m. 26,[660] n_D^{20} 1.4988,[660] d_{20} 1.1611.[660]

Me(EtOCOCH$_2$CMeH)P(O)Cl. II.[354] $b_{0.06}$ 110,[354] n_D^{20} 1.4732,[354] d^{20} 1.2066,[354] IR.[354]

ClC$_6$H$_4$(CH=CH$_2$)P(O)Cl. XXII[181] (isomer mixture). b_4 145-148,[181] n_D^{20} 1.5577,[181] d_{20} 1.3597.[181]

Ph(CHCl=CH)P(O)Cl. XXI.[692] b_2 140-141,[692] n_D^{20} 1.5821,[692] d^{20} 1.3568.[692]

Ph(CH$_2$=CH)P(O)Cl. I.[519] II.[183,210] b_3 129-130,[519] b_2 118-120,[183] b_1 110-112,[210] n_D^{20} 1.5610,[210,519] 1.5614,[183] d_4^{20} 1.2861,[210,519] kinetics of hydrolysis.[519]

ClC$_6$H$_4$(CH$_2$CH$_2$Cl)P(O)Cl. II[181] (isomer mixture). $b_{1.5-2}$ 160-162,[181] n_D^{20} 1.5762,[181] d_{20} 1.4410.[181]

Ph(ClCH$_2$CH$_2$)P(O)Cl. II.[174-176] b_3 146-148,[175] b_2 150-153,[174,176] n_D^{20} 1.5642,[174,176] d_{20} 1.3409.[174,176]

Et(4-ClC$_6$H$_4$)P(O)Cl. V.[629] $b_{0.01}$ 105.[629]

Me(2-Me-4-Cl-C$_6$H$_3$)P(O)Cl. V.[629] b_2 156.[629]

Et(Ph)P(O)Cl. I.[519] II.[209] III.[446] VIII.[364] b_4 148-150,[364] b_3 123-125,[519] $b_{0.3}$ 101,[209] n_D^{20} 1.5565,[364] 1.5494,[519] d_{20} 1.0530,[364] d_4^{20} 1.2114,[519] ^{31}P -59.0,[446,511] Cl (NQR),[522] coupl. const.,[652] kinetics of hydrolysis.[519]

Me(4-MeSC$_6$H$_4$)P(O)Cl. V.[629] m. 65.[629]

C$_8$H$_{11}$ClP(O)Cl (exo-3-chloro-endo-5-methylchlorophosphonyl-tricyclo-2.2.1.0^{2,6} heptane). XVII.[207] $b_{0.2}$ 130.[207]

(C$_4$H$_7$)$_2$P(O)Cl (iso-butenyl-). XVI.[753] m. 42-43.[753]

(EtOCH=CH)$_2$P(O)Cl. XVI.[147] b_2 152-154,[147] n_D^{20} 1.5139,[147] d^{20} 1.2186.[147]

Et(C$_6$H$_{11}$)P(O)Cl (cyclohexyl-). XII.[516,660] b_4 120-122,[516] b_3 118-119,[660] n_D^{20} 1.5002,[660] 1.4999,[516] d_{20} 1.1372.[660]

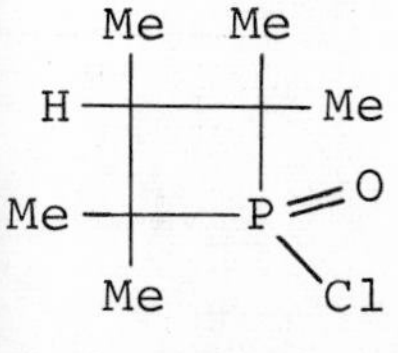

. XVI.[319,320,321] m. 74-75.[319-321]

Et(BuOCH=CH)P(O)Cl. XVII.[696] b_{10} 156-161,[696] n_D^{20} 1.4889,[696] d_{20} 1.1110.[696]

Bu$_2$P(O)Cl. I.[93,247,387,519,551] II.[334,393,587,590] V.[103,577] VII.[543] VIII.[399] X.[8] XVI.[556] XXIII.[463] b_{28} 156-157,[556] b_{15} 132-134,[556] 134,[399] b_{16} 143-145,[393] b_{12} 144-148,[551] b_{11} 142-144,[587,590] b_2 115-117,[463] 115,[519] b_1 96-97,[334] 110-112,[543] $b_{0.5}$ 103,[103] 95-100,[557] $b_{0.2}$ 85,[93] $b_{0.15}$ 80,[93] m. 49-53,[247] n_D^{20} 1.4670,[519] 1.4649,[463] 1.4643,[587,590] 1.4651,[93] 1.4652,[93] 1.4647,[334] d_4^{20} 1.0267,[519] d_{20} 1.0296,[587,590] 1.0302,[334] IR,[52] kinetics of hydrolysis.[519,535]

(i-Bu)$_2$P(O)Cl. I.[519,587,590] b_2 110-110.5,[587,590] 88-90,[519] $b_{0.2}$ 70,[93] $b_{0.15}$ 61,[93] $b_{0.1}$ 58,[93] n_D^{20} 1.4645,[519] 1.4595,[587,590] 1.4613,[93] 1.4610,[93] 1.4738,[93] d_4^{20} 1.0240,[519] d_{20} 1.0216,[587,590] IR,[52] kinetics of

hydrolysis.[519,535]
(s-Bu)$_2$P(O)Cl. IR.[52]
(t-Bu)$_2$P(O)Cl. III.[390] X.[8,673] $b_{2.7}$ 94,[390] m. 80.1-80.9,[8,673] 84-85,[390] ^{1}H,[8] mass spect.[8]
Ph(ClCOCH$_2$CH$_2$)P(O)Cl. XV.[350,351,574] $b_{0.002}$ 114-116,[350,351] $b_{0.001}$ 230,[574] n_D^{20} 1.5600,[574] 1.5595,[350,351] d^{20} 1.3676,[574] 1.3840.[350,351]
Ph(NCCH$_2$CH$_2$)P(O)Cl. XV.[362,573] $b_{0.025}$ 166,[573] $b_{0.004}$ 165-167,[362] n_D^{20} 1.5620,[573] 1.5630,[362] d^{20} 1.2864,[573] d_4^{20} 1.2824,[362] IR.[362]
Ph(C$_3$H$_6$Cl)P(O)Cl. XII[748] (isomer mixture). b_2 140,[748] n_D^{20} 1.5510,[748] d_{20} 1.2958.[748]
MeOC$_6$H$_4$(CH$_2$CH$_2$Cl)P(O)Cl. II[177] (probably isomer mixture). b_2 184,[177] n_D^{20} 1.5671,[177] d_{20} 1.2888.[177]
Ph(i-Pr)P(O)Cl. I.[519] XVI.[129] b_3 116-121,[519] $b_{0.15}$ 120-124,[129] n_D^{20} 1.5430,[519] d_4^{20} 1.1609,[519] coupl. const.,[652] kinetics of hydrolysis.[519]
Me(Me$_2$C=CCMe$_3$)P(O)Cl. V.[629] b_1 105.[629]
Ph(CH$_2$=CHOCH=CH)P(O)Cl. XVII.[690] b_1 146-148,[690] n_D^{20} 1.5820,[690] d_{20} 1.2660.[690]
4-MeC$_6$H$_4$(ClCOCH$_2$CH$_2$)P(O)Cl. XV.[353] $b_{0.005}$ 200,[353] n_D^{20} 1.5628,[353] d^{20} 1.3143.[353]
4-MeC$_6$H$_4$(MeCHCN)P(O)Cl. XV.[572] $b_{0.004}$ 170-175,[572] n_D^{20} 1.5490,[572] d_4^{20} 1.2217.[572]
4-MeC$_6$H$_4$(NCCH$_2$CH$_2$)P(O)Cl. XV.[362,573] $b_{0.04}$ 163,[573] $b_{0.004}$ 173-174,[362] n_D^{20} 1.5608,[573] 1.5615,[362] d_4^{20} 1.2634,[362] d^{20} 1.2448,[573] IR.[362]
EtOCH=CH(Ph)P(O)Cl. XVII.[11] b_2 155-156,[11] n_D^{20} 1.5613,[11] d_{20} 1.2251.[11]
Ph(s-Bu)P(O)Cl. Coupl. const.[652]
Ph(t-Bu)P(O)Cl. Coupl. const.[652]
Am$_2$P(O)Cl. I.[93] $b_{0.06}$ 97,[93] n_D^{20} 1.4655,[93] IR.[52]
(t-Am)$_2$P(O)Cl. X.[8]
4-MeC$_6$H$_4$(ClCOCH$_2$CMeH)P(O)Cl. XV.[352] $b_{0.025}$ 190,[352] $b_{0.015}$ 180,[352] n_D^{20} 1.5400,[352,353] d^{20} 1.2867.[352,353]
(ClC$_6$H$_4$)$_2$P(O)Cl. I.[107] XVII[108] (orientation of Cl not specified). b_2 205-224.[107]
(2-ClC$_6$H$_4$)$_2$P(O)Cl. X.[556] b. > 340.[556]
(3-ClC$_6$H$_4$)$_2$P(O)Cl. I.[121]
(4-ClC$_6$H$_4$)$_2$P(O)Cl. X.[556] b_{25} 280.[556]
(3-FC$_6$H$_4$)$_2$P(O)Cl. I.[607] $b_{0.02}$ 130-131,[607] n_D^{20} 1.5768,[607] Hammett const.,[607] ^{19}F.[607]
(4-FC$_6$H$_4$)$_2$P(O)Cl. I.[567,607] $b_{0.08}$ 145-147,[567] $b_{0.02}$ 130-131,[607] n_D^{20} 1.5726,[567] 1.5768,[607] Hammett const.,[567,607] ^{19}F.[567,607]
(4-NO$_2$C$_6$H$_4$)$_2$P(O)Cl. X.[556] Oil.[556]
Ph(3-ClC$_6$H$_4$)P(O)Cl. I.[121]

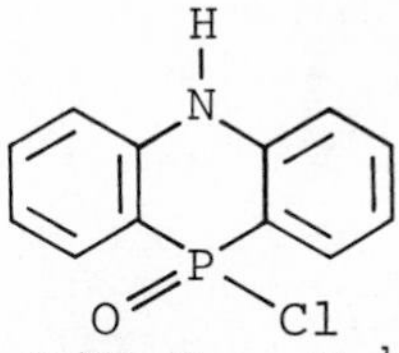 . I.[246] m. 290.[246]

$Ph_2P(O)Cl$. I.[107,111,217,392,519,551] II.[509] III.[446] VIII.[285] X.[7,159,160,484,505,614,666] XVII.[271] b_{16} 222,[484,506] b_{13} 215-218,[551] b_{10} 227-230,[392] b_1 160-165,[134] 190-194,[285] $b_{0.6}$ 167-172,[107] $b_{0.55}$ 157-158,[271] $b_{0.3}$ 132-135,[7] $b_{0.25}$ 147,[519] $b_{0.15}$ 138-139,[271] $b_{0.1}$ 135,[159] 145,[509] $b_{0.07}$ 135-136,[271] $b_{0.03-0.06}$ 137-138,[48] $b_{0.01}$ 135-140,[111] m. 68-69,[134] n_D^{20} 1.6047,[519] $n_D^{24.3}$ 1.6083,[48] d_4^{20} 1.2772,[519] IR,[114,539,666] ^{31}P -42.7,[161,446,509,612] ^{31}P signal enhancement through dynamic polarization,[566] kinetics of hydrolysis,[134,519] conductivity in liq. HCl,[538] complex: $2Ph_2P(O)Cl \cdot TiCl_4$ [m. 168-170 (dec.), IR, ^{31}P -55.9].[46]

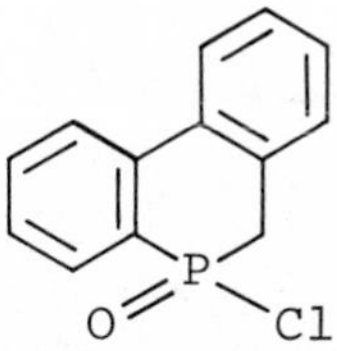 I.[436,437] m. 125-126.[436,437]

$Ph(C_6H_{11})P(O)Cl$. XII.[748] b_6 191-193,[748] n_D^{20} 1.5560,[748] d_{20} 1.2096.[748]

$(C_6H_{11})_2P(O)Cl$ (cyclohexyl-). XII.[516] m. 108-109.[516]

$(BuOCH{=}CH)_2P(O)Cl$. XVI.[147] b_3 182-183,[147] n_D^{20} 1.4988,[147] d^{20} 1.1295.[147]

$Ph(3\text{-}MeC_6H_4)P(O)Cl$. I.[172] $b_{0.2}$ 155-158.[172]

$Ph(4\text{-}MeC_6H_4)P(O)Cl$. XVI.[496] b_{25} 270-280.[496]

$Ph(3\text{-}MeOC_6H_4)P(O)Cl$. I.[273] II.[461] $b_{0.6}$ 175-178,[273] $b_{0.25}$ 171-174.[461]

$Ph(3,5\text{-}Me_2C_6H_3)P(O)Cl$. I.[172] $b_{0.1}$ 153-155.[172]

$(MeC_6H_4)_2P(O)Cl$. VII[107] (orientation of Me group not specified).

$(2\text{-}MeC_6H_4)_2P(O)Cl$. X.[556] XVI.[496] b. > 370.[496,556]

$(4\text{-}MeC_6H_4)_2P(O)Cl$. X.[556] XVI.[496] b. > 360.[496,556]

$(4\text{-}MeOC_6H_4)_2P(O)Cl$. I.[679] $b_{0.05}$ 216,[679] m. 59-63,[679] IR.[679]

$(2,4,5\text{-}Me_3C_6H_2)PhP(O)Cl$. XVI.[492,496,505] b_{10} 210-215,[492,496,505]

$(PhCCl{=}CH)_2P(O)Cl$. XVII.[150] m. 118-120.[150]

$(PhCH{=}CH)_2P(O)Cl$. XVII.[150,416] m. 79-81.[150]

$(C_8H_{17})_2P(O)Cl$. I.[725]

$Ph(C_{13}H_{11})P(O)Cl$ [(4-vinylnaphthyl-1-)methyl-]. II.[1]

$(1\text{-}C_{10}H_7)_2P(O)Cl$ (1-naphthyl-). X.[556] Undistillable oil.[556]

$(PhCH{=}CHCH{=}CH)_2P(O)Cl$. XVII.[149] m. 114-116.[149]
$Ph(Ph_3C)P(O)Cl$. XIII.[76] XIV.[76] m. 208-210,[76] IR.[76]
$[CH_2P(O)Cl(Ph)]_2$. I.[471] m. 179-181.[471]

Phosphinyl Bromides

$(CF_3)_2P(O)Br$. XXIII.[215] b. 78.3 (est.),[215] m. -35.5,[215] $\log P_{mm} = 5.0476 + 1.75 \cdot \log T - 0.003833 \cdot T - 1854/T$,[215] Trouton const. 21.4,[215] IR.[215]
$(BrCH_2)_2P(O)Br$. I.[143] $b_{0.05}$ 127-132,[143] n_D^{19} 1.6040.[143]
$Me(BrCH_2)P(O)Br$. XIV.[448] $b_{0.1}$ 97-101,[448] m. 35-36,[448] ^{31}P -52.1.[448]
$Et(BrCH_2)P(O)Br$. XIV.[706]
$C_4H_6P(O)Br$ (1-oxo-1-bromo-phospholene-3). XIX.[24,28] $b_{0.53}$ 72-74,[24,28] m. 46-48,[24,28] n_D^{20} 1.5593,[24,28] d_4^{20} 1.7037,[24,28] ^1H (double resonance).[24]
$C_5H_8P(O)Br$ (1-oxo-1-bromo-3-methyl-phospholene-2). XIX.[24] XXI.[24] $b_{0.2}$ 123-126,[24] n_D^{20} 1.5682,[24] d_4^{20} 1.6216.[24]
$C_5H_8P(O)Br$ (1-oxo-1-bromo-3-methyl-phospholene-3). XIX.[28] $b_{0.02}$ 123-125,[28] n_D^{20} 1.5682,[28] d_4^{20} 1.6216.[28]
$C_6H_{10}P(O)Br$ (1-oxo-1-bromo-3,4-dimethyl-phospholene-3). XIX.[31] $b_{0.55}$ 120-104,[28,31] m. 85-87,[28,31] ^1H.[31]
$(t\text{-}Bu)_2P(O)Br$. X.[8,673] m. 98-99,[8,673] ^1H,[8] mass spect.[8]
$Me(4\text{-}i\text{-}PrC_6H_4CH_2)P(O)Br$. I.[1]

Phosphinyl Pseudohalides

$Me_2P(O)CN$. IX.[679] XVI.[195] b. 128-130,[195] $b_{1.4}$ 74,[679] m. 40-41,[195] 53-56.[679]
$Ph_2P(O)CN$. IX.[316] $b_{0.15}$ 131-134.[316]
$Me_2P(O)NCO$. IX.[126] $b_{0.02}$ 58-60.[126] n_D^{20} 1.4610,[126] d^{20} 1.2101.[126]
$Et_2P(O)NCO$. IX.[126,679] b_1 70-72,[679] $b_{0.06}$ 56-58,[126] n_D^{20} 1.4555,[126] d^{20} 1.1200.[126]
$(i\text{-}Pr)_2P(O)NCO$. IX.[126] $b_{0.02}$ 54-56,[126] n_D^{20} 1.4605,[126] d^{20} 1.0551,[126] IR.[124]
$Bu_2P(O)NCO$. IX.[126] $b_{0.02}$ 84-85,[126] n_D^{20} 1.4520,[126] d^{20} 1.0149.[126]
$Ph_2P(O)NCO$. IX.[562,563,701,703] XXIII.[123,125,678] b_5 184-186,[125] $b_{0.5}$ 160,[678] $b_{0.3}$ 155,[678] $b_{0.15}$ 163-166,[562] 163-165,[563] $b_{0.01}$ 115,[703] 155-157,[701] m. 54-55,[701] 52-54,[678] 50-52,[703] 50-55,[562,563] IR,[562,563,703] polymerization.[562,563]
$(4\text{-}MeC_6H_4)_2P(O)NCO$. IX.[679] $b_{1.5}$ 188-190,[679] m. 45-48.[679]
$(4\text{-}MeOC_6H_4)_2P(O)NCO$. IX.[679] $b_{0.3}$ 270-275 (dec.).[679]
$Me_2P(O)NCS$. IX.[679] $b_{0.9}$ 91-92,[679] $b_{0.1}$ 73-74,[679] m. 50-52.5.[679]
$Et_2P(O)NCS$. IX.[679] $b_{0.7}$ 88-90.[679]
$Ph_2P(O)NCS$. IX.[679] $b_{0.05}$ 165-170,[679] $b_{0.008}$ 156-162,[679] m. 34.5-36.5,[679] IR.[679]
$Et_2P(O)N_3$. IX.[154] b_8 97,[154] n_D^{20} 1.4679,[154] d^{20} 1.1116.[154]
$Me(Ph)P(O)N_3$. IX.[42a] $b_{0.3\mu}$ 65,[42a] IR.[42a]

$(4-ClC_6H_4)_2P(O)N_3$. IX.[42]
$Ph_2P(O)N_3$. IX.[42,532,534,717] $b_{0.05}$ 137-140,[42] m. 71-72 (dec.),[532] molecular dist. at 0.1 μ,[49] dark yellow oil.[533,717]
$(MeC_6H_4)_2P(O)N_3$. IX[42] (orientation of Me group not specified).

Thionophosphinyl Fluorides

$Me_2P(S)F$. IX.[615,624] b_{20} 60.[624] b_{25} 63.5-64.5 [615], $n_D^{23.9}$ 1.4875 [615].
$Et_2P(S)F$. IX.[376 624] XVIII.[381] b_{19} 72.5-73,[376] b_{18} 71-73,[381] b_{12} 75,[624] n_D^{18} 1.4815,[381] n_D^{20} 1.4822,[376] d_{18} 1.0829,[381] d_4^{20} 1.0838,[376] ^{31}P -102.0.[612]
$Me(4-ClC_6H_4)P(S)F$. IX.[624] b_1 92.[624]
$Me(Ph)P(S)F$. IX.[624] b_1 68.[624]
$Me(2-Cl-4-MeC_6H_3)P(S)F$. IX.[624] b_1 100.[624]
$Me(4-MeSC_6H_4)P(S)F$. IX.[624] b_1 134.[624]
$(4-ClC_6H_4)_2P(S)F$. IX.[385] $b_{0.02}$ 158-159.[385]
$(4-MeC_6H_4)_2P(S)F$. IX.[385] $b_{0.1}$ 170-173.[385]
$Ph_2P(S)F$. IX.[615] $b_{1.5}$ 168-172 [615]. n_D^{20} 1.6250 [615].

Thionophosphinyl Chlorides

$(CF_3)_2P(S)Cl$. VIII.[198] XI.[198] XVIII.[198] b. 61,[198] m. -21.7 to -21.5,[198] $\log P_{mm} = 5.6678 + 1.75\cdot\log T - 0.0050\cdot T - 1849/T$,[198] Trouton const. 21.3,[198] IR.[198]
$Me(CCl_3)P(S)Cl$. XVIII.[131,340] b_2 100,[131,340] m. 90.[131,340]
$(ClCH_2)_2P(S)Cl$. VI.[187,536] XII.[510] b_{10} 104-106,[536] b_6 92.5-93.5,[187] n_D^{20} 1.5890,[536] 1.5872,[187] d_4^{20} 1.5580,[536] 1.5483,[187] ^{31}P -86.6.[510]
$Me(ClCH_2)P(S)Cl$. VI.[222] b_{3-4} 68-70,[222] n_D^{25} 1.568,[222] ^{31}P -85.0,[511] -85 ± 1.[222]
$Me_2P(S)Cl$. III.[446,454] V.[102,200,442,557,629] VI.[474a] VIII.[39,224] XI.[382] XII.[37,38,223] XVIII.[303,341] XXIII.[377] b_{50} 100-103,[223] b_{16} 82.5-83.5,[341] 82-83,[303,382] 107-107.5,[474a] b_{15} 72,[200] 76-78,[557] b_{12} 73.5,[102,629] 72-75,[442] b_{10} 58-61,[38,224] $b_{9.5}$ 65-67,[446] m. 22.5,[102,629] 21,[523] 25,[557] 24-25,[474a] n_D^{20} 1.5460,[446] 1.544,[61,341] 1.542,[303,382] 1.5451,[442] n_D^{25} 1.5413,[523] d_4^{20} 1.250,[61] 1.2560,[341] d_{20} 1.2651,[303] 1.2624,[303] 1.2625,[382] d_4^{25} 1.2278,[523] IR,[140,200] Raman,[140,200] dipole moment 4.01,[61] ^{31}P -87.3,[446,511] Cl (NQR),[523,632] kinetics of hydrolysis.[523]
$Et(CCl_3)P(S)Cl$. XVIII.[131,340] b_2 110,[131,340] n_D^{20} 1.5410,[131,340] d_4^{20} 1.6524.[131,340]
$Et(ClCH_2)P(S)Cl$. ^{31}P -95.6.[511]
$Me(Et)P(S)Cl$. III.[446,449] V.[442,629] XI.[382,556] XVIII.[303] b_{15} 96-99,[556] b_{12} 73.5,[629] b_{11} 90-91,[303,382,523] b_{10} 85-86,[442] 85-87,[449] n_D^{20} 1.5359,[442] 1.5379,[449] 1.5360,[303,523] 1.531,[382] d_4^{20} 1.1992,[523] d_{20} 1.1817,[303] 1.1815,[382] ^{31}P -97,[449] -98.0,[446,511] Cl (NQR),[523]

kinetics of hydrolysis.[523]

$(CH_2{=}CH)_2P(S)Cl$. III.[454] b_{740} 220-221.[454]

$Et_2P(S)Cl$. I.[65,474] III.[446,449,454] V.[102,401,537] XI.[23,65,382,397,556] XVIII.[303] b_{740} 225-227,[446] 195-200,[454] b_{15} 117-120,[556] b_{14} 99-100,[401] b_{11} 99-100,[303] 94-95,[523] b_9 94-95,[65] b_8 90-90.5,[23] b_7 84-86,[397] b_6 93-94,[382] b_5 86.5,[449] b_4 60-61,[474] b_1 70,[102] 80-85,[537] n_D^{20} 1.5326,[449] 1.5325,[303,397] 1.5289,[523] 1.5294,[23] 1.5286,[382] 1.5281,[474] n_D^{25} 1.5292,[65] d_4^{20} 1.1544,[397] 1.1427,[523] d_{20} 1.1438,[23] 1.1542,[303] 1.1549,[382] 1.1435,[474] ^{31}P -108,[449] -108.3,[446,454,511] Cl (NQR),[523,632] kinetics of hydrolysis,[523] kinetics of solvolysis.[523]

Me(Pr)P(S)Cl. V.[442] b_{10} 105-108,[442] b_3 71-75,[523] n_D^{20} 1.5167,[442] 1.5270,[523] d_4^{20} 1.1464,[523] ^{31}P -95.3,[511] kinetics of hydrolysis.[523]

$C_5H_8P(S)Cl$ (1-thio-1-chloro-2-methyl-phospholene-3). XXI.[374] b_2 103-105,[374] n_D^{20} 1.5695,[374] d_4^{20} 1.2306.[374]

$C_5H_8P(S)Cl$ (1-thio-1-chloro-3-methyl-phospholene-3). XXI.[374] b_2 118-120,[374] m. 40.4-41.[374]

Me(Bu)P(S)Cl. V.[442] b_{10} 110-112,[442] n_D^{20} 1.5132,[442] ^{31}P -96.5.[511]

$(CF_3CH_2CH_2)_2P(S)Cl$. I.[407]

$C_6H_{10}P(S)Cl$ (1-thio-1-chloro-3,4-dimethyl-phospholene-2). IR,[722] UV,[722] ^1H.[722]

$Pr_2P(S)Cl$. I.[474] V.[102,401] XI.[556] b_{15} 131-135,[556] $b_{11.5}$ 120-121,[401] $b_{2.5}$ 73.5-74,[474] b_1 96,[102] n_D^{20} 1.5182,[474] d_{20} 1.0859,[474] Cl (NQR).[523]

$(i\text{-}Pr)_2P(S)Cl$. I.[474] XII.[518] $b_{2.5}$ 58.5-59,[474] $b_{0.35}$ 38,[518] n_D^{20} 1.5232,[474] d_{20} 1.0978,[474] Cl (NQR).[523]

$Ph(CCl_3)P(S)Cl$. XVIII.[131,340] b_2 125,[131,340] m. 100.[131,340]

$Ph(ClCH_2)P(S)Cl$. III.[446] XII.[510] ^{31}P -75.7,[510] -77.0.[446,511]

Me(Ph)P(S)Cl. III.[446] IV.[450,455] V.[422,537,610] XI.[556] b_{15} 140-150,[556] b_3 124-126,[537] b_1 106-107,[610] $b_{0.16}$ 99,[523] $b_{0.03}$ 93-94,[450,455] $b_{0.01}$ 91-94,[442] n_D^{20} 1.6175,[450] 1.6117,[442] 1.6169,[523] d_4^{20} 1.2541,[523] ^{31}P -81.0,[446,450,511] Cl (NQR),[523] kinetics of hydrolysis.[523]

$Me(C_6H_{11})P(S)Cl$. V.[629] VI.[322] b_3 110-112,[629] b_2 89-90,[322] n_D^{20} 1.5570,[322] d_{20} 1.1713.[322]

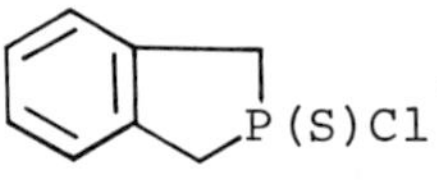

. XI.[43] m. 125-126.[43]

Et(Ph)P(S)Cl. I[557] [mixture with Et(Ph)P(O)Cl]. III.[446] V.[557,610,629] XI.[209,210,556] b_{15} 175-190,[556] b_2 126,[610,629] $b_{0.5}$ 110,[209] $b_{0.2}$ 102-108,[557] $b_{0.1}$ 100-103,[557] ^{31}P -93.7,[446,511] -94.1.[557]

$Bu_2P(S)Cl$. I.[474] V.[102] XI.[556] XII.[518] b_{15} 155-159,[556]

b_5 120,[518] b_2 84-85,[474] b_1 116,[102] n_D^{20} 1.5160,[474] d_{20} 1.0570.[474]

$(i\text{-}Bu)_2P(S)Cl$. III.[449] $b_{0.15}$ 69-72,[449] n_D^{20} 1.5108,[449] ^{31}P -100.[449]

$(ClC_6H_4)_2P(S)Cl$. I.[109] IV[455] (orientation of Cl not specified). $b_{0.8}$ 220,[109] $b_{0.1}$ 180-210,[455] solid.[109]

$(2\text{-}ClC_6H_4)_2P(S)Cl$. XI.[556] m. 41-42.[556]

$(4\text{-}ClC_6H_4)_2P(S)Cl$. IV[450] (mixture with 2-isomer). XI.[556] b_{15} 290-300,[556] $b_{0.1}$ 180-210,[450] m. 58-60,[556] 100-101,[450] ^{31}P -75.1.[450]

$(FC_6H_4)_2P(S)Cl$. IV[455] (orientation of F not specified). $b_{0.08}$ 122-127.[455]

$(3\text{-}FC_6H_4)_2P(S)Cl$. VI.[607] $b_{0.05}$ 135,[607] m. 53-55,[607] Hammett const.,[607] ^{19}F.[607]

$(4\text{-}FC_6H_4)_2P(S)Cl$. IV.[450] VI.[567] $b_{0.02}$ 140,[567] $b_{0.005}$ 100-110,[450] m. 57-58,[450] 52-54,[567] Hammet const.,[567] ^{31}P -77.7,[450] ^{19}F.[567]

$(4\text{-}NO_2C_6H_4)_2P(S)Cl$. XI.[556] m. 172-175.[556]

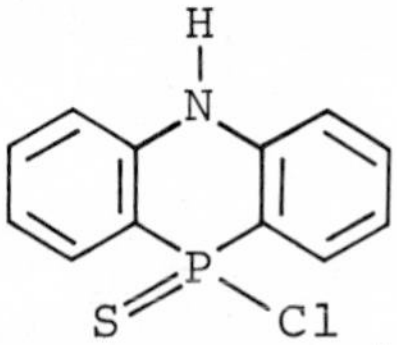

. XXIII.[479a]

$Ph_2P(S)Cl$. I.[109,271] III.[446,454] IV.[450,455] VI.[473] X.[159,160] XI.[45,537,556] b_{15} 275-280,[556] b_7 107-109,[473] b_4 200-210,[271] b_3 190-210,[109] $b_{1.5}$ 186-200,[537] b_1 195-199,[159] $b_{0.55}$ 152-155,[318] $b_{0.5}$ 160-163,[446,454] $b_{0.3}$ 155-160,[271] $b_{0.2}$ 155-159,[45] $b_{0.005}$ 147-148,[450] n_D^{20} 1.6618,[450] 1.6628,[45] 1.6563,[318] d_{20} 1.0845,[473] IR,[744] ^{31}P -79.5,[161,511,612] -79.3,[450] -79.6,[446,454] ^{31}P signal enhancement through dynamic polarization,[566] complexes: $Ph_2P(S)Cl \cdot AlCl_3$ (^{31}P -90.4, conductivity);[453] $Ph_2P(S)Cl \cdot 2AlCl_3$ (^{31}P -90.4, conductivity).[453]

$Ph_2{}^{32}P(S)Cl$.[668]

$Ph(4\text{-}MeC_6H_4)P(S)Cl$. XI.[556] b_{15} 265-270.[556]

$Bz_2P(S)Cl$. I.[474] XI.[43] b_4 160-165,[43] m. 129-131,[43] 130.5-131.5.[474]

$(MeC_6H_4)_2P(S)Cl$. IV[455] (orientation of Me group not specified). $b_{0.2}$ 180-235,[455] m. 95.[455]

$(2\text{-}MeC_6H_4)_2P(S)Cl$. XI.[556] b_{15} 225-130,[556] m. 18-19.[556]

$(4\text{-}MeC_6H_4)_2P(S)Cl$. I.[109] IV.[450] XVIII.[496] $b_{0.2}$ 180-235,[450] m. 95,[450] 96,[496] ^{31}P -79.7.[450]

$(4\text{-}Me_2NC_6H_4)_2P(S)Cl$. XI.[556] b_{15} 270-285.[556]

$(1\text{-}C_{10}H_7)_2P(S)Cl$ (1-naphthyl-). XI.[556] b_{15} 290-310,[556] m. 51-53.[556]

$(C_6H_{13}C_6H_3Me)_2P(S)Cl$. IV.[450] ^{31}P -75.2.[450]

Thionophosphinyl Bromides

$Me_2P(S)Br$. V.[102,258,442,537] XI.[447] b_{718} 205-207,[442] b_{15} 76-78,[537] b_{14} 87-88,[258] b_{13} 90,[102] b_9 82-85,[447] m. 32-34,[442,447] 34,[102] n_D^{35} 1.5482,[442] 1.5815,[447] IR,[140] Raman,[140] ^{31}P -63.2.[511]
Me(Et)P(S)Br. V.[442,444] b_{10} 92-94,[442] b_3 75-77,[442] n_D^{20} 1.5783,[442] ^{31}P -85.0.[444,511]
$Et_2P(S)Br$. III.[449] V.[400,401,444] $b_{12.5}$ 112.5,[400,401] b_{10} 92-93,[449] ^{31}P -96,[444] -98.2,[449] -98.3.[511]
Me(Pr)P(S)Br. V.[442] b_{10} 107-110,[442] n_D^{20} 1.5639.[442]
Me(Bu)P(S)Br. V.[442] b_{10} 118-122,[442] n_D^{20} 1.5541.[442]
Me(t-Bu)P(S)Br. ^{31}P -102.9,[386] INDOR study,[386] ^1H.[386]
$Pr_2P(S)Br$. V.[400,401,537] b_{12} 129.5-130.5,[400,401] b_7 108-110,[537] $d_4^{22.5}$ 1.304.[401]
Me(Ph)P(S)Br. V.[442] $b_{0.5}$ 113-116,[442] n_D^{20} 1.6512,[442] ^{31}P -61.0.[511]
Me(Bz)P(S)Br. V.[442] $b_{0.05}$ 150-160,[442] m. 48-50,[442] ^{31}P -73.7.[511]
$Bu_2P(S)Br$. V.[400,401] $b_{12.5}$ 159-160,[400,401] d_4^{19} 1.243,[401] ^{31}P -91.2.[511]
$(i\text{-}Bu)_2P(S)Br$. III.[449] $b_{0.05}$ 81-84,[449] n_D^{20} 1.5531,[449] ^{31}P -88.5.[449]
$Ph_2P(S)Br$. IV.[450] $b_{0.01}$ 155,[450] n_D^{20} 1.6939,[450] ^{31}P -64.4.[450]
$(4\text{-}MeC_6H_4)_2P(S)Br$. I.[109]

Thionophosphinyl Iodides

$(CF_3)_2P(S)I$. VIII.[198] m. 30 (dec.).[198]

Thionophosphinyl Pseudohalides

$Ph_2P(S)CN$. IX.[318] XI.[317,318] $b_{8.6}$ 213-214,[318] $b_{0.25}$ 149-151,[317,318] m. 50.0-50.2,[317] n_D^{20} 1.6393,[318] 1.6114,[317] 1.6414,[318] d^{25} 1.2033,[317] d_4^{25} 1.2023,[318] IR.[318]
$(C_6F_5)_2P(S)NCS$. XI.[154a] ^{31}P +4.2.[154a]
$Ph_2P(S)NCS$. IX.[613] $b_{0.01}$ 176-178,[613] m. 48.[613]
$Me_2P(S)N_3$. IX.[622] m. 68.[622]
$Et_2P(S)N_3$. IX.[622] Oil.[622]
$Ph_2P(S)N_3$. IX.[42,717]

Selenophosphinyl Halide

$Et_2P(Se)Cl$. XI.[464] b_{15} 113-114,[464] n_D^{25} 1.5677.[464]

(received October 27, 1971)

REFERENCES

1. Abramo, J. G., E. C. Chapin, and A. Y. Garner, U.S.

Patent 3,143,569 (1964); C.A. 61, 13346h (1964).
2. Akerfeldt, S., and L. Fagerlind, J. Med. Chem., 10, 115 (1967).
3. Aksnes, G., and S. I. Snaprud, Acta Chem. Scand., 15, 457 (1961).
4. Aleksandrov, V. N., and V. I. Emel'yanov, Zh. Obshch. Khim., 37, 2714 (1967); C.A. 69, 2982v (1968).
5. Aleksandrova, I. A., and G. M. Vinokurova, USSR Patent 226,600 (1968); C.A. 70, 47593f (1969).
6. Aliev, M. I., F. A. Mamedov, S. Z. Israfilova, and D. R. Israelyan, Azerb. Khim. Zh., 1968, 50; C.A. 70, 68476k (1969).
7. Amonoo-Neizer, E. H., S. K. Ray, R. A. Shaw, and B. C. Smith, J. Chem. Soc., 1965, 4296.
8. Angstadt, H. P., J. Amer. Chem. Soc., 86, 5040 (1964).
9. Anisimov, K. N., Izv. Akad. Nauk SSSR, Otd. Khim. Nauk, 1954, 803; C.A. 49, 13074e (1955).
10. Anisimov, K. N., and G. K. Fedorova, Dopov. Akad. Nauk Ukr. RSR, 1960, 1245; C.A. 56, 499i (1962).
11. Anisimov, K. N., and N. E. Kolobova, Izv. Akad. Nauk SSSR, Otd. Khim. Nauk, 1962, 444; C.A. 57, 12529c (1962).
12. Anisimov, K. N., and N. E. Kolobova, Izv. Akad. Nauk SSSR, Otd. Khim. Nauk, 1956, 923; C.A. 51, 4933g (1957).
13. Anisimov, K. N., N. E. Kolobova, and A. N. Nesmeyanov, Izv. Akad. Nauk SSSR, Otd. Khim. Nauk, 1954, 796; C.A. 49, 13074a (1955).
14. Anisimov, K. N., N. E. Kolobova, and A. N. Nesmeyanov, Izv. Akad. Nauk SSSR, Otd. Khim. Nauk, 1954, 799; C.A. 49, 13074c (1955).
15. Anisimov, K. N., and Kopylova, Izv. Akad. Nauk SSSR, Otd. Khim. Nauk, 1961, 277; C.A. 55, 18562i (1961).
16. Anisimov, K. N., G. M. Kunitskaya, and N. A. Slovokhotova, Izv. Akad. Nauk SSSR, Otd. Khim. Nauk, 1961, 64; C.A. 55, 18562e (1961).
17. Anisimov, K. N., and A. N. Nesmeyanov, Izv. Akad. Nauk SSSR, Otd. Khim. Nauk, 1954, 610; C.A. 49, 11540i (1955).
18. Anisimov, K. N., and B. V. Raisbaum, Izv. Akad. Nauk SSSR, Otd. Khim. Nauk, 1958, 1208; C.A. 53, 4114 (1959).
19. Anschütz, L., E. Klein, and G. Cermak, Ber., 77, 726 (1944).
20. Arbuzov, A. E., and B. A. Arbuzov, J. Russ. Phys.-Chem. Soc., 61, 217 (1929); C.A. 23, 3921 (1929).
21. Arbuzov, A. E., and K. V. Nikonorov, Zh. Obshch. Khim., 17, 2139 (1947); C.A. 42, 4546b (1948).
22. Arbuzov, B. A., and N. I. Rizpolozhenskii, Izv. Akad. Nauk SSSR, Otd. Khim. Nauk, 1952, 854; C.A. 47, 9903b (1953).

23. Arbuzov, B. A., and N. I. Rizpolozhenskii, Dokl. Akad. Nauk SSSR, 89, 291 (1953); C.A. 48, 7540g (1954).
24. Arbuzov, B. A., Yu. Yu. Samitov, A. O. Vizel, and T. V. Zykova, Dokl. Akad. Nauk SSSR, 159, 1062 (1964); C.A. 62, 6371a (1965).
25. Arbuzov, B. A., V. S. Vinogradova, and N. A. Polezhaeva, Dokl. Akad. Nauk SSSR, 137, 855 (1961); C.A. 56, 3507c (1962).
26. Arbuzov, B. A., V. S. Vinogradova, and N. A. Polezhaeva, Izv. Akad. Nauk SSSR, Otd. Khim. Nauk, 1962, 71; C.A. 57, 13791a (1962).
27. Arbuzov, B. A., V. S. Vinogradova, N. A. Polezhaeva, and A. K. Shamsutdinova, Izv. Akad. Nauk SSSR, Otd. Khim. Nauk, 1963, 675; C.A. 59, 11551f (1963).
28. Arbuzov, B. A., and A. O. Vizel, Dokl. Akad. Nauk SSSR, 158, 1105 (1964); C.A. 62, 2791g (1965).
29. Arbuzov, B. A., A. O. Vizel, A. O. Raevskii, Yu. F. Tarenko, and D. F. Fazliev, Izv. Akad. Nauk SSSR, Ser. Khim., 1968, 1882; C.A. 70, 20165f (1969).
30. Arbuzov, B. A., A. O. Vizel, O. A. Raevskii, Yu. F. Tarenko, and L. E. Petrova, Izv. Akad. Nauk SSSR, Ser. Khim., 1969, 460; C.A. 70, 115250u (1969).
31. Arbuzov, B. A., A. O. Vizel, Yu. Yu. Samitov, and K. M. Ivanovskaya, Dokl. Akad. Nauk SSSR, 159, 582 (1964); C.A. 62, 6505d (1965).
32. Arbuzov, B. A., A. O. Vizel, Yu. Yu. Samitov, and Y. F. Tarenko, Izv. Akad. Nauk SSSR, Ser. Khim., 1967, 672; C.A. 67, 100193s (1967).
33. Baaz, M., V. Gutmann, L. Huebner, F. Mairinger, and T. S. West, Z. Anorg. Allg. Chem., 311, 302 (1961).
34. Baaz, M., V. Gutmann, and L. Huebner, Monatsh. Chem., 92, 272 (1961).
35. Baaz, M., V. Gutmann, and J. R. Masaguer, Monatsh. Chem., 92, 582 (1961).
36. Bai, L. I., A. Ya. Yakubovich, and L. I. Muler, Zh. Obshch. Khim., 34, 3609 (1964); C.A. 62, 9168f (1965).
37. Baker, J. W., L. C. D. Groenweghe, and R. E. Stenseth, U.S. Patent 3,457,305 (1969); C.A. 71, 81527t (1969).
38. Baker, J. W., and R. E. Stenseth, U.S. Patent 3,429,-916 (1969); C.A. 70, 106648c (1969).
39. Baker, J. W., and R. E. Stenseth, U.S. Patent 3,457,-306 (1969); C.A. 71, 91648x (1969).
40. Baldwin, R. A., J. Org. Chem., 30, 3866 (1965).
41. Baldwin, R. A., and R. M. Washburn, J. Org. Chem., 30, 3860 (1965).
42. Baldwin, R. A., and R. M. Washburn, J. Amer. Chem. Soc., 83, 4466 (1961).
42a. Baldwin, R. A., C. O. Wilson, and R. I. Wagner, J. Org. Chem., 32, 2172 (1967).
42b. Bannard, R. A. B., J. R. Gilpin, G. R. Vavasour, and A. F. McKay, J. Can. Chem., 31, 976 (1953).

43. Baranov, Yu. I., O. F. Filippov, M. A. Sokol'skii, S. L. Varshavskii, and M. I. Kabachnik, USSR Patent 221,695 (1968); C.A. 69, 106875d (1968).
44. Batuev, M. I., L. I. Shmonina, A. D. Matveeva, and M. F. Shostakovskii, Izv. Akad. Nauk SSSR, Otd. Khim. Nauk, 1961, 513; C.A. 55, 23318g (1961).
45. Becke-Goehring, M., W. Haubold, and H. P. Latscha, Z. Anorg. Allg. Chem., 333, 120 (1964).
46. Becke-Goehring, M., and A. Slawisch, Z. Anorg. Allg. Chem., 346, 295 (1966).
47. Bel'skii, V. E., I. V. Berezovskaya, B. E. Ivanov, A. R. Panteleeva, V. G. Trutnev, and I. M. Shermergorn, Dokl. Akad. Nauk SSSR, 171, 613 (1966); C.A. 66, 46075u (1967).
48. Berlin, K. D., J. G. Morgan, M. E. Peterson, and W. C. Pivonka, J. Org. Chem., 34, 1266 (1969).
49. Berlin, K. D., R. Ranganathan, and H. Haberlein, J. Heterocycl. Chem., 5, 813 (1968).
50. Biddle, P., J. Kennedy, and J. L. Willans, Chem. Ind. (London), 1957, 1481.
51. BIOS Final Report, 1808, 19, 20 (1948).
52. Blasse, R. G., Rec. Trav. Chim. Pays-Bas, 84, 267 (1965).
53. Bliznyuk, N. K., A. F. Kolomiets, and P. S. Khokhlov, USSR Patent 172,742 (1965); C.A. 64, 757a (1966).
54. Bliznyuk, N. K., Z. N. Kvasha, and A. F. Kolomiets, Zh. Obshch. Khim., 37, 1811 (1967); C.A. 69, 2985y (1968).
55. Bliznyuk, N. K., Z. N. Kvasha, and S. L. Varshavskii, USSR Patent 230,147 (1968); C.A. 71, 3471f (1969).
56. Bliznyuk, N. K., Z. N. Kvasha, S. L. Varshavskii, and G. A. Madzhara, USSR Patent 221,696 (1968); C.A. 69, 106874c (1968).
57. Bliznyuk, N. K., G. S. Levskaya, and S. L. Varshavskii, USSR Patent 237,890 (1969); C.A. 71, 61541q (1969).
58. Bliznyuk, N. K., L. D. Protasova, Z. N. Kvasha, and S. L. Varshavskii, USSR Patent 221,697 (1968); C.A. 69, 106877f (1968).
59. Bloch, H. S., U.S. Patent 2,994,715 (1961); C.A. 56, 3516e (1962).
60. Bodnarchuk, N. D., V. V. Malovik, and G. I. Derkach, Zh. Obshch. Khim., 39, 168 (1969); C.A. 70, 96857d (1969).
61. Bogolyubov, G. M., K. S. Mingaleva, and A. A. Petrov, Zh. Obshch. Khim., 35, 1566 (1965); C.A. 63, 17860f (1965).
62. Bogolyubov, G. M., and A. A. Petrov, Zh. Obshch. Khim., 37, 229 (1967); C.A. 66, 105029g (1967).
63. Bogonostseva, N. P., and T. E. Filippova, Zh. Obshch. Khim., 33, 1363 (1963); C.A. 59, 10110d (1963).
64. Bolle, J., J. C. Mileo, and L. Nicolas, Compt. Rend.,

261, 1852 (1965).
65. Borecki, C., J. Michalski, and S. Musierowicz, J. Chem. Soc., 1958, 4081.
66. Boter, H. L., and A. J. J. Ooms, Rec. Trav. Chim. Pays-Bas, 85, 21 (1966).
67. Bothner-By, A. A., and R. H. Cox, J. Phys. Chem., 73, 1830 (1969).
68. Bott, R. W., B. F. Dowden, and C. Eaborn, J. Organometal. Chem., 4, 291 (1965).
69. Bott, R. W., B. F. Dowden, and C. Eaborn, J. Chem. Soc., 1965, 4994.
70. Bott, R. W., B. F. Dowden, and C. Eaborn, Int. Symp. Organosilicon Chem., Sci. Commun., Prague, 1965, 290; C.A. 65, 10606c (1966).
71. Boyd, D. R., and G. Chignell, J. Chem. Soc., 123, 813 (1923).
72. Boyd, D. R., and F. J. Smith, J. Chem. Soc., 125, 1477 (1924).
73. Boyd, D. R., and F. J. Smith, J. Chem. Soc., 1926, 2323.
74. Brace, N. O., J. Org. Chem., 26, 3197 (1961).
75. Briggeman, D. H., A. Longacre, and C. J. Smith, U.S. Patent 3,188,281 (1965); C.A. 63, 7044d (1965).
76. Brophy, J. J., K. L. Freeman, and M. J. Gallagher, J. Chem. Soc. (C), 1968, 2760.
77. Brown, B. B., L. J. Lutz, and C. J. Smith, U.S. Patent 3,179,696 (1965); C.A. 63, 632f (1965).
78. Bryant, P. J. R., A. H. Ford-Moore, B. J. Perry, A. W. H. Wardrop, and T. F. Watkins, J. Chem. Soc., 1960, 1553.
79. Bugerenko, E. F., E. A. Chernyshev, and A. D. Petrov, Dokl. Akad. Nauk SSSR, 143, 840 (1962); C.A. 57, 3474c (1962).
80. Bugerenko, E. F., E. A. Chernyshev, L. I. Shul'gina, and A. S. Petukhova, USSR Patent 229,516 (1968); C.A. 70, 58012s (1969).
81. Burg, A. B., and A. J. Sarkis, J. Amer. Chem. Soc., 87, 238 (1965).
82. Bystrov, V. F., A. A. Neimysheva, A. U. Stepanyants, and I. L. Knunyants, Dokl. Akad. Nauk SSSR, 156, 637 (1964); C.A. 61, 6548d (1964).
83. Cade, J. A., Tetrahedron, 2, 322 (1958).
84. Cade, J. A., J. Chem. Soc., 1959, 2266.
85. Carraher, C. E., and T. Brandt, Makromol. Chem., 123, 144 (1969).
86. Chantrell, P. G., C. A. Pearce, C. R. Toyer, and R. Twaits, J. Appl. Chem. (London), 15, 460 (1965).
87. Charrier, C., and M. P. Simonnin, Compt. Rend., C 265, 1347 (1967).
88. Chen, W.-C., Hua Hsueh Hsueh Pao, 31, 29 (1965); C.A. 63, 4128e (1965).

89. Cherbuliez, E., G. Weber, and J. Rabinowitz, Helv. Chim. Acta, 45, 2665 (1962).
90. Chernyshev, E. A., Izv. Akad. Nauk SSSR, Otd. Khim. Nauk, 1958, 96; C.A. 52, 11735e (1958).
91. Chernyshev, E. A., and E. F. Bugerenko, USSR Patent 148,049 (1962); C.A. 58, 9142e (1963).
92. Chernyshev, E. A., and A. D. Petrov, Dokl. Akad. Nauk SSSR, 105, 282 (1955); C.A. 50, 11283h (1956).
93. Christen, P. J., and L. M. van der Linde, Rec. Trav. Chim. Pays-Bas, 78, 543 (1959).
94. Christen, P. J., L. M. van der Linde, and F. N. Hooge, Rec. Trav. Chim. Pays-Bas, 78, 161 (1959).
95. Christol, C., and H. Christol, J. Chim. Phys., 62, 246 (1965).
96. Christol, H., and C. Marty, Compt. Rend., C 262, 1722 (1966).
97. Clay, J. P., J. Org. Chem., 16, 892 (1951).
98. Clay, J. P., U.S. Patent 2,744,132 (1956); C.A. 51, 1246c (1957).
99. Clayton, J. O., and W. L. Jensen, J. Amer. Chem. Soc., 70, 3880 (1948).
100. Coates, H., and R. P. Carter, British Patent 734,187 (1955); C.A. 50, 7123a (1956).
101. Coates, H., and R. P. Carter, U.S. Patent 2,853,515 (1958); C.A. 53, 7988i (1959).
102. Coelln, R., German Patent 1,054,453 (1959); C.A. 55, 6375b (1961).
103. Coelln, R., and G. Schrader, German Patent 1,056,606 (1959); C.A. 56, 11621g (1962).
104. Coelln, R., and G. Schrader, British Patent 917,085 (1963); C.A. 59, 5197g (1963).
105. Copenhaver, J. W., and J. Kwiatek, U.S. Patent 2,882,310 (1959); C.A. 53, 16964i (1959).
105a. Copenhaver, J. W., and C. W. Weber, U.S. Patents 2,882,305 and 2,882,307 (1959); C.A. 53, 16964f and 16964g (1959).
106. Corbridge, D. E. G., Topics Phosphorus Chem., 6, 235 (1968).
107. Craig, W. G., U.S. Patent 2,724,726 (1955); C.A. 50, 10129e (1956).
108. Craig, W. G., and W. A. Higgins, U.S. Patent 2,727,073 (1955); C.A. 50, 9445f (1956).
109. Craig, W. G., and W. A. Higgins, U.S. Patent 2,724,725 (1955); C.A. 50, 10129f (1956).
110. Cram, D. J., R. D. Trepka, and P. S. Janiak, J. Amer. Chem. Soc., 88, 2749 (1966).
111. Crofts, P. C., I. M. Downie, and K. Wiliamson, J. Chem. Soc., 1964, 1240.
112. Crofts, P. C., and I. S. Fox, J. Chem. Soc. (B), 1968, 1416.
112a. Crofts, P. C., and G. M. Kosolapoff, J. Amer. Chem.

Soc., 75, 3379 (1953).
112b. Crofts, P. C., and G. M. Kosolapoff, J. Amer. Chem. Soc., 75, 5738 (1953).
113. Crutchfield, M. M., C. H. Dungan, and J. R. Van Wazer, Topics Phosphorus Chem., 5, 1 (1967).
114. Daasch, L. W., and D. C. Smith, Anal. Chem., 23, 853 (1951).
115. Davankov, A. B., M. I. Kabachnik, V. V. Korshak, Yu. A. Leikin, R. F. Okhovetsker, and E. N. Tsvetkov, Zh. Obshch. Khim., 37, 1605 (1967); C.A. 67, 90884b (1967).
116. Dawson, T. P., and J. W. Armstrong, U.S. Patent 2,847,469 (1958); C.A. 53, 3058b (1959).
117. Dawson, T. P., and J. Chernack, U.S. Patent 2,929,843 (1960); C.A. 54, 15245i (1960).
118. Dawson, T. P., and K. C. Kennard, J. Org. Chem., 22, 1671 (1957).
119. Dehn, H., V. Gutmann, H. Kirch, and G. Schoeber, Monatsh. Chem., 93, 1348 (1962).
120. Dehn, H., V. Gutmann, and G. Schoeber, Monatsh. Chem., 94, 312 (1963).
121. Denham, J. M., and R. K. Ingham, J. Org. Chem., 23, 1298 (1958).
122. Derkach, G. I., Zh. M. Ivanova, and N. I. Liptuga, Khim. Org. Soedin. Fosfora, Akad. Nauk SSSR, Otd. Obshch. Tekh. Khim., 1967, 72; C.A. 69, 35212x (1968).
123. Derkach, G. I., and A. V. Kirsanov, USSR Patent 150,834 (1962); C.A. 59, 5198f (1963).
124. Derkach, G. I., and A. A. Kisilenko, Zh. Obshch. Khim., 34, 3060 (1964); C.A. 62, 2692d (1965).
125. Derkach, G. I., and L. I. Samarai, Zh. Obshch. Khim., 33, 1587 (1963); C.A. 59, 12669d (1963).
126. Derkach, G. I., and E. I. Slyusarenko, Zh. Obshch. Khim., 38, 1784 (1968); C.A. 70, 78082t (1969).
127. Derkach, G. I., E. I. Slyusarenko, V. V. Doroshenko, and B. Ya. Libman, Khim. Org. Soedin. Fosfora, Akad. Nauk SSSR, Otd. Obshch. Tekh. Khim., 1967, 64; C.A. 69, 10512g (1968).
128. Derkach, G. I., E. I. Slyusarenko, B. Ya. Libman, and N. I. Liptuga, Zh. Obshch. Khim., 35, 1881 (1965); C.A. 64, 2122a (1966).
129. Dietsche, W. H., Tetrahedron, 23, 3049 (1967).
130. Dmitrieva, L. E., S. Z. Ivin, and K. V. Karavanov, Khim. Org. Soedin. Fosfora, Akad. Nauk SSSR, Otd. Obshch. Tekh. Khim., 1967, 155; C.A. 68, 114709x (1968).
131. Dmitrieva, L. E., K. V. Karavanov, and S. Z. Ivin, USSR Patent 207,909 (1967); C.A. 69, 67531s (1968).
132. Dmitrieva, L. E., K. V. Karavanov, and S. Z. Ivin, Zh. Obshch. Khim., 38, 1547 (1968); C.A. 69, 59350s (1968).

133. Doak, G. O., and L. D. Freedman, J. Amer. Chem. Soc., 76, 1621 (1954).
134. Dostrovsky, I., and M. Halmann, J. Chem. Soc., 1953, 502.
135. Drake, L. R., and C. S. Marvel, J. Org. Chem., 2, 387 (1937).
136. Drozd, G. I., S. Z. Ivin, and M. A. Sokal'skii, Zh. Obshch. Khim., 39, 1177 (1969); C.A. 71, 50072m (1969).
137. Drozd, G. I., S. Z. Ivin, and O. G. Strukov, Zh. Obshch. Khim., 39, 1418 (1969); C.A. 71, 81475z (1969).
138. Durig, J. R., F. Block, and I. W. Levin, Spectrochim. Acta, 21, 1105 (1965).
139. Durig, J. R., B. R. Mitchell, J. S. DiYorio, and F. Block, J. Phys. Chem., 70, 3190 (1966).
140. Durig, J. R., D. W. Wertz, B. R. Mitchell, F. Block, and J. M. Greene, J. Phys. Chem., 71, 3815 (1967).
141. Dzhurinskaya, N. G., S. A. Mikhailyants, and V. P. Evdakov, Zh. Obshch. Khim., 37, 2278 (1967); C.A. 68, 87361u (1968).
142. Dzhurinskaya, N. G., S. A. Mikhailyants, and V. P. Evdakov, Zh. Obshch. Khim., 38, 1267 (1968); C.A. 69, 77380f (1968).
143. Edmundson, R. S., and E. W. Mitchell, J. Chem. Soc. (C), 1966, 1096.
144. Evdakov, V. P., N. G. Dzhurinskaya, and S. A. Mikhailyants, USSR Patent 197,585 (1967); C.A. 69, 19311y (1968).
145. Evtikhov, Z. L., N. A. Razumova, and A. A. Petrov, Dokl. Akad. Nauk SSSR, 177, 108 (1967); C.A. 69, 36218r (1968).
146. Fedorova, G. K., and A. V. Kirsanov, Zh. Obshch. Khim., 30, 4044 (1960); C.A. 55, 23401h (1961).
147. Fedorova, G. K., and A. V. Kirsanov, Zh. Obshch. Khim., 35, 1483 (1965); C.A. 63, 14901a (1965).
148. Fedorova, G. K., R. N. Ruban, and A. V. Kirsanov, Zh. Obshch. Khim., 39, 1471 (1969); C.A. 71, 113058j (1969).
149. Fedorova, G. K., and Ya. P. Shaturskii, Zh. Obshch. Khim., 36, 1262 (1966); C.A. 65, 15419b (1966).
150. Fedorova, G. K., Ya. P. Shaturskii, and A. V. Kirsanov, Probl. Organ. Sinteza, Akad. Nauk SSSR, Otd. Obshch. i Tekhn. Khim., 1965, 263; C.A. 64, 8228e (1966).
151. Fedorova, G. K., Ya. P. Shaturskii, L. S. Moskalevskaya, Yu. S. Grushin, and A. V. Kirsanov, Zh. Obshch. Khim., 37, 2686 (1967); C.A. 69, 67490c (1968).
152. Ferron, J. L., Can. J. Chem., 41, 2299 (1963).
153. Feshchenko, N. G., T. I. Alekseeva, L. F. Irodionova, and A. V. Kirsanov, Zh. Obshch. Khim., 37, 473 (1967); C.A. 67, 32745w (1967).

154. Filatova, I. M., E. L. Zaitseva, A. P. Simonov, and A. Ya. Yakubovich, Zh. Obshch. Khim., 38, 1304 (1968); C.A. 69, 87091b (1968).
154a. Fild, M., Z. Naturforsch., 23b, 604 (1968).
155. Fild, M., and R. Schmutzler, J. Chem. Soc. (A), 1969, 840.
156. Finegold, H., Ann. N.Y. Acad. Sci., 70, 765 (1958).
157. Fischer, M. H., and H. H. Incho, S. African Patent, 67 30,259 (1968); C.A. 71, 3470e (1969).
158. Fischer, M. H., H. H. Incho, P. E. Drummond, and R. E. Montgomery, French Patent 1,531,732 (1968); C.A. 71, 70736j (1969).
159. Fluck, E., German Patent 1,271,089 (1968); C.A. 69, 96872j (1968).
160. Fluck, E., and H. Binder, Angew. Chem., 77, 381 (1965).
161. Fluck, E., and H. Binder, Z. Anorg. Allg. Chem., 354, 139 (1967).
162. Flurry, R. L., and C. E. Boozer, J. Org. Chem., 31, 2076 (1966).
163. Foss, V. L., V. V. Kudinova, G. B. Postnikova, and I. F. Lutsenko, Dokl. Akad. Nauk SSSR, 146, 1106 (1962); C.A. 58, 7968h (1963).
164. Fossek, W., Monatsh. Chem., 7, 20 (1886).
165. Frank, A. W., Can. J. Chem., 46, 3573 (1968).
166. Frank, A. W., J. Org. Chem., 31, 1521 (1966).
167. Frank, A. W., and I. Gordon, Can. J. Chem., 44, 2593 (1966).
168. Fries, F. A., and E. Steinlechner, German Patent 961,886 (1957); C.A. 53, 11272d (1959).
169. Fujii, S., J. Sci. Hiroshima Univ., Ser. A-II, 31, 89 (1967); C.A. 69, 36576f (1968).
170. Funatsukuri, G., H. Yoshioka, and M. Ueda, Japanese Patent 2322 (1962); C.A. 58, 7976f (1963).
171. Furukawa, G. T., M. L. Reilly, J. H. Riccirelli, and M. Tenenbaum, J. Res. Natl. Bur. Std. (U.S.), A 68, 367 (1964); C.A. 61, 10103h (1964).
172. Gallagher, M. J., E. C. Kirby, and F. G. Mann, J. Chem. Soc., 1963, 4846.
173. Gamrath, H. R., U.S. Patent 2,764,609 (1956); C.A. 51, 5823g (1957).
174. Gefter, E. L., Zh. Vses. Khim., Obshch. D. I. Mendeleeva, 5, 479 (1960); C.A. 55, 1492c (1961).
175. Gefter, E. L., USSR Patent 130,898 (1960); C.A. 55, 7356f (1961).
176. Gefter, E. L., Zh. Obshch. Khim., 31, 952 (1961); C.A. 55, 23399d (1961).
177. Gefter, E. L., Zh. Obshch. Khim., 33, 3548 (1963); C.A. 60, 8057a (1964).
178. Gefter, E. L., and M. I. Kabachnik, Plast. Massy, 1961, 63; C.A. 56, 501g (1962).

179. Gefter, E. L., and I. A. Rogacheva, Zh. Obshch. Khim., 31, 955 (1961); C.A. 55, 23399f (1961).
180. Gefter, E. L., and I. A. Rogacheva, Zh. Obshch. Khim., 32, 964 (1962); C.A. 58, 2469c (1963).
181. Gefter, E. L., and I. A. Rogacheva, Zh. Obshch. Khim., 32, 3962 (1962); C.A. 59, 658b (1963).
182. Gefter, E. L., and I. A. Rogacheva, Zh. Obshch. Khim., 34, 88 (1964); C.A. 60, 10709g (1964).
183. Gefter, E. L., and I. A. Rogacheva, Zh. Obshch. Khim., 36, 79 (1966); C.A. 64, 14208f (1966).
184. Geiseler, G., F. Asinger, and M. Fedtke, Chem. Ber., 93, 765 (1960).
185. Geiseler, G., K. O. Binernagel, and J. Fruwert, Ber. Bunsenges. Phys. Chem., 71, 478 (1967).
186. Geiseler, G., K. Quitsch, and M. Kockert, Z. Phys. Chem., 218, 367 (1961).
187. Genkina, G. K., and V. A. Gilyarov, Izv. Akad. Nauk SSSR, Ser. Khim., 1969, 185; C.A. 70, 115265c (1969).
188. Gerding, H., and J. W. Maarsen, Rec. Trav. Chim. Pays-Bas, 76, 481 (1957).
189. Gevorkyan, A. A., and A. A. Manukyan, Arm. Khim. Zh., 21, 817 (1968); C.A. 71, 39076v (1969).
190. Gibson, C. S., and J. D. A. Johnson, J. Chem. Soc., 1928, 92.
191. Gilbert, A. R., and F. Precopio, U.S. Patent 2,835,651 (1958); C.A. 52, 15125b (1958).
192. Ginsberg, V. A., and A. Ya. Yakubovich, Zh. Obshch. Khim., 30, 3979 (1960); C.A. 55, 22099e (1961).
193. Gladshtein, B. M., and L. N. Shitov, Zh. Obshch. Khim., 37, 2586 (1967); C.A. 69, 27494a (1968).
194. Gladshtein, B. M., and L. Z. Soborovskii, USSR Patent 130,513 (1960); C.A. 55, 6375i (1961).
195. Gladshtein, B. M., and V. M. Zimin, Zh. Obshch. Khim., 37, 2055 (1967); C.A. 68, 49687x (1968).
196. Golubski, Z. E., and B. Glowiak, Bull. Acad. Polon. Sci., Ser. Sci. Chim., 12, 471 (1964); C.A. 62, 1311d (1965).
197. Gore, R. C., Discuss. Faraday Soc., 9, 138 (1950).
198. Gosling, K., and A. B. Burg, J. Amer. Chem. Soc., 90, 2011 (1968).
199. Gottlieb, H. B., J. Amer. Chem. Soc., 54, 748 (1932).
200. Goubeau, J., and D. Koettgen, Z. Anorg. Allg. Chem., 360, 182 (1968).
201. Graf, R., Chem. Ber., 85, 9 (1952).
202. Graf, R., German Patent 910,649 (1954); C.A. 52, 11117b (1958).
203. Gramstad, T., Spectrochim. Acta, 19, 497 (1963).
204. Granoth, I., A. Kalir, Z. Pelah, and E. D. Bergmann, Chem. Commun., 1969, 260.
205. Green, B. S., D. B. Sowerby, and K. J. Wihksne, Chem. Ind. (London), 1960, 1306.

206. Green, M., Proc. Chem. Soc., 1963, 177.
207. Green, M., J. Chem. Soc., 1965, 541.
208. Green, M., and R. F. Hudson, Proc. Chem. Soc., 1962, 217.
209. Green, M., and R. F. Hudson, J. Chem. Soc., 1963, 540.
210. Green, M., and R. F. Hudson, Proc. Chem. Soc., 1961, 145.
211. Greenwood, R. A., M. Scalera, and H. Z. Lecher, U.S. Patent 2,814,645 (1957); C.A. 52, 5464h (1958).
212. Griffin, C. E., and M. Gordon, J. Amer. Chem. Soc., 89, 4427 (1967).
213. Griffiths, J. E., Spectrochim. Acta, A 24, 115 (1968).
214. Griffiths, J. E., Spectrochim. Acta, A 24, 303 (1968).
215. Griffiths, J. E., and A. B. Burg, J. Amer. Chem. Soc., 84, 3442 (1962).
216. Grishina, O. N., and L. M. Bezzubova, Izv. Akad. Nauk SSSR, Ser. Khim., 1965, 1619; C.A. 64, 3587c (1966).
217. Grishina, O. N., and L. M. Bezzubova, Izv. Akad. Nauk SSSR, Ser. Khim., 1966, 1617; C.A. 66, 76095t (1967).
218. Grishina, O. N., and M. I. Potekhina, Neftekhimiya, 8, 111 (1968); C.A. 69, 2980t (1968).
219. Grishina, O. N., and R. Z. Sabirova, Neftekhimiya, 1, 796 (1961); C.A. 57, 15147a (1962).
220. Grishina, O. N., and R. Z. Sabirova, Sintezi Svoistva Monomerov, Akad. Nauk SSSR, Inst. Neftekhim. Sinteza, Sb. Rabot 12-oi (Devenadtsatoi) Konf. po. Vysokomolekul. Soedin., 1962, 80 (Publ. 1064); C.A. 62, 6508a (1965).
221. Groenweghe, L. C. D., U.S. Patent 3,024,278 (1962); C.A. 75, 7311c (1962).
222. Groenweghe, L. C. D., U.S. Patent 3,206,442 (1965); C.A. 63, 15007f (1965).
223. Groenweghe, L. C. D., U.S. Patent 3,457,307 (1969); C.A. 71, 124665f (1969).
224. Groenweghe, L. C. D., U.S. Patent 3,457,308 (1969); C.A. 71, 91646v (1969).
225. Groenweghe, L. C. D., and J. H. Payne, J. Amer. Chem. Soc., 83, 1811 (1961).
226. Gruzdev, V. G., S. Z. Ivin, and K. V. Karavanov, USSR Patent 185,910 (1966); C.A. 66, 115796u (1967).
227. Gruzdev, V. G., K. V. Karavanov, S. Z. Ivin, I. S. Mazel, and V. V. Tarasov, Zh. Obshch. Khim., 37, 450 (1967); C.A. 67, 43870j (1967).
228. Gryszkiewicz-Trochimowski, E., Bull. Soc. Chim. (France), 1967, 2232.
229. Gryszkiewicz-Trochimowski, E., Bull. Soc. Chim.

(France), 1967, 4289.
230. Gryszkiewicz-Trochimowski, E., J. Quinchon, and M. Bousquet, Mem. Poudres, 44, 119 (1962).
231. Gryszkiewicz-Trochimowski, E., J. Quinchon, and O. Gryszkiewicz-Trochimowski, Bull. Soc. Chim. (France), 1960, 1794.
232. Gryszkiewicz-Trochimowski, E., J. Quinchon, O. Gryszkiewicz-Trochimowski, and M. LeSech, Bull. Soc. Chim. (France), 1961, 739.
233. Gubnitskaya, E. S., and G. I. Derkach, Zh. Obshch. Khim., 38, 1530 (1968); C.A. 70, 87914t (1969).
234. Guichard, F., Ber., 32, 1572 (1899).
235. Guseinov, I. I., B. V. Lopatin, G. S. Vasil'ev, L. V. Orlova, and M. F. Shostakowski, Izv. Akad. Nauk SSSR, Otd. Khim. Nauk, 1962, 1550; C.A. 58, 4060e (1963).
236. Gutmann, V., U.S. Dept. Comm. Rept. PB 152,848 (1960); C.A. 58, 969e (1963).
237. Gutmann, V., Oesterr. Chem.-Ztg., 62, 326 (1961).
238. Gutmann, V., D. E. Hagen, and K. Utvary, Monatsh. Chem., 93, 627 (1962).
239. Gutmann, V., P. Heilmayer, and K. Utvary, Monatsh. Chem., 92, 196 (1961).
240. Gutmann, V., J. Imhof, and F. Mairinger, Monatsh. Chem., 99, 1615 (1968).
241. Gutmann, V., G. Moertl, and K. Utvary, Monatsh. Chem., 93, 1114 (1962).
242. Gutmann, V., J. Scherzer, and G. Schoeber, Monatsh. Chem., 95, 390 (1964).
243. Gutmann, V., and G. Schoeber, Monatsh. Chem., 93, 1353 (1962).
244. Gutmann, V., and E. Wychera, Monatsh. Chem., 96, 828 (1965).
245. Gutmann, V., E. Wychera, and F. Mairinger, Monatsh. Chem., 97, 1265 (1966).
246. Haering, M., Helv. Chim. Acta, 43, 1826 (1960).
247. Hagens, W., H. J. T. Bos, W. Voskuil, and J. F. Arens, Rec. Trav. Chim. Pays-Bas, 88, 71 (1969).
248. Hall, H. K., J. Org. Chem., 21, 248 (1956).
249. Halmann, M., J. Chem. Soc., 1959, 305.
250. Halmann, M., L. Kugel, and S. Pinchas, J. Chem. Soc., 1961, 3542.
251. Hamilton, L. A., U.S. Patent 2,365,466 (1944); C.A. 39, 46193 (1945).
252. Hamilton, L. A., U.S. Patent 2,382,309 (1945).
252a. Hanford, W. E., U.S. Patent 2,882,316 (1959); C.A. 53, 16965e (1959).
253. Hardy, D. V. N., and H. H. Hatt, J. Chem. Soc., 1952, 3778.
254. Hartle, R. J., J. Org. Chem., 31, 4288 (1966).
255. Harwood, H. J., M. L. Becker, and R. R. Smith, J.

Org. Chem., 32, 3882 (1967).
256. Harwood, H. J., and K. A. Pollart, U.S. Patent 3,082,256 (1963); C.A. 59, 10121g (1963).
257. Harwood, H. J., and K. A. Pollart, U.S. Patent 3,104,259 (1963); C.A. 60, 557c (1964).
258. Harwood, H. J., and K. A. Pollart, J. Org. Chem., 28, 3430 (1963).
259. Hasserodt, U., K. Hunger, and F. Korte, Tetrahedron, 19, 1563 (1963).
260. Hatt, H. H., J. Chem. Soc., 1933, 776.
261. Hatt, H. H., J. Chem. Soc., 1929, 2412.
262. Haven, A. C., J. Amer. Chem. Soc., 78, 842 (1956).
263. Haven, A. C., U.S. Patent 2,835,652 (1958); C.A. 53, 221g (1959).
264. Hays, H. R., and G. M. Kosolapoff, U.S. Patent 3,445,522 (1969); C.A. 71, 81520k (1969).
265. Helferich, B., and U. Curtius, Justus Liebigs Ann. Chem., 655, 59 (1962).
266. Henning, H. G., Z. Chem., 5, 103 (1965).
267. Henglein, A., Int. J. Appl. Radiat. and Isotopes, 8, 156 (1960).
268. Herail, F., Compt. Rend., C 262, 1493 (1966).
269. Herail, F., and V. Viossat, Compt. Rend., 259, 4629 (1964).
270. Heuck, C., H. Vilczek, and F. Rochlitz, German Patent 1,123,667 (1962); C.A. 57, 3486b (1962).
271. Higgins, W. A., P. W. Vogel, and W. G. Craig, J. Amer. Chem. Soc., 77, 1864 (1955).
272. Hignett, T. P., L. B. Hein, A. B. Phillips, and R. D. Young, U.S. Patent 2,875,245 (1959); C.A. 53, 12177b (1959).
273. Hinton, R. C., F. G. Mann, and D. Todd, J. Chem. Soc., 1961, 5454.
274. Hoffmann, F. W., and A. M. Reeves, J. Org. Chem., 26, 3040 (1961).
275. Hoffmann, F. W., D. H. Wadsworth, and H. D. Weiss, J. Amer. Chem. Soc., 80, 3945 (1958).
276. Hofmann, A. W., Ber., 6, 303 (1873).
277. Hogben, M. G., and W. A. G. Graham, J. Amer. Chem. Soc., 91, 283 (1969).
278. Hooge, F. N., and P. J. Christen, Rec. Trav. Chim. Pays-Bas, 77, 911 (1958).
279. Hsieh, H. L., U.S. Patent 3,147,313 (1964); C.A. 61, 12175b (1964).
280. Hudson, R. F., and L. Keay, J. Chem. Soc., 1960, 1859.
281. Hudson, R. F., and G. E. Moss, J. Chem. Soc., 1964, 1040.
282. Huffman, C. W., and M. Hamer, U.S. Patent 3,149,144 (1964); C.A. 61, 14712e (1964).
283. Huffman, C. W., and M. Hamer, U.S. Patent 3,149,137

(1964); C.A. 62, 4053b (1965).
284. Hunger, K., U. Hasserodt, and F. Korte, Tetrahedron, 20, 1593 (1964).
285. Hunt, B. B., and B. C. Saunders, J. Chem. Soc., 1957, 2413.
286. Ignat'ev, V. M., B. I. Ionin, and A. A. Petrov, Zh. Obshch. Khim., 36, 1505 (1966); C.A. 66, 11010c (1967).
287. Ignat'ev, V. M., B. I. Ionin, and A. A. Petrov, Zh. Obshch. Khim., 37, 1898 (1967); C.A. 68, 29796d (1968).
288. Inukai, K., T. Ueda, and H. Muramatsu, J. Org. Chem., 29, 2224 (1964).
289. Ionin, B. I., V. M. Ignat'ev, and V. B. Lebedev, Zh. Obshch. Khim., 37, 1863 (1967); C.A. 68, 64404w (1968).
290. Isbell, A. F., and F. T. Wadsworth, J. Amer. Chem. Soc., 78, 6042 (1956).
291. Ishida, H., T. Yoneda, and B. Sekine, Japanese Patent 15,405 (1968); C.A. 70, 68504t (1969).
292. Israilov, D., and L. A. Rodivilova, Vysokomolekul. Soedin., 7, 2089 (1965); C.A. 64, 11325e (1966).
293. Issleib, K., and A. Brack, Z. Anorg. Allg. Chem., 277, 258 (1954).
294. Ivanov, B. E., A. R. Panteleeva, R. R. Shagidullin, and I. M. Shermergorn, Zh. Obshch. Khim., 37, 1856 (1967); C.A. 68, 29797e (1968).
295. Ivanov, B. E., V. G. Trutnev, and I. M. Shermergorn, USSR Patent 187,779 (1966); C.A. 67, 11588g (1967).
296. Ivanova, Zh. M., and A. V. Kirsanov, Zh. Obshch. Khim., 31, 3991 (1961); C.A. 57, 8605i (1962).
297. Ivanova, Zh. M., and A. V. Kirsanov, Zh. Obshch. Khim., 32, 2592 (1962); C.A. 58, 9122g (1963).
298. Ivanova, Zh. M., and A. V. Kirsanov, Zh. Obshch. Khim., 35, 1974 (1965); C.A. 64, 6680e (1966).
299. Ivanova, Zh. M., S. K. Mikhailik, and G. I. Derkach, Zh. Obshch. Khim., 38, 1334 (1968); C.A. 69, 106852u (1968).
300. Ivanova, Zh. M., S. K. Mikhailik, and G. I. Derkach, Zh. Obshch. Khim., 39, 1037 (1969); C.A. 71, 61478z (1969).
301. Ivashchenko, Ya. M., L. S. Sologub, S. D. Moshchitskii, and A. V. Kirsanov, Zh. Obshch. Khim., 39, 1695 (1969); C.A. 71, 124600f (1969).
302. Ivasyuk, N. V., and I. M. Shermergorn, Izv. Akad. Nauk SSSR, Ser. Khim., 1969, 481; C.A. 71, 13174a (1969).
303. Ivin, S. Z., and K. V. Karavanov, Zh. Obshch. Khim., 28, 2958 (1958); C.A. 53, 9035b (1959).
304. Ivin, S. Z., K. V. Karavanov, and V. V. Lysenko, USSR Patent 163,617 (1964); C.A. 61, 16095g (1964).

305. Ivin, S. Z., K. V. Karavanov, and V. V. Lysenko, USSR Patent 163,615 (1964); C.A. 61, 16096a (1964).
306. Ivin, S. Z., V. N. Pastushkov, Yu. A. Kondrat'ev, K. F. Oglobin, and V. V. Tarasov, Zh. Obshch. Khim., 38, 2069 (1968); C.A. 70, 11759a (1969).
307. Ivin, S. Z., and V. K. Promonenkov, USSR Patent 163,616 (1964); C.A. 63, 4333c (1965).
308. Ivin, S. Z., and V. K. Promonenkov, Khim. Org. Soedin. Fosfora, Akad. Nauk SSSR, Otd. Obshch. Tekh. Khim., 1967, 139; C.A. 68, 114701p (1968).
309. Ivin, S. Z., and V. K. Promonenkov, Zh. Obshch. Khim., 37, 489 (1967); C.A. 67, 43873n (1967).
310. Ivin, S. Z., V. K. Promonenkov, and E. A. Fokin, Zh. Obshch. Khim., 39, 1058 (1969); C.A. 71, 61485z (1969).
311. Ivin, S. Z., and I. D. Shelakova, Zh. Obshch. Khim., 31, 4052 (1961); C.A. 57, 9878b (1962).
312. Jean, H., Bull. Soc. Chim. (France), 1956, 569.
313. Jensen, W. L., U.S. Patent 2,662,917 (1953); C.A. 48, 13711g (1954).
314. Jensen, W. L., and J. O. Clayton, U.S. Patent 2,683,168 (1954); C.A. 49, 11001d (1955).
315. Jensen, W. L., and J. O. Clayton, U.S. Patent 2,795,609 (1957); C.A. 51, 16535c (1957).
316. Johns, I. B., U.S. Patent 3,330,628 (1967); C.A. 67, 73676t (1967).
317. Johns, I. B., U.S. Patent 3,410,809 (1968); C.A. 71, 113665e (1969).
318. Johns, I. B., and H. R. Di Pietro, J. Org. Chem., 29, 1970 (1964).
319. Jungermann, E., and R. C. Clutter, U.S. Patent 3,259,671 (1966); C.A. 65, 8961d (1966).
320. Jungermann, E., and J. J. McBride, J. Org. Chem., 26, 4182 (1961).
321. Jungermann, E., J. J. McBride, R. Clutter, and A. Mais, J. Org. Chem., 27, 606 (1962).
322. Kabachnik, M. I., and N. N. Godovikov, Dokl. Akad. Nauk SSSR, 110, 217 (1956); C.A. 51, 4982g (1957).
323. Kabachnik, M. I., T. A. Mastryukova, and T. A. Melent'eva, Zh. Obshch. Khim., 33, 382 (1963); C.A. 59, 1678a (1963).
324. Kabachnik, M. I., and T. Ya. Medved, Izv. Akad. Nauk SSSR, Otd. Khim. Nauk, 1950, 635; C.A. 45, 8444 (1951).
325. Kabachnik, M. I., and T. Ya. Medved, Izv. Akad. Nauk SSSR, Otd. Khim. Nauk, 1959, 2142; C.A. 54, 10834f (1960).
326. Kabachnik, M. I., and T. Ya. Medved, Izv. Akad. Nauk SSSR, Otd. Khim. Nauk, 1961, 604; C.A. 55, 23319h (1961).
327. Kabachnik, M. I., and P. A. Rossiiskaya, Izv. Akad.

Nauk SSSR, Otd. Khim. Nauk, 1946, 403; C.A. 40, 4688[5] (1946).
328. Kabachnik, M. I., and P. A. Rossiiskaya, Izv. Akad. Nauk SSSR, Otd. Khim. Nauk, 1946, 515; C.A. 40, 88d (1947).
329. Kabachnik, M. I., and E. S. Shepeleva, Dokl. Akad. Nauk SSSR, 75, 219 (1950); C.A. 45, 6569i (1951).
330. Kabachnik, M. I., and E. S. Shepeleva, Izv. Akad. Nauk SSSR, Otd. Khim. Nauk, 1951, 185; C.A. 45, 10191b (1951).
331. Kabachnik, M. I., and E. S. Shepeleva, Izv. Akad. Nauk SSSR, Otd. Khim. Nauk, 1949, 56; C.A. 43, 5739h (1949).
331a. Kabachnik, M. I., and E. S. Shepeleva, Izv. Akad. Nauk SSSR, Otd. Khim. Nauk, 1950, 39; C.A. 44, 7257f (1950).
332. Kabachnik, M. I., and E. S. Shepeleva, Izv. Akad. Nauk SSSR, Otd. Khim. Nauk, 1953, 862; C.A. 49, 843e (1955).
333. Kabachnik, M. I., and E. N. Tsvetkov, Zh. Obshch. Khim., 30, 3227 (1960); C.A. 55, 21067d (1961).
334. Kabachnik, M. I., and E. N. Tsvetkov, Dokl. Akad. Nauk SSSR, 135, 323 (1960); C.A. 55, 14288f (1961).
335. Kabachnik, M. I., and E. N. Tsvetkov, Izv. Akad. Nauk SSSR, Otd. Khim. Nauk, 1961, 1896; C.A. 56, 8739a (1962).
336. Kalabina, A. V., and N. A. Dubovik, Izv. Fiz.-Khim. Nauchn.-Issled. Inst. Pri Irkutskom Gos. Univ., 5, 131 (1961); C.A. 61, 683g (1964).
337. Kalabina, A. V., E. F. Grechkin, T. I. Bychkova, A. K. Filippova, N. A. Tyukavkina, and L. T. Ermakova, Sintezi Svoistva Monomerov, Akad. Nauk SSSR, Inst. Neftekhim. Sinteza, Sb. Rabot 12-oi (Dvenadtsatoi) Konf. po Vysokomolekul. Soedin., 1962, 267 (Publ. 1964); C.A. 62, 6418g (1965).
338. Kamai, G., V. S. Tsivunin, and L. A. Panina, Tr. Kazansk. Khim.-Tekhnol. Inst., No. 30, 11 (1962); C.A. 60, 4180d (1964).
339. Kamai, G., and V. S. Tsivunin, Dokl. Akad. Nauk SSSR, 128, 543 (1959); C.A. 54, 7538h (1960).
340. Karavanov, K. V., L. E. Dmitrieva, and S. Z. Ivin, Zh. Obshch. Khim., 38, 1547 (1968); C.A. 69, 96823u (1968).
341. Karavanov, K. V., and S. Z. Ivin, Zh. Obshch. Khim., 35, 78 (1965); C.A. 62, 13175f (1965).
342. Karavanov, K. V., S. Z. Ivin, and V. G. Gruzdev, Khim. Org. Soedin. Fosfora, Akad. Nauk SSSR, Otd. Obshch. Tekh. Khim. 1967, 157; C.A. 68, 114700n (1968).
343. Karavanov, K. V., S. Z. Ivin, and V. V. Lysenko, Zh. Obshch. Khim., 35, 737 (1965); C.A. 63, 4327e (1965).

344. Karavanov, K. V., S. Z. Ivin, V. V. Lysenko, and G. I. Drozd, Probl. Organ. Sinteza, Akad. Nauk SSSR, Otd. Obshch. Tekhn. Khim., 1965, 291; C.A. 64, 6682b (1966).
345. Karavanov, K. V., S. Z. Ivin, and F. I. Ponomarenko, USSR Patent 223,095 (1968); C.A. 70, 29055c (1969).
346. Karavanov, K. V., S. Z. Ivin, and F. I. Ponomarenko, USSR Patent 226,601 (1968); C.A. 70, 47591d (1969).
347. Karavanov, K. V., S. Z. Ivin, and F. I. Ponomarenko, USSR Patent 229,505 (1968); C.A. 70, 58020t (1969).
348. Karelsky, M., and K. H. Pausacker, Aust. J. Chem., 11, 336 (1958).
349. Keat, R., Chem. Ind. (London), 1968, 1362.
350. Khairullin, V. K., USSR Patent 173,763 (1965); C.A. 64, 2127e (1966).
351. Khairullin, V. K., Dokl. Akad. Nauk SSSR, 162, 827 (1965); C.A. 63, 7040e (1965).
352. Khairullin, V. K., G. V. Dmitrieva, and A. N. Pudovik, Zh. Obshch. Khim., 37, 1838 (1967); C.A. 68, 13088p (1968).
353. Khairullin, V. K., G. V. Dmitrieva, and A. N. Pudovik, Khim. Org. Soedin. Fosfora, Akad. Nauk SSSR, Otd. Obshch. Tekh. Khim., 1967, 40; C.A. 69, 67486f (1968).
354. Khairullin, V. K., G. V. Dmitrieva, and A. N. Pudovik, Izv. Akad. Nauk SSSR, Ser. Khim., 1969, 1166; C.A. 71, 50121b (1969).
355. Khairullin, V. K., R. M. Kondrat'eva, and A. N. Pudovik, Izv. Akad. Nauk SSSR, Ser. Khim., 1967, 2097; C.A. 68, 39721q (1968).
356. Khairullin, V. K., R. M. Kondrat'eva, and A. N. Pudovik, Zh. Obshch. Khim., 38, 288 (1968); C.A. 69, 106816k (1968).
357. Khairullin, V. K., R. M. Kondrat'eva, and A. N. Pudovik, Zh. Obshch. Khim., 38, 858 (1968); C.A. 69, 52218w (1968).
358. Khairullin, V. K., and A. N. Pudovik, Zh. Obshch. Khim., 36, 494 (1966); C.A. 65, 738f (1966).
359. Khairullin, V. K., and A. N. Pudovik, Zh. Obshch. Khim., 37, 2742 (1967); C.A. 69, 67483c (1968).
360. Khairullin, V. K., A. N. Pudovik, and N. I. Kharitonova, Zh. Obshch. Khim., 39, 608 (1969); C.A. 71, 39082u (1969).
361. Khairullin, V. K., T. I. Sobchuk, and A. N. Pudovik, Zh. Obshch. Khim., 37, 710 (1967); C.A. 67, 54222a (1967).
361a. Khairullin, V. K., T. I. Sobchuk, and A. N. Pudovik, Zh. Obshch. Khim., 36, 296 (1966); C.A. 64, 15915c (1966).
362. Khairullin, V. K., T. I. Sobchuk, and A. N. Pudovik, Zh. Obshch. Khim., 38, 585 (1968); C.A. 69, 59347w (1968).
363. Khairullin, V. K., M. A. Vasyanina, and A. N. Pudovik,

Zh. Obshch. Khim., 38, 2071 (1968); C.A. 70, 11762w (1969).
364. Kharrasova, F. M., and G. Kamai, Zh. Obshch. Khim., 34, 2195 (1964); C.A. 61, 10705f (1964).
365. Kharrasova, F. M., G. Kamai, and G. I. Matveeva, Zh. Obshch. Khim., 38, 1262 (1968); C.A. 69, 55759j (1969).
366. Kinnear, A. M., and E. A. Perren, J. Chem. Soc., 1952, 3437.
367. Kirsanov, A. V., and G. K. Fedorova, Dopov. Akad. Nauk Ukr. RSR, 1960, 1086; C.A. 56, 15143a (1962).
368. Knapsack A. G., French Patent 1,549,019 (1968); C.A. 71, 124664e (1969).
369. Kodolov, V. I., G. A. Semerveva, and S. S. Spasskii, Tr. Inst. Khim. Akad. Nauk SSSR, Ural. Filial., No. 13, 99 (1966); C.A. 68, 109631m (1968).
370. Koehler, H., Ber., 13, 463 (1880).
371. Koehler, H., Ber., 13, 1623 (1880).
372. Koehler, H., Ber., 13, 1626 (1880).
373. Koehler, H., and A. Michaelis, Ber., 9, 1053 (1876).
374. Kolomiets, A. F., Z. N. Kvasha, and N. K. Bliznyuk, USSR Patent 172,797 (1965); C.A. 64, 756c (1966).
375. Kolomiets, A. F., Z. N. Kvasha, and N. K. Bliznyuk, USSR Patent 173,768 (1965); C.A. 64, 1702a (1966).
376. Komkov, I. P., USSR Patent 199,145 (1967); C.A. 68, 95965j (1968).
377. Komkov, I. P., USSR Patent 199,146 (1967); C.A. 68, 95962f (1968).
378. Komkov, I. P., USSR Patent 202,132 (1967); C.A. 68, 114739g (1968).
379. Komkov, I. P., S. Z. Ivin, and K. V. Karavanov, Metody Poluch. Khim. Reak. Prep., No. 12, 20 (1965); C.A. 65, 5478h (1966).
380. Komkov, I. P., S. Z. Ivin, and K. V. Karavanov, Zh. Obshch. Khim., 28, 2960 (1958); C.A. 53, 9035e (1959).
381. Komkov, I. P., S. Z. Ivin, K. V. Karavanov, and L. E. Smirnov, Zh. Obshch. Khim., 32, 301 (1962); C.A. 57, 16649a (1962).
382. Komkov, I. P., K. V. Karavanov, and S. Z. Ivin, Zh. Obshch. Khim., 28, 2963 (1958); C.A. 53, 9035h (1959).
383. Kondrat'ev, Yu. A., Ya. S. Arbisman, O. G. Strukov, S. S. Dubov, V. V. Tarasov, and S. Z. Ivin, Zh. Obshch. Khim., 39, 709 (1969); C.A. 71, 70696w (1969).
384. Kondrat'ev, Yu. A., E. S. Vdovina, Ya. S. Arbisman, V. V. Tarasov, O. G. Strukov, S. S. Dubov, and S. Z. Ivin, Zh. Obshch. Khim., 38, 2859 (1968); C.A. 71, 22153v (1969).
385. Konishi, K., Japanese Patent 14,537 (1967); C.A. 69, 19308c (1968).
386. Kosfeld, R., G. Haegele, and W. Kuchen, Angew. Chem., 80, 794 (1968).

387. Kosolapoff, G. M., J. Amer. Chem. Soc., 71, 369 (1949).
388. Kosolapoff, G. M., and A. D. Brown, J. Chem. Soc. (C), 1966, 757.
389. Kosolapoff, G. M., and A. D. Brown, J. Chem. Soc. (C), 1967, 1789.
390. Kosolapoff, G. M., and A. D. Brown, Chem. and Ind., 1969, 1272.
391. Kosolapoff, G. M., and R. F. Struck, J. Chem. Soc., 1957, 3739.
392. Kosolapoff, G. M., and R. F. Struck, J. Chem. Soc., 1959, 3950.
393. Kosolapoff, G. M., and R. M. Watson, J. Amer. Chem. Soc., 73, 4101 (1951).
394. Kosolapoff, G. M., and R. M. Watson, J. Amer. Chem. Soc., 73, 5466 (1951).
395. Kosolapov, S. N., V. I. Dubravin, L. Z. Soborovskii, S. L. Varshavskii, Yu. M. Zinov'ev, and V. E. Zhigachev, USSR Patent 212,257 (1968); C.A. 69, 67535w (1968).
396. Kozlov, E. S., and S. N. Gaidamaka, Zh. Obshch. Khim., 39, 933 (1969); C.A. 71, 50085t (1969).
397. Krasil'nikova, E. A., A. M. Potapov, and A. I. Razumov, Zh. Obshch. Khim., 38, 1098 (1968); C.A. 69, 67471x (1968).
398. Kreshkov, A. P., V. A. Drozdov, and N. A. Kolchina, Zh. Analit. Khim., 19, 1177 (1964); C.A. 62, 2242d (1965).
399. Kuchar, M., Chem. Prumysl, 13, 191 (1963); C.A. 59, 15305c (1963).
400. Kuchen, W., and H. Buchwald, Angew. Chem., 71, 162 (1959).
401. Kuchen, W., H. Buchwald, K. Strolenberg, and J. Metten, Justus Liebigs Ann. Chem., 652, 28 (1962).
402. Kulakova, V. N., Yu. M. Zinov'ev, and L. Z. Soborovskii, Zh. Obshch. Khim., 39, 838 (1969); C.A. 71, 61491y (1969).
403. Kuznetsov, E. V., R. K. Valetdinov, and Ts. Ya. Roitburd, Zh. Obshch. Khim., 33, 150 (1963); C.A. 59, 655g (1963).
404. Kwiatek, J., U.S. Patent 2,882,303 (1959); C.A. 53, 16964f (1959).
404a. Kwiatek, J., U.S. Patent 2,882,308, 2,882,312 (1959); C.A. 53, 16964h, 16965a (1959).
404b. Kwiatek, J., U.S. Patent 2,882,309, 2,882,311 (1959); C.A. 53, 16964h, 16964i (1959).
405. Kwiatek, J., and J. W. Copenhaver, U.S. Patent 2,882,313 (1959); C.A. 53, 16965b (1959).
406. Landau, M. A., V. V. Sheluchenko, and S. S. Dubov, Dokl. Akad. Nauk SSSR, 182, 134 (1968); C.A. 70, 23046d (1969).

407. Larionova, M. A., A. L. Klebanskii, and V. A. Bartashev, Zh. Obshch. Khim., 33, 265 (1963); C.A. 59, 656h (1963).
408. Larsson, L., Ark. Kemi., 13, 259 (1958); C.A. 53, 14919d (1959).
409. Lecher, H. Z., and R. A. Greenwood, U.S. Patent 2,870,204 (1959); C.A. 53, 11306b (1959.
410. Lecher, H. Z., R. A. Greenwood, K. C. Whitehouse, and T. H. Chao, J. Amer. Chem. Soc., 78, 5018 (1956).
411. Lecher, H. Z., and E. Kuh, U.S. Patent 2,654,738 (1953); C.A. 48, 10053c (1954).
412. Lequan, R. M., and M. P. Simonnin, Compt. Rend., C 268, 1400 (1969).
413. Lesfauries, P., and P. Rumpf, Bull. Soc. Chim. (France), 1950, 542.
414. Letcher, J. H., and J. R. Van Wazer, Topics Phosphorus Chem., 5, 75 (1967).
415. Levin, B. B., I. N. Fetin, and L. A. Astakhova, USSR Patent 165,458 (1964); C.A. 62, 6432g (1965).
416. Levin, Ya. A., V. S. Galeev, and N. V. Evdokimova, USSR Patent 232,973 (1968); C.A. 70, 106644y (1969).
417. Levin, Ya. A., V. S. Galeev, and E. K. Trutneva, Zh. Obshch. Khim., 37, 1872 (1967); C.A. 68, 29801b (1968).
418. Levin, Ya. A., and L. K. Gazizova, USSR Patent 210,860 (1968); C.A. 69, 36253y (1968).
419. Levin, Ya. A., and R. I. Pyrkin, USSR Patent 217,394 (1968); C.A. 69, 77488x (1968).
420. Levin, Ya. A., E. K. Trutneva, S. K. Moiseenko, and R. A. Komlev, Izv. Akad. Nauk SSSR, Ser. Khim., 1967, 1881; C.A. 68, 29786a (1968).
421. Lewis, A. H., and R. D. Stayner, U.S. Patent 2,587,340 (1952); C.A. 46, 8144h (1952).
422. Lewis, A. H., and R. D. Stayner, U.S. Patent 2,670,367 (1954); C.A. 49, 3240d (1955).
423. Lindner, J., O. Brugger, A. Jenkner, and L. Tschemernigg, Monatsh. Chem., 53/4, 263 (1929).
424. Lindner, J., and M. Strecker, Monatsh. Chem., 53/4, 274 (1929).
425. Lindner, J., W. Wirth, and B. Zaunbauer, Monatsh. Chem., 70, 1 (1937).
426. Lobanov, D. I., E. N. Tsvetkov, and M. I. Kabachnik, Zh. Obshch. Khim., 39, 841 (1969); C.A. 71, 61495c (1969).
427. Lora-Thamayo, M., E. F. Alvarez, and M. Andreu, Bull. Soc. Chim. Biol., 44, 501 (1962).
428. Lorenz, W., and G. Schrader, German Patent 1,067,017 (1958); C.A. 56, 1482d (1962).
429. Lucken, E. A. C., and M. A. Whitehead, J. Chem. Soc., 1961, 2459.
430. Lutsenko, I. F., and M. Kirilov, Dokl. Akad. Nauk

SSSR, 128, 89 (1959); C.A. 54, 1288i (1960).
431. Lutsenko, I. F., and M. Kirilov, Dokl. Akad. Nauk SSSR, 132, 842 (1960); C.A. 54, 20842a (1960).
432. Lutsenko, I. F., and M. Kirilov, Zh. Obshch. Khim., 31, 3594 (1961); C.A. 57, 8606c (1962).
433. Lutsenko, I. F., and M. Kirilov, Godishnik Sofii. Univ., Fiz.-Mat. Fak. Khim., 55, 135 and 165 (1960/61); C.A. 59, 6434a (1963).
434. Lutsenko, I. F., M. Kirilov, and G. A. Ovchinnikova, Zh. Obshch. Khim., 31, 2028 (1961); C.A., 55, 27021f (1961).
435. Lutsenko, I. F., M. Kirilov, and G. B. Postnikova, Zh. Obshch. Khim., 31, 2034 (1961); C.A. 55, 27021i (1961).
436. Lynch, E. R., J. Chem. Soc., 1962, 3729.
437. Lynch, E. R., British Patent 933,800 (1963); C.A. 60, 1796a (1964).
438. Lysenko, V. V., I. D. Shelakova, K. V. Karavanov, and S. Z. Ivin, Zh. Obshch. Khim., 36, 1507 (1966); C.A. 66, 11005m (1967).
439. Maguire, M. H., and G. Shaw, J. Chem. Soc., 1957, 311.
440. Maier, L., Angew. Chem., 71, 574 (1959).
441. Maier, L., Chem. Ber., 94, 3043 (1961).
442. Maier, L., Chem. Ber., 94, 3051 (1961).
443. Maier, L., Chem. Ber., 94, 3056 (1961).
444. Maier, L., J. Inorg. Nucl. Chem., 24, 275 (1962).
445. Maier, L., U.S. Patent 3,075,017 (1963); C.A. 58, 13995c (1963).
446. Maier, L., J. Inorg. Nucl. Chem., 24, 1073 (1962).
447. Maier, L., Helv. Chim. Acta, 46, 2026 (1963).
448. Maier, L., Helv. Chim. Acta, 46, 2667 (1963).
449. Maier, L., Helv. Chim. Acta, 47, 27 (1964).
450. Maier, L., Helv. Chim. Acta, 47, 120 (1964).
450a. Maier, L., Helv. Chim. Acta, 53, 1948 (1971).
451. Maier, L., Helv. Chim. Acta, 48, 133 (1965).
452. Maier, L., Helv. Chim. Acta, 48, 1190 (1965).
453. Maier, L., Z. Anorg. Allg. Chem., 345, 29 (1966).
454. Maier, L., U.S. Patent 3,321,557 (1967); C.A. 67, 73684u (1967).
455. Maier, L., German Patent 1,238,024 (1967); C.A. 67, 90934t (1967).
456. Maier, L., Helv. Chim. Acta, 52, 1337 (1969).
457. Maier, L., and R. Gredig, Helv. Chim. Acta, 52, 827 (1969).
458. Maier-Bode, H., and G. Koetz, East German Patent 10,881 (1955); C.A. 53, 6518i (1959).
459. Malatesta, L., Gazz. Chim. Ital., 77, 509 (1947).
460. Manatt, S. L., G. L. Juvinall, R. I. Wagner, and D. D. Elleman, J. Amer. Chem. Soc., 88, 2689 (1966).
461. Mann, F. G., B. P. Tong, and V. P. Wystrach, J. Chem. Soc., 1963, 1155.

462. Mark, V., C. H. Dungan, M. M. Crutchfield, and J. R. Van Wazer, Topics Phosphorus Chem., 5, 227 (1967).
463. Markowska, A., Bull. Acad. Pol. Sci., Ser. Sci. Chim., 15, 153 (1967); C.A. 67, 90891b (1967).
464. Markowska, A., and J. Michalski, Rocz. Chem., 34, 1675 (1960); C.A. 56, 7346a (1962).
465. Martin, G., and A. Besnard, Compt. Rend. 257, 898 (1963).
466. Martin, G., and G. Mavel, Compt. Rend., 253, 644 (1961).
467. Martin, G., and G. Mavel, Compt. Rend., 253, 2523 (1961).
468. Mashlyakovskii, L. N., and B. I. Ionin, Zh. Obshch. Khim., 35, 1577 (1965); C.A. 63, 18143b (1965).
469. Mashlyakovskii, L. N., B. I. Ionin, V. B. Lebedev, A. A. Petrov, and I. S. Okhrimenko, Khim. Org. Soedin. Fosfora, Akad. Nauk SSSR, Otd. Obshch. Tekh. Khim., 1967, 238; C.A. 68, 114704s (1968).
470. Mashlyakovskii, L. N., B. I. Ionin, I. S. Okhrimenko, and A. A. Petrov, Zh. Obshch. Khim., 37, 1307 (1967); C.A. 68, 22017s (1968).
471. Mastalerz, P., Rocz. Chem., 39, 33 (1965); C.A. 62, 16292c (1965).
472. Mastalerz, P., and Z. E. Golubski, Rocz. Chem., 39, 951 (1965); C.A. 64, 3591a (1966).
473. Mastryukova, T. A., A. E. Shipov, and M. I. Kabachnik, Zh. Obshch. Khim., 29, 1450 (1959); C.A. 54, 9729f (1960).
474. Mastryukova, T. A., A. E. Shipov, and M. I. Kabachnik, Zh. Obshch. Khim., 31, 507 (1961); C.A. 55, 22101f (1961).
474a. Mastryukova, T. A., A. E. Shipov, and M. I. Kabachnik, Zh. Obshch. Khim., 32, 3579 (1962); C.A. 58, 11394h (1963).
475. Mavel, G., Progr. NMR Spectrosc., 1, 250 (1966).
476. Mavel, G., and G. Martin, J. Phys. Radium, 24, 108 (1963).
477. Mavel, G., and G. Martin, Compt. Rend., 257, 1703 (1963).
478. Maynard, J. A., and J. M. Swan, Aust. J. Chem., 16, 596 (1963).
479. Mayo, F. R., Amer. Chem. Soc. Div. Petrol. Chem. Reprints, 5, C47 (1960); C.A. 57, 15143h (1962).
479a. McHattie, G. V., British Patent 860,629 (1961); C.A. 57, 2256d (1962).
480. McIvor, R. A., G. A. Grant, and C. E. Hubley, Can. J. Chem., 34, 1611 (1956).
481. McIvor, R. A., and C. E. Hubley, Can. J. Chem., 37, 869 (1959).
482. McIvor, R. A., C. E. Hubley, G. A. Grant, and A. A. Gray, Can. J. Chem., 36, 820 (1958).
483. Medved, T. Ya., and M. I. Kabachnik, Izv. Akad. Nauk

SSSR, Otd. Khim. Nauk, 1961, 270; C.A. 55, 20922a (1961).
484. Meisenheimer, J., J. Casper, M. Hoering, W. Lauter, L. Lichtenstadt, and W. Samuel, Justus Liebigs Ann. Chem., 449, 213 (1926).
485. Meisters, A., and J. M. Swan, Aust. J. Chem., 18, 155 (1965).
486. Melchiker, P., Ber., 31, 2915 (1898).
487. Melton, T. M., U.S. Patent 3,458,569 (1969); C.A. 71, 91652u (1969).
488. Metzger, S. H., O. H. Basedow, and A. F. Isbell, J. Org. Chem., 29, 627 (1964).
489. Metzger, S. H., and A. F. Isbell, J. Org. Chem., 29, 623 (1964).
490. Michaelis, A., Ber., 6, 816 (1873).
491. Michaelis, A., Justus Liebigs Ann. Chem., 181, 265 (1876).
492. Michaelis, A., Ber., 10, 627 (1877).
493. Michaelis, A., Ber., 13, 2174 (1880).
494. Michaelis, A., Justus Liebigs Ann. Chem., 293, 193 (1896).
495. Michaelis, A., Justus Liebigs Ann. Chem., 294, 1 (1896).
496. Michaelis, A., Justus Liebigs Ann. Chem., 315, 43 (1901).
497. Michaelis, A., Justus Liebigs Ann. Chem., 407, 290 (1915).
498. Michaelis, A., and J. Ananoff, Ber., 7, 1688 (1874).
499. Michaelis, A., and T. Becker, Ber., 30, 1003 (1897).
500. Michaelis, A., and A. Flemming, Ber., 34, 1291 (1901).
501. Michaelis, A., and F. Kammerer, Ber., 8, 1306 (1875).
502. Michaelis, A., and H. Koehler, Ber., 9, 519 (1876).
503. Michaelis, A., and H. Koehler, Ber., 9, 1053 (1876).
504. Michaelis, A., and W. LaCoste, Ber., 18, 2109 (1885).
505. Michaelis, A., and A. Link, Justus Liebigs Ann. Chem., 207, 193 (1881).
506. Michaelis, A., and C. Panek, Justus Liebigs Ann. Chem., 212, 203 (1882).
507. Michaelis, A., and F. Rothe, Ber., 25, 1747 (1892).
508. Michalski, J., and A. Skowronska, Rocz. Chem., 34, 1381 (1960); C.A. 55, 19842c (1961).
509. Moedritzer, K., J. Amer. Chem. Soc., 83, 4381 (1961).
510. Moedritzer, K., U.S. Patent 3,360,556 (1967); C.A. 68, 49755t (1968).
511. Moedritzer, K., L. Maier, and L. C. D. Groenweghe, J. Chem. Eng. Data, 7, 307 (1962).
512. Monard, C., and J. Quinchon, Bull. Soc. Chim. (France), 1961, 1084.
513. Monard, C., and J. Quinchon, Bull. Soc. Chim. (France), 1961, 1086.
514. Monsanto Chemical Co., French Patent 1,347,066

(1963); C.A. 60, 12055d (1964).
515. Moskva, V. V., V. M. Ismailov, and A. I. Razumov, Zh. Obshch. Khim., 38, 2587 (1968); C.A. 71, 3440v (1969).
516. Mueller, E., and H. G. Padeken, Chem. Ber., 100, 521 (1967).
517. Muller, N., P. C. Lauterbur, and J. Goldenson, J. Amer. Chem. Soc., 78, 3557 (1956).
518. Nagel, R. M., U.S. Patent 3,461,189 (1969); C.A. 71, 102013q (1969).
519. Neimysheva, A. A., and I. L. Knunyants, Zh. Obshch. Khim., 36, 1090 (1966); C.A. 65, 12068b (1966).
520. Neimysheva, A. A., and I. L. Knunyants, Dokl. Akad. Nauk SSSR, 177, 856 (1967); C.A. 68, 64396v (1968).
521. Neimysheva, A. A., and I. L. Knunyants, Zh. Obshch. Khim., 38, 595 (1968); C.A. 69, 66629t (1968).
522. Neimysheva, A. A., V. A. Pal'm, G. K. Semin, N. A. Loshadkin, and I. L. Knunyants, Zh. Obshch. Khim., 37, 2255 (1967); C.A. 69, 2386d (1968).
523. Neimysheva, A. A., V. I. Savchuk, M. V. Ermolaeva, and I. L. Knunyants, Izv. Akad. Nauk SSSR, Ser. Khim., 1968, 2222; C.A. 70, 56923r (1969).
524. Neimysheva, A. A., V. I. Savchuk, and I. L. Knunyants, Zh. Obshch. Khim., 36, 500 (1966); C.A. 65, 2104c (1966).
525. Nishiwaki, T., Tetrahedron, 21, 3043 (1965).
526. Nixon, J. F., and R. Schmutzler, Spectrochim. Acta, 20, 1835 (1964).
527. Nixon, J. F., and R. Schmutzler, Spectrochim. Acta, 22, 565 (1966).
528. Novitskii, N. A., K. I. Razmova, and M. V. Kivi, Zh. Obshch. Khim., 37, 1136 (1967); C.A. 68, 29800a (1968).
529. Nylen, P., Ber., 57, 1023 (1924).
530. Nyquist, R. A., Appl. Spectrosc., 22, 452 (1968).
531. Okhlobystin, O. Yu., and L. I. Zakharkin, Izv. Akad. Nauk SSSR, Otd. Khim. Nauk, 1958, 1006; C.A. 53, 1122i (1959).
532. Paciorek, K. L., Inorg. Chem., 3, 96 (1964).
533. Paciorek, K. L., and R. H. Kratzer, Inorg. Nucl. Chem. Lett., 2, 39 (1966).
534. Paciorek, K. L., and R. H. Kratzer, U.S. Patent 3,272,846 (1966); C.A. 65, 18618g (1966).
535. Pal'm, V., A. A. Neimysheva, and I. L. Knunyants, Reakts. Sposobnost Org. Soedin., 4, 38 (1967); C.A. 69, 76489m (1968).
536. Panteleeva, A. R., and I. M. Shermergorn, Izv. Akad. Nauk SSSR, Ser. Khim., 1968, 1644; C.A. 69, 87112j (1968).
536a. Passino, H. J., U.S. Patent 2,882,315 (1959); C.A. 53, 16965d (1959).

537. Patel, N. K., and H. J. Harwood, J. Org. Chem., 32, 2999 (1967).
538. Peach, M. E., and T. C. Waddington, J. Chem. Soc., 1962, 2680.
539. Peach, M. E., and T. C. Waddington, J. Chem. Soc., 1962, 3450.
540. Peach, M. E., and T. C. Waddington, J. Chem. Soc., 1963, 799.
541. Perren, E. A., and A. M. Kinnear, British Patent 707,961 (1954); C.A. 49, 7588f (1955).
542. Peterlein, K., German Patent 1,196,195 (1965); C.A. 63, 14907g (1965).
543. Peters, G., J. Org. Chem., 27, 2198 (1962).
544. Petrov, A. A., N. A. Razumova, and Zh. L. Evtikhov, Zh. Obshch. Khim., 37, 1410 (1967); C.A. 68, 2949u (1968).
545. Petrov, K. A., N. K. Bliznyuk, M. A. Korshunov, F. L. Maklyaev, and A. N. Voronkov, Zh. Obshch. Khim., 29, 3407 (1959); C.A. 54, 17245i (1960).
546. Petrov, K. A., A. I. Gavrilova, and V. P. Korotkova, Zh. Obshch. Khim., 32, 1978 (1962); C.A. 58, 4595h (1963).
547. Petrov, K. A., and L. V. Khorkhoyanu, Khim. Org. Soedin. Fosfora, Akad. Nauk SSSR, Otd. Obshch. Tekh. Khim., 1967, 176; C.A. 69, 67480z (1968).
548. Petrov. K. A., F. L. Maklyaev, and N. K. Bliznyuk, Zh. Obshch. Khim., 30, 1602 (1960); C.A. 55, 1414d (1961).
549. Petrov, K. A., E. E. Nifant'ev, and L. P. Sinogeinika, USSR Patent 143,799 (1962); C.A. 57, 5583h (1962).
550. Petrov, K. A., V. A. Parshina, and G. L. Daruze, Zh. Obshch. Khim., 30, 3000 (1960); C.A. 55, 23400a (1961).
551. Petrov, K. A., M. A. Raksha, and V. L. Vinogradov, USSR Patent 174,627 (1965); C.A. 64, 2127d (1966).
552. Petrov, K. A., M. A. Raksha, and V. L. Vinogradov, USSR Patent 179,314 (1966); C.A. 65, 2297a (1966).
553. Petrov, K. A., M. A. Raksha, and V. L. Vinogradov, Zh. Obshch. Khim., 36, 715 (1966). C.A. 65, 8951f (1966).
554. Petrov, K. A., V. V. Smirnov, and V. I. Emel'yanov, Zh. Obshch. Khim., 31, 3027 (1961); C.A. 56, 12934d (1963).
555. Pianfetti, J. A., and L. D. Quin, J. Amer. Chem. Soc., 84, 851 (1962).
556. Plets, V. M., Dissertation, Kazan, 1938.
557. Pollart, K. A., and H. J. Harwood, J. Org. Chem., 27, 4444 (1962).
558. Ponomarenko, F. I., S. Z. Ivin, and K. V. Karavanov, Zh. Obshch. Khim., 39, 382 (1969); C.A. 71, 39087z (1969).

559. Ponomarev, V. V., S. A. Golubtsov, K. A. Andrianov, and E. A. Chuprova, Izv. Akad. Nauk SSSR, Ser. Khim., 1969, 1551; C.A. 71, 113060d (1969).
560. Ponomarev, V. V., A. S. Shapatin, and S. A. Golubtsov, USSR Patent 186,477 (1966); C.A. 66, 85853r (1967).
561. Popoff, I. C., J. Polym. Sci. (Pt. B), 1, 245 (1963).
562. Popoff, I. C., and B. I. D'Iorio, U.S. Dept. Comm. Rept. AD 264,773 (1961); C.A. 59, 3947f (1963).
563. Popoff, I. C., and J. P. King, J. Polym. Sci. (Pt. B), 1, 247 (1963).
564. Popov, E. M., T. A. Mastryukova, N. P. Rodionova, and M. I. Kabachnik, Zh. Obshch. Khim., 29, 1998 (1959); C.A. 54, 8282c (1960).
565. Popov, E. M., E. N. Tsvetkov, J.-Y. Chang, and T. Ya. Medved, Zh. Obshch. Khim., 32, 3255 (1962); C.A. 58, 5165d (1963).
566. Potenza, J. A., E. H. Poindexter, P. J. Caplan, and R. A. Dwek, J. Amer. Chem. Soc., 91, 4356 (1969).
567. Prikoszovich, W., and H. Schindlbauer, Chem. Ber., 102, 2922 (1969).
568. Protopopov, I. S., and M. Ya. Kraft, Zh. Obshch. Khim., 34, 1446 (1964); C.A. 61, 5685f (1964).
569. Pudovik, A. N., I. M. Aladzheva, and L. V. Spirina, Zh. Obshch. Khim., 37, 700 (1967); C.A. 67, 21970q (1967).
570. Pudovik, A. N., R. D. Gareev, and L. I. Kuznetsova, Zh. Obshch. Khim., 39, 1536 (1969); C.A. 71, 113049g (1969).
571. Pudovik, A. N., and E. A. Ishmaeva, Zh. Obshch. Khim., 35, 358 (1965); C.A. 62, 14721c (1965).
572. Pudovik, A. N., and V. K. Khairullin, Zh. Obshch. Khim., 39, 1724 (1969); C.A. 71, 124579f (1969).
573. Pudovik, A. N., V. K. Khairullin, and G. V. Dmitrieva, Dokl. Akad. Nauk SSSR, 174, 372 (1967); C.A. 68, 2959x (1968).
574. Pudovik, A. N., V. K. Khairullin, and V. N. Eliseenkov, Zh. Obshch. Khim., 37, 455 (1967); C.A. 67, 32744v (1967).
575. Pudovik, A. N., V. K. Khairullin, Yu. Yu. Samitov, and R. R. Shagidullin, Zh. Obshch. Khim., 37, 865 (1967); C.A. 68, 105305k (1968).
576. Pudovik, A. N., and I. V. Konovalova, Zh. Obshch. Khim., 31, 1693 (1961); C.A. 55, 24540b (1961).
577. Pudovik, A. N., I. V. Konovalova, and E. A. Ishmaeva, Zh. Obshch. Khim., 33, 2509 (1963); C.A. 60, 1788b (1964).
578. Pudovik, A. N., and R. I. Tarasova, Zh. Obshch. Khim., 34, 1151 (1964); C.A. 61, 1888g (1964).
579. Pudovik, A. N., G. E. Yastrebova, V. I. Nikitina, and Yu. Yu. Samitov, Zh. Obshch. Khim., 38, 292 (1968); C.A. 69, 106815j (1968).

580. Quin, L. D., in "1,4-Cycloaddition Reactions," J. Hamer, Ed., Academic Press, New York, 1967, p. 47.
581. Quinchon, J., M. LeSech, and E. Gryszkiewicz-Trochimowski, Bull. Soc. Chim. (France), 1961, 735.
582. Rafikov, S. R., and M. E. Ergebekov, Zh. Obshch. Khim., 34, 2230 (1964); C.A. 61, 10705b (1964).
583. Rakshys, J. W., R. W. Taft, and W. A. Sheppard, J. Amer. Chem. Soc., 90, 5236 (1968).
584. Ramsden, H. E., U.S. Patent 3,109,851 (1963); C.A. 60, 3015f (1964).
585. Rattenbury, K. H., U.S. Patent 2,993,929 (1961); C.A. 56, 505d (1962).
586. Razumov, A. I., B. G. Liorber, M. B. Gazizov, and Z. M. Khammatova, Zh. Obshch. Khim., 34, 1851 (1964); C.A. 61, 8334g (1964).
587. Razumov, A. I., O. A. Mukhacheva, I. V. Zaikonnikova, N. N. Godovnikov, and N. I. Rizpolozhenskii, Khim. i Primen. Fosfororgan. Soedin., Akad. Nauk SSSR, Tr. 1-oi Konfer., 1955, 205 (Publ. 1957); C.A. 52, 293i (1958).
588. Razumov, A. I., O. A. Mukhacheva, and Sim-Do-Khen, Izv. Akad. Nauk SSSR, Otd. Khim. Nauk, 1952, 894; C.A. 47, 10466b (1953).
589. Razumov, A. I., O. A. Mukhacheva, and Sim-Do-Khen, Tr. Kazan Khim.-Teknol. Inst. S. M. Kirova, 1952, 151; C.A. 51, 6504a (1957).
590. Razumov, A. I., O. A. Mukhacheva, and I. V. Zaikonnikova, Zh. Obshch. Khim., 27, 754 (1957); C.A. 51, 16332e (1957).
591. Razumova, N. A., Zh. L. Evtikhov, L. I. Zubtsova, and A. A. Petrov, Zh. Obshch. Khim., 38, 2342 (1968); C.A. 70, 47546t (1969).
592. Razvodovskaya, L. V., A. F. Grapov, and N. N. Mel'nikov, Zh. Obshch. Khim., 39, 1260 (1969); C.A. 71, 70692s (1969).
593. Regel, E. K., and M. F. Botts, U.S. Patent 3,416,912 (1968); C.A. 71, 22185g (1969).
594. Reinhardt, H., D. Bianchi, and D. Moelle, Chem. Ber., 90, 1656 (1957).
595. Retcofsky, H. L., and C. E. Griffin, Tetrahedron Lett., 1966, 1975.
595a. Richard, J. J., K. E. Burke, J. W. O'Laughlin, and C. V. Banks, J. Amer. Chem. Soc., 83, 1722 (1961).
596. Robinson, M. M., J. Amer. Chem. Soc., 80, 5481 (1958).
597. Rochlitz, F., and H. Vilcsek, German Patent 1,108,687 (1961); C.A. 57, 3484f (1962).
598. Rochlitz, F., and H. Vilcsek, German Patent 1,138,770 (1962); C.A. 58, 9143g (1963).
599. Rochlitz, F., and H. Vilcsek, Angew. Chem., 74, 970 (1962).
600. Roesky, H. W., Chem. Ber., 101, 3679 (1968).

601. Rossiiskaya, P. A., and M. I. Kabachnik, Izv. Akad. Nauk SSSR, Otd. Khim. Nauk, 1947, 389; C.A. 42, 2924b (1948).
602. Rozen, A. M., and Z. I. Nikolotova, Zh. Neorgan. Khim., 9, 1725 (1964); C.A. 61, 10102a (1964).
603. Rozinov, V. G., and E. F. Grechkin, Zh. Obshch. Khim., 39, 934 (1969); C.A. 71, 50098z (1969).
604. Rozinov, V. G., A. L. Taskina, and E. F. Grechkin, Zh. Obshch. Khim., 39, 1647 (1969); C.A. 71, 91585z (1969).
605. Sachs, H., Ber., 25, 1514 (1892).
605a. Schimmelknecht. K., and W. Denk, German Patent 1,020,019 (1956); C.A. 1959, 11767.
606. Schindlbauer, H., Chem. Ber., 100, 3432 (1967).
607. Schindlbauer, H., and W. Prikoszovich, Chem. Ber., 102, 2914 (1969).
608. Schliebs, R., German Patent 1,119,862 (1961); C.A. 58, 6862b (1963).
609. Schliebs, R., German Patent 1,210,834 (1966); C.A. 64, 17639a (1966).
610. Schloer, H., and G. Schrader, German Patent 1,067,021 (1961); C.A. 56, 2474d (1962).
611. Schloer, H., and G. Schrader, German Patent 1,114,359 (1958); C.A. 56, 10191f (1962).
612. Schmidpeter, A., and H. Brecht, Z. Naturforsch., B 23, 1592 (1968).
613. Schmidpeter, A., and H. Groeger, Chem. Ber., 100, 3052 (1967).
614. Schmidt, M., and H. Bipp, Sitzber. Ges. Befoerder. Ges. Naturwiss, Marburg, 83/84, 523 (1961/62); C.A. 59, 6436b (1963).
615. Schmutzler, R., J. Inorg. Nucl. Chem., 25, 335 (1963).
616. Schmutzler, R., Inorg. Chem., 3, 410 (1964).
617. Schmutzler, R., Org. Synth., 45, 99 (1965).
618. Schmutzler, R., J. Chem. Soc., 1964, 4551.
619. Schmutzler, R., Org. Synth., 46, 21 (1966).
620. Schmutzler, R., and G. S. Reddy, Z. Naturforsch., 20b, 832 (1965).
621. Schrader, G., unpublished work, quoted in: Houben-Weyl, Methoden der Organischen Chemie, 4.Aufl. Band XII/1; Georg Thieme Verlag, Stuttgart 1963.
622. Schrader, G., German Patent 1,058,056 (1959); C.A. 55, 7290b (1961).
623. Schrader, G., German Patent 1,099,532 (1961); C.A. 56, 3515e (1962).
624. Schrader, G., German Patent 1,085,874 (1959); C.A. 56, 4796e (1962).
625. Schrader, G., Belgian Patent 608,802 (1962); C.A. 57, 7311d (1962).
626. Schrader, G., German Patent 1,146,882 (1963); C.A. 59, 10121g (1963).
627. Schrader, G., German Patent 1,158,509 (1963); C.A. 60, 6870d (1964).

628. Schrader, G., U.S. Patent 3,184,465 (1965); C.A. 63, 11578d (1965).
629. Schrader, G., W. Lorenz, R. Coelln, and H. Schloer, U.S. Patent 3,232,830 (1966); C.A. 64, 15923a (1966).
629a. Schwarzenbach, G., H. Ackermann, and P. Ruckstuhl, Helv. Chim. Acta, 32, 1175 (1949).
630. Seel, F., K. Ballreich, and R. Schmutzler, Chem. Ber., 95, 199 (1962).
631. Semerneva, G. A., and V. I. Kodolov, Tr. Inst. Khim., Akad. Nauk SSSR, Ural. Filial, 1968, No. 15, 19; C.A. 70, 48001y (1969).
632. Semin, G. K., and T. A. Babushkina, Teor. Eksp. Khim., 4, 835 (1968); C.A. 71, 44253e (1969).
633. Semin, G. K., and E. V. Bryuchova, Chem. Commun., 1968, 605.
634. Sennewald, K., A. Ohorodnik, D. Kirstein, and H. J. Hardel, U.S. Patent 3,397,122 (1968); C.A. 69, 87182g (1968).
635. Shaturskii, Ya. P., Y. S. Grushin, G. K. Fedorova, and A. V. Kirsanov, Zh. Obshch. Khim., 39, 1467 (1969); C.A. 71, 113048f (1969).
636. Sheluchenko, V. V., M. A. Landau, S. S. Dubov, A. A. Neimysheva, and I. L. Knunyants, Dokl. Akad. Nauk SSSR, 177, 376 (1967); C.A. 68, 44610g (1968).
637. Shevchenko, V. I., A. S. Shtepanek, and A. V. Kirsanov, Zh. Obshch. Khim., 31, 3062 (1961); C.A. 56, 15541c (1962).
638. Shikhiev, I. A., M. I. Aliev, S. Z. Israfilova, and D. R. Israelyan, Azerb. Khim. Zh., 1960, 29; C.A. 55, 22167h (1961).
639. Shokol, V. A., V. V. Doroshenko, and G. I. Derkach, Zh. Obshch. Khim., 39, 938 (1969); C.A. 71, 50087v (1969).
640. Shokol, V. A., V. F. Gamaleya, and G. I. Derkach, Zh. Obshch. Khim., 39, 856 (1969); C.A. 71, 61497e (1969).
641. Shokol, V. A., V. F. Gamaleya, and G. I. Derkach, Zh. Obshch. Khim., 37, 2528 (1967); C.A. 68, 87352s (1968).
642. Shokol, V. A., V. F. Gamaleya, and G. I. Derkach, Zh. Obshch. Khim., 38, 1081 (1968); C.A. 69, 59340p (1968).
643. Shokol, V. A., V. F. Gamaleya, and G. I. Derkach, Zh. Obshch. Khim., 38, 1104 (1968); C.A. 69, 59336s (1968).
644. Shokol, V. A., V. F. Gamaleya, and G. I. Derkach, Zh. Obshch. Khim., 39, 1703 (1969); C.A. 71, 124593f (1969).
645. Shokol, V. A., V. F. Gamaleya, L. I. Molyavko, and G. I. Derkach, Khim. Org. Soedin. Fosfora, Akad. Nauk SSSR, Otd. Obshch. Tekh. Khim., 1967, 109; C.A.

69, 36225r (1968).
646. Short, J. N., U.S. Patent 2,871,263 (1959); C.A. 53, 11225f (1959).
647. Shostakovskii, M. F., I. I. Guseinov, L. I. Shmonina, G. S. Vasil'ev, and B. V. Lopatin, Zh. Obshch. Khim., 30, 2836 (1960); C.A. 55, 17487h (1961).
648. Shostakovskii, M. F., I. I. Guseinov, and G. S. Vasil'ev, Zh. Obshch. Khim., 30, 2832 (1960); C.A. 55, 17487e (1961).
649. Shostakovskii, M. F., L. I. Shmonina, and I. M. Tikhomirova, Izv. Akad. Nauk SSSR, Ser. Khim., 1963, 2193; C.A. 60, 9137d (1964).
650. Shtepanek, A. S., Yu. V. Piven, and G. I. Derkach, Khim. Org. Soedin. Fosfora, Akad. Nauk SSSR, Otd. Obshch. Tekh. Khim., 1967, 61; C.A. 68, 114695q (1968).
651. Siddall, T. H., and C. A. Prohaska, J. Org. Chem., 28, 2908 (1963).
652. Siddall, T. H., and C. A. Prohaska, J. Amer. Chem. Soc., 84, 2502 (1962).
653. Siddall, T. H., C. A. Prohaska, and W. E. Shuler, Nature, 190, 903 (1961).
654. Slovokhotova, N. A., K. N. Anisimov, G. M. Kunitskaya, and N. E. Kolobova, Izv. Akad. Nauk SSSR, Otd. Khim. Nauk, 1961, 71; C.A. 55, 23049b (1961).
655. Smirnov, E. A., Yu. A. Kondrat'ev, V. A. Petrunin, and Y. M. Zinov'ev, USSR Patent 196,818 (1967); C.A. 68, 69126s (1968).
656. Smirnov, E. A., Yu. M. Zinov'ev, and V. A. Petrunin, Zh. Obshch. Khim., 38, 1551 (1968); C.A. 69, 96827y (1968).
657. Smirnov, O. K., and N. I. Grineva, Neftekhimiya, 2, 237 (1962); C.A. 59, 656g (1963).
658. Smith, W. C., U.S. Patent 2,950,306 (1960); C.A. 55, 2569i (1961).
659. Smith, W. C., J. Amer. Chem. Soc., 82, 6176 (1960).
660. Soborovskii, L. Z., and Yu. M. Zinov'ev, Zh. Obshch. Khim., 24, 516 (1954); C.A. 49, 6086i (1955).
661. Soborovskii, L. Z., Yu. M. Zinov'ev, and M. A. Englin, Dokl. Akad. Nauk SSSR, 73, 333 (1950); C.A. 45, 2854c (1951).
662. Soborovskii, L. Z., Yu. M. Zinov'ev, and M. A. Englin, Dokl. Akad. Nauk SSSR, 67, 293 (1949); C.A. 44, 1401h (1950).
663. Soborovskii, L. Z., Yu. M. Zinov'ev, and L. I. Muler, Dokl. Akad. Nauk SSSR, 109, 98 (1956); C.A. 51, 1825i (1957).
664. Soborovskii, L. Z., Yu. M. Zinov'ev, and T. G. Spiridonova, Zh. Obshch. Khim., 29, 1139 (1959); C.A. 54, 1270c (1960).
665. Sorokin, M. F., and I. Manoviciu, Mater. Plast., 3,

136 (1966); C.A. 65, 17131c (1966).
666. Sosnovsky, G., and D. J. Rawlinson, J. Org. Chem., 33, 2325 (1968).
667. Sowerby, D. B., J. Inorg. Nucl. Chem., 22, 205 (1961).
668. Spence, R. A., J. M. Swan, W. G. R. M. DeBoer, T. Ghose, R. C. Nairn, J. M. Rolland, H. A. Ward, and S. H. B. Wright, Clin. Exp. Immunol., 3, 865 (1968).
669. Steger, E., and M. Kuntze, Spectrochim. Acta, A 23, 2189 (1967).
670. Steger, E., J. Rehak, and H. Faltus, Z. Physik. Chem. (Leipzig), 229, 110 (1965).
671. Stetter, H., and W. D. Last, Chem. Ber., 102, 3364 (1969).
672. Steyermark, P. R., J. Org. Chem., 28, 586 (1963).
673. Sun Oil Co., Netherlands Appl. 6,501,808 (1965); C.A. 64, 6693c (1966).
674. Tarasov, V. V., Ya. S. Arbisman, Yu. A. Kondrat'ev, and S. Z. Ivin, Zh. Obshch. Khim., 38, 130 (1968); C.A. 69, 58713a (1968).
675. Tarasov, V. V., Yu. A. Kondrat'ev, S. S. Dubov, O. G. Strukov, and S. Z. Ivin, Zh. Vses. Khim. Obshch., 14, 116 (1969); C.A. 70, 114405m (1969).
676. Timofeeva, T. N., B. I. Ionin, Yu. L. Kleiman, N. V. Morkovin, and A. A. Petrov, Zh. Obshch. Khim., 38, 1255 (1968); C.A. 69, 77353z (1968).
677. Titov, A. I., M. V. Sizova, and P. O. Gitel, Dokl. Akad. Nauk SSSR, 159, 385 (1964); C.A. 62, 6509h (1965).
678. Tomaschewski, G., and B. Breitfeld, J. Prakt. Chem., 311, 256 (1969).
679. Tomaschewski, G., A. Otto, and D. Zanke, Arch. Pharm. (Weinheim), 301, 520 (1968).
680. Toy, A. D. F., J. Amer. Chem. Soc., 70, 186 (1948).
681. Toy, A. D. F., U.S. Patent 2,482,810 (1949); C.A. 44, 658f (1960).
682. Toy, A. D. F., U.S. Patent 2,425,765 (1947); C.A. 42, 596d (1948).
683. Toy, A. D. F., U.S. Patent 2,425,766 (1947); C.A. 42, 596h (1948).
684. Toy, A. D. F., and K. H. Rattenbury, U.S. Patent 3,244,745 (1966); C.A. 64, 17638f (1966).
685. Toy, A. D. F., and E. H. Uhing, U.S. Patent 3,110,727 (1963); C.A. 60, 4185c (1964).
686. Traise, T. P., and E. N. Walsh, U.S. Patent 2,924,560 (1960); C.A. 54, 11993 (1960).
687. Tsivunin, V. S., S. V. Fridland, T. V. Zykova, and G. Kamai, Zh. Obshch. Khim., 36, 1424 (1966); C.A. 66, 28843p (1967).
688. Tsivunin, V. S., and G. Kamai, Dokl. Akad. Nauk SSSR, 131, 1113 (1960); C.A. 54, 20843d (1960).
689. Tsivunin, V. S., G. Kamai, and S. V. Fridland, Zh.

Obshch. Khim., 33, 2146 (1963); C.A. 59, 12837g (1963).
690. Tsivunin, V. S., G. K. Kamai, and S. V. Fridland, Zh. Obshch. Khim., 36, 436 (1966); C.A. 65, 741d (1966).
691. Tsivunin, V. S., G. Kamai, and R. S. Khisamutdinova, Dokl. Akad. Nauk SSSR, 164, 594 (1965); C.A. 63, 18143g (1965).
692. Tsivunin, V. S., G. Kamai, R. S. Khisamutdinova, and E. M. Smirnov, Zh. Obshch. Khim., 35, 1231 (1965); C.A. 63, 11608b (1965).
693. Tsivunin, V. S., G. K. Kamai, and V. V. Kormachev, USSR Patent 187,784 (1966); C.A. 67, 82264j (1967).
694. Tsivunin, V. S., G. K. Kamai, and V. V. Kormachev, Zh. Obshch. Khim., 36, 1663 (1966); C.A. 66, 55550x (1967).
695. Tsivunin, V. S., G. Kamai, and G. K. Makeeva, Dokl. Akad. Nauk SSSR, 135, 1157 (1960); C.A. 55, 12271d (1961).
696. Tsivunin, V. S., G. Kamai, and D. B. Sultanova, Zh. Obshch. Khim., 33, 2149 (1963); C.A. 59, 12839b (1963).
697. Tsvetkov, E. N., D. I. Lobanov, and M. I. Kabachnik, Zh. Obshch. Khim., 38, 2285 (1968); C.A. 71, 13176c (1969).
698. Tsvetkov, E. N., G. K. Semin, T. A. Babushkina, D. I. Lobanov, and M. I. Kabachnik, Izv. Akad. Nauk SSSR, Ser. Khim., 1967, 2375; C.A. 68, 34550y (1968).
699. Tullock, C. W., and D. D. Coffman, J. Org. Chem., 25, 2016 (1960).
700. Utvary, K., Inorg. Nucl. Chem. Lett., 1, 77 (1965).
701. Utvary, K., E. Freundlinger, and V. Gutmann, Monatsh. Chem., 97, 348 (1966).
702. Utvary, K., E. Freundlinger, and V. Gutmann, Monatsh. Chem., 97, 679 (1966).
703. Utvary, K., and R. Hagenauer, Monatsh. Chem., 94, 797 (1963).
704. Van Wazer, J. R., C. F. Callis, J. N. Shoolery, and R. C. Jones, J. Amer. Chem. Soc., 78, 5715 (1956).
705. Van Wazer, J. R., and J. H. Letcher, Topics Phosphorus Chem., 5, 169 (1967).
706. Van Winkle, J. L., U.S. Patent 3,014,964 (1961); C.A. 56, 10191e (1962).
707. Vasil'ev, G. S., I. Guseinov, and M. F. Shostakovskii, Zh. Obshch. Khim., 34, 1216 (1964); C.A. 61, 1890a (1964).
708. Vasil'ev, G. S., E. N. Prilezhaeva, V. F. Bystrov, and M. F. Shostakovskii, Zh. Obshch. Khim., 35, 1350 (1965); C.A. 63, 14899b (1965).
709. Vilcsek, H., and F. Rochlitz, German Patent 1,103,922 (1961); C.A. 56, 1481i (1962).

710. Vilkov, L. V., N. I. Sadova, and I. Yu. Zil'berg, Zh. Strukt. Khim., 8, 528 (1967); C.A. 67, 112116g (1967).
711. Vilkov, L. V., V. P. Spiridonov, and N. I. Sadova, Zh. Strukt. Khim., 9, 187 (1968); C.A. 69, 69730e (1968).
712. Vinokurova, G. M., and I. A. Aleksandrova, Izv. Akad. Nauk SSSR, Ser. Khim., 1969, 884; C.A. 71, 50116d (1969).
713. Viout, M. P., and P. Rumpf, Bull. Soc. Chim. (France), 1957, 768.
714. Walsh, E. N., U.S. Patent 2,685,603 (1954); C.A. 49, 10358g (1955).
715. Walsh, E. N., T. M. Beck, and A. D. F. Toy, J. Amer. Chem. Soc., 78, 4455 (1956).
716. Walsh, E. N., T. M. Beck, and W. H. Woodstock, J. Amer. Chem. Soc., 77, 929 (1955).
717. Washburn, R. M., U.S. Patent 3,212,844 (1965); C.A. 64, 3601c (1966).
718. Weber, C. W., U.S. Patent 2,822,304 (1959); C.A. 53, 16964f (1959).
718a. Weber, C. W., U.S. Patent 2,882,306 (1959); C.A. 53, 16964g (1959).
719. Weber, C. W., U.S. Patent 2,882,242 (1959); C.A. 53, 17904h (1959).
719a. Weber, C. W., U.S. Patent 2,882,314 (1959); C.A. 53, 16965c (1959).
720. Weeks, M. H., N. P. Musselman, P. P. Yevich, K. H. Jacobson, and F. W. Oberst, Amer. Ind. Hyg. Assoc. J., 25, 470 (1964).
721. Weilmuenster, E. A., and J. J. Minnick, U.S. Patent 3,179,695 (1965); C.A. 63, 632e (1965).
722. Weitkamp, H., and F. Korte, Z. Anal. Chem., 204, 245 (1964).
723. Weller, J., Ber., 20, 1718 (1887).
724. Weller, J., Ber., 21, 1492 (1888).
725. Williams, R. H., and L. A. Hamilton, J. Amer. Chem. Soc., 74, 5418 (1952).
726. Wismer, M., U.S. Patent 3,294,745 (1966); C.A. 66, 56695k (1967).
727. Woodstock, W. H., U.S. Patent 2,471,472 (1949); C.A. 43, 7499e (1949).
728. Woodstock, W. H., U.S. Patent 2,495,799 (1950); C.A. 44, 3517h (1950).
729. Yagupolskii, L. M., and Zh. M. Ivanova, Zh. Obshch. Khim., 29, 3766 (1959); C.A. 54, 19553c (1960).
730. Yagupolskii, L. M., and Zh. M. Ivanova, Zh. Obshch. Khim., 30, 1284 (1960); C.A. 55, 429d (1961).
731. Yagupolskii, L. M., and Zh. M. Ivanova, Zh. Obshch. Khim., 30, 4026 (1960); C.A. 55, 22196e (1961).
732. Yagupolskii, L. M., and P. A. Yufa, Zh. Obshch. Khim., 30, 1294 (1960); C.A. 55, 431h (1961).

733. Yakubovich, A. Ya., and V. A. Ginsburg, Dokl. Akad. Nauk SSSR, 82, 273 (1952); C.A. 47, 2685f (1953).
734. Yakubovich, A. Ya., and V. A. Ginsburg, Zh. Obshch. Khim., 24, 1465 (1954); C.A. 49, 10834g (1955).
735. Yakubovich, A. Ya., V. A. Ginsburg, and S. P. Makarov, Dokl. Akad. Nauk SSSR, 71, 303 (1950); C.A. 44, 8319h (1950).
736. Yakubovich, A. Ya., L. Z. Soborovskii, L. I. Muler, and V. S. Faermark, Zh. Obshch. Khim., 28, 317 (1958); C.A. 52, 13613d (1958).
737. Yoke, J. T., and R. G. Laughlin, U.S. Patent 3,304,330 (1967); C.A. 67, 34100f (1967).
738. Yokoyama, M., E. Akagi, and K. Minami, Kogyo Kagaku Zasshi, 68, 460 (1965); C.A. 63, 7037e (1965).
739. Yoshioka, H., and S. Horie, Japanese Patent 11,823 (1962); C.A. 59, 10125d (1963).
740. Yoshioka, H., and S. Horie, Japanese Patent 11,824 (1962); C.A. 59, 10125d (1963).
741. Yuldashev, A., M. A. Askarov, and F. Ibragimov, Khim. Fiz.-Khim. Prirodn. Sintetich. Polomerov, Akad. Nauk Uz. SSSR, Inst. Khim. Polomerov, No. 2, 149 (1964); C.A. 61, 10702f (1964).
742. Zeffert, B. M., P. B. Coulter, and H. Tannenbaum, J. Amer. Chem. Soc., 82, 3843 (1960).
743. Zhmurova, I. N., and I. Yu. Voitsekhovskaya, Zh. Obshch. Khim., 34, 1171 (1964); C.A. 61, 1889c (1964).
744. Zingaro, R. A., Inorg. Chem., 2, 192 (1963).
745. Zinov'ev, Yu. M., V. N. Kulakova, and L. Z. Soborovskii, Zh. Obshch. Khim., 28, 1551 (1958); C.A. 53, 1117g (1959).
746. Zinov'ev, Yu. M., and L. Z. Soborovskii, Zh. Obshch. Khim., 29, 615 (1959); C.A. 54, 340g (1960).
747. Zinov'ev, Yu. M., and L. Z. Soborovskii, Zh. Obshch. Khim., 29, 2643 (1959); C.A. 54, 10836c (1960).
748. Zinov'ev, Yu. M., and L. Z. Soborovskii, Zh. Obshch. Khim., 26, 3030 (1956); C.A. 51, 8662b (1957).
749. Zinov'ev, Yu. M., and L. Z. Soborovskii, Zh. Obshch. Khim., 30, 1571 (1960); C.A. 55, 1415b (1961).
750. Zinov'ev, Yu. M., and L. Z. Soborovskii, USSR Patent 139,319 (1960); C.A. 56, 8749f (1962).
751. Zinov'ev, Yu. M., and L. Z. Soborovskii, Zh. Obshch. Khim., 34, 929 (1964); C.A. 60, 15904d (1964).
752. Zinov'ev, Yu. M., and L. Z. Soborovskii, Khim. Org. Soedin. Fosfora, Akad. Nauk SSSR, Otd. Obshch. Tekh. Khim., 1967, 200; C.A. 69, 36213k (1968).
753. Zinsmeister, R., and R. Wirtz, German Patent 1,183,080 (1964); C.A. 62, 9173h (1965).
754. Zyablikova, T. A., I. M. Magdeev, and I. M. Shermergorn, Izv. Akad. Nauk SSSR, Ser. Khim., 1968, 397; C.A. 69, 59345u (1968).

Chapter 10. Phosphonous Acids (Thio-, Seleno Analogs) and Derivatives

A. W. FRANK

United States Department of Agriculture,
New Orleans, Louisiana

In this chapter we consider all those compounds which have in common a single carbon-phosphorus bond and a phosphorus atom in a lower state of oxidation (i.e. RPX_2 where X = H, Cl, OH, NR_2, etc.). The only exceptions are the primary phosphines (RPH_2) and the phosphonous dihalides ($RPCl_2$, etc.), which were considered in Chapters 1 and 8, respectively.

The nomenclature of the compounds containing a free acid function presents a problem, because they can be named either as derivatives of phosphonous acid (1) or of phosphinic acid (2).[149]

$(HO)_2PH$ (1) $HOP(O)H_2$ (2)

The evidence favors structure (2) for the acids and thioacids, but structure (1) for the amides formed by the replacement of OH by NHR (see p. 309). The policy adopted throughout this chapter is to name all compounds according to the form in which they normally exist. The ester MePH(O)OEt, for example, is named as a phosphinate ester rather than a phosphonite ester: ethyl methylphosphinate.

A. SYNTHETIC ROUTES

A.1. Primary Phosphine Oxides (and Sulfides)

Primary phosphine oxides are prepared by the oxidation of primary phosphines with hydrogen peroxide at 0° (method 1)[96] or by the acid-catalyzed addition of phosphine to ketones (method 2).[95]

$$RPH_2 + H_2O_2 \longrightarrow RP(O)H_2 + H_2O \quad (1)$$

$$PH_3 + R_2CO \longrightarrow R_2CHP(O)H_2 \quad (2)$$

Method 2 is limited to primary phosphine oxides that contain secondary alkyl substituents. Simple ketones do not react with phosphine in the presence of catalytic amounts of acid; in strongly acidic media such as concentrated HCl, however, addition takes place under mild conditions, giving a mixture of the primary phosphine oxide and its ketone adduct $R_2CHPH(O)CR_2OH$. The reaction rate and the ratio of the products are determined by steric factors. The adduct predominates with less hindered ketones such as cyclohexanone, and the primary phosphine oxide with more hindered ketones such as 4-heptanone.[95]

Primary phosphine sulfides are prepared by the reaction of primary phosphines with sulfur at 50° (method 3).[342,402]

$$RPH_2 + S \longrightarrow RP(S)H_2 \qquad (3)$$

Only one compound of this class is known (R = Ph).

A.2. Phosphonous Acids (Thio-, Seleno Analogs)

A.2.1. From Hypophosphorous Acid

A.2.1.1. Addition to Olefins. Hypophosphorous acid[234] and its salts[698,712] add to olefins in the presence of free radical catalysts, giving phosphinic acids in almost quantitative yield (method 1a).

$$RCH{=}CH_2 + H_3PO_2 \xrightarrow{t\text{-}Bu_2O_2} RCH_2CH_2PH(O)OH \qquad (1a)$$

Further addition to the remaining phosphorus-hydrogen bond is possible.[234,523,712] The α,β-unsaturated esters, such as diethyl maleate, are reported to react with hypophosphorous acid in the absence of a catalyst upon prolonged heating at 100° (method 1b),[598] but other investigators have been unable to confirm this.[364]

A.2.1.2. Reaction with Carbinols. Hypophosphorous acid reacts with triphenylcarbinol, dinaphthopyranol, and Michler's hydrol giving products with carbon-phosphorus bonds (method 2).[187,188] Substantial amounts of the corres-

$$Ph_3COH + H_3PO_2 \longrightarrow Ph_3CPH(O)OH + H_2O \qquad (2)$$

ponding hydrocarbons may also be formed, depending on the reaction conditions.

A.2.1.3. Addition to Carbonyl Compounds. Crystalline hypophosphorous acid adds to aldehydes or ketones upon prolonged heating on the water bath, forming α-hydroxyalkylphosphinic acids (method 3;[736] kinetics[266]).

$$R_2CO + H_3PO_2 \longrightarrow R_2C(OH)PH(O)OH \qquad (3)$$

The products are capable of reacting further with the aldehyde or with acetone, forming bis(α-hydroxyalkyl)phosphinic acids. This eventuality is avoided by maintaining the hypophosphorous acid in excess.[736] Ketones higher than acetone do not form disubstitution products.[420,421]

Reactions with aromatic α-diketones are complicated by

redox reactions. The products are those of the addition of hypophosphorous acid to the carbonyl group of the resulting benzoin or desoxybenzoin.[560]

A.2.1.4. Addition to Schiff Bases. The addition of hypophosphorous acid to aldehydes or ketones gives α-aminoalkylphosphinic acids if a primary amine is added to the reaction mixture (method 4).[672-674] The ketone may be

$$RNH_2 + R_2^1CO + H_3PO_2 \longrightarrow RNHCR_2^1PH(O)OH + H_2O \quad (4)$$

present in excess as the solvent,[672] but aldehydes are best used in the form of Schiff bases, to avoid side reactions.[363,385] α-Hydroxyalkylphosphinic acids are not intermediates, for they do not react with primary amines under these conditions.[672]

Secondary amines do not participate in this reaction[672] unless the carbonyl compound is formaldehyde (method 5).[406]

$$R_2NCH_2OH + H_3PO_2 \longrightarrow R_2NCH_2PH(O)OH + H_2O \quad (5)$$

Enamines derived from primary amines and 1,3-dicarbonyl compounds react in a normal manner, but those derived from secondary amines are unreactive. Unsubstituted enamines give products that evolve ammonia, giving α,β-unsaturated phosphinic acids.[364]

A.2.1.5. Reaction with Acid Anhydrides. Potassium hypophosphite is acetylated by acetic anhydride at 100°, giving the monoacetyl derivative in 15% yield (method 6).[322]

$$Ac_2O + KH_2PO_2 \longrightarrow AcPH(O)OK + AcOH \quad (6)$$

Other acid anhydrides react similarly.[322,323]

A.2.1.6. Reaction with Diazonium Salts and Arylhydrazines. Aromatic phosphinic acids are said to be obtained by the reaction of sodium hypophosphite with diazonium salts (method 7)[558,648] or arylhydrazines (method 8).[559] The validity of these methods has been questioned.[347a,737,746]

A.2.2. From White Phosphorus or Phosphorus Trichloride and Olefins

When solutions of white phosphorus in benzene or petroleum ether contain olefins such as cyclohexene, pinene, and oleic acid, they rapidly absorb oxygen, giving crystalline "phosphorates" (3), which contain two phosphorus atoms per olefin unit. Hydrolysis of (3) gives an α,β-unsaturated phosphinic acid (method 9).[617,756]

$$RCH=CH_2 \xrightarrow{P_4,\ O_2} \underset{(3)}{RC_2H_3P_2O_4} \xrightarrow[-H_3PO_4]{H_2O} RCH=CHPH(O)OH \quad (9)$$

Similar products are obtained by the addition of phosphorus trichloride to olefins in the presence of acetic anhydride. The resulting "phosphorites" (4), or their oxidation products the "phosphorates" (3), are both hydrolyzed to α,β-unsaturated phosphinic acids (method 10).[706,707] The structures of (3) and (4) are discussed elsewhere.[191b,740]

$$RCH=CH_2 \xrightarrow{PCl_3,\ Ac_2O} \underset{(4)}{RC_2H_3P_2O_3} \xrightarrow[-H_3PO_3]{H_2O} RCH=CHPH(O)OH \quad (10)$$

A.2.3. From Phosphonous Dihalides

A.2.3.1. Hydrolysis. The phosphonous dihalides are readily hydrolyzed to phosphinic acids by water [methods 11 (X = Cl),[231,447,614] 12 (X = F),[280] 13 (X = Br),[448] and 14 (X = I)[162,262]].

$$RPX_2 + 2H_2O \longrightarrow RPH(O)OH + 2HX \quad (11\text{-}14)$$

Hydrolysis with aqueous alkali may be used with those compounds (R = p-$Me_2NC_6H_4$,[66,451,661] Ph_3C[248]), which give acid-sensitive phosphinic acids.

Partial hydrolysis to a phosphinic chloride has been found to be possible in only one instance (R = 2,4,6-t-$Bu_3C_6H_2$). Steric hindrance prevents hydrolysis of the second phosphorus-chlorine group (method 15).[120]

$$RPCl_2 + H_2O \longrightarrow RPH(O)Cl + HCl \quad (15)$$

A.2.3.2. Alcoholysis. The products of aqueous hydrolysis are often contaminated by a repulsive phosphine-like odor, attributable to the reducing action of the hydrogen halide (HI>>HBr>HCl).[348,448] If phosphonous dichlorides are first added to alcohol and then heated briefly to reflux with water, the yields are improved and the products are odorless (method 16).[159,344,348]

$$RPCl_2 \xrightarrow{R'OH} RPH(O)OR' \xrightarrow[HCl]{H_2O} RPH(O)OH \qquad (16)$$

Similar results are obtained with the phosphonous dibromides (method 17)[100] and diiodides (method 18).[342]

A.2.4. From Phosphonous Diesters

Phosphinic acids may be prepared from phosphonous diesters by prolonged refluxing with water (method 19a).[136,443,450] The reaction is catalyzed by both acids (method 19b)[658] and bases (method 19c).[658]

$$RP(OR')_2 + H_2O + NaOH \longrightarrow RPH(O)ONa + 2R'OH \qquad (19c)$$

The phosphinic esters, which are the products of partial hydrolysis under acid conditions (p. 272), are also converted to phosphinic acids by hydrolysis. Hydrolysis is slow in a neutral medium and is catalyzed by both acids and bases (methods 20a and 20b, respectively).[113,136,658]

$$RPH(O)OR' + H_2O \xrightarrow{HCl} RPH(O)OH + R'OH \qquad (20a)$$

Anhydrous conditions favor more extensive cleavage of the ester groups. Thus phosphinic acids, rather than esters, are obtained upon treatment of phosphonous diesters with hydrogen bromide (method 21)[100] and upon alcoholysis of the sodium salts of phosphinic esters (method 22).[11]

A.2.5. From Phosphonous Diamides

The phosphonous diamides are hydrolyzed to phosphinic acids by refluxing with hydrochloric acid (method 23).[452,734] In this case, partial hydrolysis products are more

$$RP(NR_2)_2 + H_2O\ (HCl) \longrightarrow RPH(O)OH + 2R_2NH \cdot HCl \qquad (23)$$

difficult to prepare (see p. 280).

Hydrolysis of the iminophosphines RP=NNHAr with hot water yields the arylhydrazine salts of the phosphinic acids (method 24).[449]

$$RP{=}NNHAr + 2H_2O \longrightarrow RPH(O)OH \cdot NH_2NHAr \qquad (24)$$

A.2.6. From Primary Phosphines

The primary phosphines are highly sensitive to oxidation, but with careful work it is possible to oxidize them either to primary phosphine oxides (method 1, A.1) or to phosphinic acids (method 25).

$$RPH_2 + O_2 \longrightarrow RPH(O)OH \tag{25}$$

Air oxidation[342,473] yields some phosphoric acid and phosphonic acid, in addition to the phosphinic acid, regardless of the oxygen concentration.[93,378] Mild oxidants such as hydrogen peroxide[93] or dinitrogen tetroxide[93] produce less overoxidation than air or oxygen. The preferred oxidants are mercuric chloride[137] or 30% hydrogen peroxide.[93,258]

Hydrogen peroxide is also used to oxidize primary phosphine oxides to phosphinic acids (method 26).[95,96]

$$RP(O)H_2 + H_2O_2 \longrightarrow RPH(O)OH + H_2O \tag{26}$$

A.2.7. From Phosphinous Acids and their Derivatives

Trifluoromethylphosphinic acid $CF_3PH(O)OH$ is formed upon aqueous hydrolysis of bis(trifluoromethyl)phosphinous acid and its derivatives [$(CF_3)_2PX$(X = OH,[67] Cl,[67] I,[67,162] NH_2[238])](method 27). The product is stable to water, but

$$(CF_3)_2POH + H_2O \longrightarrow CF_3PH(O)OH + CHF_3 \tag{27}$$

not to aqueous base. It liberates fluoroform quantitatively on treatment with alkali.[67]

The acid may also be prepared by shaking the tertiary phosphine $(CF_3)_3P$ with 1 mole of aqueous base (method 28).[67,161]

$$(CF_3)_3P + H_2O + KOH \longrightarrow CF_3PH(O)OK + 2CHF_3 \tag{28}$$

Other perfluoro derivatives behave similarly.[62]

A.3. Phosphonous Esters (and Thio- Analogs)

A.3.1. From Hypophosphite Esters and Olefins or Carbonyl Compounds

Hypophosphite esters, like the acid itself, add to olefins,[495] aldehydes,[493] or ketones.[182] Butyl hypophosphite reacts with cyclohexene when these compounds are heated in an autoclave in the presence of di-t-butyl peroxide catalyst, giving butyl cyclohexylphosphinate (method 1).[495]

$$H_2P(O)OR + C_6H_{10} \longrightarrow C_6H_{11}PH(O)OR \qquad (1)$$

Hypophosphorous acid is esterified rapidly by acetals or ketals, forming products that interact slowly (20 days for ketals, 5 months for acetals), giving the phosphinic esters in high yield (method 2).[182] (See also Chapter 14).

$$H_3PO_2 + Me_2C(OMe)_2 \longrightarrow [H_2P(O)OMe + Me_2CO] \longrightarrow$$

$$HOCMe_2PH(O)OMe \qquad (2)$$

A.3.2. From Phosphite Esters

A.3.2.1. Reaction with Grignard Reagents. Under carefully controlled conditions, trialkyl phosphites react with Grignard reagents, giving phosphonous diesters in 30 to 50% yields (method 3a).[400,658]

$$(RO)_3P + R'MgX \longrightarrow R'P(OR)_2 + Mg(OR)X \qquad (3a)$$

The conditions to be used which vary with the Grignard reagent, should be as mild as possible to avoid further substitution.[658] Similar results are obtained with the trialkyl phosphorotrithioites (method 4).[658]

If one of the ester groups in the trialkyl phosphite is replaced by a chlorine atom, the reaction with the Grignard reagent takes place at ice-cold temperatures (<10°), displacing only the chlorine atom (method 5a).[301,310,658]

$$(RO)_2PCl + R'MgCl \longrightarrow R'P(OR)_2 + MgCl_2 \qquad (5a)$$

Yields of 60 to 70% are common. A disadvantage of this method is the difficulty of preparing pure dialkyl phosphorochloridites, which may account for the difference between physical properties observed in products prepared by this method and those prepared by other methods.[40] One way to overcome this difficulty is through the use of aromatic phosphorochloridites, which are easier to prepare than the aliphatic compounds.[310,658]

Phosphonous diesters are easily hydrolyzed. If the reaction mixtures are decomposed with 5% aqueous ammonium chloride, as commonly done in Grignard reactions, phosphinic esters are obtained in 50 to 60% yields (method 6).[305]

$$(RO)_2PCl + R'MgCl \xrightarrow{H_2O} R'PH(O)OR + ROH + MgCl_2 \quad (6)$$

The Grignard reagents may be replaced by organolithium or organosodium compounds in the reactions with trialkyl phosphites (methods 3b[436] and 3c[250]) and dialkyl phosphorochloridites (methods 5b[302,303] and 5c[164]). Organocadmium reagents give lower yields (method 5d).[658]

Sodium salts of nitroparaffins react with dialkyl phosphorochloridites giving products of the decomposition of the nitronic esters $(RO)_2PON(O){=}CHR$, but in one instance a phosphonous diester was isolated in 13% yield.[475]

The reaction of dialkyl phosphorochloridites with α-stannylacetate esters gives the phosphonous diesters in 74 to 79% yield (method 5e).[568] α-Stannylketones tend to

$$(RO)_2PCl + R_3SnCH_2CO_2R \longrightarrow (RO)_2PCH_2CO_2R + R_3SnCl \quad (5e)$$

give mixtures of β-ketophosphonites and vinyl phosphites.[553]

Phosphonous diesters are also obtained in the reaction of dialkyl phosphorothioites with α-mercuriacetates (method 7).[513]

$$(RO)_2P(S)H + Hg(CH_2CO_2R)_2 \longrightarrow$$

$$(RO)_2PCH_2CO_2R + HgS + MeCO_2R \quad (7)$$

A.3.2.2. Reaction with Ketene Dimer. Trialkyl phosphites react with dimethylketene dimer in the dione form (but not in the lactone form), giving phosphonous diesters after 30 days at 100° (method 8).[70]

$$(RO)_3P + \begin{array}{c} Me_2C - C{=}O \\ | \quad\quad | \\ O{=}C - CMe_2 \end{array} \longrightarrow (RO)_2CCMe_2COCMe_2P(OR)_2 \quad (8)$$

Ring opening is accompanied by an apparent alkoxyl group migration. Diphenylketene, however, forms 2:1 adducts of a different type.[476]

A.3.3. From Phosphenimidous Esters

Phosphinic esters are formed in the condensation of active methylene compounds such as $CH_2(CN)_2$, $NCCH_2CO_2Et$, or $CH_2(CO_2Et)_2$ with ethyl phenylphosphenimidite in the presence of p-toluenesulfonic acid (method 9).[477] The reaction is

$$ROP{=}NAr + CH_2(CN)_2 \longrightarrow (NC)_2CHPH(O)OR \quad (9)$$

believed to proceed through a metaphosphite intermediate [ROP=O].[477]

A.3.4. From Phosphonous Dichlorides

A.3.4.1. Esterification. Phosphinic esters are obtained when phosphonous dichlorides are added to alcohols at ice-cold temperatures (method 10a).[41,346,610] If the

$$RPCl_2 + 2R'OH \longrightarrow RPH(O)OR' + R'Cl + HCl \qquad (10a)$$

temperature is not controlled, the second ester group may also be cleaved (p. 263). A tertiary amine (1 equiv.) may be added to the reaction mixture to bind the by-product HCl if the product is acid-sensitive (method 10b)[23,236] or if the acid is HBr (method 11).[403] With 2 equiv. of a tertiary amine, the product is the phosphonous diester (method 12b).[36,40,629]

$$RPCl_2 + 2R'OH + 2B \longrightarrow RP(OR')_2 + 2B{\cdot}HCl \qquad (12b)$$

The tertiary amines most commonly used are pyridine, dimethylaniline, and triethylamine. Anhydrous ammonia may be used if it is added at a rate just sufficient to bind the hydrogen chloride as it is formed[237] or if it is added to the phosphonous dichloride/alcohol complex at temperatures below -40° (method 12c).[209] Secondary amines are similarly used.[668]

The acid acceptor can be omitted when phosphonous dichlorides are esterified with phenols, since the aromatic esters are not attacked by acids (method 12a).[35,432]

A third type of product, the phosphonochloridous ester RP(OR')Cl, results if the ratio of phosphonous dichloride to alcohol and tertiary amine is adjusted to 1:1:1 (method 13).[261,293]

$$RPCl_2 + R'OH + B \longrightarrow RP(OR')Cl + B{\cdot}HCl \qquad (13)$$

The phosphonous difluorides form adducts with alcohols even in the presence of tertiary amines such as dibutylaniline.[152] Hydrogen fluoride is split out by more basic tertiary amines such as triethylamine, but not by potassium fluoride. The preferred method is to treat the adduct with ammonia at -40 to -50° (method 14).[152]

$$RPF_2 \xrightarrow{R'OH} RPH(OR')F_2 \xrightarrow{NH_3} RP(OR')F \qquad (14)$$

Other acid-binding agents used occasionally in place

of the tertiary amines are the alkoxides (method 12d)[127] and the epoxides (methods 12e[79] and 10c[78]). The phosphonous dichlorides also react with alkoxides in dry ether (method 15a).[28,344]

$$RPCl_2 + 2R'ONa \longrightarrow RP(OR')_2 + 2NaCl \qquad (15a)$$

This method, now seldom used, was employed recently to prepare a series of phosphonofluoridous esters (method 16).[367]

The reaction of phenylphosphonous dichloride with mercuribis(acetaldehyde) in the presence of a tertiary amine gives the divinyl ester in 43% yield (method 15b).[397]

$$PhPCl_2 + Hg(CH_2CHO)_2 \xrightarrow{R_3N} PhP(OCH{=}CH_2)_2 + HgCl_2 \qquad (15b)$$

A.3.4.2. Reaction with Epoxides. The reaction of phosphonous dichlorides with epoxides gives either the phosphonochloridous ester (method 17)[281,319] or the phosphonous diester (method 18),[201,556,699] depending on the molar ratio used. No base is required.

$$RPCl_2 + 2R'\overset{O}{\widehat{CH-CH_2}} \longrightarrow RP(OCHR'CH_2Cl)_2 \qquad (18)$$

The products are prone to undergo self-isomerization if heated to 150° or above during distillation.[643]

A.3.4.3. Reaction with Mercaptans. Several of the methods described previously have been adopted successfully for the synthesis of sulfur-containing esters. Phosphonodithious diesters are obtained by the reaction of phosphonous dichlorides with sodium mercaptide (method 19a)[32] or lead mercaptide (method 19b).[395] Phosphonodithious diesters and phosphonochloridothious esters are prepared by the reaction of phosphonous dichlorides with alkylene sulfides (method 20)[285,516] or with mercaptans in the presence of tertiary amines (methods 21b[14,45] and 22b[15,678]). Methods 21b and 22b are also applicable to the phosphonous difluorides (method 23),[151] dibromides (method 24),[352] and diiodides (method 25).[99]

Recently, it has been learned that it is not necessary to use a base as an acid acceptor in these reactions, since the thioesters are not attacked by hydrogen chloride. The phosphonochloridothious esters (method 22a)[16] and the phosphonodithious diesters (method 21a)[585] are obtained by the reaction of phosphonous dichlorides with mercaptans in chlorocarbon solvents such as methylene chloride or chloroform.

The only novel method is the reaction of phosphonous dichlorides with xanthate esters, which takes place at 220° giving the phosphonodithious ester in 27% yield (method 26).[144]

$$RPCl_2 + 2ROC(S)SR \longrightarrow RP(SR)_2 + 2COS + 2RCl \qquad (26)$$

A.3.4.4. Reaction with Phosphonous Diesters. Ester interchange between the phosphonous dichlorides and the phosphonous diesters takes place at ice-cold temperatures, giving phosphonochloridous esters in 70 to 80% yield (method 27).[708]

$$RPCl_2 + RP(OR)_2 \longrightarrow 2RP(OR)Cl \qquad (27)$$

Similar methods are used to prepare the phosphonochloridothious esters (method 28)[15] and the analogous bromine compounds (method 29).[352] The phosphonodithious diesters also undergo ester exchange with the phosphonothioic dichlorides $RP(S)Cl_2$ at 150° (method 30).[353]

The members of the three-component mixture of phosphonous dichloride, aldehyde, and phosphorus(III) ester interact to give phosphonous diesters with phosphorus(V)-containing ester groups (method 31).[74] The phosphorus(III)

$$RPCl_2 + R'CHO + (RO)_3P \longrightarrow$$

$$RP[OCHR'P(O)(OR)_2]_2 + 2RCl \qquad (31)$$

ester, which is isomerized in the process, may be a compound of the type $(RO)_3P$ or $RP(OR)_2$.[74]

A.3.4.5. Reaction with Silyl Esters. Phosphonous diesters may be prepared by the cleavage of the silicon-oxygen bond in silyl esters (method 32).[694]

$$RPCl_2 + R_2Si(OR)_2 \longrightarrow RP(OR)_2 + R_2SiCl_2 \qquad (32)$$

The driving force for this reaction is the displacement of the more volatile halosilane. Phosphonodithious diesters are similarly obtained by the cleavage of the silicon-sulfur bond in alkylthiosilanes (method 33).[1]

A.3.5. From Phosphonochloridous Esters

A.3.5.1. Hydrolysis. Phosphinic esters and their thio analogs may be prepared by the reaction of phosphonochloridous esters with water (method 34a)[326] or with hydrogen sulfide (method 35a).[270] These preparations are best carried

$$RP(OR)Cl + H_2S \longrightarrow RPH(S)OR + HCl \quad (35a)$$

out with pyridine present as an acid acceptor (methods 34b[163] and 35b[452,457,682]).

A.3.5.2. Esterification. Phosphonochloridous esters react with alcohols, giving phosphonous diesters if a tertiary amine is present (method 36)[681,708] and phosphinic esters if the amine is absent (method 37).[603]

Mixed O,S-diesters may be prepared by the reaction of phosphonochloridothious esters with epoxides (method 38),[15] or in one of the following ways (method 39[328,708] or method 40[15,440,759]).

$$RP(OR)Cl + R'SH \xrightarrow{B} RP(OR)SR' + B{\cdot}HCl \quad (39)$$

$$RP(SR)Cl + R'OH \xrightarrow{B} RP(OR')SR + B{\cdot}HCl \quad (40)$$

Mixed dithioesters of the type RP(SR)SR' are obtained by the reaction of phosphonochloridothious esters with mercaptans, either in the presence (method 41b)[15,678] or absence (method 41a)[358] of base. The results are the same because the products are not attacked by acids.[358,585]

Phosphonodithious diesters are also obtained in low yield by the reaction of phosphonochloridothious esters with dialkyl sulfides in sealed tubes at 150° (method 42).[355,358]

$$RP(SR)Cl + R_2S \longrightarrow RP(SR)_2 + RCl \quad (42)$$

A.3.5.3. Halogen Exchange. The phosphonochloridothious esters react with hydrogen bromide, liberating hydrogen chloride (method 43).[352]

$$RP(SR)Cl + HBr \longrightarrow RP(SR)Br + HCl \quad (43)$$

This method cannot be used with the phosphonochloridous esters because of the sensitivity of the ester group to cleavage by acid. The phosphonochloridous esters do, however, react with antimony trifluoride in ether, giving the fluoridous esters in 20 to 25% yield (method 44).[367]

$$RP(OR)Cl \xrightarrow{SbF_3} RP(OR)F \quad (44)$$

A.3.6. From Phosphinic Acids

Acid-catalyzed esterification of phosphinic acids gives high yields of the phosphinic esters, provided the by-product water is removed by azeotropic distillation (method 45).[113,236,541]

$$RPH(O)OH + R'OH \overset{H^+}{\rightleftharpoons} RPH(O)OR' + H_2O \qquad (45)$$

Aliphatic phosphinic acids that contain a hydroxyl group in the γ- or δ-position form cyclic esters (phostones) on heating.[698] Phosphinic acids are also esterified by epoxides (method 46),[706,707] by diazomethane (method 47),[707] and by dialkyl sulfates (method 48).[248]

The reaction of the silver salts of the phosphinic acids with alkyl halides is a classical method (method 49a).[414 416] Sodium salts have been used in the preparation of phosphinic esters from β-aminoalkyl halides[146] and chlorotrialkylsilanes or -germanes[87] (method 49b).

$$RPH(O)ONa + R_3SiCl \longrightarrow RPH(O)OSiR_3 + NaCl \qquad (49b)$$

The esterifications of phosphonous anhydrides (method 50)[228,491] or their thio analogs (method 51)[464] are new methods. Their utility has not yet been fully explored.

$$[RPO]_n + R'OH \longrightarrow RPH(O)OR' \qquad (50)$$

Ester interchange between the phosphinic acids and the phosphonous diesters (method 52)[215,741] or the tertiary phosphite esters (method 53)[215] takes place when the compounds are warmed to 150°. The former method is preferred,

$$RPH(O)OH + RP(OR')_2 \longrightarrow 2RPH(O)OR' \qquad (52)$$

since it gives phosphinic esters free from by-products.

A.3.7. From Phosphonous Diesters

A.3.7.1. Hydrolysis. Phosphonous diesters are hydrolyzed slowly in a neutral or basic medium, but hydrolysis proceeds rapidly in an acidic medium.[113,136,658] This was demonstrated in the Grignard synthesis discussed earlier (p. 266). It is sufficient to treat the phosphonous diester with an excess of water containing one drop of a mineral acid (method 54).[658]

Phosphinic esters may also be prepared by dealkylation of phosphonous diesters under anhydrous conditions with

$$RP(OR')_2 + H_2O \xrightarrow{H^+} RPH(O)OR' + R'OH \qquad (54)$$

acids such as hydrogen chloride[203,474] or acetic acid[474] (but not HBr[100]) (method 55).

$$RP(OR')_2 + HCl \longrightarrow RPH(O)OR' + R'Cl \qquad (55)$$

Ester interchange with the phosphinic acids was discussed in the preceding section. Other P-acids such as the phosphorothioic or -dithioic diesters react similarly (method 56).[594,595]

Phosphonous diesters are attacked by lithium[23] or magnesium[659] salts in polar solvents such as 2-ethoxyethanol, liberating alkyl halide (method 57). The products yield phosphinic esters upon acidification.[23]

A.3.7.2. Transesterification. Esters of phosphonous acids may be prepared by the reaction of phosphonous diesters with alcohols or glycols, preferably with the displacement of a low-boiling alcohol, such as ethanol, by a high-boiling alcohol.[260] The esters are transesterified at 160° in the presence of sodium catalyst (method 58b).[260,547,576] Unsymmetrical esters are obtained when the amount of alcohol or phenol is insufficient.[260] In the absence

$$RP(OR')_2 \xrightarrow{R''OH} RP(OR')OR'' \xrightarrow{R''OH} RP(OR'')_2 \qquad (58b)$$

of a catalyst, the reaction is slower and the yields are not as high (method 58a).[260]

The phosphonodithious diesters react similarly, in the presence of sodium catalyst, giving phosphonous diesters (method 59).[539]

A.3.7.3. Redistribution. Mixtures of phosphonous diesters tend to equilibrate on standing (method 60).[260]

$$RP(OR')_2 + RP(OR'')_2 \rightleftharpoons 2RP(OR')OR'' \qquad (60)$$

The composition of the resulting mixture, which depends on the molar ratio of the reactants, is readily calculated. The mixed esters (forward reaction) can be isolated by distillation.[260] Conversely, the mixed esters tend to disproportionate (reverse reaction), sometimes on standing[260] and sometimes during preparation.[534] The ester $MeP(OEt)OC_8H_{17}$-n, for example, disproportionates to the extent of 48% after six months on the shelf.[260]

A.3.7.4. Reaction with Acid Chlorides. Phosphonodithious diesters react with acetyl chloride at 80° or with benzoyl chloride at 130 to 140°, giving phosphonochloridothious esters in 45 to 50% yield (method 61).[14] If suffi-

$$RP(SR)_2 + AcCl \longrightarrow RP(SR)Cl + AcSR \qquad (61)$$

cient benzoyl chloride is present, the products react further, giving phosphonous dichlorides.[14]

A.3.8. From Phosphinic Esters

A.3.8.1. Transesterification. The phosphinic esters are transesterified by primary or secondary (but not tertiary) alcohols[537] or glycols[500,538] at 150 to 160° in the presence of a small amount of sodium catalyst (method 62b). Yields are of the order of 50 to 70% (mechanism[576]).

$$RPH(O)OR + R'OH \xrightarrow{Na} RPH(O)OR' + ROH \qquad (62b)$$

Acid catalysts such as phosphoric acid (method 62c)[537,540] and aluminum chloride[537] give poorer yields. In the absence of catalysts, the reaction is slow and the yields are not as high (method 62a).[537] A phosphinothioic ester, however, has been transesterified in 85% yield, and no catalyst was used (method 63).[526]

A.3.8.2. Reaction with Phosphorus Pentasulfide. Phosphinothioic esters may be prepared in 35 to 40% yield by the reaction of phosphinic esters with phosphorus pentasulfide (method 64).[392,526] The reaction is carried out

$$RPH(O)OR \xrightarrow{P_2S_5} RPH(S)OR \qquad (64)$$

at boiling water-bath temperature.

A.3.9. From Phosphonous Diamides

A.3.9.1. Alcoholysis. Alcohols displace secondary amines from phosphonous diamides under relatively mild conditions, giving phosphonous diesters in almost quantitative yield (method 65).[403,546,669] This reaction, first

$$RP(NR_2)_2 + 2R'OH \longrightarrow RP(OR')_2 + 2R_2NH \qquad (65)$$

reported in 1961, has developed rapidly into a useful method for preparing many other phosphonous acid derivatives. Reaction with phenols or thiophenols gives the aromatic diesters or dithioesters (methods 65[532,533,546] and 66[530]). Aliphatic mercaptans, with the exception of benzyl mercaptan, are simply desulfurized (p. 289). Partial substitution products of the types $RP(NR_2)OR$ and $RP(NR_2)SR$ are obtained by reaction with 1 mole of an alcohol, phenol, or thiophenol (p. 281).

Alcoholysis of a phosphinic amide with a sugar derivative is reported to give the sugar ester (method 67).[548]

$$RPH(O)NR_2 + R'OH \longrightarrow RPH(O)OR' + R_2NH \qquad (67)$$

A.3.9.2. Reaction with Carbon Disulfide. Phosphonous diamides react exothermally with carbon disulfide, giving the products of insertion into the phosphorus-nitrogen bonds (method 68).[446,517] The products lose a mole of carbon disulfide upon repeated recrystallization (p. 282).

$$RP(NR_2)_2 + 2\ CS_2 \longrightarrow RP(SC[S]NR_2)_2 \qquad (68)$$

A.3.10. From Phosphonamidous Esters

The ease of cleavage of the phosphorus-nitrogen bond in the phosphonamidous esters has been used to advantage in the preparation of several phosphonous acid derivatives.

Hydrolysis with water gives the phosphinic ester (method 69),[533] and treatment with hydrogen sulfide produces the phosphinothioic ester (method 70).[535]

$$RP(NR_2)OR' + H_2O \longrightarrow RPH(O)OR' + R_2NH \qquad (69)$$

Reaction with alcohols or phenols gives the mixed phosphonous diester (method 71).[533,534] Reaction with benzyl mercaptan gives the mixed O,S-diester (method 72),[530] but other mercaptans are simply desulfurized (p. 289).

$$RP(NR_2)OR' + R''OH \longrightarrow RP(OR')OR'' + R_2NH \qquad (71)$$

The phosphonamidothious ester $PhP(NEt_2)SPr$ reacts with acetyl chloride in 2 hr at 20 to 50°, giving the phosphonochloridothious ester in 57% yield (method 73).[641]

$$RP(NR_2)SR + AcCl \longrightarrow RP(SR)Cl + AcNR_2 \qquad (73)$$

A.3.11. From Primary Phosphines

Primary phosphines react with sulfenyl chlorides (method 74)[551] or disulfides (method 75)[198,219] under mild conditions, giving 80 to 90% yields of the phosphonodithious diesters.

$$RPH_2 + 2R'SCl \longrightarrow RP(SR')_2 + 2HCl \qquad (74)$$

$$RPH_2 + 2R'SSR' \longrightarrow RP(SR')_2 + 2R'SH \qquad (75)$$

Method 75 requires a radical inhibitor. In its absence, the products interact giving the phosphonotrithioic diester $RP(S)(SR)_2$.[219] Primary phosphines that contain a γ-hydroxyl substituent form cyclic thioesters (1,2-oxaphospholanes), owing to transesterification of the initially formed dithioesters.[218] (See also Chapter 1).

The sodium derivative of methylphosphine reacts with diethyl disulfide in benzene, giving the phosphonodithious diester in 41% yield (method 76).[94] The stannyl derivative gives either the dithioester or the partial ester, depending on the reactant ratio (method 77).[94]

$$RP(SnR_3)_2 \xrightarrow{R'SSR'} RP(SnR_3)SR' \xrightarrow{R'SSR'} RP(SR')_2 \qquad (77)$$

A.3.12. From Cyclotetraphosphines or Cyclopentaphosphines

The thermal decomposition of cyclotetraphosphines (R = Ph or CF_3) in the presence of dialkyl disulfides gives phosphonodithious diesters in 60 to 90% yields (method 78).[125,675] Disulfides are also used to trap the carbenoid

$$[RP]_4 + 4R'SSR' \longrightarrow 4RP(SR')_2 \qquad (78)$$

phosphinidene group (RP:) generated in other ways.[675] An attempt to prepare the silyl ester by splitting the silicon-silicon bond in hexamethyldisilane was unsuccessful.[125]

Phenyl cyclopentaphosphine reacts when heated with α-mercurated carbonyl compounds, giving vinyl esters in 53.5% yield (method 79).[566] (See Chapter 2).

$$[PhP]_5 + 5\ Hg(CH_2COR)_2 \longrightarrow 5PhP(OCR=CH_2)_2 + 5Hg \qquad (79)$$

A.3.13. From Phosphonothioic Esters

The phosphonothioic acid ester MeP(S)(OPr-i)OH is desulfurized by Raney nickel in ethanol at reflux (method 80).[638]

$$RP(S)(OR)OH + H_2\ (Ni) \longrightarrow RPH(O)OR + H_2S \qquad (80)$$

The reaction is stereospecific.

A.3.14. From Phosphinous Esters

Fluoroform is displaced from phosphinous esters containing trifluoromethyl substituents by mild alcoholysis in the presence of base (method 81).[210]

$$RP(CF_3)OR' + R'OH \xrightarrow{RONa} RP(OR')_2 + CHF_3 \qquad (81)$$

A.4. Phosphonous Amides

A.4.1. From Phosphorus Trichloride and 2-Aminobiphenyl

The heterocyclic compound 10-chloro-9,10-dihydro-9,10-azaphosphaphenanthrene (6) is obtained in good yield as a crystalline solid when phosphorus trichloride is treated with 2-aminobiphenyl in dry benzene; the intermediate, presumably (5), is cyclized with aluminum chloride at 210 to 220° (method 1).[243] Similar reactions occur with

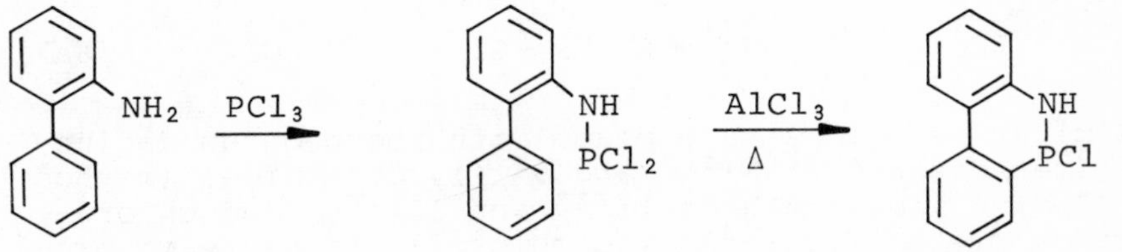

(1)

(5) (6)

substituted 2-aminobiphenyls.[107]

A.4.2. From Phosphorodiamidous Chlorides and Grignard Reagents

Phosphonous diamides may be prepared in 50 to 60% yields by the reaction of phosphorodiamidous chlorides with Grignard reagents in ether (method 2a).[102,169,710,734]

$$(R_2N)_2PCl + R'MgCl \longrightarrow R'P(NR_2)_2 + MgCl_2 \qquad (2a)$$

The dichlorides R_2NPCl_2 give phosphonamidous chlorides by partial substitution if the Grignard reagent is R = cyclohexyl, but with other Grignard reagents there were separation problems (method 3).[278,686]

Organolithium, -aluminum, or -lead compounds may be used in place of the Grignard reagents (methods 2b,[169,509] 2c[404] and 2d[412] respectively). The yields are usually higher because there is less cleavage of the phosphorus-nitrogen bond than with the Grignard reagents.[404,408]

A.4.3. From Phosphorous Triamides and Perfluoroalkyl Halides

The reaction of hexamethylphosphorous triamide with perfluoroalkyl iodides takes place at room temperature or above, giving phosphonous diamides in 70% yield (method 5).[25]

$$2(R_2N)_3P + R_fI \longrightarrow R_fP(NR_2)_2 + (R_2N)_4PI \qquad (5)$$

A.4.4. From Phosphonous Dichlorides

A.4.4.1. Reaction with Primary or Secondary Amines. Phosphonous dichlorides react violently with primary or secondary amines if no solvent is used. The reaction with secondary amines takes place smoothly in ether or benzene, giving phosphonamidous chlorides (method 6a)[44,170,263,644] or phosphonous diamides (method 7a),[170,446,744] depending on how much secondary amine is employed.

$$RPCl_2 + 2R_2NH \longrightarrow RP(NR_2)Cl + R_2NH{\cdot}HCl \qquad (6a)$$

Similar reactions take place with phosphonous difluorides (methods 8[154,156,507] and 9[221]), dibromides (method 10),[403] diiodides (methods 11[227] and 12[118]), and chlorofluorides (method 13).[156] The chlorofluorides react preferentially at the chlorine atom.

A tertiary amine may be added as an acid acceptor if the use of an excess of the secondary amine is undesirable or impractical (methods 6b[130,686] and 7b[514,703]). Alternatively, the reaction may be carried out with the sodium salt of the secondary amine (method 14).[257]

The use of primary amines in these reactions is less common, although there appears to be no inherent reason for this. Primary amines have been used in synthesizing phosphonamidous chlorides (method 6)[239,663] and fluorides (method 8)[156] from the respective phosphonous dihalide, and phosphonous diamides (method 7)[372,700] from phosphonous dichlorides. Phosphonamidous fluorides are also obtained from the adducts $RPH(NHR)F_2$ by treatment with a tertiary amine (method 15).[155]

If a tertiary amine is present, the reaction of phosphonous dichlorides with primary amines gives iminophosphines, e.g., $[PhP{-}NPh]_3$ (method 16b).[289] The reactions

$$RPCl_2 + R'NH_2 \xrightarrow{R_3N} [RP{=}NR']_n + 2R_3N{\cdot}HCl \qquad (16b)$$

in which R = Me, R' = t-Bu[663] and R = CF_3, R' = NMe_2[253] are exceptions.

Unsubstituted phosphonous diamides having the structure $RP(NH_2)_2$ have been isolated in only two instances (method 7a, R = t-Bu[664] or CF_3[247,253]).

A.4.4.2. Reaction with Hydrazines. The reaction of phosphonous dichlorides (or diiodides, method 12)[525] with unsym-dimethylhydrazine gives products having the structures $RP(NHNMe_2)Cl$ (method 6b)[487] or $RP(NHNMe_2)_2$ (method 7a),[487] but phenylhydrazine gives iminophosphines RP=NNHPh (method 16a).[449] A small amount of the dihydrazide was isolated in one experiment.[449]

A.4.4.3. Reaction with Phosphonous Diamides. Phosphonamidous chlorides and bromides may be prepared by the redistribution of equimolar mixtures of phosphonous dichlorides (or dibromides, method 18)[733] and phosphonous diamides (method 17).[403,733] The reaction takes place

$$RPCl_2 + RP(NR_2)_2 \longrightarrow 2RP(NR_2)Cl \qquad (17)$$

exothermally and quantitatively as soon as the reagents are mixed.

A.4.4.4. Reaction with Imino Esters. The reaction of phosphonous dichlorides with oxalimidic esters gives the phosphodiazole derivatives; one example is structure (7) (method 19).[142,150]

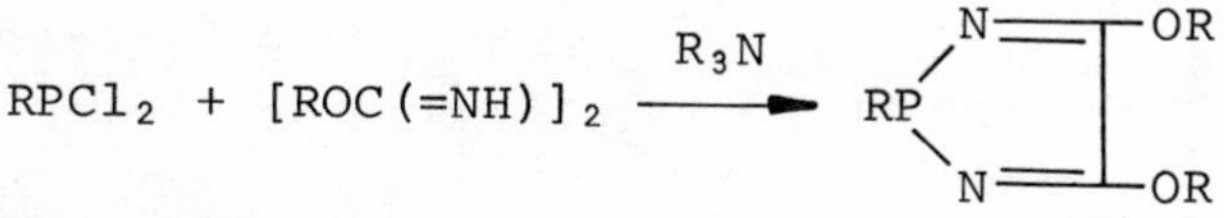

(19)

(7)

A.4.4.5. Reaction with Tertiary Phosphine Imides. Phosphonous dibromides add to the phosphorus-nitrogen double bond in tertiary phosphine imides giving bis(phosphonium salts) (method 20).[410]

$$RPBr_2 + 2Ph_3P{=}NPh \longrightarrow RP(N(Ph)PPh_3^{+}Br^{-})_2 \qquad (20)$$

A.4.4.6. Reaction with Azides. Phenylphosphonous dichloride reacts with the N-Grignard reagent resulting from the interaction of methyl azide with a Grignard reagent, giving the 1,3-triazene in 40% yield (method 21).[89]

$$RMgCl \xrightarrow{RN_3} [R_2N_3MgCl] \xrightarrow{PhPCl_2} PhP(N_3R_2)_2 \qquad (21)$$

The product explodes if heated above 120°.

A.4.4.7. Reaction with Aminosilanes. Phosphonous diamides are readily prepared by the cleavage of the silicon-nitrogen bonds in aminosilanes (method 22).[3,62] This

$$RPCl_2 + 2R'_3SiNR_2 \longrightarrow RP(NR_2)_2 + 2\ R'_3SiCl \qquad (22)$$

method is useful in cases such as $C_6F_5PCl_2$, where direct aminolysis is complicated by substitution on the ring.[62,399]

A.4.4.8. Reaction with N-Silyllithium Compounds. Silyl derivatives of the phosphonaminous chlorides (method 23)[664] or the phosphonous diamides (method 24)[180,664] are prepared by the reaction of phosphonous dichlorides with 1 or 2 equiv. of an N-silyllithium compound.

$$RPCl_2 + RN(SiR_3)Li \longrightarrow RP(NRSiR_3)Cl + LiCl \qquad (23)$$

Silyl compounds of the type $R_2Si(NRLi)_2$ give 2,4-diaza-3-phosphasiletidines.[179,180]

A.4.5. From Phosphonochloridous Esters

Phosphonochloridous esters react with primary or secondary amines, giving phosphonamidous esters (method 25a).[328,533,535,708]

$$RP(OR')Cl + 2R_2NH \longrightarrow RP(OR')NR_2 + R_2NH{\cdot}HCl \qquad (25a)$$

A tertiary amine may be added as an acid acceptor if the use of a primary amine for this purpose is undesirable (method 25b).[246,328,686]

The phosphonochloridothious esters react similarly with secondary amines in the absence of a tertiary amine (method 26a)[15,17] and with secondary amines or hydrazines in the presence of a tertiary amine (method 26b).[7,17]

A.4.6. From Phosphonous Diamides

A.4.6.1. Hydrolysis. Partial hydrolysis of a phosphonous diamide was stated in one instance to give the phosphinic amide, although the product was not isolated (method 27).[548]

$$RP(NR_2)_2 + H_2O \longrightarrow RPH(O)NR_2 + R_2NH \qquad (27)$$

Other attempts have not been successful,[88,198,269,372] perhaps because the products are more easily hydrolyzed than the starting materials.[198]

A.4.6.2. Alcoholysis. Partial alcoholysis of phosphonous diamides takes place on heating with 1 mole of an alcohol to 100°, giving phosphonamidous esters in high yield (method 28).[404,533,534]

$$RP(NR_2)_2 + R'OH \longrightarrow RP(NR_2)OR' + R_2NH \qquad (28)$$

Similar reactions take place with phenols (method 28)[533] and thiophenols (method 29).[530] Ethanolamine derivatives give cyclic phosphonamidous esters called 1,3,2-oxaza or -thiazaphospholidenes.[462,463,498,499]

A.4.6.3. Transamidation. The reaction known as transamidation, which can be considered to be the counterpart of transesterification in the ester series, occurs when a phosphonous diamide is heated with a primary or secondary amine to 100 to 120°, preferably with the displacement of a volatile secondary amine (method 30).[532] Unsymmetrical

$$RP(NR_2)_2 + 2R'_2NH \longrightarrow RP(NR'_2)_2 + 2R_2NH \qquad (30)$$

diamides of the type $RP(NR_2)(NR'_2)$ are obtained when the amount of secondary amine is insufficient.[532]

Transamidation also takes place in the reaction of phosphonamidous esters with primary or secondary amines, in preference to displacement of alcohol (method 31).[494]

A.4.6.4. Reaction with Phosphorus(III) Halides. The reaction of phosphonous diamides with phosphorus trifluoride or phosphorus oxyfluoride gives the phosphonamidous fluoride, together with unidentified solids (method 32).[92]

$$RP(NR_2)_2 + PF_3 \longrightarrow RP(NR_2)F + R_2NPF_2 \qquad (32)$$

Other phosphorus(III) halides tend to split both phosphorus-nitrogen bonds. The reaction with phosphonous dichlorides is discussed on page 279.

A.4.6.5. Reaction with Acid Fluorides. A weakly exothermic reaction takes place when a phosphonous diamide is mixed with benzoyl fluoride in an NMR tube (method 33).[91]

$$RP(NR_2)_2 + PhCOF \longrightarrow RP(NR_2)F + PhCONR_2 \qquad (33)$$

Excess benzoyl fluoride displaces the second amide group.[91]

A.4.6.6. Reaction with CO_2, CS_2, Isocyanates, and Isothiocyanates. The phosphonous diamides undergo exothermic reactions with carbon dioxide and with alkyl isothiocyanates, giving the products of insertion into one of the phosphorus-nitrogen bonds (methods 34[517] and 35[517]).

$$RP(NR_2)_2 + CO_2 \longrightarrow RP(NR_2)OC(O)NR_2 \qquad (34)$$

$$RP(NR_2)_2 + R'NCS \longrightarrow RP(NR_2)NR'C(S)NR_2 \qquad (35)$$

Carbon disulfide forms a 2:1 adduct initially, but this loses a mole of carbon disulfide upon repeated recrystallization (method 68, A.3).[446] Phenyl isocyanate also gives a 2:1 adduct (method 36).[130]

$$RP(NR_2)_2 + R'NCO \longrightarrow RP[NR'C(O)NR_2]_2 \qquad (36)$$

A.4.6.7. Reaction with Halosilanes. t-Butylphosphonous diamide reacts with chlorotrimethylsilane, giving the silyl derivative in 70% yield (method 37).[664]

$$RP(NH_2)_2 + 2R_3SiCl \longrightarrow RP(NHSiR_3)_2 + 2HCl \qquad (37)$$

The product is very sensitive to air and moisture.

A.4.7. From Phosphonamidous Chlorides

A.4.7.1. Hydrolysis. Hydrolysis of the phosphonamidous chloride (6) by shaking with water in methylene chloride gives the crystalline phosphinic amide (method 38).[143]

$$RP(NHR')Cl + H_2O \longrightarrow RPH(O)NHR' + HCl \qquad (38)$$

A.4.7.2. Esterification. The esterification of phosphonamidous chlorides with alcohols or phenols takes place readily in ether solution in the presence of tertiary amines, giving phosphonamidous esters in 60 to 70% yield (method 39).[44,483]

$$RP(NR_2)Cl + R'OH \xrightarrow{R_3N} RP(NR_2)OR' + R_3N{\cdot}HCl \qquad (39)$$

In the absence of a tertiary amine, the products are phosphonous diesters (method 40).[44,130]

A.4.7.3. Reaction with Primary or Secondary Amines. The reaction of phosphonamidous chlorides with primary or secondary amines is a useful method of preparing phosphonous diamides, particularly mixed amides (method 41).[170,253,509]

$$RP(NR_2)Cl + 2R_2NH \longrightarrow RP(NR_2)_2 + R_2NH{\cdot}HCl \qquad (41)$$

Phosphonamidous fluorides react similarly (method 42).[221,507]

The phosphonamidous chloride $MeP(NMe_2)Cl$ reacts with the sodium salt of p-toluenesulfonamide in ether at -70°, giving the mixed amide (20) (method 43).[671] The structure

$$RP(NR_2)Cl + ArSO_2NHNa \longrightarrow \underset{\underline{20}}{RPH(=NSO_2Ar)NR_2} + NaCl \qquad (43)$$

of the product (20), is discussed on page 309.

A.4.7.4. Reaction with Isocyanates. Phenyl isocyanate gives insertion products with phosphonamidous chlorides (method 44).[130]

$$RP(NR_2)Cl + R'NCO \longrightarrow RP[NR'C(O)NR_2]Cl \qquad (44)$$

A.4.7.5. Halogen Exchange. The phosphonamidous chlorides react with sodium fluoride in tetramethylenesulfone at 100° or with antimony trifluoride in ether at 25 to 30°, giving phosphonamidous fluorides in 50 to 60% yield (method 45).[156,507] In the absence of a solvent, the product may

$$RP(NR_2)Cl \xrightarrow{SbF_3} RP(NR_2)F \qquad (45)$$

be fluorinated further to the trifluorophosphorane $RP(NR_2)F_3$.[677] Antimony pentafluoride also gives the phosphorus(III) fluoride under these conditions (method 46).[156]

Phosphonamidous iodides are prepared from phosphonamidous chlorides by exchange with lithium iodide in ether (method 47).[735]

A.4.8. From Diaminodiphosphines

Diphosphines of the type $RP(NR_2)P(NR_2)R$, which are prepared by the reaction of phosphonamidous chlorides with sodium/potassium alloy in ether, are cleaved by bromine or iodine in benzene solution giving phosphonamidous bromides in 65 to 70% yield (methods 48[686] and 49[684]).

$$RP(NR_2)P(NR_2)R + Br_2 \longrightarrow 2RP(NR_2)Br \qquad (48)$$

The iodides are also obtained by cleavage with alkyl iodides (method 50).[684]

$$RP(NR_2)P(NR_2)R + 2RI \longrightarrow RP(NR_2)I + [R_3PNR_2]I \quad (50)$$

With the stoichiometric quantity of hydrogen chloride, cleavage gives the phosphonamidous chlorides (method 51).[684]

$$RP(NR_2)P(NR_2)R + 2HCl \longrightarrow RP(NR_2)Cl + \frac{1}{n}[RP]_n + R_2NH \cdot HCl \quad (51)$$

A.4.9. From Phosphonous Diesters

Phenol is displaced from aromatic phosphonous diesters by aromatic diamines at 150° (method 52).[557]

$$RP(OAr)_2 + 2ArNH_2 \longrightarrow RP(NHAr)_2 + 2ArOH \quad (52)$$

Similar reactions take place with the aminodiphosphazenes $Ar_2P(NH_2){=}NP({=}NH)Ar_2$, giving the cyclic phosphonous diamides (21) (method 53).[557] The structures of the products are discussed on page 309.

A.4.10. From Primary Phosphines

The reaction of 1,1,2,2-tetrafluoroethylphosphine with dimethylamine in ether gives a 64% yield of the phosphonamidous fluoride, together with 2% of the corresponding diamide (methods 54 and 55).[221] This is another example

$$RCF_2PH_2 \xrightarrow{R_2NH} RCH_2P(NR_2)F \quad (54)$$

of the phosphaalkene reaction discussed on page 296. See also Chapter 1.

The phosphine reacts with aniline in a similar manner, giving the iminophosphine $RCH_2P{=}NPh$.[221] With ammonia, the product is a polymer having the composition $[(RCH_2P)_3N_2]_n$.[210]

A.4.11. From Tertiary Phosphines

Perfluoroalkyl groups are displaced in a stepwise manner from tertiary perfluoroalkylphosphines by N-chloramines at 60 to 70° (method 56).[24]

$$(R_f)_3P \xrightarrow{R_2NCl} (R_f)_2PNR_2 \xrightarrow{R_2NCl} R_fP(NR_2)_2 \quad (56)$$

A.5. Phosphonous Anhydrides (and Thio- Analogs)

A.5.1. From Phosphonous Dichlorides

A.5.1.1. Hydrolysis. Phosphonous anhydrides of the type $[RPO]_n$ are obtained when phosphonous dichlorides are hydrolyzed with the minimal amount of water or formic acid at 0 to 10° (method 1a).[228,464]

$$RPCl_2 + H_2O \longrightarrow [RPO]_n + 2HCl \qquad (1a)$$

The products are described as undistillable oils,[491,492] glassy substances,[464] or crystalline solids,[228] depending on the subsequent workup. Similar substances are obtained when a tertiary amine is added as an acid acceptor (method 1b).[464,491,492]

Only in the case of the sterically hindered phosphonous dichloride $RPCl_2$ [R = 2,4,6-(t-Bu)$_3C_6H_2$] is it possible to isolate the phosphinic chloride RPH(O)Cl. This substance is converted to a phosphinic anhydride of the structure $[RPH(O)]_2O$ upon attempted oxidation with potassium ferricyanide (method 2).[120]

$$2RPH(O)Cl + H_2O \longrightarrow [RPH(O)]_2O + 2HCl \qquad (2)$$

A.5.1.2. Reaction with Hydrogen Sulfide. Treatment of phenylphosphonous dichloride with hydrogen sulfide for 3 hr at 140 to 150° gives a crystalline phosphonothious anhydride having the structure $[PhPS]_n$ (method 3).[464]

$$RPCl_2 + H_2S \longrightarrow [RPS]_n + 2HCl \qquad (3)$$

In the early literature, a mixed anhydride structure $R_2PSP{=}S$ was proposed for this compound to explain the formation of diphenylphosphinic acid upon oxidation.[343]

Phosphonous dichlorides react with disulfan in ether at room temperature, giving alkyl- or arylthionophosphine sulfides $[RPS_2]_2$.[64] With a 2:1 ratio, products having the composition of mixed phosphorus(III)/phosphorus(V) thioanhydrides can be isolated as pale yellow oils (method 4).[64]

$$2RPCl_2 + H_2S_2 \longrightarrow RP(Cl)SP(S)(Cl)R + 2HCl \qquad (4)$$

A.5.1.3. Reaction with Alkylthiosilanes. Cleavage of the silicon-sulfur bonds in $(R_3Si)_2S$ or $[R_2SiS]_n$ (n = 2 or 3) by phenylphosphonous dichloride gives the crystalline phosphonothious anhydride $[PhPS]_3$ (method 5).[1] (structure discussed in Chapter 2).

$$RPCl_2 + (R_3Si)_2S \longrightarrow [RPS]_n + 2R_3SiCl \qquad (5)$$

A.5.1.4. Reaction with Phosphorus Esters. Treatment of phosphonous dichlorides with dialkyl phosphites (method 6) or their sodium salts (method 7) under conditions normally used for esterification results in the formation of mixed phosphonous-phosphorous anhydrides in 40 to 50% yield.[45]

$$RPCl_2 + 2(RO)_2P(O)H + 2R_3N \longrightarrow RP[OP(OR)_2]_2 + 2R_3N{\cdot}HCl \quad (6)$$

Similar reactions with phosphinic esters result in mixed anhydrides of the type $RP[OP(OR)R]_2$ (method 8).[45] Mixed phosphonous-phosphinous anhydrides $RP(OR)OPR_2$ occur when phosphinous chlorides interact with the sodium salts of phosphinic esters (method 9).[46]

$$RP(OR)ONa + R_2PCl \longrightarrow RP(OR)OPR_2 + NaCl \quad (9)$$

A.5.1.5. Acetylation. Anhydrides of the type $CF_3P(OAc)_2$ have been prepared by the reaction of phosphonous dichlorides (or diiodides method 11)[524] with silver acetate (method 10)[524] or acetic anhydride (method 12).[524] The

$$RPCl_2 + 2Ac_2O \longrightarrow RP(OAc)_2 + 2AcCl \quad (12)$$

reaction with the anhydride appears to proceed through the intermediate chloroacetoxyphosphine $CF_3P(OAc)Cl$.[524]

A similar reaction occurs between the phosphonochloridous ester MeP(OEt)Cl and potassium acetate in ether at -10°, giving the anhydride ester MeP(OEt)OAc in 27% yield (method 13).[550]

A.5.2. From Phosphinic Acids

Phenylphosphinic anhydride has been detected by ^1H NMR in $CDCl_3$ solutions of phenylphosphinic acid and dicyclohexylcarbodiimide (method 14).[198]

$$ArPH(O)OH + (RN{=})_2C \longrightarrow [ArPH(O)]_2O + (RNH)_2CO \quad (14)$$

A.5.3. From Phosphinic Esters

The reaction involving phosphinic esters is encountered as a side reaction in the Michaelis-Becker synthesis (page 303). Phosphinic esters react with their sodium salts in toluene at 50°, giving the phosphonous-phosphonic anhydride (method 15).[11] If little or no solvent is present, however,

$$ArPH(O)OR + ArP(OR)ONa \longrightarrow ArP(OR)OP(O)(ONa)Ar + RH \quad (15)$$

the reaction gives a mixture of ArPH(O)ONa and ArRP(O)OR.[11]

A.5.4. From Phosphonous Diamides

A.5.4.1. Hydrolysis. Hydrolysis of phosphonous diamides with the minimal amount of water (method 16)[269,464] gives the same product as method 1. Attempts to stop this

$$RP(NR_2)_2 + H_2O \longrightarrow [RPO]_n + 2R_2NH \qquad (16)$$

reaction at the phosphinic amide stage $RPH(O)NR_2$ have not been successful (page 280).

A.5.4.2. Reaction with Phosphonic Anhydrides. Heating phosphonous diamides till dissolution with phosphonic anhydrides at 115 to 125° gives the mixed phosphonous-phosphonic anhydride (method 17).[465]

$$RP(NR_2)_2 + RPO_2 \longrightarrow RP(NR_2)OP(O)(NR_2)R \qquad (17)$$

Similar anhydrides with phosphoric acid esters are prepared by reaction with the acyl phosphates $AcOP(X)(OR)_2$ (X = O or S)(method 18).[167]

$$RP(NR_2)_2 + AcOP(O)(OR)_2 \longrightarrow RP(NR_2)OP(O)(OR)_2 + AcNR_2 \qquad (18)$$

A.5.5. From Phosphonamidous Esters

The phosphonamidous esters react with acetic anhydride at 20 to 30°, giving the products of displacement of the amino group (method 19).[168]

$$RP(NR_2)OR' + Ac_2O \longrightarrow RP(OAc)OR' + AcNR_2 \qquad (19)$$

Reaction with the acyl phosphorothioate $AcOP(S)(OBu)_2$ gives the mixed anhydride $RP(OR)OP(S)(OBu)_2$.[167]

B. BASIC CHEMISTRY

The dominant feature of the compounds of this chapter is their state of oxidation, which lies between the primary phosphines, discussed in Chapter 1, and the phosphonic acids, discussed in Chapter 12. Compared with compounds of a similar state of oxidation, the phosphonous acids are generally speaking more reactive than the phosphorous acid derivatives of Chapter 15, but less reactive than the phosphinous acid derivatives of Chapter 11. The reactivity of the ester toward the Michaelis-Arbuzov reaction, for example, increases in the following order:[620,629]

$$(RO)_3P < R'P(OR)_2 < R'_2POR$$

With these two broad generalizations in mind, let us consider the basic chemistry of this class of compounds in detail. The order of presentation is, for convenience, the same as that of the introductory chapter.

B.1. Addition to the Phosphorus(III) Group

B.1.1. Oxidation

The compounds of this chapter are all oxidizable to some extent. Air exposure suffices with most of the phosphonous diesters,[40,629] phosphonous diamides,[272,487] phosphonofluoridous esters,[367] and phosphonamidous chlorides.[44] The phosphonodithious diesters do not react at room temperature, and they explode on warming.[753] Ultraviolet catalysis has been reported in the oxidation of phosphonamidous esters at 80°.[534]

$$RP(NR_2)OR \xrightarrow[{[O]}]{h\nu} RP(O)(NR_2)OR$$

Hydrogen peroxide has been employed with the phosphonodithious diesters[678,758] and the phosphonous diamides,[111] and 50% nitric acid was used with the phosphonous diamides;[110] however, nonaqueous oxidants are generally preferred. These include active manganese dioxide [for $RP(OR)_2$[293,567] and $RP(NHR)_2$[372,700]], ozone [for $RP(NHR)_2$[372]], diethyl peroxide [for $RP(NR_2)_2$[140]], and dinitrogen tetroxide [for $RP(OR)_2$,[549,642] $RP(SR)_2$,[539] $RP(NR_2)_2$,[142] and $RP(NR_2)SR$[17]].

The oxidation of compounds having the P(O)H or P(S)H structure is discussed on page 295.

B.1.2. Addition of Sulfur, Selenium, or Tellurium

The phosphonous diesters add sulfur readily, giving phosphonothionic diesters in 70 to 80% yield.[40,629] No catalyst is required.

$$RP(OR)_2 + S \longrightarrow RP(S)(OR)_2$$

Similar additions take place with the dithioesters,[32] diamides,[403,446,530,700] anhydrides,[46,464] and thioanhydrides,[464] and with the mixed function compounds RP(OR)SR,[678,759] $RP(NR_2)OR$,[44] $RP(NR_2)SR$,[17,752] and

RP(OR)Cl.[261,281] The halogen-containing compounds RP(SR)Cl,[16] $RP(NR_2)Cl$,[44] and probably RP(OR)F,[367] RP(SR)F,[151] and $RP(NR_2)F$[145] require a little aluminum chloride catalyst. In the absence of the catalyst, there is no reaction, even at 120°.[16,44]

Most of these compounds also add selenium and tellurium in a similar manner. Selenium adducts have been reported for the phosphonous diesters,[229,389,624,625,629,752] the phosphonodithious diesters,[753] the phosphonous diamides,[372] the phosphonamidous chlorides,[515] and the phosphonamidothious esters.[752] Tellurium adducts have been reported for the phosphonous diesters[229,625] and the phosphonous diamides.[220]

Aliphatic mercaptans are desulfurized by phosphonodithious diesters in a reaction that is radical because it does not take place when a radical inhibitor is present.[219]

$$RP(SR)_2 + R'SH \longrightarrow RP(S)(SR)_2 + R'H$$

Similar reactions occur with the phosphonous diamides[530] and the phosphonamidous esters.[530]

Sulfur adducts are occasionally obtained by exchange processes with other phosphorus compounds, such as the phosphinothioic chlorides.[357]

$$RP(SR)_2 + R_2P(S)Cl \longrightarrow RP(S)(SR)_2 + R_2PCl$$

B.1.3. Formation of Complexes

The phosphonous diesters, in common with other phosphorus(III) compounds, react with cuprous iodide and other copper(I) halides forming crystalline 1:1 or 1:2 adducts, which are often useful for characterization of the esters.[29,40,310,505] (See also Chapter 3B).

$$4CuI + 4L \longrightarrow [CuIL]_4 \quad \text{where } L = RP(OR)_2$$

Complexes are also formed with other transition metal halides (or pseudohalides), such as nickel(II) cyanide[112,704] or osmium(IV) chloride,[147] and with other phosphorus(III) compounds such as the phosphonous diamides,[170,373,511,685,687,700] the phosphonodithious diesters,[32,45] the phosphonofluoridothious esters,[151] and even the phosphonous anhydrides, $[RPO]_n$.[228]

The divalent iron and ruthenium complexes of the phosphonous diesters are reduced by sodium borohydride to hydride complexes, e.g., FeH_2L_4.[373,437]

Transition metal carbonyl complexes are prepared by displacement of carbon monoxide from transition metal carbonyls by phosphonous diesters.[58,84,148,226,288,481,720]

$$Mo(CO)_6 + L \longrightarrow Mo(CO)_5L + CO$$

Phosphonous diamides[720] or phosphonamidous fluorides[677] react similarly.

Products of a higher degree of substitution, such as the Ni(0) or Pd(0) complexes NiL_4 or PdL_4, are prepared by reduction of the transition metal halides with an excess of the ligand,[520] or, better, by the displacement of olefins from π complexes such as nickelocene[721] or cycloheptatrienechromium tricarbonyl.[288]

$$Ni(C_5H_5)_2 + 4L \longrightarrow NiL_4 + 2C_5H_5$$

The latter method has been used with the phosphonous diamides[274] and the phosphonamidous fluorides,[677] as well as the phosphonous diesters.[288,721]

Borane complexes of the phosphonous diesters,[705,765] diamides,[129,510] phosphonamidous esters,[375] or phosphonamidous chlorides[129] are obtained by reaction of the ligand with diborane[510] or decaborane[705] or by the displacement

$$B_{10}H_{14} + 2L \longrightarrow B_{10}H_{12}L_2 + H_2$$

of a weaker ligand (e.g. L = MeCN, R_2S, or R_3As) from its borane complex.[129,375,765] The phosphorus-containing

$$B_{10}H_{12}L_2 + 2L' \longrightarrow B_{10}H_{12}L'_2 + 2L$$

ligands are themselves displaced by stronger ligands such as the tertiary amines or tertiary phosphines.[264,765]

Borane complexes are also obtained by the reaction of phosphonous diesters with sodium borohydride in the presence of carbon dioxide.[635]

$$NaBH_4 + L + CO_2 \longrightarrow BH_3L + HCO_2Na$$

The phosphonous diamides form 1:1 complexes with trialkylboranes at 0° or below.[264,265] At higher temperatures, exchange reactions take place, giving products with boron-nitrogen bonds. Complexes of phosphonous diesters[119] or phosphonous diamides[703] with aluminum trialkyls are similarly prepared.

B.1.4. Reaction with Azides

Phosphonous diesters react with phenyl azide[294,296] and with acetyl azide[295] to give phosphonimidic diesters in 60 to 70% yield (kinetics[716]).

$$RP(OR')_2 + PhN_3 \longrightarrow RP(=NPh)(OR')_2 + N_2$$

Organic azides also react in this manner with phosphonous diamides,[142,767] phosphonofluoridous esters,[366] phosphonothious diesters,[366] phosphonamidous chlorides,[263] and phosphonamidous esters,[208] but not with phosphonous dichlorides.[263,359]

Sulfonyl derivatives of the phosphonimidic diamides are obtained by the reaction of phosphonous diamides with Chloramine-B.[150]

$$RP(NR_2)_2 + ArSO_2NNaCl \longrightarrow RP(=NSO_2Ar)(NR_2)_2 + NaCl$$

B.1.5. Reaction with Chloramine

The chloramination of phosphonous diamides with gaseous chloramine gives crystalline aminophosphonium salts.[239]

$$RP(NR_2)_2 + ClNH_2 \longrightarrow [RP(NR_2)_2NH_2]Cl$$

Owing to the presence of ammonia in the chloramine gas, however, the same reaction with phosphonamidous chlorides gives $[RP(NR_2)(NH_2)_2]Cl$.

B.1.6. Reaction with Halogens

With chlorine, the phosphonous diamides form solid 1:1 adducts which are easily hydrolyzed to phosphonic diamides.[446]

$$RP(NR_2)_2 \xrightarrow{Cl_2} RP(NR_2)_2Cl_2 \xrightarrow{H_2O} RP(O)(NR_2)_2$$

Similar adducts are formed with the phosphonamidous fluorides[158] and with aromatic phosphonous diesters such as $PhP(OPh)_2$.[484] Aliphatic ester groups are attacked by the halogens (page 305).

B.1.7. Reaction with Alkyl Halides

The phosphonous diamides undergo neither the Michaelis-Arbuzov reaction[446] (page 301) nor the Perkow reaction[272] (page 302). Phosphonium salts are formed--always at phosphorus, not nitrogen[170,372,700]--and the salts are stable to heat[597] and to solvolysis with alcohols.[272]

$$RP(NR_2)_2 + R'X \longrightarrow [RR'P(NR_2)_2]X$$

Like tertiary phosphines, the phosphonous diamides form

betaines with p-benzoquinone[396] and ylids with quaternary ammonium salts.[272]

B.1.8. Addition to o-Quinones and α-Diketones

The phosphonous diesters react with o-quinones[76,751] and α-diketones,[76] forming cyclic 1:1 adducts of pentacovalent structure.

$$RP(OR)_2 + R'COCOR' \longrightarrow \begin{array}{c} RO \\ R-P \\ RO \end{array} \begin{array}{c} O \\ O \end{array} \begin{array}{c} R' \\ R' \end{array}$$

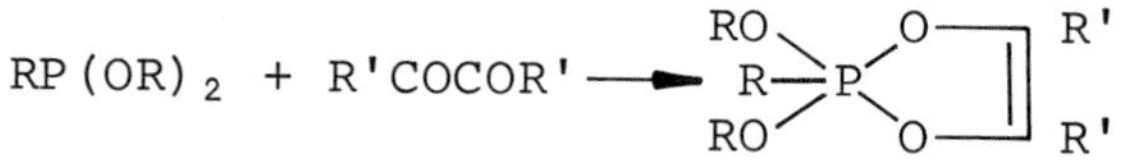

Similar adducts are formed with phosphonamidous esters[466] and phosphonamidous fluorides[153] and with other conjugated substances such as the 1,3-dienes[634] and the α,β-unsaturated ketones.[217] (See Chapter 5A).

The reaction of diethyl phenylphosphonite with diethyl peroxide gives the adduct PhP(OEt)$_4$, which is cleaved by benzoic acid to PhP(O)(OEt)$_4$.[140] The phosphonous diamide PhP(NEt$_2$)$_2$, however, is simply oxidized by diethyl peroxide.[140]

B.1.9. Addition to Carbon Disulfide

The phosphonous diesters form deep red solutions with carbon disulfide. This distinguishes them from the tertiary phosphines, which form crystalline adducts with carbon disulfide, and the tertiary phosphite esters, which give a negative test.[658]

B.2. Reactions at the P-X Bond

The compounds of this chapter that contain the P-X bond are the phosphonochloridous esters, the phosphonochloridothious esters, the phosphonamidous chlorides, and the corresponding fluorine, bromine, and iodine compounds. The reactions of these substances with water, alcohols, and amines are discussed in Section A.

B.2.1. Reaction with Grignard Reagents

The phosphonochloridous esters react with Grignard reagents at -60°, giving phosphinous esters.[293,300,708]

$$PhP(OR)Cl + R'MgCl \longrightarrow PhR'POR + MgCl_2$$

Similar reactions take place with the phosphonamidous chlorides[62,107,143] and iodides.[227]

The use of organolithium compounds in place of the Grignard reagents has been reported for the phosphonamidous chlorides.[107,169] An organotin compound α-triethylstannylacetone was used with a phosphonochloridous ester.[512]

B.2.2. Coupling by Sodium

Coupling of the phosphonamidous chlorides is accomplished by reaction with sodium[735] or sodium-potassium alloy[686] in ether or benzene.

$$2RP(NR_2)Cl + 2Na \longrightarrow RP(NR_2)P(NR_2)R + 2NaCl$$

An attempt to couple a phosphonamidous iodide with mercury failed because the product was disproportionated by the by-product mercuric iodide.[735]

B.2.3. Thermal Decomposition

The phosphonochloridous esters decompose under relatively mild conditions (R' = alkyl at 50°, aryl at 90°) with the evolution of alkyl chloride.[708] The structure of the residue has been given as $[RPO]_n$,[550] RP(OR')OPRCl,[528] or RP(O)(OR')PRCl.[183] The corresponding fluoridous esters are stable at low temperatures, and they are distillable.[152] By contrast, the phosphonochloridothious esters are stable to heat.[15]

Further study is obviously needed to resolve these discrepancies.

B.2.4. Reaction with Aldehydes

The reaction of phosphonous diesters with aldehydes and phosphorus(III) chlorides containing at least one phosphorus-chlorine group gives products such as the following:[74]

$$3RP(OR)_2 + 3R'CHO + PCl_3 \longrightarrow P[OCHR'P(O)(OR)R]_3 + 3RCl$$

The phosphorus(III) halide may be phosphorus trichloride, as shown, or a compound of the types $RPCl_2$, $ROPCl_2$, $(RO)_2PCl$, or $(R_2N)_2PCl$.[74]

If the amount of ester introduced is less than 1:1 relative to the other two components, the product is a polymer.[75]

B.2.5. Valency Expansion Reactions

The phosphonochloridous esters react with α,β-unsaturated amides with the evolution of heat and hydrogen chloride, giving esters of β-cyanoethylphosphinic acids in 20 to 50% yield.[330,586]

$$RP(OR')Cl + CH_2{=}CHCONH_2 \longrightarrow RP(O)(OR')CH_2CH_2CN + HCl$$

The phosphonochloridous esters react with sulfenamides under mild conditions, giving esters of phosphonamidothiolic acids.[528]

$$RP(OR')Cl + RSNR_2 \longrightarrow RP(O)(NR_2)SR + R'Cl$$

B.3. Reactions at the Phosphorus-Hydrogen Bond

B.3.1. Deuterium Exchange

The phosphorus-bound hydrogen in phenylphosphinic acid may be exchanged for deuterium with D_2O,[54] ethanol-d,[640] or DCl.[640] The exchange is catalyzed by both acids and bases, but it is negligible in buffered neutral solution.[54,640] No ^{18}O exchange is detectable under these conditions with water enriched in ^{18}Oxygen. The kinetics of the exchange process is consistent in both cases with a mechanism in which the rate-determining step is the rupture of the phosphorus-hydrogen bond.[640]

Deuterium exchange with methanol-d has been performed with the compounds $ClCH_2PH(O)OH$[225] and MePH(O)OPr-i.[638] Exchange in the acid is slow compared with the competing rearrangement (page 296).

B.3.2. Reaction with Halogens

The phosphinic acids react violently with chlorine, evolving hydrogen chloride and bursts of flame.[447,450] The action of aqueous bromine[418,419] or iodine[162,248] is more moderate and results in oxidation to the phosphonic acids.

The phosphinic esters react with chlorine at 0° or below, giving phosphonochloridates.[23,286,626,741]

$$RPH(O)OR' + Cl_2 \longrightarrow RP(O)(OR')Cl + HCl$$

Chlorine may be replaced by sulfuryl chloride in the preparation of phosphonochloridates from phosphinic esters[286,456,552] and phosphinothioic esters[457,526] and by N-chlorosuccinimide in the chlorination of phosphinic esters.[638]

B.3.3. The Atherton-Todd Reaction

The phosphinic esters, like the dialkyl phosphites, react with carbon tetrachloride in the presence of bases, giving phosphonochloridates or their derivatives.

$$RPH(O)OR + CCl_4 \xrightarrow{R_3N} RP(O)(OR)Cl + CHCl_3$$

If the base is ammonia[268] or a primary[23] or secondary[545] amine, the product is a phosphonamidic ester $RP(O)(OR)NR_2$. If the base is a tertiary amine, as shown, the product can be converted to a phosphonic diester by reaction with an epoxide.[139]

Similar reactions take place with the phosphinothioic esters[393] and probably also with the phosphinic acids.[189]

Phosphonous diamides that contain a free NH group react with carbon tetrachloride at 0°, giving phosphonamidimidic chlorides (or bromides with CBr_4).[664]

$$RP(NHR)_2 + CCl_4 \longrightarrow RP(=NR)(NHR)Cl + CHCl_3$$

Phosphinic esters are also converted to phosphonamidic esters by aqueous chloramine at -5° to 30°.[552]

$$RPH(O)OR + ClNH_2 \longrightarrow RP(O)(NH_2)OR + HCl$$

B.3.4. Oxidation

The primary phosphine oxides are readily oxidized to phosphonic acids by hydrogen peroxide, with the exception of two compounds (R = CHMePh and CH_2CH_2CN), which give the phosphinic acids (method 26, A.2).[95,96]

$$RP(O)H_2 \xrightarrow{H_2O_2} RPH(O)OH \xrightarrow{H_2O_2} RP(O)(OH)_2$$

The phosphinic acids usually require stronger oxidizing agents. Useful oxidants include concentrated hydrogen peroxide,[98,162] nitric acid,[93,231,445] alkaline permanganate,[161,248] mercuric chloride,[137,161,417,661] aqueous bromine,[175,418,419] and iodine.[162,248] The aromatic phosphinic acids may suffer ring nitration if nitric acid is employed.[52,450] Phosphinic acids reduce ammoniacal silver nitrate and other metallic salts to the free metals.[191]

Phosphinic esters may be oxidized by air, preferably under UV catalysis,[531] or by oxidizing agents such as dinitrogen tetroxide[541] or ozone.[490] Hydrogen peroxide has been employed with the phosphinothioic esters.[678]

B.3.5. Reaction with Sulfur and its Compounds

The phosphinic esters add sulfur in basic solvents such as dioxane[529] or in the presence of a base such as triethylamine[529] or ammonia.[522]

$$RPH(O)OR + S \longrightarrow RP(O)(OR)SH$$

In the form of their sodium salts, the phosphinic esters react readily with both sulfur[85] and selenium.[426] The phosphinothioic esters behave similarly.[360,457]

Phosphinic esters react with sulfenyl chlorides,[392,660,] sulfenamides,[528] disulfides,[394,527] and isothiocyanates,[391,542,593] giving O,S-phosphonothiolic diesters and HCl, R_2NH, RSH, or HCN, respectively.

$$RPH(O)OR + R'SNR_2 \longrightarrow RP(O)(OR)SR' + R_2NH$$

Basic (Na)[394,527] or radical (UV)[471] catalysts have been employed with the disulfides, but these are not always necessary.[527] The isothiocyanates require an equivalent amount (i.e., the sodium salt[391]). Similar reactions with the phosphinothioic esters give the O,S-phosphonodithiolic diesters RP(S)(OR)SR.[391,392,394,535,660]

The reaction of phosphinic esters with P-sulfenyl chlorides gives P-O-P anhydrides, owing to isomerization of the intermediate P-S-P anhydrides (see page 305).[453]

B.3.6. 1,2-Hydride Shifts

Chloromethylphosphinic acid (8) reacts rapidly with aqueous alkali at 100°, but only a minor amount (ca. 12%) of the anticipated hydrolysis product (9) is obtained.[727] The major product is methylphosphinic acid (10).

$$ClCH_2PH(O)OH \xrightarrow{OH^-} HOCH_2PH(O)OH + MeP(O)(OH)_2$$

(8) (9) (10)

Weaker bases, such as calcium or magnesium hydroxide, lead to a higher proportion of (9).[727] Later work showed that this reaction is common to other phosphonous acid derivatives containing an α-chlorine atom, such as $ClCH_2PH(O)OMe$ (11)[213] or $ClCH_2PH_2$,[185] and to their reactions with a variety of strong bases such as the alkoxides, primary and secondary amines, and even Grignard reagents.

Deuterium exchange measurements have indicated that the rearrangement is predominantly (95%) intramolecular.[225] A mechanism involving a 1,2-hydride shift has been suggested.[225]

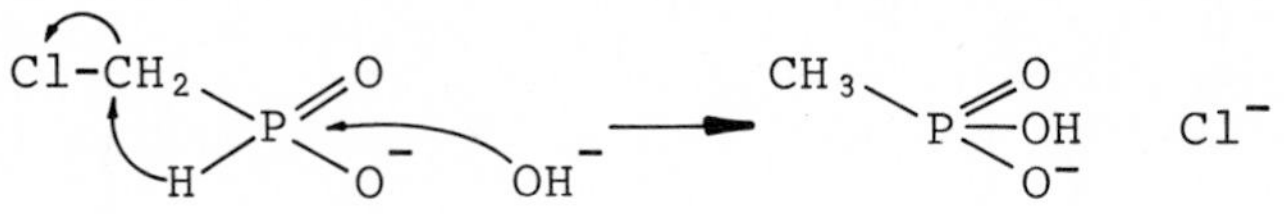

Alternative mechanisms involving phosphaalkene intermediates [e.g., CH_2=P(O)OH] were ruled out in this instance,[225] but they may be operative in the reaction of (11) with Grignard reagents.[213]

B.3.7. Addition to Olefins

The phosphinic acids and their derivatives add to olefins in the presence of radical catalysts, giving compounds containing a new carbon-phosphorus bond. The phosphinic acids react at 125°.[523,712]

$$RPH(O)ONa + R'CH{=}CH_2 \longrightarrow R(R'CH_2CH_2)P(O)ONa$$

Similar reactions take place with phosphinic esters[587,712] and primary phosphine oxides.[96] The reactivity of the phosphorus compound decreases in the following order:[587]

$$(RO)_2P(S)H > RPH(O)OR > (RO)_2P(O)H$$

Base-catalyzed addition is preferred with α,β-unsaturated esters,[80,96,610] amides,[96] nitriles,[80,96,457,601] ketones,[605] and sulfones,[607] with 1,3-dienes,[588] and with other active olefins such as vinyl acetate[589,593] and 2-vinylpyridine.[428]

$$RPH(O)OR + CH_2{=}CHCN \xrightarrow[ROH]{RONa} RP(O)(OR)CH_2CH_2CN$$

Phosphinic esters also add to the triple bond in alkynes, giving products that contain vinyl substituents. The addition to simple alkynes such as 1-heptyne takes place upon prolonged heating in the presence of a radical catalyst at 80 to 90°.[592] Further addition requires a

$$RPH(O)OR + R'C{\equiv}CH \longrightarrow RP(O)(OR)CH{=}CHR'$$

higher temperature (140-180°), resulting in a 2:1 adduct.[606]

Phosphinic esters also add to α,β-acetylenic esters, if a basic catalyst is used.[611] α,β-Acetylenic aldehydes and ketones, however, react preferentially at the carbonyl group.[574]

Phosphinic esters add to the carbon-carbon double bond in ketene[600] and ketene acetal.[213] The acetyl derivatives formed in the ketene reaction usually react further with the ketene, forming ultimately α-acetoxyvinylphosphonate esters.[600]

$$RPH(O)OR \xrightarrow{CH_2=C=O} RP(O)(OR)COCH_3 \xrightarrow{CH_2=C=O} RP(O)(OR)OC(=CH_2)OAc$$

B.3.8. Addition to Aldehydes or Ketones

Primary phosphine oxides add to aldehydes or ketones in the presence of an acid catalyst (HCl) giving either secondary or tertiary phosphine oxides, depending on the structure of the carbonyl compound.[95,96] Dialdehydes and ketones displace one phosphorus-hydrogen group, but aldehydes displace both.[96]

$$RP(O)H_2 + R'_2CO \longrightarrow RPH(O)CR'_2OH$$

The phosphinic acids add to aldehydes, but not to ketones other than acetone.[363,423,608] The phosphinic esters (and phosphinothioic esters[6,457]) add to both aldehydes and ketones under mild conditions.[9,299] The use of basic catalysts, such as sodium ethoxide,[9,299] contributes little to the reaction rate and is actually detrimental with compounds that give products that undergo base-catalyzed rearrangement (e.g., the α-halocarbonyl compounds or the 1,2-dicarbonyl compounds). α-Halocarbonyl compounds such as chloral[9,59,486,502] or the α-haloketones[5b,486] give secondary phosphinic esters that rearrange to phosphonic esters under the influence of base. Fluoroketones

$$RP(O)(OR)CHOHCCl_3 \xrightarrow{OH^-} RP(O)(OR)OCH{=}CCl_2$$

yield exclusively the rearrangement products.[282,283]

1,2-Dicarbonyl compounds such as the α-diketones[579] or the α-ketoesters[580,590,591] react with phosphinic esters at 100° or below, giving secondary phosphinic esters; but if the temperature is allowed to rise above 100°, the products rearrange to phosphonic esters.

$$RP(O)(OR)CR_2OH \xrightarrow{\Delta} RP(O)(OR)OCHR_2$$

The α-ketophosphonate esters behave similarly,[5b,579] as do the products of addition to other phosphorus-hydrogen compounds, including the phosphinic acids[581] and the phosphinothioic esters.[583,584]

Phosphinic esters have been reported to add to the carbon-sulfur double bond in thiourea, forming products having structure (12).[4] It seems more likely, however, that these products (like those of the corresponding dialkyl phosphites[460]) are S-alkyl isothiuronium salts (13).

$RP(O)(OR)C(NH_2)_2SH$ (12) $[RPH(O)O]^-[RSC(NH_2)_2]^+$ (13)

B.3.9. Addition to the C=N Bond

Phosphinic esters add to aldehydes or ketones in alcoholic ammonia solution at 100°, giving α-aminoalkylphosphinic esters in 20 to 30% yield.[299] Some hydrolysis of the ester group occurs.

$$RPH(O)OR + R'_2CO + NH_3 \longrightarrow RP(O)(OR)CR'_2NH_2$$

Primary amines react similarly, preferably in the form of Schiff bases and in the presence of an alkoxide catalyst.[570]

$$RPH(O)OR + R'CH{=}NR \xrightarrow{NaOR} RP(O)(OR)CHR'NHR$$

Secondary amines may be used in the form of their methylenediamine derivatives.[533] Phosphinic acids[536,608]

$$RPH(O)OR + CH_2(NR_2)_2 \longrightarrow RP(O)(OR)CH_2NR_2 + R_2NH$$

and phosphinothioic esters[535] react similarly.

Primary phosphine oxides[96] and phosphinic esters[593] add to the carbon-nitrogen double bond in isocyanates, forming P-carbamoyl derivatives. Phenylphosphinic acid

$$RPH(O)OR + R'NCO \xrightarrow{NaOR} RP(O)(OR)CONHR'$$

reacts similarly if an excess of triethylamine is present,[189] otherwise the acid is converted to phenylphosphonic anhydride.[190]

Isothiocyanates react in a different manner (see page 296).

B.3.10. Reaction with Orthoesters

Phosphinic esters react with orthoformate esters in sealed tubes at 182°, giving phosphorylated formals. Conversions are low, and some decomposition of the products is observed.[472,627]

$$RPH(O)OR + CH(OR)_3 \longrightarrow RP(O)(OR)CH(OR)_2 + ROH$$

B.3.11. Formation of Cyclotetraphosphines

Cyclopolyphosphines are the end products of several reactions involving phosphorus-hydrogen compounds. The primary phosphine oxides are dehydrated by pumping down at 60° (1mm).[252] Yields are low (15-20%), owing to the

$$4RP(O)H_2 \longrightarrow [RP]_4 + 2H_2O$$

concomitant disproportionation. (See also Chapter 2).

Several attempts have been made to prepare mixed compounds of the type RPHCl or RPHBr, but the compounds, if formed, disproportionate with the loss of HX.[276,277,750] Disproportionation also occurs following attempts to prepare compounds of the type $RPHNR_2$, either by reduction of the phosphonamidous chlorides with lithium aluminum hydride[279] or by the thermal decomposition of phosphonous diamides.[374]

$$4RPHNHR \longrightarrow [RP]_4 + 4RNH_2$$

B.4. Reactions at the P-OR (or -SR) Bond

The esters and thioesters comprise by far the largest group of compounds in this chapter. They include all the compounds in Section A.3, and some of the compounds in Sections A.4 and A.5. Reactions with water, alcohols and amines are discussed under Section A.

B.4.1. Reaction with Grignard Reagents

Grignard reagents react with phosphonous diesters to form unsymmetrical tertiary phosphines.[306,658] It is not

$$RP(OR)_2 + 2R'MgCl \longrightarrow RR'_2P + 2MgClOR$$

possible to stop the reaction at the phosphinous ester stage RR'POR.[658] The phosphinic esters, however, react with Grignard reagents, giving secondary phosphine oxides in yields ranging from 25 to 75%.[163] The reaction initially produces the magnesium salts of phosphinous acids, which, if not hydrolyzed, can be brought into reaction with sulfur to give phosphinothioic acids.[297,432,618]

$$RPH(O)OR \xrightarrow{R'MgCl} RR'POMgCl \xrightarrow{S} RR'P(S)OH$$

B.4.2. Reaction with Phosphorus(III) Chlorides

Phosphonous diesters react with phosphorus trichloride

in ether at 5°, giving phosphonous dichlorides in 50% yield.[658] With excess PCl_3, the by-product is $ROPCl_2$.

$$RP(OR)_2 + PCl_3 \longrightarrow RPCl_2 + (RO)_2PCl$$

Phosphonous dichlorides are also obtained from phosphonodithious diesters and benzoyl chloride at 130 to 140°.[14]

$$RP(SR)_2 + 2PhCOCl \longrightarrow RPCl_2 + 2PhCOSR$$

The preparation of intermediate species such as RP(OR)Cl, RP(SR)Cl, and the analogous bromine compounds by this method is described on page 270.

B.4.3. Reduction with Lithium Aluminum Hydride

Phosphonous diesters are reduced to primary phosphines by lithium aluminum hydride.[658]

$$RP(OR)_2 \xrightarrow{LiAlH_4} RPH_2$$

B.4.4. The Michaelis-Arbuzov Reaction

Phosphonous diesters undergo the Michaelis-Arbuzov reaction with a wide variety of alkyl halides, giving esters of secondary phosphinic acids[28] (kinetics;[19] review[240]). The reaction is usually complete in 2 to 3 hr

$$RP(OR)_2 + R'X \longrightarrow RR'P(O)OR + RX$$

at 100 to 120°, compared with the 5 to 6 hr at 160° required for the trialkyl phosphites. The phosphonous diesters are also more susceptible to self-isomerization than the trialkyl phosphites, but in truly pure state they isomerize with difficulty. Esters of aromatic phosphonous acids isomerize at 250°, but the aliphatic compounds are stable to 300°.[596]

The scope of this reaction encompasses α-halo ethers,[36] α-halo esters,[30,40] α-halo acetals,[630] carbamoyl chlorides,[40] chloroformate esters,[235] acid chlorides,[205,621] and some α-halo ketones,[573] and polyhalogen compounds such as CCl_4,[309] CCl_3Br,[309] $CF_3CF{=}CF_2$,[284,340] and the alkylene dihalides.[199,244,430] Epichlorohydrin gives a variety of products.[10]

Secondary and tertiary alkyl halides, with few exceptions, give poor results. Some exceptions are Ph_3CBr,[31] $MeCHBrCO_2Et$,[30,254] $EtCHBrCO_2Et$,[80] $BrCH(CO_2Et)_2$,[33] cyanuric chloride,[255] and PhCCl=NP(O)(OPh)$_2$.[141] Catalysts such as the tertiary amines[34,362] are seldom necessary.

Alkyltin halides were at first believed to give products with phosphorus-tin bonds,[38] but the products were later shown to be salts of phosphinic acids.[37,599]

With alkyl halides, aromatic esters of phosphonous acids form stable quasi-phosphonium salts that are not decomposed under the conditions of the Michaelis-Arbuzov reaction.[307] The salts can, however, be decomposed by drastic heating,[310] by alcoholysis,[35] or by alkaline hydrolysis.[444] The thermal method is sometimes complicated by redistribution of the alkyl and phenoxy groups.[35,485]

Crystalline quasi-phosphonium salts have also been prepared by the addition of methyl iodide to unsaturated phosphonous diesters containing branched ester radicals, such as $CH_2=CHCH_2P(OBu\text{-}i)_2$.[386,387] The adducts decompose to secondary phosphinic esters on mild heating.

The phosphonodithious diesters also undergo the Michaelis-Arbuzov reaction,[32] but the sulfur atoms compete with the phosphorus for the alkyl halide and mixtures of products are usually obtained.[45,356] (See also Chapter 7).

$$RP(SR')_2 \xrightarrow{R''X} [RR''P(SR')_2]\,X \xrightarrow{\text{normal}} RR''P(S)SR' + R'X$$

$$RP(SR')_2 \xrightarrow{R''X} [RP(SR')SR'R'']\,X \xrightarrow{\text{abnormal}} RP(SR')X + R'SR''$$

Further interaction of the products gives rise to abnormal products such as tertiary phosphine sulfides[45,358] and trialkylsulfonium halides.[356] Normal rearrangement products are said to be obtained from α-halo esters[32] and chloroformates.[235] The phosphonodithious diesters differ in this respect from the esters $(RS)_3P$, which give only the products of addition to sulfur under the conditions of the Michaelis-Arbuzov reaction.

Phosphonous diamides do not take part in this reaction (see page 291), but the phosphonamidous esters[483] and thioesters[17] do. The products are phosphinic or phosphinothioic amides.

The phosphonofluoridous esters also participate in this reaction, even though the other ester halides do not.[368]

$$RP(OR)F + R'I \longrightarrow RR'P(O)F + RI$$

B.4.5. The Perkow Reaction

The use of α-haloketones in the Michaelis-Arbuzov reaction gives either the normal rearrangement products or enol phosphonates, depending on the halogen employed and the reaction temperature.[71,573,748] The formation of the normal product is more favorable than in the case of the tertiary phosphite esters, because of the milder conditions needed (review[383,738]).

Enol phosphonates are formed exclusively when the alkyl halide is an α-haloaldehyde such as chloral.[22,371,632]

$$RP(OR)_2 + CCl_3CHO \longrightarrow RP(O)(OR)OCH=CCl_2 + RCl$$

Similar reactions take place with the phosphonamidous esters[644,680] and the phosphonothious diesters,[680] but the products of the phosphonodithious diesters have not been identified.[14]

Esters of trichloroacetic acid also give enol phosphonates.[633]

B.4.6.   The Michaelis-Becker Reaction

The Michaelis-Arbuzov reaction is sometimes modified by replacing the phosphonous diester with the sodium salt of a phosphinic ester[43,346] or thioester.[682]

$$RP(OR)ONa + R'X \longrightarrow RR'P(O)OR + NaX$$

The alkyl halide may be replaced by a p-toluenesulfonate ester[131] or by an α-haloester[254,682] or salt.[726] Carbamoyl chlorides give the P-carbamoyl derivatives.[42]

The use of α-haloketones in this reaction, as in the Michaelis-Arbuzov reaction, leads to abnormal products. The products are either epoxyalkylphosphinates (14),[50] vinylphosphonates (15),[51] or both.

$$RP(OR)ONa + MeCOCH_2Cl \Big\langle \begin{array}{l} RP(O)(OR)CMeCH_2 \\ \qquad\qquad\qquad \diagdown O \diagup \\ \qquad\qquad (14) \\ \\ RP(O)(OR)OCMe=CH_2Cl \end{array} + NaCl$$

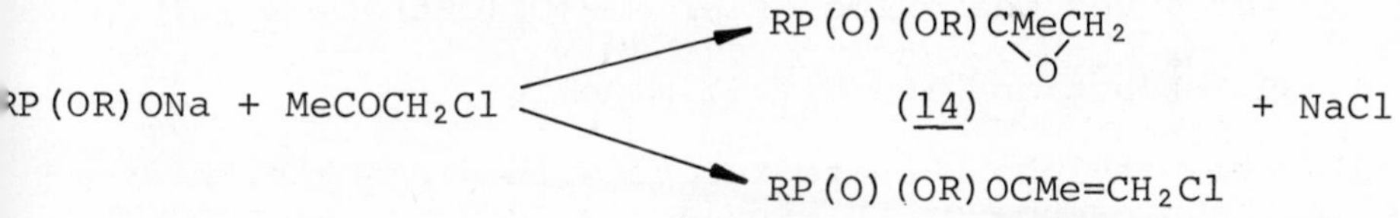

$$(15)$$

The "normal" products [the β-ketophosphinic esters, RP(O)(OR)CH$_2$COR] are not obtained at all by this method, although they are accessible through the enol ethers of the α-haloketones.[714]

Acid chlorides react with the sodium salts of phosphinic esters giving α-ketophosphonate esters, but these react further, adding another mole of the phosphinic ester (see page 298) and then undergoing a base-catalyzed rearrangement. The ultimate products are phosphonic esters of α-hydroxyalkylphosphinates.[582,609]

$$2RP(OR)ONa + R'COCl \longrightarrow RP(O)(OR)OCHR'P(O)(OR)R + 2NaCl$$

B.4.7. Reaction with α,β-Unsaturated Acids

Phosphonous diesters react with α,β-unsaturated acids such as acrylic or methacrylic acid by addition to the double bond and transfer of an ester radical to the carboxylic acid.[318]

$$RP(OR')_2 + CH_2=CHCO_2H \longrightarrow RP(O)(OR')CH_2CH_2CO_2R'$$

Similar reactions take place with the phosphonochloridous esters[329,585] and thioesters,[16] giving the corresponding carboxylic acid chlorides [e.g., $RP(O)(OR')CH_2CH_2COCl$].
  Phosphonous diesters also react with α,β-unsaturated aldehydes[318] and ketones,[243] giving vinyl ethers.

$$RP(OR')_2 + CH_2=CHCHO \longrightarrow RP(O)(OR')CH_2CH=CHOR'$$

B.4.8. Reaction with Nitro Compounds

Diethyl methylphosphonite displaces a nitro group from o-dinitrobenzene, giving ethyl methyl(o-nitrophenyl)phosphinate[105] (kinetics[104]).

$$MeP(OEt)_2 + o\text{-}C_6H_4(NO_2)_2 \longrightarrow Me(o\text{-}NO_2C_6H_4)P(O)OEt + EtNO_2$$

In other cases the ester acts as a reducing agent, being more effective in this regard than triethyl phosphite or $(Et_2N)_3P$.[106] Reaction with the α-halonitro compound $Me_2C(NO_2)Br$ gives the oxime ester $MeP(O)(OEt)ON=CMe_2$.[21]

B.4.9. Reaction with Nitrilimines

Dibutyl vinylphosphonite reacts with diphenylnitrilimine in the presence of triethylamine, giving the product of 1,3-addition.[349]

$$CH_2=CHP(OBu)_2 + PhNHN=CClPh \xrightarrow{\quad Et_3N \quad} PhN \underset{N=}{\overset{\diagup\!\!\!\!\diagup}{\diagdown}} P(O)OBu + BuCl$$

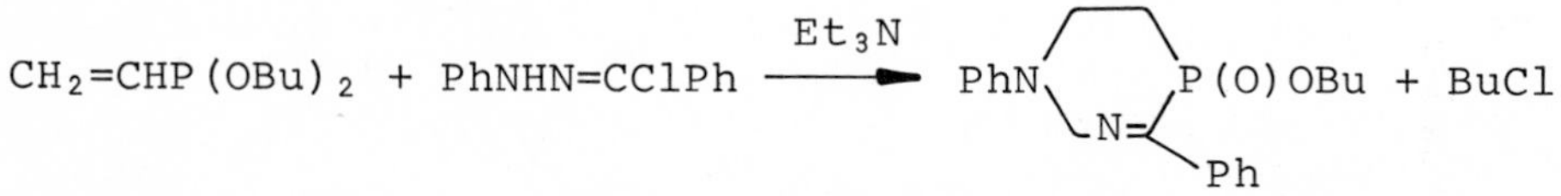

B.4.10. Reaction with Sulfur Compounds

Phosphonous diesters react with sulfenyl chlorides,[470] disulfides,[241] and isothiocyanates,[459] giving O,S-phosphonothiolate esters and RCl, R_2S, and RCN, respectively.

$$RP(OR')_2 + RSCl \longrightarrow RP(O)(OR')SR + R'Cl$$

These reactions resemble those of the phosphinic esters (page 296) but require no catalyst. The disulfide reaction, in fact, may be carried out in the presence of a free radical inhibitor (hydroquinone).[241]

Similar reactions take place with the phosphonamidous esters, giving phosphonamidothiolic esters $RP(O)(NR_2)SR$.[649]

The reaction of phosphonous diesters with P-sulfenyl chlorides[454,455] or disulfides[458] yields the expected anhydrides, but these rearrange on distillation to the more stable P-O-P isomers.

$$RP(OR')_2 + (RO)_2P(O)SCl \longrightarrow RP(O)(OR')SP(O)(OR)_2 + R'Cl$$

$$\downarrow \Delta$$

$$RP(O)(OR')OP(S)(OR)_2$$

B.4.11. Reaction with N-Haloamides

The reaction of phosphonous diesters with N-chloro- or N-bromosuccinimide proceeds with displacement of alkyl halide in a manner similar to the Michaelis-Arbuzov reaction, giving phosphonamidic esters in 55 to 75% yield.[724,768]

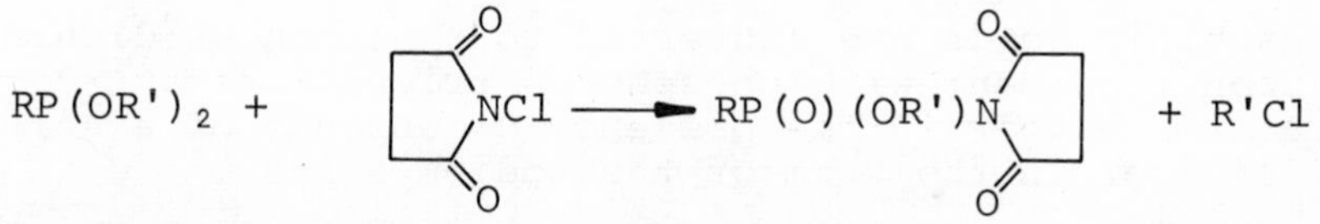
RP(OR')2 +
NCl
RP(O)(OR')N
+ R'Cl

B.4.12. Reaction with Halogens

The phosphonous diesters react with chlorine,[203,293] bromine,[629] and other mild halogenating agents, giving alkyl phosphonohalidates in good yield.

$$RP(OR')_2 + Cl_2 \longrightarrow RP(O)(OR')Cl + R'Cl$$

Similar reactions take place with the phosphonofluoridous esters[157] and thioesters.[151]

The reaction of phosphonous diesters with halogens takes a different course when an alcohol is present. The product is a phosphonic diester.[194]

$$RP(OR')_2 + Cl_2 + R'OH \longrightarrow RP(O)(OR')_2 + R'Cl + HCl$$

Similar "oxidations" occur with other positive halogen

sources, such as carbon tetrachloride[103] or α-haloketones,[272] when the reaction is carried out in the presence of an alcohol.

Reactions with phosphorus(III) halides result in the replacement of ester radicals without valency expansion (see Section B.4.2). Phosphorus pentachloride converts the phosphonous diesters to phosphonic dichlorides $RP(O)Cl_2$.[297]

B.5. Reactions at the P-OH Bond

B.5.1. Preparation of Sodium Derivatives

Phosphinic esters react with sodium in dry ether or benzene, forming the sodium derivatives.[42,43,346] The

$$RPH(O)OR' + Na \longrightarrow RP(OR')ONa + H_2$$

products are soluble in ether, gasoline, benzene, and other organic solvents.[43,346] The corresponding thio derivatives are prepared by the reaction of phosphinothioic esters with sodium alkoxides.[360,682]

$$RPH(S)OR' + NaOR' \longrightarrow RP(OR')SNa + R'OH$$

B.5.2. Reaction with Phosphorus(III) Chlorides

Phosphinic acids are converted to phosphonous dichlorides upon treatment with phosphorus trichloride at room temperature, either in the presence or absence of a solvent[193,248] or in the form of the sodium salts.[491]

$$3RPH(O)OH + 2PCl_3 \longrightarrow 3RPCl_2 + 2H_3PO_3$$

Phosphorus pentachloride gives the phosphonic dichloride $RP(O)Cl_2$.[447]

B.5.3. Reduction with Lithium Aluminum Hydride

The phosphinic acids are difficult to reduce; classical reducing agents are ineffective. Phenylphosphinic acid poisons Raney nickel catalyst[267] and is reduced to phenylphosphine in only 13% yield by lithium aluminum hydride.[745]

B.6. Reactions at the Phosphorus-Nitrogen Bond

The amides comprise all the compounds of Section A-4. Reactions with water, alcohols, mercaptans, and amines are discussed in Section A.

B.6.1. Cleavage with Acids

The phosphorus-nitrogen bond in phosphonous diamides is cleaved easily and quantitatively by hydrogen chloride under anhydrous conditions, at room temperature or below.[102,169,170,710]

$$RP(NR_2)_2 + 4HCl \longrightarrow RPCl_2 + 2R_2NH\cdot HCl$$

Similar reactions take place with the phosphonamidous chlorides[278] and phosphonamidous fluorides.[156,221,507] The RPClF from the latter are stable if R = phenyl,[156] but if R = alkyl they disproportionate rapidly to $RPCl_2$ and RPF_2.[156,221,507]

The phosphorus-nitrogen bond in the phosphonous diamides is also split by hydrogen bromide and hydrogen iodide,[102] but with hydrogen fluoride the phosphonamidous fluorides form crystalline adducts, e.g., $[R_2NH_2][RPHF_4]$.[109,154]

Exchange reactions also take place with boron halides (BF_3 or BCl_3),[265] phosphorus halides (PF_3,[92] POF_3,[92] or PCl_3[733]), and even acid fluorides such as benzoyl fluoride.[91] Partial reactions of this nature are discussed on page 281.

The phosphonous diamide $CF_3P(NMe_2)_2$ reacts with $CF_3P(PMe_2)_2$ at 132°, presumably giving $CF_3P(NMe_2)PMe_2$; however, the product could not be separated from the starting materials.[125]

B.6.2. Reaction with Aldehydes or Ketones

Insertion of aldehydes or ketones into the phosphorus-nitrogen bond has been observed with the phosphonamidous esters[466] and thioesters.[8]

$$RP(NR_2)OR + R'CHO \longrightarrow RP(O)(OR)CHR'NR_2$$

Yields are low (20 to 50%).

B.7. Reactions at the Phosphorus-Carbon Bond

B.7.1. The Haloform Reaction

Trifluoromethylphosphinic acid liberates fluoroform quantitatively when treated with aqueous base at room temperature[67,100] or when heated in water to 100°.[67,161]

$$CF_3PH(O)OH + H_2O \longrightarrow CHF_3 + H_3PO_3$$

Similar reactions take place with phosphinic acids

containing other fluoroalkyl substituents[98,162] and with derivatives of these acids such as the phosphonodithious diesters,[99] the phosphonamidous chlorides,[253] and the phosphonous diiodides,[118] diamides,[221,253,507] and diacetates.[524]

B.7.2. Other Scissions

Other phosphinic acids that suffer carbon-phosphorus bond scission under relatively mild conditions are the acids RPH(O)OH (R = Me_2COH,[416] p-$R_2NC_6H_4$,[661] p-$NH_2C_6H_4$,[737] 2-Me-4-$Me_2NC_6H_3$,[413] 2,4-$(MeO)_2C_6H_3$,[569] and Ph_3C[248]), which are hydrolyzed by aqueous acid but not by aqueous alkali. Also included in this category are the acids in which R = Me_2COH,[416] PhCHOH,[736] p-$Me_2NC_6H_4$,[661] and CCl_3,[506] which are cleaved upon heating to 100° or above in preference to the usual path of thermal decomposition, disproportionation.

$$HOCMe_2PH(O)OH \xrightarrow{\Delta} Me_2CO + H_3PO_2$$

The β-ketoesters $RCOCH_2P(OR)_2$ and their enolic derivatives $RC(OR')=CHP(OR)_2$ are prone to hydrolytic cleavage under neutral or acidic conditions.[567]

$$RC(OR)=CHP(OR)_2 + 2H_2O \longrightarrow (RO)_2P(O)H + RCOCH_3 + 2ROH$$

B.8. Reactions at the P-O-P Bond

The phosphonous anhydrides are sensitive to air and moisture and must be handled in a dry, inert atmosphere. They react vigorously with alkyl halides, forming quaternary phosphonium salts.[45,46] In mixed anhydrides, such as the phosphonous-phosphinous anhydride shown below, the phosphorus atom bearing the smaller number of phosphorus-oxygen bonds is the one that is quaternized.[46]

$$RP(OR)OPR'_2 + 2R'Br \longrightarrow R'_4PBr + RP(O)(OR)Br$$

Bromination of the thioanhydride $[PhPS]_n$ with bromine at 60° gives phenylphosphonothioic dibromide $PhP(S)Br_2$ in 73% yield.[353] Under the same conditions, sulfuryl chloride gives phenylphosphonic dichloride $PhP(O)Cl_2$.[353]

The mixed anhydrides $RP(OAc)_2$ are easily cleaved by anhydrous hydrogen chloride, regenerating the phosphonous dichloride.[524]

C. GENERAL DISCUSSION OF PHYSICAL PROPERTIES

C.1. Tautomerism

The phosphonous acids, in common with other lower oxyacids of phosphorus, exist predominantly in the "keto" form (17) of the tautomeric pair:

$$RP(OH)_2 \rightleftarrows RPH(O)OH$$

(16) (17)

The acids are monobasic[377,613] and show both P-H and P=O absorption in the IR spectrum.[134] Even the trifluoromethyl derivative ($R = CF_3$) has this form.[100] The "enol" form (16) cannot be detected spectroscopically[198,287] owing to its low concentration, but its existence has been established by kinetic methods. The proportion of "enol" in phenylphosphinic acid (R = Ph), estimated from deuterium exchange measurements, probably does not exceed that found in aqueous acetone solutions.[54,640]

The stability of the "keto" form in the phosphinic acids is a consequence of the high energy of the phosphorus-oxygen double bond, which in turn owes its stability to p_π-d_π bonding, electrostatic factors, or both.[271a] This effect also appears in the phosphorus-sulfur and the phosphorus-nitrogen double bonds, although to a more limited extent. Phosphonodithious acids are unknown, but the partial esters RPH(S)OR show P-H and P=S absorption in the IR and ^1H and ^{31}P NMR spectra which support the "keto" structure (19).

$$RP\begin{cases} OR \\ SH \end{cases} \rightleftarrows RPH(S)OR$$

(18) (19)

The phosphonous diamides, on the other hand, favor the "enol" form $RP(NHR)_2$ almost exclusively.[62,372,700] The IR spectrum of $PhP(NHBu\text{-}t)_2$, for example, shows no evidence of P-H (2440-2350 cm^{-1}) or P=N (1300 cm^{-1}) linkages.[700] Only in those compounds in which the basicity of the nitrogen is drastically reduced, relative to the phosphorus, does the "keto" or phosphinimidic amide form RPH(=NR)NHR predominate. One such compound is (20), in which the N-substituent is the strongly electronegative tosyl group.[671] In others (21), the N-P(H)=N group forms part of a phosphazene ring.[670]

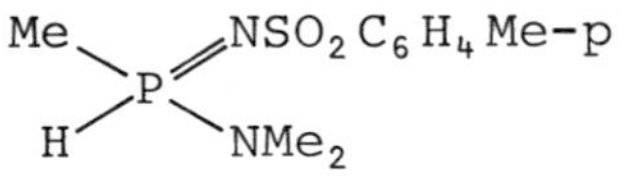

(20)

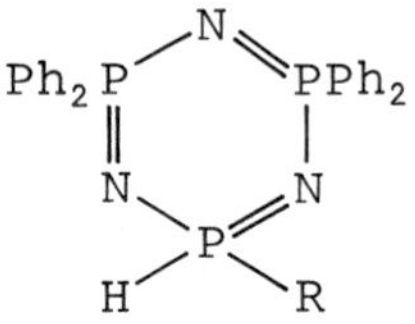

(21)

The "enol" form is also dominant in the phosphonous dihydrazides $RP(NHNR_2)_2$[487,525] and in mixed amides of the types RP(NHR)F,[156] RP(NHR)Cl,[239,663] RP(NHR)OR[328,499] (with the possible exception of one glucose derivative[494]), and RP(NHR)SR.[7,498]

C.2. Acid Dissociation Constants

Acid dissociation constants have been determined for about 28 phosphinic acids by potentiometric,[290,291,377,613] spectroscopic,[290,555] or other[338] methods. Acid strengths have been determined not only in water but also in methanol[290,361] and other alcohols,[361] ethylene glycol,[121] and dimethyl sulfoxide.[121] The phosphinic acids are stronger than the corresponding phosphonic acids by about 0.3 pK unit.[291,377,555]

Plots of pK_a versus Hammett σ constants for series of m- or p-substituted arylphosphinic acids $XC_6H_4PH(O)OH$ [X = H, Cl, Br, CH_3, OCH_3, $N(CH_3)_2$, CO_2H, CN] are linear, indicating a good fit to the Hammett equation.[608,613] The reaction constant ρ is 0.830, slightly larger than the value of 0.755 reported for phosphonic acids.[613]

An estimate has been made of the magnitude of the Hammett substituent constant for the phosphinic acid group itself, -PH(O)OH. The estimate, > +0.37, is based on the position of the methyl signal in the ^1H spectrum of p-tolylphosphinic acid.[665]

The basicity of the phosphoryl oxygen in the esters PhPH(O)OR (R = $SiMe_3$ or $GeEt_3$) has been measured by a hydrogen-bonding method, using the shift of the γ_{OH} band of phenol in carbon tetrachloride.[87]

C.3. Disproportionation

The phosphinic acids undergo a characteristic redox reaction when heated by themselves to about 150°.[447] One mole of a primary phosphine is formed for every 2 moles of phosphonic acid:

$$3RPH(O)OH \xrightarrow{\Delta} RPH_2 + 2RP(O)(OH)_2$$

Aliphatic phosphinic acids, contrary to an earlier report,[231] decompose at about the same temperature as the aromatic phosphinic acids.[658] The only known exceptions are acids that suffer carbon-phosphorus bond scission at temperatures below 150° (see page 308).

Considering the antiquity of the disproportionation reaction (first noted by Michaelis in 1874[447]), surprisingly little is known about it. Kosolapoff and Powell,[348] observing that the phosphinic acids exist in a state of trimeric aggregation, suggested that the disproportionation was the result of a rearrangement of hydrogen and oxygen atoms within the cyclic trimer. The reaction, however, takes place at a temperature where dissociation to the monomer would be favored. Trifluoromethylphosphinic acid, for example, dissociates from dimer to monomer at about 134°.[100] More recently, Gallagher and Jenkins[198] have suggested the intermediacy of the carbenelike phosphinidene [PhP], which could be formed by thermal decomposition of the phosphinic anhydride $[PhPH(O)]_2O$. The formation of phenylphosphine is explained as an abstraction of hydrogen by [PhP] from either phenylphosphinic acid or its anhydride.[198]

The only other compounds in this chapter that undergo a comparable redox reaction are the primary phosphine oxides, which disproportionate at about room temperature to primary phosphine and phosphinic acid (method 1).[95,96]

$$2RP(O)H_2 \longrightarrow RPH_2 + RPH(O)OH \qquad (1)$$

The primary phosphine oxides are more stable in polar solvents than in the solid state, and considerably more stable in strongly acidic media. This suggests that they are more stable in the form of the conjugate acids $RP(OH)H_2^+$ than as the free bases.[96]

Phenylphosphine sulfide, a viscous yellow liquid,[342] decomposes rapdily at room temperature, evolving phenylphosphine and hydrogen sulfide and depositing the crystalline cyclotetraphosphine monosulfide.[402] (Structure see Chapters 1 and 2).

$$5RP(S)H_2 \longrightarrow [RP]_4S + RPH_2 + 4H_2S$$

The catechol ester (<u>22</u>) disproportionates when heated at 100°, especially in the presence of acid catalysts, giving the crystalline phosphorane (<u>23</u>) and an unidentified phosphorus-rich residue.[751]

Phosphonous diamides with primary amino radicals $PhP(NHR)_2$ lose amine at 140 to 160°, yielding pentaphenyl cyclopentaphosphine, some monophosphazene, and other high-molecular-weight products.[372,374] (See Chapter 2.)

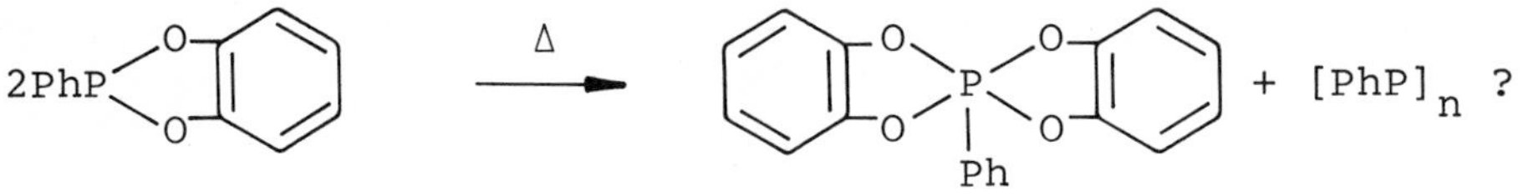

(22) (23)

$$2PhP(NHR)_2 \longrightarrow PhP(=NR)(NHR)_2 + \tfrac{1}{5}[PhP]_5 + RNH_2$$

Under similar conditions, the phosphonous dihydrazide $PhP(NHNMe_2)_2$ yields primarily the P=N polymer $[PhP(NMe_2)=N]_n$.[760]

The acetoxyphosphines $CF_3P(OAc)_2$ and $CF_3P[OC(O)CF_3]_2$ disproportionate rapidly at room temperature or below, giving the acid anhydride and impure $[CF_3PO]_n$ polymer (method 2).[524]

$$2RP(OAc)_2 \longrightarrow [RPO]_n + Ac_2O \qquad (2)$$

Trimethylamine seems to promote polymerization.[524]

C.4. Acetylene-Allene Rearrangement

The use of α-acetylenic alcohols such as propargyl alcohol in the synthesis of phosphonous diesters from phosphonous dichlorides (method 12b, A.3) gives products that rearrange spontaneously to allenic phosphinates, even at 25°.[424,571,829]

$$RP(OCH_2C{\equiv}CH)_2 \longrightarrow RP(O)(OCH_2C{\equiv}CH)CH{=}C{=}CH_2$$

A cyclic mechanism (S_Ni') was proposed in which the phosphorus lone pair attacks one of the ethynyl groups, forming a new phosphorus-carbon bond and simultaneously cleaving the carbon-oxygen bond. The structures reported in the original paper[311] are incorrect.

C.5. Spectroscopy

The use of spectroscopy as an aid in the identification and characterization of organophosphorus compounds has grown enormously in the two decades since the first edition of this book. In this section we shall examine only the use of spectroscopic methods relating to the phosphonous acids or their derivatives as a whole. References to spectral data for individual substances are given in Section D.

C.5.1. Ultraviolet (UV)

Ultraviolet spectra have been reported for 12 phosphinic acids[287,554,655,746] and for a few other compounds of the types $RP(OR)_2$,[112] RPH(O)NHR,[143] and $RP(OAc)_2$.[524]

The UV spectra of PhPH(O)OH[287] and CH_2=CHP$(OBu)_2$[562] show no evidence of resonance interaction between the phosphorus atom and the benzene ring or vinyl group. The spectra of PhPH(O)OH and PhP(O)$(OH)_2$ are almost identical.[287]

C.5.2. Infrared (IR)

Infrared spectra have been reported for about half of the compounds in this chapter. The most important band assignments are given in Table 1.

Table 1. Infrared Absorption Band Assignments

Group	Stretching Frequency (cm^{-1})	Number of Compounds
P-H	2440-2280 (ν_{P-H})	67
	990-940 (δ_{P-H})	3
P-D	1750-1650	4
P=O	1286-1170	19
P=S	654,645	2
P=N	1140	1
P-O-C	1050-960 ($\nu_{(P)-O-C}$)	13
	805-775 ($\nu_{P-O-(C)}$)	4
P-S-C	515-497 ($\nu_{P-S-(C)}$)	3
P-N-C	1070-953 ($\nu_{(P)-N-C}$)	8
	775,730 ($\nu_{P-N-(C)}$)	2
P-O-H	2900-2500 ($\nu_{(P)-O-H}$)	7
P-F	833, 790	2
P-Cl	490	1

The phosphorus-hydrogen stretching frequency has been investigated extensively in a study of intermolecular association in the phosphinic esters.[761] The intensity of the band correlates well with the electronegativities of the phosphorus substituents.[657] The phosphorus-hydrogen band appears as a doublet in the esters ArPH(O)OR, owing to the presence of rotational isomers;[215,270,762] it also appears as a doublet in the primary phosphine oxide n-$C_8H_{17}P(O)H_2$.[96]

Efforts to establish a phosphorus-carbon stretching assignment have not been too successful. t-Butyl[133] and trichloromethyl[506] compounds absorb in the 780 to 620 cm^{-1} region, which has been tentatively assigned to ν_{P-C}.[122a] In CF_3 compounds, however, the ν_{P-C} band is given as 481, 439 cm^{-1}.[99]

Other bands have been found to be useful, in particular the carbon-fluorine deformation band at 770 to 730 cm^{-1} in P-CF_3 compounds[67,161,524] and the two sharp (ring-breathing) bands at 1003 to 1000 and 1445 to 1440 cm^{-1} in P-C_6H_5 compounds.[73,134] Thiophene compounds[69] absorb at 1470 cm^{-1} and ferrocene compounds[701,702] absorb at 1319, 1310 cm^{-1}.

C.5.3. Raman

Raman spectra have been reported for the esters $RP(OBu)_2$ and RPH(O)OBu (R = vinyl,[562] allyl,[562] or cyclopentadienyl[303]) and for the diaziridides $RP(\overline{NCH_2CH_2})_2$ (R=Me or Ph).[499] The phosphinic esters contain bands in the P-H region (2387-2340 cm^{-1}) and in the P=O region (1239-1216 cm^{-1}) that support the assignments made for these bands in the IR spectrum (Table 1).

There is no evidence for resonance interaction (conjugation) between the double bond and the phosphorus atom in any of these compounds.[562]

C.5.4. Nuclear Magnetic Resonance (NMR)

^{31}P Spectra have been reported for about 100 of the compounds in this chapter.[425] The chemical shifts vary from +1 to -65 ppm for the quadruply connected compounds and from -56 to -218 ppm for the triply connected compounds (Table 2). Only one positive chemical shift has been reported.[671]

The ^{31}P spectra of the primary phosphine oxides $RP(O)H_2$ show a 1:2:1 triplet with a large coupling constant (J_{PH} 470-510 Hz).[95,96] The phosphinic acids, and their derivatives containing a single phosphorus-hydrogen bond, all show a 1:1 doublet with a large coupling constant (J_{PH} 470-710 Hz). When the acid is neutralized, as in PhPH(O)ONa, the ^{31}P chemical shift becomes more positive by about 4 ppm and the coupling constant decreases by about 50 Hz.[467]

Smaller coupling constants, usually of the order of 10 to 20 Hz, have been reported for 2-bond splittings (J_{PCH}, J_{PNH}) and 3-bond splittings (J_{PCCH}, J_{POCH}, J_{PSCH}, J_{PNCH}). The ranges are reported in Table 3. A few 4-bond splittings have been reported also: J_{PCCCH}, 0.5 to 5.0 Hz;[111,696] J_{POCCH}, 0.5 Hz;[692,755] J_{PNCCH}, 1.0 to 1.3 Hz;[61,376,663] and J_{PNSiCH}, 0.65 to 2.05 Hz.[664]

Phosphorus-fluorine coupling constants are usually somewhat larger than the corresponding phosphorus-hydrogen

Table 2. ^{31}P Chemical Shifts

Class	Range (ppm[a])	Number of Compounds
Quadruply connected		
$RP(O)H_2$	-6 to -22	5
$RPH(O)OH$	-6 to -44	10
$RPH(O)OR$	-20 to -45	16
$RPH(S)OR$	-61 to -65	2
$RPH(=NR)NR_2$	+1 to -12	3
Triply connected		
$RP(OR)_2$	-146 to -201	15
$RP(OR)SR$	-79	1
$RP(SR)_2$	-56 to -91	3
$RP(OR)F$	-214 to -218	3
$RP(SR)F$	-139 to -214	8
$RP(NR_2)_2$	-76 to -107	16
$RP(NR_2)Cl$	-118 to -151	6
$RP(NR_2)F$	-135 to -170	9
$RP(NR_2)Br$	-161	1
$RP(NR_2)OR$	-132.5 to -154	2

[a]Relative to 85% H_3PO_4

coupling constants. The ranges for the phosphorus-fluorine coupling constants are included in Table 3 for comparison purposes.

In addition to the ^{31}P spectra discussed earlier, ^{1}H spectra have been reported for 78 compounds, ^{13}C spectra for one compound,[639] and ^{19}F spectra for 25 compounds.[435] Most of the data compiled in Table 3 were taken from these spectra, rather than the ^{31}P spectra.

The ^{31}P chemical shifts predicted by Maslov[429] with the aid of his "equation of weighted averages" have not been borne out by subsequent data and are therefore omitted from this study.

C.5.5. Mass Spectra

Fragmentation patterns have been examined in the mass spectra of a few phosphinic esters,[97] phosphonous diamides,[350,525] and phosphonamidous fluorides.[221] In some, only the parent ion (M^+) is reported, giving the molecular weight.[24,62,92,677]

C.6. Chromatographic Methods

Chromatographic methods are gaining in usage in an index

Table 3. Phosphorus-Hydrogen and Phosphorus-Fluorine Coupling Constants

Coupling	Range (Hz)	Number of Compounds	Coupling	Range (Hz)	Number of Compounds
J_{PH}	470-710	45	J_{PF}	873-1130	20
J_{PCH}	7-19	30	J_{PCF}	51-150	7
J_{PNH}	6-20	5			
J_{PCCH}	0.5-18	24	J_{PCCF}	17-39	3
J_{POCH}	3-13	17			
J_{PSCH}	11	1			
J_{PNCH}	4-16	33	J_{PNCF}	23	1

Table 4. Other Physical Measurements

Measurement	Number of Compounds	Ref.
Phosphinic acids, RPH(O)OH		
X-ray analysis	6	323,709
Ionic conductivity	3	132,161,162
Surface tension (γ)	10	385,656
Paper electrophoresis	5	175,233,337,697,763
Thermogravimetric analysis (TGA)	2	406
Differential thermal analysis (DTA)	1	651a
Phosphinic esters, RPH(O)OR		
Magnetic rotation, $[\rho]_M$	6	370
Surface tension (γ) and parachor (P)	10	49
Dipole moment	10	47
Polarographic reduction	2	381
Atomic refraction, phosphorus	16	603
Phosphonous diesters, $RP(OR)_2$		
Magnetic rotation, $[\rho]_M$	11	370,739
Surface tension (γ) and parachor (P)	1	48
Dipole moment	1	434
Atomic refraction, phosphorus	27	313,332,333,628
Viscosity	1	469
Thermogravimetric analysis (TGA)	6	596
Phosphonous diamides, $RP(NR_2)_2$		
Differential thermal analysis (DTA)	1	597
Heat of formation (ΔH_f)	1	733
Phosphonamidous Esters, $RP(NR_2)OR$		
Magnetic rotation, $[\rho]_M$ and magnetic susceptibility (Y_M)	1	375

of purity. Paper chromatography (PC) has been used for this purpose with the phosphinic acids,[175,196,233,322,336,341,697,727,763] the phosphinic esters,[230,500] and the phosphonodithious diesters.[675] Thin-layer chromatography (TLC) has been used with the phosphinic acids,[216,488,769] the phosphinic esters,[488,497,548,766] the phosphonous diesters,[488,494,496,499] the phosphonodithious diesters,[355] the phosphonochloridothious esters,[357] and the phosphonamidous esters.[230,488,494,497,548] Gas chromatography (GC) has been used with one phosphonous diester.[72]

C.7. Other Physical Measurements

Physical measurements other than those already discussed are summarized in Table 4.

D. LIST OF COMPOUNDS

D.1. Primary Phosphine Oxides (and Sulfides)

D.1.1. Primary Phosphine Oxides

C_3H_6NOP

$NCCH_2CH_2P(O)H_2$. ^{31}P -9.[96]

$C_4H_{11}OP$

i-BuP(O)H_2. ^{31}P -6.[96]

$C_5H_{11}OP$

c-$C_5H_9P(O)H_2$. 2.[95] Yellow oil.[95]

$C_5H_{13}OP$

$Et_2CHP(O)H_2$. 2.[95] Oil, dec. when distilled.[95]

C_6H_7OP

PhP(O)H_2. ^{31}P -7.[96]

$C_6H_{13}OP$

c-$C_6H_{11}P(O)H_2$. 2.[95] ^{31}P -22.[425a]

$C_7H_{17}OP$

s-$C_7H_{15}P(O)H_2$. 2.[95] Oil.[95]
$Pr_2CHP(O)H_2$. 2.[95] m. -17 to -14°, dec. at room temp.[95]

$C_8H_{19}OP$

n-$C_8H_{17}P(O)H_2$. 1.[96] m. 46-48°,[96] IR,[96] ^{31}P -10.[96]

D.1.2. Primary Phosphine Sulfides

C_6H_7PS

$PhP(S)H_2$. 3.[342,402] Unstable, viscous yellow liq.,[342,402] IR.[402]

D.2. Phosphonous Acids (Thio-, Seleno Analogs)

D.2.1. Phosphinic Acids

$CH_2Cl_3O_2P$

$CCl_3PH(O)OH$. 11.[506] m. 34°,[506] IR,[506] ^1H 6.85.[506]
-, Na, K salts.[506]

$CH_2F_3O_2P$

$CF_3PH(O)OH$. 16.[67] 17.[100] 18.[67] 21.[100] 27.[67,238] 28.[67,161] b_{extrap} 217°,[100] $\log P_{mm} = 3.9810 + 1.75 \log T - 0.0017706T - 2419/T$[100], dimer in liq. or vap. state, dissociating to monomer above 134° ($\Delta F° = 16.00 - 0.0247T$),[100] IR,[100] pK_a 1.01.[161]
-, Na salt,[67,161] IR,[67,161] ionic conduct. 29.1;[161] NH_3 salt, IR;[238] K salt.[161]

CH_4ClO_2P

$ClCH_2PH(O)OH$. 11.[727] 20b.[213] Viscous liq.,[727] n_D^{25} 1.4905,[727] d_{25}^{25} 1.5369,[727] ^1H 3.67 (CH_2), 6.68 (PH).[225]
-, Na salt, d. 124°.[727]

CH_5O_2P

$MePH(O)OH$. 16,[175] 19a;[136] $ClCH_2PH_2$, aq. NaOH,[185] oil,[136,175] ^{31}P -35.0,[175] ^1H 1.45 (CH_3), 7.04 (PH), 13.13 (POH),[136] other ^1H data,[175,185] PC,[175] paper electrophoresis,[175] pK_a 2.3.[175]

CH_5O_3P

$HOCH_2PH(O)OH$. 3.[266,693,727] ^1H 3.97 (CH_2), 7.25 (PH),[225] PC.[727]
-, Ca salt, crystals.[693]

$C_2H_4ClO_2P$

ClCH=CHPH(O)OH. 11.[719] Phenylhydrazine salt, m. 163°.[719]

$C_2H_4Cl_3O_2P$

CCl_3CHOHPH(O)OH. 3.[165,503] m. 142°,[165] 147-147.5° dec.,[503] IR,[503] highly toxic, LD_{50} 515 mg/kg in mice.[504]

$C_2H_5O_3P$

MeCOPH(O)OH. 6.[322] Oil,[322] pK_a 1.93.[323]
-, K salt IR,[323] PC;[322] Ni salt;[323] x-ray powder diagrams on Li, Na, K, Ba, and Zn salts.[323]

$C_2H_7O_2P$

EtPH(O)OH. 11.[231,445] Oil, d^{19} 1.2952.[231]

$C_2H_7O_3P$

MeCHOHPH(O)OH. 3.[419,736] MeCOPH(O)OH, H_2 (Raney Ni),[322] oil, dec. 130°,[736] ^{31}P -34.3,[182] PC.[322]
-, Ba salt;[736] NH_3 salt, TLC.[769]

$C_3H_2F_7O_2P$

n-C_3F_7PH(O)OH. 11.[162] 14.[118,162] 27.[162] b_1 90°,[118] $b_{0.7}$ 71°,[162] conduct. 363.[162]

$C_3H_6NO_2P$

$NCCH_2CH_2$PH(O)OH. 26.[96] ^{31}P -44.[425b]
-, aniline salt, m. 107-108°.[96]

$C_3H_7O_3P$

$MeCOCH_2$PH(O)OH. Possible product of the reaction of acetone with phosphorus and iodine.[479] [See Ref. 347b.]
-, Ba salt.[479]
EtCOPH(O)OH. 6.[322] pK_a 2.16.[323]
-, K salt, IR,[323] PC;[322] x-ray powder diagrams on Na, K, and Ba salts.[323]

$C_3H_8ClO_2P$

$MeCHClCH_2$PH(O)OH. 11.[719] syrup.[719]
-, phenylhydrazine salt, m. 185°.[719]

$C_3H_9O_2P$

PrPH(O)OH. 11.[231] Oil, d^{13} 1.1418.[231]
i-PrPH(O)OH. 11.[231] d^{19} 1.1891.[231]

$C_3H_9O_3P$

$HOCMe_2PH(O)OH$. 3.[415,416] m. 52°,[414] dec. 110-120°.[416]
-Pb salt;[415] Cu, Ni, Co, Ag salts.[416]

$C_3H_{10}NO_2P$

$Me_2NCH_2PH(O)OH$. 5.[406] m. 56-58°,[406] ^{31}P -9.6.[406]

$C_3H_{15}B_{10}O_2P$

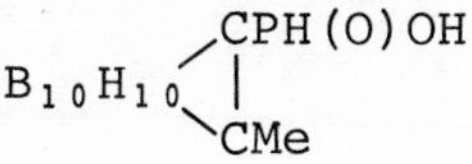

11.[764] 23.[764] m. 237-238°.[764]

$C_4H_5O_2PS$

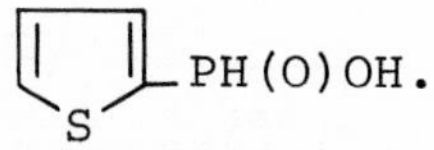

11.[650] m. 70°.[650]

$C_4H_9O_3P$

PrCOPH(O)OH. 6.[322] pK_a 2.22.[323]
-, K salt, IR,[323] PC;[322] Zn salt;[323] x-ray powder diagrams on Li, Na, K, Ba, and Ni salts.[323]

$C_4H_{11}O_2P$

BuPH(O)OH. 19b.[658] 19c.[658] 20a.[658] 20b.[658] Viscous oil, dec. 180°,[658] IR.[658]
i-BuPH(O)OH. 1a.[491] 11.[231] Oil, d^{23} 1.0740.[231]
-, Na salt, crystals.[491]
t-BuPH(O)OH. 11.[133] m. 86°,[133] IR,[133] ^1H 1.15 (C_4H_9), 6.76 (PH), 12.45 (OH).[133]

$C_4H_{11}O_3P$

HOCMeEtPH(O)OH. 3[414,420] Oil.[420]
-, Pb, Cu, Ag salts.[420]

$C_4H_{12}NO_2P$

$MeNHCMe_2PH(O)OH$. 4.[363] m. 221°.[363]

$C_4H_{13}N_4O_2P$

$NH_2C(=NH)NHNHCMe_2PH(O)OH$. 4.[672,673] m. 186°.[672,673]

$C_5H_9O_3P$

MeCOCH=CMePH(O)OH. 4.[364] Oil.[364]

$C_5H_{12}NO_3P$

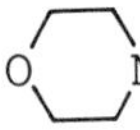$NCH_2PH(O)OH$. 5.[406] m. 203-205° (hydrate),[406] IR,[406] ^{31}P -8.6,[406] TGA.[406]

$C_5H_{13}O_2P$

i-AmPH(O)OH. 11.[231] Oil, d^{23} 1.0613.[231]
-, NH_3, Fe salts.[231]

$C_5H_{13}O_3P$

i-BuCHOHPH(O)OH. 3.[419,736] Oil, dec. 170°.[736]
-, Ba salt.[736]
MePrCOHPH(O)OH. 3.[414,421] Oil, dec. 100°.[421]
-, Pb salt.[421]
$Et_2COHPH(O)OH$. 3.[414,422] Oil.[422]
-, Pb salt.[422]
$BuOCH_2PH(O)OH$. From $ClCH_2PH(O)OH$, BuOH (NaOH),[728] n_D^{25} 1.4480.[728]

$C_5H_{14}NO_2P$

$Et_2NCH_2PH(O)OH$. 5.[406] Viscous yellow oil,[406] ^{31}P -6.6.[406]

$C_6H_2F_5O_2P$

$C_6F_5PH(O)OH$. 11.[399] 23.[62] 27.[62] m. 81-83°,[399] ^1H 7.71 (PH), 13.2 (OH),[399] pK_a 4.40.[399]
-, t-$BuNH_2$ salt, m. 211°,[62] IR;[62] Na salt, IR.[399]

$C_6H_6BrO_2P$

m-$BrC_6H_4PH(O)OH$. 11.[614] m. 97-98°,[614] pK_a 1.39.[613]
p-$BrC_6H_4PH(O)OH$. 8.[559] 11.[441] 24.[441] 25.[441] m. 142°,[377] 143°,[441,746] 145°,[559] UV,[746] pK_a 1.25 (H_2O),[377] ∿3.0 (EtOH).[608]
- aniline salt, prisms;[441] phenylhydrazine salt, m. 181° dec.;[441] NH_3, K, Cu, Ca, Ba, Pb salts.[441]

$C_6H_6ClO_2P$

o-$ClC_6H_4PH(O)OH$. 11.[614] 16.[725] m. 128-129°,[83] 128.5-129°.[614]
m-$ClC_6H_4PH(O)OH$. 11.[614] 16.[390] m. 90.5-91.5°,[614] 91-92°,[390] UV,[655b] IR,[653d] pK_a 1.35.[613]
p-$ClC_6H_4PH(O)OH$. 11.[441] 12.[280] 14.[174] 16.[83] 24.[441] 25.[441] m. 129-130°,[280] 130-131°,[60,174,441] 130.5-131.5°,[725] 131°,[52] 131-132°,[83,614] IR,[608] pK_a 1.57 (H_2O),[613] ∿3.0 (EtOH).[608]
-, phenylhydrazine salt, m. 169°;[441] NH_3, Cu, Ba salts.[441]

$C_6H_6FO_2P$

m-$FC_6H_4PH(O)OH$. 11.[666] m. 60°.[666]

$C_6H_6NO_4P$

o-$NO_2C_6H_4PH(O)OH$. 7.[558,648] m. 155-157°,[648] 157°.[558]
-, Ag salt.[648]
p-$NO_2C_6H_4PH(O)OH$. 7.[558] 8.[559] m. 134°.[558]

$C_6H_7O_2P$

PhPH(O)OH. 7.[558] 8.[559] 11.[447] 13.[448] 14.[262] 16.[159,215,348] 18.[342] 19b.[400] 20a.[344] 20b.[299] 22.[11] 23.[198] 24.[449] 25.[137,342] 28.[65] 1 (C.3).[96] m. 70°,[447] 70-71°,[640] 70.5°,[558] 78-80°,[137] 80°,[215,377] 81-82°,[745] 83-84°,[262,400] 84°,[159] 86°,[348] 86-87°,[193] UV,[287,655a,745,746] IR,[117d,134,173,215,608,652a,653a,c] far IR,[232] ^{31}P -20 (acetone),[732] -23 (H_2O),[732] ^1H 7.64 (PH),[193] other ^1H data,[640] MW trimer (C_6H_6),[348] monomer (HOAc),[195] pK_a (H_2O) 1.1,[338] 1.35,[377] 1.53,[291] 1.54,[290] 1.75,[613] pK_a (alcohols),[290,291,361,523,608] acidity function (H_0),[121] PC,[336,697] TLC,[216] paper electrophoresis.[337,697]
-, cyclohexylamine salt, m. 198-200°;[137] aniline salt, m. 101°,[746] 103-104°;[96] hexamethylenediamine salt;[90] phenylhydrazine salt, m. 134° dec.[159] m. 135°;[449] p-tolylhydrazine salt, m. 148°;[449] benzylphenylhydrazine salt, m. 108°;[449] S-p-chlorobenzylthiuronium salt, m. 180-182°;[242] Na salt, UV,[287] IR,[54,173,653e] ^{31}P -17.4;[467] NH_3 salt,[299,477] m. 175°;[299] Fe salt,[57,447] m. 275-280° dec.;[57] $TiCl_2$ and $Ti(OR)_3$ (R = i-Pr or Bu) salts;[172] K, Ca, Pb and Ba salts;[447] lanthanide salts.[478]

$C_6H_8NO_2P$

p-$NH_2C_6H_4PH(O)OH$. From p-$BrC_6H_4PH(O)OH$, NH_4OH (Cu_2O),[339] m. 169°,[339] 210° (hydrate),[737] pK_a 3.68,[339] 3.7.[737]

$C_6H_{11}O_3P$

$EtO_2CCH{=}CMePH(O)OH$. 4.[364] Oil.[364]

$C_6H_{13}O_2P$

$c\text{-}C_6H_{11}PH(O)OH$. 1a.[234,491,523] 11.[746] Viscous liq.,[234,523] pK_a (EtOH) 4.55,[523] acidity function (H_0),[121] TLC.[488]
 -, aniline salt, m. 106-110°;[746] Na salt, col. crystals.[491]

$C_6H_{13}O_7P$

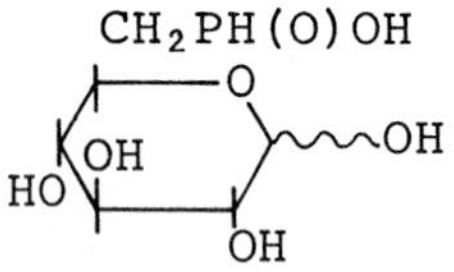

3 (from levoglucosan).[341]
 -, NH_3 salt, PC;[341] Na salt.[341]

$C_6H_{14}NO_2P$

$NCH_2PH(O)OH$ 5.[406] m. 146-147° (hydrate),[406] IR,[406] ^{31}P -9.0,[406] TGA.[406]
 -, HCl salt, m. 137-147°.[406]

$C_6H_{14}NO_4P$

$HO_2CCH_2C(NHEt)MePH(O)OH$. From the ethyl ester (aq. KOH),[364] m. 168-170).[364]

$C_6H_{15}O_2P$

$n\text{-}C_6H_{13}PH(O)OH$. 1a.[491,712] Na salt, crystals.[491,712]

$C_6H_{16}NO_2P$

$PrNHCMe_2PH(O)OH$. 4.[672,673] m. 227° dec.[672,673]

$C_7H_6Cl_3O_3P$

$2,4,5\text{-}Cl_3C_6H_2OCH_2PH(O)OH$. 25.[401] m. 129-130°,[401] UV,[554] pK_a (H_2O) 0.94.[555]

$C_7H_6NO_2P$

$p\text{-}NCC_6H_4PH(O)OH$. 11.[614] m. 166-167°,[614] pK_a (H_2O) 1.19.[613]

$C_7H_7Cl_2O_3P$

2,4-$Cl_2C_6H_3OCH_2PH(O)OH$. 25.[401] m. 100-101°,[401] UV,[554] pK_a (H_2O) 0.98.[555]
-, p-toluidine salt, m. 145°.[401]

$C_7H_7O_4P$

p-$HO_2CC_6H_4PH(O)OH$. By hydrolysis of p-$NCC_6H_4PH(O)OH$,[613] pK_a (H_2O) 1.32 (PO_2H_2), 4.06 (CO_2H).[613]

$C_7H_8ClO_2P$

3-Cl-4-$MeC_6H_3PH(O)OH$. 11.[614] m. 97.5-98.5.[614]
4-Cl-3-$MeC_6H_3PH(O)OH$. 11.[439] m. 70°.[439] Structure corrected by Quin and Freedman.[614]
-, phenylhydrazine salt, m. 156.5°;[439] NH_3, Ba salts.[439]

$C_7H_8ClO_3P$

o-$ClC_6H_4OCH_2PH(O)OH$. 25.[401] p-toluidine salt, m. 124-127° (hemihydrate).[401]
p-$ClC_6H_4OCH_2PH(O)OH$. 25.[401] m. 122-124°,[401] UV,[554] pK_a (H_2O) 1.00.[555]

$C_7H_9O_2P$

$PhCH_2PH(O)OH$. 11.[746] 25.[378] Syrupy liq.[378] or oil.[746]
-, Mg, Ca, Ba, Zn, Cd, and Pb salts.[378]
o-$MeC_6H_4PH(O)OH$. 7.[558] 11.[442,450] Oil,[442,450] m. 115°.[558]
-, aniline salt, m. 94°,[746] Cu, Ca, Pb, Ba, Fe, and Ag salts.[450]
m-$MeC_6H_4PH(O)OH$. 11.[442] Syrup.[442]
-, phenylhydrazine salt, m. 131° dec.[442] NH_3, K, Ba salts.[442]
p-$MeC_6H_4PH(O)OH$. 7.[558] 11.[450] 19a.[450] 24.[449] 25.[450] m. 104-105°,[345,450,614] 104-106°,[746] 105.5°,[159] 106°,[558] UV,[746] IR,[134,608] ^1H 2.383 (CH_3),[665] x-ray,[709] pK_a (H_2O) 1.83,[613] pK_a (EtOH) $\sim$3.3.[608]
-, phenylhydrazine salt, m. 161°;[449] NH_3, K, Cu, Ba, and Pb salts.[450]

$C_7H_9O_2PS$

$PhSCH_2PH(O)OH$. From $ClCH_2PH(O)OH$, PhSH (NaOH),[729] n_D^{25} 1.6046.[729]

$C_7H_9O_3P$

$PhCHOHPH(O)OH$. 3.[256,266,414,736] m. 108°,[414] 110°,[377,737] dec. 140°,[736] UV,[655d] IR,[652c,653b] x-ray,[709] pK_a (H_2O)

1.4,[377] DTA.[651a]
-, aniline salt, m. 158°;[672] Na salt, UV,[655c] IR;[652b, 653e] Ca salt, dec. 140°;[645] Ag,[414] Ba,[736] and Pb[736] salts.

$PhOCH_2PH(O)OH$. From $ClCH_2PH(O)OH$, PhOH (NaOH).[728] Oil,[728] UV.[554]

o-$MeOC_6H_4PH(O)OH$. 11.[614] m. 100-102°.[614]

p-$MeOC_6H_4PH(O)OH$. 11.[308,441] m. 112°,[377,441] 114-114.5°,[308] pK_a (H_2O) 1.75,[377] pK_a (EtOH) ∿3.5.[608]
-, aniline salt, m. 97-98°;[308] p-toluidine salt, m. 99-100°;[308] phenylhydrazine salt, m. 116°;[441] Ca and Pb salts.[308]

$C_7H_9O_4P$

o-$HOC_6H_4CHOHPH(O)OH$. 3.[736] Resin.[736]

$C_7H_{13}O_3P$

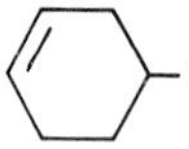-CHOHPH(O)OH. 3.[646,647] Na salt, m. 236° dec.;[646] ω-aminoacetanilide and other amine salts.[647]

$C_7H_{16}NO_3P$

$MeCOCH_2C(NHEt)MePH(O)OH$. 4.[364] m. 122°.[364]

$C_7H_{16}NO_4P$

$EtO_2CCH_2C(NHMe)MePH(O)OH$. 4.[364] m. 176-177°.[364]

$C_7H_{17}O_2P$

n-$C_7H_{15}PH(O)OH$. 1a.[536] 25.[743] Oil,[743] $b_{0.0001}$ 140°,[536] n_D^{20} 1.4547,[536] d_4^{20} 0.9917.[536]

$C_7H_{17}O_3P$

n-$C_6H_{13}CHOHPH(O)OH$. 3.[256,736] m. 55-57°,[256,736] dec. 120°.[736]
-, Pb salt.[256]

$C_7H_{18}NO_2P$

$BuNHCMe_2PH(O)OH$. 4.[363] m. 219°.[363]

$C_7H_{18}NO_3P$

$MeO(CH_2)_3NHCMe_2PH(O)OH$. 4.[363] m. 191°.[363]

$C_8H_6MnO_5P$

$(OC)_3MnC_5H_4PH(O)OH$ (cyclopentadiene deriv.). 11.[690] m. >295°,[690] IR.[690]

$C_8H_9O_2P$

PhCH=CHPH(O)OH. 11.[742] m. 74-75°.[742]
p-CH_2=$CHC_6H_4PH(O)OH$. 23.[615] m. 74-75.5°.[615]

$C_8H_{11}O_2P$

PhCHMePH(O)OH. 26.[95] Aniline salt, m. 114-116°.[95]
p-$EtC_6H_4PH(O)OH$. 11.[192,442] 24.[442] 25.[442] m. 62°,[746] 63-64°,[345,442] 67-68°,[192] UV,[746] IR.[134]
-, phenylhydrazine salt, m. 133°;[442] Na salt, white crystals;[192] NH_3, Cu, Ba salts.[442]

$C_8H_{11}O_2P$

2,3-$Me_2C_6H_3PH(O)OH$. 11.[442] m. 43°.[442]
2,4-$Me_2C_6H_3PH(O)OH$. 11.[442] m. 100°.[442]
2,5-$Me_2C_6H_3PH(O)OH$. 11.[747] Crystallizes with difficulty.[747]

$C_8H_{11}O_3P$

MePhCOHPH(O)OH. 3.[414,422] m. 85°,[414] dec. 90-100°.[414]
-, Pb salt.[414,422]
p-$EtOC_6H_4PH(O)OH$. 11.[441] m. 115°.[441]

$C_8H_{11}O_4P$

2,4-$(MeO)_2C_6H_3PH(O)OH$. 11.[569] m. 160-161°.[569]

$C_8H_{12}NO_2P$

PhNHCHMePH(O)OH. 4.[672,674] m. 190° dec.[672,674]
p-$Me_2NC_6H_4PH(O)OH$. 11.[66,661,737] 16.[86,451,619] m. 162°,[377,619,648,661] 163°,[86] pK_a (H_2O) 4.1,[377] pK_a (EtOH) ~4.0.[608]
-, HCl salt, crystals;[451] Na, Pb salts;[66,619,451,737] Cu,[66,451] K,[451] Ba,[619] and Ca[66] salts.

$C_8H_{13}N_4O_2P$

NH_2C(=NH)NHNHCHPhPH(O)OH. 4.[674] m. 152°.[674]

$C_8H_{15}O_6P$

$EtO_2CCH_2CH(CO_2Et)PH(O)OH$. 1b.[598] m. 250-251°.[598]

$C_8H_{18}NO_4P$

$EtO_2CCH_2C(NHEt)MePH(O)OH$. 4.[364] m. 186-188°.[364]

$C_8H_{19}O_2P$

n-$C_8H_{17}PH(O)OH$. 1a,[519,712] 25.[473] Oil.[473]
 -, Ag salt.[473]
i-$C_8H_{17}PH(O)OH$. 11.[742] Oil.[742]
t-$C_8H_{17}PH(O)OH$. 25.[93]
$BuCHEtCH_2PH(O)OH$. 1a.[234] Mixture with $(BuCHEtCH_2)_2P(O)OH$.[234]

$C_9H_{11}O_4P$

PhCHOAcPH(O)OH. -.[736] Yellow solid.[736]

$C_9H_{13}N_2O_4P$

m-$NO_2C_6H_4NHCMe_2PH(O)OH$. 4.[363] m. 149-150°.[363]
p-$NO_2C_6H_4NHCMe_2PH(O)OH$. 4.[363] m. 131-132°.[363]

$C_9H_{13}O_2P$

p-(i-Pr)$C_6H_4PH(O)OH$. 11.[443] Oil.[443]
 -, phenylhydrazine salt, m. 161° (1:1),[443] m. 135° (1:2);[443] Ba salt.[443]
2,4,5-$Me_3C_6H_2PH(O)OH$. 11.[443] 19a.[443] 25.[443] m. 128°.[443]
 -, phenylhydrazine salt, m. 180°;[443] K, Ba, Cu, and Pb salts.[443]
2,4,6-$Me_3C_6H_2PH(O)OH$. 11.[132,443] 24.[443] 25.[443] m. 146°,[88] 147°,[443] 147.35°,[132] conductance (H_2O) 376.[132]
 -, aniline salt, needles;[443] phenylhydrazine salt, m. 132° dec.;[443] Na salt;[132] NH_3, K, Cu, Ca, and Ba salts.[443]

$C_9H_{14}NO_2P$

$PhNHCMe_2PH(O)OH$. 4.[672,673] m. 214° dec.[672,673]
2-Me-4-$Me_2NC_6H_3PH(O)OH$. 11.[66] Na salt,[66,732] ^{31}P -20;[732] K, Cu, Ca, and Fe salts.[66]

$C_9H_{14}NO_3P$

o-$HOC_6H_4NHCMe_2PH(O)OH$. 4.[363] m. 188-189°.[363]
p-$HOC_6H_4NHCMe_2PH(O)OH$. 4.[363,754] m. 205°,[754] 230-231°.[363]

$C_9H_{15}N_2O_2P$

$PhNHNHCMe_2PH(O)OH$. 4.[672,673] m. 165°.[672,673]
 -, Na salt.[673]

$C_9H_{15}N_2O_2PS$

p-$NH_2SO_2C_6H_4NHCMe_2PH(O)OH$. 4.[672,673] m. 200° dec.[672,673]
-, Na salt, col. powder.[672,673]

$C_9H_{17}O_6P$

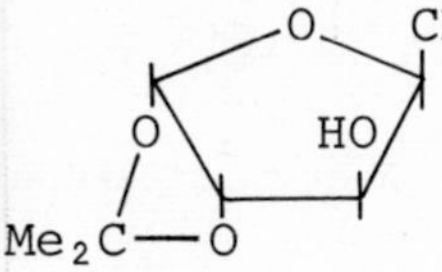

$CH_2CH_2PH(O)OH$. 25.[749] Cyclohexylamine salt, m. 166°,[749] IR,[749] ^1H 7.24 (PH) and other data,[749] $[\alpha]_D^{25}$ -10.9.[749]

$C_9H_{19}O_4P$

n-$C_6H_{13}CH(OAc)PH(O)OH$. From n-$C_6H_{13}CHOHPH(O)OH$, AcCl,[736] Oil.[736]

$C_9H_{20}NO_2P$

c-$C_6H_{11}NHCMe_2PH(O)OH$. 4.[672,673] m. 217° dec.[672,673]

$C_9H_{20}NO_3P$

$MeCOCH_2C(NHBu)MePH(O)OH$. 4.[364] m. 124-125°.[364]

$C_9H_{21}O_2PS$

n-$C_8H_{17}SCH_2PH(O)OH$. From $ClCH_2PH(O)OH$, n-$C_8H_{17}SH$ (NaOH).[729] n_D^{25} 1.4898.[729]

$C_9H_{22}NO_2P$

i-$C_6H_{13}NHCMe_2PH(O)OH$. 4.[673] m. 217°.[673]

$C_{10}H_9O_2P$

1-$C_{10}H_7PH(O)OH$. 7.[558] 11.[324,746] 23.[452,734] m. 122-124°,[746] 122.5-123.5°,[734] 125-126°,[324] d_4 (solid) 1.377,[683] UV.[746]
2-$C_{10}H_7PH(O)OH$. 7.[558] 11.[384,746] m. 135-136°,[345] 137°,[384,746] 138°,[746] UV,[746] IR.[134]

$C_{10}H_{11}FeO_2P$

$C_5H_5FeC_5H_4PH(O)OH$ (ferrocenyl). 11.[701] m. 137-138°,[701] IR,[701] pK_a (H_2O) 3.32.[701]

$C_{10}H_{13}O_2P$

2-$C_{10}H_{11}PH(O)OH$ ($C_{10}H_{11}$ = 5,6,7,8-tetrahydronaphthyl).

20a.[268] m. 95-97°,[268] IR.[268]

$C_{10}H_{14}NO_4P$

o-$HO_2CC_6H_4NHCMe_2PH(O)OH$. From the CO_2Me ester (alc. KOH).[363] Oil.[363]
p-$HO_2CC_6H_4NHCMe_2PH(O)OH$. From the CO_2Et ester (alc. KOH).[363] m. 160°.[363]

$C_{10}H_{15}O_2P$

2-Me-5-(i-Pr)$C_6H_3PH(O)OH$. (or 5,2-isomer). 11.[443] Liq.[443]
-, Ba salt.[443]

$C_{10}H_{15}O_3P$

p-(i-Pr)$C_6H_4CHOHPH(O)OH$. 3.[736] m. 105°.[736]

$C_{10}H_{16}NO_2P$

$PhCH_2NHCMe_2PH(O)OH$. 4.[363,673] m. 215-216°,[363] 220° dec.[673]
o-$MeC_6H_4NHCMe_2PH(O)OH$. 4.[363] m. 110°.[363]
p-$MeC_6H_4NHCMe_2PH(O)OH$. 4.[363] m. 197-198°.[363]

p-$Et_2NC_6H_4PH(O)OH$. 11.[86,451] Oil.[86,451]

$C_{10}H_{17}N_2O_4PS$

p-$(NH_2SO_2)C_6H_4CH_2NHCMe_2PH(O)OH$. 4.[672,673] m. 232° dec. (hydrate).[672,673]
-, Na salt, crystals.[673]

$C_{10}H_{17}O_2P$

1-$C_{10}H_{15}PH(O)OH$ ($C_{10}H_{15}$ = adamantyl). 11.[711] 25.[711] m. 167-168°,[711] 171-173°.[711]

$C_{10}H_{19}O_2P$

2-$C_{10}H_{17}PH(O)OH$. ($C_{10}H_{17}$ = pinanyl). 1a.[612] n_D^{20} 1.4851,[612] d^{20} 1.030,[612] $[\alpha]_{578}$ 84.7°,[612] $[\alpha]_{546}$ -97.2°,[612] dispersion (D_{578}^{546}) 1.15.[612]
-, Na salt, m. 220°.[612]

$C_{10}H_{21}O_2P$

n-$C_8H_{17}CH{=}CHPH(O)OH$. 10.[706,707] Syrup,[706,707] IR.[707]
-, Na salt.[706,707]

$C_{10}H_{22}NO_4P$

$EtO_2CCH_2C(NHBu)MePH(O)OH$. 4.[364] m. 182-183°.[364]

$C_{10}H_{23}O_2P$

n-$C_{10}H_{21}PH(O)OH$. 11.[656] Oil,[656] surface tension,[656] pK_a 4.5.[656]

$C_{11}H_{11}O_2P$

1-$C_{10}H_7CH_2PH(O)OH$. 23.[734] m. 136.5-137.5°.[734]

$C_{11}H_{16}ClN_2O_3P$

3-Cl-4-$AcNHC_6H_3NHCMe_2PH(O)OH$. 4.[672,673] m. 200° dec. (dihydrate),[672,673]

$C_{11}H_{16}NO_3P$

$MeCOCH_2C(NHPh)MePH(O)OH$. 4.[364] m. 198-200°.[364]

$C_{11}H_{16}NO_4P$

o-$MeO_2CC_6H_4NHCMe_2PH(O)OH$. 4.[363] Oil.[363]

$C_{11}H_{17}N_2O_3P$

p-$AcNHC_6H_4NHCMe_2PH(O)OH$. 4.[672,673] m. 216° dec.,[673] 216° dec. (dihydrate).[672]
-, Na salt, solid.[672,673]

$C_{11}H_{17}O_3P$

p-(t-Bu)$C_6H_4OCH_2PH(O)OH$. From $ClCH_2PH(O)OH$, p-(t-Bu)C_6H_4OH (NaOH).[728] m. 104-106°.[728]

$C_{11}H_{18}NO_2P$

2,3-$Me_2C_6H_3NHCMe_2PH(O)OH$. 4.[363] m. 134-135°.[363]
2-Me-4-$Et_2NC_6H_3PH(O)OH$. 11.[66] Na, Cu, Ca salts.[66]

$C_{11}H_{18}NO_3P$

p-$EtOC_6H_4NHCMe_2PH(O)OH$. 4.[672,673] m. 189° dec.[672,673]
-, Na salt.[673]

$C_{11}H_{19}N_2O_2P$

p-$Me_2NC_6H_4NHCMe_2PH(O)OH$. 4.[363] m. 224-225°.[363]

$C_{11}H_{19}O_4P$

3-($C_{10}H_{15}O$)CHOHPH(O)OH ($C_{10}H_{15}O$ = camphoryl). 3.[256] Na salt, crystals.[256]

$C_{11}H_{22}NO_3P$

$MeCOCH_2C(NHC_6H_{11}$-c)MePH(O)OH. 4.[364] m. 137-138°.[364]

$C_{12}H_9O_3P$

2-($C_{12}H_7O$)PH(O)OH ($C_{12}H_7O$ = dibenzofuran-). 11.[138] m. 125°.[138]

$C_{12}H_{11}O_2P$

p-PhC_6H_4PH(O)OH. 11.[444] crystal powder.[444]
-, K, Ca, and Ba salts.[444]

$C_{12}H_{16}N_3O_4PS_2$

p-(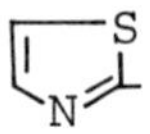 $NHSO_2)C_6H_4NHCMe_2PH$(O)OH. 4.[672,673] m. 184° dec.[672,673] (from sulfathiazole).
-, diethylamine salt.[673]

$C_{12}H_{18}NO_3P$

$MeCOCH_2C(NHCH_2Ph)MePH$(O)OH. 4.[364] m. 133-134°.[364]
$MeCOCH_2C(NHC_6H_4Me$-p)MePH(O)OH. 4.[364] m. 194-195°.[364]

$C_{12}H_{18}NO_4P$

$EtO_2CCH_2C(NHPh)MePH$(O)OH. 4.[364] m. 136-137°.[364]
p-$EtO_2CC_6H_4NHCMe_2PH$(O)OH. 4.[363] m. 131°.[363]

$C_{12}H_{19}N_2O_3P$

p-$AcNHC_6H_4NHCMeEtPH$(O)OH. 4.[672,673] m. 142° dec. (dihydrate).[672,673]

$C_{12}H_{25}O_2P$

n-$C_{10}H_{21}CH$=CHPH(O)OH. 10.[706,707] Viscous oil,[706] syrup.[707]

$C_{12}H_{27}O_2P$

n-$C_{12}H_{25}PH$(O)OH. 1a.[234,519] 11.[656] m. 43-44.5°,[656] 45-50°,[234] surface tension,[656] pK_a 4.8.[656]
-, Ca salt, m. 198-242°.[234]

$t\text{-}C_{12}H_{25}PH(O)OH$. 25.[93 258] Pale yellow syrup.[258]

$C_{13}H_{10}Cl_4NO_2P$

$3,4\text{-}Cl_2C_6H_3NHCH(C_6H_3Cl_2\text{-}3,4)PH(O)OH$. 4.[385] m. 75-80°.[385]

$C_{13}H_{11}Cl_3NO_2P$

$2,4\text{-}Cl_2C_6H_3NHCH(C_6H_4Cl\text{-}o)PH(O)OH$. 4.[385] m. 84-87°.[385]
$3,4\text{-}Cl_2C_6H_3NHCH(C_6H_4Cl\text{-}p)PH(O)OH$. 4.[385] m. 76-80°.[385]

$C_{13}H_{12}Cl_2NO_2P$

$p\text{-}ClC_6H_4NHCH(C_6H_4Cl\text{-}o)PH(O)OH$. 4.[385] m. 87-90°.[385]
$p\text{-}ClC_6H_4NHCH(C_6H_4Cl\text{-}p)PH(O)OH$. 4.[385] m. 82-87°.[385]

$C_{13}H_{12}Cl_2NO_3P$

$3,4\text{-}Cl_2C_6H_3NHCH(C_6H_4OH\text{-}o)PH(O)OH$. 4.[385] m. 105-108°.[385]
$3,4\text{-}Cl_2C_6H_3NHCH(C_6H_4OH\text{-}p)PH(O)OH$. 4.[385] m. 147-149°.[385]

$C_{13}H_{13}O_2P$

$p\text{-}PhCH_2C_6H_4PH(O)OH$. 11.[444] m. 84°.[444]
-, phenylhydrazine salt, m. 171°;[444] NH_3, Na, K, Cu, Ba, and Pb salts.[444]

$C_{13}H_{13}O_3P$

$Ph_2COHPH(O)OH$. 3.[414,421] m. 150-151°.[421]
-, Ag and Pb salts.[421]

$C_{13}H_{14}NO_2P$

PhNHCHPhPH(O)OH. 4.[672,674] m. 149-150°,[385] 150°.[672,674]
-, Na salt.[674]
$p\text{-}(PhMeN)C_6H_4PH(O)OH$. 11.[451] m. 150.5°.[451]
-, Na salt, m. 265°.[451]

$C_{13}H_{14}NO_3P$

$o\text{-}HOC_6H_4CH(NHPh)PH(O)OH$. 4.[363] m. 172-173°.[363]

$C_{13}H_{20}NO_4P$

$EtO_2CCH_2C(NHCH_2Ph)MePH(O)OH$. 4.[364] m. 193°.[364]
$EtO_2CCH_2C(NHC_6H_4Me\text{-}p)MePH(O)OH$. 4.[364] m. 128-129°.[364]

$C_{13}H_{20}N_2O_2P$

p-( N)$C_6H_4NHCMe_2PH(O)OH$. 4.[754] m. 198°.[754]

$C_{13}H_{23}N_2O_2P$

p-$Et_2NC_6H_4NHCMe_2PH(O)OH$. 4.[754] m. 218°.[754]

$C_{13}H_{23}N_2O_3P$

p-$(HOCH_2CH_2NEt)C_6H_4NHCMe_2PH(O)OH$. 4.[754] m. 216°.[754]

$C_{13}H_{29}O_2P$

n-$C_{12}H_{25}SCH_2PH(O)OH$. From $ClCH_2PH(O)OH$, n-$C_{12}H_{25}SH$ (NaOH).[729] m. 35-36°.[729]

$C_{14}H_{13}Br_2O_4P$

p-$BrC_6H_4CHOHC(OH)(C_6H_4Br$-p$)PH(O)OH$. 3.[560] m. 125° dec.[560]

$C_{14}H_{13}O_4P$

PhCH(OBz)PH(O)OH. From PhCHOHPH(O)OH.[414] m. 93°.[414]

$C_{14}H_{15}O_2P$

p-$(PhCH_2CH_2)C_6H_4PH(O)OH$. 11.[444] m. 156-157°.[444]
-, phenylhydrazine salt, crystals;[444] NH_3, Na, K, Cu, and Pb salts.[444]

$C_{14}H_{15}O_3P$

$PhCH_2C(OH)PhPH(O)OH$. 3.[560] m. 135-137°.[560]

$C_{14}H_{15}O_4P$

PhCHOHC(OH)PhPH(O)OH. 3.[560] m. 150° dec.[560]

$C_{14}H_{16}NO_2P$

p-$(PhCH_2NMe)C_6H_4PH(O)OH$. 11.[451] m. 96°.[451]
-, Na salt, m. 233°.[451]

$C_{14}H_{16}NO_3P$

$PhCH_2NHCH(C_6H_4OH$-o$)PH(O)OH$. 4.[672] m. 225°.[672]

$C_{14}H_{17}N_2O_5PS$

p-$(NH_2SO_2)C_6H_4CH_2NHCH(C_6H_4OH$-o$)PH(O)OH$. 4.[674] m. 190.5° dec.[674]
-, NH_3, Na salts.[674]

$C_{14}H_{18}N_3O_4PS$

p-(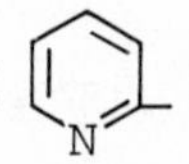 $NHSO_2)C_6H_4NHCMe_2PH(O)OH$. 4[673] (from sulfapyridine). m. 180° dec.[673]
-, Na salt, crystals.[673]

$C_{14}H_{20}N_3O_3P$

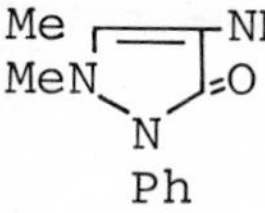

$NHCMe_2PH(O)OH$ 4[672,673] (from aminoantipyrine). m. 185° dec.[672,673]
-, Na, Ca salts.[673]

$C_{14}H_{23}O_3P$

3,5-(t-Bu)$_2$-4-$HOC_6H_2PH(O)OH$ 11.[135] m. 145-146°,[135] IR.[135]
-, Na salt, crystals.[135]

$C_{14}H_{25}N_2O_2P$

p-$Et_2NC_6H_4NHCMeEtPH(O)OH$ 4.[754] m. 208°.[754]

$C_{14}H_{27}O_6P$

i-$AmO_2CCH_2CH(CO_2Am$-i$)PH(O)OH$. 1b.[598] m. 252-254°.[598]

$C_{14}H_{29}O_2P$

n-$C_{12}H_{25}CH{=}CHPH(O)OH$. 10.[706,707] Syrup.[706,707]

$C_{14}H_{31}O_2P$

n-$C_{14}H_{29}PH(O)OH$. 1a.[519,712] 11.[656] m. 54.5-55.5°,[656] surface tension,[656] pK_a 4.8.[656]

$C_{15}H_{17}O_5P$

p-$MeOC_6H_4CHOHC(OH)PhPH(O)OH$ (or p'-MeO isomer). 3.[560] m. 150° dec.[560]

$C_{15}H_{18}N_3O_2P$

$PhN{=}NC_6H_4NHCMe_2PH(O)OH$. 4.[363] m. 168-169°.[363]

$C_{15}H_{19}N_2O_4PS$

p-$(p\text{-}NH_2C_6H_4SO_2)C_6H_4NHCMe_2PH(O)OH$. From the N-acetyl deriv. (NaOH).[673] m. 192°,[673]
-, Na salt.[673]

$C_{15}H_{23}O_4P$

2,6-$(t\text{-}Bu)_2$-4-$(HO_2C)C_6H_2PH(O)OH$. By oxidation of 2,4,6-$(t\text{-}Bu)_3C_6H_2PH(O)Cl$ with alkaline permanganate.[120] m. 226-227°,[120] IR.[120]

$C_{15}H_{27}N_2O_3P$

2-EtO-4-$Et_2NC_6H_3NHCMe_2PH(O)OH$. 4.[754] Hygroscopic.[754]

$C_{16}H_{19}O_5P$

o-$MeOC_6H_4CH_2C(OH)(C_6H_4OMe\text{-}o)PH(O)OH$. 3.[560] m. 190° dec.[560]

$C_{16}H_{19}O_6P$

m-$MeOC_6H_4CHOHC(OH)(C_6H_4OMe\text{-}m)PH(O)OH$. 3.[560] m. 175° dec.[560]

$C_{16}H_{20}NO_3P$

p-$EtOC_6H_4NHCMePhPH(O)OH$. 4.[672,673] m. 167° dec.[672,673]

$C_{16}H_{29}N_2O_3P$

p-$Et_2NC_6H_4NHCMe(CH_2CMe_2OH)PH(O)OH$. 4.[754] m. 210°.[754]

$C_{16}H_{35}O_2P$

n-$C_{16}H_{33}PH(O)OH$. 1a.[519] 11.[656] m. 62.5-63.5°,[656] surface tension,[656] pK_a 4.9.[656]

$C_{17}H_{23}N_2O_2P$

$(p\text{-}Me_2NC_6H_4)_2CHPH(O)OH$. 2.[187] m. < 110°.[187]

$C_{17}H_{27}Cl_2N_2O_2P$

2-Me-4-$(ClCH_2CH_2)_2NC_6H_3CH(NHC_5H_9\text{-}c)PH(O)OH$. 4.[563] m. 210-211°.[563]

$C_{18}H_{37}O_2P$

Oleyl-PH(O)OH. la.[234] Mixture with $(oleyl)_2P(O)OH$.[234]

$C_{18}H_{39}O_2P$

n-$C_{18}H_{37}PH(O)OH$. la.[234,519] ll.[656] m. 65-68°,[234] 69.3-69.5°,[656] IR,[461] surface tension.[656]

$C_{18}H_{40}NO_2P$

n-$C_{15}H_{31}NHCMe_2PH(O)OH$. 4.[363] m. 175-177°.[363]

$C_{18}H_{40}NO_3P$

n-$C_{12}H_{25}O(CH_2)_3NHCMe_2PH(O)OH$. 4.[363] m. 180°.[363]

$C_{19}H_{17}O_2P$

$Ph_3CPH(O)OH$. 2.[187] ll.[248] m. 220-222°,[709] 245-248°,[248] x-ray.[709]
-, Ag salt.[248]

$C_{19}H_{32}Cl_2NO_2P$

3,4-$Cl_2C_6H_3CH(NHC_{12}H_{25}$-n)PH(O)OH. 4.[385] m. 197-199°,[385] surface tension.[385]

$C_{19}H_{33}ClNO_2P$

p-$ClC_6H_4CH(NHC_{12}H_{25}$-n)PH(O)OH. 4.[385] m. 205-206°,[385] surface tension.[385]

$C_{19}H_{34}NO_2P$

n-$C_{12}H_{25}NHCHPhPH(O)OH$. 4.[385] m. 197-198°,[385] surface tension.[385]

$C_{21}H_{15}O_3P$

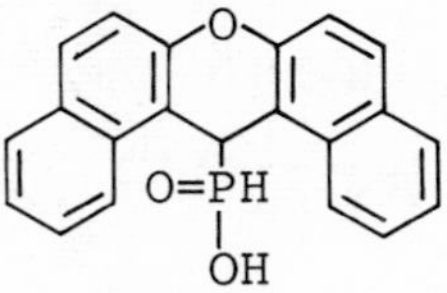

2.[188] Col. crystals.[188]
-, Na, Ba salts.[188]

$C_{21}H_{46}NO_2P$

n-$C_{18}H_{37}NHCMe_2PH(O)OH$. 4.[363,673] m. 200-201°,[363] 208°.[673]

-, Na salt, crystals.[673]

$C_{25}H_{25}Cl_2N_2O_2P$

p-$(ClCH_2CH_2)_2NC_6H_4CH(NHC_{14}H_9)PH(O)OH$ ($C_{14}H_9$ = 2-anthryl). 4.[563] m. 155-175°.[563]

$C_{25}H_{42}Cl_2NO_2P$

3,4-$Cl_2C_6H_3CH(NHC_{18}H_{35})PH(O)OH$ ($C_{18}H_{35}$ = oleyl). 4.[385] m. 194-195°,[385] surface tension.[385]

$C_{25}H_{46}NO_2P$

n-$C_{18}H_{37}NHCHPhPH(O)OH$. 4.[385] m. 186-188°,[385] surface tension.[385]

$C_{29}H_{53}Cl_2N_2O_2P$

p-$(ClCH_2CH_2)_2NC_6H_4CH(NHC_{18}H_{37}$-n$)PH(O)OH$. 4.[563] m. 80-115°.[563]
-, H_3PO_2 salt, m. 65-75°.[563]

Compounds with two Phosphorus Atoms

$C_4H_{12}O_4P_2$

$HOPH(O)(CH_2)_4PH(O)OH$. 19b.[659] m. 122-124°.[659]

$C_5H_{14}O_4P_2$

$HOPH(O)(CH_2)_5PH(O)OH$. 19b.[659] m. 61-63°.[659]

$C_6H_8O_4P_2$

p-$C_6H_4[PH(O)OH]_2$. 11.[60] 23.[110] m. 216.5-217.5°,[60] 217°,[110] IR.[110]

$C_6H_{16}N_2O_4P_2$

$HOPH(O)CH_2N$$NCH_2PH(O)OH$. 5.[406] Also from $ClCH_2PH(O)OH$, piperazine (NaOH).[722] m. 238-242° (dihydrate),[406] IR,[406] ^{31}P -9.5.[406]

$C_6H_{16}O_4P_2$

$HOPH(O)(CH_2)_6PH(O)OH$. 19b.[659] m. 84-86°.[659]

$C_8H_{12}O_4P_2$

p-C_6H_4[CH_2PH(O)OH]$_2$. 11.[60] m. 223-224°.[60]

$C_8H_{12}O_6P_2$

m-C_6H_4[CHOHPH(O)OH]$_2$. 3.[256] Na salt, yellow crystal/ powder.[256]

$C_{12}H_{12}O_4P_2$

p-[C_6H_4PH(O)OH]$_2$. 7.[558] m. 167°.[558]

$C_{12}H_{22}N_2O_4P_2$

p-C_6H_4[NHCMe$_2$PH(O)OH]$_2$. 4.[363] m. 195-196°.[363]

$C_{13}H_{24}N_2O_4P_2$

p-(o-MeC_6H_3)[NHCMe$_2$PH(O)OH]$_2$. 4.[363] m. 160°.[363]

$C_{16}H_{22}N_2O_4P_2$

[-CH_2NHCHPhPH(O)OH]$_2$. 4.[363,438] m. 246-248°,[438] dec. 240°,[363] pK_1 4.32, pK_2 7.58.[438]

$C_{16}H_{22}N_2O_6P_2$

[-CH_2NHCH(C_6H_4OH-o)PH(O)OH]$_2$. 4.[438] m. 219.5-220°,[438] pK_1 4.61, pK_2 7.57, pK_3 10.84, pK_4 11.25.[438]

$C_{16}H_{22}N_2O_6P_2S$

p,p'-SO_2[C_6H_4NHCHMePH(O)OH]$_2$. 4.[672,674] Na salt, pale yellow powder (dihydrate).[674]

$C_{18}H_{26}As_2N_2O_6P_2$

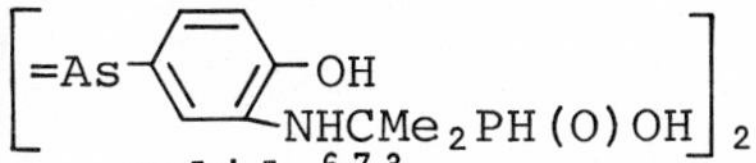

. 4.[672,673] Na salt, yellow solid.[673]

$C_{18}H_{26}N_2O_6P_2S$

p,p'-SO_2[C_6H_4NHCMe$_2$PH(O)OH]$_2$. 4.[673] m. 204° dec.[673]
-, Na salt, white powder.[673]

$C_{30}H_{46}Cl_4N_4O_4P_2$

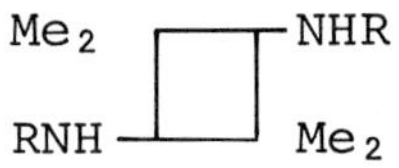

(R = CH[PH(O)OH]$C_6H_4N(CH_2CH_2Cl)_2$-p). 4.[563] m. 210°.[563]

Deuterium-Labeled Acids

CH_4ClO_2P

$ClCH_2PD(O)OD$. 11 (using D_2O).[225]

$C_6H_7O_2P$

PhPD(O)OD. 11 (using D_2O).[54] Also from PhPH(O)OH, EtOD (DCl).[640] m. 70-71°,[640] IR,[134] ^1H.[640]
-, Na salt, IR.[54,653f]

^{32}P-Labeled Acids

$CH_2Cl_3O_2P$

$CCl_3{}^{32}PH(O)OH$. By neutron irradiation of P_4 in CCl_4.[196] PC.[196]

CH_5O_2P

$Me^{32}PH(O)OH$. By neutron irradiation of Me_3P.[233] PC,[233] paper electrophoresis.[233]

$C_3H_9O_4P$

$HOCH_2CHOHCH_2{}^{32}PH(O)OH$. By neutron irradiation of glycerol and H_3PO_4.[763]

$C_6H_7O_2P$

$Ph^{32}PH(O)OH$. By neutron irradiation of PhPH(O)OH and other P-Ph compounds.[336]

D.2.2. Phosphinic Chlorides

$C_{18}H_{30}ClOP$

2,4,6-(t-Bu)$_3C_6H_2$PH(O)Cl. 15.[120] m. 133-134°,[120] IR.[120]

D.3. Phosphonous Esters (and Thio Analogs)

D.3.1. Phosphinic Esters

$C_2H_6ClO_2P$

$ClCH_2PH(O)OMe$. 10a.[213] $b_{0.2}$ 55-57°,[213] ^1H 3.57 (CH_3), 3.60 (CH_2), 6.55 (PH).[213]

$C_2H_7O_2P$

MePH(O)OMe. 10b.[657] n_D^{20} 1.422,[657] d_4^{20} 1.123,[657] IR,[657,761] ^{31}P -32.0.[761]

$C_3H_8ClO_2P$

$MePH(O)OCH_2CH_2Cl$. 62c.[540] b_5 102-104°,[540] n_D^{20} 1.4575,[540] d_4^{20} 1.2483.[540]

$C_3H_9O_2P$

MePH(O)OEt. 10b.[392,529] 11.[403] 50.[228] 54.[136] 69.[533] b_6 58-59°,[228] b_8 60-61°,[533] b_{15} 70°,[529] other bp data,[392,403] n_D^{20} 1.4220,[228,533] 1.4221,[529] 1.4225,[657] d_4^{20} 1.0511,[529] 1.0531,[228] 1.057,[657] IR,[403,657,761] ^{31}P -32.6,[178,761] ^1H 1.25 ($CH_2C\underline{H}_3$), 1.51 (PCH_3), 4.08 (CH_2), 6.87 (PH),[136] TLC.[488]

$C_3H_9O_2P$

EtPH(O)OMe. 10a.[41,456] b_{11} 65-66°,[41] b_{16} 71-72°,[456] n_D^{20} 1.4273,[41] n_D^{25} 1.4269,[456] d_0^{20} 1.0795,[49] d_4^{20} 1.0796,[41] surface tension,[49] parachor,[49] dipole moment 3.17 D.[47]

$C_3H_9O_3P$

$MePH(O)OCH_2CH_2OH$. 62b.[500] $b_{0.0001}$ 115-135° (bath),[500] n_D^{20} 1.4565,[500] d_4^{20} 1.2093,[500] PC.[500]

$C_3H_{10}NO_2P$

$MePH(O)OCH_2CH_2NH_2$. 62c.[540] n_D^{20} 1.4954.[540]
-, picrate, m. 126-128°.[540]

$C_4H_9O_2P$

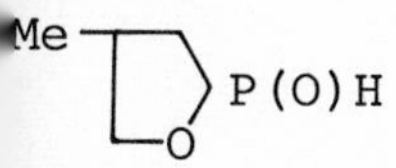

45.[698] $b_{0.2}$ 100-104°.[698]

$C_4H_{10}ClO_2P$

$MePH(O)OCH_2CH_2CH_2Cl$. 62c.[540] b_5 114-115°,[540] n_D^{20} 1.4560,[540] d_4^{20} 1.1997.[540]
$EtPH(O)OCH_2CH_2Cl$. 10b.[59] b_9 118-119°,[59] n_D^{20} 1.4670,[59] d_4^{20} 1.2360.[59]

$C_4H_{11}O_2P$

MePH(O)OPr. 10b.[529] 54.[209] b_{12} 80-81°,[209] b_{15} 84°,[529] n_D^{20} 1.4261,[209] 1.4265,[529] d_4^{20} 1.0298,[209] d_4^{20} 1.0305,[529] ^{31}P -31.5.[178]
MePH(O)OPr-i. 10b.[529] 54.[209] 80.[638] b_{11} 69-70°,[209,529] n_D^{20} 1.4205,[209] 1.4209,[529] d_4^{20} 1.0090,[209] 1.0117.[529]
-, (R)-(-)-isomer, b_7 77°,[638] $[\alpha]_D$.[638]
-, (S)-(+)-isomer.[638]
EtPH(O)OEt 10a.[41,456] 10b.[392] 52.[215] b_9 68-69°,[41] b_{15} 77.5°,[626] b_{16} 80-81°,[456] other bp data,[392] n_D^{20} 1.4242,[41] 1.4262,[626] 1.4268,[215,657] n_D^{25} 1.4238,[456] d_0^{20} 1.0195,[49] d_4^{20} 1.0170,[215,657] 1.0197,[41] 1.0779,[626] IR,[215,657] surface tension,[49] parachor,[49] dipole moment, 3.48 D.[47]

$C_4H_{11}O_3P$

$MePH(O)OCH_2CH_2CH_2OH$. 62b.[500] $b_{0.001}$ 100-120° (bath),[500] n_D^{20} 1.4555,[500] d_4^{20} 1.2503,[500] PC.[500]
$MePH(O)OCH_2CH_2OMe$. 62b.[540] b_5 96-98°,[540] n_D^{20} 1.4369,[540] d_4^{20} 1.1155.[540]
MeCHOHPH(O)OEt. 2.[182] ^{31}P -40.9.[182]
$HOCMe_2PH(O)OMe$. 2.[182] 49a.[416] Clear, col. oil,[182] n_D^{16} 1.462,[416] d^{16} 1.212,[416] ^{31}P -45.0,[182] ^1H 1.30 (CMe_2), 3.78 (OMe), 6.03 (COH), 6.42 (PH).[182]

$C_5H_7N_2O_2P$

$(NC)_2CHPH(O)OEt$. 9.[477] b_4 60-64°,[477] n_D^{15} 1.4091,[477] IR.[477]

$C_5H_7O_2PS$

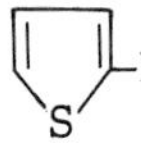
-PH(O)OMe. 10a.[69] $b_{0.2}$ 100-101.5°,[69] n_D^{14} 1.5535,[69] IR.[69]

$C_5H_{11}O_2P$

$EtPH(O)OCH_2CH=CH_2$. 50.[228] $b_{1.3}$ 41-43°,[228] n_D^{20} 1.4500,[228] d_4^{20} 1.0424.[228]

$C_5H_{11}O_4P$

$MeO_2CCH_2PH(O)OEt$. 54.[398] b_2 92°,[398] n_D^{20} 1.4510,[398] d_4^{20} 1.1934,[398] IR,[398] ^1H 3.07 (PCH_2), 3.72 (OCH_3).[553]

$C_5H_{13}O_2P$

MePH(O)OBu. 6.[305] 10a.[77] 10b.[529] 50.[228] 54.[209] 62b.[537] b_2 47-48°,[305] b_{10} 88-89°,[529] b_{18} 101-103°,[537] n_D^{20} 1.4263,[537] 1.4321,[305] 1.4332,[529] d_4^{20} 0.9959,[305] 0.9967,[529] 0.9978,[537] add'l data,[77,209,228,657] IR.[657,761]
MePH(O)OBu-i. 10b.[529] 54.[209] b_8 74-75°,[209] b_8 75-76°,[529] n_D^{20} 1.4268,[209] 1.4270,[529] d_4^{20} 0.9871,[209] 0.9880.[529]
EtPH(O)OPr. 10a.[41,456] 56.[594,595] $b_{0.6}$ 50-51°,[456] b_9 80°,[49] b_{10} 81-82°,[41] n_D^{20} 1.4280,[41] 1.4285,[49] n_D^{25} 1.4276,[456] d_0^{20} 0.9866,[49] d_4^{20} 0.9937,[41] add'l data,[594,595] surface tension,[49] parachor,[49] dipole moment, 3.37 D.[47]
EtPH(O)OPr-i. 10a.[41] b_{10} 69-70°,[41] n_D^{20} 1.4233,[41] d_0^{20} 0.9820,[49] d_4^{20} 0.9822,[41] surface tension,[49] parachor,[49] dipole moment, 3.34 D.[47]
BuPH(O)OMe. 10b.[5a] b_{10} 85-86°,[5a] n_D^{20} 1.4355,[5a] d_4^{20} 1.0140.[5a]

$C_5H_{13}O_3P$

$HOCMe_2PH(O)OEt$. 49a.[416] Oil,[416] $n_D^{18.5}$ 1.452,[414] $d^{22.5}$ 1.122.[414]

$C_6H_9O_2PS$

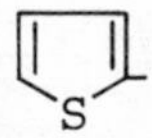
PH(O)OEt. 10a.[69] $b_{0.6}$ 113°,[69] n_D^{14} 1.5369,[69] d^{27} 1.3377,[69] IR.[69]

$C_6H_9O_3P$

$MePH(O)OCH_2$-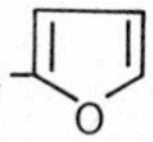 . 62b.[540] b_2 121-123°,[540] n_D^{20} 1.4674,[540] d_4^{20} 1.1870.[540]

$C_6H_{11}O_2P$

HC≡CPH(O)OBu. 54.[293] $b_{1.5}$ 65-66°,[293] n_D^{20} 1.4492,[293] d_4^{20} 1.0322.[293]

$C_6H_{13}O_2P$

CH_2=CHPH(O)OBu. 54.[293] b_2 50-51°,[293] n_D^{20} 1.4479,[293] d_4^{20} 1.0040,[293] Raman.[562]

$C_6H_{13}O_3P$

MePH(O)OCH$_2$- 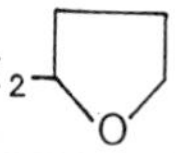 . 62b.[540] b_7 129-131°,[540] n_D^{20} 1.4638,[540] d_4^{20} 1.1525.[540]

$C_6H_{13}O_4P$

EtO_2CCH_2PH(O)OEt. 54.[398] b_1 98°,[398] n_D^{20} 1.4434,[398] d_4^{20} 1.1411,[398] IR,[398] ^1H 3.07 (PCH_2).[553]

$C_6H_{14}ClO_2P$

BuPH(O)OCH_2CH_2Cl. 10b.[5a] 10c.[78] b_2 114-116°,[78] b_6 136-137°,[5a] n_D^{20} 1.4612,[78] 1.4624,[5a] d_4^{20} 1.1390,[5a] 1.1441.[78]

$C_6H_{15}O_2P$

MePH(O)OAm-i. 10b.[529] b_{15} 106-107°,[529] n_D^{20} 1.4330,[529] d_4^{20} 0.9825.[529]
MePH(O)OAm-s. 10b.[59] b_{10} 91-92°,[59] n_D^{20} 1.4315,[59] d_4^{20} 0.9766.[59]
EtPH(O)OBu. 6.[305] 10a.[41,456] $b_{0.5}$ 68-69°.[456] $b_{1.5}$ 68°,[49] b_9 92-93°,[41] b_{10} 94-95°,[305] n_D^{20} 1.4311,[41] 1.4340,[49] 1.4350,[305] n_D^{25} 1.4298,[456] d_0^{20} 0.9753,[49] d_4^{20} 0.9759,[41] 0.9769,[305] surface tension,[49] parachor,[49] dipole moment, 3.41 D.[47]
EtPH(O)OBu-i. 10a.[41] $b_{0.7}$ 63-64°,[456] b_9 86-87°,[41] n_D^{20} 1.4300,[41] n_D^{25} 1.4292,[456] d_0^{20} 0.9688,[49] d_4^{20} 0.9688,[41] surface tension,[49] parachor,[49] dipole moment, 3.38 D.[47]
BuPH(O)OEt. 6.[305] 10b.[5a] 54.[163,658] $b_{1.5}$ 49-49.5°,[305] b_{15} 105°,[658] n_D^{20} 1.4350,[305] n_D^{22} 1.4291,[163] 1.4302,[658] d_4^{20} 0.9834,[5a,305] add'l data,[5a,163] IR.[658]
i-BuPH(O)OEt. 6.[305] 10b.[491] 50.[491] b_6 76-77°,[305] b_9 78-79°,[491] b_{10} 80-81°,[491] n_D^{20} 1.4310,[305] 1.4320,[491] 1.4325,[491] d_4^{20} 0.9501,[491] 0.9505,[491] 0.9730.[305]
t-BuPH(O)OEt. 10a.[133] b_{10} 70°,[133] n_D^{20} 1.4360,[133] IR,[133] ^1H 1.10 (t-Bu), 1.32 ($CH_2C\underline{H}_3$), 4.14 (CH_2), 6.79 (PH).[133]

$C_6H_{15}O_3P$

i-BuPH(O)OCH_2CH_2OH. 62b.[500] $b_{0.001}$ 90-110° (bath),[500] n_D^{20} 1.4575,[500] d_4^{20} 1.1121,[500] PC.[500]

$C_7H_6Cl_3O_2P$

$MePH(O)OC_6H_2Cl_3$-2,4,6. 10b.[59] m. 131-132°.[59]

$C_7H_7Cl_2O_2P$

$MePH(O)OC_6H_3Cl_2$-2,4. 10b.[59] b_4 120-121°,[59] n_D^{20} 1.5880,[59] d_4^{20} 1.6740.[59]

$C_7H_8BrO_2P$

m-$BrC_6H_4PH(O)OMe$. 47.[613] $b_{0.15}$ 123-124°,[613] IR.[613]

$C_7H_8ClO_2P$

$MePH(O)OC_6H_4Cl$-m. 10b.[59] b_4 112-113°,[59] n_D^{20} 1.5690,[59] d_4^{20} 1.5470.[59]
m-$ClC_6H_4PH(O)OMe$. 47.[613] $b_{0.25}$ 88°,[613] IR.[613]
p-$ClC_6H_4PH(O)OMe$. 47.[613] $b_{0.5}$ 101-102°.[613]

$C_7H_8NO_4P$

$MePH(O)OC_6H_4NO_2$-p. 10b.[59] m. 81-82°.[59]

$C_7H_9O_2P$

MePH(O)OPh. 10b.[59] b_9 115-116°,[59] n_D^{20} 1.5645,[59] d_4^{20} 1.4260.[59]
PhPH(O)OMe. 10a.[215,610] 37.[603] 47.[613] 56.[595] b_1 91-93°,[610,613] b_2 102°,[215] b_{10} 132.5°,[603] n_D^{20} 1.5322,[610] 1.5327,[603] 1.5340,[215] d_4^{20} 1.1663,[215] 1.1722,[603] 1.1770,[610] add'l data,[370,595] IR,[215,613,761] ^{31}P -25.2,[761] mass spect.,[97] magn. rotation.[370]

$C_7H_{11}O_2PS$

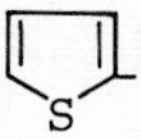PH(O)OPr. 10a.[69] $b_{0.2}$ 106°,[69] n_D^{14} 1.5308,[69] d^{27} 1.3137,[69] IR.[69]

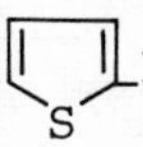PH(O)OPr-i. 10a.[69] $b_{0.25}$ 100°,[69] n_D^{14} 1.5271,[69] d^{28} 1.3184,[69] IR.[69]

$C_7H_{13}O_6P$

MePH(O)OR (R = 1,4:3,6-dianhydrosorbit-2-yl). 62a.[548] $b_{0.001}$ 160-170° (bath).[548]

$C_7H_{15}O_2P$

$MePH(O)OC_6H_{11}$-c. 62b.[537] b_{10} 117-118°,[537] n_D^{20} 1.4645,[537] d_4^{20} 1.6422.[537]
$CH_2{=}CHCH_2PH(O)OBu$. 54.[293] $b_{1.5}$ 68-68.5°,[293] n_D^{20} 1.4495,[293] d_4^{20} 0.9862,[293] Raman.[562]
c-$C_6H_{11}PH(O)OMe$. n_D^{20} 1.4670,[370] d_4^{20} 1.0822,[370] IR,[657] magn. rotation.[370]

$C_7H_{15}O_4P$

$PrPH(O)OCH_2CH_2OAc$. 55.[474] $b_{0.15}$ 72-74°,[474] IR.[474]

$C_7H_{17}O_2P$

$MePH(O)OC_6H_{13}$-n. 10b.[529] 54.[209] 62a.[537] 62b.[537] 62c.[537] b_1 82-83°,[209] b_6 103-104°,[529] b_{12} 117-119°,[537] n_D^{20} 1.4320,[537] 1.4351,[209] 1.4371,[529] d_4^{20} 0.9592,[529] 0.9616,[209] 0.9700.[537]
MePH(O)OCHMeBu-t. 62b.[537] b_{10} 92-94°,[537] n_D^{20} 1.4320,[537] d_4^{20} 0.9728.[537]
EtPH(O)OAm. 10b.[59] b_9 112-113°,[59] n_D^{20} 1.4421,[59] d_4^{20} 0.9858.[59]
EtPH(O)OAm-s. 10b.[59] b_9 100-101°,[59] n_D^{20} 1.4385,[59] d_4^{20} 0.9658.[59]
PrPH(O)OBu. 6.[305] b_2 67-67.5°,[305] n_D^{20} 1.4347,[305] d_4^{20} 0.9635.[305]
i-PrPH(O)OBu. 6.[305] $b_{2.5}$ 58.5-60.2°,[305] n_D^{20} 1.4321,[305] d_4^{20} 0.9581.[305]
BuPH(O)OPr. 10b.[5a] b_{10} 108-109°,[5a] n_D^{20} 1.4365,[5a] d_4^{20} 0.9674.[5a]
BuPH(O)OPr-i. 10b.[5a] b_7 91-92°,[5a] n_D^{20} 1.4300,[5a] d_4^{20} 0.9588.[5a]

$C_8H_{10}ClO_2P$

$MePH(O)OCH_2C_6H_4Cl$-o. 10b.[59] b_1 146-147°,[59] n_D^{20} 1.5000,[59] d_4^{20} 1.3370.[59]
$PhPH(O)OCH_2CH_2Cl$. 10c.[78] 52.[741] 55.[203,474] $b_{0.2}$ 132-136°,[474] $b_{1.5}$ 149-151°,[78] n_D^{20} 1.5412,[78] n_D^{25} 1.5510,[741] d_4^{20} 1.2626,[78] IR.[474]
3-Cl-4-$MeC_6H_3PH(O)OMe$. 47.[613] $b_{0.4}$ 120-121°.[613]

$C_8H_{11}O_2P$

EtPH(O)OPh. 56.[595] $b_{0.08}$ 62-64°,[595] n^{20} 1.5190.[595]
PhPH(O)OEt. 10a.[159,346,610] 10c.[78] 34b.[163] 37.[603] 45.[113] 52.[215] 53.[215] 54.[344,658] 56.[595] $b_{0.2}$ 102-103°,[163] $b_{0.5}$ 78-80°,[215] 109°,[658] b_1 94-95°,[610] 98-99°,[346] b_9 138°,[603] n_D^{20} 1.5180,[215] 1.5210,[658] 1.5220,[603] 1.5231,[610] n_D^{24} 1.5196,[163] d_4^{20} 1.1259,[603] 1.1275,[215]

1.1291,[610] add'l data,[78,113,299,370,657] IR,[134,163,215,224,761] ^{31}P -23.5,[425h] magn. rotation.[370]
p-$MeC_6H_4PH(O)OMe$. 37.[603] b_9 146°,[603] n_D^{20} 1.5210,[603] d_4^{20} 1.0966.[603]

$C_8H_{11}O_2PS$

$PhSCH_2PH(O)OMe$. From $ClCH_2PH(O)OMe$, PhSNa.[213] Viscous liq.,[213] ^1H 3.18 (CH_2), 3.38 (OCH_3), 6.53 (PH).[213]

$C_8H_{11}O_3P$

PhCHOHPH(O)OMe. 49.[414] m. 99°.[414]

$C_8H_{13}O_2PS$

PH(O)OBu. 10a.[69] $b_{0.15}$ 112°,[69] n_D^{14} 1.5242,[69] d^{26} 1.2670,[69] IR.[69]

$C_8H_{17}O_2P$

c-$C_6H_{11}PH(O)OEt$. 10b.[491] b_3 105-106°,[491] n_D^{20} 1.4583,[370] 1.4690,[491] d_4^{20} 1.0307,[491] 1.0437,[370] IR,[657] ^{31}P -40.3,[97,761] magn. rotation.[370]

$C_8H_{17}O_3P$

EtC(OEt)=CHPH(O)OEt. 54.[567] b_1 93-95°,[567] n_D^{20} 1.4622,[567] d_4^{20} 1.0438,[567] IR.[567]
c-$C_6H_{11}PH(O)OCH_2CH_2OH$. 62b.[500] $b_{0.001}$ 125° (bath).[500] n_D^{20} 1.4686,[500] d_4^{20} 1.0894,[500] PC.[500]

$C_8H_{17}O_4P$

$MePH(O)OCH_2CMe-CH_2$. 62a.[545] $b_{0.001}$ 160-170°,[545] n_D^{20} 1.4590,[545] d_4^{20} 1.1458.[545]

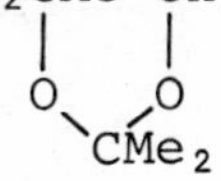

$C_8H_{18}ClO_2P$

n-$C_6H_{13}PH(O)OCH_2CH_2Cl$. 10c.[78] b_2 131-132°,[78] n_D^{20} 1.4597,[78] d_4^{20} 1.0785.[78]

$C_8H_{19}O_2P$

EtPH(O)OC_6H_{13}-n. 10a.[41] $b_{1.5}$ 87-88°,[41] n_D^{20} 1.4374,[41] d_0^{20} 0.9507,[49] d_4^{20} 0.9511,[41] surface tension,[49] parachor,[49] dipole moment, 3.36 D.[47]

BuPH(O)OBu. 10b.[5a] b_9 109-110°,[5a] n_D^{20} 1.4374,[5a] d_4^{20} 0.9523.[5a]
BuPH(O)OBu-i. 10b.[5a] b_8 110-111°,[5a] n_D^{20} 1.4350,[5a] d_4^{20} 0.9510.[5a]
BuPH(O)OBu-s. 10b.[5a] b_7 106-107°,[5a] n_D^{20} 1.4348,[5a] d_4^{20} 0.9454.[5a]

$C_8H_{19}O_2PS$

$EtSCH_2CH_2PH(O)OBu$. From $CH_2{=}CHPH(O)OBu$, EtSH (Et_3N).[293] b_2 118°,[293] n_D^{20} 1.4810,[293] d_4^{20} 1.0451.[293]

$C_9H_{11}Cl_2O_2P$

p-$ClC_6H_4CH_2PH(O)OCH_2CH_2Cl$. 10c.[78] n_D^{20} 1.5513,[78] d_4^{20} 1.3485.[78]

$C_9H_{11}Cl_2O_3P$

$MePH(O)OCH_2CH_2OC_6H_3Cl_2$-2,4. 10a.[77] 10c.[78] $b_{0.5}$ 178-180°,[78] b_1 186-190°,[77] n_D^{20} 1.5542,[77] 1.5559,[78] phytotoxic.[77]

$C_9H_{11}O_2P$

PhCH=CHPH(O)OMe. 10a.[380] $b_{0.04}$ 138°,[380] n_D^{20} 1.5886,[380] d_4^{20} 1.1541,[380] IR.[380]
$PhPH(O)OCH_2CH{=}CH_2$. 10b.[236] 37.[603] b_1 107°,[603] $b_{1.5}$ 121-123°,[236] n_D^{20} 1.5320,[603] d_4^{20} 1.1229.[603]

$C_9H_{12}ClO_2P$

$EtPH(O)OCH_2C_6H_4Cl$-o. 10b.[59] b_1 155-156°,[59] n_D^{20} 1.5383,[59] d_4^{20} 1.3680.[59]
$PhCH_2PH(O)OCH_2CH_2Cl$. 10c.[78] b_2 165-168°,[78] n_D^{20} 1.5390,[78] d_4^{20} 1.2425.[78]
$PhPH(O)OCH_2CH_2CH_2Cl$. 55.[474] $b_{0.015}$ 117-120°,[474] IR.[474]

$C_9H_{12}ClO_3P$

$MePH(O)OCH_2CH_2OC_6H_4Cl$-p. 10a.[77] 10c.[78] b_1 173-175°,[78] $b_{1.5}$ 176-180°,[77] n_D^{20} 1.5396,[78] 1.5435,[77] d_4^{20} 1.3002,[78] 1.3036,[77] phytotoxic.[77]

$C_9H_{12}FO_3P$

$MePH(O)OCH_2CH_2OC_6H_4F$-p. 10a.[77] $b_{1.5}$ 155-159°,[77] n_D^{20} 1.5105,[77] d_4^{20} 1.2535,[77] phytotoxic.[77]

$C_9H_{13}O_2P$

EtPH(O)OCH$_2$Ph. 10b.[23] $b_{0.1}$ 90-95° (mol. still).[23]
PhCH$_2$PH(O)OEt. 54.[658] b_3 119°,[658] n_D^{21} 1.5241.[658]
PhPH(O)OPr. 37.[603] 45.[113] 52.[215] 53.[215] $b_{0.3}$ 95-99°,[113] $b_{0.5}$ 90-94°,[215] b_{15} 158°,[603] n_D^{20} 1.515,[215] 1.5151,[603] d_4^{20} 1.0949,[603] 1.0962,[215] IR,[215,761] ^{31}P -26.2,[425h] mass spect.[97]
PhPH(O)OPr-i. 10a.[610] 37.[603] 53.[215] $b_{0.5}$ 80°,[215] 82°,[215] b_1 106-107°,[610] b_{10} 146°,[603] n_D^{20} 1.5075,[215] 1.5111,[610] 1.5154,[603] d_4^{20} 1.0862,[215] 1.0869,[215] 1.0922,[610] 1.0942,[603] IR,[215,761] ^{31}P -20.8.[425h]
2,5-Me$_2$C$_6$H$_3$PH(O)OMe. 10b.[762] $b_{0.05}$ 100°,[762] n_D^{20} 1.5322,[762] d_4^{20} 1.1271,[762] IR,[761,762] ^{31}P -27.5.[762]
3,4-Me$_2$C$_6$H$_3$PH(O)OMe. 54.[270] n_D^{20} 1.5368,[270] d_4^{20} 1.1273,[270] IR.[270,761]

$C_9H_{13}O_3P$

MePH(O)OCH$_2$CH$_2$OPh. 10a.[77] $b_{1.5}$ 153-156°,[77] n_D^{20} 1.5295,[77] d_4^{20} 1.1901.[77]

$C_9H_{19}O_2P$

c-C$_6$H$_{11}$PH(O)OPr-i. IR.[657] ^{31}P -37.4.[425b]

$C_9H_{19}O_3P$

c-C$_6$H$_{11}$PH(O)OCH$_2$CH$_2$CH$_2$OH. 62b.[500] $b_{0.008}$ 120-130° (bath),[500] n_D^{20} 1.4814,[500] d_4^{20} 1.1085,[500] PC.[500]

$C_9H_{19}O_4P$

MePH(O)OCH$_2$ 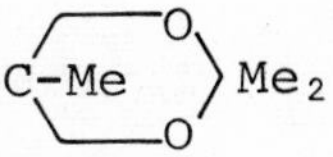 . 62b.[545] $b_{0.001}$ 160-170° (bath),[545] n_D^{20} 1.4590,[545] d_4^{20} 1.1458.[545]

$C_9H_{21}O_2P$

MePH(O)OC$_8$H$_{17}$-n. 54.[209] 62b.[537] b_2 108-111°,[537] 116-117°,[209] n_D^{20} 1.4378,[537] 1.4430,[209] d_4^{20} 0.9359,[537] 0.9444.[209]
MePH(O)OC$_8$H$_{17}$-s. 62b.[537] $b_{0.5}$ 90-92°,[537] n_D^{20} 1.4334,[537] d_4^{20} 0.9359.[537]
EtPH(O)OC$_7$H$_{15}$-n. 10a.[41] $b_{3.5}$ 102°,[49] b_4 99-100°,[41] n_D^{20} 1.4395,[41,49] d_0^{20} 0.9383,[49] d_4^{20} 0.9386,[41] surface tension,[49] parachor,[49] dipole moment, 3.31 D.[47]
BuPH(O)OAm. 10b.[5a] b_7 131-132°,[5a] n_D^{20} 1.4405,[5a] d_4^{20} 0.9482.[5]
BuPH(O)OCHEt$_2$. 10b.[5a] b_7 119-120°,[5a] n_D^{20} 1.4385,[5a] d_4^{20}

0.9393.[5a]

n-$C_7H_{15}PH(O)OEt$. 10b.[536] b_1 104-106°,[536] n_D^{20} 1.4310,[536] d_4^{20} 0.9658,[536] TLC.[488]

$C_{10}H_{13}O_2P$

PhCH=CHPH(O)OEt. 10a.[380] $b_{0.03}$ 137-138°,[380] $b_{0.08}$ 140-141°,[381] n_D^{20} 1.5665,[380,381] d_4^{20} 1.1241,[380] 1.1249,[381] IR.[380]

$C_{10}H_{13}O_4P$

$PhPH(O)OCH_2CH_2OAc$. 55.[474] $b_{0.04}$ 139-141°,[474] IR.[474]

$C_{10}H_{15}O_2P$

EtPH(O)OCHMePh. 10b.[59] b_1 115-116°,[59] n_D^{20} 1.5330,[59] d_4^{20} 1.2480.[59]

$EtPH(O)OC_6H_4Et$-o. 10b.[59] b_1 171-172°,[59] n_D^{20} 1.5520,[59] d_4^{20} 1.2880.[59]

BuPH(O)OPh. 10b.[5a] b_6 179-180°,[5a] n_D^{20} 1.5495,[5a] d_4^{20} 1.2781.[5a]

PhPH(O)OBu. 6.[305] 10a.[77,346] 10c.[78] 37.[603] 45.[541] 53.[215] $b_{0.1}$ 89-90°,[131] $b_{0.12}$ 90.5-92.0°,[286] $b_{0.5}$ 105°,[215] b_1 99.2-100°,[305] b_3 149°,[346] b_{11} 157°,[603] n_D^{20} 1.5077,[299] 1.5116,[215] 1.5144,[305] n_D^{25} 1.5081,[286] n_D^{29} 1.5045,[346] d_4^{20} 1.0627,[215] 1.0758,[305] d_4^{29} 1.0695,[346] add'l data,[77,78,215,370,541,603] IR,[215,761] ^{31}P -23.8,[425h] MW (C_6H_6) monomer,[348] magn. rotation.[370]

PhPH(O)OBu-i. 10a.[610] 37.[603] 53.[215] $b_{0.5}$ 85-87°,[215] b_1 112-113°,[610] b_9 149°,[603] n_D^{20} 1.5075,[603] 1.5080,[215] 1.5081,[610] d_4^{20} 1.0650,[603] 1.0673,[215] 1.0675,[610] IR,[117e,215,761]

p-$MeC_6H_4PH(O)OPr$. 37.[603] b_9 157°,[603] n_D^{20} 1.5159,[603] d_4^{20} 1.0729.[603]

p-$MeC_6H_4PH(O)OPr$-i. 37.[603] b_{10} 156°,[603] n_D^{20} 1.5152,[603] d_4^{20} 1.0699.[603]

2,5-$Me_2C_6H_3PH(O)OEt$. 54.[762] $b_{0.05}$ 95°,[762] n_D^{20} 1.5194,[762] 1.521,[762] d_4^{20} 1.0852,[762] 1.0892,[762] IR.[761,762]

2,6-$Me_2C_6H_3PH(O)OEt$. 54.[88] $b_{0.1}$ 92-93°,[88] m. 40°,[88] n_D^{20} 1.5316,[88] d_4^{20} 1.1068.[88]

3,4-$Me_2C_6H_3PH(O)OEt$. IR.[761]

$C_{10}H_{21}O_2P$

c-$C_6H_{11}PH(O)OBu$. 1.[495] b_5 111-112°,[495] n_D^{20} 1.4581,[370] 1.4625,[495] d_4^{20} 0.9760,[495] 0.9995,[370] IR,[657] ^{31}P -40.8,[425b] magn. rotation.[370]

$C_{10}H_{23}O_2P$

$MePH(O)OC_9H_{19}$-n. 62b.[537] b_1 123-127°,[537] n_D^{20} 1.4391,[537] d_4^{20} 0.9305.[537]
$EtPH(O)OC_8H_{17}$-n. 10a.[41] b_1 102-103°,[41] n_D^{20} 1.4440,[41] d_0^{20} 0.9316,[49] d_4^{20} 0.9320,[41] surface tension,[49] parachor,[49] dipole moment, 3.27 D.[47]

$C_{10}H_{24}NO_2P$

$Et_2NCH_2CH_2PH(O)OBu$. From CH_2=CHPH(O)OBu, Et_2NH.[306] b_1 107.5-108.0°,[306] n_D^{20} 1.4520,[306] d_4^{20} 0.9578.[306]

$C_{11}H_{15}O_2P$

PhCH=CHPH(O)OPr. 10a.[380] $b_{0.05}$ 143-145°,[380] n_D^{20} 1.5649,[380] d_4^{20} 1.0960,[380] IR.[380]
PhCH=CHPH(O)OPr-i. 10a.[380] $b_{0.03}$ 126-127°,[380] n_D^{20} 1.5545,[380] d_4^{20} 1.0898,[380] IR.[380]
2-$C_{10}H_{11}$PH(O)OMe. ($C_{10}H_{11}$ = 5,6,7,8-tetrahydronaphthyl). 54.[268] $b_{0.1}$ 125-130°,[268] n_D^{20} 1.5567,[268] d_4^{20} 1.1614,[268] IR.[268,761]

$C_{11}H_{15}O_4P$

$PhPH(O)OCH_2CH_2O_2CEt$. 55.[474] $b_{0.18}$ 125-126°,[474] IR.[474]
$PhPH(O)OCH_2CH_2CH_2OAc$. 55.[474] $b_{0.01}$ 160-162°,[474] IR.[474]

$C_{11}H_{17}O_2P$

$PhCH_2PH(O)OBu$. 6.[305] b_2 113-115.2°,[305] n_D^{20} 1.5160,[305] d_4^{20} 1.0646.[305]
PhPH(O)OAm-i. 10b.[552] b_3 138-139°,[552] n_D^{20} 1.5055,[552] d_4^{20} 1.0502.[552]
p-$MeC_6H_4PH(O)OBu$. 37.[603] b_{10} 167°,[603] n_D^{20} 1.5102,[603] d_4^{20} 1.0490.[603]
p-$MeC_6H_4PH(O)OBu$-i. 37.[603] b_9 160°,[603] n_D^{20} 1.5074,[603] d_4^{20} 1.0438.[603]
2,5-$Me_2C_6H_3PH(O)OPr$. 54.[762] n_D^{22} 1.516,[762] d_4^{22} 1.0648,[762] IR,[761 762] ^{31}P -25.1.[425h]
2,5-$Me_2C_6H_3PH(O)OPr$-i. 54.[762] n_D^{20} 1.5153,[762] d_4^{20} 1.0594,[762] IR,[761,762] ^{31}P -1.6.[762]
3,4-$Me_2C_6H_3PH(O)OPr$-i. 54.[270] n_D^{20} 1.5170,[270] d_4^{20} 1.0610,[270] IR.[270,761]

$C_{11}H_{18}NO_3P$

o-$MeC_6H_4NHCMe_2PH(O)OMe$. 47.[363] Undistillable liq.[363]

$C_{11}H_{23}O_2P$

$MePH(O)OC_{10}H_{19}$ [$C_{10}H_{19}$ = (-)-menthyl]. 34a.[68] Epimers separated by fractional cryst'n. form Ia, m. 42°,[68] $[\alpha]_D^{25}$ -96.6,[68] ^1H 7.31 (PH);[68] form Ib, ^1H 7.25 (PH),[68] Other ^1H data.[68]
$n\text{-}C_8H_{17}PH(O)OCH_2CH{=}CH_2$. 45.[236] b_1 134°,[236] n_D^{20} 1.4536.[236]

$C_{11}H_{24}NO_2P$

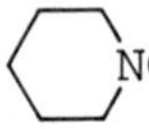$NCH_2CH_2PH(O)OBu$. 54.[306] Also from $CH_2{=}CHPH(O)OBu$, piperidine.[306] b_1 126-126.5°,[306] $b_{1.5}$ 125-126°,[306] n_D^{20} 1.4730,[306] d_4^{20} 1.0032,[306] 1.0056.[306]

$C_{11}H_{25}O_2P$

$MePH(O)OC_{10}H_{21}\text{-}n$. 62b.[537] b_2 132-135°,[537] n_D^{20} 1.4421,[537] d_4^{20} 0.9208.[537]
$EtPH(O)OC_9H_{19}\text{-}n$. 10a.[41] b_4 122-123°,[41] n_D^{20} 1.4431,[41] d_0^{20} 0.9216,[49] d_4^{20} 0.9230,[41] add'l data,[49] surface tension,[49] parachor,[49] dipole moment, 3.34 D.[47]

$C_{12}H_{11}O_2P$

PhPH(O)OPh. 52.[741] n_D^{25} 1.5924.[741]

$C_{12}H_{14}Cl_3O_2P$

PhPH(O)O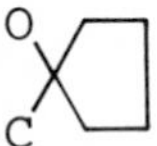 . 34a.[326] m. 101-102.5°,[326] IR.[326]
Cl_3C

$C_{12}H_{17}O_2P$

$PhCH{=}CHPH(O)OBu$. 10a.[380] $B_{0.05}$ 149-150°,[380] n_D^{20} 1.5578,[380] d_4^{20} 1.0709,[380] IR.[380]
$PhCH{=}CHPH(O)OBu\text{-}i$. 10a.[380] $b_{0.03}$ 137-138°,[380] n_D^{20} 1.5469,[380] d_4^{20} 1.0684,[380] IR.[380]
$PhCH{=}CHPH(O)OBu\text{-}s$. 10a.[380] $b_{0.03}$ 148-149°,[380] n_D^{20} 1.5518,[380] d_4^{20} 1.0790,[380] IR.[380]

$C_{12}H_{19}O_2P$

$PhPH(O)OC_6H_{13}\text{-}n$. 10a.[610] 37.[603] b_1 139°,[610] b_9 160°,[603] n_D^{20} 1.5015,[603] 1.5030,[610] d_4^{20} 1.0337,[603] 1.0388.[610]

$C_{12}H_{25}O_2P$

c-C_6H_{11}PH(O)OC_6H_{13}-n. 62b.[491] b_3 117-118°,[491] n_D^{20} 1.4690,[491] d_4^{20} 0.9850.[491]

$C_{12}H_{27}O_2P$

EtPH(O)O$C_{10}H_{21}$-n. 10a.[41] $b_{1.5}$ 129-130°,[41] n_D^{20} 1.4458,[41] d_4^{20} 0.9296,[41] dipole moment, 3.42 D.[47]
i-C_8H_{17}PH(O)OBu-i. IR.[117f]

$C_{13}H_{13}O_2P$

PhPH(O)OCH_2Ph. 10b.[23] 57.[23] $b_{0.05}$ 90-95°.[23]

$C_{13}H_{19}O_2P$

PhCH=CHPH(O)OAm. 10a.[380] $b_{0.03}$ 155-156°,[380] n_D^{20} 1.5420,[380] d_4^{20} 1.0334,[380] IR.[380]
PhCH=CHPH(O)OAm-i. 10a.[380] $b_{0.03}$ 158-159°,[380] n_D^{20} 1.5390,[380] d_4^{20} 1.0414,[380] IR.[380]
2-$C_{10}H_{11}$PH(O)OPr-i. ($C_{10}H_{11}$ = 5,6,7,8-tetrahydronaphthyl). 10b.[268] $b_{0.05}$ 154-155°,[268] n_D^{20} 1.5347,[268] d_4^{20} 1.0963,[268] IR.[270,761]

$C_{13}H_{21}O_2P$

PhPH(O)OC_7H_{15}-n. 10a.[610] b_1 150°,[610] n_D^{20} 1.4996,[610] d_4^{20} 1.0187.[610]
p-MeC_6H_4PH(O)OC_6H_{13}-n. 37.[603] b_1 132-133°,[603] n_D^{20} 1.5051,[603] d_4^{20} 1.0237.[603]

$C_{13}H_{23}O_2PSi$

PhPH(O)O$(CH_2)_4$SiMe_3. 10b.[325] b_1 142-148°.[325]

$C_{13}H_{23}O_7P$

MePH(O)OR (R = 1,2:5,6-di-O-isopropylideneglucos-3-yl). 69.[497] $b_{0.01}$ 170-180°,[497] TLC.[497]
MePH(O)OR (R = 1,2:3,4-di-O-isopropylidenegalactos-6-yl). 62a.[548] 67.[548] 69.[548] $b_{0.01}$ 160-165° (bath),[548] TLC.[548]

$C_{14}H_{13}Cl_2O_3P$

PhPH(O)OCH_2CH_2O$C_6H_3Cl_2$-2,4. 10a.[77] Viscous mass,[77] n_D^{20} 1.5912.[77]

$C_{14}H_{14}ClO_3P$

$PhPH(O)OCH_2CH_2OC_6H_4Cl$-p. 10a.[77] Viscous mass,[77] n_D^{20} 1.5815.[77]

$C_{14}H_{14}FO_3P$

$PhPH(O)OCH_2CH_2OC_6H_4F$-p. 10a.[77] $b_{0.5}$ 205-208°,[77] n_D^{20} 1.5460,[77] d_4^{20} 1.2249.[77]

$C_{14}H_{15}O_3P$

$PhPH(O)OCH_2CH_2OPh$. 10a.[77] Viscous mass,[77] n_D^{20} 1.5745.[77]

$C_{14}H_{23}O_2P$

$PhPH(O)OC_8H_{17}$-n. 10a.[610] b_1 155°,[610] n_D^{20} 1.4982,[610] d_4^{20} 1.0079.[610]
$PhPH(O)OC_8H_{17}$-i. 52.[741] n_D^{25} 1.4988.[741]
p-$MeC_6H_4PH(O)OC_7H_{15}$-n. 37.[603] b_1 143-144°,[603] n_D^{20} 1.5017,[603] d_4^{20} 1.0095.[603]

$C_{15}H_{25}O_2P$

$PhPH(O)OC_9H_{19}$-n. 10a.[610] b_1 158-160°,[610] n_D^{20} 1.4900,[610] d_4^{20} 0.9843.[610]
p-$MeC_6H_4PH(O)OC_8H_{17}$-n. 37.[603] b_1 145-146°,[603] n_D^{20} 1.5006,[603] d_4^{20} 1.0005.[603]

$C_{16}H_{25}O_2P$

$PhPH(O)OC_{10}H_{19}$ [$C_{10}H_{19}$ = (-)-menthyl]. 10b.[163] $b_{0.1}$ 165° (mol. still),[163] IR,[163] ^1H 7.42 (PH).[163]

$C_{16}H_{29}O_7P$

i-BuPH(O)OR. (R = 1,2:3,4-di-O-isopropylidenegalactos-6-yl). 50.[492] $b_{0.0003}$ 120-121°,[492] $[\alpha]_D^{20}$ -31.8,[492] TLC.[492]
i-BuPH(O)OR (R = 2,3:5,6-di-O-isopropylidenemannos-1-yl). 50.[230] m. 110-111°,[230] $[\alpha]_D^{20}$ +20,[230] TLC.[230]

$C_{16}H_{33}O_2P$

n-$C_{10}H_{21}CH{=}CHPH(O)O(CH_2CH_2O)_2H$. 46.[706,707] Oil.[706,707]

$C_{19}H_{42}NO_3P$

n-$C_{12}H_{25}O(CH_2)_3NHCMe_2PH(O)OMe$. 47.[363] Undistillable liq.[363]

$C_{20}H_{19}O_2P$

$Ph_3CPH(O)OMe$. 48.[248] m. 163-164°.[248]

$C_{21}H_{21}O_2P$

$Ph_3CPH(O)OEt$. 48.[248] 49a.[248] m. 118-121.5°.[248]

$C_{25}H_{30}NO_2P$

$Ph_3CPH(O)OCH_2CH_2NEt_2$. 49b.[146] HCl salt, m. 139-140°;[146] methobromide, cryst.[146]

$C_{26}H_{53}O_8P$

$n\text{-}C_{12}H_{25}CH{=}CHP(O)O(CH_2CH_2O)_6H$. 46.[706,707] Liq.[706,707]

$C_{27}H_{34}NO_2P$

$Ph_3CPH(O)OCH_2CH_2N(Pr\text{-}i)_2$. 49b.[146] m. 86-87°.[146]

Compounds with Two Phosphorus Atoms

$C_4H_{11}ClO_4P_2$

$MePH(O)OCH_2CH_2OP(O)(Cl)Me$. From $[MePH(O)OCH_2\text{-}]_2$, Cl_2.[542] n_D^{20} 1.4855,[542] d_4^{20} 1.4610.[542]

$C_4H_{12}O_4P_2$

$MePH(O)OCH_2CH_2OPH(O)Me$. 62a.[501] 62b,[500,538,766] $b_{0.0001}$ 85-88°,[538] n_D^{20} 1.4705,[538] 1.4725,[766] 1.4730,[501] d_4^{20} 1.2850,[766] 1.2855,[501] 1.2902,[538] TLC.[488,766]

$C_5H_{14}O_4P_2$

$MePH(O)O(CH_2)_3OPH(O)Me$. 62b.[538] $b_{0.0001}$ 95-97°,[538] n_D^{20} 1.4608,[538] d_4^{20} 1.2147.[538]

$C_6H_{16}O_4P_2$

$MePH(O)OCH_2CHMeOPH(O)Me$. 62b.[538] $b_{0.0001}$ 98-101°,[538] n_D^{20} 1.4627,[538] d_4^{20} 1.1940.[538]
$EtPH(O)OCH_2CH_2OPH(O)Et$. 62a.[602] 62b.[604] $b_{0.05}$ 94-95°,[602] n_D^{20} 1.4578,[602] d_4^{20} 1.2207.[602]
$MeOPH(O)(CH_2)_4PH(O)OMe$. 54.[659] b_1 108-115°,[659] n_D^{22} 1.4732.[659]

$C_6H_{16}O_5P_2$

$[MePH(O)OCH_2CH_2]_2O$. 62b.[538] $b_{0.0001}$ 113-115°,[538] n_D^{20}

1.4694,[538] d_4^{20} 1.2437.[538]

$C_7H_{18}O_4P_2$

$MePH(O)O(CH_2)_5OPH(O)Me$. 62b.[766] $b_{0.0001}$ 140-155° (bath),[766] n_D^{20} 1.4995,[766] d_4^{20} 1.1675,[766] TLC.[766]
$EtPH(O)O(CH_2)_3OPH(O)Et$. 62b.[604] $b_{0.04}$ 128-130°,[604] n_D^{20} 1.4663, d_4^{20} 1.1630.[604]
$EtPH(O)OCH_2CHMeOPH(O)Et$. 62b.[604] $b_{0.04}$ 116-119°,[604] n_D^{20} 1.4612,[604] d_4^{20} 1.1580.[604]

$C_8H_{16}O_6P_2$

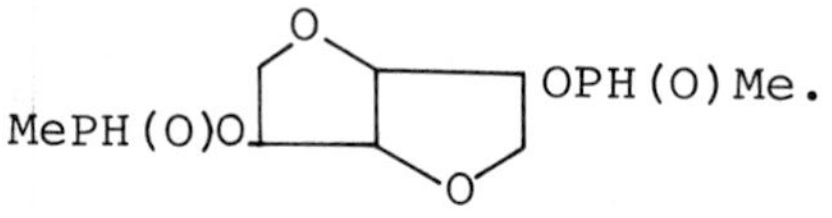

62b.[538] 69.[497] $b_{0.0001}$ 124-127°,[538] $b_{0.001}$ 190-195°,[497] n_D^{20} 1.5077,[497] 1.5087,[538] TLC.[497]

$C_8H_{20}O_4P_2$

$EtPH(O)OCH_2CH_2CHMeOPH(O)Et$. 62b.[604] $b_{0.08}$ 143-144°,[604] n_D^{20} 1.4640,[604] d_4^{20} 1.1335.[604]
$EtPH(O)OCHMeCHMeOPH(O)Et$. 62b.[604] $b_{0.04}$ 128-129°,[604] n_D^{20} 1.4626,[604] d_4^{20} 1.1360.[604]
$EtOPH(O)(CH_2)_4PH(O)OEt$. 54.[659] Col. liq.[659]

$C_8H_{20}O_5P_2$

$[EtPH(O)OCH_2CH_2]_2O$. 62b.[604] $b_{0.06}$ 160-162°,[604] n_D^{20} 1.4696,[604] d_4^{20} 1.1815.[604]

$C_9H_{22}O_4P_2$

$EtPH(O)O(CH_2)_5OPH(O)Et$. 62b.[604] $b_{0.08}$ 161-163°,[604] n_D^{20} 1.4672,[604] d_4^{20} 1.1162.[604]

$C_9H_{23}NO_4P_2$

$MePH(O)OCH_2CH_2OP(O)(Me)CH_2NEt_2$. From $[MePH(O)OCH_2-]_2$, $CH_2(NEt_2)_2$,[542] $b_{0.0001}$ 98-99°,[542] n_D^{20} 1.4742,[542] d_4^{20} 1.1360.[542]
-, methiodide, m. 43-45°.[542]

$C_{10}H_{16}O_4P_2$

$p-C_6H_4[CH_2OPH(O)Me]_2$. 62b.[538] $b_{0.0001}$ 118-123°,[538] n_D^{20} 1.4840,[538] d_4^{20} 1.1892.[538]

$C_{10}H_{24}O_4P_2$

$EtOPH(O)(CH_2)_6PH(O)OEt$. 54.[659] Col. oil.[659]

$C_{10}H_{24}O_5P_2$

$[EtPH(O)OCHMeCH_2]_2O$. 62b.[604] $b_{0.04}$ 144-147°,[604] n_D^{20} 1.4613,[604] d_4^{20} 1.1219.[604]

$C_{12}H_{28}O_7P_2$

$EtPH(O)(OCH_2CH_2)_4OPH(O)Et$. 62b.[604] $b_{0.03}$ 152-155°,[604] n_D^{20} 1.4661,[604] d_4^{20} 1.1618.[604]

Compounds with Three or More Phosphorus Atoms

$C_8H_{12}O_6P_3$

$MeC[CH_2OPH(O)Me]_3$. 62b.[766] $b_{0.0001}$ 180-190° (bath),[766] n_D^{20} 1.4860,[766] TLC.[766]

$C_9H_{24}O_8P_4$

$C[CH_2OPH(O)Me]_4$. 62b.[766] $b_{0.0001}$ 190-200° (bath),[766] n_D^{20} 1.4930.[766]

Silicon and Germanium Esters

$C_9H_{15}O_2PSi$

$PhPH(O)OSiMe_3$. 49b.[87] $b_{0.07}$ 91°,[87] n_D^{20} 1.4902,[87] d_4^{20} 1.054,[87] IR,[87] ^{31}P -10.1,[87] ^1H 0.26 (CH_3), 7.42 (PH).[87]

$C_{12}H_{21}GeO_2P$

$PhPH(O)OGeEt_3$. 49b.[87] $b_{0.01}$ 128-129°,[87] n_D^{20} 1.518,[87] d_4^{20} 1.2193,[87] IR,[87] ^{31}P -14.6,[87] ^1H 1.06 (Et), 7.44 (PH).[87]

Deuterium-labeled Esters

$C_2H_6ClO_2P$

$ClCH_2PD(O)OMe$. 10a.[213] ^1H.[213]

$C_4H_{11}O_2P$

MePD(O)OPr-i. From MePH(O)OPr-i, MeOD.[638] $[\alpha]_D$,[638] ^1H.[638]

$C_9H_{13}O_2P$

PhPD(O)OPr. 54.[215] $b_{0.5}$ 94-95°,[215] n_D^{20} 1.5168,[215] d_4^{20} 1.1044,[215] IR.[215]

D.3.2. Phosphonofluoridous Esters

C_2H_6FOP

MeP(OMe)F. 16.[367] b_{750} 56-58°,[367] d_4^{20} 1.0150.[367]

C_3H_8FOP

MeP(OEt)F. 14,[152] 16.[367] b_{750} 60-62°,[367] b_{760} 63-65°,[152] n_D^{20} 1.3935,[367] 1.3967,[152] d_4^{20} 1.0396,[152] 1.0436,[367] ^{31}P -218,[152] ^{19}F (F_2 ref.) 528.0.[689]

$C_4H_{10}FOP$

MeP(OPr)F. b_{120} 26-28°,[152] n_D^{20} 1.3990,[152] ^{31}P -214,[152] ^{19}F (F_2 ref.) 528.4.[689]

MeP(OPr-i)F. 16.[367] b_{750} 67-69°,[367] n_D^{20} 1.3820,[367] d_4^{20} 0.9300.[367]

$C_5H_{12}FOP$

MeP(OBu)F. 44.[367] b_{65} 43-44°,[367] n_D^{20} 1.4000,[367] d_4^{20} 0.9701.[367]

MeP(OBu-i)F. 14.[152] 44.[367] b_{100} 37-39°,[152] 47-48°,[367] n_D^{20} 1.3907,[152] 1.3990,[367] d_4^{20} 0.9529,[367] IR,[152] ^{31}P -216,[152] ^{19}F (F_2 ref.) 530.4.[689]

MeP(OBu-s)F. 16.[367] b_{125} 49-50°,[367] n_D^{20} 1.4112,[367] d_4^{20} 0.9702.[367]

EtP(OPr-i)F. ^{31}P -29 reported[732] for this compound is actually that of the phosphonofluoridic ester, EtP(O)(OPr-i)F.[676]

$C_6H_{14}FOP$

BuP(OEt)F. 16.[367] b_{12} 30-32°,[367] n_D^{20} 1.4120,[367] d_4^{20} 0.9203.[367]

C_7H_8FOP

MeP(OPh)F. 16.[367] b_{14} 58-59°,[367] n_D^{20} 1.5020,[367] d_4^{20} 1.1320.[367]

$C_7H_{14}FOP$

MeP(OC_6H_{11}-c)F. 16.[367] b_{26} 53-55°,[367] n_D^{20} 1.4502,[367] d_4^{20} 1.0483.[367]

$C_8H_{16}FOP$

EtP(OC_6H_{11}-c)F. 16.[367] b_{25} 75°,[367] n_D^{20} 1.4500,[367] d_4^{20} 1.0332.[367]

D.3.3. Phosphonochloridous Esters

$C_3H_7Cl_2OP$

$MeP(OCH_2CH_2Cl)Cl$. 17.[281] b_{12} 59-60°,[281] n_D^{18} 1.495,[281] d_4^{18} 1.2699.[281]

$C_4H_9Cl_2OP$

$MeP(OCHMeCH_2Cl)Cl$. 17.[281] b_{6-7} 56-57°,[281] n_D^{18} 1.483,[281] d_4^{18} 1.1985.[281]
$EtP(OCH_2CH_2Cl)Cl$. 13.[679] 17.[281,319] b_6 46°,[319] b_{15} 75°,[281] n_D^{18} 1.490,[281] n_D^{20} 1.4860,[319] d_4^{18} 1.2152,[281] d_4^{20} 1.2200.[319]

$C_4H_{10}ClOP$

$EtP(OEt)Cl$. 27.[708] b_{18} 35-36°,[708] n_D^{20} 1.4554.[708]

$C_5H_{10}Cl_3OP$

$EtP(OCH[CH_2Cl]_2)Cl$. 17.[329] $b_{0.15}$ 100°,[329] n_D^{20} 1.4995,[329] d_4^{20} 1.2962.[329]

$C_5H_{11}Cl_2OP$

$EtP(OCHMeCH_2Cl)Cl$. 17.[281] b_7 80°,[281] n_D^{18} 1.4810,[281] d_4^{18} 1.1900.[281]

$C_6H_{10}Cl_2OP$

$EtP(OCH_2CHClCH{=}CH_2)Cl$. 17.[577] $b_{0.08}$ 50-51°,[577] n_D^{20} 1.4895,[577] d_4^{20} 1.1580.[577]

$C_8H_9Cl_2OP$

$PhP(OCH_2CH_2Cl)Cl$. 13.[585] 17.[300] $b_{0.10}$ 122°,[585] b_1 99-100°,[300] n_D^{20} 1.5645,[300] 1.5670,[585] d_4^{20} 1.2896,[300] 1.2934.[585]

$C_8H_{10}ClOP$

$PhP(OEt)Cl$. 27.[708] $b_{0.2}$ 74-75°,[708] n_D^{20} 1.5664.[708]

$C_8H_{13}Cl_4OP$

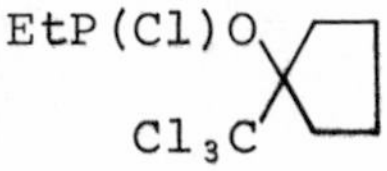

13.[331] $b_{0.15}$ 107°,[331] n_D^{20} 1.5250,[331] d_4^{20} 1.3645.[331]

$C_8H_{16}Cl_2OP$

EtP(O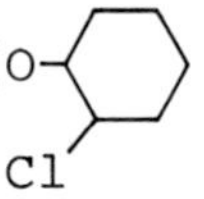)Cl. 17.[281] b_{10} 132°,[281] n_D^{13} 1.5080,[281] d_4^{13} 1.200.[281]

$C_9H_{10}Cl_3OP$

PhP(OCH[CH_2Cl]$_2$)Cl. 17.[585] $b_{0.12}$ 118-119°,[585] n_D^{20} 1.5670,[585] d_4^{20} 1.3442.[585]

$C_{10}H_{11}Cl_4OP$

PhP($OCMe_2CCl_3$)Cl. 13.[369] $b_{0.1}$ 112-113°,[369] n_D^{20} 1.5633,[369] d_4^{20} 1.3754.[369]

$C_{10}H_{13}Cl_2OP$

EtP($OCH_2CHClPh$)Cl. 17.[577] b_1 110-112°,[577] n_D^{20} 1.5504,[577] d_4^{20} 1.2040.[577]

$C_{10}H_{14}ClOP$

PhP(OBu)Cl. 13.[293] b_2 77-78°,[293] n_D^{20} 1.5352,[293] d_4^{20} 1.1049.[293]

$C_{11}H_{13}Cl_4OP$

p-$MeC_6H_4P(OCMe_2CCl_3)$Cl. 13.[328] $b_{0.15}$ 132-133°,[328] n_D^{20} 1.5630,[328] d_4^{20} 1.3408.[328]

$C_{12}H_{10}ClOP$

PhP(OPh)Cl. IR.[115]

$C_{12}H_{13}Cl_4OP$

PhP(Cl)O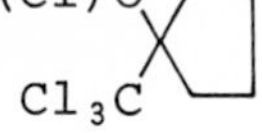 . 13.[326] $b_{0.1}$ 145-147°,[326] n_D^{20} 1.5782,[326] d_4^{20} 1.3989.[326]

$C_{13}H_{15}Cl_4OP$

PhP(Cl)O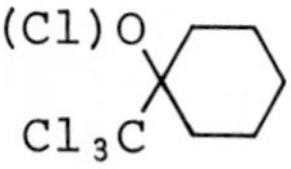 . 13.[327] $b_{0.1}$ 150-151°,[327] m. 59-60°.[327]

D.3.4. Phosphonous Diesters

$C_3H_4F_3O_2P$

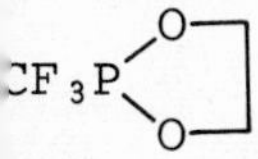

. 12a.[100] b. 113° (extrap),[100] log P_{mm}= 5.0948 + 1.75 log T - 0.003702T - 2051/T,[100] m. -33.0°,[100] IR.[100]

$C_3H_6F_3O_2P$

$CF_3P(OMe)_2$. 12a.[100] b. 88.8° (extrap),[100] log P_{mm} = 5.1731 + 1.75 log T - 0.00355 T - 1985/T,[100] IR.[100]

$C_3H_7F_2O_2P$

$CHF_2P(OMe)_2$. 81.[210] b. 107° (extrap).[210]

$C_3H_8ClO_2P$

$ClCH_2P(OMe)_2$. 12b.[200] n_D^{25} 1.453.[200]

$C_3H_9O_2P$

$MeP(OMe)_2$. 65.[403,669] b_{295} 61°,[669] b_{300} 62-65°,[403] n_D^{20} 1.4172,[403] IR,[403] ^{31}P -200.8,[403] ^1H 1.14 (PCH_3), 3.53 (OCH_3).[731]

$C_4H_{11}O_2P$

$EtP(OMe)_2$. 12b.[40,382] b_{210} 73-77°,[382] b_{225} 73.5-74.5°,[40] n_D^{20} 1.4210,[40] d_4^{20} 0.9515,[40] inflames in air.[40]

$C_5H_{10}Cl_3O_2P$

$CCl_3P(OEt)_2$. 12b.[53] b_8 86-87°.[53]

$C_5H_{11}O_2P$

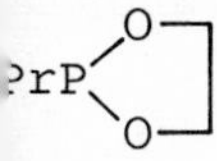

. 5a.[474] b_{20} 68-69°,[474] IR.[474]

$C_5H_{13}O_2P$

$MeP(OEt)_2$. 12b.[237,259,260,636] 12c.[209,237] b_{50} 47°,[259] b_{90} 55-57°,[209] b. 120-122°,[237] n_D^{20} 1.4155,[209] n_D^{25} 1.4168,[259] n_D^{26} 1.4513,[237] d_4^{20} 0.905,[237] 0.9051,[209] add'l data,[260,636] IR,[718] ^1H [136,435a] 1.12 (PCH_3), 1.19 ($CH_2C\underline{H}_3$), 3.81 (CH_2).[136]

$C_6H_9O_2PS$

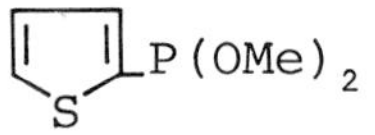 12b.[69] $b_{0.9}$ 57°,[69] n_D^{20} 1.5444,[69] d^{25} 1.1859,[69] IR.[69]

$C_6H_{11}Cl_2O_2P$

CH_2=CHP(OCH_2CH_2Cl)$_2$. 5a.[293,298] b_1 71-71.5°,[293] b_{22} 93°,[298] n_D^{20} 1.4945,[298] 1.4947,[293] d_4^{20} 1.2308,[298] 1.2322.[293]

$C_6H_{11}O_2P$

CH≡CP(OEt)$_2$. 5a.[13] b_{20} 75-80°,[13] IR,[13] ^{1}H 1.29 (CH_3), 3.09 (CH), 3.6-4.3 (CH_2),[13] C.A. Registry No. 20505-16-2.

$C_6H_{11}O_4P$

$MeCO_2$CH=CHP(OMe)$_2$. 12b.[565] $b_{1.5}$ 76°,[565] n_D^{20} 1.4680,[565] d_4^{20} 1.1411.[565]

$C_6H_{12}ClO_4P$

AcOCHClCH_2P(OMe)$_2$. 12b.[564] b_1 86-87°,[564] n_D^{20} 1.4590,[564] d_4^{20} 1.2127.[564]

$C_6H_{13}Cl_2O_2P$

EtP(OCH_2CH_2Cl)$_2$. 18.[319] b_5 92°,[319] n_D^{20} 1.4795,[319] d_4^{20} 1.2030.[319]

$C_6H_{13}O_2P$

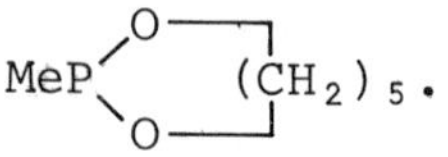 65.[543] b_4 46-48°,[543] polymerizes on standing.[543]

$C_6H_{13}O_3P$

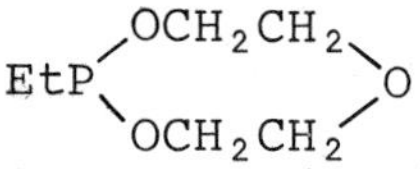 58a.[576] b_{12} 73-74°,[576] n_D^{20} 1.4745,[576] d_4^{20} 1.1081,[576] becomes viscous on standing.[576]

EtOCH=CHP(OMe)$_2$. 12b.[186] $b_{15.5}$ 81-82°,[186] n_D^{20} 1.4618,[186] d_4^{20} 1.0261.[186]

$C_6H_{15}O_2P$

$EtP(OEt)_2$. 3a.[658] 5a.[215,658] 12b.[40,260,620,628,629] 12c.[209] 65.[669] 40 (A.4.).[44] b_{10} 35.5-36.5°,[620] b_{32} 55-56°,[44] b_{38} 58-60°,[40] b_{55} 71°,[669] b_{760} 137-139°,[629] n_D^{20} 1.4212,[40] 1.4215,[44] 1.4222,[629] 1.4230,[620] d_0^{20} 0.9064,[629] d_4^{20} 0.9065,[620] 0.9200,[44] 0.9275,[40] add'l data,[209,215,260,596,628,658] IR,[215,718] TGA.[596]
$BuP(OMe)_2$. 3c.[250] b_{15} 89-92°.[250]

$C_7H_7O_2P$

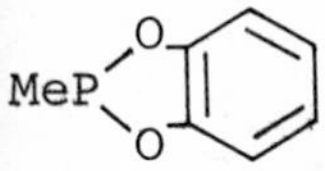 . 12b.[752] b_{10} 76°,[752] m. 2°,[752] ^1H 1.18 (CH_3), 7.05 (C_6H_4).[752]

$C_7H_{13}O_2P$

$MeP(OCH_2CH{=}CH_2)_2$. 12c.[209] b_{35} 63-64°,[209] n_D^{20} 1.4558,[209] d_4^{20} 0.9598.[209]
$CH_2{=}C{=}CHP(OEt)_2$. ^1H 4.736 ($CH_2{=}C$), 5.443 (CH).[696]

$C_7H_{13}O_4P$

$EtCO_2CH{=}CHP(OMe)_2$. 12b.[565] b_1 80°,[565] n_D^{20} 1.4675,[565] d_4^{20} 1.1247.[565]

$C_7H_{14}ClO_4P$

$EtCO_2CHClCH_2P(OMe)_2$. 12b.[564] $b_{0.2}$ 84-86°,[564] n_D^{20} 1.4580,[564] d_4^{20} 1.1716.[564]

$C_7H_{15}Cl_2O_2P$

$MeP(OCH_2CHClMe)_2$. 18.[556] n_D^{25} 1.4638,[556] d^{35} 1.137.[556]

$C_7H_{15}O_2P$

$CH_2{=}CHCH_2P(OEt)_2$. 5a.[625] 12b.[625] b_{11} 47-49°,[625] b_{13} 53-55°,[625] n_D^{20} 1.4380,[625] 1.4428,[625] d_4^{20} 0.9189,[625] 0.9314.[625]

$C_7H_{15}O_3P$

$i\text{-}PrOCH{=}CHP(OMe)_2$. 12b.[186] b_8 77-78°,[186] n_D^{20} 1.4604,[186] d_4^{20} 1.0050.[186]

$C_7H_{15}O_4P$

$MeO_2CCH_2P(OEt)_2$. 5e,[553,568] 7.[513] b_8 84-85°,[568] b_{8-9} 87°,[513] n_D^{20} 1.4420,[513,568] d_4^{20} 1.0662,[568] 1.0678,[513] IR,[513] ^1H 3.16 (PCH_2), 3.60 (OCH_3).[513,553]

$C_7H_{16}NO_4P$

$EtCH(NO_2)P(OEt)_2$. 5c.[475] $b_{0.5}$ 85-87°,[475] IR.[475]

$C_7H_{17}O_2P$

$MeP(OPr)_2$. 12b.[259] 12c.[209] b_{33} 76°,[259] b_{35} 63-64°,[209] n_D^{20} 1.4248,[209] n_D^{25} 1.4243,[259] d_4^{20} 0.9031.[209]
$MeP(OPr\text{-}i)_2$. 12b.[259,430] 12c.[209] b_{30} 51-54°,[430] b_{36} 55°,[259] b_{40} 57-58°,[209] n_D^{20} 1.4168,[209] n_D^{25} 1.4157,[259] d_4^{20} 0.8887,[209] ^{31}P -173.0,[376] ^1H 1.10 (PCH_3), 1.20 (OCH), 4.20 ($CHCH_3$).[376]
$PrP(OEt)_2$. 12b.[628] b_{12} 52-53°,[628] n_D^{20} 1.4275,[628] d_4^{20} 0.9029.[628]

$C_7H_{17}O_2PS$

$MeP(OEt)OCH_2CH_2SEt$. 12a.[259] $b_{0.040}$ 42-43°,[259] n_D^{25} 1.4782,[259] d_4^{25} 0.9963.[259]

$C_8H_6F_5O_2P$

$C_6F_5P(OMe)_2$. 12b.[177] $b_{0.1}$ 44°,[177] IR.[177]

$C_8H_9O_2P$

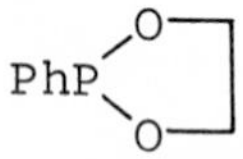 12b.[245,474,634] 58a.[576] $b_{0.8}$ 79-80°,[474] b_1 88-89°,[576] $b_{1.6}$ 86-89°,[245] b_4 101-102°,[634] cryst. below room temp.,[245] n_D^{20} 1.5656,[245] 1.5682,[634] 1.5789,[576] d_4^{20} 1.2090,[634] 1.2453,[576] becomes viscous on standing,[474,576] IR.[245,474]

$C_8H_{10}BrO_2P$

$p\text{-}BrC_6H_4P(OMe)_2$. 12b.[334] b_{16} 134.5-135.5°,[334] n_D^{20} 1.5660,[334] d_4^{20} 1.4350.[334]

$C_8H_{10}ClO_2P$

$p\text{-}ClC_6H_4P(OMe)_2$. 12b.[314,480] b_8 107-109°,[480] b_{11} 113-115°,[314] n_D^{20} 1.5430,[314] d_0^{20} 1.2030,[314] d_4^{20} 1.2095.[480]

$C_8H_{10}FO_2P$

m-$FC_6H_4P(OMe)_2$. 12c.[616] $b_{0.7}$ 45°,[616] ^{19}F (C_6H_5F ref.) -0.03.[616]

p-$FC_6H_4P(OMe)_2$. 12c.[616] $b_{2.7}$ 75°,[616] ^{19}F (C_6H_5F ref.) -1.94.[616]

$C_8H_{11}O_2P$

$EtP(OCH_2C{\equiv}CH)_2$. 12b.[311] Structure revised.[82,424,571]

$PhP(OMe)_2$. 12b.[36,127,241,244,309,636] 15.[29] b_7 77-79°,[713] b_{13} 94.5°,[36] b_{15} 101-102°,[29] n_D^0 1.5118,[309] n_D^{20} 1.5264,[370] 1.5270,[596] 1.5280,[29] n_D^{25} 1.5261,[244] 1.5278,[241] d_0^0 1.0972,[36,309] 1.1022,[29] d_0^{20} 1.0849,[29] d_0^{24} 1.0732,[36] d_4^{20} 1.0820,[370] 1.0839,[596] add'l data,[36,116,119,127,241,244,596,636] IR,[116] ^{31}P -159,[425f] magn. rotation,[370] GC,[72] TGA,[596] available commercially.[713]

$C_8H_{13}O_2PS$

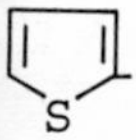$P(OEt)_2$. 12b.[69] b_3 81-83°,[69] n_D^{25} 1.5208,[69] d^{25} 1.1033,[69] IR.[69]

$C_8H_{15}O_2P$

$EtP(OCH_2CH{=}CH_2)_2$. 12b.[572,644] b_{10} 67-67.5°,[572] b_{11} 65-67°,[644] n_D^{20} 1.4550,[572] 1.4553,[644] d_4^{20} 0.9396,[644] 0.9403.[572]

$C_8H_{15}O_4P$

$EtP(OCH_2\overset{O}{CH{-}CH_2})_2$. 12b.[642] $b_{0.5}$ 91-93°,[642] n_D^{20} 1.4700,[642] d_4^{20} 1.1477.[642]

$MeCO_2CH{=}CHP(OEt)_2$. 12b.[565] b_1 82-83°,[565] n_D^{20} 1.4600,[565] d_4^{20} 1.0980,[565] IR.[565]

$PrCO_2CH{=}CHP(OMe)_2$. 12b.[565] b_1 91°,[565] n_D^{20} 1.4627,[565] d_4^{20} 1.0829.[565]

$C_8H_{16}ClO_4P$

$AcOCHClCH_2P(OEt)_2$. 12b.[564] b_1 95-97°,[564] n_D^{20} 1.4517,[564] d_4^{20} 1.1286.[564]

$PrCO_2CHClCH_2P(OMe)_2$. 12b.[564] $b_{0.5}$ 90-91°,[564] n_D^{20} 1.4565,[564] d_4^{20} 1.1358.[564]

$C_8H_{17}Cl_2O_2P$

$EtP(OCH_2CH_2CH_2Cl)_2$. 12b.[578] n_D^{20} 1.4688,[578] d_4^{20} 1.1630.[578]

$BuP(OCH_2CH_2Cl)_2$. 5a.[292] b_2 100-101°,[292] n_D^{20} 1.4777,[292] d_4^{20} 1.1372.[292]

$C_8H_{17}O_2P$

$MeCH{=}CHCH_2P(OEt)_2$. 12b.[334] $b_{0.3}$ 54-56°,[334] n_D^{20} 1.4380,[334] d_4^{20} 0.9145.[334]
$Me_2C{=}CHP(OEt)_2$. 12b.[320] b_{100} 66-68°,[320] n_D^{20} 1.4540,[320] d_4^{20} 0.9445.[320]
$c\text{-}C_6H_{11}P(OMe)_2$. n_D^{20} 1.4640,[370] d_4^{20} 0.9921,[370] magn. rotation.[370]

$C_8H_{17}O_3P$

$EtOCH{=}CHP(OEt)_2$. 12b.[26,186] b_2 113°,[26] $b_{11.5}$ 93-93.3°,[186] n_D^{20} 1.4567,[186] 1.4570,[26] d_4^{20} 0.9764,[26] 0.9788.[186]
$BuOCH{=}CHP(OMe)_2$. 12b.[186] b_{11} 103-104°,[186] n_D^{20} 1.4635,[186] d_4^{20} 0.9940.[186]
$EtC(OEt){=}CHP(OMe)_2$. 12b.[567] b_7 83-84°,[567] n_D^{20} 1.4637,[567] d_4^{20} 0.9977.[567]

$C_8H_{17}O_4P$

$EtO_2CCH_2P(OEt)_2$. 5b.[568] b_9 95-96°,[568] n_D^{20} 1.4380,[568] d_4^{20} 1.0349.[568]

$C_8H_{18}ClO_2P$

$Cl(CH_2)_4P(OEt)_2$. 12b.[251] $b_{0.05}$ 70-71° [mixture with $Cl(CH_2)_8Cl$].[251]

$C_8H_{19}O_2P$

$EtP(OPr)_2$. 5a.[301] 12b.[40] 12c.[209] b_7 56-56.5°,[301] b_{10} 64-66°,[596] b_{11} 65-66°,[40] n_D^{20} 1.4278,[40] 1.4280,[596] 1.4318,[301] d_4^{20} 0.8935,[301] 0.9017,[596] 0.9021,[40] add'l data,[209] TGA.[596]
$EtP(OPr\text{-}i)_2$. 12b.[40,430] b_{12} 50-52°,[430] b_{35} 73.5-74.0°,[40] n_D^{20} 1.4169,[40] d_0^{20} 0.8844.[40]
$EtP(OEt)OBu$. 36.[708] b_{12} 71°,[708] n_D^{20} 1.4310.[708]
$BuP(OEt)_2$. 3a.[658] 5a.[658] 5d.[658] 12b.[628] b_{12} 68.5-70°,[628] b_{20} 78°,[658] n_D^{20} 1.4308,[628] 1.4310,[658] d_4^{20} 0.8963.[628]
$t\text{-}BuP(OEt)_2$. 12b.[133] b_{10} 85°,[133] n_D^{20} 1.4162,[133] IR,[133] ^1H 1.10 (t-Bu), 1.26 ($CH_2C\underline{H}_3$), 4.06 (CH_2).[133]

$C_8H_{19}O_2PS$

$MeP(OPr)OCH_2CH_2SEt$. 58a.[259] $b_{0.025}$ 53-56°,[259] n_D^{25} 1.4715.[259]
$MeP(OPr\text{-}i)OCH_2CH_2SEt$. 58a.[259] $b_{0.100}$ 42-46°,[259] n_D^{25} 1.4678.[259]

$C_9H_{11}O_2P$

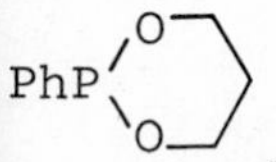 12b.[245,474,549] $b_{0.15}$ 72-74°,[474] $b_{0.3}$ 96-98°,[245] b_2 86-89°,[549] n_D^{20} 1.5460,[549] n_D^{25} 1.5659,[245] d_4^{20} 1.1524,[549] IR,[474] becomes viscous on standing.[474]

$C_9H_{11}O_2P$

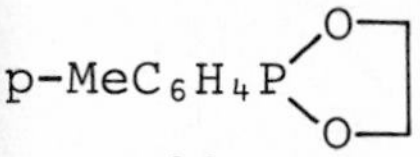 12b.[312] b_{12} 128-131°,[312] n_D^{20} 1.5719,[312] d_4^{20} 1.1986.[312]

$C_9H_{12}ClO_2P$

$MeP(OEt)OC_6H_4Cl$-p. 71.[533] b_8 107-109°,[533] n_D^{20} 1.5110,[533] d_4^{20} 1.0424.[533]

$C_9H_{13}O_2P$

$PrP(OCH_2C{\equiv}CH)_2$. 12b.[311] Structure revised.[82,424,571]
MeP(OEt)OPh. 71.[533] b_{11} 91-92°,[533] n_D^{20} 1.5002,[533] d_4^{20} 1.0234.[533]
p-$MeC_6H_4P(OMe)_2$. 12b.[335] b_{14} 107-109°,[335] n_D^{10} 1.5325,[335] n_D^{20} 1.5280,[313] d_0^0 1.0709,[335] d_0^{10} 1.0427,[335] d_4^{20} 1.0630.[313]

$C_9H_{15}O_2P$

$CH_2{=}CHCH_2P(OCH_2CH{=}CH_2)_2$. 12b.[389] b_{10} 80-82°,[389] n_D^{20} 1.4698,[389] d_4^{20} 0.9547.[389]

$C_9H_{17}O_2P$

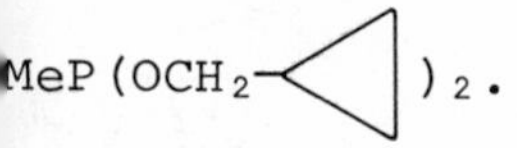 12c.[209] b_{17} 70-71°,[209] n_D^{20} 1.4612,[209] d_4^{20} 0.9679.[209]

$C_9H_{17}O_4P$

$EtCO_2CH{=}CHP(OEt)_2$. 12b.[565] $b_{1.5}$ 87°,[565] n_D^{20} 1.4575,[565] d_4^{20} 1.0519.[565]

$C_9H_{18}ClO_4P$

$EtCO_2CHClCH_2P(OEt)_2$. 12b.[564] b_1 106°,[564] n_D^{20} 1.4570,[564] d_4^{20} 1.1232.[564]

$C_9H_{19}O_2P$

CH_2=$CHCH_2P(OPr)_2$. 5a.[625] 12b.[625] b_8 72-72.5°,[625] b_{13} 81-83°,[625] n_D^{20} 1.4415,[625] 1.4445,[625] d_4^{20} 0.9080,[625] 0.9111.[625]
CH_2=$CHCH_2P(OPr\text{-}i)_2$. 12b.[625] b_{15} 66-67°,[625] n_D^{20} 1.4375,[625] d_4^{20} 0.9031.[625]

$C_9H_{19}O_3P$

i-PrOCH=CHP(OEt)$_2$. 12b.[186] $b_{11.5}$ 98.5°,[186] n_D^{20} 1.4547,[186] d_4^{20} 0.9648.[186]
MeC(OBu)=CHP(OMe)$_2$. 12b.[567] b_9 104-106°,[567] n_D^{20} 1.4680,[567] d_4^{20} 0.9878.[567]

$C_9H_{21}O_2P$

MeP(OBu)$_2$. 2c.[209] 5a.[301] b_1 39-40°,[301] b_2 57-58°,[209] n_D^{20} 1.4312,[209] 1.4348,[301] d_4^{20} 0.8902,[301] 0.8957.[209]
MeP(OBu-i)$_2$. 12c.[209] b_1 45-46°,[209] n_D^{20} 1.4278,[209] d_4^{20} 0.8941.[209]
PrP(OPr)$_2$. 12b.[628] b_{11} 81-82°,[628] n_D^{20} 1.4223,[628] d_4^{20} 0.8917.[628]
PrP(OPr-i)$_2$. 12b.[430] b_{10} 61-62°.[430]
i-PrP(OPr-i)$_2$. 12b.[631] $b_{12.5}$ 105.5-107.5°,[629] b_{14} 59.5-60.5°,[631] n_D^{20} 1.4150,[631] d_0^{20} 0.8794.[631]

$C_9H_{21}O_2PS_2$

MeP(OCH$_2$CH$_2$SEt)$_2$. 58a.[259] $b_{0.025}$ 89-94°,[259] n_D^{25} 1.5067,[259] d_4^{25} 1.0675.[259]

$C_{10}H_{10}F_5O_2P$

$C_6F_5P(OEt)_2$. 12b.[177] $b_{0.2}$ 58°,[177] IR.[177]

$C_{10}H_{11}O_2P$

PhP(OCH=CH$_2$)$_2$. 15b.[397] b_2 76-78°,[397] n_D^{20} 1.5385,[397] d_4^{20} 1.0633.[397]

$C_{10}H_{12}Cl_3O_2P$

p-ClC$_6$H$_4$P(OCH$_2$CH$_2$Cl)$_2$. 18.[204] n_D^{20} 1.5534,[204] d_4^{20} 1.3600,[204] contains some m-isomer.[204]

$C_{10}H_{13}Cl_2O_2P$

$PhP(OCH_2CH_2Cl)_2$. 18.[201] b_3 138-140°,[201] n_D^{20} 1.5420,[201] d_4^{20} 1.2620.[201]

$C_{10}H_{13}O_2P$

$PhCH{=}CHP(OMe)_2$. 12b.[379] $b_{0.06}$ 88-90°,[379] n_D^{20} 1.5692,[379] d_4^{20} 1.0865,[379] IR.[379]

$C_{10}H_{13}O_2P$

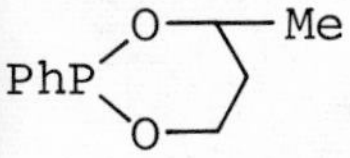

. 12b.[244,549] b_1 93-95°,[549] $b_{1.5}$ 100°,[244] m. 82-83°,[244] n_D^{20} 1.5498,[549] d_4^{20} 1.1388.[549]

$C_{10}H_{13}O_3P$

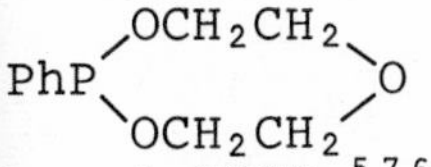

. 58a.[576] $b_{0.5}$ 109°,[576] n_D^{20} 1.5502,[576] d_4^{20} 1.2027,[576] becomes viscous on standing.[576]

$C_{10}H_{14}BrO_2P$

p-$BrC_6H_4P(OEt)_2$. 12b.[334] $b_{11.5}$ 149-150°,[334] n_D^{20} 1.5432,[334] d_4^{20} 1.3180,[334] IR.[334]

$C_{10}H_{14}ClO_2P$

m-$ClC_6H_4P(OEt)_2$. 12b.[390] b_3 81°,[390] n_D^{20} 1.5232,[390] d_4^{20} 1.1265.[390]
p-$ClC_6H_4P(OEt)_2$. 12b.[315,596,725] b_4 100-101°,[725] b_9 132-133°,[596] b_{11} 129-130.5°,[315] n_D^{20} 1.5252,[315,596,725] d_0^{20} 1.1210,[315] d_4^{20} 1.1084,[596] 1.1258,[725] TGA.[596]

$C_{10}H_{15}O_2P$

$BuP(OCH_2C{\equiv}CH)_2$. 12b.[311] Structure revised.[82,424,571]
$MeP(OEt)OC_6H_4Me$-p. 71.[533] b_9 108-109°,[533] n_D^{20} 1.5040,[533] d_4^{20} 1.0105.[533]
EtP(OEt)OPh. 58b.[260] $b_{0.080}$ 51°,[260] n_D^{25} 1.5055.[260]
$PhP(OEt)_2$. 3a.[658] 5a.[301,658] 5b.[302] 12b.[36,114,255,309,668] 15a.[28,223,344] 65.[372,669] b_1 63-65°,[302] b_4 99-100°,[114] b. 235°,[344] 235-237°,[48] n_D^0 1.5063,[309] n_D^{20} 1.5113,[302] 1.5118,[114] 1.5120,[48] d_0^0 1.0405,[309] d^{16} 1.032,[344] d_0^{20} 1.0247,[48] d_4^{20} 1.0235,[302] 1.0243,[114] add'l data,[19,223,255,301,362,370,372,587,596,658,668,669,713] UV,[112,123] IR,[134,215] ^{31}P -153.5,[721] -154.0,[425f] -156,[140] -158.5,[425f] ^1H 1.17,[692] 1.261[755] (CH_3),

3.80,[692] 3.954[755] (CH[1]), fig.,[717a] ^{13}C,[639] magn. rotation,[370] TGA,[596] surface tension,[48] available commercially.[713]

2,5-$Me_2C_6H_3P(OMe)_2$. 12b.[762] $b_{0.05}$ 80°,[762] n_D^{20} 1.525,[762] d_4^{20} 1.052.[762]

3,4-$Me_2C_6H_3P(OMe)_2$. 12b.[270] n_D^{20} 1.5382,[270] d_4^{20} 1.0551.[270]

$C_{10}H_{16}NO_2P$

p-$Me_2NC_6H_4P(OMe)_2$. 12d.[127,222] $b_{0.4}$ 125-126°,[222] $b_{0.8}$ 119-120°.[127]

$C_{10}H_{17}Cl_2O_2P$

$EtP(OCH_2CHClCH{=}CH_2)_2$. 18.[577] $b_{0.08}$ 94-95°,[577] n_D^{20} 1.4873,[577] d_4^{20} 1.1290.[577]

$C_{10}H_{17}O_2P$

$MeCH{=}CHCH_2P(OCH_2CH{=}CH_2)_2$. 12b.[334] $b_{0.1}$ 72-74°,[334] n_D^{20} 1.4710,[334] d_4^{20} 0.9505.[334]

$C_{10}H_{17}O_2PS$

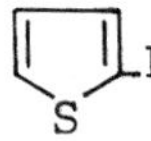

$P(OPr)_2$. 12b.[69] $b_{0.3}$ 80-82°,[69] n_D^{24} 1.5092,[69] d^{25} 1.0698,[69] IR.[69]

$P(OPr\text{-}i)_2$. 12b.[69] $b_{0.5}$ 70°,[69] n_D^{25} 1.5030,[69] d^{25} 1.0559,[69] IR.[69]

$C_{10}H_{19}O_2P$

$EtP(OCH_2CH{=}CHMe)_2$. 12b.[572] b_2 70.5-71°,[572] n_D^{20} 1.4628,[572] d_4^{20} 0.9315.[572]

$EtP(OCHMeCH{=}CH_2)_2$. 12b.[572] b_3 50-51°,[572] n_D^{20} 1.4470,[572] d_4^{20} 0.9059.[572]

$BuP(OCH_2CH{=}CH_2)_2$. 12b.[311] b_{11} 98-100°,[311] n_D^{20} 1.4550,[311] d_0^{20} 0.9340.[311]

$HC{\equiv}CP(OBu)_2$. 5a.[293] b_2 58.8-60°,[293] n_D^{20} 1.4520,[293] d_4^{20} 0.9289.[293]

$C_{10}H_{19}O_4P$

$PrCO_2CH{=}CHP(OEt)_2$. 12b.[565] $b_{2.5}$ 103°,[565] n_D^{20} 1.4590,[565] d_4^{20} 1.0448.[565]

$C_{10}H_{20}ClO_4P$

$PrCO_2CHClCH_2P(OEt)_2$. 12b.[564] $b_{0.2}$ 91-92°,[564] n_D^{20} 1.4545,[564] d_4^{20} 1.0872.[564]

$C_{10}H_{21}Cl_2O_2P$

$EtP(O[CH_2]_4Cl)_2$. 12b.[578] n_D^{20} 1.4698,[578] d_4^{20} 1.1280.[578]

$C_{10}H_{21}Cl_2O_4P$

$EtP([OCH_2CH_2]_2Cl)_2$. 12b.[578] n_D^{20} 1.4698,[578] d_4^{20} 1.1280[578] (same data as the preceding compound).

$C_{10}H_{21}O_2P$

CH_2=CHP(OBu)$_2$. 5a.[293] b_2 49-50°,[293] n_D^{20} 1.4471,[293] d_4^{20} 0.9039,[293] add'l data,[293] UV,[562] Raman.[562]
MeCH=CHCH$_2$P(OPr)$_2$. 12b.[387] $b_{0.1}$ 65-67°,[387] n_D^{20} 1.4435,[387] d_4^{20} 0.9029.[387]
MeCH=CHCH$_2$P(OPr-i)$_2$. 12b.[334] $b_{0.1}$ 56-58°,[334] n_D^{20} 1.4380,[334] d_4^{20} 0.8931,[334] IR.[334]
c-$C_6H_{11}P(OEt)_2$. n_D^{20} 1.4624,[370] d_4^{20} 0.9521,[370] magn. rotation.[370]

$C_{10}H_{21}O_3P$

BuOCH=CHP(OEt)$_2$. 12b.[186] b_{11} 116-116.5°,[186] n_D^{20} 1.4584,[186] d_4^{20} 0.9584.[186]
EtC(OEt)=CHP(OEt)$_2$. 12b.[567] b_7 96-97°,[567] n_D^{20} 1.4588,[567] d_4^{20} 0.9639.[567]
EtC(OBu)=CHP(OMe)$_2$. 12b.[567] b_{10} 112-113°,[567] n_D^{20} 1.4649,[567] d_4^{20} 0.9738.[567]

$C_{10}H_{23}O_2P$

EtP(OBu)$_2$. 5a.[301] 12b.[40] 12c.[209] 36.[708] b_1 47-58°,[301] b_3 63-63.5°,[40] b_{12} 94-95°,[708] n_D^{20} 1.4353,[40] 1.4354,[708] 1.4370,[301] d_4^{20} 0.8871,[301] 0.8977,[40] add'l data.[209]
EtP(OBu-i)$_2$. 12c.[209] b_5 60-62°,[209] n_D^{20} 1.4285,[209] d_4^{20} 0.8901.[209]
EtP(OBu-s)$_2$. 12b.[633] b_{12} 78-82°,[633] n_D^{20} 1.4280,[633] d_4^{20} 0.8870.[633]
BuP(OPr)$_2$. 12b.[628] b_{14} 99.5-101°,[628] n_D^{20} 1.4355,[628] d_4^{20} 0.8961.[628]
BuP(OPr-i)$_2$. 12b.[430,518] b_{12} 80-82°,[430] b_{18} 85-87°.[518]

$C_{11}H_{14}Cl_3O_2P$

PhP(OMe)OCMe$_2$CCl$_3$. 36.[369] $b_{0.12}$ 113°,[369] n_D^{20} 1.5440,[369] d_4^{20} 1.3384.[369]

p-$ClCH_2C_6H_4P(OCH_2CH_2Cl)_2$. 18.[304] n_D^{20} 1.5620,[304] d_4^{20} 1.3328.[304]

$C_{11}H_{15}Cl_2O_2P$

$MeC_6H_4P(OCH_2CH_2Cl)_2$. 18.[206] n_D^{20} 1.5381,[206] d_4^{20} 1.2311.[206]

$C_{11}H_{15}Cl_2O_3P$

$MeOC_6H_4P(OCH_2CH_2Cl)_2$. 18.[202] n_D^{20} 1.5462,[202] d_4^{20} 1.2996,[202] isomerizes when distilled.[202]

$C_{11}H_{15}O_2P$

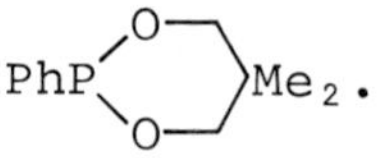

. 12b.[197] cryst.,[197] ^1H 0.6 ($CH_{3eq.}$), 1.3 ($CH_{3ax.}$), 3.25-3.83 (CH_2), 7.35 (C_6H_5).[197]

$C_{11}H_{15}O_2PS$

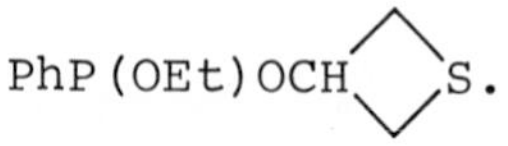

. 71.[39] $b_{0.06}$ 98-100°,[39] n_D^{20} 1.5600,[39] d_4^{20} 1.1464.[39]

$C_{11}H_{17}O_2P$

$PhCH_2P(OEt)_2$. 3a.[658] 5a.[658] 12c.[209] b_2 92-94°,[209] b_3 88-90°,[658] n_D^{20} 1.5021,[209] 1.5032,[658] d_4^{20} 1.0161.[209]
o-$MeC_6H_4P(OEt)_2$. ^{31}P -178.[425g]
p-$MeC_6H_4P(OEt)_2$. 12b.[335] 15a.[450] b_9 123-125°,[335] 134-136°,[596] b. 280°,[450] n_D^{18} 1.5138,[335] n_D^{20} 1.5138,[596] 1.5141,[313] d_0^{10} 1.0380,[335] d_0^{18} 1.0210,[335] d_4^{20} 1.0190,[313] 1.0256,[596] TGA.[596]

$C_{11}H_{17}O_3P$

p-$MeOC_6H_4P(OEt)_2$. 12b.[307] b_{13} 136-138°,[307] n_D^{16} 1.4986,[307] d_0^0 1.0529,[307] d_0^{16} 1.0433.[307]

$C_{11}H_{21}O_5P$

$MeO_2CCMe_2COCMe_2P(OMe)_2$. 8.[70] $b_{0.03}$ 76-77°,[70] IR,[70] ^1H 1.13 ($PCMe_2$), 1.27 (CMe_2CO_2), 3.55 ($POCH_3$), 3.63 (CO_2CH_3).[70]

$C_{11}H_{23}O_2P$

$CH_2{=}CHCH_2P(OBu)_2$. 5a.[293,622] $b_{1.5}$ 59-59.5°,[293] b_{10} 102.5-103°,[622] n_D^{20} 1.4465,[622] 1.4500,[293] d_4^{20} 0.8981,[293]

0.9006,[622] Raman.[562]
CH_2=$CHCH_2P(OBu\text{-}i)_2$. 12b.[386] b_{13} 94-96°,[386] n_D^{20} 1.4428,[386] d_4^{20} 0.8920.[386]
CH_2=$CHCH_2P(OBu\text{-}s)_2$. 12b.[624] b_8 87-89°,[624] n_D^{20} 1.4428,[624] d_4^{20} 0.8931.[624]

$C_{11}H_{23}O_3P$

$MeCOCH_2P(OBu)_2$. 5e.[553] ^1H 1.75 ($COCH_3$), 3.4 (PCH_2).[553]
MeC(OBu)=CHP$(OEt)_2$. 12b.[567] b_{10} 118-120°,[567] n_D^{20} 1.4600,[567] d_4^{20} 0.9568.[567]
PrC(OBu)=CHP$(OMe)_2$. 12b.[567] $b_{1.5-2}$ 99-101°,[567] n_D^{20} 1.4637,[567] d_4^{20} 0.9335.[567]

$C_{11}H_{24}NO_2P$

MeP(OEt)O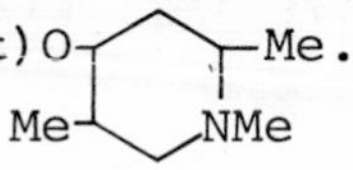. 71.[534] $b_{0.06}$ 43-45°,[534] n_D^{20} 1.4733,[534] d_4^{20} 0.9350.[534]

$C_{11}H_{25}O_2P$

MeP(OEt)OC_8H_{17}-n. 58a.[260] 60.[260] $b_{0.1}$ 65°,[260] n_D^{25} 1.4372.[260]
MeP$(OAm\text{-}i)_2$. 12c.[209] b_1 62-63°,[209] n_D^{20} 1.4311,[209] d_4^{20} 0.8854.[209]
PrP$(OBu)_2$. 5a.[301,432] 5b.[302] b_1 59-60°,[302] b_7 93.5-95°,[432] n_D^{20} 1.4375,[302] 1.4400,[432] d_4^{20} 0.8836,[302] 0.8844,[432] add'l data.[301]
AmP$(OPr\text{-}i)_2$. 12b.[431] b_{15} 99-101°.[431]

$C_{12}H_9O_2P$

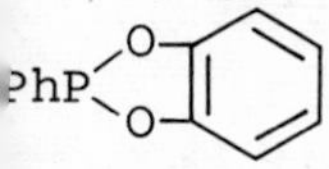 12a.[73,103a] 12b,[751] 58a.[207] $b_{0.1}$ 91°,[751] b_{2-3} 180-183°,[207] f. 28°,[751] m. 140-145°,[73] 168-169°,[103a] n_D^{20} 1.5804,[207] IR.[73,103a]

$C_{12}H_{13}O_2P$

1-$C_{10}H_7P(OMe)_2$. 12b.[222,315] $b_{0.15}$ 101°,[222] b_4 137-138°,[315] n_D^{20} 1.6096,[315] 1.6127,[222] d_0^{20} 1.1550.[315]

$C_{12}H_{14}ClO_2P$

p-$ClC_6H_4P(OCH_2CH$=$CH_2)_2$. 12b.[317,469] b_1 116-117°,[572] b_2 125-126°,[313] b_3 126-127°,[317] n_D^{20} 1.5376,[317] 1.5410,[313],

[572] d_0^0 1.1685,[317] d_0^{20} 1.1490,[317] d_4^{20} 1.1372,[572] 1.1379.[313]

$C_{12}H_{15}O_2P$

$PhP(OCH_2CH=CH_2)_2$. 12b.[236,317,469,572] $b_{0.3}$ 100°,[469] $b_{0.5}$ 92-93°,[572] b_3 116-117°,[317] n_D^{20} 1.5240,[317] 1.5300,[572] d_0^0 1.0620,[317] d_0^{20} 1.0443,[317] d_4^{20} 1.0443,[572] pour-point < -65°F,[469] viscosity.[469]

$C_{12}H_{15}O_4P$

$BzOCMe=CHP(OMe)_2$. 12b.[564] $b_{0.025}$ 105-107°,[564] n_D^{20} 1.5285,[564] d_4^{20} 1.1613.[564]

$PhP(OCH_2CH\overset{O}{-}CH_2)_2$. 12b.[642,699] $b_{0.007}$ 129-130°,[642] $b_{0.5}$ 99°,[699] n_D^{20} 1.5213,[642] n_D 1.5579,[699] d_4^{20} 1.2042.[642]

$C_{12}H_{16}Cl_3O_2P$

$PhP(OEt)OCMe_2CCl_3$. 36.[369] $b_{0.18}$ 123°,[369] n_D^{20} 1.5352,[369] d_4^{20} 1.2772.[369]

$C_{12}H_{17}Cl_2O_2P$

$PhP(OCH_2CHClMe)_2$. 18.[556] n_D^{25} 1.5308,[556] d^{35} 1.080.[556]

$C_{12}H_{17}O_2P$

$PhCH=CHP(OEt)_2$. 12b.[379] $b_{0.03}$ 96-98°,[379] n_D^{20} 1.5482,[379] d_4^{20} 1.0322,[379] IR.[379]

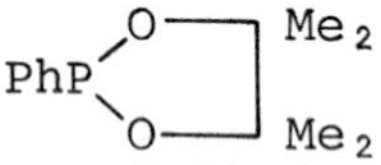

. 12b.[184] m. 103-104°,[184] ^1H 0.98 (CH_3), 7.25-7.81 (C_6H_5).[184]

$p-CH_2=CHC_6H_4P(OEt)_2$. 5a.[293] b_2 96.5-97°,[293] n_D^{20} 1.5398,[293] d_4^{20} 1.0251.[293]

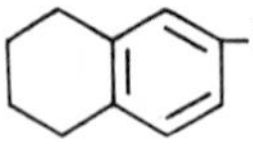
-P(OMe)₂. 12b.[268] $b_{0.1}$ 100°,[268] n_D^{20} 1.5532,[268] d_4^{20} 1.0997,[268] ^{31}P -160.6.[425f]

$C_{12}H_{17}O_3P$

$PhOCH=CHP(OEt)_2$. 12b.[186] $b_{2.5}$ 107-108°,[186] n_D^{20} 1.5228,[186] d_4^{20} 1.0635.[186]

$C_{12}H_{18}BrO_2P$

$p-BrC_6H_4P(OPr)_2$. 12b.[334] b_4 135-135.5°,[334] n_D^{20} 1.5327,[334]

d_4^{20} 1.2580.[334]

$C_{12}H_{18}ClO_2P$

p-$ClC_6H_4P(OPr)_2$. 12b.[315] b_{12} 153°,[315] n_D^{20} 1.5179,[315] d_0^{20} 1.0990.[315]
p-$ClC_6H_4P(OPr\text{-}i)_2$. 12b.[315] b_{11} 138.5-139°,[315] n_D^{20} 1.5108,[315] d_0^{20} 1.0880.[315]

$C_{12}H_{19}O_2P$

$PhP(OPr)_2$. 5a.[301] 12b.[36,215,309] b_1 73-74°,[301] b_{10-13} 132.5-133.5°,[309] b_{15} 137°,[36] n_D^0 1.4939,[309] n_D^{20} 1.5026,[370] 1.5072,[301] n_D^{25} 1.4939,[36] d_0^0 1.0123,[36,309] d_4^{20} 0.9994,[370] 1.0000,[301] 1.0039,[215] d_0^{23} 0.9925,[36] IR,[215] ^1H 0.88 (CH_3), 1.49 ($\underline{CH_2}CH_3$), 3.61 (OCH_2), 7.37 (C_6H_5),[717b] magn. rotation.[370]
$PhP(OPr\text{-}i)_2$. 12b.[34] $b_{0.5}$ 110°,[505] b_{10} 121-122°,[34] n_D^{18} 1.5021,[34] n_D^{25} 1.5000,[505] d_0^0 1.0103,[34] d_0^{17} 0.9952,[34] ^{31}P -151.5,[425f] ^1H 1.20 (CH_3), 4.21 (CH),[692] fig.[717c]
PhP(OEt)OBu. 36.[708] $b_{0.2}$ 105-107°,[708] n_D^{20} 1.5039.[708]
p-$EtC_6H_4P(OEt)_2$. 12b.[630] $b_{0.014}$ 74-75°,[630] n_D^{20} 1.5078,[630] d_4^{20} 1.0037.[630]
2,5-$Me_2C_6H_3P(OEt)_2$. 12b.[762] $b_{0.05}$ 75°,[762] n_4^{20} 1.5144,[762] d_4^{20} 1.0139,[762] add'l data.[762]
2,6-$Me_2C_6H_3P(OEt)_2$. 5a.[88] $b_{0.1}$ 72-73°,[88] $b_{0.2}$ 86-87°,[88] n_D^{20} 1.5135,[88] 1.5145,[88] d_4^{20} 1.0157,[88] 1.0185.[88]

$C_{12}H_{21}O_2PS$

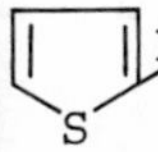
$P(OBu)_2$. 12b.[69] $b_{0.4}$ 96-97°,[69] n_D^{20} 1.5042,[69] d^{25} 1.0355,[69] IR.[69]

$C_{12}H_{23}O_4P$

$EtP(OCH_2$ 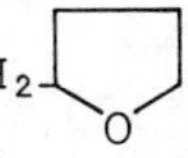 $)_2$. 12b.[633] $b_{0.35}$ 131-134°,[633] n_D^{20} 1.4790,[633] d_4^{20} 1.0839.[633]

$C_{12}H_{25}O_2P$

$MeCH{=}CHCH_2P(OBu)_2$. 12b.[387] $b_{0.06}$ 104-106°,[387] n_D^{20} 1.4448,[387] d_4^{20} 0.8963.[387]
$MeCH{=}CHCH_2P(OBu\text{-}i)_2$. 12b.[387] $b_{0.04}$ 83-85°,[387] n_D^{20} 1.4430,[387] d_4^{20} 0.8880.[387]
c-$C_6H_{11}P(OPr\text{-}i)_2$. ^{31}P -179.6.[425f]

$C_{12}H_{25}O_3P$

EtC(OBu)=CHP(OEt)$_2$. 12b.[567] b_8 120-122°,[567] n_D^{20} 1.4570,[567] d_4^{20} 0.9486.[567]

$C_{12}H_{26}ClO_2P$

Cl(CH$_2$)$_4$P(OBu)$_2$. 12b.[251] $b_{0.05}$ 99°.[251]

$C_{12}H_{27}O_2P$

EtP(OAm)$_2$. 12b.[644] b_2 81-82°,[644] n_D^{20} 1.4390,[644] d_4^{20} 0.8834.[644]

EtP(OEt)OC$_8$H$_{17}$-n. 58a.[260] $b_{0.3}$ 72°,[260] n_D^{25} 1.4387.[260]

BuP(OBu)$_2$. 3a.[658] 5a.[301] 5b.[302] 12b.[628] 58b.[658] $b_{1.5}$ 70.5-71.5°,[302] b_{10} 116.5-118°,[628] b_{15} 125-128°,[658] n_D^{20} 1.4390,[658] 1.4410,[628] 1.4421,[302] d_4^{20} 0.8814,[302] 0.8883,[628] add'l data,[301,658,739] magn. rotation.[739]

i-BuP(OBu-i)$_2$. 12b.[629] $b_{12.5}$ 105.5-107.5°,[629] n_D^{20} 1.4290,[629] d_0^{20} 0.8677.[629]

n-C$_6$H$_{13}$P(OPr-i)$_2$. 12b.[431] b_{15} 112-116°.[431]

n-C$_8$H$_{17}$P(OEt)$_2$. 12c.[237] $b_{0.12}$ 58-60°,[237] n_D^{25} 1.4384,[237] d_4^{20} 0.927.[237]

$C_{13}H_{12}ClO_2P$

ClCH$_2$P(OPh)$_2$. ^{31}P -153.6.[468]

$C_{13}H_{13}O_2P$

MeP(OPh)$_2$. 5a.[310] 12a.[544] 65.[532,546] $b_{0.06}$ 88-89°,[670] b_1 105-106°,[532] b_9 144-148°,[546] b_{10} 212-213°,[310] n_D^{20} 1.5560,[546] 1.5671,[532] 1.5825,[310] d_0^0 1.1904,[310] d_0^{20} 1.1742,[310] d_4^{20} 1.1420,[546] add'l data,[532,544] ^{31}P -178.5,[670] ^{1}H 0.88 (CH$_3$).[670]

$C_{13}H_{16}Cl_3O_2P$

PhP(OMe)O  Cl$_3$C . 36.[326] $b_{0.15}$ 139°,[326] n_D^{20} 1.5583,[326] d_4^{20} 1.3230.[326]

$C_{13}H_{17}O_2P$

p-MeC$_6$H$_4$P(OCH$_2$CH=CH$_2$)$_2$. 12b.[572] $b_{0.3}$ 97-98°,[572] n_D^{20} 1.5288,[572] d_4^{20} 1.0262.[572]

1-C$_9$H$_7$P(OEt)$_2$. (C$_9$H$_7$ = indenyl). 5a.[302] 5b.[302] $b_{1.5}$ 99-99.5°,[302] n_D^{20} 1.5491,[302] d_4^{20} 1.0655.[302]

$C_{13}H_{18}Cl_3O_2P$

$PhP(OPr)OCMe_2CCl_3$. 36.[369] $b_{0.13}$ 125°,[369] n_D^{20} 1.5320,[369] d_4^{20} 1.2547.[369]
$PhP(OPr\text{-}i)OCMe_2CCl_3$. 36.[369] $b_{0.12}$ 105-107°,[369] n_D^{20} 1.5290,[369] d_4^{20} 1.2416.[369]

$C_{13}H_{21}O_2P$

$PhCH_2P(OPr\text{-}i)_2$. 12b.[63] b_{15} 128-131°.[63]
$p\text{-}MeC_6H_4P(OPr)_2$. 12b.[335] b_6 129-130°,[335] n_D^{16} 1.5070,[335] n_D^{20} 1.5045,[313] d_0^0 1.0091,[335] d_0^{16} 0.9937,[335] d_4^{20} 0.9925.[313]
$p\text{-}MeC_6H_4P(OPr\text{-}i)_2$. 12b.[312] b_{12} 129-131°,[312] n_D^{20} 1.5003,[312] d_4^{20} 0.9977.[312]
$p\text{-}(i\text{-}Pr)C_6H_4P(OEt)_2$. 12b.[315] b_{11} 132-134°,[315] n_D^{20} 1.5055,[315] d_0^{20} 0.9886.[315]
$2,4,5\text{-}Me_3C_6H_2P(OEt)_2$. 15a.[443] b_{100} 232-233°,[443] n_D^{15} 1.505,[443] d^{15} 1.048.[443]

$C_{13}H_{23}O_2P$

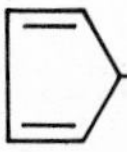—P(OBu)$_2$. 5a.[301,303] 5c.[303] b_1 80-81°,[303] b_2 81.5-83°,[301] n_D^{20} 1.4822,[303] 1.4827,[301] d_4^{20} 0.9610,[303] Raman.[303] Dimer (on standing) n_D^{20} 1.4950,[303] d_4^{20} 1.0110.[303]

$C_{13}H_{25}O_2P$

$MeP(OC_6H_{11})_2$. 59.[539] b_{12} 145-150°,[539] n_D^{20} 1.4480,[539] d_4^{20} 0.9461.[539]

$C_{13}H_{27}O_2P$

$CH_2{=}CHCH_2P(OAm)_2$. 12b.[625] b_9 125-127°,[625] n_D^{20} 1.4502,[625] d_4^{20} 0.8922.[625]
$CH_2{=}CHCH_2P(OAm\text{-}neo)_2$. 12b.[386] $b_{0.02}$ 70-72°,[386] n_D^{20} 1.4375,[386] d_4^{20} 0.8752.[386]
$c\text{-}C_5H_9P(OBu)_2$. 5a.[296] Also from $c\text{-}C_5H_5P(OBu)_2$, H_2 (Ni).[303] b_1 77-78°,[296] $b_{1.5}$ 83-83.5°,[303] n_D^{20} 1.4595,[296] 1.4620,[303] d_4^{20} 0.9284,[296] 0.9314.[303]

$C_{13}H_{27}O_3P$

$PrC(OBu){=}CHP(OEt)_2$. 12b.[567] b_1 97-98°,[567] n_D^{20} 1.4598,[567] d_4^{20} 0.9392.[567]

$C_{13}H_{29}O_2P$

$MeP(OC_6H_{13}\text{-}n)_2$. 12c.[209] 59.[539] b_1 97-98°,[209] b_8 129-133°,[539] n_D^{20} 1.4382,[209] 1.4454,[539] d_4^{20} 0.8793,[209] 0.9404.[539]

$MeP(OEt)OC_{10}H_{21}\text{-}n$. 58a.[260] $b_{0.25}$ 85°,[260] n_D^{25} 1.4414.[260]

$C_{14}H_{15}O_2P$

$EtP(OPh)_2$. 5a.[310,547] 12b.[40] 65.[514] b_1 113-115°,[547] 115-115.5°,[40] b_{12} 223-225°,[310] n_D^{15} 1.5912,[310] n_D^{20} 1.5623,[547] 1.5660,[40] d_0^{15} 1.1923,[310] d_4^{20} 1.1000,[547] 1.1050,[40] IR,[514] add'l data.[307,514]

$C_{14}H_{17}O_2P$

$1\text{-}C_{10}H_7P(OEt)_2$. 12b.[315] b_{10} 167-168°,[315] n_D^{20} 1.5848,[315] d_0^{20} 1.1000.[315]

$C_{14}H_{18}Cl_3O_2P$

PhP(OMe)O  Cl_3C . 36.[327] $b_{0.2}$ 145-146°,[327] m. 63-64°.[327]

PhP(OEt)O  Cl_3C . 36.[326] $b_{0.09}$ 127°,[326] n_D^{20} 1.5510,[326] d_4^{20} 1.2957.[326]

$C_{14}H_{19}O_2P$

$PhP(OCEt{=}CH_2)_2$. 79.[566] $b_{1.5}$ 119-120°,[566] n_D^{20} 1.5210,[566] d_4^{20} 1.0183,[566] IR.[566]

$C_{14}H_{19}O_4P$

$BzOCMe{=}CHP(OEt)_2$. 12b.[564] $b_{0.025}$ 120-121°,[564] n_D^{20} 1.5220,[564] d_4^{20} 1.1248.[564]

$C_{14}H_{19}O_6P$

$PhP(OCH_2CO_2Et)_2$. 58b.[540] $b_{0.01}$ 200-220° (bath),[540] n_D^{20} 1.5073,[540] d_4^{20} 1.2100.[540]

$C_{14}H_{20}Cl_3O_2P$

$PhP(OBu)OCMe_2CCl_3$. 36.[369] $b_{0.16}$ 124-125°,[369] n_D^{20} 1.5281,[369] d_4^{20} 1.2239.[369]

PhP(OBu-i)OCMe$_2$CCl$_3$. 36.[369] $b_{0.10}$ 114-115°,[369] n_D^{20} 1.5270,[369] d_4^{20} 1.2209.[369]
PhP(OBu-s)OCMe$_2$CCl$_3$. 36.[369] $b_{0.15}$ 135-137°,[369] n_D^{20} 1.5290,[369] d_4^{20} 1.2356.[369]

$C_{14}H_{20}NO_2P$

p-Me$_2$NC$_6$H$_4$P(OCH$_2$CH=CH$_2$)$_2$. 12b.[572] $b_{0.03}$ 114-115°,[572] n_D^{20} 1.5645,[572] d_4^{20} 1.0563.[572]

$C_{14}H_{21}O_2P$

PhCH=CHP(OPr)$_2$. 12b.[379] $b_{0.03}$ 108-109°,[379] n_D^{20} 1.5369,[379] d_4^{20} 1.0058,[379] IR.[379]
PhCH=CHP(OPr-i)$_2$. 12b.[379] $b_{0.04}$ 92-94°,[379] n_D^{20} 1.5310,[379] d_4^{20} 0.9990,[379] IR.[379]

$C_{14}H_{22}BrO_2P$

p-BrC$_6$H$_4$P(OBu)$_2$. 12b.[334] b_4 169-170°,[334] n_D^{20} 1.5248,[334] d_4^{20} 1.2110.[334]

$C_{14}H_{22}ClO_2P$

p-ClC$_6$H$_4$P(OBu)$_2$. 12b.[315] b_{14} 172-173°,[315] n_D^{20} 1.5096,[315] d_0^{20} 1.0700.[315]

$C_{14}H_{23}O_2P$

PhP(OBu)$_2$. 5c.[301] 12b.[316] 12e.[79] 65.[407] 71.[533] 40 (A.4).[130] b_1 97.5-98.5°,[301] b_3 116-117°,[533] b_8 139-140.5°,[316] b_{28} 171-172°,[713] n_D^{20} 1.4975,[316] 1.4993,[301] 1.4995,[533] d_4^{20} 0.9769,[301] 0.9770,[316] add'l data,[79,370,407] IR,[117b] ^{31}P -157.3,[407] -178,[425f] ^1H 3.78 (OCH$_2$), 7.46 (C$_6$H$_5$),[407] magn. rotation,[370] available commercially.[27,713]
PhP(OBu-i)$_2$. 12b.[309,362] 15a.[30] $b_{0.8}$ 104°,[362] b_7 134.5°,[30] b_{10-13} 144-145°,[309] n_D^0 1.4658,[309] d_0^0 1.0060.[309]
2,4-Me$_2$C$_6$H$_3$P(OPr-i)$_2$. ^{31}P -146.9.[425g]
2,5-Me$_2$C$_6$H$_3$P(OPr)$_2$. 12b.[762] $b_{0.05}$ 103-105°,[762] n_D^{22} 1.5068,[762] d_4^{22} 0.9874.[762]
2,5-Me$_2$C$_6$H$_3$P(OPr-i)$_2$. 12b.[762] $b_{0.05}$ 100-104°,[762] n_D^{20} 1.5015,[762] d_4^{20} 0.9776,[762] ^{31}P -146.4.[762]
3,4-Me$_2$C$_6$H$_3$P(OPr-i)$_2$. 12b.[270] n_D^{20} 1.5048,[270] d_4^{20} 0.9775.[270]

$C_{14}H_{23}O_4P$

PhP(OCH$_2$CH$_2$OEt)$_2$. 12b.[482] $b_{0.02}$ 142-145°,[482] n_D^{20} 1.5000,[482] d_4^{20} 1.0656.[482]

$C_{14}H_{27}O_2P$

$EtP(OC_6H_{11}\text{-}c)_2$. 12b.[633] $b_{0.2}$ 128°,[633] n_D^{20} 1.4840,[633] d_4^{20} 0.9887.[633]

$C_{14}H_{29}NO_2P$

$PhP(OEt)OCH_2CH_2NEt_2$. 36.[681] $b_{0.01}$ 90°.[681]

$C_{14}H_{29}O_2P$

$c\text{-}C_6H_{11}P(OBu)_2$. n_D^{20} 1.4617,[370] d_4^{20} 0.9440,[370] ^{31}P -184.2,[425f] magn. rotation.[370]

$C_{14}H_{31}O_2P$

$EtP(OC_6H_{13}\text{-}n)_2$. 12b.[40] b_2 100-101°,[40] n_D^{20} 1.4435,[40] d_4^{20} 0.8875.[40]
$EtP(OEt)OC_{10}H_{21}\text{-}n$. 58a.[260] $b_{0.06}$ 100°,[260] n_D^{25} 1.4426.[260]
$n\text{-}C_8H_{17}P(OPr\text{-}i)_2$. 12b.[431] b_4 107-112°.[431]

$C_{14}H_{32}NO_2P$

$Et_2NCH_2CH_2P(OBu)_2$. From $CH_2{=}CHP(OBu)_2$, Et_2NH.[306] $b_{1.5}$ 93.5-95.0°,[306] n_D^{20} 1.4520,[306] d_4^{20} 0.9003.[306]

$C_{15}H_{11}Cl_6O_2P$

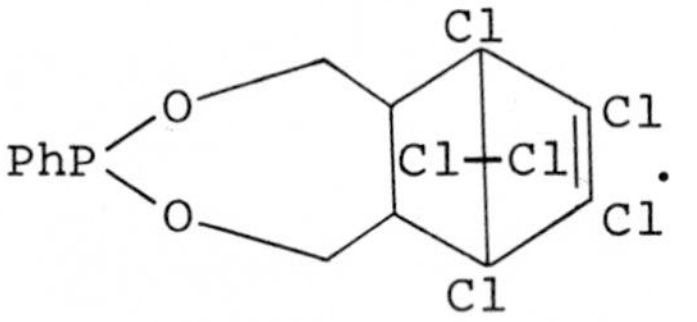

. 12b.[56] m. 150°,[56] IR,[56] ^1H.[56]

$C_{15}H_{17}O_2P$

$i\text{-}PrP(OPh)_2$. 5a.[310] b_{11} 212-214°,[310] n_D^{17} 1.5782,[310] d_0^{17} 1.1625.[310]

$C_{15}H_{20}Cl_3O_2P$

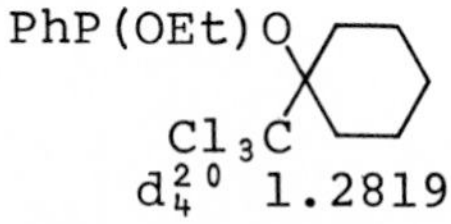

. 36.[327] $b_{0.18}$ 158-159°,[327] n_D^{20} 1.5489,[327] d_4^{20} 1.2819.[327]

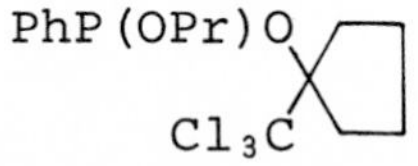

. 36.[326] $b_{0.1}$ 133-134°,[326] n_D^{20} 1.5460,[326]

d_4^{20} 1.2667.[326]

PhP(OPr-i)O 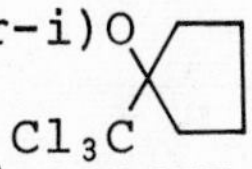

. 36.[326] $b_{0.12}$ 136-137°,[326] n_D^{20} 1.5445,[326] d_4^{20} 1.2655.[326]

$C_{15}H_{22}Cl_3O_2P$

$PhP(OAm\text{-}i)OCMe_2CCl_3$. 36.[369] $b_{0.16}$ 130-132°,[369] n_D^{20} 1.5225,[369] d_4^{20} 1.2012.[369]

$C_{15}H_{25}O_2P$

$PhCH_2P(OBu)_2$. 5a.[301] b_1 103-104°,[301] n_D^{20} 1.4972,[301] d_4^{20} 0.9742.[301]
$PhCH_2P(OBu\text{-}i)_2$. 12c.[209] b_3 137-138°,[209] n_D^{20} 1.5080,[209] d_4^{20} 0.9998.[209]
$p\text{-}MeC_6H_4P(OBu)_2$. 12b.[335] b_{13} 170-171°,[335] n_D^{16} 1.5024,[335] n_D^{20} 1.5003,[313] d_0^0 0.9899,[335] d_0^{16} 0.9776,[335] d_4^{20} 0.9753.[313]
$p\text{-}MeC_6H_4P(OBu\text{-}i)_2$. 12b.[335] b_{13} 155-156°,[335] n_D^{16} 1.4987,[335] d_0^0 0.9807,[335] d_0^{16} 0.9667.[335]
$p\text{-}(i\text{-}Pr)C_6H_4P(OPr)_2$. 12b.[315] b_{12} 163-165°,[315] n_D^{20} 1.5009,[315] d_0^{20} 0.9802.[315]

$C_{15}H_{27}O_2P$

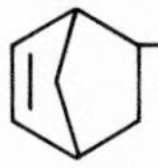
$-P(OBu)_2$. From $CH_2=CHP(OBu)_2$, cyclopentadiene.[293]

$b_{1.5}$ 99.7-101°,[293] n_D^{20} 1.4792,[293] d_4^{20} 0.9696.[293]

$C_{15}H_{27}O_3PSi$

$Me_2Si(OEt)CH_2C_6H_4P(OEt)_2$. 12b.[561] $b_{0.5}$ 140°,[561] n_D^{20} 1.4990,[561] d_4^{20} 1.0009,[561] IR.[561]

$C_{15}H_{28}NO_7P$

$MeP(OR)OCH_2CH_2NH_2$ (R = 1,2:5,6-di-O-isopropylideneglucos-3-yl). 71.[499] $b_{0.5}$ 152-154°,[499] n_D^{20} 1.4815,[499] TLC.[499]
$MeP(OR)OCH_2CH_2NH_2$ (R = 1,2:3,4-di-O-isopropylidenegalactos-6-yl). 71.[499] $b_{0.5}$ 145-148°,[499] n_D^{20} 1.4818,[499] TLC.[499]

$C_{15}H_{31}O_2P$

$CH_2=CHCH_2P(OC_6H_{13}\text{-}n)_2$. 5a.[623] $b_{0.25}$ 110-112°,[623] n_D^{20}

1.4500,[623] d_4^{20} 0.8889.[623]

$C_{15}H_{31}O_3P$

MeC(OBu)=CHP(OBu)$_2$. 12b.[567] $b_{1.5}$ 125-127°,[567] n_D^{20} 1.4617,[567] d_4^{20} 0.9321,[567] IR.[567]

$C_{15}H_{32}NO_2P$

NCH$_2$CH$_2$P(OBu)$_2$. From CH$_2$=CHP(OBu)$_2$, piperidine.[306] b_1 118-118.5°,[306] $b_{1.5}$ 122-122.5°,[306] n_D^{20} 1.4694,[306] 1.4710,[306] d_4^{20} 0.9378,[306] 0.9399.[306]

$C_{15}H_{33}O_2P$

AmP(OAm)$_2$. 12b.[739] $b_{0.5}$ 87°,[739] n_D^{20} 1.442,[739] d_4^{20} 0.8791,[739] magn. rotation,[739] dipole moment 2.67 D.[434]

$C_{16}H_{15}O_4P$

PhP(OCH$_2$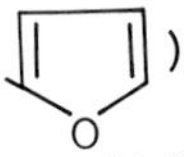)$_2$. 58b.[540] $b_{0.0001}$ 180-200° (bath),[540] n_D^{20} 1.5220,[540] d_4^{20} 1.1525.[540]

$C_{16}H_{16}Cl_3O_2P$

PhP(OPh)OCMe$_2$CCl$_3$. 36.[369] m. 44-46°.[369]

$C_{16}H_{19}O_2P$

EtP(OCH$_2$Ph)$_2$. 12b.[644] $b_{1.5}$ 140-141°,[644] n_D^{20} 1.5499,[644] d_4^{20} 1.0761.[644]

EtP(OC$_6$H$_4$Me-p)$_2$. 12b.[273] $b_{0.1-0.3}$ 130-209°,[273] n_D^{28} 1.5583.[273]

BuP(OPh)$_2$. 3a.[250] 5a.[310] b_1 135-140°,[250] b_8 225-226°,[310] n_D^{15} 1.5878,[310] d_0^0 1.1870,[310] d_0^{15} 1.1744.[310]

$C_{16}H_{21}O_2P$

1-C$_{10}$H$_7$P(OPr)$_2$. 12b.[315] b_{12} 188-189.5°,[315] n_D^{20} 1.5672,[315] d_0^{20} 1.0630.[315]

1-C$_{10}$H$_7$P(OPr-i)$_2$. 12b.[315] b_{12} 176-178°,[315] n_D^{20} 1.5648,[315] d_0^{20} 1.0671.[315]

$C_{16}H_{22}Cl_3O_2P$

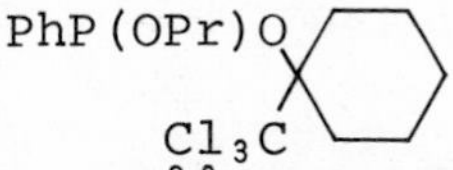

. 36.[327] $b_{0.17}$ 159-160°,[327] n_D^{20} 1.5495,[327] d_4^{20} 1.2598.[327]

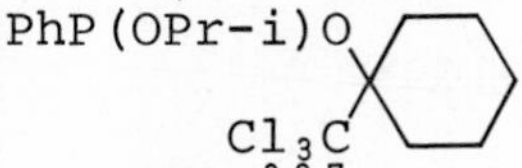

. 36.[327] $b_{0.22}$ 153-154°,[327] m. 57-58°.[327]

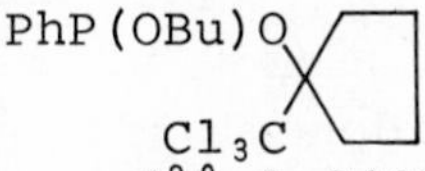

. 36.[326] $b_{0.14}$ 149-151°,[326] n_D^{20} 1.5412,[326] d_4^{20} 1.2401.[326]

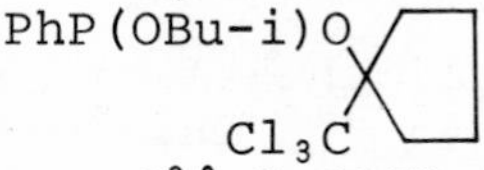

. 36.[326] $b_{0.1}$ 134-135°,[326] n_D^{20} 1.5405,[326] d_4^{20} 1.2379.[326]

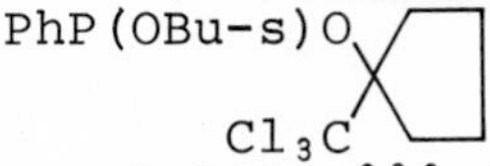

. 36.[326] $b_{0.13}$ 138-140°,[326] n_D^{20} 1.5395,[326] d_4^{20} 1.2473.[326]

$C_{16}H_{23}O_4P$

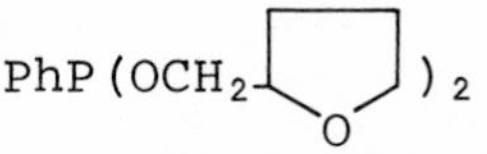

. 58b.[540] $b_{0.0001}$ 160-180° (bath),[540] n_D^{20} 1.5290,[540] d_4^{20} 1.1505.[540]

$C_{16}H_{25}O_2P$

PhCH=CHP(OBu)$_2$. 12b.[379] $b_{0.06}$ 123-125°,[379] n_D^{20} 1.5287,[379] d_4^{20} 0.9875,[379] IR.[379]

PhCH=CHP(OBu-i)$_2$. 12b.[379] $b_{0.03}$ 108-109°,[379] n_D^{20} 1.5245,[379] d_4^{20} 0.9780,[379] IR.[379]

$C_{16}H_{25}O_2P$

2-$C_{10}H_{11}$P(OPr)$_2$ ($C_{10}H_{11}$ = 5,6,7,8-tetrahydronaphthyl). n_D^{20} 1.5271,[653f] d_4^{20} 1.029,[653f] IR.[652d,653f]

2-$C_{10}H_{11}$P(OPr-i)$_2$. 12b.[268] b_1 165-170°,[268] n_D^{20} 1.5213,[268] d_4^{20} 1.0169.[268]

$C_{16}H_{26}BrO_2P$

p-BrC_6H_4P(OAm)$_2$. 12b.[334] b_3 168.5-169.5°,[334] n_D^{20} 1.5191,[334] d_4^{20} 1.1718.[334]

$C_{16}H_{29}O_3PSi$

$Me_2Si(OEt)CH_2CH_2C_6H_4P(OEt)_2$. 12b.[561] $b_{0.5}$ 137-140°,[561] n_D^{20} 1.4955,[561] d_4^{20} 0.9962,[561] IR.[561]

$C_{16}H_{29}O_4PSi$

$MeSi(OEt)_2CH_2C_6H_4P(OEt)_2$. 12b.[561] $b_{0.5}$ 138-140°,[561] n_D^{20} 1.4922,[561] d_4^{20} 1.0512,[561] IR.[561]

$C_{16}H_{30}NO_7P$

$MeP(OR)OCH_2CH_2NHMe$ (R = 1,2:5,6-di-O-isopropylideneglucos-3-yl). 71.[499] $b_{0.5}$ 160-163°,[499] n_D^{20} 1.4792,[499] IR,[499] TLC.[499]

$MeP(OR)OCH_2CH_2NHMe$ (R = 1,2:3,4-di-O-isopropylidenegalactos-6-yl). 71.[499] $b_{0.5}$ 155-157°,[499] n_D^{20} 1.4803,[499] IR,[499] TLC.[499]

$C_{17}H_{19}Cl_2O_4P$

$MeP(OCH_2CH_2OC_6H_4Cl\text{-}p)_2$. 12e.[79] n_D^{20} 1.5538,[79] d_4^{20} 1.2899.[79]

$C_{17}H_{19}O_2P$

$9\text{-}C_{13}H_9P(OEt)_2$ ($C_{13}H_9$ = fluorenyl). 5b.[302] b_2 148.5-149°,[302] m. 67.5-70°,[302] subl. 2 mm.[302]

$C_{17}H_{24}Cl_3O_2P$

PhP(OBu)O 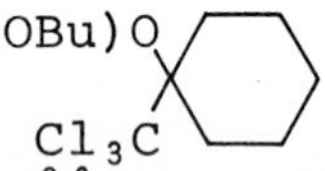 . 36.[327] $b_{0.18}$ 162-163°,[327] n_D^{20} 1.5410,[327] d_4^{20} 1.2377.[327]

PhP(OBu-i)O 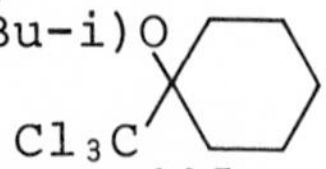 . 36.[327] $b_{0.2}$ 156-158°,[327] m. 56-58.5°.[327]

PhP(OBu-s)O 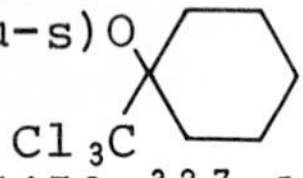 . 36.[327] $b_{0.2}$ 151-152°,[327] n_D^{20} 1.5450,[327] d_4^{20} 1.2583.[327]

$C_{17}H_{29}O_2P$

$p\text{-}MeC_6H_4P(OAm)_2$. 12b.[312] b_4 159°,[313] b_{12} 181-184°,[312] n_D^{20} 1.4943,[312] 1.4962,[313] d_4^{20} 0.9611,[313] 0.9692.[312]

$C_{17}H_{31}O_4PSi$

$MeSi(OEt)_2CH_2CH_2C_6H_4P(OEt)_2$. 12b.[561] $b_{0.5}$ 138-140°,[561]

n_D^{20} 1.4922,[561] d_4^{20} 1.0512.[561]

$C_{17}H_{31}O_5PSi$

$(EtO)_3SiCH_2C_6H_4P(OEt)_2$. 12b.[561] $b_{0.5}$ 161°,[561] n_D^{20} 1.4932,[561] d_4^{20} 1.0620.[561]

$C_{17}H_{32}NO_7P$

$EtP(OR)OCH_2CH_2NHMe$ (R = 1,2:5,6-di-O-isopropylideneglucos-3-yl). 71.[499] $b_{0.5}$ 172-177°,[499] n_D^{20} 1.4774,[499] TLC.[499]

$EtP(OR)OCH_2CH_2NHMe$ (R = 1,2:3,4-di-O-isopropylidenegalactos-6-yl). 71.[499] $b_{0.5}$ 170-175°,[499] n_D^{20} 1.4770,[499] TLC.[499]

$C_{17}H_{35}N_2O_2P$

MeP(O 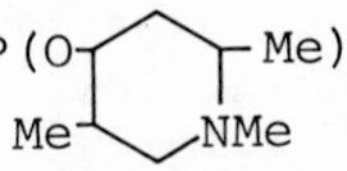

)$_2$. 65.[534] $b_{0.06}$ 82-84°,[534] n_D^{20} 1.4834,[534] d_4^{20} 0.9842.[534]

$C_{17}H_{35}O_3P$

$PrC(OBu)=CHP(OBu)_2$. 12b.[567] b_2 148-150°,[567] n_D^{20} 1.4612,[567] d_4^{20} 0.9229.[567]

$C_{17}H_{37}O_2P$

$MeP(OC_8H_{17}\text{-}n)_2$. 58a.[260] 59.[539] $b_{0.3}$ 125-130°,[260] b_1 128-135°,[539] n_D^{20} 1.4582,[539] n_D^{25} 1.4460,[260] d_4^{20} 0.9333,[539] TLC.[488]

$MeP(OCH_2CHEtBu)_2$. 65.[546] $b_{0.7}$ 150-152°,[546] n_D^{20} 1.4365,[546] d_4^{20} 0.9142.[546]

$MeP(OC_8H_{17}\text{-}s)_2$. 59.[539] b_1 110-115°,[539] n_D^{20} 1.4618,[539] d_4^{20} 0.9386.[539]

$C_{18}H_{10}F_5O_2P$

$C_6F_5P(OPh)_2$. 12b.[177] $b_{0.1}$ 145°,[177] IR.[177]

$C_{18}H_{13}Cl_2O_2P$

$PhP(OC_6H_4Cl\text{-}p)_2$. 12a.[433] $b_{3.1}$ 223.5-226°.[433]

$C_{18}H_{13}O_2P$

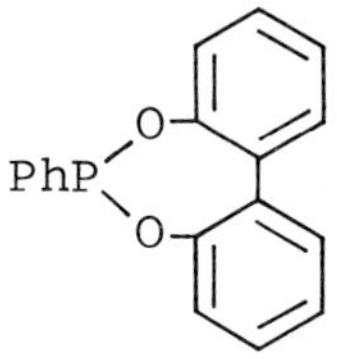

. 32.[694] b_1 215°,[694] solid at room temp.[694]

$C_{18}H_{14}BrO_2P$

p-$BrC_6H_4P(OPh)_2$. 12b.[334] b_5 218-220°,[334] n_D^{20} 1.6283,[334] d_4^{20} 1.3798.[334]

$C_{18}H_{15}O_2P$

$PhP(OPh)_2$. 5a.[310] 12a.[35,544] 12b.[35] 65.[533] $b_{0.5}$ 160°,[117c] b_1 175-177°,[544] b_{10} 223-224°,[310] b_{14} 230°,[35] n_D^{20} 1.6025,[310] 1.6109,[544] $n_D^{24.5}$ 1.6104,[35] n_D^{25} 1.6080,[117c] d_0^0 1.1784,[35] 1.1960,[310] d_0^{20} 1.1810,[310] $d_0^{24.5}$ 1.1517,[35] add'l data,[35,307,533] IR,[117c] ^{31}P -164.9.[468]

$C_{18}H_{18}Cl_3O_2P$

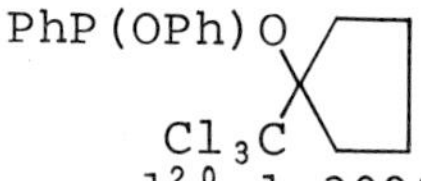

. 36.[326] $b_{0.12}$ 161-162°,[326] n_D^{20} 1.5855,[326] d_4^{20} 1.3093.[326]

$C_{18}H_{21}O_2P$

c-$C_6H_{11}P(OPh)_2$. 5a.[310] b_{10} 233-235°,[310] n_D^{20} 1.5705,[310] d_0^0 1.1499,[310] d_0^{20} 1.1360.[310]

$C_{18}H_{25}O_2P$

1-$C_{10}H_7P(OBu)_2$. 12b.[315] b_4 175-176°,[315] n_D^{20} 1.5575,[315] d_0^{20} 1.0401.[315]

$C_{18}H_{29}O_2P$

$PhCH{=}CHP(OAm)_2$. 12b.[379] $b_{0.03}$ 138-139°,[379] n_D^{20} 1.5225,[379] d_4^{20} 0.9757,[379] IR.[379]
$PhCH{=}CHP(OAm\text{-}i)_2$. 12b.[379] $b_{0.04}$ 128-129°,[379] n_D^{20} 1.5204,[379] d_4^{20} 0.9753,[379] IR.[379]

$C_{18}H_{31}O_2P$

$PhP(OC_6H_{13}\text{-}n)_2$. 12c.[237] $b_{0.3}$ 123-126°,[237] n_D^{25} 1.4880,[237] d_4^{20} 0.954.[237]

$C_{18}H_{33}O_5PSi$

$(EtO)_3SiCH_2CH_2C_6H_4P(OEt)_2$. 12b.[561] $b_{0.5}$ 167°,[561] n_D^{20} 1.4920,[561] d_4^{20} 1.0599,[561] IR.[561]

$C_{18}H_{33}O_9P$

$MeP(OR)OCH_2CHOMeCH_2OMe$ (R = 1,2:5,6-di-O-isopropylideneglucos-3-yl). 71.[494] $b_{0.01}$ 130-140°,[494] n_D^{20} 1.4697,[494] TLC.[494]

$C_{18}H_{37}O_2P$

$c\text{-}C_6H_{11}P(OC_6H_{13}\text{-}n)_2$. 65.[491] b_4 156-157°,[491] n_D^{20} 1.4620,[491] d_4^{20} 0.9190.[491]

$C_{18}H_{39}O_2P$

$EtP(OC_8H_{17}\text{-}n)_2$. 12b.[40,273] 12c.[237] 58a.[260] $b_{0.05}$ 123°,[260] 180°,[273] $b_{0.15}$ 119-124°,[237] b_3 142-144°,[40] n_D^{20} 1.4489,[40] n_D^{25} 1.4459,[237] 1.4475,[260] n_D^{28} 1.4443,[273] d_4^{20} 0.878,[237] 0.8809.[40]

$n\text{-}C_{12}H_{25}P(OPr\text{-}i)_2$. 12b.[431] b_8 165-170°.[431]

$C_{19}H_{20}Cl_3O_2P$

PhP(OPh)O 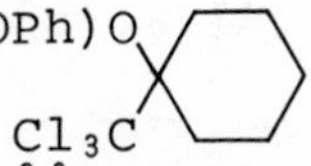 . 36.[327] $b_{0.25}$ 172-173°,[327] n_D^{20} 1.5745,[327] d_4^{20} 1.2863.[327]

$C_{19}H_{26}Cl_3O_2P$

$PhP(OC_6H_{11}\text{-}c)O$ 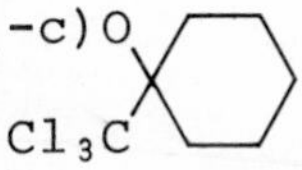 . 36.[327] m. 69-72°.[327]

$C_{19}H_{33}O_2P$

$p\text{-}MeC_6H_4P(OC_6H_{13}\text{-}n)_2$. 12b.[312] b_{12} 194-195°,[312] n_D^{20} 1.4929,[312] d_4^{20} 0.9596.[312]

$C_{19}H_{36}NO_7P$

$MeP(OR)OCH_2CH_2NEt_2$ (R = 1,2:5,6-di-O-isopropylideneglucos-3-yl). 71.[494] b_1 143-146°,[494] n_D^{20} 1.4625,[494] TLC.[494]

$C_{20}H_{19}O_2P$

$PhP(OCH_2Ph)_2$. 12b.[23] $b_{0.1}$ 65-70° (mol. distillation).[23]

$PhP(OC_6H_4Me\text{-}o)_2$. 12a.[35] 12b.[35] b_{10} 229-231°,[35] n_D^{22}

1.6022,[35] d_0^0 1.1452,[35] d_0^{22} 1.1290.[35]

$C_{20}H_{33}O_2P$

$PhCH{=}CHP(OC_6H_{13}\text{-}n)_2$. 12b.[379] $b_{0.02}$ 159-160°,[379] n_D^{20} 1.5150,[379] d_4^{20} 0.9611,[379] IR.[379]

$C_{20}H_{35}O_2P$

$PhP(OC_7H_{15}\text{-}n)_2$. 12e.[79] n_D^{20} 1.4900,[79] d_4^{20} 0.9480.[79]

$C_{21}H_{21}O_2P$

$2,4,5\text{-}Me_3C_6H_2P(OPh)_2$. 15a.[443] b_{40} 283° dec.,[443] m. 59°,[443] n_D^{15} 1.5085,[443] d^{15} 1.144.[443]

$C_{21}H_{45}O_2P$

$MeP(OC_{10}H_{21}\text{-}n)_2$. 58a.[260] 58b.[547] $b_{0.075}$ 138°,[260] b_1 192-195°,[547] n_D^{20} 1.4523,[547] n_D^{25} 1.4493,[260] d_4^{20} 0.8698.[547]
$n\text{-}C_7H_{15}P(OC_7H_{15}\text{-}n)_2$. 12b.[739] $b_{0.8}$ 155°,[739] n_D^{20} 1.451,[739] d_4^{20} 0.8724,[739] magn. rotation.[739]

$C_{22}H_{17}O_2P$

$1\text{-}C_{10}H_7P(OPh)_2$. 5a.[310] b_{10} 245-247°,[310] n_D^{15} 1.6178,[310] d_0^0 1.2055,[310] d_0^{15} 1.1912.[310]

$C_{22}H_{20}Cl_3O_2P$

$PhP(OC_{10}H_7\text{-}1)O$ 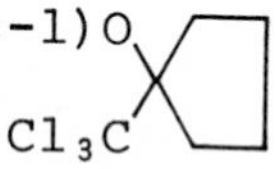 . 36.[326] n_D^{20} 1.6248,[326] d_4^{20} 1.3090.[326]

$C_{22}H_{21}Cl_2O_4P$

$PhP(OCH_2CH_2OC_6H_4Cl\text{-}p)_2$. 12e.[79] n_D^{20} 1.5886,[79] d_4^{20} 1.2985.[79]

$C_{22}H_{25}N_2O_2P$

$PhP(OCH_2CH_2NHPh)_2$. 65.[462] 71.[462] $b_{0.001}$ 145-147°,[462] $b_{0.03}$ 124-126°,[462] IR.[462]
$PhP(OC_6H_4NMe_2\text{-}m)_2$. 12b.[181] $b_{2.3\mu}$ 150°,[181] n_D^{25} 1.6280.[181]

$C_{22}H_{37}O_2P$

$PhCH{=}CHP(OC_7H_{15}\text{-}n)_2$. 12b.[379] $b_{0.04}$ 170-171°,[379] n_D^{20} 1.5105,[379] d_4^{20} 0.9528,[379] IR.[379]

$C_{22}H_{39}O_2P$

$PhP(OC_8H_{17}\text{-}n)_2$. 58a.[207,489] 58b.[547] b_2 164-167°,[489] b_3 162-164°,[207] 178-181°,[547] n_D^{20} 1.4925,[489] 1.4930,[207] 1.5010,[547] d_4^{20} 0.9434,[489] 0.9626,[547] 1.0129.[207]
$PhP(OCH_2CHEtBu)_2$. 58a.[207] $b_{0.5}$ 148-150°,[117a] $b_{3-3.5}$ 172-173°,[207] n_D^{20} 1.4908,[207] n_D^{25} 1.48995,[117a] d_4^{20} 0.9465,[207] IR.[117a]

$C_{22}H_{45}O_2P$

$c\text{-}C_6H_{11}P(OC_8H_{17}\text{-}n)_2$. 65.[491] b_3 150-151°,[491] n_D^{20} 1.4660,[491] d_4^{20} 0.9190.[491]

$C_{22}H_{47}O_2P$

$EtP(OC_{10}H_{21}\text{-}n)_2$. 58a.[260] 58b.[547] $b_{0.05}$ 150°,[260] b_3 195-200°,[547] n_D^{20} 1.4558,[547] n_D^{25} 1.4509,[260] d_4^{20} 0.8764.[547]

$C_{23}H_{22}Cl_3O_2P$

$PhP(OC_{10}H_7\text{-}1)O$ 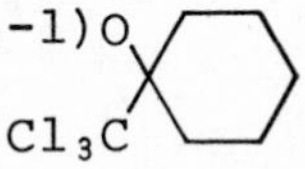

. 36.[327] m. 89-91°,[327] n_D^{20} 1.6250.[327]

$C_{24}H_{43}O_2P$

$PhP(OC_9H_{19}\text{-}n)_2$. 58a.[207] $b_{2-2.5}$ 198-200°,[207] n_D^{20} 1.4890,[207] d_4^{20} 0.9498.[207]

$C_{25}H_{41}O_{12}P$

$MeP(OR)_2$ (R = 1,2:5,6-di-O-isopropylideneglucos-3-yl). 65.[496] m. 46-49°,[496] TLC.[496]
$MeP(OR')_2$ (R = 1,2:3,4-di-O-isopropylidenegalactos-6-yl). 65.[496] m. 37-41°,[496] TLC.[496]
$MeP(OR)OR'$ (R and R' as in the two preceding compounds). 71.[496] m. 40-43°,[496] TLC.[496]

$C_{26}H_{47}O_2P$

$PhP(OC_{10}H_{21}\text{-}n)_2$. 58a.[207] b_{2-3} 204-212°,[207] n_D^{20} 1.4860,[207] d_4^{20} 0.9381.[207]

$C_{28}H_{35}O_2P$

$PhP(OC_6H_4[Am\text{-}t]\text{-}p)_2$. 12a.[433] $b_{3.0}$ 225-228°.[433]

$C_{30}H_{23}O_4P$

$PhP(OC_6H_4[OPh]\text{-}m)_2$. 12b.[108] $b_{0.45}$ 260-275°.[108]

Compounds with Two Phosphorus Atoms

$C_6H_{22}B_{10}O_4P_2$

$m\text{-}B_{10}H_{10}C_2[P(OMe)_2]_2$. 12a.[20] n_D^{26} 1.5420.[20]

$C_8H_{20}O_4P_2$

$(MeO)_2P(CH_2)_4P(OMe)_2$. 3a.[659] 5a.[659] b_1 87-89°,[659] n_D^{22} 1.4693.[659]

$C_{10}H_{16}O_4P_2$

$p\text{-}C_6H_4[P(OMe)_2]_2$. 12b.[55] $b_{0.1-0.2}$ 94-97°,[55] n_D^{20} 1.5479,[55] IR.[55]

$C_{10}H_{24}O_4P_2$

$(MeO)_2(CH_2)_6P(OMe)_2$. 3a.[659] $b_{1.5}$ 128-132°,[659] n_D^{21} 1.4704.[659]

$C_{12}H_{28}O_4P_2$

$(EtO)_2P(CH_2)_4P(OEt)_2$. 5a.[659] b_2 135-137°,[659] n_D^{20} 1.4604.[659]

$C_{13}H_{30}O_4P_2$

$(EtO)_2P(CH_2)_5P(OEt)_2$. 5a.[659] b_3 154-160°,[659] n_D^{21} 1.4568.[659]

$C_{14}H_{24}O_4P_2$

$p\text{-}C_6H_4[P(OEt)_2]_2$. 12b.[60] b_2 146-148°,[60] n_D^{25} 1.5181,[60] d_4^{25} 1.1207.[60]

$C_{14}H_{32}O_4P_2$

$(EtO)_2P(CH_2)_6P(OEt)_2$. 3a.[659] 5a.[659] $b_{0.02}$ 115-120°,[659] b_2 160-163°,[659] n_D^{15} 1.4620,[659] n_D^{20} 1.4590.[659]

$C_{16}H_{36}O_4P_2$

$(PrO)_2P(CH_2)_4P(OPr)_2$. 58a.[659] b_1 137-140°,[659] n_D^{16} 1.4630.[659]
$(i\text{-}PrO)_2P(CH_2)_4P(OPr\text{-}i)_2$. 3a.[659] 58a.[659] b_1 132-135°,[659] n_D^{21} 1.4515.[659]

$C_{18}H_{32}O_4P_2$

p-C_6H_4[P(OPr)$_2$]$_2$. 12b.[60] b_2 165-167°,[60] n_D^{25} 1.5048,[60] d_4^{25} 1.0464.[60]

$C_{18}H_{40}O_4P_2$

(PrO)$_2$P(CH_2)$_6$P(OPr)$_2$. 58a.[659] $b_{0.5}$ 155-160°,[659] n_D^{22} 1.4595.[659]
(i-PrO)$_2$P(CH_2)$_6$P(OPr-i)$_2$. 58a.[659] b_2 155-157°,[659] n_D^{22} 1.4515.[659]

$C_{20}H_{44}O_4P_2$

(BuO)$_2$P(CH_2)$_4$P(OBu)$_2$. 58a.[659] b_1 188-193°,[659] n_D^{20} 1.4610.[659]

$C_{22}H_{40}O_4P_2$

p-C_6H_4[P(OBu)$_2$]$_2$. 12b.[60] $b_{0.02}$ 170-171°,[60] n_D^{25} 1.4981,[60] d_4^{25} 1.0195.[60]

$C_{22}H_{48}O_4P_2$

(BuO)$_2$P(CH_2)$_6$P(OBu)$_2$. 3a.[659] $b_{0.1}$ 192-195°,[659] n_D^{25} 1.4575.[659]

$C_{29}H_{30}O_4P_2$

(PhO)$_2$P(CH_2)$_5$P(OPh)$_2$. 5a.[12] $b_{0.02}$ 158-160°.[12]

Compounds with Three Phosphorus Atoms

$C_{20}H_{33}Cl_4O_8P_3$

PhP[OCHEtP(O)(OCH_2CH_2Cl)$_2$]$_2$. 31.[74] n_D^{25} 1.5170.[74]

$C_{26}H_{33}O_6P_3$

PhP[OCHMeP(O)(OEt)Ph]$_2$. 31.[74] n_D^{25} 1.5583.[74]

$C_{44}H_{69}O_8P_3$

PhP[OCHPhP(O)(OC_6H_{13}-n)$_2$]$_2$. 31.[74] n_D^{25} 1.5068.[74]

D.3.5. Phosphinothioic Esters

C_3H_9OPS

MePH(S)OEt. 64.[657] n_D^{20} 1.4908,[657] d_4^{20} 1.0695,[657] IR.[657]

$C_4H_{11}OPS$

MePH(S)OPr. 51.[166] 64.[526] 70.[535] b_5 70-71°,[526] b_6 69-71°,[535] n_D^{20} 1.4840,[535] 1.4864,[526] d_4^{20} 1.0348.[535]
MePH(S)OPr-i. ^{31}P -61.3.[178]
EtPH(S)OEt. 35b.[457] 64.[392] b_3 58°,[392] b_{19} 84°,[457] n_D^{20} 1.4894,[457] n_D^{24} 1.4903.[392]

$C_5H_{13}OPS$

MePH(S)OBu. 64.[526] 70.[535] b_9 96-98°,[526] b_{10} 92-93°,[535] n_D^{20} 1.4840,[526,535] 1.4885,[657] d_4^{20} 0.9932,[535] 1.0147,[657] IR.[657]
EtPH(S)OPr. 35b.[457] b_{16} 96-97°,[457] n_D^{25} 1.4870.[457]
EtPH(S)OPr-i. 35b.[457] b_{12} 79°,[457] n_D^{25} 1.4742.[457]

$C_6H_{15}OPS$

EtPH(S)OBu. 35b.[452,457] b_{15} 107-108°,[452] 108°,[457] n_D^{20} 1.4840,[452] n_D^{25} 1.4830,[457] d_4^{20} 0.9030.[452]

$C_7H_{17}OPS$

BuPH(S)OPr. 64.[6] b_2 118-120°,[6] n_D^{20} 1.4780,[6] d_4^{20} 0.9705.[6]

$C_8H_{11}OPS$

PhPH(S)OEt. 35a.[270] 35b.[457] 64.[526] 70.[405] $b_{0.1}$ 105°,[682] $b_{0.2}$ 120-127°,[405] $b_{0.4}$ 82-83°,[457] n_D^{20} 1.5700,[270] 1.5720,[526] 1.5732,[405] n_D^{25} 1.5432,[457] d_4^{20} 1.1006,[526] 1.1392,[270] add'l data,[657] IR,[270,405,657,761] ^{31}P -64.4,[425c,657] -65.3.[405]

$C_9H_{13}OPS$

PhPH(S)OPr. 51.[464] 64.[526] 70.[535] $b_{0.03}$ 60°,[535] $b_{0.3}$ 99-101°,[464] n_D^{20} 1.5509,[464] 1.5613,[526] 1.5632,[535] d_4^{20} 1.1082.[535]

$C_{10}H_{15}OPS$

PhPH(S)OBu. 70.[535] $b_{0.03}$ 75°,[535] n_D^{20} 1.5530,[535] d_4^{20} 1.0894.[535]

$C_{12}H_{19}OPS$

PhPH(S)OC$_6$H$_{13}$-n. 63.[526] $b_{0.03}$ 150-160° (bath),[526] n_D^{20} 1.5186,[526] d_4^{20} 1.0419.[526]

D.3.6. Phosphonofluoridothious Esters

$C_5H_{12}FPS$

MeP(SBu)F. 23.[151] b_{55} 72-74°,[151] n_D^{20} 1.4745,[151] d_4^{20} 1.0650,[151] ^{31}P -214,[151] ^{19}F (F_2 ref.) 578.[151]

$C_8H_{10}FPS$

PhP(SEt)F. ^{31}P -196.[689]

D.3.7. Phosphonochloridothious Esters

C_2H_6ClPS

MeP(SMe)Cl. 22b.[678] b_{12} 45°.[678]

$C_3H_7Cl_2PS$

MeP(SCH_2CH_2Cl)Cl. 20.[285] b_{15} 102-103°,[285] n_D^{20} 1.5660,[285] d_4^{20} 1.2830.[285]

C_3H_8ClPS

MeP(SEt)Cl. 22a.[691] 22b.[678] $b_{0.8}$ 40-41°,[691] b_1 45°,[678] n_D^{20} 1.5321,[691] d_4^{20} 1.1270,[691] ^{31}P -151.[691]

$C_4H_9Cl_2PS$

MeP($SCH_2CHMeCl$)Cl. 20.[285] b_{16} 110-111°,[285] n_D^{20} 1.5225,[285] d_4^{20} 1.2105.[285]
EtP(SCH_2CH_2Cl)Cl. 20.[285] b_{26} 125°,[285] n_D^{20} 1.5508,[285] d_4^{20} 1.2240.[285]

$C_4H_{10}ClPS$

EtP(SEt)Cl. 22a.[16] 22b.[15,678] 28.[15] 30.[353] 61.[14] b_1 50°,[678] b_7 53-54°,[14] b_8 54-56°,[15] b_{14} 64°,[353] n_D^{20} 1.5280,[353] 1.5308,[15] 1.5309,[14] d_4^{20} 1.0820,[353] 1.1022,[15] 1.1038,[14] add'l data,[14-16,357] TLC.[357]

$C_5H_{11}Cl_2PS$

EtP($SCH_2CHClMe$)Cl. 20.[285,516] b_3 94-95°,[285] b_{10} 104-106°,[516] n_D^{20} 1.5160,[285] 1.5389,[516] d_4^{20} 1.1805,[285] 1.2133.[516]

$C_5H_{12}ClPS$

MeP(SBu)Cl. 22a.[691] b_1 58-59°,[691] n_D^{20} 1.5198,[691] d_4^{20} 1.0891,[691] ^{31}P -153.[691]

EtP(SPr)Cl. 22a.[16] 28.[15] b_8 70-72°,[15] b_{10} 76-77°,[16] n_D^{20} 1.5250,[16] 1.5252,[15] d_4^{20} 1.0734,[15] 1.0743.[16]
EtP(SPr-i)Cl. 22a.[16] 28.[15] b_{10} 65-67°,[15] 66.5-68°,[16] n_D^{20} 1.5200,[16] 1.5202,[15] d_4^{20} 1.0613,[16] 1.0620.[15]

$C_6H_{12}ClPS$

Me_2C=CHP(SEt)Cl. 22b.[678] b_1 76°.[678]

$C_6H_{14}ClPS$

EtP(SBu)Cl. 22a.[16,691] 28.[15] 30.[353] b_8 83-85°,[15] b_9 96-97°,[353] n_D^{20} 1.5160,[353] 1.5191,[15] d_4^{20} 1.0511,[15] 1.0578,[353] add'l data,[16,357,691] ^{31}P -163.[691]
EtP(SBu-i)Cl. 22a.[16] 28.[15] b_8 77-79°,[15] b_{10} 82-83°,[16] n_D^{20} 1.5163,[15] 1.5176,[16] d_4^{20} 1.0474,[15] 1.0476.[16]

$C_7H_7Cl_2PS$

MeP(SC_6H_4Cl-p)Cl. 22a.[691] b_1 120-122°,[691] n_D^{20} 1.6282,[691] d_4^{20} 1.3415.[691]

C_7H_8ClPS

MeP(SPh)Cl. 22a.[691] b_2 111-112°,[691] n_D^{20} 1.6164,[691] d_4^{20} 1.2321,[691] ^{31}P -148.[691]

$C_8H_9Cl_2PS$

PhP(SCH_2CH_2Cl)Cl. 20.[516] b_1 115.5-116°,[516] n_D^{20} 1.6211,[516] d_4^{20} 1.3138.[516]
p-ClC_6H_4P(SEt)Cl. 22b.[678] b_1 122°.[678]

$C_8H_{10}ClPS$

MeP(SC_6H_4Me-p)Cl. 22a.[691] b_1 115-116°,[691] n_D^{20} 1.6888,[691] d_4^{20} 1.2100,[691] ^{31}P -150.[691]
PhP(SEt)Cl. 22a.[691] 22b.[678] b_1 92°,[678] $b_{1.5}$ 106-108°,[691] n_D^{20} 1.6041,[691] d_4^{20} 1.2122,[691] ^{31}P -139.[691]

$C_9H_{11}Cl_2PS$

PhP(SCH_2CHClMe)Cl. 20.[516] 22a.[516] b_1 110-111°,[516] 111-112°,[516] n_D^{20} 1.6120,[516] d_4^{20} 1.2737,[516] 1.2749.[516]

$C_9H_{12}ClPS$

PhP(SPr)Cl. 22b.[641] 73.[641] $b_{0.07}$ 83-84°,[641] 83-85°,[641] n_D^{20} 1.5986,[641] 1.5995,[641] d_4^{20} 1.1627,[641] 1.1654.[641]

$C_{10}H_{20}ClPS$

Me_2C=C(Bu-t)P(SEt)Cl. 22b.[678] b_1 96°.[678]

D.3.8. Phosphonobromidothious Esters

$C_4H_{10}BrPS$

EtP(SEt)Br. 24.[357] 29.[352] b_8 74-75°,[357] b_{10} 80-82°,[352] n_D^{20} 1.5620,[352] 1.5650,[357] d_4^{20} 1.3998,[352] 1.4108.[357]

$C_6H_{14}BrPS$

EtP(SBu)Br. 24.[357,691] $b_{0.4}$ 69-70°,[691] b_9 118-119°,[357] n_D^{20} 1.5472,[691] 1.5560,[357] d_4^{20} 1.2841,[691] 1.3220,[357] ^{31}P -163.[691]

$C_7H_{16}BrPS$

EtP(SAm)Br. 24.[352] 43.[352] b_9 121-122°,[352] n_D^{20} 1.5384,[352] d_4^{20} 1.2680.[352]

D.3.9. Phosphonothious Esters

$C_6H_{15}OPS$

EtP(OEt)SEt. 40.[15] b_7 58-60°,[15] n_D^{20} 1.4865,[15] d_4^{20} 0.9782.[15]

C_7H_7OPS

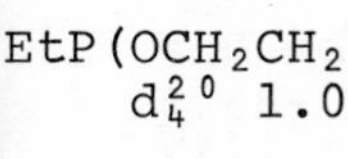

From $MePCl_2$, o-mercaptophenol (R_3N).[752] b_2 88°,[752] ^1H 1.42 (CH_3), 6.69 (C_6H_4).[752]

$C_7H_{16}ClOPS$

EtP(OCH_2CH_2Cl)SPr-i. 38.[15] $b_{0.005}$ 51-52°,[15] n_D^{20} 1.5055,[15] d_4^{20} 1.0779.[15]

$C_9H_{11}OPS$

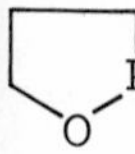

PSPh. 75.[218] $b_{0.2}$ 116°,[218] IR,[218] ^{31}P -79.0,[218] ^1H 1.8 (CH_2), 4.18 (OCH_2), 7.3 (C_6H_5).[218]

$C_{10}H_{15}OPS$

$MeP(OEt)SCH_2Ph$. 72.[530] $b_{0.5}$ 109-111°,[530] n_D^{20} 1.5597,[530] d_4^{20} 1.0821.[530]
$PhP(OEt)SEt$. 39.[708] b_{17} 172-174°,[708] n_D^{20} 1.5619.[708]

$C_{15}H_{20}Cl_3OPS$

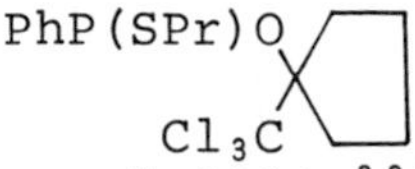

39.[328] $b_{0.2}$ 157-159°,[328] n_D^{20} 1.5720,[328] d_4^{20} 1.2834.[328]

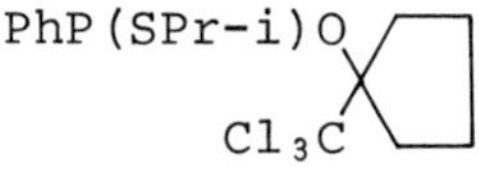

. 39.[328] n_D^{20} 1.5720,[328] d_4^{20} 1.2906.[328]

$C_{16}H_{22}Cl_3OPS$

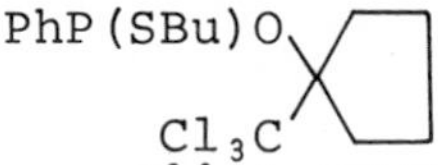

. 39.[328] $b_{0.003}$ 149-150°,[328] n_D^{20} 1.5670,[328] d_4^{20} 1.2577.[328]

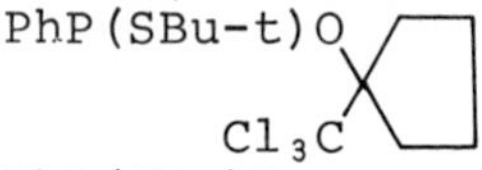

. 39.[328] n_D^{20} 1.5676,[328] d_4^{20} 1.2712.[328]

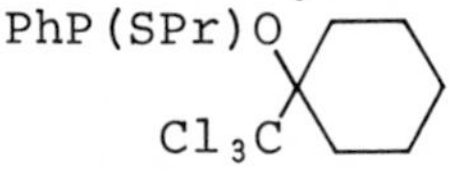

. 39.[327] n_D^{20} 1.5680,[327] d_4^{20} 1.2974.[327]

Compounds with Two Phosphorus Atoms

$C_{14}H_{24}O_2P_2S_2$

$p-C_6H_4[OP(SPr)Me]_2$. 40.[440] $b_{0.025}$ 171°,[440] n_D^{27} 1.5650,[440] d^{20} 1.22.[440]

D.3.10. Phosphonodithious Diesters

$C_3H_6F_3PS_2$

$CF_3P(SMe)_2$. 25.[99] 78.[125] b. 168.4° extrap.,[99] $\log P = 5.7827 + 1.75 \log T - 0.00400\ T - 2545/T$,[99] m. -65.3 to -64.7°,[99] IR,[99] ^1H,[125] ^{19}F ($CFCl_3$ ref.) 60.4.[125]

$C_3H_7PS_2$

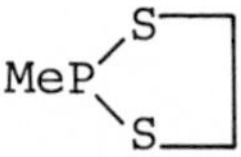

. 21b.[753] b_5 90°,[753] m. -5°.[753]

$C_4H_9PS_2$

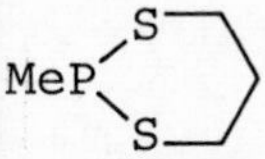
. 21b.[753] b_5 102°.[753]

$C_5H_{13}PS_2$

MeP(SEt)$_2$. 19a.[539] 76.[94] 77.[94] b_8 83-85°,[539] b_9 85-87°,[94] n_D^{20} 1.5575,[94] 1.5577,[539] d_4^{20} 1.0378,[539] 1.0574,[94] add'l data.[94]

$C_6H_{15}PS_2$

EtP(SEt)$_2$. 21a.[18] 21b.[45] 41a.[358] $b_{0.04}$ 44-45.5°,[18] b_{10} 98-100°,[45,356] n_D^{20} 1.5460,[356] 1.5490,[45] 1.5501,[18] d_4^{20} 1.0200,[45] 1.0201,[356] 1.0217,[18] TLC.[357]

$C_7H_{15}N_2PS_4$

MeP(SC[S]NMe$_2$)$_2$. 68.[517] m. 97-98°.[517]

$C_8H_9PS_2$

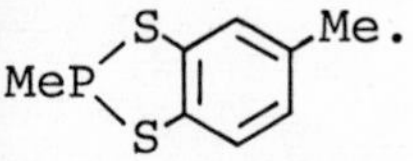
21b.[753] b_2 110-114°,[753] ^1H 1.5 (PCH_3), 2.3 (CCH_3), 7.1 (C_6H_3).[752]

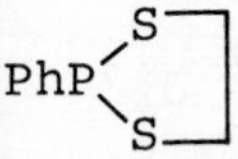
. 33.[2] $b_{0.2}$ 136-137°,[2] n_D^{23} 1.6880,[2] ^1H 2.95,3.10 (CH_2), 7.10-7.75 (C_6H_5).[654a]

$C_8H_{15}PS_2$

EtP(SCH$_2$CH=CH$_2$)$_2$. 21a.[356] $b_{0.1}$ 129-131°,[356] n_D^{20} 1.5690,[356] d_4^{20} 1.0503.[356]

$C_8H_{19}PS_2$

EtP(SEt)SBu-i. 41b.[15] $b_{0.008}$ 53-55°,[15] n_D^{20} 1.5359,[15] d_4^{20} 0.9883.[15]
EtP(SPr)$_2$. 21b.[14] $b_{0.1}$ 86-88°,[14] n_D^{20} 1.5362,[14] d_4^{20} 0.9901.[14]
EtP(SPr-i)$_2$. 21b.[14] b_9 95-97°,[14] n_D^{20} 1.5289,[14] d_4^{20} 0.9783.[14]

$C_9H_{20}ClPS_2$

ClCH$_2$P(SBu)$_2$. 21b.[723] b_1 128-130°,[723] n_D^{25} 1.5400.[723]

$C_9H_{21}PS_2$

$MeP(SBu)_2$. 19a.[758] $b_{0.2}$ 90-95°.[758]

$C_{10}H_{15}PS_2$

$PhP(SEt)_2$. 19a.[32] 21b.[198] 33.[1] 75.[198] 78.[675] $b_{0.001}$ 92°,[1] $b_{0.1}$ 109-112°,[675] $b_{0.8}$ 122°,[198] $b_{3.5}$ 143-144°,[32] n_D^{20} 1.6175,[1] n_D^{22} 1.6165,[198] d_0^0 1.1417,[32] ^1H 1.23 (CH_3), 2.68 (CH_2), 7.25-7.73 (C_6H_5).[198]

$C_{10}H_{23}PS_2$

$EtP(SBu)_2$. 21b.[14] 41a.[358] 42.[355] $b_{0.02}$ 114-115°,[351] $b_{0.07}$ 92-94°,[14] b_8 148-149°,[355] n_D^{20} 1.5240,[355] 1.5263,[14] 1.5275,[351] d_4^{20} 0.9695,[14] 0.9726,[351] 0.9752,[355] add'l data,[356] TLC.[355]
$EtP(SBu\text{-}i)_2$. 21b.[14] b_{120} 154-157°,[14] n_D^{20} 1.5237,[14] d_4^{20} 0.9638.[14]
$BuP(SPr)_2$. 4.[658] $b_{2.5}$ 110-112°,[658] n_D^{20} 1.5270.[658]

$C_{12}H_{17}N_2PS_4$

$PhP(SC[S]NMe_2)_2$. 68.[130] m. 169-170°,[130] IR.[130]

$C_{12}H_{27}PS_2$

$EtP(SAm)_2$. 21a.[356] $b_{0.1}$ 119-120°,[356] n_D^{20} 1.5162,[356] d_4^{20} 0.9540.[356]
$EtP(SAm\text{-}i)_2$. 21a.[356] $b_{0.06}$ 82-84°,[356] n_D^{20} 1.5214,[356] d_4^{20} 0.9605.[356]
$BuP(SBu)_2$. 75.[219] $b_{0.25}$ 96-110°,[219] IR.[219]

$C_{13}H_{13}PS_2$

$MeP(SPh)_2$. 66.[530] $b_{0.007}$ 115-120° (bath),[530] n_D^{20} 1.6701,[530] d_4^{20} 1.1952.[530]

$C_{14}H_{23}PS_2$

$PhP(SBu)_2$. 21a.[637] 26.[144] 74.[551] $b_{0.05}$ 145-155°,[637] $b_{0.2}$ 152-154°,[144] b_4 173-176°,[551] n_D^{20} 1.5824,[551] 1.5850,[144] n_D^{26} 1.5548,[637] d_4^{20} 1.0510.[551]
$PhP(SBu\text{-}i)_2$. 19a.[32] $b_{12.5}$ 191-192°,[32] d_0^0 1.0637.[32]

$C_{15}H_{16}NPS_2$

p-$Me_2NC_6H_4$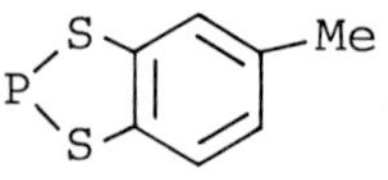

. 21b.[107] m. 125-126°.[107]

$C_{15}H_{17}PS_2$

$MeP(SCH_2Ph)_2$. 66.[530] $b_{0.002}$ 150-160° (bath),[530] n_D^{20} 1.6328,[530] d_4^{20} 1.1553.[530]

$C_{16}H_{19}PS_2$

$EtP(SCH_2Ph)_2$. 21a.[356] $b_{0.03}$ 162-163°,[356] n_D^{20} 1.6330,[356] d_4^{20} 1.1508.[356]

$BuP(SPh)_2$. 75.[219] $b_{0.30}$ 175-180°.[219]

$C_{18}H_5Cl_{10}PS_2$

$PhP(SC_6Cl_5)_2$. 19b.[395] m. 249-250°.[395]

$C_{18}H_{15}PS_2$

$PhP(SPh)_2$. 21b;[521] also from thermal dec. of $PhP(SPh)F_3$.[521] $b_{0.1}$ 221-223°,[521] ^{31}P -90.8.[521]

$C_{18}H_{25}N_2PS_4$

PhP(SC[S]N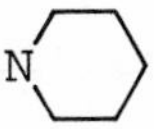)$_2$. 68.[446] m. 144°.[446]

$C_{18}H_{27}PS_2$

$c\text{-}C_6H_{11}P(SPh)_2$. 75.[219] $b_{0.15}$ 183-189°,[219] IR.[219]

D.4. Phosphonous Amides

D.4.1. Phosphinic Amides

$C_5H_{14}NOP$

$MePH(O)NEt_2$. 27.[548] Not isolated.[548]

$C_{12}H_{10}NOP$

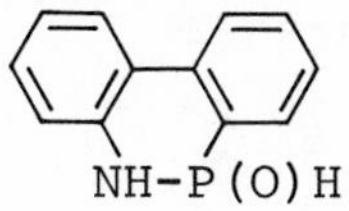

. 38.[143] m. 193-194°,[143] UV.[143]

D.4.2. Phosphonamidous Fluorides

Primary Amine Derivatives

C_4H_9FNP

$MeP(NHCH_2CH{=}CH_2)F$. 8.[156] ^{31}P -151.[156]

$C_4H_{11}FNP$

MeP(NHPr-i)F. 15.[155] b_{90} 31-32°,[155] n_D^{20} 1.4254,[155] d_4^{20} 0.9776,[155] ^{31}P -169,[155] ^{19}F (F_2 ref.) 534.[689]

$C_5H_{13}FNP$

MeP(NHBu-i)F. 15.[155] b_{60} 36-38°,[155] n_D^{20} 1.4214,[155] d_4^{20} 0.9650,[155] ^{19}F (F_2 ref.) 533.[689]

$C_{14}H_{23}FNP$

PhP(NHBu-i)F. 8.[156] ^{31}P -135.8.[156]

Secondary Amine Derivatives

$C_3H_6F_4NP$

$CF_3P(NMe_2)F$. 8.[507] 45.[507] b. 75.5°(extrap.),[507] log P = 8.024 - 1794/T,[507] ΔH_{vap}. 8210 cal/mol,[507] IR,[507] ^1H 2.87 (CH_3),[507] ^{19}F ($CFCl_3$ ref.) 70.6 (CF_3), 134.3 (PF).[507]

C_3H_9FNP

$MeP(NMe_2)F$. 32.[92] 45.[156,677] b_{100} 29-30°,[156] b_{760} 85-86°,[677] n_D^{20} 1.4200,[156] d_4^{20} 0.9473,[156] IR,[92] ^{31}P -165.9,[156] -168.9,[677] -170.4,[425h] ^1H 1.22 (PCH_3), 2.75 (NMe_2),[677] ^{19}F ($CFCl_3$ ref.) 117.5,[677] mass spect.[92,677]

$C_4H_6F_4NP$

CF_2=$CFP(NMe_2)F$. From CF_2=CFMgI, $(Me_2N)_2PCl$ (by-product).[126] Unstable, condenses at -45°,[126] ^1H 2.56,[126] ^{19}F ($CFCl_3$ ref.) 90.1, 113.0 (CF_2), 190.0 (CF), 126.3 (PF).[126]

$C_4H_7Cl_2F_3NP$

$CHCl_2CF_2P(NMe_2)F$. 45.[214] $b_{0.1}$ 40-42°.[214] ^1H 2.85 (CH_3), 5.85 (CH),[214] ^{19}F ($CFCl_3$ ref.) 115.6 (CF_2), 135.4 (PF).[214]

$C_4H_7F_5NP$

$CHF_2CF_2P(NMe_2)F$. 45.[214] ^1H 2.88 (CH_3), 5.80 (CH),[214] ^{19}F ($CFCl_3$ ref.) 126.1 (PCF_2), 134.7 (CHF_2), 136.5 (PF).[214]

$C_4H_9F_3NP$

$CHF_2CH_2P(NMe_2)F$. 8.[221] 54.[221] Volatile liq.,[221] IR,[221] ^1H 2.40 (CH_2), 2.78 (CH_3), 5.95 (CH),[160] ^{19}F ($CFCl_3$ ref.) 110.2 (CHF_2), 119.5 (PF),[221] mass spect.[221]

C_5H_7FNP

MeP(N 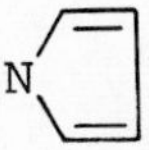)F. 45.[156] b_6 67-69°,[156] n_D^{20} 1.4765,[156] d_4^{20} 1.1316.[156]

$C_5H_{13}FNP$

MeP(NEt_2)F. 8.[154] 45.[156] 46.[156] b_{45} 29-30°,[154,156] b_{80} 52-54°,[156] n_D^{20} 1.4135,[156] 1.4175,[156] d_4^{20} 0.9420,[154] 0.9425,[156] 0.9460,[156] ^{31}P -165.2,[156] ^{19}F (CF_3CO_2H ref.) 32.2.[156]

$C_6H_{13}FNP$

MeP(N 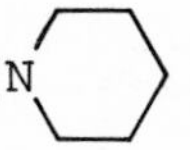)F. 45.[156] b_{20} 50-52°,[156] n_D^{20} 1.4450,[156] d_4^{20} 1.0340,[156] ^{31}P -165.1,[156] ^{19}F (CF_3CO_2H ref.) 35.4.[156]

$C_8H_{11}FNP$

PhP(NMe_2)F. 33.[91] 45.[677] b_{15} 88-90°,[677] ^{31}P -158,[91] -159.8,[677] ^1H 2.60 (CH_3),[677] ^{19}F ($CFCl_3$ ref.) 128.5,[677] 129.5,[91] mass spect.[677]

$C_9H_{21}FNP$

MeP(NBu_2)F. 8.[154,156] 13.[156] $b_{3.5}$ 50-53°,[156] b_4 56-57°,[154,156] n_D^{20} 1.4069,[156] 1.4103,[154,156] d_4^{20} 0.8929,[154,156] 0.8945,[156] IR,[156] ^{31}P -166.9.[156]

MeP(NBu-i_2)F. 8.[154,156] b_3 49-50°,[156] n_D^{20} 1.4075,[156] d_4^{20} 0.8876.[156]

$C_{10}H_{15}FNP$

PhP(NEt_2)F. 13.[156] 45.[156,677] 46.[156] $b_{0.25}$ 54-56°,[677] b_1 78-80°,[156] n_D^{20} 1.4993,[156] d_4^{20} 1.0370,[156] add'l data,[156] ^{31}P -154.9,[156] -156.0,[677] ^1H 0.97 (CH_3), 2.97 (CH_2),[677] ^{19}F 45.5 (CF_3CO_2H ref.),[156] 125.7 ($CFCl_3$ ref.),[677] mass spect.[677]

$C_{10}H_{23}FNP$

$EtP(NBu_2)F$. 8.[156] $b_{1.5}$ 52-54°,[156] n_D^{20} 1.4167.[156]

D.4.3. Phosphonamidous Chlorides

Primary Amine Derivatives

$C_2H_4ClF_3NP$

$CF_3P(NHMe)Cl$. 17.[253] b. 101.6°(extrap.),[253] log P = 5.0210 + 1.75 log T - 0.0033 T - 2077/T,[253] IR.[253]

$C_5H_{13}ClNP$

MeP(NHBu-t)Cl. 6a.[663] b_5 48-52°,[663] ^1H.[663]

$C_8H_{19}ClNP$

t-BuP(NHBu-t)Cl. 6a.[663] $b_{0.1}$ 35-36°,[663] ^1H.[663]

$C_{10}H_{15}ClNP$

PhP(NHBu-t)Cl. 6a.[239] $b_{0.1}$ 78-80°,[239] IR,[239] ^{31}P -118,[239] ^1H 0.98 (CH_3), 3.02 (NH), 7.16 (C_6H_5).[239]

$C_{12}H_9ClNP$

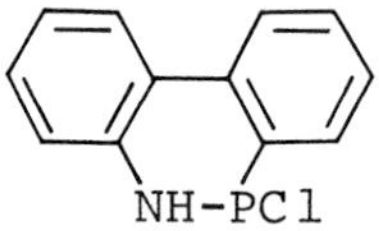

. 1.[107,143] m. 132-134°,[143] $subl._{0.05}$ 180-190°.[143]

Secondary Amine Derivatives

$C_3H_6ClF_3NP$

$CF_3P(NMe_2)Cl$. 6a.[101,507] b. 115.0°(extrap),[101] log P = 5.3231 + 1.75 log T - 0.0038 T - 2134/T,[101] IR.[101]

$C_3H_8Cl_2NP$

$ClCH_2P(NMe_2)Cl$. 6a.[677] b_{12} 64°,[677] ^{31}P -123.0,[677] ^1H 2.73 (CH_3), 3.82 (CH_2),[677] CH_2 signal split at -20°.[211]

C_3H_9ClNP

$MeP(NMe_2)Cl$. 6a.[169,677] 17.[403,733] b_{12} 64°,[677] b_{720} 138-144°,[403] b. 142.0-142.5°,[169] IR,[169] ^{31}P -150.2,[677] -150.7,[403] -151,[733] ^1H 1.65 (PCH_3), 2.66 (NMe_2).[677]

$C_4H_7ClF_4NP$

$CHF_2CF_2P(NMe_2)Cl$. 6a.[214] ^1H 2.88 (CH_3), 5.83 (CH),[214] ^{19}F ($CFCl_3$ ref.) 121.6 (PCF_2), 134.9 (CHF_2).[214]

$C_4H_7Cl_3F_2NP$

$CHCl_2CF_2P(NMe_2)Cl$. 6a.[214] $b_{0.01}$ 55-57°,[214] ^1H 2.85 (CH_3), 5.90 (CH),[214] ^{19}F ($CFCl_3$ ref.) 112.6.[214]

$C_4H_{11}ClNP$

$EtP(NMe_2)Cl$. 6a.[644] b_{16} 51-52°,[644] n_D^{20} 1.4855,[644] d_4^{20} 1.0271.[644]

$C_5H_{10}ClF_3NP$

$CF_3P(NEt_2)Cl$. 6a.[227] b_{28} 61-63°,[227] n_D^{20} 1.4176,[227] d_4^{20} 1.2257.[227]

$C_5H_{12}Cl_2NP$

$ClCH_2P(NEt_2)Cl$. ^1H,[211] CH_2 signal split at -40°.[211]

$C_5H_{13}ClNP$

$MeP(NEt_2)Cl$. 6a.[533,686] 17.[733] b_6 57-59°,[533] b_9 55°,[81] b_{17} 71-72°, b_{740} 182°,[733] b. 184°,[686] n_D^{20} 1.4870,[533] 1.4879,[81] d_4^{20} 1.007,[533] 1.0161,[81] ^{31}P -143,[733] -143.0.[468]
i-$PrP(NMe_2)Cl$. ^1H.[212]

$C_6H_{15}ClNP$

$EtP(NEt_2)Cl$. 6a.[44,686] 51.[684] b_3 52-53°,[44] b_{15} 81-82°,[686] b_{18} 83°,[684] b. 196°,[686] n_D^{20} 1.4836,[44] d_4^{20} 0.9964.[44]

$C_8H_6ClF_5NP$

$C_6F_5P(NMe_2)Cl$. 3.[62] $b_{0.1}$ 75-78°,[62] IR,[62] ^1H 2.69,[61] ^{19}F ($CFCl_3$ ref.) 129.0 (o-F), 151.0 (p-F), 162.5 (m-F),[61] mass spect.[62]

$C_8H_{11}ClNP$

$PhP(NMe_2)Cl$. 6a.[169,275,509] 6b.[130] $b_{0.1}$ 68-70°,[275] 118°,[130] $b_{0.25}$ 47-50°,[124] b_2 78°,[169] $b_{2.5}$ 79°,[509] n_D^{20} 1.5440,[130] 1.5749,[509] d_4^{19} 1.135,[509] IR,[169] ^{31}P -141.0,[677] ^1H (CH_3) 2.46,[275] 2.50,[677] 2.60,[124] 2.62,[694] split at -80° into two doublets,[694] $\Delta F^{\dagger}$(-50°) 10.9 kcal/mole,[124] $\Delta G^{\ddagger}$(-50°) 10.8 kcal/mole.[275]

$C_8H_{19}ClNP$

t-BuP(NEt_2)Cl. 6a.[662] b_{12} 86°,[662] ^1H 0.51 ($CH_2C\underline{H}_3$), 0.59 (t-Bu), 2.42 (NCH_2).[662]

$C_{10}H_{15}ClNP$

PhP(NEt_2)Cl. 6a.[170,246,263,535,686] 51.[684] $b_{0.05}$ 82-84°,[170,275] b_2 146-148°,[124] b_6 128-130°,[535] b_{12} 142-143°,[686] 144°,[684] b_{14} 140-143°,[246,263] n^{20} 1.5560,[535] d_4^{20} 1.0953,[535] IR,[239] ^{31}P -118,[239] -140.4,[677] ^1H 0.88 (CH_3), 2.73 (CH_2), 7.24 (C_6H_5),[239] add'l ^1H data;[124,275,677] CH_3 signal split at low temp.,[124,275] $\Delta F^\dagger$(-57°) 10.8 kcal/mole,[124] $\Delta G^\ddagger$(-50°) 10.0 kcal/mole.[275]

$C_{10}H_{21}ClNP$

c-C_6H_{11}P(NEt_2)Cl. 3.[278] 6a.[686] 51.[684] b_{13} 143-144°,[684] b_{16} 143-144°.[278]

$C_{11}H_{17}ClNP$

PhP(NMeBu)Cl. 6a.[239] $b_{0.15}$ 103-106°,[239] IR,[239] ^1H 0.64 ($CH_2C\underline{H}_3$), 1.08 (NCH_2), 2.04 (CH_2), 2.18 (NCH_3), 7.24 (C_6H_5).[239]

$C_{12}H_{19}ClNP$

PhP(NPr_2)Cl. 6a.[275] $b_{0.15}$ 136-140°,[275] ^1H 0.7 (CH_3),[275] $\Delta G^\ddagger$(-55°) 10.8 kcal/mole.[275]

PhP(NPr-i_2)Cl. 6a.[124,275] $b_{0.05}$ 87-88°,[124] $b_{0.2}$ 122°,[275] ^1H 1.09, 1.29 (CH_3, nonequiv.),[124] add'l data,[275] $\Delta F^\dagger$(-15°) 12.8 kcal/mole,[124] $\Delta G^\ddagger$(-5°) 13.2 kcal/mole.[275]

$C_{13}H_{13}ClNP$

PhP(NMePh)Cl. 6a.[275] $b_{0.15}$ 178-181°,[275] ^1H 2.71.[275]

$C_{13}H_{15}ClNP$

PhP(NMe[CH_2Ph])Cl. 6a.[275] $b_{0.1}$ 154-160°,[275] ^1H 2.29 (CH_3),[275] $\Delta G^\ddagger$(48°) 11.1 kcal/mole.[275]

$C_{13}H_{19}ClNP$

c-C_6H_{11}P(NMePh)Cl. 6a.[686] b_{17} 193-196°.[686]

$C_{14}H_{15}ClNP$

EtP(NPh_2)Cl. 6b.[686] b_{17} 209-211°.[686]

PhP(NEtPh)Cl. 6a.[275] $b_{0.1}$ 180-185°,[275] ^1H 0.82.[275]

$C_{14}H_{23}ClNP$

$PhP(NBu_2)Cl$. 6a.[275] $b_{0.1}$ 145-148°,[275] ^1H 0.76 (CH_3),[275] $\Delta G^\ddagger$(-60°) 10.7 kcal/mole.[275]
$PhP(NBu\text{-}s_2)Cl$. 6a.[275] $b_{0.2}$ 187°,[275] ^1H 1.3 (CH_3),[275] $\Delta G^\ddagger$(15°) 14.6 kcal/mole.[275]

$C_{15}H_{16}ClN_2OP$

$PhP(NPhCONMe_2)Cl$. 44.[130] m. 67-69°,[130] IR.[130]

$C_{16}H_{19}ClNP$

$PhP(N[Pr\text{-}i]CH_2Ph)Cl$. 6a.[275] $b_{0.1}$ 180°,[275] ^1H 1.20 (CH_3).[275]

$C_{20}H_{19}ClNP$

$PhP(N[CH_2Ph]_2)Cl$. 6a.[124] $b_{0.05}$ 170-175°,[124] m. 47-49°,[124] ^1H 4.09 (CH_2).[124]

Hydrazine Derivatives

$C_8H_{12}ClN_2P$

$PhP(NHNMe_2)Cl$. 6a.[487] Viscous liq.[487]

Bis(phosphonochloridous) Imides

$C_2H_3Cl_2F_6NP_2$

$(CF_3PCl)_2NMe$. 17.[253] b. 148°extrap.,[253] log P = 4.915 + 1.75 log T - 0.003 T - 2260/T,[253] m.~ -50°,[253] IR.[253]

D.4.4. Phosphonamidous Bromides

C_3H_9BrNP

$MeP(NMe_2)Br$. 18.[733] ^{31}P -161,[733] -161.1.[468]

$C_6H_{11}BrNP$

$EtP(NEt_2)Br$. 48.[686] b_{12} 96-97°.[686]

$C_{10}H_{15}BrNP$

$PhP(NEt_2)Br$. 48.[686] b_{13} 156-157°.[686]

$C_{10}H_{21}BrNP$

$c\text{-}C_6H_{11}P(NEt_2)Br$. 48.[686] b_{15} 152-154°.[686]

D.4.5. Phosphonamidous Iodides

$C_5H_{10}F_3INP$

$CF_3P(NEt_2)I$. 11.[227] b_{17} 88-90°,[227] n_D^{20} 1.5046,[227] d_4^{20} 1.6858.[227]

$C_6H_{15}INP$

$EtP(NEt_2)I$. 50.[684] b_{15} 120-122°.[684]

$C_7H_{10}F_7INP$

$C_3F_7P(NEt_2)I$. 11.[227] b_6 71°,[227] n_D^{20} 1.4480,[227] d_4^{20} 1.7093.[227]

$C_8H_{11}INP$

$PhP(NMe_2)I$. 47.[735] Reddish-brown oil.[735]

$C_{10}H_{15}INP$

$PhP(NEt_2)I$. 50.[684] b_{12} 176-178°.[684]

$C_{10}H_{21}INP$

c-$C_6H_{11}P(NEt_2)I$. 49.[684] 50.[684] b_{11} 170-171°.[684]

D.4.6. Phosphonamidous Esters

Primary Amine Derivatives

C_3H_8NOP

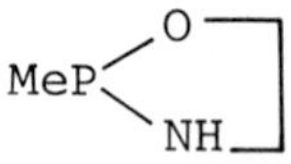

. 28.[499] Viscous oil (heptamer),[499] IR.[499]

$C_8H_{10}NOP$

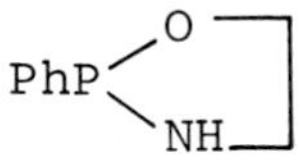

. 28.[463] b_{4-5} 115-118°.[463]

$C_{11}H_{18}NO_2P$

$EtP(NHC_6H_4OMe$-p$)OEt$. 25b.[208] b_2 122-125°,[208] n_D^{20} 1.5413,[208] d_4^{20} 1.072.[208]

$C_{12}H_{20}NO_2P$

$EtP(NHC_6H_4OMe\text{-}p)OPr\text{-}i$. 25b.[208] b_2 124-126°,[208] n_D^{20} 1.5332,[208] d_4^{20} 1.0468.[208]

$C_{13}H_{17}Cl_3NOP$

$PhP(NHCH_2CH{=}CH_2)OCMe_2CCl_3$. 25a.[328] $b_{0.16}$ 145-147°,[328] n_D^{20} 1.5610,[328] d_4^{20} 1.2725.[328]

$C_{13}H_{19}Cl_3NOP$

$PhP(NHPr)OCMe_2CCl_3$. 25a.[328] n_D^{20} 1.5475,[328] d_4^{20} 1.2425.[328]

$C_{14}H_{19}Cl_3NOP$

$p\text{-}MeC_6H_4P(NHCH_2CH{=}CH_2)OCMe_2CCl_3$. 25a.[328] $b_{0.14}$ 146°,[328] n_D^{20} 1.5570,[328] d_4^{20} 1.2446.[328]

$C_{14}H_{21}Cl_3NOP$

$PhP(NHBu)OCMe_2CCl_3$. 25a.[328] n_D^{20} 1.5370,[328] d_4^{20} 1.2347.[328]

$C_{15}H_{18}NO_2P$

$PhP(NHC_6H_4OMe\text{-}p)OEt$. 25b.[208] $b_{0.02}$ 143-146°,[208] n_D^{20} 1.5950,[208] d_4^{20} 1.1434.[208]

$C_{15}H_{19}Cl_3NOP$

$PhP(NHCH_2CH{=}CH_2)O$ 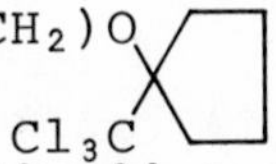

. 25a.[328] $b_{0.2}$ 140°,[328] n_D^{20} 1.5380,[328] d_4^{20} 1.2776.[328]

$C_{15}H_{21}Cl_3NOP$

$PhP(NHPr)O$ 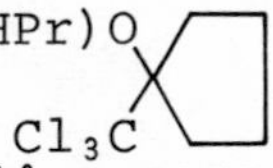

. 25a.[328] $b_{0.1}$ 145-146°,[328] n_D^{20} 1.5635,[328] d_4^{20} 1.2153.[328]

$C_{15}H_{28}NO_7P$

$MePH({=}NCH_2CH_2OH)OR$ (R = 1,2:5,6-di-O-isopropylideneglucos-3-yl). 31.[494] b_1 132-135°,[494] n_D^{20} 1.4718,[494] IR,[494] TLC.[494]

$C_{16}H_{23}Cl_3NOP$

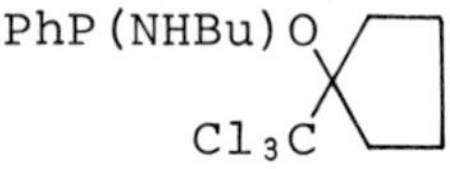

. 25a.[328] n_D^{20} 1.5552,[328] d_4^{20} 1.2392.[328]

$C_{17}H_{25}Cl_3NOP$

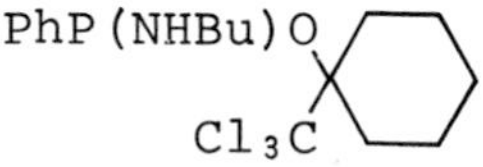

. 25a.[327] n_D^{20} 1.5450,[327] d_4^{20} 1.2321.[327]

$C_{17}H_{30}NO_8P$

MeP($NHCH_2CO_2Et$)OR (R = 1,2:5,6-di-O-isopropylideneglucos-3-yl). 31.[494] $b_{0.005}$ 120-125°,[494] n_D^{20} 1.4793,[494] TLC.[494]

$C_{22}H_{21}Cl_3NOP$

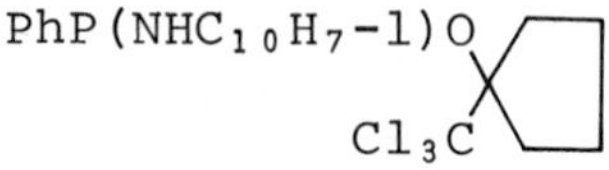

. 25b.[326] m. 107-110°.[326]

Compounds with Two Phosphorus Atoms

$C_{32}H_{32}Cl_6N_2O_2P_2$

p-[PhP($OCMe_2CCl_3$)NHC_6H_4-]$_2$. 25b.[328] s. 82-90°.[328]

$C_{34}H_{36}Cl_6N_2O_2P_2$

p-[p-MeC_6H_4P($OCMe_2CCl_3$)NHC_6H_4-]$_2$. 25b.[328] s. 89-101°.[328]

$C_{36}H_{36}Cl_6N_2O_2P_2$

p-[PhPNHC_6H_4-]$_2$. 25b.[328] n_D^{20} 1.5640.[328]

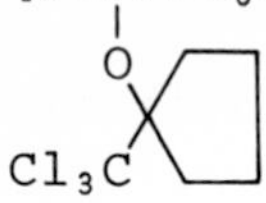

Secondary Amine Derivatives

$C_4H_{10}NOP$

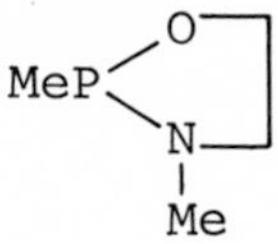

. 28.[499] Dimer, b_5 137-140°,[499] n_D^{20} 1.4995,[499] d_4^{20} 1.0790,[499] IR.[499]

$C_5H_{12}NOP$

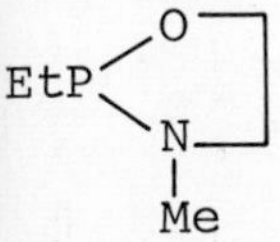

. 28.[499] Dimer, $b_{0.5}$ 135-138°,[499] n_D^{20} 1.0502,[499] IR.[499]

$C_6H_{15}N_2O_2P$

$MeP(NMe_2)OC(O)NMe_2$. 34.[517] b_{13} 103°,[517] n_D^{20} 1.4710.[517]

$C_7H_{14}NOP$

$CH_2=C=CHP(NMe_2)OEt$. ^1H 4.639 ($CH_2=C$), 5.252 (CH).[696]

$C_7H_{18}NOP$

$MeP(NEt_2)OEt$. 25a.[533] 39.[533] b_9 46-48°,[533] b_{11} 49-50°,[533] n_D^{20} 1.4380,[533] d_4^{20} 0.8827.[533]
$EtP(NMe_2)OPr$-i. 39.[644] b_{27} 71-73°,[644] n_D^{20} 1.4377,[644] d_4^{20} 0.8932.[644]
$EtP(NEt_2)OMe$. 39.[44] b_{10} 49-50°,[44] n_D^{20} 1.4450,[44] d_4^{20} 0.8954.[44]

$C_8H_{20}NOP$

$MeP(NEt_2)OPr$. 25a.[535] b_{11} 62-63°,[535] n_D^{20} 1.4366,[535] d_4^{20} 0.9260.[535]
$MeP(NEt_2)OPr$-i. $b_{0.1}$ 41°,[375] n_D^{20} 1.448,[375] d_4^{20} 0.8672,[375] ^{31}P -132.5,[376] ^1H 1.05 ($CH_2\underline{CH_3}$), 1.09 (PCH_3), 1.18 ($CH\underline{CH_3}$), 3.03 (NCH_2), 3.90 ($\overline{O}CH$),[376] magn. rotation,[375] magn. susceptibility.[375]
$EtP(NEt_2)OEt$. 39.[44] b_9 53-54°,[44] n_D^{20} 1.4418,[44] d_4^{20} 0.8847.[44]

$C_9H_{12}NOP$

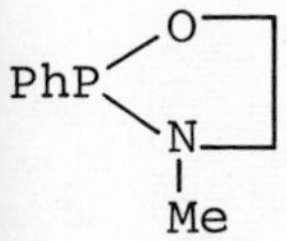

. 28.[462,463] $b_{0.02}$ 88-91°,[463] $b_{0.03}$ 64-66°.[462]

$C_9H_{22}NOP$

$MeP(NEt_2)OBu$. 25a.[535] b_7 72-73°,[535] n_D^{20} 1.4430,[535] d_4^{20} 0.8969.[535]
$EtP(NEt_2)OPr$. 39.[44] b_{10} 73-74°,[44] n_D^{20} 1.4439,[44] d_4^{20} 0.8820.[44]
$EtP(NEt_2)OPr$-i. 39.[44] b_9 63-65°,[44] n_D^{20} 1.4400,[44] d_4^{20} 0.8730.[44]

$C_{10}H_{14}NOP$

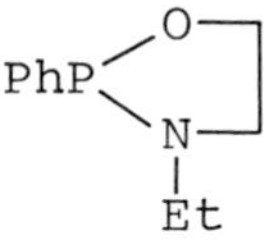 . 28.[463] b_{1-2} 79-81°.[463]

$C_{10}H_{15}ClNOP$

p-$ClC_6H_4P(NMe_2)OEt$. -.[680] b_2 102°.[680]

$C_{10}H_{16}NOP$

$PhP(NMe_2)OEt$. 28.[404] $b_{0.02}$ 74-80°,[404] n_D^{20} 1.5263,[404] IR,[404] ^{31}P -154.4.[404]

$C_{10}H_{24}NOP$

$EtP(NEt_2)OBu$. 39.[44] b_{10} 88-89°,[44] n_D^{20} 1.4450,[44] d_4^{20} 0.8800.[44]

$EtP(NEt_2)OBu$-i. 39.[44] b_{10} 79-80°,[44] n_D^{20} 1.4423,[44] d_4^{20} 0.8739.[44]

$C_{11}H_{18}NOP$

$MeP(NEt_2)OPh$. 28.[533] b_8 118-120°,[533] n_D^{20} 1.5220,[533] d_4^{20} 0.9991.[533]

$C_{12}H_{18}NOP$

EtP(N)OPh. 28.[514] $b_{0.06}$ 60-64°,[514] n_D^{20} 1.5182,[514] d_4^{20} 1.0123,[514] IR.[514]

$C_{12}H_{20}NOP$

$PhP(NEt_2)OEt$. 25a.[708] $b_{0.7}$ 83-85°,[708] n_D^{20} 1.5189.[708]

$C_{12}H_{28}NOP$

$EtP(NEt_2)OC_6H_{13}$-n. 39.[44] b_1 80-81°,[44] n_D^{20} 1.4500,[44] d_4^{20} 0.8787.[44]

$C_{13}H_{20}NOPS$

$PhP(NEt_2)OCH$S. 28.[39] $b_{0.07}$ 105°,[39] n_D^{20} 1.5600,[39] d_4^{20} 1.0910.[39]

$C_{13}H_{22}NOP$

PhP(NEt$_2$)OPr. 25a.[535] b_4 116-118°,[535] n_D^{20} 1.5132,[535] d_4^{20} 1.0049.[535]

$C_{13}H_{29}N_2OP$

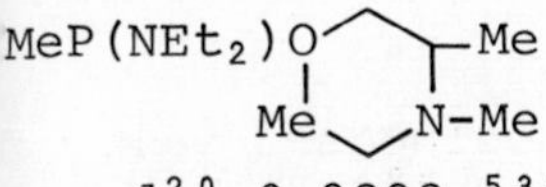

. 28.[534] $b_{0.05}$ 54-55°,[534] n_D^{20} 1.4750,[534] d_4^{20} 0.9393.[534]

$C_{13}H_{30}NOP$

EtP(NEt$_2$)OC$_7$H$_{15}$-n. 39.[44] b_1 101-102°,[44] n_D^{20} 1.4513,[44] d_4^{20} 0.8794.[44]

$C_{14}H_{13}ClNOP$

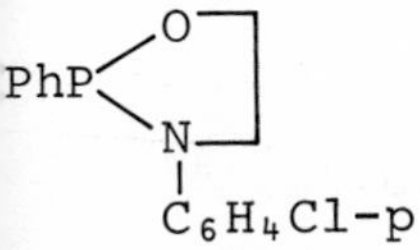

. 28.[463] $b_{0.08}$ 154-162°,[463] m. 53-55°.[463]

$C_{14}H_{14}NOP$

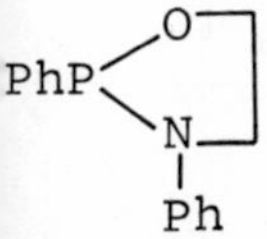

. 28.[462,463] $b_{0.03}$ 130-132°,[462,463] m. 75-76°,[463] 75-77°.[462]

$C_{14}H_{24}NOP$

PhP(NEt$_2$)OBu. 25a.[535] 28.[533] b_3 110-111°,[533] b_5 136-138°,[535] n_D^{20} 1.5090,[535] 1.5150,[533] d_4^{20} 0.9504,[533] 0.9778.[535]

$C_{14}H_{32}NOP$

EtP(NEt$_2$)OC$_8$H$_{17}$-n. 39.[44] b_2 115-116°,[44] n_D^{20} 1.4528,[44] d_4^{20} 0.8794.[44]

$C_{15}H_{16}NOP$

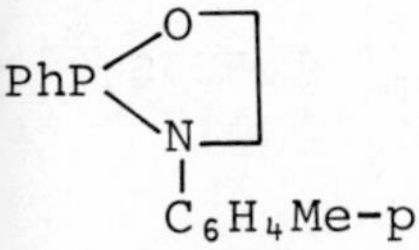

. 28.[463] $b_{0.1}$ 163-165°,[463] m. 62-64°.[463]

$C_{15}H_{16}NO_2P$

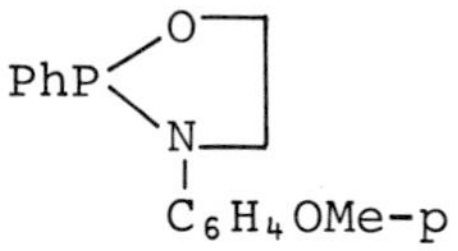

. 28.[463] $b_{0.05}$ 158-162°,[463] $b_{0.1-0.15}$ 174-178°,[463] m. 41-44°,[463] 56-58°.[463]

$C_{15}H_{28}NO_6P$

$MeP(NMe_2)OR$ (R = 1,2:5,6-di-O-isopropylideneglucos-3-yl). 28.[494] b_1 160-170°,[494] n_D^{20} 1.4755,[494] TLC.[494]
$MeP(NMe_2)OR'$ (R' = 2,3:5,6-di-O-isopropylidenemannos-1-yl). 28.[230] $b_{0.001}$ 110-120° (bath),[230] m. 60-62°,[230] TLC.[230]

$C_{15}H_{34}NOP$

$EtP(NEt_2)OC_9H_{19}$-n. 39.[44] b_2 121-122°,[44] n_D^{20} 1.4550,[44] d_4^{20} 0.8800.[44]

$C_{16}H_{20}NOP$

$PhP(NEt_2)OPh$. 39.[483] $b_{0.06}$ 135-137°,[483] n_D^{20} 1.5708,[483] d_4^{20} 1.0670.[483]

$C_{16}H_{36}NOP$

$EtP(NEt_2)OC_{10}H_{21}$-n. 39.[44] b_2 134-135°,[44] n_D^{20} 1.4569,[44] d_4^{20} 0.8806.[44]

$C_{17}H_{30}NO_7P$

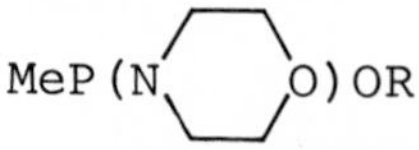

(R = 1,2:5,6-di-O-isopropylideneglucos-3-yl). 31.[494] $b_{0.01}$ 130-135°,[494] n_D^{20} 1.4765,[494] TLC.[494]

$C_{17}H_{32}NO_6P$

$MeP(NEt_2)OR'$ (R' = 1,2:3,4-di-O-isopropylidenegalactos-6-yl). 28.[548] $b_{0.01}$ 130-140° (bath),[548] n_D^{20} 1.4768,[548] TLC.[488,548]

$C_{19}H_{33}N_2OP$

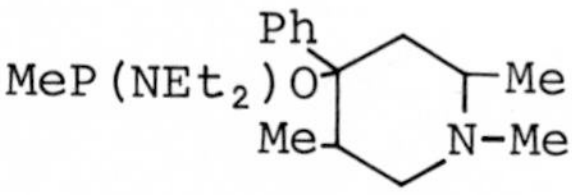

. 28.[534] $b_{0.03}$ 73-76°,[534] n_D^{20} 1.5250,[534] d_4^{20} 1.0350.[534]

$C_{21}H_{33}Cl_3NOP$

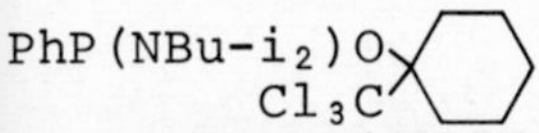

. 25a.[327] m. 87-88°.[327]

Compounds with Two Phosphorus Atoms

$C_{16}H_{34}N_2O_4P_2$

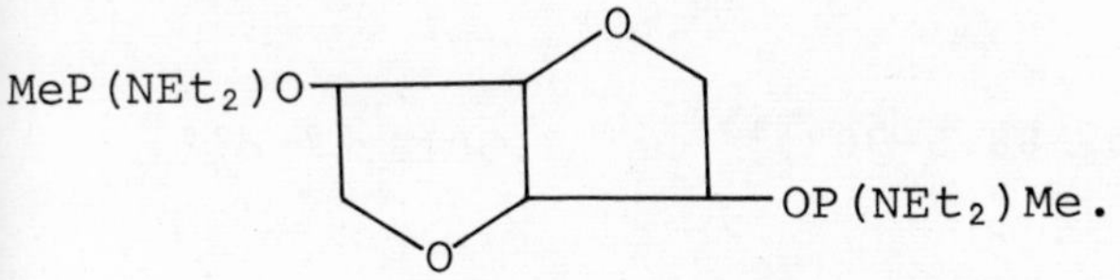

28.[497]

-, 1,4:3,6-dianhydroglucitol deriv., $b_{0.1}$ 140-150°,[497] n_D^{25} 1.4994,[497] TLC.[497]
-, 1,4:3,6-dianhydromannitol deriv., $b_{0.1}$ 145-155°,[497] n_D^{25} 1.4985,[497] TLC.[497]

D.4.7. Phosphonamidothious Esters

Primary Amine Derivatives

C_3H_8NPS

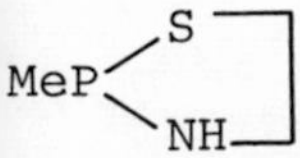

. 29.[498] b_2 59-60°,[498] n_D^{20} 1.5974,[498] d_4^{20} 1.1810,[498] IR.[498]

C_7H_8NPS

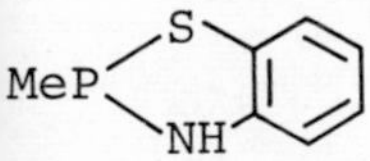

. From $MePCl_2$, o-$NH_2C_6H_4SH$ (R_3N).[752] b_2 122-124°,[742] m. 120°,[752] ^1H 1.22 (CH_3), 7.01 (C_6H_4).[752]

$C_8H_{10}NPS$

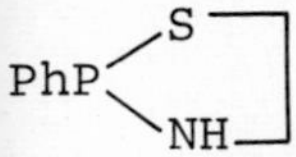

. 29.[498] $b_{2.5}$ 134-135°,[498] n_D^{20} 1.5985,[498] d_4^{20} 1.1012.[498]

Secondary Amine Derivatives

$C_5H_{12}NPS$

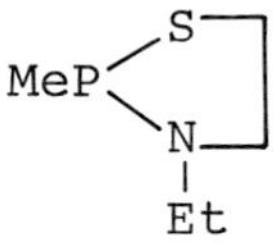

. 29.[365] $b_{0.5}$ 43-44°,[365] n_D^{20} 1.5500,[365] d_4^{20} 1.1112.[365]

$C_8H_{20}NPS$

EtP(NEt$_2$)SEt. 26a.[17] b_9 88.5-90°,[17] n_D^{20} 1.5018,[17] d_4^{20} 0.9357.[17]

$C_9H_{22}NPS$

EtP(NEt$_2$)SPr. 26a.[15] b_7 98.5-100°,[15] n_D^{20} 1.4992,[15] d_4^{20} 0.9312.[15]

EtP(NEt$_2$)SPr-i. 26a.[17] b_{10} 96-97°,[17] n_D^{20} 1.4954,[17] d_4^{20} 0.9226.[17]

$C_{10}H_{24}NPS$

EtP(NEt$_2$)SBu. 26b.[17] b_9 115-117°,[17] n_D^{20} 1.4950,[17] d_4^{20} 0.9213.[17]

$C_{11}H_{18}NPS$

MeP(NEt$_2$)SPh. 29.[530] $b_{0.02}$ 70-75°,[530] n_D^{20} 1.5462,[530] d_4^{20} 1.1119.[530]

$C_{17}H_{25}N_2PS_2$

. 68 (A.3.).[446] m. 137°.[446]

Hydrazine Derivatives

$C_{10}H_{17}N_2PS$

EtP(NHNHPh)SEt. 26b.[7] $b_{0.5}$ 130°,[7] n_D^{20} 1.5915,[7] d_4^{20} 1.0904,[7] IR.[7]

$C_{11}H_{19}N_2PS$

MeP(NHNHPh)SBu. 26b.[7] $b_{0.5}$ 140°,[7] n_D^{20} 1.5810,[7] d_4^{20} 1.0767,[7] IR.[7]

EtP(NHNHPh)SPr-i. 26b.[7] $b_{0.5}$ 135°,[7] n_D^{20} 1.5780,[7] d_4^{20} 1.0634,[7] IR.[7]

D.4.8. Phosphonous Diamides

Ammonia Derivatives

$CH_4F_3N_2P$

$CF_3P(NH_2)_2$. 7a.[247,253] m. >0°.[253]

$C_4H_{13}N_2P$

t-BuP$(NH_2)_2$. 7a.[664] m. 77-79°,[664] subl.$_{12}$ 75-85°,[664] ^1H 0.45 (CH_3).[664]

Primary Amine Derivatives

$C_3H_8F_3N_2P$

$CF_3P(NHMe)_2$. 7a,[253] 41.[253] b. 134.5°(extrap),[253] log P = 7.1770 + 1.75 log T - 0.0058 T - 2650/T,[253] m. -29.5 to -28.5°,[253] IR.[253]

$C_6H_{17}N_2P$

t-BuP$(NHMe)_2$. 7a.[664] b_{12} 70-72°,[664] m. -26°,[664] ^1H 0.61 (CCH_3), 2.23 (NCH_3).[664]

$C_8H_8F_5N_2P$

$C_6F_5P(NHMe)_2$. 7a.[62] Col. oil, dec. 70°,[62] IR.[62]

$C_8H_{13}N_2P$

PhP$(NHMe)_2$. 7a.[372] $b_{0.4}$ 70-80°,[372] IR.[372]

$C_{10}H_{17}N_2O_2PS$

MePH$(=NSO_2C_6H_4Me\text{-}p)NMe_2$. 43.[671] IR,[671] ^{31}P + 1.4,[671] ^1H 1.66 (NCH_3), 2.53 (PCH_3), 7.07 (PH).[671]

$C_{10}H_{17}N_2P$

PhP$(NHEt)_2$. 7a.[372] $b_{0.001}$ 53-56°,[372] n_D^{22} 1.5486,[372] m. 6.5-7.5°,[372] IR,[372] ^1H 1.04 (CH_3), 2.22 (NH), 2.82 (CH_2), 7.35-7.65 (C_6H_5).[372]

$C_{12}H_{21}N_2P$

PhP$(NHPr)_2$. 7a.[372] $b_{0.001}$ 100°,[372] IR.[372]
PhP$(NHPr\text{-}i)_2$. 7a.[372] $b_{0.001}$ 52°,[372] m. 13.5-14.5°,[372] $n_D^{19.5}$ 1.5298,[372] IR,[372] ^1H 1.05, 1.09 $(CH_3$, nonequiv.), 1.92 (NH), 3.2 (CH), 7.1-7.6 (C_6H_5).[372]

$C_{13}H_{13}N_2P$

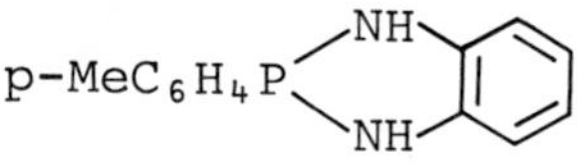 . 52.[557] m. 245-248°.[557]

$C_{13}H_{15}N_2P$

$MeP(NHPh)_2$. 30.[532] m. 154-155°.[532]

$C_{14}H_{20}F_5N_2P$

$C_6F_5P(NHBu\text{-}t)_2$. 7a.[62] m. 54°,[62] subl.,[62] IR,[62] ^{31}P,[61] ^{1}H 1.16 (CH_3), 2.64 (NH),[61] ^{19}F ($CFCl_3$ ref.) 137.7 (o-F), 155.7 (p-F), 162.1 (m-F),[61] mass spect.[62]

$C_{14}H_{25}N_2P$

$PhP(NHBu\text{-}t)_2$. 7a.[372,700,757] $b_{0.2}$ 97-100°,[757] $b_{0.4}$ 98-100°,[700] $b_{0.001}$ 64-65°,[372] m. 13.5-14.7°,[372] n_D^{22} 1.5198,[372] IR,[372,700] ^{1}H 1.22 (CH_3).[372]

$C_{18}H_{15}N_2PS_2$

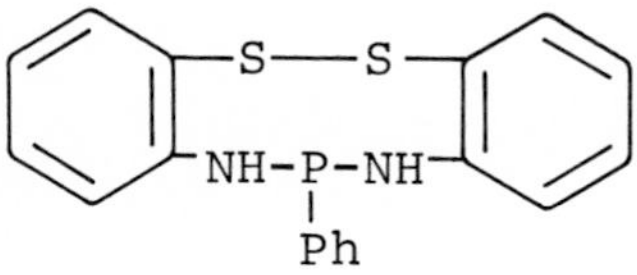 . 52.[557] m. 110-118°.[557]

$C_{19}H_{17}N_2P$

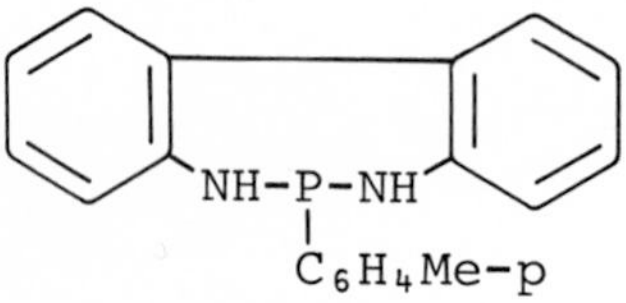 . 52.[557] m. 117-118°.[557]

$C_{19}H_{17}N_2PS_2$

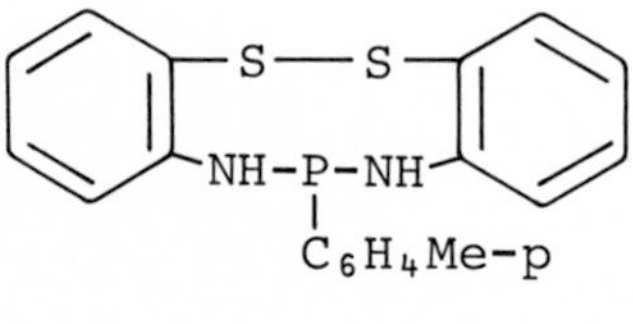 . 52.[557] m. 99-102°.[557]

$C_{20}H_{16}F_5N_2P$

$C_6F_5P(NHCH_2Ph)_2$. 7a.[62] m. 78-79°,[62] IR.[62]

$C_{24}H_{49}N_2O_6PSi_2$

$PhP(NHCH_2CH_2CH_2Si[OEt]_3)_2$. 7b.[171] cryst.,[171] IR.[171]

Compounds with Three Phosphorus Atoms

$C_{25}H_{24}N_3P_3$

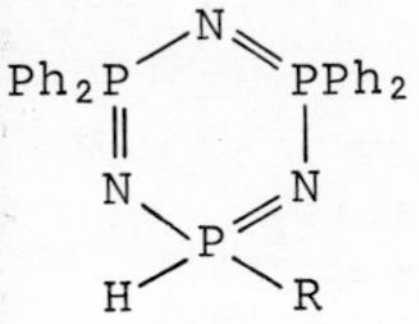

R = Me: 53.[670] m. 152°,[670] IR,[670] ^{31}P -6.9 (PMe), -12.9 (PPh_2),[670] ^1H 1.57 (CH_3), 7.65 (PH).[670]

$C_{26}H_{26}N_3P_3$

R = Et: 53.[670] m. 127-129°,[670] IR,[670] ^{31}P -12.5 (PEt), -15.5 (PPh_2),[670] ^1H 1.03 (CH_3), 1.72 (CH_2), 7.58 (PH).[670]

Secondary Amine Derivatives

$C_5F_{15}N_2P$

$CF_3P(N[CF_3]_2)_2$. 56.[24] b. 93°(extrap),[24] log P = 7.450 - 1670/T,[24] IR,[24] ^{19}F ($CFCl_3$ ref.) 51.5 (NCF_3), 58.6 (PCF_3),[24] mass spect.[24]

$C_5H_9N_2O_2P$

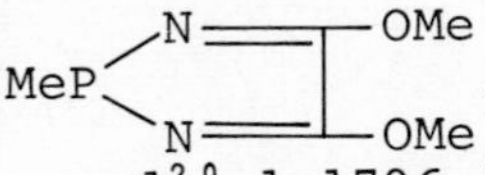

. 19.[142] $b_{0.7}$ 53-54°,[142] n_D^{20} 1.5158,[142] d_4^{20} 1.1796.[142]

$C_5H_{11}N_2P$

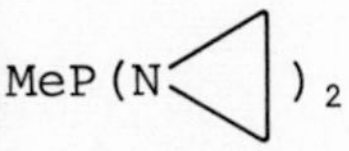

. Raman.[688]

$C_5H_{12}F_3N_2P$

$CF_3P(NMe_2)_2$. 5.[25] 7a.[507] 42.[507] b. 105°(extrap),[25]

135°(extrap),[507] log P = 7.09 - 1844/T,[25] 6.766 + 1.75 log T - 0.005 T - 2618/T,[507] $\Delta H_{vap.}$ 9430 cal/mole,[507] IR,[25,507] [1]H 2.75,[125] 2.76,[507] 2.85,[25] [19]F (CFCl$_3$ ref.) 61.5,[125] 62.0.[507]

$C_5H_{15}N_2P$

MeP(NMe$_2$)$_2$.   2a.[169]   2d.[412]   2c.[411]   7a.[265,403,532,669] 10.[403]   b$_{10}$ 39-40°,[532] b$_{49-50}$ 64-67°,[169] b$_{86}$ 77°,[669] b$_{755}$ 141°,[265] vap. press.$_{0.1-20}$,[265] m. -52°,[265] n$_D^{20}$ 1.4620,[532] 1.4630,[403,411] d$_4^{20}$ 0.8814,[532] add'l data,[403,] [411,412] IR,[169] [31]P -86.4,[403] [1]H 1.12 (PCH$_3$), 2.62 (NCH$_3$),[731] add'l data.[264,427]

$C_6H_{12}F_3N_2P$

CF$_2$=CFP(NMe$_2$)$_2$.   2a.[126]   Unstable,[126] [1]H 2.45,[126] [19]F (CFCl$_3$ ref.) 89.4, 107.8 (CF$_2$), 181.3 (CF).[126]

$C_6H_{13}N_2P$

EtP(N◁)$_2$.   7b.[514]   b$_8$ 63-64°,[514] n$_D^{20}$ 1.5116,[514] d$_4^{20}$ 0.9988.[514]
CH≡CP(NMe$_2$)$_2$.   2a.[111]   b$_{15}$ 58°,[111] [1]H 2.70 (CH$_3$), 2.98 (CH).[111]

$C_6H_{15}F_2N_2P$

CHF$_2$CH$_2$P(NMe$_2$)$_2$.   7a.[221]   9.[221]   42.[221]   55.[221]   Col. viscous oil,[221] IR,[221] [1]H 2.39 (CH$_2$), 2.72 (NCH$_3$), 5.93 (CH).[160]

$C_6H_{15}N_2P$

EtP(NMe$_2$)$_2$.   2c.[404]   7a.[669]   b$_{14}$ 45°,[669] b$_{15}$ 52°,[350] b$_{720}$ 153-158°,[404] n$_D^{20}$ 1.4618,[350] 1.4632,[404] IR,[404] [31]P -99.9,[404] [1]H 1.0 (CH$_2$CH$_3$), 2.51 (NCH$_3$),[350] mass spect.[350]

$C_6H_{23}B_{10}N_2P$

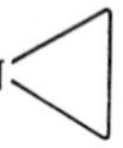

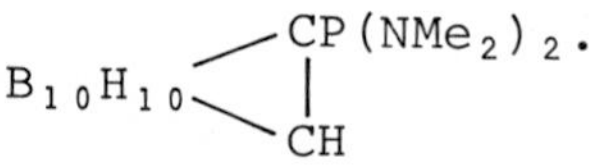

B$_{10}$H$_{10}$ —CP(NMe$_2$)$_2$ / CH.   2b.[764]   m. 69-70°.[764]

$C_7H_{12}F_7N_2P$

C$_3$F$_7$P(NMe$_2$)$_2$.   12.[118]   b. 27-29°.[118]
i-C$_3$F$_7$P(NMe$_2$)$_2$.   5.[25]   b. 143°(extrap),[25] log P = 6.41 -

1960/T,[25] IR,[25] ^1H 2.78.[25]

$C_7H_{13}N_2O_2P$

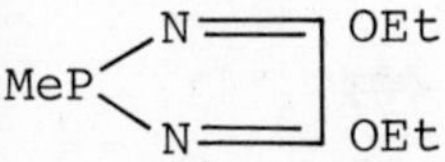

. 19.[142] $b_{0.7}$ 91-93°,[142] n_D^{20} 1.4971,[142] d_4^{20} 1.1004.[142]

$C_7H_{15}N_2P$

CH_2=C=CHP$(NMe_2)_2$. ^1H 4.59 (CH_2), 5.15 (CH).[696]
MeC≡CP$(NMe_2)_2$. 2a.[111] b_{14} 84°,[111] ^1H 2.02 (CCH_3), 2.675 (NCH_3).[111]

$C_7H_{17}N_2O_2P$

$MeO_2CCH_2P(NMe_2)_2$. ^1H 2.80 (CH_2), 3.71 (OCH_3).[553]

$C_7H_{18}N_3PS$

$MeP(NMe_2)NMeC(S)NMe_2$. 35.[517] $b_{0.15}$ 120-122°,[517] n_D^{20} 1.5520.[517]

$C_7H_{19}N_2P$

$PrP(NMe_2)_2$. 2a.[102]

$C_7H_{25}B_{10}N_2P$

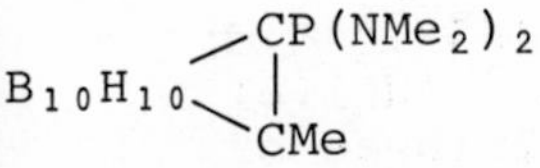

. 2b.[764] m. 76-78°.[764]

$C_8H_{15}N_2O_2P$

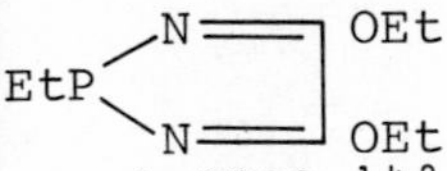

. 19.[142] $b_{0.5}$ 84-85°,[142] n_D^{20} 1.4968,[142] d_4^{20} 1.0830.[142]

$C_8H_{17}N_2P$

CH_2=C=CMeP$(NMe_2)_2$. ^1H 1.60 (CCH_3), 4.51 (CH_2).[696]

$C_8H_{19}N_2P$

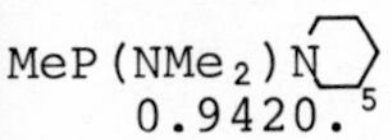

. 30.[532] b_2 50°,[532] n_D^{20} 1.4950,[532] d_4^{20} 0.9420.[532]

$C_8H_{21}N_2P$

$BuP(NMe_2)_2$. 2a.[269] 2b.[509] b_2 61°,[269] b_{11} 75°,[509] b_{722} 190°,[509] n_D^{20} 1.4598,[269] 1.4600,[509] d_4^{20} 0.8639,[269] 0.867.[509]
i-$BuP(NMe_2)_2$. 2c.[404] b_{720} 185-191°,[404] n_D^{20} 1.4615,[404] IR,[404] ^{31}P -92.4.[404]

$C_9H_{17}N_2O_2P$

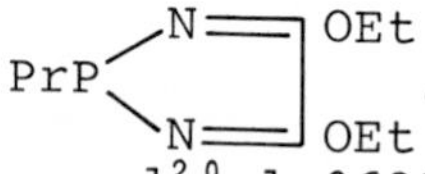
. 19.[142] $b_{0.7}$ 104-106°,[142] n_D^{20} 1.4932,[142] d_4^{20} 1.0683.[142]

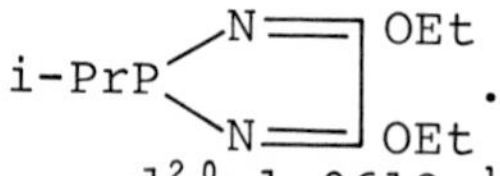
. 19.[142] $b_{0.5}$ 96-97°,[142] n_D^{20} 1.4922,[142] d_4^{20} 1.0613.[142]

$C_9H_{19}N_2O_2P$

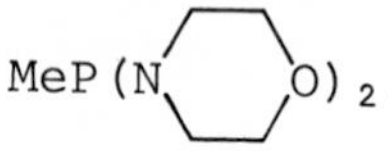
. 10.[403] $b_{0.15}$ 87-88°,[403] m. 45-47°,[403] ^{31}P -82.5.[403]

$C_9H_{23}N_2P$

$MeP(NEt_2)_2$. 7a.[496,546] 10.[403,733] b_5 78°,[403] b_{10} 80-82°,[496] b_{11} 82-85°,[546] n_D^{20} 1.4655,[496] 1.4658,[546] d_4^{20} 0.8990,[546] ^{31}P -79.5,[376] -80,[733] -80.4,[403] ^1H 1.01 ($CH_2C\underline{H}_3$), 1.18 (PCH_3), 2.96 (CH_2),[376] ΔH_f (heat of formation).[733]

$C_{10}H_{11}N_2O_2P$

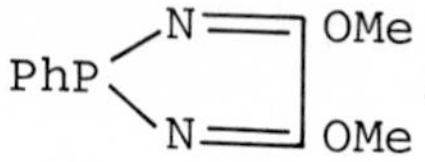
. 19.[150] $b_{0.2}$ 145-146°,[150] m. 61-65°.[150]

$C_{10}H_{12}F_5N_2P$

$C_6F_5P(NMe_2)_2$. 2a.[62] 7a.[62] 22.[62] $b_{0.1}$ 50-52°,[62] 54°,[399] IR,[62,399] ^1H 3.17,[61] ^{19}F ($CFCl_3$ ref.) 140.2 (o-F), 157.1 (p-F), 164.1 (m-F),[61] mass spect.[62]

$C_{10}H_{13}N_2P$

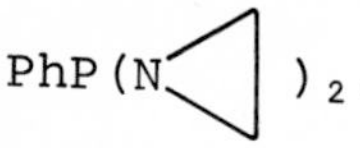
. 7b.[514] $b_{0.03}$ 67-68°,[514] n_D^{20} 1.5912,[514] d_4^{20}

1.0986,[514] Raman.[688]

$C_{10}H_{16}ClN_2P$

m-$ClC_6H_4P(NMe_2)_2$. 7a.[390] b_3 97°,[390] m. 42-44°,[390] n_D^{20} 1.5583,[390] d_4^{20} 1.1055.[390]
p-$ClC_6H_4P(NMe_2)_2$. 7a.[725] b_5 116°,[725] n_D^{20} 1.5602,[725] d_4^{20} 1.1022.[725]

$C_{10}H_{16}FN_2P$

m-$FC_6H_4P(NMe_2)_2$. 7a.[616] $b_{0.2}$ 87°,[616] ^{19}F (C_6H_5F ref.) 0.63.[616]
p-$FC_6H_4P(NMe_2)_2$. 7a.[616] $b_{1.3}$ 83°,[616] ^{19}F (C_6H_5F ref.) 2.33.[616]

$C_{10}H_{17}N_2P$

$PhP(NMe_2)_2$. 2b.[102,169] 7a.[130,170,669] 10.[403] 41.[509] $b_{0.02}$ 40-43°,[169] $b_{0.05}$ 75-76°,[403] $b_{0.2}$ 59.0-59.5°,[170] b_3 80°,[509] b_{12} 124°,[669] n_D^{20} 1.5476,[130] 1.5479,[509] 1.5510,[403] d_4^{21} 1.001,[509] add'l data,[91,130,170,272] IR,[169] ^{31}P -100.3,[403] -101.5,[91] ^1H 2.78 (CH_3).[508]

$C_{10}H_{19}N_2O_2P$

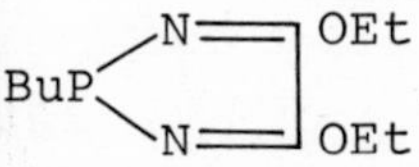

. 19.[150] $b_{0.2}$ 101-102°.[150]

$C_{10}H_{20}F_3N_2P$

$CF_2{=}CFP(NEt_2)_2$. 2a.[710] b_5 75-77°,[710] b_{11} 89-90°,[710] n_D^{20} 1.4470,[710] d_4^{20} 1.054.[710]

$C_{10}H_{21}N_2P$

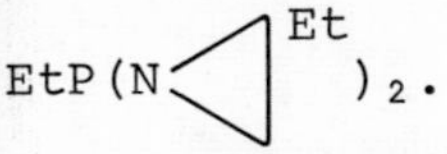

. 7b.[514] b_{10} 97-98°,[514] n_D^{20} 1.4782,[514] d_4^{20} 0.9192.[514]

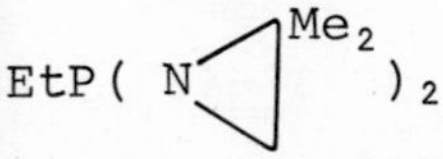

. 7b.[514] b_{10} 87-88°,[514] n_D^{20} 1.4779,[514] d_4^{20} 0.9153,[514] IR.[514]

$C_{10}H_{25}N_2P$

$EtP(NEt_2)_2$. 2a.[710] b_3 72-73°,[710] n_D^{20} 1.4680,[710] d_4^{20} 0.885,[710] DTA.[597]

$C_{11}H_{23}N_2P$

MeP(N$\bigcirc$)$_2$.   10.[403]   30.[532] $b_{0.13}$ 80-82°,[403] b_1 106-108°,[532] n_D^{20} 1.5164,[532] d_4^{20} 0.9843.[532]
$C_5H_{11}C\equiv CP(NMe_2)_2$.   2a.[111] b_{14} 125°.[111]

$C_{12}H_{14}ClN_2O_2P$

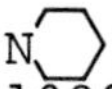

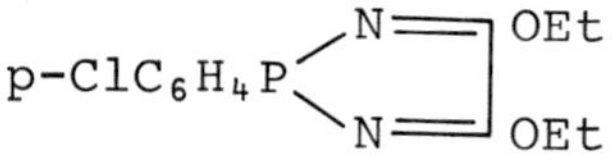

p-ClC$_6$H$_4$P$\begin{array}{c}N=\!=\!=OEt\\N=\!=\!=OEt\end{array}$.   19.[150] $b_{0.25}$ 152-153°.[150]

$C_{12}H_{15}N_2O_2P$

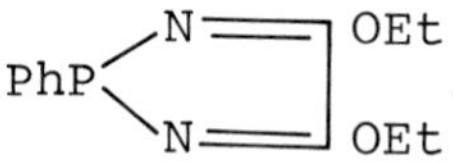

PhP$\begin{array}{c}N=\!=\!=OEt\\N=\!=\!=OEt\end{array}$.   19.[150] $b_{0.25}$ 149-150°,[150] m. 62-65°.[150]

$C_{12}H_{21}N_2P$

PhP(NMe$_2$)NEt$_2$.   41.[170] $b_{0.1}$ 72.5°.[170]
2,5-Me$_2$C$_6$H$_3$P(NMe$_2$)$_2$.   7a.[88] $b_{0.07}$ 86-87°,[88] n_D^{20} 1.5432,[88] d_4^{20} 0.9815.[88]
3,4-Me$_2$C$_6$H$_3$P(NMe$_2$)$_2$.   7a.[88] $b_{0.05}$ 73-75°,[88] n_D^{20} 1.5479,[88] d_4^{20} 0.9868.[88]

$C_{12}H_{27}B_{10}N_2P$

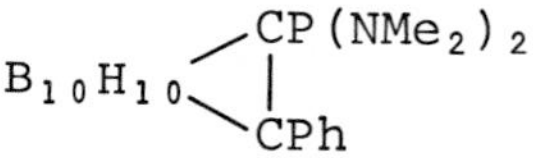

B$_{10}$H$_{10}$$\begin{array}{c}CP(NMe_2)_2\\|\\CPh\end{array}$.   2b.[764]   m. 74-75°.[764]

$C_{12}H_{29}N_2P$

BuP(NEt$_2$)$_2$. ^{31}P -99.8.[425e]
s-BuP(NEt$_2$)$_2$.   2a.[269] $b_{0.8}$ 64-65°,[269] n_D^{20} 1.4676,[269] d_4^{20} 0.8808.[269]

$C_{13}H_{17}N_2O_2P$

p-MeC$_6$H$_4$P$\begin{array}{c}N=\!=\!=OEt\\N=\!=\!=OEt\end{array}$.   19.[150] $b_{0.3}$ 158-160°,[150] m. 37-42°.[150]

$C_{13}H_{23}N_2P$

MeP(N[CH$_2$CH=CH$_2$]$_2$)$_2$.   10.[409] $b_{0.02}$ 80-85°.[409]

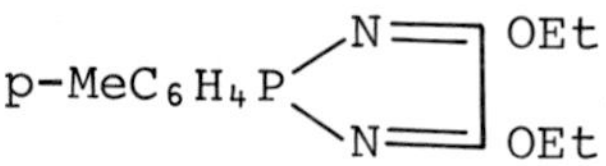

$C_{14}H_{20}F_5N_2P$

$C_6F_5P(NEt_2)_2$. 7b.[176] $b_{0.7}$ 100-102°,[176] ^{31}P -79.2.[176]

$C_{14}H_{21}N_2O_2P$

PhP(N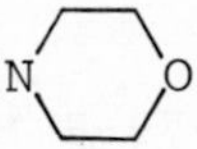O)$_2$. 7a.[198] 7b.[272] m. 113-115°.[198,272]

$C_{14}H_{21}N_2P$

PhP(N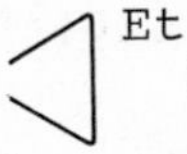)$_2$. 7b.[514] $b_{0.05}$ 91-92°,[514] n_D^{20} 1.5389,[514] d_4^{20} 1.0021.[514]

PhP(N)$_2$. 7b.[514] $b_{0.05}$ 82-83°,[514] n_D^{20} 1.5406,[514] d_4^{20} 1.0068.[514]

$C_{14}H_{24}ClN_2P$

o-$ClC_6H_4P(NEt_2)_2$. 2a.[667] b_{10} 164-167°.[667]
m-$ClC_6H_4P(NEt_2)_2$. 2a.[667] b_{11} 168-172°.[667]
p-$ClC_6H_4P(NEt_2)_2$. 2a.[667] b_{10} 163-168°.[667]

$C_{14}H_{24}FN_2P$

m-$FC_6H_4P(NEt_2)_2$. 2a.[666] b_{10} 147°,[666] n_D^{20} 1.5199.[666]
p-$FC_6H_4P(NEt_2)_2$. 2a.[667] b_9 148-151°.[667]

$C_{14}H_{25}N_2P$

$PhP(NEt_2)_2$. 7a.[170] 7b.[462] 22.[3] $b_{0.005}$ 80°,[3] $b_{0.1}$ 91.5°,[170] $b_{0.2}$ 115-117°,[407] 123-125°,[462] b_{10} 150-151°,[483] n_D^{20} 1.5338,[483] 1.5339,[3] IR,[239] ^{31}P -60.8,[407] -95.6,[425e] -98,[239] ^1H 0.94 (CH_3), 3.06 (CH_2), 7.38 (C_6H_5),[407] add'l ^1H data.[239]

$C_{14}H_{31}N_2P$

c-$C_6H_{11}P(NEt_2)_2$. 2a.[269] 7a;[269,491] also from c-$C_6H_{11}P(NEt_2)Cl$, Na.[686] $b_{0.1}$ 89-90°,[269] b_3 125-126°,[491] b_6 133-135°,[686] n_D^{20} 1.4890,[269] 1.4900,[269,491] d_4^{20} 0.9248,[269] 0.9270,[491] 0.9276,[269] ^{31}P -107.1.[425e]

$C_{15}H_{19}N_2P$

$MeP(NMePh)_2$. 10.[403] $b_{0.2}$ 136-138°,[403] ^{31}P -76.2.[403]

$C_{15}H_{27}N_2P$

o-$MeC_6H_4P(NEt_2)_2$. 2a.[667] b_{10} 155-157°.[667]
m-$MeC_6H_4P(NEt_2)_2$. 2a.[667] b_{12} 159-161°.[667]
p-$MeC_6H_4P(NEt_2)_2$. 2a.[667] b_{12} 154-156°.[667]

$C_{15}H_{31}N_2P$

$MeP(NMeC_6H_{11}$-c$)_2$. 10.[403] $b_{0.13}$ 116-119°,[403] ^{31}P -79.6.[403]

$C_{15}H_{35}N_2P$

n-$C_7H_{15}P(NEt_2)_2$. 7a.[536] b_1 145-148°,[536] n_D^{20} 1.4620,[536] d_4^{20} 0.9156.[536]

$C_{16}H_{24}ClN_2P$

p-ClC_6H_4P(N$)_2$. 7a.[446] m. 95°.[446]

$C_{16}H_{25}N_2P$

PhP(N$)_2$. 7a.[159,446] m. 76°,[744,746] 78°,[446] 79.0-80.5°,[274] m. 80.5°,[159] 82-83°,[193] C.A. Registry No. 22979-14-2.

$C_{16}H_{29}N_2P$

$PhP(NEt_2)NPr_2$. 41.[170] $b_{0.05}$ 110-114°.[170]
2,5-$Me_2C_6H_3P(NEt_2)_2$. 7a.[88] $b_{0.25}$ 108°,[88] n_D^{20} 1.5325,[88] d_4^{20} 0.9635,[88] ^{31}P -92.8.[88]
2,6-$Me_2C_6H_3P(NEt_2)_2$. 2a.[88] $b_{0.07}$ 104-105°,[88] n_D^{20} 1.5365,[88] d_4^{20} 0.984,[88] ^{31}P -92.8.[425e]
3,4-$Me_2C_6H_3P(NEt_2)_2$. 7a.[88] $b_{0.1}$ 113.5°,[88] n_D^{20} 1.534,[88] d_4^{20} 0.962.[88]

$C_{17}H_{27}N_2OP$

p-$MeOC_6H_4P$(N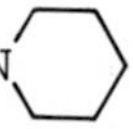$)_2$. 7a.[446] m. 69°.[446]

$C_{17}H_{27}N_2P$

p-MeC_6H_4P(N$)_2$. 7a.[446] m. 80°,[446] m. 85°.[744]

$C_{18}H_{29}N_2OP$

p-$EtOC_6H_4P$(N 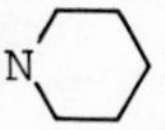)$_2$. 7a.[446] m. 84°.[446]

$C_{18}H_{33}N_2P$

$PhP(NPr_2)_2$. 7a.[170] $b_{0.1}$ 122°,[170] IR,[239] ^{31}P -98°,[239] ^1H 0.40 (CH_3), 1.33 ($\underline{CH_2}CH_3$), 2.92 (NCH_2), 7.09 (C_6H_5).[239]

$C_{20}H_{29}FeN_2P$

$C_5H_5FeC_5H_4P$(N 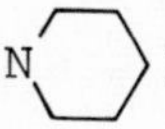)$_2$ (ferrocene deriv.). 7a.[702] m. 106-107°,[702] IR.[702]

$C_{22}H_{33}N_2P$

PhP(N )$_2$. 7a.[700] m. 136-137°.[700]

$C_{22}H_{40}FN_2P$

m-$FC_6H_4P(NBu_2)_2$. 2a.[666] $b_{0.45}$ 200°,[666] n_D^{20} 1.5223.[666]

$C_{22}H_{41}N_2P$

$PhP(NBu_2)_2$. 7a.[239] $b_{0.14}$ 141°,[239] IR.[239]

$C_{24}H_{25}N_2P$

$PhP(NC_9H_{10})_2$ (C_9H_{10} = tetrahydroquinolyl). 7a.[446] m. 150°.[446]

$C_{24}H_{27}N_4O_2P$

$PhP(NPhC(O)NMe_2)_2$. 36.[150] m. 57-59°,[150] IR.[150]

$C_{25}H_{27}N_2P$

p-$MeC_6H_4P(NC_9H_{10})_2$ (C_9H_{10} = tetrahydroquinolyl). 7a.[446] m. 140°.[446]

$C_{26}H_{31}N_4O_6PS_2$

PhP(N[$SO_2C_6H_4Me$-p]$CONMe_2$)$_2$. 36.[128] Not described.

$C_{29}H_{31}N_2P$

$MeP(N[CH_2Ph]_2)_2$. 10.[409] $b_{0.13}$ 115-117°.[409]

$C_{30}H_{25}N_2P$

$PhP(NPh_2)_2$. 14.[257] m. 139-140°.[257]

Compounds with Two Phosphorus Atoms

$C_{10}H_{34}B_{10}N_4P_2$

m-$B_{10}H_{10}C_2[P(NMe_2)_2]_2$. 7a.[20] m. 73-75°.[20]

$C_{14}H_{28}N_4P_2$

p-$C_6H_4[P(NMe_2)_2]_2$. 2b.[102,169] m. 146.5-147.7°,[169] IR.[169]

$C_{22}H_{44}N_4P_2$

p-$C_6H_4[P(NEt_2)_2]_2$. 2a.[110] m. 100-101°,[110] IR.[110]

Compounds with Three Phosphorus Atoms

$C_{34}H_{35}N_2P_3$

$PhP(NEtPPh_2)_2$. From $PhP(NHEt)_2$, Ph_2PCl (Et_3N).[372] m. 144-145°.[372]

$C_{49}H_{43}Br_2N_2P_3$

$MeP(NPhPPh_3{}^+ Br^-)_2$. 19.[410] m. 196-198°,[410] reacts with $HgBr_2$, $NaBPh_4$, and $Cr(SCN)_4(NH_3)_2$, forming salts; BPh_4 salt, m. 172-173°.[410]

Deuterium-Labeled Compounds

$C_{14}H_{25}N_2P$

p-$DC_6H_4P(NEt_2)_2$. 2b.[249] b_2 132-134°.[249]

Hydrazine Derivatives

$C_5H_{14}F_3N_4P$

$CF_3P(NHNMe_2)_2$. 12.[525] b. 184°$_{extrap.}$,[525] unstable > 100°,[525] $\log P = 5.9707 + 1.75 \log T - 0.002678\ T - 2985/T$,[525] m. 49.20°,[525] ^1H 2.41 (CH_3), 4.08 (NH),[525] ^{19}F ($CFCl_3$ ref.) 65.4,[525] mass spect.[525]

$C_{10}H_{19}N_4P$

$PhP(NHNMe_2)_2$. 7a.[487] m. 61.5-63.0°,[487] m. 68-69.5°,[760] IR.[487,760]

$C_{12}H_{23}N_4P$

$PhP(NMeNMe_2)_2$. 7b.[321] $b_{0.5}$ 120-133°,[321] IR,[321] ^{31}P -80.2,[321] ^1H 2.34 (NMe_2), 2.40 (NMe), 2.48 (NMe), 7.00 (C_6H_5).[321]

$C_{18}H_{19}N_4P$

$PhP(NHNHPh)_2$. 7a.[449] White, cryst. powder.[449]

Compounds with Two Phosphorus Atoms

$C_{16}H_{22}N_4P_2$

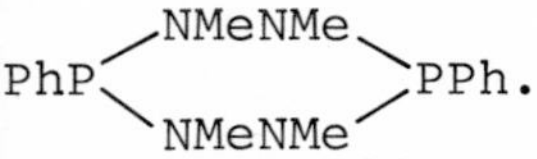
7b.[703] m. 222-223°,[703] IR,[703] ^{31}P -29.9,[703] ^1H 3.04 (CH_3), 7.56 (C_6H_5).[703]

Triazene Derivatives

$C_{10}H_{17}N_6P$

$PhP(NMeN{=}NMe)_2$. 21.[89] $b_{0.2}$ 109°,[89] explodes > 120°,[89] ^1H 3.78 (terminal CH_3), 4.17 (middle CH_3).[89]

D.4.9. Iminophosphines

Primary Amine Derivatives

$C_8H_8F_2NP$

$CHF_2CH_2P{=}NPh$. 55.[221] m. 160°,[221] IR.[221]

$C_{12}H_{10}NP$

PhP=NPh. 16b.[289] Trimer: b. 447° (extrap),[289] log P = 8.44 - 4000/T,[289] m. 264-265°,[289] d. 352°.[289]

Hydrazine Derivatives

$C_{12}H_{10}BrN_2P$

$p\text{-}BrC_6H_4P{=}NNHPh$. 16a.[441] m. 160°.[441]

$C_{12}H_{10}ClN_2P$

p-ClC_6H_4P=NNHPh. 16a.[441] m. 161° dec.[441]

$C_{12}H_{11}N_2P$

PhP=NNHPh. 16a.[449] m. 152°.[449]

$C_{13}H_{13}N_2P$

PhP=NNHC_6H_4Me-p. 16a.[449] m. 162°.[449]

$C_{14}H_{15}N_2P$

p-EtC_6H_4P=NNHPh. 16a.[442] m. 139°.[442]

$C_{15}H_{17}N_2P$

2,4,6-$Me_3C_6H_2$P=NNHPh. 16a.[443] m. 135°.[443]

$C_{19}H_{17}N_2P$

PhP=NNPhCH_2Ph. 16a.[449] m. 141°.[449]

D.4.10. N-Silyl Derivatives

Phosphonamidous Chloride Derivatives

$C_8H_{21}ClNPSi$

t-Bu(NMeSiMe_3)Cl. 23.[664] $b_{0.1}$ 45-48°,[664] m. -4°,[664] ^1H -0.26 (SiCH_3), 0.72 (t-Bu), 2.12 (NCH_3).[664]

Phosphonous Diamide Derivatives

$C_8H_{23}N_2PSi$

t-BuP(NHMe)NHSiMe_3. 41.[664] $b_{0.1}$ 42-44°,[664] ^1H -0.24 (SiCH_3), 0.49 (t-Bu), 2.08 (NCH_3).[664]

$C_9H_{25}N_2PSi$

t-BuP(NHMe)NMeSiMe_3. 41.[664] $b_{0.1}$ 41-43°,[664] ^1H -0.11 (SiCH_3), 0.70 (t-Bu), 1.98, 2.13 (NCH_3).[664]

$C_{10}H_{29}N_2PSi_2$

t-BuP(NHSiMe_3)$_2$. 37.[664] $b_{0.1}$ 51-55°,[664] m. -1°,[664] ^1H 0.30 (t-Bu), -0.33 (SiCH_3).[664]

$C_{14}H_{29}N_2PSi_3$

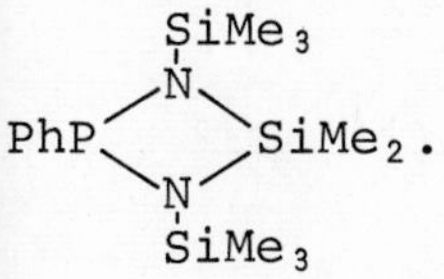

24.[180] b_4 107°,[180] n_D^{20} 1.5029.[180]

$C_{30}H_{25}N_2PSi$

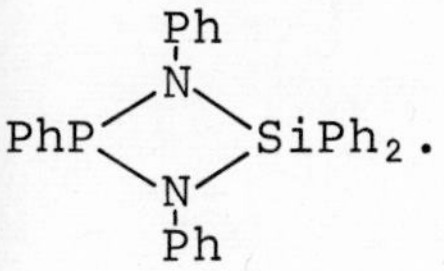

24.[179] m. 212°.[179]

D.5. Phosphonous Anhydrides

D.5.1. Phosphinic Anhydrides

$C_{12}H_{12}O_3P_2$

$[PhPH(O)]_2O$. 14.[198] ^1H 7.4-7.9 (C_6H_5), 8.05 (PH).[198]

$C_{36}H_{60}O_3P_2$

$[2,4,6\text{-}(t\text{-}Bu)_3C_6H_2PH(O)]_2O$. 2.[120] m. 259.5-260.5°,[120] IR.[120]

D.5.2. Phosphonochloridous Anhydrides

Anhydrides with Carboxylic Acids

$C_3H_3ClF_3OP$

$CF_3P(OAc)Cl$. 12.[524] Vap. press. 14 mm (0°).[524]

D.5.3. Phosphonous Ester Anhydrides

Anhydrides with Phosphinous Acids

$C_8H_{20}O_2P_2$

$EtP(OEt)OPEt_2$. 9.[46] $b_{1.5}$ 91-93°,[46] n_D^{20} 1.4868,[46] d_4^{20} 1.0042.[46]

Anhydrides with Phosphorothionic Diesters

$C_{17}H_{30}O_4P_2S$

$PhP(OPr)OP(S)(OBu)_2$. 19.[167] $b_{0.001}$ 101-102°,[167] n_D^{20} 1.5003,[167] d_4^{20} 1.1010.[167]

Anhydrides with Carboxylic Acids

$C_5H_{11}O_3P$

MeP(OEt)OAc. 13.[550] b_{11} 89-91°,[550] n_D^{20} 1.4320,[550] d_4^{20} 1.0970.[550]

$C_{11}H_{15}O_3P$

PhP(OPr)OAc. 19.[168] b_1 101-103°,[168] n_D^{20} 1.5130,[168] d_4^{20} 1.0870.[168]

D.5.4. Phosphonamidous Anhydrides

Anhydrides with Phosphoric Diesters

$C_{18}H_{33}NO_4P_2$

$PhP(NEt_2)OP(O)(OBu)_2$. 18.[167] $b_{0.0001}$ 111-112°,[167] n_D^{20} 1.4557,[167] IR.[730]

Anhydrides with Phosphorothionic Diesters

$C_{18}H_{33}NO_3P_2S$

$PhP(NEt_2)OP(S)(OBu)_2$. 18.[167] $b_{0.0001}$ 99-100°,[167] n_D^{20} 1.5130,[167] d_4^{20} 1.0639,[167] IR.[730]

Anhydrides with Phosphonamidic Acids

$C_{15}H_{28}N_2O_2P$

$PhP(NEt_2)OP(O)(NEt_2)Me$. 17.[465] $b_{0.05}$ 93°,[465] n_D^{20} 1.5130,[465] d_4^{20} 1.0820.[465]

D.5.5. Phosphonochloridothious Anhydrides

Anhydrides with Phosphonochloridothionic Acids

$C_4H_{10}Cl_2P_2S_2$

EtP(Cl)SP(S)(Cl)Et. 4.[64] n_D^{20} 1.5777,[64] IR.[64]

$C_{12}H_{10}Cl_2P_2S_2$

PhP(Cl)SP(S)(Cl)Ph. 4.[64] n_D^{20} 1.6530,[64] IR.[64]

D.5.6. Phosphonous Anhydrides

CF_3OP

$[CF_3PO]_n$. 2 (C.3.).[524] red-brown waxy solid.[524]

CH_3OP

[MePO]$_n$. 1a.[228] 1b.[492] col. dense oil[492] or cryst.,[228] polymerizes rapidly to amorphous powder.[492]

C_2H_5OP

[EtPO]$_n$. 1a.[228] cryst.[228]

C_4H_9OP

[i-BuPO]$_n$. 1b.[491] thick col. oil,[491] becomes viscous on standing.[491]

C_6H_5OP

[PhPO]$_n$. 1a.[464] 1b.[464] 16.[464] glassy solid, s. ~ 100° (n = 5-7).[464]

$C_6H_{11}OP$

[c-C_6H_{11}PO]$_n$. 16.[269] cryst., m. 252°.[269]

Anhydrides with Phosphorous Diesters

$C_6H_{17}O_6P_3$

EtP[OP(OMe)$_2$]$_2$. 6.[45] Explodes during distillation (140°).[45]

$C_{10}H_{25}O_6P_3$

EtP[OP(OEt)$_2$]$_2$. 6.[45] 7.[45] $b_{0.5}$ 126-128°,[45] b_1 128-130°,[45] 1.4768,[45] 1.4772,[45] d_4^{20} 1.1524,[45] 1.1531.[45]

$C_{14}H_{33}O_6P_3$

EtP[OP(OPr)$_2$]$_2$. 7.[45] b_1 145-147°,[45] n_D^{20} 1.4662,[45] d_4^{20} 1.0785.[45]

EtP[OP(OPr-i)$_2$]$_2$. 7.[45] b_1 130-132°,[45] n_D^{20} 1.4622,[45] d_4^{20} 1.0693.[45]

$C_{18}H_{41}O_6P_3$

EtP[OP(OBu-i)$_2$]$_2$. 6.[45] b_2 165-167°,[45] n_D^{20} 1.4643,[45] d_4^{20} 1.0420.[45]

Anhydrides with Phosphonous Esters

$C_{12}H_{29}O_4P_3$

EtP[OP(OPr-i)Et]$_2$. 8.[45] b_1 144-146°,[45] n_D^{20} 1.4928,[45] d_4^{20}

1.0778.[45]

Anhydrides with Carboxylic Acids

$C_5F_9O_4P$

$CF_3P(OC(O)CF_3)_2$. 10.[524] 11.[524] b. 111°$_{extrap}$,[524] log P = 6.738 + 1.75 log T - 0.005935 T - 2342/T,[524] m. - 23.7°,[524] IR.[524]

$C_5H_6F_3O_4P$

$CF_3P(OAc)_2$. 10.[524] 12.[524] Vap. press. 0.4 mm (22°),[524] UV,[524] IR.[524]

D.5.7. Phosphonothious Anhydrides

C_6H_5PS

$[PhPS]_n$. 3.[464] 5.[1] m. 126-133° (n = 5),[464] m. 148° (n = 3),[1] IR.[1]

(received March 2, 1971)

REFERENCES

1. Abel, E. W., D. A. Armitage, and R. P. Bush, J. Chem. Soc., 1964, 5584.
2. Abel, E. W., D. A. Armitage, and R. P. Bush, J. Chem. Soc., 1965, 7098.
3. Abel, E. W., D. A. Armitage, and G. R. Willey, J. Chem. Soc., 1965, 57.
4. Abramov, V. S., Dokl. Akad. Nauk SSSR, 117, 811 (1957).
5a. Abramov, V. S., and V. I. Barabanov, Khim. Org. Soedin. Fosfora, 1967, 135; C.A. 69, 67469c (1968).
5b. Abramov, V. S., and V. I. Barabanov, Zh. Obshch. Khim., 36, 1830 (1966).
6. Abramov, V. S., V. I. Barabanov, and Z. Ya. Sazonova, Zh. Obshch. Khim., 39, 1543 (1969).
7. Abramov, V. S., R. Sh. Chenborisov, and V. V. Markin, Zh. Obshch. Khim., 38, 2588 (1968).
8. Abramov, V. S., R. Sh. Chenborisov, and V. V. Markin, Zh. Obshch. Khim., 39, 464 (1969).
9. Abramov, V. S., and M. I. Kashirskii, Zh. Obshch. Khim., 28, 3059 (1958).
10. Abramov, V. S., and R. N. Savintseva, Zh. Obshch. Khim., 39, 1967 (1969).
11. Abramov, V. S., and L. A. Tarasov, Zh. Obshch. Khim., 38, 674 (1968).
12. Adrova, N. A., L. K. Prokhorova, and M. M. Koton, Izv. Akad. Nauk SSSR, Ser. Khim., 1966, 1824.
13. Aguiar, A. M., J. R. S. Irelan, C. J. Morrow, J. P.

John, and G. W. Prejean, J. Org. Chem., 34, 2684 (1969).
14. Akamsin, V. D., and N. I. Rizpolozhenskii, Izv. Akad. Nauk SSSR, Ser. Khim., 1966, 493.
15. Akamsin, V. D., and N. I. Rizpolozhenskii, Izv. Akad. Nauk SSSR, Ser. Khim., 1967, 825.
16. Akamsin, V. D., and N. I. Rizpolozhenskii, Izv. Akad. Nauk SSSR, Ser. Khim., 1967, 1976.
17. Akamsin, V. D., and N. I. Rizpolozhenskii, Izv. Akad. Nauk SSSR, Ser. Khim., 1967, 1983.
18. Akamsin, V. D., and N. I. Rizpolozhenskii, Izv. Akad. Nauk SSSR, Ser. Khim., 1967, 1987.
19. Aksnes, G., and D. Aksnes, Acta Chem. Scand., 18, 38 (1964).
20. Alexander, R. P., and H. Schroeder, Inorg. Chem., 5, 493 (1966).
21. Allen, J. F., J. Amer. Chem. Soc., 79, 3071 (1957).
22. Allen, J. F., and O. H. Johnson, J. Amer. Chem. Soc., 77, 2871 (1955).
23. Anand, N., and A. R. Todd, J. Chem. Soc., 1951, 1867.
24. Ang, H. G., and H. J. Emeléus, J. Chem. Soc., A, 1968, 1334.
25. Ang, H. G., G. Manoussakis, and Y. O. El-Nigumi, J. Inorg. Nucl. Chem., 30, 1715 (1968).
26. Anisimov, K. N., and N. E. Kolobova, Izv. Akad. Nauk SSSR, Otd. Khim. Nauk, 1962, 442; C.A. 57, 13790f (1962).
27. Arapahoe Chemicals, Boulder, Colo. 80302; Chem. Week, March 4, 1970, p. 75.
28. Arbuzov, A. E., J. Russ. Phys.-Chem. Soc., 42, 395 (1910); Chem. Zentralbl., 1910, II, 453.
29. Arbuzov, A. E., Zh. Obshch. Khim., 4, 898 (1934); C.A. 29, 2146 (1935).
30. Arbuzov, A. E., and B. A. Arbuzov, J. Russ. Phys.-Chem. Soc., 61, 1599 (1929); C.A. 24, 5289 (1930).
31. Arbuzov, A. E., and I. A. Arbuzova, J. Russ. Phys.-Chem. Soc., 61, 1905 (1929); C.A. 24, 5289 (1930).
32. Arbuzov, A. E., and G. Kamai, J. Russ. Phys.-Chem. Soc., 61, 2037 (1929); C.A. 24, 5736 (1930).
33. Arbuzov, A. E., and G. Kamai, Zh. Obshch. Khim., 17, 2149 (1947); C.A. 42, 4523g (1948).
34. Arbuzov, A. E., G. Kamai, and O. N. Belorossova, Zh. Obshch. Khim., 15, 766 (1945); C.A. 41, 105f (1947).
35. Arbuzov, A. E., G. Kamai, and L. V. Nesterov, Tr. Kazan. Khim. Tekhnol. Inst., 16, 17 (1951); C.A. 51, 5720f (1957).
36. Arbuzov, A. E., and A. I. Razumov, Izv. Akad. Nauk SSSR, Otd. Khim. Nauk, 1945, 167; C.A. 40, 3411 (1946).
37. Arbuzov, B. A., and N. P. Grechkin, Izv. Akad. Nauk SSSR, Otd. Khim. Nauk, 1956, 440.
38. Arbuzov, B. A., and N. P. Grechkin, Zh. Obshch. Khim.,

50, 107 (1950).
39. Arbuzov, B. A., and O. A. Nuretdinova, Izv. Akad. Nauk SSSR, Ser. Khim., 1969, 1314.
40. Arbuzov, B. A., and N. I. Rizpolozhenskii, Izv. Akad. Nauk SSSR, Otd. Khim. Nauk, 1952, 854; C.A. 47, 9903b (1953).
41. Arbuzov, B. A., and N. I. Rizpolozhenskii, Izv. Akad. Nauk SSSR, Otd. Khim. Nauk, 1952, 956; C.A. 47, 9904c (1953).
42. Arbuzov, B. A., and N. I. Rizpolozhenskii, Izv. Akad. Nauk SSSR, Otd. Khim. Nauk, 1954, 631; C.A. 49, 11541h (1955).
43. Arbuzov, B. A., and N. I. Rizpolozhenskii, Izv. Akad. Nauk SSSR, Otd. Khim. Nauk, 1955, 253.
44. Arbuzov, B. A., N. I. Rizpolozhenskii, and M. A. Zvereva, Izv. Akad. Nauk SSSR, Otd. Khim. Nauk, 1955, 1021.
45. Arbuzov, B. A., N. I. Rizpolozhenskii, and M. A. Zvereva, Izv. Akad. Nauk SSSR, Otd. Khim. Nauk, 1957, 179.
46. Arbuzov, B. A., N. I. Rizpolozhenskii, and M. A. Zvereva, Izv. Akad. Nauk SSSR, Otd. Khim. Nauk, 1958, 706.
47. Arbuzov, B. A., and T. G. Shavsha-Tolkacheva, Izv. Akad. Nauk SSSR, Otd. Khim. Nauk, 1954, 812.
48. Arbuzov, B. A., and V. S. Vinogradova, Izv. Akad. Nauk SSSR, Otd. Khim. Nauk, 1947, 459.
49. Arbuzov, B. A., and V. S. Vinogradova, Izv. Akad. Nauk SSSR, Otd. Khim. Nauk, 1954, 622.
50. Arbuzov, B. A., V. S. Vinogradova, and M. A. Zvereva, Izv. Akad. Nauk SSSR, Otd. Khim. Nauk, 1960, 1772.
51. Arbuzov, B. A., V. S. Vinogradova, and M. A. Zvereva, Izv. Akad. Nauk SSSR, Otd. Khim. Nauk, 1960, 1981.
52. Arnold, G. B., and C. S. Hamilton, J. Amer. Chem. Soc., 63, 2637 (1941).
53. Atkinson, R. E., J. I. G. Cadogan, and J. Dyson, J. Chem. Soc., C. 1967, 2542.
54. Bailey, W. J., and R. B. Fox, J. Org. Chem., 29, 1013 (1964).
55. Baldwin, R. A., C. O. Wilson, Jr., and R. I. Wagner, J. Org. Chem., 32, 2172 (1967).
56. Ballschmiter, K., and H. Singer, Chem. Ber., 101, 7 (1968).
57. Banks, J. E., Anal. Chim. Acta, 19, 331 (1958).
58. Bannister, W. D., B. L. Booth, M. Green, and R. N. Haszeldine, J. Chem. Soc., A, 1969, 698.
59. Barabanov, V. I., and V. S. Abramov, Zh. Obshch. Khim., 35, 2225 (1965).
60. Baranov, Yu. I., O. F. Filippov, S. L. Varshavskii, and M. I. Kabachnik, Dokl. Akad. Nauk SSSR, 182, 337 (1968); C.A. 70, 11756 x (1969).

61. Barlow, M. G., M. Green, R. N. Haszeldine, and H. G. Higson, J. Chem. Soc., B, 1966, 1025.
62. Barlow, M. G., M. Green, R. N. Haszeldine, and H. G. Higson, J. Chem. Soc., C, 1966, 1592.
63. Batkowski, T., P. Mastalerz, M. Michalewska, and B. Nitka, Rocz. Chem., 41, 471 (1967); C.A. 67, 32738 w (1967).
64. Baudler, M., and H.-W. Valpertz, Z. Naturforsch., 22 B, 222 (1967).
65. Beg, M. A. A., and H. C. Clark, Can. J. Chem., 39, 564 (1961).
66. Benda, L., and W. Schmidt, German Patent 379,813 (1924); Chem. Zentralbl., 1924, II, 1271.
67. Bennett, F. W., H. J. Emeléus, and R. N. Haszeldine, J. Chem. Soc., 1954, 3896.
68. Benschop, H. P., D. H. J. M. Platenburg, F. H. Meppelder, and H. L. Boter, Chem. Commun., 1970, 33.
69. Bentov, M., L. David, and E. D. Bergmann, J. Chem. Soc., 1964, 4750.
70. Bentrude, W. G., and E. R. Witt, J. Amer. Chem. Soc., 85, 2522 (1963).
71. Beriger, E., and R. Sallmann, U.S. Patent 2,908,605 (1959); C.A. 54, 15189a (1960).
72. Berlin, K. D., T. H. Austin, M. Nagabhushanam, M. Peterson, J. Calvert, L. A. Wilson, and D. Hopper, J. Gas Chromatogr., 3, 256 (1965).
73. Berlin, K. D., and M. Nagabhushanam, J. Org. Chem., 29, 2056 (1964).
74. Birum, G. H., U.S. Patent 3,014,944 (1961); C.A. 56, 11622h (1962).
75. Birum, G. H., U.S. Patent 3,014,954 (1961); C.A. 56, 12946h (1962).
76. Birum, G. H., and J. L. Dever, U.S. Patent 2,961,455 (1960); C.A. 55, 8292g (1961).
77. Bliznyuk, N. K., A. F. Kolomiets, Z. N. Kvasha, G. S. Levskaya, and V. V. Antipina, Zh. Obshch. Khim., 36, 475 (1966).
78. Bliznyuk, N. K., Z. N. Kvasha, P. S. Khokhlov, and A. F. Kolomiets, Zh. Obshch. Khim., 37, 884 (1967).
79. Bliznyuk, N. K., Z. N. Kvasha, and A. F. Kolomiets, Zh. Obshch. Khim., 37, 888 (1967).
80. Bochwic, B., and J. Michalski, Rocz. Chem., 26, 593 (1952); C.A. 49, 2345g (1955).
81. Bogolyubov, G. M., and A. A. Petrov, Zh. Obshch. Khim., 37, 229 (1967).
82. Boisselle, A. P., and N. A. Meinhardt, J. Org. Chem., 27, 1828 (1962).
83. Bokanov, A. I., and V. A. Plakhov, Zh. Obshch. Khim., 35, 350 (1965).
84. Booth, B. L., and R. N. Haszeldine, J. Chem. Soc., A, 1966, 157.

85. Borecki, C., J. Michalski, and S. Musierowitz, J. Chem. Soc., 1958, 4081.
86. Bourneuf, M., Bull. Soc. Chim. (France), [4], 33, 1808 (1923).
87. Brazier, J.-F., D. Houalla, and R. Wolf, Bull. Soc. Chim. (France), 1970, 1089.
88. Brazier, J.-F., F. Mathis, and R. Wolf, C.R. Acad. Sci. Paris, Ser. C, 262, 1393 (1966).
89. Brinckman, F. E., H. S. Haiss, and R. A. Robb, Inorg. Chem., 4, 936 (1965).
90. Brinkman, Jr., G. H., and D. L. Elbert, U.S. Patent 3,235,534 (1966); C.A. 58, 8081h (1962).
91. Brown, C., M. Murray, and R. Schmutzler, J. Chem. Soc., C, 1970, 878.
92. Brown, D. H., K. D. Crosbie, G. W. Fraser, and D. W. A. Sharp, J. Chem. Soc., A, 1969, 551.
93. Brown, H. C., U.S. Patent 2,584,112 (1952); C.A. 46, 9580h (1952).
94. Bruker, A. B., L. D. Balashova, and L. Z. Soborovskii, Zh. Obshch. Khim., 36, 75 (1966).
95. Buckler, S. A., and M. Epstein, Tetrahedron, 18, 1211 (1962).
96. Buckler, S. A., and M. Epstein, Tetrahedron, 18, 1221 (1962).
97. Budzikiewicz, H., and Z. Pelah, Monatsh. Chem., 96, 1739 (1965).
98. Burch, G. M., H. Goldwhite, and R. N. Haszeldine, J. Chem. Soc., 1964, 572.
99. Burg, A. B., and K. Gosling, J. Amer. Chem. Soc., 87, 2113 (1965).
100. Burg, A. B., and J. E. Griffiths, J. Amer. Chem. Soc., 83, 4333 (1961).
101. Burg, A. B., K. K. Joshi, and J. F. Nixon, J. Amer. Chem. Soc., 88, 31 (1966).
102. Burg, A. B., and R. I. Wagner, U.S. Patent 2,934,564 (1960); C.A. 54, 18437b (1960).
103. Burn, A. J., and J. I. G. Cadogan, J. Chem. Soc., 1963, 5788.
103a. Butcher, F. K., W. Gerrard, M. Howarth, E. F. Mooney, and H. A. Willis, Spectrochim. Acta, 20, 79 (1964).
104. Cadogan, J. I. G., and D. T. Eastlick, J. Chem. Soc., B, 1970, 1314.
105. Cadogan, J. I. G., D. J. Sears, and D. M. Smith, J. Chem. Soc., C, 1969, 1314.
106. Cadogan, J. I. G., and M. J. Todd, J. Chem. Soc., C, 1969, 2808.
107. Campbell, I. G. M., and J. K. Way, J. Chem. Soc., 1960, 5034.
108. Campbell, J. R., and R. E. Hatton, U.S. Patent 3,071,609 (1963); C.A. 58, 11401f (1963).
109. Cavell, R. G., and J. F. Nixon, Proc. Chem. Soc.,

1964, 229.
110. Chantrell, P. G., C. A. Pearce, C. R. Toyer, and R. Twaits, J. Appl. Chem., 14, 563 (1964).
111. Charrier, C., and M.-P. Simonnin, C. R. Acad. Sci. Paris, Ser. C, 264, 995 (1967).
112. Chastain, B. B., E. A. Rick, R. L. Pruett, and H. B. Gray, J. Amer. Chem. Soc., 90, 3994 (1968).
113. Cherbuliez, E., G. Weber, and J. Rabinowitz, Helv. Chim. Acta, 46, 2464 (1963).
114. Chernyshev, E. A., E. F. Bugerenko, N. A. Nikolaeva, and A. D. Petrov, Dokl. Akad. Nauk SSSR, 147, 117 (1962).
115. Chittenden, R. A., and L. C. Thomas, Spetrochim. Acta, 21, 861 (1965).
116. Christol, H., and C. Marty, C.R. Acad. Sci. Paris, Ser. C, 262, 1722 (1966).
117. Coblentz Society, "Infra-red Spectra," Sadtler Research Laboratories, Philadelphia, Pa., 1959 ff: (a) no. 613, (b) no. 614, (c) no. 615, (d) no. 1545, (e) no. 4368, (f) no. 4369.
118. Codell, M., J. Chem. Eng. Data, 8, 460 (1963).
119. Cohen, B. M., A. R. Cullingworth, and J. D. Smith, J. Chem. Soc., A, 1969, 2193.
120. Cook, A. G., J. Org. Chem., 30, 1262 (1965).
121. Cook, A. G., and G. W. Mason, J. Inorg. Nucl. Chem., 28, 2579 (1966).
122. Corbridge, D. E. C., Topics Phosphorus Chem., 6, 235 (1969); (a) p. 298.
123. Coskran, K. J., J. M. Jenkins, and J. G. Verkade, J. Amer. Chem. Soc., 90, 5437 (1968).
124. Cowley, A. H., M. J. S. Dewar, W. R. Jackson, and W. B. Jennings, J. Amer. Chem. Soc., 92, 5206 (1970).
125. Cowley, A. H., and D. S. Dierdorf, J. Amer. Chem. Soc., 91, 6609 (1969).
126. Cowley, A. H., and M. W. Taylor, J. Amer. Chem. Soc., 91, 1929 (1969).
127. Coyne, D. M., W. E. McEwen, and C. A. VanderWerf, J. Amer. Chem. Soc., 78, 3061 (1956).
128. Cragg, R. H., Chem. Ind. (London), 1967, 1751.
129. Cragg, R. H., M. S. Fortuin, and N. N. Greenwood, J. Chem. Soc., A, 1970, 1817.
130. Cragg, R. H., and M. F. Lappert, J. Chem. Soc., A, 1966, 82.
131. Cram, D. J., R. D. Trepka, and P. S. Janiak, J. Amer. Chem. Soc., 88, 2749 (1966).
132. Creighton, H. J. M., J. Phys. Chem., 30, 1209 (1926).
133. Crofts, P. C., and D. M. Parker, J. Chem. Soc., C, 1970, 332.
134. Daasch, L. W., and D. C. Smith, Anal. Chem., 23, 853 (1951).
135. Dannels, B. F., and A. F. Shepard, U.S. Patent

3,402,196 (1968); C.A. 70, 4275 m (1969).
136. Daugherty, K. E., W. A. Eychanes, and J. I. Stevens, Appl. Spectrosc. 22, 95 (1968).
137. Davies, J. H., Chem. Ind. (London), 1964, 1755.
138. Davies, W. C., and C. W. Othen, J. Chem. Soc., 1936, 1236.
139. Demarcq, M. C., and J. Sleziona, French addn. 86,531 (1966); C.A. 65, 13762h (1966).
140. Denney, D. B., D. Z. Denney, B. C. Chang, and K. L. Marsi, J. Amer. Chem. Soc., 91, 5243 (1969).
141. Derkach, G. I., and A. V. Kirsanov, Zh. Obshch. Khim., 29, 1815 (1959).
142. Derkach, G. I., and Yu. V. Piven, Zh. Obshch. Khim., 36, 1087 (1966).
143. Dewar, M. J. S., and V. P. Kubha, J. Amer. Chem. Soc., 82, 5685 (1960).
144. Dietsche, W. H., Tetrahedron, 23, 3049 (1967).
145. Dobbie, R. C., L. F. Doty, and R. G. Cavell, J. Amer. Chem. Soc., 90, 2015 (1968).
146. Dornfeld, C. A., U.S. Patent 2,924,615 (1960); C.A. 54, 12072g (1960).
147. Douglas, P. G., and B. L. Shaw, J. Chem. Soc., A, 1970, 334.
148. Douglas, P. G., and B. L. Shaw, J. Chem. Soc., A, 1970, 1556.
149. Drake, L. R., Committee Report, Chem. Eng. News, 30, 4515 (1952).
150. Dregval, G. F., and G. I. Derkach, Zh. Obshch. Khim., 33, 2952 (1963).
151. Drozd, G. I., and S. Z. Ivin, Zh. Obshch. Khim., 39, 1417 (1969).
152. Drozd, G. I., S. Z. Ivin, V. N. Kulakova, and V. V. Sheluchenko, Zh. Obshch. Khim., 38, 576 (1968).
153. Drozd, G. I., S. Z. Ivin, and V. V. Sheluchenko, Zh. Obshch. Khim., 38, 1906 (1968).
154. Drozd, G. I., S. Z. Ivin, and V. V. Sheluchenko, Zh. Vses. Khim. Obshchest., 12, 472 (1967); C.A. 67, 116917s (1967).
155. Drozd, G. I., S. Z. Ivin, V. V. Sheluchenko, and M. A. Landau, Zh. Obshch. Khim., 38, 1653 (1968).
156. Drozd, G. I., S. Z. Ivin, V. V. Sheluchenko, B. I. Tetel'baum, G. M. Luganskii, and A. D. Varshavskii, Zh. Obshch. Khim., 37, 1631 (1967).
157. Drozd, G. I., S. Z. Ivin, and O. G. Strukov, Zh. Obshch. Khim., 39, 1418 (1969).
158. Drozd, G. I., M. A. Sokal'skii, V. V. Sheluchenko, M. A. Landau, and S. Z. Ivin, Zh. Obshch. Khim., 39, 936 (1969).
159. Dye, Jr., W. T., Naval Research Lab. Report P-3044 (1946).
160. Dyer, J., and J. Lee, J. Chem. Soc., B, 1970, 409.

161. Emeléus, H. J., R. N. Haszeldine, and R. C. Paul, J. Chem. Soc., 1955, 563.
162. Emeléus, H. J., and J. D. Smith, J. Chem. Soc., 1959, 375.
163. Emmick, T. L., and R. L. Letsinger, J. Amer. Chem. Soc., 90, 3459 (1968).
164. Engelke, E. F., U.S. Patent 2,377,870 (1945); C.A. 39, 4099 (1945).
165. Ettel, V., and J. Horak, Collect. Czech. Chem. Commun., 26, 2087 (1961); C.A. 56, 5994 i (1962).
166. Evdakov, V. P., L. I. Mizrakh, and G. P. Sizova, USSR Patent 162,846 (1963); C.A. 61, 9528h (1964).
167. Evdakov, V. P., and E. K. Shlenkova, Dokl. Akad. Nauk SSSR, 168, 1323 (1966); C.A. 65, 13525g (1966).
168. Evdakov, V. P., and E. K. Shlenkova, Zh. Obshch. Khim., 35, 739 (1965).
169. Evleth, Jr., E. M., L. D. Freeman, and R. I. Wagner, J. Org. Chem., 27, 2192 (1962).
170. Ewart, G., D. S. Payne, A. L. Porte, and A. P. Lane, J. Chem. Soc., 1962, 3984.
171. Fekete, F., U.S. Patent 3,203,924 (1965); C.A. 64, 3597e (1966).
172. Feld, R., J. Chem. Soc., 1964, 3963.
173. Ferraro, J. R., and C. M. Andrejasich, J. Inorg. Nucl, Chem., 26, 377 (1964).
174. Feshchenko, N. G., T. V. Kovaleva, and A. V. Kirsanov, Zh. Obshch, Khim., 39, 2184 (1969).
175. Fiat, D., M. Halmann, L. Kugel, and J. Reuben, J. Chem. Soc., 1962, 3837.
176. Fild, M., O. Glemser, and I. Hollenberg, Z. Naturforsch., 21 B, 920 (1966); C.A. 66, 65593c (1967).
177. Fild, M., I. Hollenberg, and O. Glemser, Naturwiss., 54, 89 (1967); C.A. 67, 21966t (1967).
178. Finegold, H., Ann. N.Y. Acad. Sci., 70, 875 (1958).
179. Fink, W., Angew. Chem., Int. Ed. Engl., 5, 760 (1966).
180. Fink, W., Chem. Ber., 96, 1071 (1963).
181. Fitch, H. M., U.S. Patent 2,759,961 (1956); C.A. 51, 482b (1957).
182. Fitch, S. J., J. Amer. Chem. Soc., 86, 61 (1964).
183. Fluck, E., and H. Binder, Inorg. Nucl. Chem. Lett., 3, 307 (1967).
184. Fontal, B., and H. Goldwhite, Tetrahedron, 22, 3275 (1966).
185. Fontal, B., H. Goldwhite, and D. G. Rowsell, J. Org. Chem., 31, 2424 (1966).
186. Foss, V. L., V. V. Kudinova, G. B. Postnikova, and I. F. Lutsenko, Dokl. Akad. Nauk SSSR, 146, 1106 (1962); C.A. 58, 7968e (1963).
187. Fosse, R., Bull. Soc. Chim. (France), [4] 7, 231 (1910).

188. Fosse, R., Bull. Soc. Chim. (France), [4] 7, 357 (1910).
189. Fox, R. B., and W. J. Bailey, J. Org. Chem., 25, 1447 (1960).
190. Fox, R. B., and W. J. Bailey, J. Org. Chem., 26, 2542 (1961).
191. Frank, A. W., Chem. Rev., 61, 389 (1961); (a) p. 411, (b) pp. 391-392.
192. Frank, A. W., J. Org. Chem., 24, 966 (1959).
193. Frank, A. W., J. Org. Chem., 26, 850 (1961).
194. Frank, A. W., and C. F. Baranauckas, J. Org. Chem., 31, 872 (1966).
195. Freedman, L. D., and G. O. Doak, J. Org. Chem., 21, 1533 (1956).
196. Gabov, N. I., and A. I. Shafiev, Radiokhimiya, 8, 330 (1966); C.A. 65, 13070e (1966).
197. Gagnaire, D., J. B. Robert, and J. Verrier, Bull. Soc. Chim. (France), 1968, 2392.
198. Gallagher, M. J., and I. D. Jenkins, J. Chem. Soc., C, 1966, 2176.
199. Garner, A. Y., U.S. Patent 2,916,510 (1959); C.A. 54, 5571b (1960).
200. Garner, A. Y., U.S. Patent 3,161,607 (1964); C.A. 62, 5405e (1965).
201. Gefter, E. L., Zh. Obshch. Khim., 31, 949 (1961).
202. Gefter, E. L., Zh. Obshch. Khim., 33, 3548 (1963).
203. Gefter, E. L., and I. A. Rogacheva, Zh. Obshch. Khim., 32, 964 (1962).
204. Gefter, E. L., and I. A. Rogacheva, Zh. Obshch. Khim., 32, 3962 (1962).
205. Gefter, E. L., and I. A. Rogacheva, Zh. Obshch. Khim., 33, 1177 (1963).
206. Gefter, E. L., and I. A. Rogacheva, Zh. Obshch. Khim., 34, 88 (1964).
207. Gefter, E. L., and I. A. Rogacheva, Zh. Obshch. Khim., 34, 92 (1964).
208. Genkina, G. K., V. A. Gilyarov, and M. I. Kabachnik, Zh. Obshch. Khim., 38, 2513 (1968).
209. Gladshtein, B. M., and L. N. Shitov, Zh. Obshch. Khim., 39, 1951 (1969).
210. Goldwhite, H., R. N. Haszeldine, and D. G. Rowsell, J. Chem. Soc., 1965, 6875.
211. Goldwhite, H., and D. G. Rowsell, Chem. Commun., 1968, 1665.
212. Goldwhite, H., and D. G. Rowsell, Chem. Commun., 1969, 713.
213. Goldwhite, H., and D. G. Rowsell, J. Amer. Chem. Soc., 88, 3572 (1966).
214. Goldwhite, H., and D. G. Rowsell, J. Mol. Spectrosc., 27, 364 (1968).
215. Gonçalves, H., F. Mathis, and R. Wolf, Bull. Soc.

Chim. (France), 1961, 1595.
216. Gonnet, C., and A. Lamotte, Bull. Soc. Chim. (France), 1969, 2932.
217. Gorenstein, D., and F. H. Westheimer, J. Amer. Chem. Soc., 92, 634 (1970).
218. Grayson, M., and C. E. Farley, Chem. Commun., 1967, 830.
219. Grayson, M., and C. E. Farley, J. Org. Chem., 32, 236 (1967).
220. Grechkin, N. P., I. A. Nuretdinov, and N. A. Buina, Izv. Akad. Nauk SSSR, Ser. Khim., 1969, 168.
221. Green, M., R. N. Haszeldine, B. R. Iles, and D. G. Rowsell, J. Chem. Soc., 1965, 6879.
222. Green, M., and R. F. Hudson, J. Chem. Soc., 1958, 3129.
223. Green, M., and R. F. Hudson, J. Chem. Soc., 1963, 540.
224. Griffin, C. E., Chem. Ind. (London), 1960, 1058.
225. Griffin, C. E., E. H. Uhing, and A. D. F. Toy, J. Amer. Chem. Soc., 87, 4757 (1965).
226. Grim, S. O., P. R. McAllister, and R. M. Singer, Chem. Commun., 1969, 38.
227. Grinblat, M. P., A. L. Klebanskii, and V. N. Prons, Zh. Obshch. Khim., 39, 172 (1969).
228. Grudzev, V. G., K. V. Karavanov, and S. Z. Ivin, Zh. Obshch. Khim., 38, 1548 (1968).
229. Gryszkiewicz-Trochimowski, E., J. Quinchon, and O. Gryszkiewicz-Trochimowski, Bull. Soc. Chim. (France), 1960, 1794.
230. Gudkova, I. P., I. K. Golovnikova, and E. E. Nifant'ev, Zh. Obshch. Khim., 38, 1340 (1968).
231. Guichard, F., Ber., 32, 1572 (1899).
232. Hadzi, D., J. Chem. Phys., 34, 1445 (1961).
233. Halmann, M., and L. Kugel, J. Inorg. Nucl. Chem., 25, 1343 (1963).
234. Hamilton, L. A., and R. H. Williams, U.S. Patent 2,957,931 (1960); C.A. 55, 10317h (1961).
235. Harman, D., U.S. Patent 2,629,731 (1953); C.A. 48, 1418d (1954).
236. Harman, D., and A. R. Stiles, U.S. Patent 2,659,714 (1953); C.A. 48, 12168d (1954).
237. Harowitz, C. L., U.S. Patent 2,903,475 (1959); C.A. 54, 2169f (1960).
238. Harris, G. S., J. Chem. Soc., 1958, 512.
239. Hart, W. A., and H. H. Sisler, Inorg. Chem., 3, 617 (1964).
240. Harvey, R. G., and E. R. De Sombre, Topics Phosphorus Chem., 1, 57 (1964).
241. Harvey, R. G., H. I. Jacobson, and E. V. Jensen, J. Amer. Chem. Soc., 85, 1618 (1963).
242. Harvey, R. G., and E. V. Jensen, J. Org. Chem., 28,

470 (1963).
243. Harvey, R. G., and E. V. Jensen, Tetrahedron Lett., 1963, 1801.
244. Harwood, H. J., and D. W. Grisley, Jr., J. Amer. Chem. Soc., 82, 423 (1960).
245. Harwood, H. J., and N. K. Patel, Macromolecules, 1, 233 (1968).
246. Hassel, G., Dissertation, University of Mainz, 1959, cited in Refs. 263 and 509.
247. Haszeldine, R. N., H. Goldwhite, and D. G. Rowsell, British Patent 1,069,201 (1967); C.A. 67, 44304q (1967).
248. Hatt, H. H., J. Chem. Soc., 1933, 776.
249. Hawes, W., and S. Trippett, J. Chem. Soc., C, 1969, 1465.
250. Hechenbleikner, I., and K. R. Molt, U.S. Patent 3,316,333 (1967); C.A. 68, 49765w (1968).
251. Helferich, B., and E. Aufderhaar, Ann., 658, 100 (1962).
252. Henderson, Jr., W. A., M. Epstein, and F. S. Seichter, J. Amer. Chem. Soc., 85, 2462 (1963).
253. Heners, J., and A. B. Burg, J. Amer. Chem. Soc., 88, 1677 (1966).
254. Henning, H.-G., and G. Hilgetag, J. Prakt. Chem., [4] 29, 86 (1965).
255. Hewertson, W., R. A. Shaw, and B. C. Smith, J. Chem. Soc., 1963, 1670.
256. Hirschmann, H., U.S. Patent 2,370,903 (1945); C.A. 39, 4892 (1945).
257. Hnoosh, M. H., and R. A. Zingaro, Can. J. Chem., 47, 4679 (1969).
258. Hoff, M. C., and P. Hill, J. Org. Chem., 24, 356 (1959).
259. Hoffmann, F. W., and T. R. Moore, J. Amer. Chem. Soc., 80, 1150 (1958).
260. Hoffmann, F. W., R. G. Roth, and T. C. Simmons, J. Amer. Chem. Soc., 80, 5937 (1958).
261. Hoffmann, F. W., D. H. Wadsworth, and H. D. Weiss, J. Amer. Chem. Soc., 80, 3945 (1958).
262. Hoffmann, H., and R. Grünewald, Chem. Ber., 94, 186 (1961).
263. Hoffmann H., R. Grünewald, and L. Horner, Chem. Ber., 93, 861 (1960).
264. Holmes, R. R., and R. P. Carter, Jr., Inorg. Chem., 2, 1146 (1963).
265. Holmes, R. R., and R. P. Wagner, J. Amer. Chem. Soc., 84, 357 (1962).
266. Horak, J., and V. Ettel, Collect. Czech. Chem. Commun., 26, 2410 (1961); C.A. 56, 5995e (1962).
267. Horner, L., H. Reuter, and E. Herrmann, Ann. Chem., 660, 1 (1962).

268. Houalla, D., R. Miquel, and R. Wolf, Bull. Soc. Chim. (France), 1963, 1152.
269. Houalla, D., M. Sanchez, and R. Wolf, Bull. Soc. Chim. (France), 1965, 2368.
270. Houalla, D., and R. Wolf, C.R. Acad. Sci. Paris, 259, 180 (1964); C.A. 61, 8161d (1964).
271. Hudson, R. F., Structure and Mechanism in Organo-Phosphorus Chemistry, Academic Press, New York, 1965. (a) p. 83.
272. Hudson, R. F., P. A. Chopard, and G. Salvadori, Helv. Chim. Acta, 47, 632 (1964).
273. Huffman, C. W., and M. Hamer, U.S. Patent 3,149,145 (1964); C.A. 61, 14712g (1964).
274. Hussey, A. S., and Y. Takeuchi, J. Org. Chem., 35, 643 (1970).
275. Imbery, D., and H. Friebolin, Z. Naturforsch., 23 B, 759 (1968); C.A. 69, 56081u (1968).
276. Issleib, K., and G. Döll, Chem. Ber., 94, 2664 (1961).
277. Issleib, K., and D. Jacob, Chem. Ber., 94, 107 (1961).
278. Issleib, K., and W. Seidel, Chem. Ber., 92, 2681 (1959).
279. Issleib, K., and H. Weichmann, Chem. Ber., 97, 721 (1964).
280. Ivanova, Zh. M., and A. V. Kirsanov, Zh. Obshch. Khim., 31, 3991 (1961).
281. Ivin, S. Z., and K. V. Karavanov, Zh. Obshch. Khim., 29, 3456 (1959).
282. Ivin, S. Z., V. K. Promonenkov, and E. A. Fokin, Zh. Obshch. Khim., 37, 1642 (1967).
283. Ivin, S. Z., V. K. Promonenkov, and E. A. Fokin, Zh. Obshch. Khim., 37, 2511 (1967).
284. Ivin, S. Z., V. K. Promonenkov, and E. A. Fokin, Zh. Obshch. Khim., 39, 1058 (1969).
285. Ivin, S. Z., and I. D. Shelakova, Zh. Obshch. Khim., 35, 1220 (1965).
286. Jackson, H. L., U.S. Patent 2,722,538 (1955); C.A. 50, 4218c (1956).
287. Jaffe, H. H., and L. D. Freedman, J. Amer. Chem. Soc., 74, 1069 (1952).
288. Jenkins, J. M., J. R. Moss, and B. L. Shaw, J. Chem. Soc., A, 1969, 2796.
289. Johns, I. B., E. A. McElhill, and J. O. Smith, J. Chem. Eng. Data, 7, 277 (1962).
290. Juillard, J., Bull. Soc. Chim. (France), 1966, 1727.
291. Juillard, J., and N. Simonet, Bull. Soc. Chim. (France), 1968, 1883.
292. Kabachnik, M. I., G. A. Balueva, T. Ya. Medved, E. N. Tsvetkov, and Chung-Yu Chang, Kinet. Katal., 6, 212 (1965); C.A. 63, 7039b (1965).

293. Kabachnik, M. I., Chung-Yu Chang, and E. N. Tsvetkov, Zh. Obshch. Khim., 32, 3351 (1962).
294. Kabachnik, M. I., and V. A. Gilyarov, Dokl. Akad. Nauk SSSR, 96, 991 (1954); C.A. 49, 8842a (1955).
295. Kabachnik, M. I., V. A. Gilyarov, and E. M. Popov, Zh. Obshch. Khim., 32, 1598 (1962).
296. Kabachnik, M. I., V. A. Gilyarov, and E. N. Tsvetkov, Izv. Akad. Nauk SSSR, Otd. Khim. Nauk, 1959, 2135; C.A. 54, 10918g (1960).
297. Kabachnik, M. I., T. A. Mastryukova, and T. A. Melent'eva, Zh. Obshch. Khim., 32, 267 (1962).
298. Kabachnik, M. I., T. A. Mastryukova, and T. A. Melent'eva, Zh. Obshch. Khim., 33, 382 (1963).
299. Kabachnik, M. I., and T. Ya. Medved, Izv. Akad. Nauk SSSR, Otd. Khim. Nauk, 1954, 1024; C.A. 50, 219i (1956).
300. Kabachnik, M. I., T. Ya. Medved, and Yu. M. Polikarpov, Dokl. Akad. Nauk SSSR, 135, 849 (1960); C.A. 55, 14288c (1961).
301. Kabachnik, M. I., and E. N. Tsvetkov, Dokl. Akad. Nauk SSSR, 117, 817 (1957); C.A. 52, 8070c (1958).
302. Kabachnik, M. I., and E. N. Tsvetkov, Izv. Akad. Nauk SSSR, Otd. Khim. Nauk, 1960, 133.
303. Kabachnik, M. I., and E. N. Tsvetkov, Zh. Obshch. Khim., 30, 3227 (1960).
304. Kabachnik, M. I., and E. N. Tsvetkov, Zh. Obshch. Khim., 31, 684 (1961).
305. Kabachnik, M. I., E. N. Tsvetkov, and Chung-Yu Chang, Dokl. Akad. Nauk SSSR, 125, 1260 (1959); C.A. 53, 21752f (1960).
306. Kabachnik, M. I., E. N. Tsvetkov, and Chung-Yu Chang, Zh. Obshch. Khim., 32, 3340 (1962).
307. Kamai, G., Dokl. Akad. Nauk SSSR, 66, 389 (1949); C.A. 44, 127e (1950).
308. Kamai, G., Zh. Obshch. Khim., 4, 192 (1934); Chem. Zentralbl., 1935, I 2669.
309. Kamai, G., Zh. Obshch. Khim., 18, 443 (1948); C.A. 42, 7723c (1948).
310. Kamai, G., and E. A. Gerasimova, Tr. Kazan. Khim. Tekhnol. Inst., 15, 26 (1950); C.A. 51, 11273i (1957).
311. Kamai, G., and E. A. Gerasimova, Tr. Kazan. Khim. Tekhnol. Inst., 23, 138 (1957); C.A. 52, 9946a (1958).
312. Kamai, G., and R. K. Ismagilov, Zh. Obshch. Khim., 34, 439 (1964).
313. Kamai, G., and F. M. Kharrasova, Zh. Obshch. Khim., 33, 3846 (1963).
314. Kamai, G., F. M. Kharrasova, R. B. Sultanova, and S. Yu. Tukhvatullina, Izv. Vyssh. Uchebn. Zaved., Khim. Khim. Tekhnol., 5, 759 (1962); C.A. 58, 13985a (1963).

315. Kamai, G., F. M. Kharrasova, R. B. Sultanova, and S. Yu. Tukhvatullina, Zh. Obshch. Khim., 31, 3550 (1961).
316. Kamai, G., F. M. Kharrasova, and S. Yu. Tukhvatullina, Tr. Kazan. Khim. Tekhnol. Inst., 30, 18 (1962); C.A. 60, 5542g (1964).
317. Kamai, G., and V. A. Kukhtin, Zh. Obshch. Khim., 25, 1932 (1955).
318. Kamai, G., and V. A. Kukhtin, Zh. Obshch. Khim., 28, 939 (1958).
319. Kamai, G., and V. S. Tsivunin, Dokl. Akad. Nauk SSSR, 128, 543 (1959); C.A. 54, 7538g (1960).
320. Kamai, G., V. S. Tsivunin, and S. Kh. Nuretdinov, Zh. Obshch. Khim., 35, 1817 (1965).
321. Kanamueller, J. M., and H. H. Sisler, Inorg. Chem., 6, 1765 (1967).
322. Kasparek, F., Monatsh. Chem., 94, 809 (1963).
323. Kasparek, F., Z. Anorg. Allg. Chem., 362, 205 (1968).
324. Kelbe, W., Ber. 11, 1499 (1878).
325. Kerschner, P. M., and B. W. Greenwald, U.S. Patent 2,864,845 (1958); C.A. 53, 12239d (1959).
326. Khairullin, V. K., Izv. Akad. Nauk SSSR, Ser. Khim., 1965, 1792.
327. Khairullin, V. K., M. A. Kuryleva, and T. I. Sobchuk, Izv. Akad. Nauk SSSR, Ser. Khim., 1965, 1083.
328. Khairullin, V. K., M. A. Kuryleva, and T. I. Sobchuk, Izv. Akad. Nauk SSSR, Ser. Khim., 1967, 99.
329. Khairullin, V. K., M. A. Vasyanina, and A. N. Pudovik, Izv. Akad. Nauk SSSR, Ser. Khim., 1967, 950.
330. Khairullin, V. K., M. A. Vasyanina, and A. N. Pudovik, Izv. Akad. Nauk SSSR, Ser. Khim., 1967, 1603.
331. Khairullin, V. K., M. A. Vasyanina, A. N. Pudovik, and Yu. Yu. Samitov, Khim. Org. Soedin. Fosfora, 1967, 29; C.A. 69, 59335r (1968).
332. Kharrasova, F. M., and G. Kamai, Zh. Obshch. Khim., 38, 359 (1968).
333. Kharrasova, F. M., and G. Kamai, Zh. Obshch. Khim., 38, 617 (1968).
334. Kharrasova, F. M., G. Kamai, and R. R. Shagidullin, Zh. Obshch. Khim., 35, 1993 (1965).
335. Khisamova, Z. L., and G. Kamai, Zh. Obshch. Khim., 20, 1162 (1950); C.A. 45, 1531c (1951).
336. Kiso, Y., M. Kobayashi, Y. Kitaoka, K. Kawamoto, and J. Takeda, Bull. Chem. Soc. Japan, 40, 2779 (1967).
337. Kiso, Y., M. Kobayashi, Y. Kitaoka, K. Kawamoto, and J. Takeda, J. Chromatogr., 33, 561 (1968); C.A. 68, 90202y (1968).
338. Kiso, Y., M. Kobayashi, Y. Kitaoka, K. Kawamoto, and J. Takeda, J. Chromatogr., 36, 215 (1968); C.A. 69, 90093j (1968).
339. Klotz, I. M., and R. T. Morrison, J. Amer. Chem. Soc., 69, 473 (1947).

340. Knunyants, I. L., R. N. Sterlin, V. V. Tyuleneva, and L. N. Pinkina, Izv. Akad. Nauk SSSR, Otd. Khim. Nauk, 1963, 1123; C.A. 59, 8784d (1963).
341. Kochetkov, N. K., E. E. Nifant'ev, and I. P. Gudkova, Zh. Obshch. Khim., 37, 277 (1967).
342. Köhler, H., and A. Michaelis, Ber., 10, 807 (1877).
343. Köhler, H., and A. Michaelis, Ber., 10, 815 (1877).
344. Köhler, H., and A. Michaelis, Ber., 10, 816 (1877).
345. Kohn, E. J., U. E. Hanninen, and R. B. Fox, Naval Research Lab. Report, C-3180 (1947).
346. Kosolapoff, G. M., J. Amer. Chem. Soc., 72, 4292 (1950).
347. Kosolapoff, G. M., Organophosphorus Compounds, Wiley, New York, 1950. (a) p. 142, (b) p. 147.
348. Kosolapoff, G. M., and J. S. Powell, J. Amer. Chem. Soc., 72, 4291 (1950).
349. Kosovtsev, V. B., V. N. Chistokletov, and A. A. Petrov, Zh. Obshch. Khim., 39, 223 (1969).
350. Kostyanovskii, R. G., I. A. Nuretdinov, N. P. Grechkin, and I. I. Chervin, Izv. Akad. Nauk SSSR, Ser. Khim., 1969, 2588.
351. Krasil'nikova, E. A., O. I. Korol, and A. I. Razumov, Tr. Kazan. Khim. Tekhnol. Inst., 33, 171 (1964); C.A. 66, 10999m (1967).
352. Krasil'nikova, E. A., A. M. Potapov, and A. I. Razumov, Zh. Obshch. Khim., 37, 1173 (1967).
353. Krasil'nikova, E. A., A. M. Potapov, and A. I. Razumov, Zh. Obshch. Khim., 37, 1409 (1967).
354. Krasil'nikova, E. A., A. M. Potapov, and A. I. Razumov, Zh. Obshch. Khim., 37, 2365 (1967).
355. Krasil'nikova, E. A., A. M. Potapov, and A. I. Razumov, Zh. Obshch. Khim., 37, 2585 (1967).
356. Krasil'nikova, E. A., A. M. Potapov, and A. I. Razumov, Zh. Obshch. Khim., 38, 609 (1968).
357. Krasil'nikova, E. A., A. M. Potapov, and A. I. Razumov, Zh. Obshch. Khim., 38, 1098 (1968).
358. Krasil'nikova, E. A., A. M. Potapov, and A. I. Razumov, Zh. Obshch. Khim., 38, 1101 (1968).
359. Kratzer, R. H., and K. L. Paciorek, Inorg. Chem., 4, 1767 (1965).
360. Krawiecki, C., J. Michalski, and Z. Tulimowski, Chem. Ind. (London), 1965, 34.
361. Kreshkov, A. P., V. A. Drozdov, and N. A. Kolchina, Zh. Fiz. Khim., 40, 2150 (1966); C.A. 66, 6121u (1967).
362. Kreutzkamp, N., and J. Pluhatsch, Arch. Pharm. (Weinheim), 292, 159 (1959); C.A. 53, 17938d (1959).
363. Kreutzkamp, N., C. Schimpfky, and K. Storck, Arch. Pharm. (Weinheim), 300, 868 (1967); C.A. 68, 49689z (1968).
364. Kreutzkamp, N., C. Schimpfky, and K. Storck, Arch.

Pharm. (Weinheim), 301, 247 (1968).
365. Kruglyak, Yu. K., S. I. Malekin, and I. V. Martynov, Zh. Obshch. Khim., 39, 466 (1969).
366. Kulakova, V. N., Yu. M. Zinov'ev, S. P. Makarov, V. A. Shpanskii, and L. Z. Soborovskii, Zh. Obshch. Khim., 39, 385 (1969).
367. Kulakova, V. N., Yu. M. Zinov'ev, and L. Z. Soborovskii, Zh. Obshch. Khim., 39, 579 (1969).
368. Kulakova, V. N., Yu. M. Zinov'ev, and L. Z. Soborovskii, Zh. Obshch. Khim., 39, 838 (1969).
369. Kuryleva, M. A., and V. K. Khairullin, Izv. Akad. Nauk SSSR, Ser. Khim., 1965, 2133.
370. Labarre, M.-C., and J.-F. Labarre, J. Chim. Phys. Physicochim. Biol., 63, 1577 (1966); C.A. 66, 99904v (1967).
371. Ladd, E. C., and M. P. Harvey, U.S. Patent 2,631,162 (1953); C.A. 48, 7048h (1954).
372. Lane, A. P., D. A. Morton-Blake, and D. S. Payne, J. Chem. Soc., A, 1967, 1492.
373. Lane, A. P., and D. S. Payne, J. Chem. Soc., 1963, 4004.
374. Lane, A. P., and D. S. Payne, Proc. Chem. Soc., 1964, 403.
375. Laurent, J.-P., and G. Jugie, Bull. Soc. Chim. (France), 1969, 26.
376. Laurent, J.-P., G. Jugie, and R. Wolf, J. Chim. Phys. Physicochim. Biol., 66, 409 (1969).
377. Lesfauries, P., and P. Rumpf, C.R. Acad. Sci. Paris, 228, 1018 (1949).
378. Letts, E. A., and R. F. Blake, Trans. Roy. Soc. Edinburgh, 35, 527 (1889); J. Chem. Soc. Abstr., 58, 766 (1890).
379. Levin, Ya. A., and V. S. Galeev, Zh. Obshch. Khim., 37, 1327 (1967).
380. Levin, Ya. A., and V. S. Galeev, Zh. Obshch. Khim., 37, 2736 (1967).
381. Levin, Ya. A., Yu. M. Kargin, V. S. Galeev, and V. I. Sannikova, Izv. Akad. Nauk SSSR, Ser. Khim., 1968, 411.
382. Lewis, R. A., and K. Mislow, J. Amer. Chem. Soc., 91, 7009 (1969).
383. Lichtenthaler, F. W., Chem. Rev., 61, 607 (1961), tables 30 and 31.
384. Lindner, J., and M. Strecker, Monatsh. Chem., 53/54, 274 (1929).
385. Linfield, W. M., E. Jungermann, and A. T. Guttmann, J. Org. Chem., 26, 4088 (1961).
386. Liorber, B. G., Z. M. Khammatova, I. V. Berezovskaya, and A. I. Razumov, Zh. Obshch. Khim., 38, 165 (1968).
387. Liorber, B. G., Z. M. Khammatova, and A. I. Razumov, Zh. Obshch. Khim., 39, 1551 (1969).

388. Liorber, B. G., Z. M. Khammatova, A. I. Razumov, T. V. Zykova, and T. B. Borisova, Zh. Obshch. Khim., 38, 878 (1968).
389. Liorber, B. G., and A. I. Razumov, Zh. Obshch. Khim., 36, 314 (1966).
390. Lobanov, D. I., E. N. Tsvetkov, and M. I. Kabachnik, Zh. Obshch. Khim., 39, 841 (1969).
391. Lorenz, W., Belgian Patent 625,216 (1963); C.A. 60, 10600d (1964).
392. Lorenz, W., and G. Schrader, Belgian Patent 609,076 (1962); C.A. 57, 16661d (1962).
393. Lorenz, W., and G. Schrader, German Patent 1,067,017 (1959); Chem. Zentralbl., 1960, 9382.
394. Lorenz, W., and G. Schrader, U.S. Patent 3,082,240 (1963); C.A. 59, 5077g (1963).
395. Lucas, C. R., and M. E. Peach, Can. J. Chem., 48, 1869 (1970).
396. Lucken, E. A. C., J. Chem. Soc., 1963, 5123.
397. Lutsenko, I. F., and Z. S. Kraits, Dokl. Akad. Nauk SSSR, 132, 612 (1960); C.A. 54, 24346d (1960).
398. Lutsenko, I. F., Z. S. Novikova, L. Ya. Barinova, and M. V. Proskurnina, Zh. Obshch. Khim., 37, 1330 (1967).
399. Magnelli, D. D., G. Tesi, J. U. Lowe, Jr., and W. E. McQuistion, Inorg. Chem., 5, 457 (1966).
400. Maguire, M. H., and G. Shaw, J. Chem. Soc., 1955, 2039.
401. Maguire, M. H., and G. Shaw, J. Chem. Soc., 1957, 311.
402. Maier, L., Helv. Chim. Acta, 46, 1812 (1963).
403. Maier, L., Helv. Chim. Acta, 46, 2667 (1963).
404. Maier, L., Helv. Chim. Acta, 47, 2129 (1964).
405. Maier, L., Helv. Chim. Acta, 49, 1249 (1966).
406. Maier, L., Helv. Chim. Acta, 50, 1742 (1967).
407. Maier, L., Helv. Chim. Acta, 52, 858 (1969).
408. Maier, L., J. Inorg. Nucl. Chem., 24, 1073 (1962).
409. Maier, L., U.S. Patent 3,137,692 (1964); C.A. 61, 7045c (1964).
410. Maier, L., U.S. Patent 3,188,294 (1965); C.A. 63, 13318f (1965).
411. Maier, L., U.S. Patent 3,320,251 (1967); C.A. 67, 54262p (1967).
412. Maier, L., U.S. Patent 3,321,557 (1967); C.A. 67, 73684u (1967).
413. Marelli, O., Ann. Chim. Appl., 34, 149 (1944); C.A. 40, 7515 (1946).
414. Marie, C., Ann. Chim. Phys., [8] 3, 335 (1904); J. Chem. Soc. Abstr., 88, I 17 (1905).
415. Marie, C., C.R. Acad. Sci. Paris, 133, 219 (1901).
416. Marie, C., C.R. Acad. Sci. Paris, 134, 286 (1902).
417. Marie, C., C.R. Acad. Sci. Paris, 134, 847 (1902).

418. Marie, C., C.R. Acad. Sci. Paris, 135, 1118 (1902).
419. Marie, C., C.R. Acad. Sci. Paris, 136, 48 (1903).
420. Marie, C., C.R. Acad. Sci. Paris, 136, 234 (1903).
421. Marie, C., C.R. Acad. Sci. Paris, 136, 508 (1903).
422. Marie, C., C.R. Acad. Sci. Paris, 137, 124 (1903).
423. Marie, C., C.R. Acad. Sci. Paris, 138, 1707 (1904).
424. Mark, V., Tetrahedron Lett., 1962, 281.
425. Mark, V., C. H. Dungan, M. M. Crutchfield, and J. R. Van Wazer, Topics Phosphorus Chem., 5, 227 (1967). (a) p. 282, (b) p. 317, (c) p. 368, (d) p. 277, (e) p. 273, (f) p. 255, (g) p. 266, (h) p. 275.
426. Markowska, A., and J. Michalski, Rocz. Chem., 34, 1675 (1960); C.A. 56, 7345h (1962).
427. Martin, G., and G. Mavel, C.R. Acad. Sci. Paris, 255, 2095 (1962).
428. Maruszewska-Wieczorkowska, E., and J. Michalski, Bull. Acad. Pol. Sci., Ser. Sci. Chim., Geol. Geogr., 6, 19 (1958); C.A. 52, 16349 (1958).
429. Maslow, P. G., J. Phys. Chem., 72, 1424 (1968).
430. Mastalerz, P., Rocz. Chem., 38, 61 (1964); C.A. 60, 14535h (1964).
431. Mastalerz, P., and R. Tyka, Rocz. Chem., 38, 1529 (1964); C.A. 62, 9169h (1964).
432. Mastryukova, T. A., A. E. Shipov, and M. I. Kabachnik, Zh. Obshch. Khim., 29, 1450 (1959).
433. Mattson, R. W., U.S. Patent 2,769,743 (1956); C.A. 51, 3915d (1957).
434. Mauret, P., J.-P. Fayet, D. Voigt, M.-C. Labarre, and J.-F. Labarre, J. Chim. Phys. Physicochim. Biol., 65, 549 (1968); C.A. 69, 13828a (1968).
435. Mavel, G., Progress in Nucl. Magn. Resonance Spectroscopy, Vol. 1, J. W. Emsley, J. Feeney, and L. H. Sutcliffe, Eds., Pergamon Press, Oxford, 1966, pp. 251-373. (a) p. 307.
436. McBee, E. T., O. R. Pierce, and H. M. Metz, U.S. Patent 2,899,454 (1959); C.A. 54, 9765c (1960).
437. Meakin, P., L. J. Guggenberger, J. P. Jesson, D. H. Gerlach, F. N. Tebbe, W. G. Peet, and E. L. Muetterties, J. Amer. Chem. Soc., 92, 3484 (1970).
438. Medved, T. Ya., M. V. Rudomino, N. M. Dyatlova, and M. I. Kabachnik, Izv. Akad. Nauk SSSR, Ser. Khim., 1968, 1211.
439. Melchiker, P., Ber. 31, 2915 (1898).
440. Melton, T. M., U.S. Patent 3,205,252 (1965); C.A. 63, 14905h (1965).
441. Michaelis, A., Ann., 293, 193 (1896).
442. Michaelis, A., Ann., 293, 261 (1896).
443. Michaelis, A., Ann., 294, 1 (1897).
444. Michaelis, A., Ann., 315, 43 (1901).
445. Michaelis, A., Ber., 13, 2174 (1880).
446. Michaelis, A., Ber., 31, 1037 (1898).

447. Michaelis, A., and J. Ananoff, Ber., 7, 1688 (1874).
448. Michaelis, A., and H. Köhler, Ber., 7, 519 (1876).
449. Michaelis, A., and F. Oster, Ann., 270, 123 (1892).
450. Michaelis, A., and C. Paneck, Ann., 212, 203 (1882).
451. Michaelis, A., and A. Schenk, Ann., 260, 1 (1890).
452. Michalski, J., Rocz. Chem., 29, 960 (1955), C.A. 50, 10641 (1956).
453. Michalski, J., M. Mikolajczyk, and A. Ratajczak, Chem. Ind. (London), 1962, 819.
454. Michalski, J., M. Mikolajczyk, and A. Skowronska, Chem. Ind. (London), 1962, 1053.
455. Michalski, J., and A. Skowronska, J. Chem. Soc., C, 1970, 703.
456. Michalski, J., and A. Skowronska, Rocz. Chem., 30, 799 (1956); C.A. 54, 10832a (1960).
457. Michalski, J., and Z. Tulimowski, Rocz. Chem., 36, 1781 (1962); C.A. 59, 10109f (1963).
458. Michalski, J., and J. Wieczorkowski, Bull. Acad. Pol. Sci., Cl. 3, 5, 917 (1957); C.A. 52, 6157g (1958).
459. Michalski, J., and J. Wieczorkowski, Rocz. Chem., 33, 105 (1959); Chem. Zentralbl., 1961, 3292.
460. Miller, B., and T. P. O'Leary, Jr., Chem. Ind. (London), 1961, 55.
461. Miller, C. D., R. C. Miller, and W. Rogers, Jr., J. Amer. Chem. Soc., 80, 1562 (1958).
462. Mitsunobu, O., T. Ohashi, M. Kikuchi, and T. Mukaiyama, Bull. Chem. Soc. Japan, 39, 214 (1966).
463. Mitsunobu, O., T. Ohashi, M. Kikuchi, and T. Mukaiyama, Bull. Chem. Soc. Japan, 40, 2964 (1967); C.A. 69, 2931c (1968).
464. Mizrakh, L. I., and V. P. Evdakov, Zh. Obshch. Khim., 36, 469 (1966).
465. Mizrakh, L. I., V. P. Evdakov, and L. Yu. Sandalova, Zh. Obshch. Khim., 35, 1871 (1965).
466. Mizrakh, L. I., L. Yu. Sandalova, and V. P. Evdakov, Zh. Obshch. Khim., 37, 1875 (1967).
467. Moedritzer, K., Inorg. Chem., 6, 936 (1967).
468. Moedritzer, K., L. Maier, and L. C. D. Groenweghe, J. Chem. Eng. Data, 7, 307 (1962).
469. Morris, R. C., V. W. Buls, and S. A. Ballard, U.S. Patent 2,577,796 (1951); C.A. 46, 9581e (1952).
470. Morrison, D. C., J. Org. Chem., 21, 705 (1956).
471. Mosher, W. A., and R. R. Irino, J. Amer. Chem. Soc., 91, 756 (1969).
472. Moskva, V. V., A. I. Maikova, and A. I. Razumov, Zh. Obshch. Khim., 39, 2451 (1969).
473. Möslinger, W., Ber. 9, 998 (1876).
474. Mukaiyama, T., T. Fujiwara, Y. Tamura, and Y. Yokota, J. Org. Chem., 29, 2572 (1964).
475. Mukaiyama, T., and H. Nambu, J. Org. Chem., 27, 2201 (1962).

476. Mukaiyama, T., H. Nambu, and M. Okamoto, J. Org. Chem., 27, 3651 (1962).
477. Mukaiyama, T., and K. Osaka, Bull. Chem. Soc. Japan, 39, 566 (1966); C.A. 64, 19395 (1966).
478. Mukherji, A.K., Anal. Chim. Acta, 30, 591 (1964).
479. Mulder, E., J. Prakt. Chem., 91, 472 (1864).
480. Neimysheva, A. A., V. I. Savchuk, and I. L. Knunyants, Zh. Obshch. Khim., 36, 500 (1966).
481. Nesmeyanov, A. N., K. N. Anisimov, and N. E. Kolobova, Izv. Akad. Nauk SSSR, Otd. Khim. Nauk, 1962, 722.
482. Nesterov, L. V., and N. A. Aleksandrova, Zh. Obshch. Khim., 39, 931 (1969).
483. Nesterov, L. V., A. Ya. Kessel', and R. I. Mutalapova, Zh. Obshch. Khim., 39, 2453 (1969).
484. Nesterov, L. V., and R. I. Mutalapova, Zh. Obshch. Khim., 37, 1843 (1967).
485. Nesterov, L. V., and R. I. Mutalapova, Zh. Obshch. Khim., 37, 1847 (1967).
486. Newallis, P. E., J. W. Baker, and J. P. Chupp, U.S. Patent 3,070,489 (1962); C.A. 58, 9144d (1963).
487. Nielsen, R. P., and H. H. Sisler, Inorg. Chem., 2, 753 (1963).
488. Nifant'ev, E. E., Zh. Obshch. Khim., 35, 1980 (1965).
489. Nifant'ev, E. E., and I. V. Fursenko, Zh. Obshch. Khim., 35, 1882 (1965).
490. Nifant'ev, E. E., and I. V. Komlev, USSR Patent 186,469 (1966); C.A. 66, 76150g (1967).
491. Nifant'ev, E. E., and M. P. Koroteev, Zh. Obshch. Khim., 37, 1366 (1967).
492. Nifant'ev, E. E., M. P. Koroteev, N. L. Ivanova, I. P. Gudkova, and D. A. Predvoditelev, Dokl. Akad. Nauk SSSR, 173, 1345 (1967); C.A. 67, 108710d (1967).
493. Nifant'ev, E. E., and L. P. Levitan, Zh. Obshch. Khim., 35, 758 (1965).
494. Nifant'ev, E. E., S. M. Markov, A. P. Tuseev, and A. F. Vasil'ev, Sint. Prir. Soedin., Ikh Analogov Fragm., 1965, 42; C.A. 65, 10651c (1966).
495. Nifant'ev, E. E., and L. V. Matveeva, Zh. Obshch. Khim., 37, 1692 (1967).
496. Nifant'ev, E. E., and A. P. Tuseev, Sint. Prir. Soedin., Ikh Analogov Fragm., 1965, 34; C.A. 65, 5511h (1966).
497. Nifant'ev, E. E., A. P. Tuseev, and Yu. I. Koshurin, Sint. Prir. Soedin., Ikh Analogov Fragm., 1965, 38; C.A. 65, 10651f (1966).
498. Nifant'ev, E. E., A. P. Tuseev, S. M. Markov, and G. F. Lidenko, Zh. Obshch. Khim., 36, 319 (1966).
499. Nifant'ev, E. E., A. P. Tuseev, and V. V. Tarasov, Zh. Obshch. Khim., 36, 1124 (1966).
500. Nifant'ev, E. E., A. I. Zavalishina, and I. V. Komlev, Zh. Obshch. Khim., 37, 2497 (1967).

501. Nifant'ev, E. E., A. I. Zavalishina, I. S. Nasonovskii, and I. V. Komlev, Zh. Obshch. Khim., 38, 2538 (1968).
502. Nikonorov, K. V., and E. A. Gurylev, Izv. Akad. Nauk SSSR, Ser. Khim., 1965, 2136.
503. Nikonorov, K. V., E. A. Gurylev, and F. Fakhrislamova, Izv. Akad. Nauk SSSR, Ser. Khim., 1966, 1095.
504. Nikonorov, K. V., E. A. Gurylev, L. G. Urazaeva, M. N. Nazypov, R. A. Asadov, and S. D. Anisin, Izv. Akad. Nauk SSSR, Ser. Khim., 1969, 2241.
505. Nishizawa, Y., Bull. Chem. Soc. Japan, 34, 1170 (1961); C.A. 56, 11429h (1962).
506. Nixon, J. F., J. Chem. Soc., 1964, 2471.
507. Nixon, J. F., and R. G. Cavell, J. Chem. Soc., 1964, 5983.
508. Nixon, J. F., and R. Schmutzler, Spectrochim. Acta, 22, 565 (1966).
509. Nöth, H., and H.-J. Vetter, Chem. Ber., 96, 1109 (1963).
510. Nöth, H., and H.-J. Vetter, Chem. Ber., 96, 1298 (1963).
511. Nöth, H., and H.-J. Vetter, Chem. Ber., 96, 1479 (1963).
512. Novikova, Z. S., E. A. Efimova, and I. F. Lutsenko, Zh. Obshch. Khim., 38, 2345 (1968).
513. Novikova, Z. S., M. A. Krasnovskaya, and I. F. Lutsenko, Zh. Obshch. Khim., 39, 1060 (1969).
514. Nuretdinov, I. A., and N. P. Grechkin, Izv. Akad. Nauk SSSR, Ser. Khim., 1967, 436.
515. Nuretdinov, I. A., N. P. Grechkin, N. A. Buina, and L. K. Nikonorova, Izv. Akad. Nauk SSSR, Ser. Khim., 1969, 1535.
516. Nuretdinova, O. N., Izv. Akad. Nauk SSSR, Ser. Khim., 1965, 1901.
517. Oertel, G., H. Malz, and H. Holtschmidt, Chem. Ber., 97, 891 (1964).
518. Ogata, Y., and H. Tomioka, J. Org. Chem., 35, 596 (1970).
519. Okamoto, Y., and H. Sakurai, Kogyo Kagaku Zasshi, 68, 2080 (1965); C.A. 64, 14075h (1966).
520. Orio, A. A., B. B. Chastain, and H. B. Gray, Inorg. Chim. Acta, 3, 8 (1969).
521. Peake, S. C., and R. Schmutzler, J. Chem. Soc., A, 1970, 1049.
522. Pelchowicz, Z., and H. Leader, J. Chem. Soc., 1963, 3320.
523. Peppard, D. F., G. W. Mason, and C. M. Andrejasich, J. Inorg. Nucl. Chem., 27, 697 (1965).
524. Peterson, L. K., and A. B. Burg, J. Amer. Chem. Soc., 86, 2587 (1964).
525. Peterson, L. K., G. L. Wilson, and K. I. Thé, Can. J. Chem., 47, 1025 (1969).

526. Petrov, K. A., A. A. Basyuk, V. P. Evdakov, and L. I. Mizrakh, Zh. Obshch. Khim., 34, 2226 (1964).
527. Petrov, K. A., N. K. Bliznyuk, and I. Yu. Mansurov, Zh. Obshch. Khim., 31, 176 (1961).
528. Petrov, K. A., N. K. Bliznyuk, and V. A. Savostenok, Zh. Obshch. Khim., 31, 1361 (1961).
529. Petrov, K. A., N. K. Bliznyuk, Yu. N. Studnev, and A. F. Kolomiets, Zh. Obshch. Khim., 31, 179 (1961).
530. Petrov, K. A., V. P. Evdakov, G. I. Abramtsev, and A. K. Strautman, Zh. Obshch. Khim., 32, 3070 (1962).
531. Petrov, K. A., V. P. Evdakov, and K. A. Bilevich, USSR Patent 149,779 (1962); C.A. 58, 11401b (1963).
532. Petrov, K. A., V. P. Evdakov, K. A. Bilevich, and V. I. Chernykh, Zh. Obshch. Khim., 32, 3065 (1962).
533. Petrov, K. A., V. P. Evdakov, K. A. Bilevich, and Yu. S. Kosarev, Zh. Obshch. Khim., 32, 1974 (1962).
534. Petrov, K. A., V. P. Evdakov, and L. I. Mizrakh, Zh. Obshch. Khim., 33, 1246 (1963).
535. Petrov, K. A., V. P. Evdakov, L. I. Mizrakh, and V. P. Romodin, Zh. Obshch. Khim., 32, 3062 (1962).
536. Petrov, K. A., T. N. Lysenko, B. Ya. Libman, and V. V. Pozdnev. Khim. Org. Soedin. Fosfora, 1967, 181; C.A. 69, 67487g (1968).
537. Petrov, K. A., E. E. Nifant'ev, and R. G. Gol'tsova, Zh. Obshch. Khim., 31, 2367 (1961).
538. Petrov, K. A., E. E. Nifant'ev, and R. G. Gol'tsova, Zh. Obshch. Khim., 31, 2370 (1961).
539. Petrov, K. A., E. E. Nifant'ev, and R. G. Gol'tsova, Zh. Obshch. Khim., 31, 3174 (1961).
540. Petrov, K. A., E. E. Nifant'ev, and R. G. Gol'tsova, Zh. Obshch. Khim., 32, 3716 (1962).
541. Petrov, K. A., E. E. Nifant'ev, R. G. Gol'tsova, M. A. Belaventsev, and S. M. Korneev, Zh. Obshch. Khim., 32, 1277 (1962).
542. Petrov, K. A., E. E. Nifant'ev, R. G. Gol'tsova, and G. V. Gubin, Zh. Obshch. Khim., 31, 2732 (1961).
543. Petrov, K. A., E. E. Nifant'ev, R. G. Gol'tsova, and L. M. Solntseva, Vysokomol. Soedin., 5, 1691 (1963); C.A. 60, 12121c (1964).
544. Petrov, K. A., E. E. Nifant'ev, L. V. Khorkhoyanu, and R. G. Gol'tsova, Vysokomol. Soedin., 5, 1799 (1963); C.A. 60, 12121g (1964).
545. Petrov, K. A., E. E. Nifant'ev, L. V. Khorkhoyanu, and I. G. Shcherba, Zh. Obshch. Khim., 34, 70 (1964).
546. Petrov, K. A., E. E. Nifant'ev, T. N. Lysenko, and V. P. Evdakov, Zh. Obshch. Khim., 31, 2377 (1961).
547. Petrov, K. A., E. E. Nifant'ev, A. A. Shchegolev, M. M. Butilov, and I. F. Rebus, Zh. Obshch. Khim., 33, 899 (1963).
548. Petrov, K. A., E. E. Nifant'ev, A. A. Shchegolev, and A. P. Tuseev, Zh. Obshch. Khim., 34, 690 (1964).

549. Petrov, K. A., E. E. Nifant'ev, and I. I. Sopikova, Vysokomol. Soedin., 2, 685 (1960); C.A. 55, 9935f (1961).
550. Petrov, K. A., E. E. Nifant'ev, I. I. Sopikova, and V. M. Budanov, Zh. Obshch. Khim., 31, 2373 (1961).
551. Petrov, K. A., V. A. Parshina, B. A. Orlov, and G. M. Tsypina, Zh. Obshch. Khim., 32, 4017 (1962).
552. Petrov, K. A., and O. S. Urbanskaya, Zh. Obshch. Khim., 30, 1233 (1960).
553. Petrovskaya, L. I., M. V. Proskurnina, Z. S. Novikova, and I. F. Lutsenko, Izv. Akad. Nauk SSSR, Ser. Khim., 1968, 1277.
554. Phillips, J. N., Aust. J. Chem., 12, 199 (1959); C.A. 53, 14683h (1959).
555. Phillips, J. N., J. Chem. Soc., 1958, 4271.
556. Pianfetti, J. A., and G. M. Kosolapoff, U.S. Patent 2,881,202 (1959); C.A. 53, 14937f (1959).
557. Pilgram, K., and F. Korte, Tetrahedron, 19, 137 (1963).
558. Plets, V. M., Zh. Obshch. Khim., 7, 84 (1937); C.A. 31, 4965 (1937).
559. Plets, V. M., Zh. Obshch. Khim., 7, 90 (1937); C.A. 31, 4965 (1937).
560. Polonovski, M., M. Pesson, and G. Polmanss, C.R. Acad. Sci. Paris, 239, 1506 (1954); C.A. 50, 269 (1956).
561. Ponomarev, V. V., S. A. Golubtsov, K. A. Andrianov, and G. N. Kondrashova, Izv. Akad. Nauk SSSR, 1969, 1743.
562. Popov, E. M., E. N. Tsvetkov, Chung-Yu Chang, and T. Ya. Medved, Zh. Obshch. Khim., 32, 3255 (1962).
563. Popp, F. D., and W. Kirsch, J. Org. Chem., 26, 3858 (1961).
564. Postnikova, G. B., A. S. Kostyuk, and I. F. Lutsenko, Zh. Obshch. Khim., 35, 2204 (1965).
565. Postnikova, G. B., A. S. Kostyuk, and I. F. Lutsenko, Zh. Obshch. Khim., 36, 1129 (1966).
566. Postnikova, G. B., and I. F. Lutsenko, Zh. Obshch. Khim., 33, 4029 (1963).
567. Postnikova, G. B., and I. F. Lutsenko, Zh. Obshch. Khim., 37, 233 (1967).
568. Proskurnina, M. V., Z. S. Novikova, and I. F. Lutsenko, Dokl. Akad. Nauk SSSR, 159, 619 (1964); C.A. 62, 6508h (1965).
569. Protopopov, I. S., and M. Ya. Kraft, Zh. Obshch. Khim., 34, 1446 (1964).
570. Pudovik, A. N., Dokl. Akad. Nauk SSSR, 92, 773 (1953); C.A. 49, 3049i (1955).
571. Pudovik, A. N., and I. M. Aladzheva, Dokl. Akad. Nauk SSSR, 151, 1110 (1963); C.A. 59, 13798d (1963).
572. Pudovik, A. N., I. M. Aladzheva, and L. V. Spirina,

Zh. Obshch. Khim., 37, 700 (1967).
573. Pudovik, A. N., and V. P. Aver'yanova, Zh. Obshch. Khim., 26, 1426 (1956).
574. Pudovik, A. N., and O. S. Durova, Zh. Obshch. Khim., 36, 1460 (1966).
575. Pudovik, A. N., and G. I. Evstaf'ev, Dokl. Akad. Nauk SSSR, 183, 842 (1968); C.A. 70, 67280e (1969).
576. Pudovik, A. N., and G. I. Evstaf'ev, Vysokomol. Soedin., 6, 2139 (1964); C.A. 62, 9168a (1965).
577. Pudovik, A. N., and E. M. Faizullin, Zh. Obshch. Khim., 34, 882 (1964).
578. Pudovik, A. N., E. M. Faizullin, and S. V. Yakovleva, Zh. Obshch. Khim., 37, 460 (1967).
579. Pudovik, A. N., I. V. Gur'yanova, L. V. Banderova, and M. G. Zimin, Zh. Obshch. Khim., 37, 876 (1967).
580. Pudovik, A. N., I. V. Gur'yanova, S. P. Perevezentseva, and T. V. Zykova, Zh. Obshch. Khim., 37, 1313 (1967).
581. Pudovik, A. N., I. V. Gur'yanova, and G. V. Romanov, Zh. Obshch. Khim., 39, 2418 (1969).
582. Pudovik, A. N., I. V. Gur'yanova, and M. G. Zimin, Zh. Obshch. Khim., 37, 2088 (1967).
583. Pudovik, A. N., I. V. Gur'yanova, and M. G. Zimin, Zh. Obshch. Khim., 38, 1533 (1968).
584. Pudovik, A. N., I. V. Gur'yanova, M. G. Zimin, and O. E. Raevskaya, Zh. Obshch. Khim., 38, 1539 (1968).
585. Pudovik, A. N., V. K. Khairullin, and M. A. Vasyanina, Zh. Obshch. Khim., 37, 411 (1967).
586. Pudovik, A. N., V. K. Khairullin, M. A. Vasyanina, and G. F. Novikova, Izv. Akad. Nauk SSSR, Ser. Khim., 1969, 2334.
587. Pudovik, A. N., and I. V. Konovalova, Zh. Obshch. Khim., 30, 2348 (1960).
588. Pudovik, A. N., and I. V. Konovalova, Zh. Obshch. Khim., 31, 1693 (1961).
589. Pudovik, A. N., and I. V. Konovalova, Zh. Obshch. Khim., 32, 467 (1962).
590. Pudovik, A. N., I. V. Konovalova, and L. V. Banderova, Zh. Obshch. Khim., 35, 1206 (1965).
591. Pudovik, A. N., I. V. Konovalova, and L. V. Dedova, Zh. Obshch. Khim., 34, 2905 (1964).
592. Pudovik, A. N., I. V. Konovalova, and O. S. Durova, Zh. Obshch. Khim., 31, 2656 (1961).
593. Pudovik, A. N., I. V. Konovalova, and R. E. Krivonosova, Zh. Obshch. Khim., 26, 3110 (1956).
594. Pudovik, A. N., and V. K. Krupnov, Zh. Obshch. Khim., 38, 194 (1968).
595. Pudovik, A. N., and V. K. Krupnov, Zh. Obshch. Khim., 38, 305 (1968).
596. Pudovik, A. N., and V. K. Krupnov, Zh. Obshch. Khim., 38, 1287 (1968).

597. Pudovik, A. N., and V. K. Krupnov, Zh. Obshch. Khim., 39, 1890 (1969).
598. Pudovik, A. N., T. M. Moshkina, and I. V. Konovalova, Zh. Obshch. Khim., 29, 3338 (1959).
599. Pudovik, A. N., and A. A. Muritova, Dokl. Akad. Nauk SSSR, 158, 419 (1964); C.A. 62, 11646f (1965).
600. Pudovik, A. N., V. I. Nikitina, and G. F. Krupnov, Zh. Obshch. Khim., 29, 4019 (1959).
601. Pudovik, A. N., and N. G. Poloznova, Zh. Obshch. Khim., 25, 778 (1955).
602. Pudovik, A. N., and M. A. Pudovik, Zh. Obshch. Khim., 33, 3353 (1963).
603. Pudovik, A. N., and M. A. Pudovik, Zh. Obshch. Khim., 36, 1467 (1966).
604. Pudovik, A. N., and M. A. Pudovik, Zh. Obshch. Khim., 36, 1658 (1966).
605. Pudovik, A. N., R. D. Sabirova, and T. A. Tener, Zh. Obshch. Khim., 24, 1026 (1954).
606. Pudovik, A. N., and O. S. Shulyndina, Zh. Obshch. Khim., 39, 1014 (1969).
607. Pudovik, A. N., and F. N. Sitdikova, Dokl. Akad. Nauk SSSR, 125, 826 (1959); C.A. 53, 19850c (1959).
608. Pudovik, A. N., L. V. Spirina, M. A. Pudovik, Yu. M. Kargin, and L. S. Andreeva, Zh. Obshch. Khim., 39, 1715 (1969).
609. Pudovik, A. N., and R. I. Tarasova, Zh. Obshch. Khim., 34, 3946 (1964).
610. Pudovik, A. N., and D. Kh. Yarmukhametova, Izv. Akad. Nauk SSSR, Otd. Khim. Nauk, 1952, 902; C.A. 47, 10469 (1953).
611. Pudovik, A. N., and D. Kh. Yarmukhametova, Izv. Akad. Nauk SSSR, Otd. Khim. Nauk, 1954, 636; C.A. 49, 8789a (1955).
612. Quesnel, G., M. de Botton, A. Chambolle, and R. Dulou, C.R. Acad. Sci. Paris, 251, 1074 (1960); C.A. 55, 4569b (1961).
613. Quin, L. D., and M. R. Dysart, J. Org. Chem., 27, 1012 (1962).
614. Quin, L. D., and J. S. Humphrey, Jr., J. Amer. Chem. Soc., 83, 4124 (1961).
615. Rabinowitz, R., U.S. Patent 3,211,782 (1965); C.A. 64, 3600c (1966).
616. Rakshys, J. W., R. W. Taft, and W. A. Sheppard, J. Amer. Chem. Soc., 90, 5236 (1968).
617. Rankov, G., Ann. Univ. Sofia, II, Fac. Phys.-Math., 2, 32, 51 (1936); C.A. 31, 2168 (1937).
618. Ratajczak, A., Rocz. Chem., 36, 175 (1962); C.A. 57, 15147 (1962).
619. Raudnitz, H., Ber. 60, 743 (1927).
620. Razumov, A. I., Zh. Obshch. Khim., 29, 1635 (1959).
621. Razumov, A. I., and M. B. Gazizov, Zh. Obshch. Khim.,

37, 2738 (1967).
622. Razumov, A. I., and B. G. Liorber, Dokl. Akad. Nauk SSSR, 135, 1150 (1960); C.A. 55, 13298b (1961).
623. Razumov, A. I., and B. G. Liorber, Zh. Obshch. Khim., 32, 4063 (1962).
624. Razumov, A. I., B. G. Liorber, R. S. Andronova, and I. A. Milyutina, Tr. Kazan. Khim. Tekhnol. Inst., 36, 487 (1967); C.A. 70, 11753u (1969).
625. Razumov, A. I., B. G. Liorber, M. B. Gazizov, and Z. M. Khammatova, Zh. Obshch. Khim., 34, 1851 (1964).
626. Razumov, A. I., E. A. Markovich, and A. D. Reshetnikova, Zh. Obshch. Khim., 27, 2394 (1957).
627. Razumov, A. I., and V. V. Moskva, Zh. Obshch. Khim., 35, 1595 (1965).
628. Razumov, A. I., and O. A. Mukhacheva, Zh. Obshch. Khim., 26, 1436 (1956).
629. Razumov, A. I., O. A. Mukhacheva, and Sim-Do-Khen, Izv. Akad. Nauk SSSR, Otd. Khim. Nauk, 1952, 894; C.A. 47, 10466b (1953).
630. Razumov, A. I., and G. A. Savicheva, Zh. Obshch. Khim., 34, 2595 (1964).
631. Razumov, A. I., and Sim-Do-Khen, Tr. Kazan. Khim. Tekhnol. Inst., 19/20, 167 (1954/55); C.A. 51, 6503h (1957).
632. Razumov, A. I., and N. G. Zabusova, Zh. Obshch. Khim., 30, 1307 (1960).
633. Razumov, A. I., N. G. Zabusova, R. L. Poznyak, and O. I. Korol, Tr. Kazan. Khim. Tekhnol. Inst., 29, 22 (1960); C.A. 58, 544e (1963).
634. Razumova, N. A., F. V. Bagrov, and A. A. Petrov, Zh. Obshch. Khim., 39, 2368 (1969).
635. Reetz, T., U.S. Patent 3,104,253 (1963); C.A. 60, 557g (1964).
636. Reetz, T., W. A. Busch, and D. H. Chadwick, U.S. Patent 3,057,904 (1962); C.A. 58, 3316e (1963).
637. Regel, E. K., U.S. Patent 3,193,372 (1965); C.A. 63, 8976e (1965).
638. Reiff, L. P., and H. S. Aaron, J. Amer. Chem. Soc., 92, 5275 (1970).
639. Retcofsky, H. L., and C. E. Griffin, Tetrahedron Lett., 1966, 1975.
640. Reuben, J., D. Samuel, and B. L. Silver, J. Amer. Chem. Soc., 85, 3093 (1963).
641. Rizpolozhenskii, N. I., and V. D. Akamsin, Izv. Akad. Nauk SSSR, Ser. Khim., 1969, 1398.
642. Rizpolozhenskii, N. I., L. V. Boiko, and M. A. Zvereva, Dokl. Akad. Nauk SSSR, 155, 1137 (1964); C.A. 61, 1817e (1964).
643. Rizpolozhenskii, N. I., and A. A. Muslinkin, Izv. Akad. Nauk SSSR, Otd. Khim. Nauk, 1961, 1600.
644. Rizpolozhenskii, N. I., and M. A. Zvereva, Izv. Akad.

Nauk SSSR, Otd. Khim. Nauk, 1959, 358.
645. Rocamora, R. de R., Spanish Patent 259,881 (1960); C.A. 57, 7312c (1962).
646. Ruschig, H., and W. Aumüller, German Patent 1,005,063 (1957); C.A. 53, 15008h (1959).
647. Ruschig, H., and W. Aumüller, German Patent 1,009,623 (1957); C.A. 53, 16063f (1959).
648. Ryazanov, I. P., and I. P. Khazhova, Tr. Magnitogorsk. Gornomet. Inst., 13, 29 (1957); C.A. 54, 6393d (1960).
649. Rylyakova, N. S., Yu. A. Kondrat'ev, and S. Z. Ivin, Zh. Obshch. Khim., 37, 483 (1967).
650. Sachs, H., Ber. 25, 1514 (1892).
651. Sadtler Research Laboratories, Philadelphia, Pa., DTA Spectra. (a) no. 1414.
652. Sadtler Research Laboratories, Philadelphia, Pa., Infrared Grating Spectra. (a) no. 69, (b) no. 736, (c) no. 8323, (d) no. 11087.
653. Sadtler Research Laboratories, Philadelphia, Pa., Infrared Prism Spectra, 1962 ff. (a) no. 201, (b) no. 4123, (c) no. 8812, (d) no. 15165, (e) no. 20278, (f) no. 33090.
654. Sadtler Research Laboratories, Philadelphia, Pa., NMR Spectra, 1966 ff. (a) no. 1931.
655. Sadtler Research Laboratories, Philadelphia, Pa., Ultraviolet Spectra. (a) no. 2285, (b) no. 4476, (c) no. 6799, (d) no. 11464.
656. Sakurai, H., Y. Okamoto, and K. Horiuchi, Kogyo Kagaku Zasshi, 68, 961 (1965); C.A. 63, 9982d (1965).
657. Sanchez, M., R. Wolf, and F. Mathis, Spectrochim. Acta, Part A, 23, 2617 (1967).
658. Sander, M., Chem. Ber., 93, 1220 (1960).
659. Sander, M., Chem. Ber., 95, 473 (1962).
660. Schegk, E., H. Schloer, and G. Schrader, German Patent 1,138,394 (1962); C.A. 58, 6863g (1963).
661. Schenk, A., and A. Michaelis, Ber., 21, 1497 (1888).
662. Scherer, O. J., and W. Gick, Chem. Ber., 103, 71 (1970).
663. Scherer, O. J., and P. Klusmann, Angew. Chem., Int. Ed. Engl., 8, 752 (1969).
664. Scherer, O. J., and P. Klusmann, Z. Anorg. Allg. Chem., 370, 171 (1969).
665. Schiemenz, G. P., Angew. Chem., Int. Ed. Engl., 7, 544 (1968).
666. Schindlbauer, H., Chem. Ber., 102, 2914 (1969).
667. Schindlbauer, H., Monatsh. Chem., 96, 1936 (1965).
668. Schliebs, R., German Patent 1,088,955 (1960); C.A. 55, 27215d (1961).
669. Schliebs, R., German Patent 1,098,940 (1961); C.A. 56, 506c (1962).
670. Schmidpeter, A., and J. Ebeling, Angew. Chem., Int. Ed. Engl., 7, 209 (1968).

671. Schmidpeter, A., and H. Roseknecht, Angew. Chem., Int. Ed. Engl., 8, 614 (1969).
672. Schmidt, H., Chem. Ber., 81, 477 (1948).
673. Schmidt, H., German Patent 870,701 (1953); C.A. 52, 16291h (1958).
674. Schmidt, H., German Patent 875,662 (1953).
675. Schmidt, U., I. Boie, C. Osterroht, R. Schröer, and H.-F. Grützmacher, Chem. Ber., 101, 1381 (1968).
676. Schmutzler, R., J. Chem. Soc., 1964, 4551.
677. Schmutzler, R., J. Chem. Soc., 1965, 5630.
678. Schrader, G., Belgian Patent 608,802 (1962); C.A. 57, 7311d (1962).
679. Schrader, G., Belgian Patent 614,005 (1962); C.A. 58, 2472d (1963).
680. Schrader, G., German Patent 954,244 (1956); C.A. 53, 11305e (1959).
681. Schrader, G., German Patent 1,064,510 (1959); C.A. 55, 14381f (1961).
682. Schrader, G., German Patent 1,115,248 (1961); C.A. 56, 14328h (1962).
683. Schröder, H., Ber., 12, 561 (1879).
684. Seidel, W., Z. Anorg. Allg. Chem., 330, 141 (1964); C.A. 61, 7040f (1964).
685. Seidel, W., Z. Chem., 3, 429 (1963); C.A. 60, 5053h (1964).
686. Seidel, W., and K. Issleib, Z. Anorg. Allg. Chem., 325, 113 (1963); C.A. 60, 1791h (1964).
687. Seidel, W., and H. Schöler, Z. Chem., 7, 431 (1967); C.A. 68, 35452e (1968).
688. Shagidullin, R. R., and N. P. Grechkin, Zh. Obshch. Khim., 38, 150 (1968).
689. Sheluchenko, V. V., S. S. Dubov, G. I. Drozd, and S. Z. Ivin, Zh. Strukt. Khim., 9, 909 (1968); C.A. 70, 24524v (1969).
690. Shen, L. M. C., G. G. Long, and C. G. Moreland, J. Organometal. Chem., 5, 362 (1966).
691. Shitov, L. N., and B. M. Gladshtein, Zh. Obshch. Khim., 39, 1251 (1969).
692. Siddall III, T. H., and C. A. Prohaska, J. Amer. Chem. Soc., 84, 3467 (1962).
693. Siegfried, B., A.-G., Swiss Patent 311,982 (1956); C.A. 51, 10559f (1957).
694. Silcox, C. M., and J. J. Zuckerman, J. Amer. Chem. Soc., 88, 168 (1966).
695. Simonnin, M. P., J. J. Basselier, and C. Charrier, Bull. Soc. Chim. (France), 1967, 3544.
696. Simonnin, M. P., and C. Charrier, C. R. Acad. Sci. Paris, Ser. C, 267, 550 (1968).
697. Siuda, A., Nukleonika, 10, 459 (1965); C.A. 64, 16600h (1966).
698. Smith, C. W., U.S. Patent 2,648,695 (1953); C.A. 48,

8252d (1954).
699. Smith, C. W., G. B. Payne, and E. C. Shokol, U.S. Patent 2,856,369 (1958); C.A. 53, 2686i (1959).
700. Smith, N. L., J. Org. Chem., 28, 863 (1963).
701. Sollott, G. P., and E. Howard, Jr., J. Org. Chem., 27, 4034 (1962).
702. Sollott, G. P., and E. Howard, Jr., J. Org. Chem., 29, 2451 (1964).
703. Spangenberg, S. F., and H. H. Sisler, Inorg. Chem., 8, 1004 (1969).
704. Stalick, J. K., and J. A. Ibers, Inorg. Chem., 8, 1084 (1969).
705. Stanko, V. I., A. I. Klimova, and L. I. Zakharkin, Izv. Akad. Nauk SSSR, Otd. Khim. Nauk, 1962, 919.
706. Stayner, R. D., U.S. Patent 2,686,803 (1954); C.A. 49, 11,000c (1955).
707. Stayner, R. D., U.S. Patent 2,693,482 (1954); C.A., 49, 13287h (1955).
708. Steininger, E., Chem. Ber., 95, 2993 (1962).
709. Stelling, O., Z. Phys. Chem., 117, 161 (1925); Chem. Zentralbl., 1926, I 833.
710. Sterlin, R. N., R. D. Yatsenko, L. N. Pinkina, and I. L. Knunyants, Izv. Akad. Nauk SSSR, Otd. Khim. Nauk, 1960, 1991.
711. Stetter, H., and W.-D. Last, Chem. Ber., 102, 3364 (1969).
712. Stiles, A. R., and F. F. Rust, U.S. Patent 2,724,718 (1955); C.A. 50, 10124d (1956).
713. Strem Chemicals Inc., Danvers, Mass., Product Information, bull., 1970.
714. Sturtz, G. Bull. Soc. Chim. (France), 1964, 2340.
715. Tebbe, F. N., P. Meakin, J. P. Jesson, and E. L. Muetterties, J. Amer. Chem. Soc., 92, 1068 (1970).
716. Temple, R. D., and J. E. Leffler, Tetrahedron Lett., 1968, 1893.
717. Texas A & M University, College Station, Tex., "Catalog of Selected NMR Spectral Data," Thermodynamic Research Center Data Project, 1960 ff. (a) no. 159, (b) no. 165, (c) no. 65.
718. Thomas, L. C., and R. A. Chittenden, Spectrochim. Acta, 20, 489 (1964).
719. Titov, A. I., M. V. Sizova, and P. O. Gitel', Dokl. Akad. Nauk SSSR, 159, 385 (1964); C.A. 62, 6509h (1965).
720. Tolman, C. A., J. Amer. Chem. Soc., 92, 2953 (1970).
721. Tolman, C. A., J. Amer. Chem. Soc., 92, 2956 (1970).
722. Toy, A. D. F., and E. H. Uhing, U.S. Patent 3,160,632 (1961); C.A. 62, 4053f (1965).
723. Toy, A. D. F., E. N. Walsh, and J. R. Froli, Jr., U.S. Patent 3,094,405 (1963); C.A. 59, 12844e (1963).

724. Tsolis, A. K., Ph.D. Dissertation, University of Kansas, 1963; Diss. Abstr., 25, 1586 (1964).
725. Tsvetkov, E. N., D. I. Lobanov, and M. I. Kabachnik, Zh. Obshch. Khim., 38, 2285 (1968).
726. Tsvetkov, E. N., R. A. Malevannaya, and M. I. Kabachnik, Zh. Obshch. Khim., 37, 695 (1967).
727. Uhing, E., K. Rattenbury, and A. D. F. Toy, J. Amer. Chem. Soc., 83, 2299 (1961).
728. Uhing, E. H., and A. D. F. Toy, French Patent 1,346,938 (1963); C.A. 60, 12055b (1964).
729. Uhing, E. H., and A. D. F. Toy, French Patent 1,356,435 (1964); C.A. 61, 688f (1964).
730. Upadysheva, A. V., E. K. Shlenkova, M. G. Bel'skaya, and V. P. Evdakov, Khim. Org. Soedin. Fosfora, 1967, 235; C.A. 69, 2328m (1968).
731. Vahrenkamp, H., and H. Nöth, J. Organometal. Chem., 12, 281 (1968).
732. Van Wazer, J. R., C. F. Callis, J. N. Shoolery, and R. C. Jones, J. Amer. Chem. Soc., 78, 5715 (1956).
733. Van Wazer, J. R., and L. Maier, J. Amer. Chem. Soc., 86, 811 (1964).
734. Van de Westeringh, C., and H. Veldstra, Rec. Trav. Chim. Pays-Bas, 77, 1096 (1958).
735. Vetter, H.-J., and H. Nöth, Chem. Ber., 96, 1816 (1963).
736. Ville, J., Ann. Chim. Phys., [6] 23, 289 (1891); Chem. Zentralbl., 1891, II 343.
737. Viout, M. P., J. Rech. CNRS, 28, 15 (1954); C.A. 50, 7077e (1956).
738. Vladimirova, I. L., A. F. Grapov, and V. I. Lomakina, Reakts. Metody Issled. Org. Soedin., 16, 7 (1966); C.A. 66, 94646c (1967).
739. Voigt, D., M. C. Labarre, and J.-P. Jaureguy, Bull. Soc. Chim. (France), 1964, 3087.
740. Walling, C., F. W. Stacey, S. E. Jamison, and E. S. Huyser, J. Amer. Chem. Soc., 80, 4543 and 4546 (1958).
741. Walsh, E. N., U.S. Patent 2,860,155 (1958); C.A. 54, 1425a (1960).
742. Walsh, E. N., T. M. Beck, and W. H. Woodstock, J. Amer. Chem. Soc., 77, 929 (1955).
743. Watt, G. W., and R. C. Thomson, Jr., J. Amer. Chem. Soc., 70, 2295 (1948).
744. Weil, T., B. Prijs, and H. Erlenmeyer, Helv. Chim. Acta, 35, 1412 (1952).
745. Weil, T., B. Prijs, and H. Erlenmeyer, Helv. Chim. Acta, 36, 142 (1953).
746. Weil, T., B. Prijs, and H. Erlenmeyer, Helv. Chim. Acta, 36, 1314 (1953).
747. Weller, J., Ber. 21, 1492 (1888).
748. Whetstone, R. R., W. J. Raab, and W. E. Hall, U.S. Patent 2,867,646 (1959); C.A. 53, 12237h (1959).

749. Whistler, R. L., C.-C. Wang, and S. Inokawa, J. Org. Chem., 33, 2495 (1968).
750. Wiberg, E., and H. Nöth, Z. Naturforsch., 12 B, 125 (1967); C.A. 52, 290 (1958).
751. Wieber, M., and W. R. Hoos, Tetrahedron Lett., 1968, 5333.
752. Wieber, M., and J. Otto, Chem. Ber., 100, 974 (1967).
753. Wieber, M., J. Otto, and M. Schmidt, Angew. Chem., Int. Ed. Engl., 3, 586 (1964).
754. Willems, J. F., Belgian Patent 557,556 (1957); C.A. 51, 17545c (1957).
755. Williamson, M. P., and C. E. Griffin, J. Phys. Chem., 72, 4043 (1968).
756. Willstätter, R., and E. Sonnenfeld, Ber., 47, 2801 (1914).
757. Wilson, H. F., and C. E. Glassick, French Patent 1,366,248 (1964); C.A. 61, 16096g (1964).
758. Wilson, Jr., J. H., U.S. Patent 3,162,570 (1964); C.A. 62, 7799h (1965).
759. Wilson, Jr., J. H., U.S. Patent 3,261,743 (1966); C.A. 65, 9662c (1966).
760. Winyall, M., and H. H. Sisler, Inorg. Chem., 4, 655 (1965).
761. Wolf, R., D. Houalla, and F. Mathis, Spectrochim. Acta, Part A, 23, 1641 (1967).
762. Wolf, R., J. R. Miquel, and F. Mathis, Bull. Soc. Chim. (France), 1963, 825.
763. Yagi, T., S. A. El-Kinawy, and A. A. Benson, J. Amer. Chem. Soc., 85, 3462 (1963).
764. Zakharkin, L. I., A. V. Kazantsev, and M. N. Zhubekova, Izv. Akad. Nauk SSSR, Ser. Khim., 1969, 2056.
765. Zakharkin, L. I., V. I. Stanko, and A. I. Klimova, Izv. Akad. Nauk SSSR, Ser. Khim., 1964, 917.
766. Zavalishina, A. I., S. F. Sorokina, and E. E. Nifant'ev, Zh. Obshch. Khim., 38, 2271 (1968).
767. Zhmurova, I. N., R. I. Yurchenko, and A. V. Kirsanov, Zh. Obshch. Khim., 38, 2078 (1968).
768. Zinov'ev, Yu. M., V. N. Kulakova, R. I. Mironova, and L. Z. Soborovskii, Zh. Obshch. Khim., 39, 606 (1969).
769. Zyablikova, T. A., I. M. Magdeev, and I. M. Shermergorn, Izv. Akad. Nauk SSSR, Ser. Khim., 1968, 397.

Chapter 11. Phosphinous Acids and Derivatives

L. A. HAMILTON* and P. S. LANDIS

Mobil Research and Development Corporation,
Paulsboro, New Jersey

*Deceased

The compounds described in this chapter include phosphinous acids, their sulfur and selenium analogs, and esters, amides, and anhydrides derived from them. Also included are the isomerized "keto" forms of the free acids (i.e., the secondary phosphine oxides, secondary phosphine sulfides, and secondary phosphine selenides).

In all cases the substances contain the disubstituted phosphinyl group having two organic radicals (alkyl or aryl) bound directly to a phosphorus atom. Either oxygen, nitrogen, sulfur, or selenium is also attached to the phosphorus atom by a single bond (except in the case of the "keto" forms). A tabulation of the types of compounds with specific examples is given in Table 1.

A. METHODS OF PREPARATION

I. PHOSPHINOUS ACIDS

The only well-characterized, reported phosphinous acid is the bis(trifluoromethyl) derivative.[83,84] It is prepared either by reaction of the anhydride with dry hydrogen

Table 1. Phosphinous Acids and Derivatives

General Type	Specific Example
R_2POH	$(CF_3)_2POH$
Phosphinous acids	Bis(trifluoromethylphosphinous) acid
R_2PNH_2	$Ph_2PN(CH_3)_2$
Phosphinous amides	N,N-Dimethyl diphenylphosphinous amide
R_2POR	Ph_2POCH_3
Phosphinites	Methyl diphenylphosphinite
R_2PSH	$(CF_3)_2PSH$
Phosphinothious acids	Bis(trifluoromethylphosphinothious) acid
R_2PSR	Ph_2PSEt
Phosphinothioites	Ethyl diphenylphosphinothioite
$R_2PH(O)$	MeEtPH(O)
Secondary phosphine oxides	Methylethylphosphine oxide
$R_2PH(S)$	$Ph_2PH(S)$
Secondary phosphine sulfides	Diphenylphosphine sulfide
$R_2PH(Se)$	$Ph_2PH(Se)$
Secondary phosphine selenides	Diphenylphosphine selenide
R_2POPR_2	$(CF_3)_2POP(CF_3)_2$
Phosphinous anhydrides	Bis(trifluoromethylphosphinous) anhydride

chloride or by reaction of bis(trifluoromethylphosphine) with mercuric oxide. In the first synthesis the anhydride is prepared by reaction of silver carbonate with the dial-

$$2(CF_3)_2PI + Ag_2CO_3 \longrightarrow (CF_3)_2POP(CF_3)_2 + 2AgI + CO_2$$

$$(CF_3)_2POP(CF_3)_2 + HCl \longrightarrow (CF_3)_2POH + (CF_3)_2PCl$$

kyliodide at room temperature. Treatment of the anhydride with dry HCl for eight days at 80° in a high-vacuum all-glass apparatus gave essentially a quantitative yield of bis(trifluoromethylphosphinous) acid.[83]

Synthesis of the acid by reaction of the phosphine with mercuric oxide is believed to form the phosphine oxide, which rapidly equilibrates to the more stable enol form when the alkyl groups contain strong electron- withdrawing groups.

$$(CF_3)_2PH + HgO \longrightarrow [(CF_3)_2P(O)H] \longrightarrow (CF_3)_2POH + Hg$$

Attempts to make mixed methyl trifluoromethylphosphinous acid by these techniques have been unsuccessful.[26]

II. SYNTHESIS OF SECONDARY PHOSPHORINE OXIDES BY USE OF ORGANOMETALLIC REAGENTS

The most generally used synthesis of secondary phosphine oxides involves treatment of the appropriate Grignard reagent with a solution of dialkylphosphonate in ether or tetrahydrofuran. This reaction has been used to prepare symmetrical[91,121,210,211] and unsymmetrical[61] secondary phosphine oxides of low and high molecular weight.

$$(RO)_2P(O)H + 3R'MgX \longrightarrow R'_2P(O)MgX \xrightarrow{H^{\oplus}} R'_2P(O)H$$

The reaction is rapid at room temperature and yields of 40 to 80% are obtained. Three moles of Grignard reagent are required, the first mole is used to replace the active hydrogen of the phosphonate to produce an alkane and the intermediate $(RO)_2POMgX$.[91] Subsequent reaction of this intermediate with Grignard reagent, followed by hydrolysis, yields the secondary phosphine oxide. Most phosphine oxides are solid and are separated and purified by solvent extraction and recrystallization.

Unsymmetrical, optically active secondary phosphine oxides have been prepared in 25 to 75% yield by nucleophilic displacement of ethoxide from an ethyl ester of a monosubstituted phosphinic acid using a Grignard reagent.[61]

$$R-P(O)OEt + R'MgX \longrightarrow RR'P(O)H$$
$$\quad\quad |$$
$$\quad\quad H$$

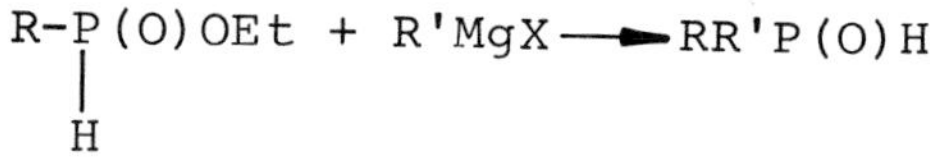

Sterically hindered secondary phosphine oxides can also be prepared by reaction of a Grignard reagent with alkyl or aryl dichlorophosphines.[19][94][120]   Only a bulky alkyl

$$RP(O)Cl_2 + t\text{-BuMgX} \xrightarrow{\quad H_3O^+ \quad} R\ t\text{-BuP(O)H}$$

group can be easily introduced by this technique, although the presence of a bulky R group on the phosphorus atom initially permits the utilization of less bulky Grignard reagents (e.g., isopropyl derivatives) to obtain the mono-substitution-reduction product. In such cases disubstitution is also observed, leading to the formation of tertiary phosphine oxides.

In actual practice, four types of products are obtained from this reaction:

$$RP(O)Cl_2 + t\text{-BuMgCl} \longrightarrow$$

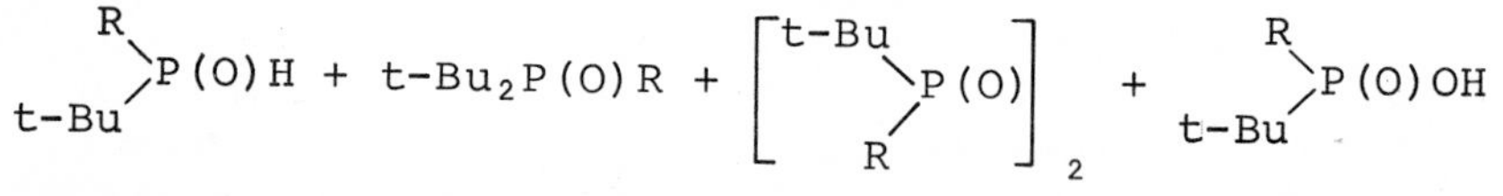

$$\underset{(\underline{1})}{\overset{R}{\underset{t\text{-Bu}}{>}}P(O)H} + \underset{(\underline{2})}{t\text{-Bu}_2P(O)R} + \underset{(\underline{3})}{\left[\overset{t\text{-Bu}}{\underset{R}{>}}P(O)\right]_2} + \underset{(\underline{4})}{\overset{R}{\underset{t\text{-Bu}}{>}}P(O)OH}$$

With aliphatic phosphonic dichlorides, t-butyl magnesium chloride reacts to form (1) in 50% yield. Yields of (1) increase with increasing chain length of the radical on the phosphorus atom. Tertiary phosphine oxides (2) are usually formed in low yields (2-7%); no type (3) products are observed, and type (4) products are usually found to the extent of 5 to 10%.

When the reaction is applied to arylphosphonic dichlorides and a t-butyl Grignard reagent, the primary product is again type (1), but appreciable yields of (3) and (4) are obtained.

These results indicate that the attachment of two bulky alkyl groups to the phosphorus atom is hindered but not prohibited. Steric hindrance in an intermediate formed in reaction of the Grignard reagent at the phosphorus atom may be the reason for the introduction of only one t-butyl group in the reaction.[120]

Aryl lithium compounds have also been used as reagents for the preparation of secondary phosphine oxides.[209] The lithium reagents are more active nucleophiles than the corresponding Grignard reagent and generally give high yields of phosphine oxide.

III. SYNTHESIS OF SECONDARY PHOSPHINE OXIDES BY WAY OF FRIEDEL-CRAFTS REACTION USING PHOSPHINOUS CHLORIDES

A simple synthesis of secondary aryl phosphine oxides has been developed using Friedel-Crafts reactions.[15,25]

$$ArH + PCl_3 \xrightarrow{AlCl_3} ArPCl_2 + Ar_2PCl \xrightarrow{H_2O} ArPH(O)OH + Ar_2P(O)H$$

Hydrolysis of the reaction mixtures from these preparations gives a mixture of phosphinic acid and secondary phosphine oxide. With durene, mesitylene, or pentamethylbenzene, the phosphine oxide is formed in 30 to 60% yield. In these reactions the phosphine oxide coprecipitates with by-product $AlCl_3 \cdot POCl_3$ and is separated by hydrolysis and extraction with benzene. Co-product phosphinic acid is separated by caustic extraction of the benzene solution.

When the diarylphosphinous chloride is available, it can be hydrolyzed by refluxing a wet benzene[140] or carbon tetrachloride[56,94] solution.

Unsymmetrical secondary aryl phosphine oxides have been prepared by a sequence of reactions involving initial reaction of an arylphosphonous dichloride with an arenediazonium fluoborate, reduction of the product with aluminum powder, and hydrolysis of the resulting diarylphosphinous chloride.[178]

$$PhN_2BF_4 + ArPCl_2 \xrightarrow{CuBr} Ph\text{-}\overset{\oplus}{P}(Ar)Cl_2\,BF_4^{\ominus} \xrightarrow[{[H]}]{Al}$$

$$PhArPCl \xrightarrow{H_2O} PhArP(O)H$$

The phosphinous chloride is separated from any phosphonous chloride by direct distillation. Yields are generally 30 to 60%. Hydrolysis of the phosphinous chloride is carried out in wet benzene in order to minimize disproportionation. Yields in the hydrolysis are between 70 and 95%.

IV. SYNTHESIS OF SECONDARY PHOSPHINE OXIDES BY REDUCTION OF PHOSPHINYL HALIDES AND PHOSPHINIC ACID ESTERS

Reduction of phosphinyl halides with lithium aluminum hydride or sodium borohydride is carried out at 0° with an excess of reducing agent. This technique is a simple method for the preparation of dialkylphosphine oxides.[211] It has a major disadvantage, however, in the limited availability of phosphinic acids, the precursors of the halides.

Reduction of esters of disubstituted phosphinic acids with lithium aluminum hydride has been used to prepare unsymmetrical, optically active secondary phosphine oxides.[61] Yields of 25 to 75% are reported.

V. SYNTHESIS OF SECONDARY PHOSPHINE OXIDES BY OXIDATION TECHNIQUES

Secondary phosphine oxides are obtained by treating a secondary phosphine with dry air in a solvent at 50 to 70°.[183-185] The reaction can be carried out in a thin film reactor using dry air without a solvent. In this case, cooling is required to moderate the reaction exothermicity. Aprotic solvents are most frequently used, and dry air is bubbled through the solution. Yields are surprisingly high (70-90%) for this reaction, which shows typical free radical characteristics.

With more powerful chemical oxidants such as peroxides, ozone, or perchlorates, it is difficult to stop the oxidation at the intermediate phosphine oxide stage so that large amounts of phosphinic acids $R_2P(O)OH$ are observed. Mercuric oxide has found limited use in the oxidation of secondary phosphines.[85]

Phosphinous amides are converted by oxidation to secondary phosphine oxides by chemical oxidants, including manganese dioxide and hydrogen peroxide.[202]

VI. SYNTHESIS OF SECONDARY PHOSPHINE OXIDES BY HYDROLYSIS OF PHOSPHINITES

Controlled hydrolysis of dialkylphosphinites in the presence of a catalytic amount of sulfuric acid produces secondary phosphine oxides in 80 to 95% yields.[110] The starting materials are available from the reaction of alkylphosphorodichloridites with Grignard reagents[111] or by reaction of an alcohol with a dialkylphosphinous chloride.[5,114,115,135,163,164,167,188]

$$ROPCl_2 + 2R'MgX \longrightarrow R'_2POR + 2MgXCl$$

$$R_2PCl + ROH \longrightarrow R_2POR + HCl$$

$$R_2POR' + H_2O \xrightarrow{H^{\oplus}} R_2P(O)H + R'OH$$

VII. SYNTHESIS OF SECONDARY PHOSPHINE OXIDES BY REACTION OF PRIMARY PHOSPHINES WITH ALDEHYDES AND KETONES

Secondary phosphine oxides can be obtained by careful control of the reaction of primary phosphines with aldehydes or ketones in the presence of hydrochloric acid.[23,24] In this reaction the yield of the product is determined by oxygen transfer, multiple addition, reaction conditions, and the nature of the substituent on the phosphorus atom (see also Chapter 7). Five separate reactions are possible.

1. Oxygen transfer

$$RPH_2 + R'_2CO \longrightarrow R\text{-}P(O)(H)CHR'_2$$

2. Oxygen transfer and addition

$$RPH_2 + 2R'_2CO \longrightarrow RP(O)(CHR'_2)(HOCR'_2)$$

3. Multiple addition

$$RPH_2 + 2R'_2CO \longrightarrow RP[C(OH)R'_2]_2$$

4. Multiple addition and phosphonium salt formation

$$RPH_2 + 2R'_2CO + HX \longrightarrow RP[CR'_2(OH)]_2 \cdot HX$$

$$RPH_2 + 3R'_2CO + HX \longrightarrow \overset{\oplus}{R}P[CR'_2(OH)]_3X^{\ominus}$$

5. Addition and cyclization

$$RPH_2 + 3R_2'CO \longrightarrow$$

$$R_2'C \underset{\underset{O}{|}}{\overset{\overset{R}{\underset{|}{P}}}{\diagup}} \underset{\underset{O}{|}}{\overset{}{\diagdown}} CR_2' + H_2O$$

with the central ring:

$$\begin{array}{c} R \\ | \\ P \\ \diagup \quad \diagdown \\ R_2'C \quad\quad CR_2' \\ | \quad\quad\quad | \\ O \quad\quad\quad O \\ \diagdown \quad \diagup \\ C \\ | \\ R_2' \end{array}$$

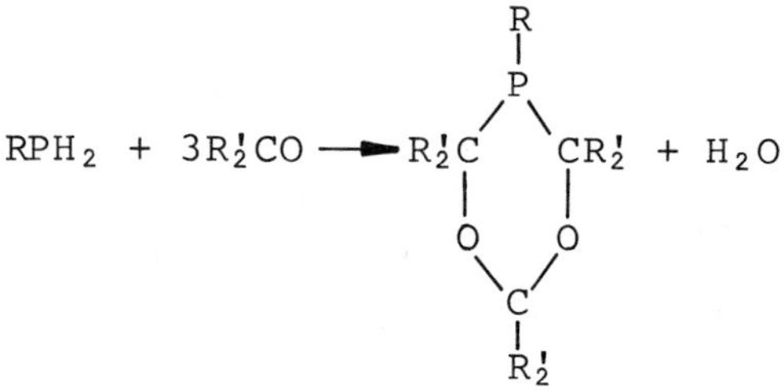

The choice between reactions 1 and 2 is determined
largely by steric factors in the phosphine and carbonyl
reagents. Reactant ratios also influence the type and
amount of each product obtained. For example, arylphos-
phines react with cyclohexanone at 1:1 and 1:2 molar ratios
to produce tertiary phosphine oxides. With less hindered
ketones such as acetophenone, the reactant ratios dictate
the course of reaction and either secondary or tertiary
phosphine oxide can be obtained. When a hindered phosphine
is reacted with bulky ketone, the secondary phosphine oxide
is the only product obtained. Excess acid is important to
suppress reactions 4 and 5 (see Chapters 7 and 4).

Aromatic aldehydes generally produce 1-hydroxy tertiary
phosphine oxides on reaction with primary phosphines. However, we know that these derivatives decompose to give
salts of secondary phosphine oxides,[95] thus providing an
indirect route to dialkylphosphine oxides.

$$RPH_2 + 2R'CHO \xrightarrow{\text{H}^{\oplus}} \underset{\underset{CH_2R'}{|}}{RP(O)CH(OH)R'} \xrightarrow[\text{2) H}^{\oplus}]{\text{1) OH}^{\oplus}}$$

$$R(R'CH_2)P(O)H + RCHO$$

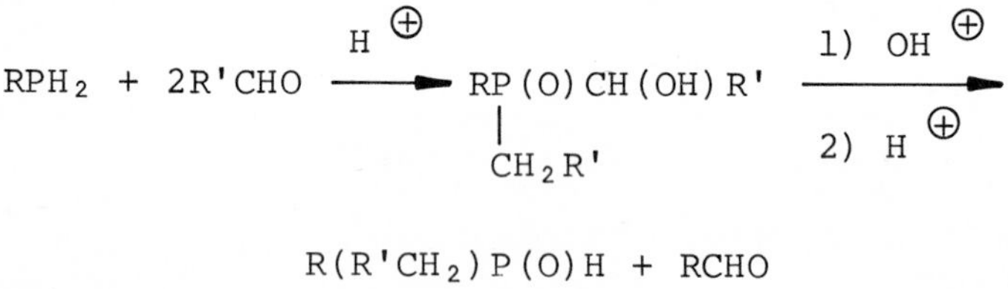

Aliphatic aldehydes generally yield the normal carbonyl
addition products, although dialdehydes including glyoxal,
succinaldehyde, adipaldehyde, terephthalaldehyde, and iso-
phthalaldehyde are reported to produce secondary diphosphine oxides.[23]

Base-catalyzed addition of primary phosphine oxides to
olefins offers another route to a variety of secondary
phosphine oxides.[24]

$$C_8H_{17}P(O)H_2 + CH_2=CHCONH_2 \xrightarrow{\text{NaOMe}} \underset{\underset{H}{|}}{RP(O)CH_2CH_2CONH_2}$$

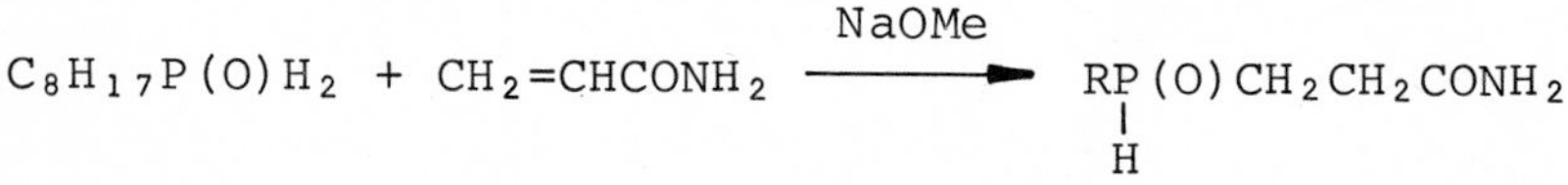

Methanol is used as a solvent and excess olefin must be avoided to suppress the formation of the tertiary phosphine oxide.

VIII. SYNTHESIS OF ESTERS OF PHOSPHINOUS ACID AND PHOSPHINITES BY ALCOHOLYSIS OF MONOCHLOROPHOSPHINES

The most satisfactory laboratory synthesis of phosphinites involves the addition of an alcohol or phenol to a disubstituted chlorophosphine in the presence of a tertiary base, usually pyridine, triethylamine, or dimethylaniline.[4,5,8,9,32,109,111,114,116,117,135,164,167,169,188,205]

$$R_2PX + R'OH + B \longrightarrow R_2POR' + B \cdot HX$$

Less satisfactory results are obtained with sodium alkoxides, since the separation of colloidal sodium halide is cumbersome and free alkali in the alkoxide promotes isomerization and salt formation.

The reaction is generally carried out in anhydrous ether or tetrahydrofuran in an inert atmosphere and at temperatures close to 0°. Temperatures above 150° must be avoided during distillation in order to minimize isomerization of reactive products to the trisubstituted phosphine oxide.[8] Aryl derivatives frequently are crystalline solids with sharp, unpleasant odors.

The wide range of alcohols that have been used as nucleophiles in this reaction includes polyhydric alcohols and unsaturated alcohols.[17] In the latter case, isomerization to the pentavalent phosphine oxide occurs at relatively low temperatures. Acetylenic esters have not been isolated, although they are postulated as intermediates.

Epoxides undergo uncatalyzed addition to chlorophosphines, producing β-chloroethyl dialkylphosphinites.[105,109,113] This reaction, which is generally applicable to dialkyl- and diarylchlorophosphines, is usually carried out in an inert atmosphere at -20°. The products are thermally and oxidatively unstable. On heating to 150°, there is rapid rearrangement to a β-chloroethylphosphine oxide. Prolonged heating results in the elimination of dichloroethane and the production of esters of ethylenediphosphinic acid.[109] Hydrogen chloride is eliminated in the presence of base, forming vinyl dialkylphosphine oxides.[109]

$$R_2POCH_2CH_2Cl \xrightarrow{150°} R_2P(O)CH_2CH_2Cl \begin{array}{c} \xrightarrow{base} R_2P(O)CH=CH_2 \\ \xrightarrow{\Delta} R_2P(O)CH_2CH_2(O)PR_2 \end{array}$$

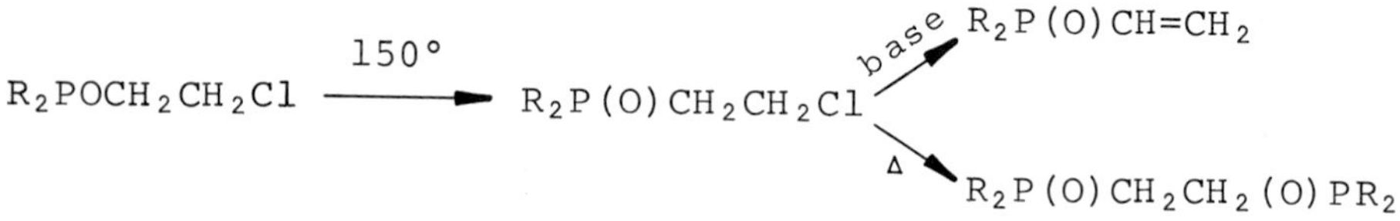

IX. SYNTHESIS OF ESTERS OF PHOSPHINOUS ACID AND PHOS-PHINITES BY TRANSESTERIFICATION

Transesterification of phosphinites proceeds under mild conditions utilizing base catalysis.[163,164] The reaction is carried out in the usual fashion at a temperature permitting the removal of the exchanged alcohol from the reactants.

In contrast to phosphonites, the phosphinites are transesterified not only by primary and secondary alcohols but also by tertiary alcohols. Also, the transesterification is more facile and produces higher yields in phosphinites than in analogous phosphonites. This is in accord with the formation of an intermediate phosphonium compound utilizing the unshared pair of electrons on phosphorus.

$$R_2\ddot{P}OR' + R''OH \longrightarrow \left[\begin{array}{c} R \diagdown \quad \diagup OR' \\ P \\ R \diagup \;\; | \diagdown OR'' \\ H \end{array} \right] \xrightarrow{-R'OH} R_2POR''$$

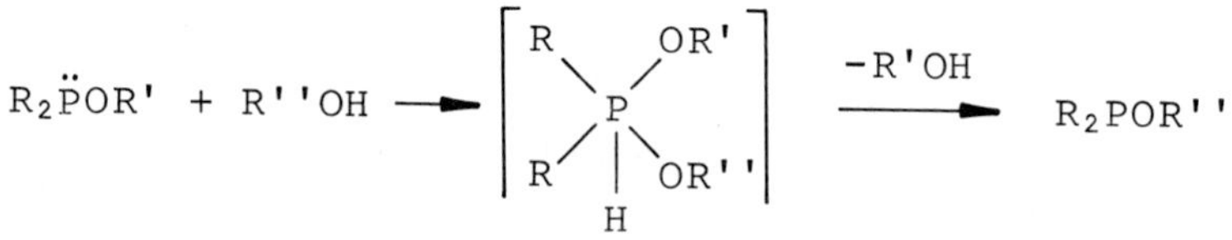

Factors that increase the electron density at the phosphorus atom facilitate the transesterification. Thus phosphinites react more readily than phosphonites or phosphites, and arylphosphinites react more sluggishly than alkyl derivatives.[163] Steric hindrance in the alcohol as well as in the phosphinite plays an important role in the reaction kinetics.

Glycols, which have also been used in the transesterification reaction, produce mono- and polysubstituted phosphinites. The stoichiometry of the reactants normally dictates the yield and distribution of products.

$$
\begin{array}{ccc}
CH_2 & CH_2 & CH_2 \\
| & | & | \\
OPR_2 & OH & OH
\end{array}
$$

$$+$$

$$
R_2POR' \;+\;
\begin{array}{ccc}
CH_2 & CH & CH_2 \\
| & | & | \\
OH & OH & OH
\end{array}
\;\longrightarrow\;
\begin{array}{ccc}
CH_2 & CH & CH_2 \\
| & | & | \\
OPR_2 & OPR_2 & OH
\end{array}
$$

$$+$$

$$
\begin{array}{ccc}
CH_2 & CH & CH_2 \\
| & | & | \\
OPR_2 & OPR_2 & OPR_2
\end{array}
$$

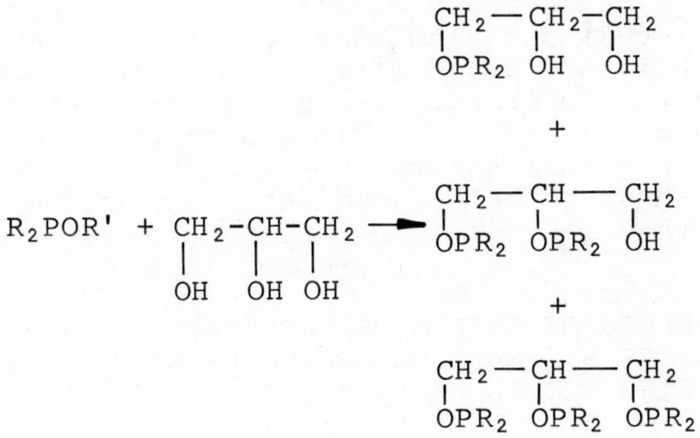

X. SYNTHESIS OF ESTERS OF PHOSPHINOUS ACID AND PHOS-PHINITES USING GRIGNARD REAGENTS

1. DISPLACEMENT OF ALKOXIDE GROUPS IN PHOSPHITES. Prior to 1950 little was known concerning the synthetic utility and the mechanism of the displacement of an alkoxide group from trivalent esters by Grignard and other organometallic reagents. A preliminary communication recorded the formation of a trisubstituted phosphine oxide from reaction of excess aryl Grignard with trimethylphosphite. This product was explained by positing intramolecular rearrangement of the alkyl diarylphosphinite, which resulted from displacement of two methoxy groups.

Subsequent work[14] has demonstrated that the reaction follows a stepwise path that can be interrupted at a point where the major product is phosphonite, phosphinite, or phosphine. The ratios of the three trivalent phosphorus

$$(RO)_3P + R'MgX \rightleftharpoons (RO)_3P \cdot R'MgX \longrightarrow (RO)_2PR' + ROMgX$$

$$(RO)_2PR' + R'MgX \rightleftharpoons (RO)_2PR' \cdot R'MgX \longrightarrow (RO)PR'_2 + ROMgX$$

$$(RO)PR'_2 + R'MgX \rightleftharpoons (RO)PR'_2 \, R'MgX \longrightarrow R'_3P + ROMgX$$

compounds are dependent on the concentration of reactants, the nature of the leaving group, the solvent, and the addition time. To optimize phosphinite formation, the reaction is carried out with 2 moles of Grignard reagent using ether as a solvent and a temperature of 0 to 20°.

2. DISPLACEMENT OF HALOGEN FROM DIHALOPHOSPHITES. Replacement of two halogens from the easily obtainable alkyl- or aryldihalophosphites (see Chapter 15) produces

phosphinites in 30 to 70% yield.[111,163,191] Low reaction temperatures (-60°) and slow addition of 1 equiv. of phosphite to the Grignard reagent serve to minimize displacement of the alkoxide group. Yields are further enhanced by the use of pyridine or 8-hydroxyquinoline, which displaces the esters from their complexes with the magnesium halide. Tetrahydrofuran is apparently a better solvent medium than ether.

XI. SYNTHESIS OF ESTERS OF PHOSPHINOUS ACID AND PHOSPHINITES BY REACTION OF PHOSPHINOUS AMIDES WITH CYANATES AND ALCOHOLS

Recently aryl cyanates have been reported to undergo reaction with N,N-disubstituted diarylphosphinous amides to produce moderate yields of diarylphosphinous esters.[118] Mechanistic details are not reported.

$$Ph_2PNMe_2 + PhCNO \longrightarrow Ph_2POPh + Me_2NCN$$

Alcoholysis of phosphinous amides is discussed on page 505.

XII. SYNTHESIS OF ESTERS OF PHOSPHINOUS ACID AND PHOSPHINITES BY REACTION OF PHOSPHINES WITH PEROXIDES

The reaction of trialkylphosphines with di-t-butyl peroxides leads to formation of t-butyl dialkylphosphinite and trialkylphosphine oxide.[20] These products suggest two modes of cleavage in the radical reaction. With alkyl

$$t\text{-BuO}\cdot + R_3P \longrightarrow t\text{-BuOPR}_2 + R\cdot \qquad (a)$$

$$t\text{-BuO}\cdot + R_3P \longrightarrow R_3PO + t\text{-Bu}\cdot \qquad (b)$$

derivatives path (a) is favored; path (b) is preferred with aryl derivatives.

XIII. SYNTHESIS OF ESTERS OF PHOSPHINOUS ACID AND PHOSPHINITES BY REACTION OF CHLOROPHOSPHINES WITH ALDEHYDES AND PHOSPHITES

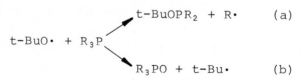

High yields of ester containing the phosphinite structure are reported in the reaction of a diarylchlorophosphine with an aldehyde and a trialkylphosphite.[16,187]

$$Ph_2PCl + EtCHO + P(OEt)_3 \longrightarrow Ph_2POCH(Et)P(OEt)_2 + EtCl$$

XIV. ACETOXYPHOSPHINES

The only known acetoxydialkylphosphines are the type $(CF_3)_2POR$ where R is acetyl and trifluoromethyl. Two synthetic routes, 1[159] and 2,[26] to these compounds have been used. Apparently the strong electron-withdrawing trifluoromethyl group is necessary to stabilize the trivalent oxyphosphorus product. Generally, attempts to pre-

1. $(CF_3)_2PI + CF_3COOAg \longrightarrow (CF_3)_2POCOCF_3 + AgI$

2. $(CF_3)_2POP(CF_3)_2 + (CH_3CO)_2O \longrightarrow 2(CF_3)_2POCOCH_3$

pare such trivalent compounds lead to phosphorus valence expansion and formation of a phosphine oxide. Thus reaction of methyl diphenylphosphinite with acetyl chloride produces acetyl diphenylphosphine oxide.[8]

XV. PHOSPHINOTHIOUS ACIDS

The only reported phosphinothious acids are bis(trifluoromethylphosphinothious) acid and its deutero analog. Although hydrogen sulfide does not directly react with $(CF_3)_2PCl$, rapid reaction occurs at room temperature in the presence of trimethylamine. Yields are limited to 50%, since the mercaptan further condenses with the chlorophosphine to produce di(bistrifluoromethylphosphino)sulfide.

$$(CF_3)_2PCl + H_2S \xrightarrow{Me_3N} (CF_3)_2PSH$$

$$(CF_3)_2PSH + (CF_3)_2PCl \xrightarrow{Me_3N} (CF_3)_2PSP(CF_3)_2$$

Since the phosphinothious acid forms a salt with trimethylamine, yields from the foregoing reaction can be enhanced by using a twofold excess of amine, thus effectively blocking further condensation of the thioacid with unreacted halophosphine. Anhydrous hydrogen bromide is used to generate the free thioacid.

An alternative route to the same product involves

reversible cleavage of di(bistrifluoromethylphosphino)sulfide by anhydrous hydrogen bromide at 150°.[29,41]

The parent compound H_2PSH has been detected in a flow system by mass spectrometry and low-temperature IR spectroscopy. Phosphine and hydrogen sulfide at 0.1 mm pressure are passed through a glow discharge from a high-frequency generator and the products are condensed at -196°. Conversion at flow rates of 0.5 mmole/min are 0.5 to 1.0%.

XVI. SYNTHESIS OF PHOSPHINOTHIOITES BY DISPLACEMENT OF HALOGEN FROM DIALKYLPHOSPHINOUS CHLORIDES BY MERCAPTANS AND XANTHATES

The simplest and most economical synthesis of phosphinothioites R_2PSR' involves the slow addition of mercaptans to dialkyl- or diarylphosphinous chlorides in methylene chloride or chloroform solution. The reaction typically provides 50 to 80% yields of the thioester.[1,29,41] It is often convenient to carry out the reaction in the presence of pyridine or a tertiary base to remove by-product hydrogen chloride.[32] This technique provides nearly quantitative yields of thioester, minimizing several unwanted side reactions, including isomerization of sensitive esters to the tertiary phosphine sulfide,[3,41,205] (see Chapter 7).

Reaction of the phosphinous halides with an alkali metal mercaptide is also a satisfactory route to phosphinothioites.[5,7]

A convenient technique for thioalkylation of phosphinous halides involves reaction of equimolar quantities of a xanthate with a monohalophosphine at 160 to 180°.[58] The reaction is usually complete in several hours, as evidenced by the cessation of evolution of by-product gases, carbonylsulfide, and ethyl chloride.

$$R_2PCl + R'OC(S)SEt \longrightarrow R_2PSR' + COS + EtCl$$

XVII. SYNTHESIS OF PHOSPHINOTHIOITES BY REACTION OF SECONDARY PHOSPHINES WITH DISULFIDES.

In the presence of free radical inhibitors such as hydroquinone, dialkyl- or diarylphosphines react with disulfides to form phosphinothioites in surprisingly high yields.[53,80] The reaction is carried out in dilute benzene solution using an inert atmosphere. Mercaptan is a by-product. In the absence of a radical inhibitor, further abstraction of sulfur occurs to give phosphinodithioites $R_2P(S)SR$. The formation of both products presumably occurs by S_N2 displacement at the disulfide bond in manner analo-

$$R_2PH + R'SSR' \xrightarrow{\text{Inhibitor}} R_2PSR' + R'SH$$

$$R_2PH + R'SSR' \longrightarrow R_2P(S)SR'$$

gous to the reaction of tertiary phosphines with disulfides to form tertiary phosphine sulfides.[131]

In rather sharp contrast, it has recently been reported[198a] that diphosphines, on irradiation with UV light, react with disulfides, diselenides, ditellurides, or hydrazines to give the respective dialkylphosphinohetero ester. Nearly quantitative yields are reported.

$$R_2P\text{—}PR_2 + R'E\text{—}ER' \longrightarrow 2R_2P\text{—}ER'$$

Where E = sulfur, selenium, tellurium or nitrogen.

XVIII. SYNTHESIS OF SECONDARY PHOSPHINE SULFIDES AND SELENIDES BY THE USE OF ORGANOMETALLIC REAGENTS

Dialkylphosphonothionates react with excess Grignard reagent to produce secondary phosphine sulfides.[131,132] This reaction has not been explored as extensively as the

$$(EtO)_2P(S)H + 3ArMgX \longrightarrow Ar_2P(S)H + 2EtOMgX$$

analogous formation of secondary phosphine oxides, but it appears that reaction conditions, solvents, and techniques are similar. Yields are frequently greater than 90%.

It is often more convenient to use an organolithium reagent because the secondary phosphine sulfides are easier to separate from the reaction products.[131,132]

XIX. SYNTHESIS OF SECONDARY PHOSPHINE SULFIDES AND SELENIDES BY REACTION OF SECONDARY PHOSPHINES WITH SULFUR OR SELENIUM

In the absence of air, addition of a stoichiometric amount of sulfur or selenium to a solution of a secondary phosphine in an inert solvent produces nearly quantitative yields of secondary phosphine sulfides[157,182] or selenides.[131a] Liquid products are purified by distillation, although disproportionation and decomposition often occur. The higher molecular weight derivatives, which are crystalline solids, are purified by recrystallization. Careful control of the reaction including slow addition of the sulfur is used in order to prevent further sulfurization of the phosphine sulfide to the phosphinodithioic acid $R_2P(S)SH$.

Disecondary alkylene diphosphine disulfides have also been prepared by this technique.[101,102]

$$R(H)P(CH_2)_nP(H)R + 2S \longrightarrow R(H)P(S)(CH_2)_n(S)P(H)R$$

Elemental sulfur will also react with phosphinous amides under carefully controlled conditions to produce secondary phosphine sulfides.[130,200,201] Only two secondary phosphine selenides have been prepared so far; they are both crystalline solids.[131a]

XX. SYNTHESIS OF SECONDARY PHOSPHINE SULFIDES AND SELENIDES BY HYDROLYTIC CLEAVAGE OF DIPHOSPHINE DISULFIDES

Aliphatic diphosphine disulfides are cleaved by aqueous alkali to produce equimolar quantities of dialkylphosphine sulfide and dialkylphosphinothioic acid.[47] The sulfide is

$$R_2P(S){-}P(S)R_2 + NaOH \longrightarrow R_2P(S)ONa + R_2P(S)H$$

easily separated from the acid by virtue of its solubility in base. This cleavage reaction is apparently limited to aliphatic diphosphine disulfides, since aryl derivatives do not react with caustic at elevated temperatures.

XXI. SYNTHESIS OF SECONDARY SULFIDES AND SELENIDES BY REACTION OF PHOSPHINOUS AMIDES WITH HYDROGEN SULFIDE

Reaction of phosphinous amides with anhydrous hydrogen sulfide in an inert solvent provides high yields of secondary phosphine sulfides.[132] Because of the ease of prepa-

$$R_2PNR'_2 + H_2S \longrightarrow R_2P(S)H + R'_2NH\cdot H_2S$$

ration, the high yields, the generality for either alkyl or aryl derivatives, and the easy accessibility of the starting materials, this method is the preferred synthesis of secondary phosphine sulfides.

XXII. DIALKYL- AND DIARYLPHOSPHINOUS AMIDES BY

REACTION OF DIALKYLCHLOROPHOSPHINES WITH AMMONIA AND AMINES

Dialkylphosphinous amides having no substituents on the nitrogen atom are stable only when the phosphorus substituents are highly electronegative or are very bulky. Thus bis(trifluoromethylphosphinous) amide[86] and di-t-butylphosphinous amide[195] have been prepared in high yield by reaction of ammonia with the appropriate dialkylphosphinous chloride. The reaction is usually carried out in dry ether at -50° and the products separated and purified by vacuum distillation.

Bis(trifluoromethylphosphine) has been reported to undergo an elimination-addition reaction with ammonia to produce difluoromethyltrifluoromethylphospinous amide in 64% yield.[79]

$$(CF_3)_2PH + NH_3 \xrightarrow{-NH_4F} CF_3P{=}CF_2 \xrightarrow{NH_3} CF_3CHF_2PNH_2$$

All attempts to prepare $(CH_3)_2PNH_2$ have led to $[(CH_3)_2P]_2NH$, which itself is unstable and decomposes on heating to $[(CH_3)_2P]_3N$.[34] Apparently the stability of dimethylphosphinous amide is governed by nitrogen hybridization, and π-bond character dictates the reactivity of the trifluoromethyl derivative. The chemistry of methyltrifluoromethylphosphinous amide is intermediate.[26]

A wide variety of N-substituted phosphinous amides have been prepared by treatment of a disubstituted phosphinous halide with a primary or secondary amine in ether or benzene at ice bath temperatures.[34,65,87,103,130,170,200,201] Generally these reactions proceed smoothly, with precipitation of the amine hydrochloride. However, with some diarylamines, it has been found necessary to prepare the sodium salt of the diarylamine and subsequently react it with a diarylphosphinous chloride.[65]

$$R_2PX + 2R'_2NH \longrightarrow R_2PNR'_2 + R'_2NH{\cdot}HX$$

$$Ar_2PX + Ar'_2NNa \longrightarrow Ar_2PNAr'_2 + NaX$$

XXIII. SYNTHESIS OF DIALKYL- AND DIARYLPHOSPHINOUS AMIDES BY ANIONIC DISPLACEMENT OF HALOGEN FROM DIHALOPHOSPHINOUS AMIDES

Alkyl anions from organometallic reagents have been used to displace halogen from dihalophosphinous amides to produce dialkylphosphinous amides.[35] Although this reaction

$$Me_2NPCl_2 + 2MeMgX \longrightarrow Me_2NPMe_2 + 2MgXCl$$

has found occasional use,[35,103] the yields of desired product are often below 50% because the Grignard reagent attacks the phosphorus-nitrogen bond almost as readily as the phosphorus-chlorine bond. Inverse addition of the Grignard reagent to the phosphine is often used to minimize such phosphorus-nitrogen bond attack. An interesting exception is the synthesis of the aromatic heterocyclic (5), wherein addition of the Grignard reagent occurs in high yield with no observable phosphorus-nitrogen bond cleavage.[57]

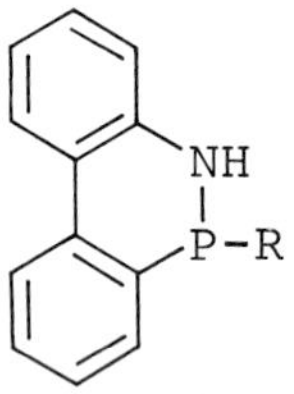

(5)

Organolithium[42,63,153,207a] and aluminum alkyl[129] reagents react with dihalophosphinous amides in a manner analogous to that of Grignard reagents forming phosphinous amides.

$$R_2NPCl_2 + 2CH_3C{\equiv}CLi \longrightarrow R_2NP(C{\equiv}CCH_3)_2 + 2LiCl$$

$$3R_2NPCl_2 + 2AlR'_3 \longrightarrow 3R_2NPR'_2 + 2AlCl_3$$

XXIV. SYNTHESIS OF DIALKYL- AND DIARYLPHOSPHINOUS AMIDES BY REACTION OF CHLOROPHOSPHINES WITH HYDRAZINES

A series of diarylphosphinous hydrazides has been synthesized by displacement of halogen from diarylphosphinous chloride with hydrazine or its derivatives.[149,160] In some

$$2Ph_2PCl + 3H_2NNH_2 \longrightarrow Ph_2PNHNHPPh_2 + 2H_2NNH_2{\cdot}HCl$$

cases the solubility of hydrazinium halide by-product prevents precipitation and renders the method inconvenient. This difficulty is overcome by the use of a hydrogen halide acceptor such as triethylamine.

It is possible to replace more than one proton in a hydrazine with the diphenylphosphinol group. The first proton may be substituted at room temperature, but temperatures above 100° are required to substitute additional protons. Apparently a steric factor contributes to the greater stability of the phosphorus-substituted hydrazine over the parent hydrazine, since it is well known that the attachment of phosphorus to nitrogen reduces the electron density and the basic character of the latter[34,65,200,201] and thus should promote its proton-donor character.

With an unsymmetrical compound such as methylhydrazine, bis(trifluoromethylphosphinous) iodide reacts at both nitrogen atoms with equal preference giving the isomeric compounds $(CF_3)_2PNHNHMe$ and $(CF_3)_2PNMeNH_2$. Apparently the increased basicity of the alkylamino group is compensated by steric interactions.

The products are potential bifunctional chelating agents with phosphorus-nitrogen dative bonding modifying the chelating ability as well as the character of the nitrogen-nitrogen bond.

The hydrazine derivatives are characterized by ^1H, ^{31}P, and ^{19}F NMR, IR spectra and by quantitative cleavage of the phosphorus-nitrogen bond with hydrogen chloride.[161]

$$(CF_3)_2PNHNMe_2 + 2HCl \longrightarrow (CF_3)_2PCl + Me_2NNH_2 \cdot HCl$$

XXV. MISCELLANEOUS SYNTHESES OF DIALKYL- AND DIARYL-PHOSPHINOUS AMIDES

Bis(phosphinous) amides are prepared by the reaction of phosphinous amides with anhydrous ammonia[35] or by reaction of ammonium bromide with a sodium amide-dimethylphosphine adduct.[208] Better characterized, more stable bis-

$$Me_2PNMe_2 + 2NH_3 \longrightarrow Me_2PNHPMe_2$$

(phosphinous) amides are obtained using trifluoromethyl derivatives.

Several phosphinous amides of type (6) have been prepared in good yield (a) by reaction of hexamethyldisilazane and diphenylchlorophosphine in the presence of triethylamine

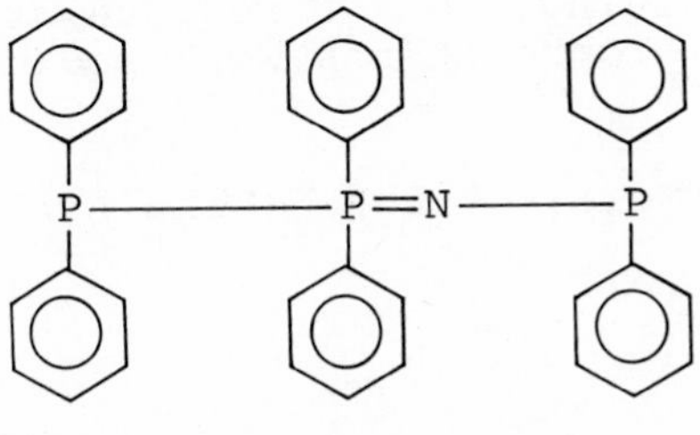

(6)

or (b) from tris(diphenylphosphino)immonium chloride and triethylamine.[67]

XXVI. COMPOUNDS CONTAINING $R_2P-O-Si-$, $R_2P-N-Si$, R_2P-O-B LINKAGES

The volatile, stable compound $Me_3SiOP(CF_3)_2$ has been prepared by three different quantitative processes.[27]

$$[(CF_3)_2P]_2O + (Me_3Si)_2O \longrightarrow 2(CF_3)_2POSiMe_3$$

$$[(CF_3)_2P]_2O + Me_3SiCl \longrightarrow (CF_3)_2POSiMe_3 + (CF_3)_2PCl$$

$$(CF_3)_2PCl + (Me_3Si)_2O \longrightarrow (CF_3)_2POSiMe_3 + Me_3SiCl$$

High-vacuum techniques are used for monitoring these syntheses, and fractional condensation is the customary technique for isolating products.

The analogous compound $H_3SiOP(CF_3)_2$ is less stable because the SiH structure cleaves the phosphorus-oxygen bond.

Similar preparative techniques have been used for the polyphosphinoxysilanes including $HSi[OP(CF_3)_2]_3$ and $Si[OP(CF_3)_2]_4$. All the products are surprisingly stable against disproportionation to P-O-P and Si-O-Si bonds.

Three volatile boron phosphinites have been prepared by cleavage of bis(trifluoromethylphosphinous) acid anhydride with the corresponding boron chloride.[28] The

$$3(CF_3)_2POP(CF_3)_2 + BCl_3 \longrightarrow [(CF_3)_2PO]_3B + 3(CF_3)_2PCl$$

$$2(CF_3)_2POP(CF_3)_2 + MeBCl_2 \longrightarrow [(CF_3)_2PO]_2BMe + 2(CF_3)_2PCl$$

$$(CF_3)_2POP(CF_3)_2 + Me_2BCl \longrightarrow (CF_3)_2POBMe_2 + (CF_3)_2PCl$$

reactions are carried out in sealed tubes at 50° and the products are separated by high-vacuum fractionation.

Silylamines containing the R_2P residue are prepared by reaction of diarylchlorophosphines with sodium bis(trimethylsilyl)amines.[67]

$$Ph_2PCl + NaN(SiMe_3)_2 \longrightarrow Ph_2PN(SiMe_3)_2 + NaCl$$

B. BASIC CHEMISTRY AND PROPERTIES

B.1. Phospinous Acids

Phosphinous acids R_2POH were listed as acids provisionally until 1968, when direct evidence on their isolation and properties was obtained. Only trifluoromethylphosphinous acid and its deuterium analog have been isolated.

The tautomeric equilibrium $R_2POH \rightleftharpoons R_2P(O)H$ has long been cited because of the structural relation to phosphorus acid, dialkylphosphonates, and arylphosphinates.[189] This equilibrium was the basis for the Arbuzov postulate that a true hydroxyl cannot exist on a trivalent phosphorus atom but must rearrange to the keto form. It has been pointed out that the facile ketonization is the result of the formation of stronger bonds.[134]

Direct evidence for this equilibrium in phosphinous acids include the following observations: formation of an insoluble silver salt R_2POAg is slow when silver nitrate is added to diphenylphosphine oxide;[99] IR spectroscopy clearly shows predominance of the keto form;[49] in reactions of diarylphosphine oxides with disulfides, acceleration of the rate by electron-withdrawing groups on the aromatic rings attached to phosphorus support, as an initial reaction step, a rapid equilibrium between keto and reactive trivalent enol forms;[81] rapid hydrogen-deuterium exchange is observed on treatment of secondary phosphine oxides with methanol-d.[61]

In the case of the trifluoromethyl derivative, the R_2POH structure is permanently stable and, indeed, ditrifluoromethylphosphine oxide easily converts to the phosphinous acid.[83] It appears that the $(CF_3)_2P$ group is weakly basic, so that the attached oxygen atom preferentially retains the proton. Also the electron-acceptor character of this $(CF_2)_3P$-O group can support a strong π bond between the lone pair of electrons over oxygen and the two pairs of 3d orbitals of the phosphorus atom. With alkyl derivatives neither of these stabilizing effects applies (i.e., the phosphorus is considerably more basic and less receptive to π bonding).

Bis(trifluoromethylphosphinous) acid is thermally stable.[83] Excess mercuric oxide converts it to the phosphinic acid in high yield. Alkaline hydrolysis degrades the molecule, producing fluoroform in quantitative yield.[84] Reaction with trimethylamine gives a white, hygroscopic salt characterized by the disappearance of the phosphorus-hydroxyl stretching band at 854 cm^{-1} and the appearance of a terminal P=O stretching band at 1305 cm^{-1}.[84]

B.2. Secondary Phosphine Oxides

Evidence for the existence of an equilibrium between secondary phosphine oxides and phosphinous acids has been presented earlier. Considerable amounts of spectroscopic data have been used to support the preponderance of the "keto" form and, indeed, potentiometric titrations have demonstrated that secondary phosphine oxides are neutral.[210]

With the exception of the ethyl, isopropyl, and trifluoromethyl derivatives, all the reported dialkyl- and diarylphosphine oxides are solid. At one time the lower dialkyl derivatives were thought to be too unstable to exist, but recently they have been isolated in good yield by careful distillation and by use of appropriate solvents.[91] A plot of the melting points of the di-n-alkylphosphine oxides versus the number of carbon atoms in the alkyl group shows a steady increase in melting point with increasing chain length, with the usual exception of the ethyl derivative.

Dialkylphosphine oxides that are insoluble in water are generally recrystallized by benzene, hexane, or acetone or reacted with one of these compounds at elevated temperatures.[91] Benzene or alcohol is preferred as solvent for diarylphosphine oxides.

Most secondary phosphine oxides are relatively unstable, decomposing when heated above their melting points or when dried in a vacuum at 100°.[210] The usual decomposition leads to the formation of a dialkylphosphinic acid and a dialkylphosphine by disproportionation. Lower alkyl deri-

$$2R_2P(O)H \longrightarrow R_2P(O)OH + R_2PH$$

vatives are more stable than the higher homologs. Dimethyl- and diethylphosphine oxide are not decomposed by heating in aqueous solution at 96° for 36 hr.

B.2.1. Oxidation

Although many dialkylphosphine oxides are easily oxidized to the dialkylphosphinic acid,[69,94,99,143,209-211] certain derivatives with bulky groups are quite resistant to oxidation, presumably because of steric hindrance.[19,69]

Hydrogen peroxide (30% aqueous) is the most commonly used chemical oxidant for the lower molecular weight dialkylphosphine oxides. With higher molecular weight derivatives, solubility problems render the reaction sluggish. Peracetic acid is a better solvent and a more useful oxidant in these cases.

Diaryl- and aryl alkylphosphine oxides are similarly oxidized to diaryl- and aryl alkylphosphinic acids with hydrogen peroxide or peracetic acid.[61,99,121,209]

Dimesityl-, diduryl-, and dipentamethylphenylphosphine oxides are exceptions.[69] Alkaline ferricyanide or permanganate are necessary to convert these hindered compounds to the phosphinic acids.

Reaction with phosphorus pentachloride in dry benzene followed by hydrolysis has also been used as a technique to produce phosphinic acids.[61,211]

Oxygen or air in basic methanolic solution is commonly used as an oxidation medium for converting secondary phosphine oxides to phosphinic acids. Oxidation in air is relatively slow at room temperature (several days are required for complete oxidation), but the yields are surprisingly high, often between 80 and 90%.[61]

Sulfur reacts readily with secondary phosphine oxides in benzene solution to produce phosphinothioic acids $R_2P(S)OH$.[94,110]

B.2.2. Addition to Unsaturated Compounds

Secondary phosphine oxides can be added to a wide variety of unsaturated compounds.[94,110,142,185] Thus dibenzylphosphine oxide adds to acrylonitrile in the presence of basic catalysts to give dibenzylphosphinylpropionitrile in 66% yield (see also Chapter 6). Similar additions to

$$(PhCH_2)_2P(O)H + CH_2{=}CHCN \xrightarrow{NaOEt} (PhCH_2)_2P(O)CH_2CH_2CN$$

α,β-unsaturated amides, esters, and ketones are reported. A base-catalyzed chain mechanism has been proposed for the following reaction:

$$R_2P(O)H + R'ONa \longrightarrow R_2P(O)Na + R'OH$$

$$R_2P(O)Na + CH_2{=}CHCN \longrightarrow [R_2P(O)CH_2\overset{\ominus}{C}HCN]\ Na^{\oplus}$$

$$[R_2P(O)CH_2\overset{\ominus}{C}HCN]\ Na^{\oplus} + R_2P(O)H \longrightarrow R_2P(O)CH_2CH_2CN + R_2P(O)Na$$

When this reaction is carried out with diphenylphosphine oxide and tetracyclone, adducts are formed by attack of tervalent phosphorus only at the carbon alpha to the carbonyl group. The final site of protonation, however, depends on reaction conditions.[139] Thus the formation of (8) appears to be favored in acidic solution even though (7) formed in basic solution, is quite stable to acid.

Ketones, aldehydes, and anils undergo condensation with dialkylphosphine oxides to give dialkylphosphinyl alcohols[110,142,185] (see Chapter 6). It is also possible to

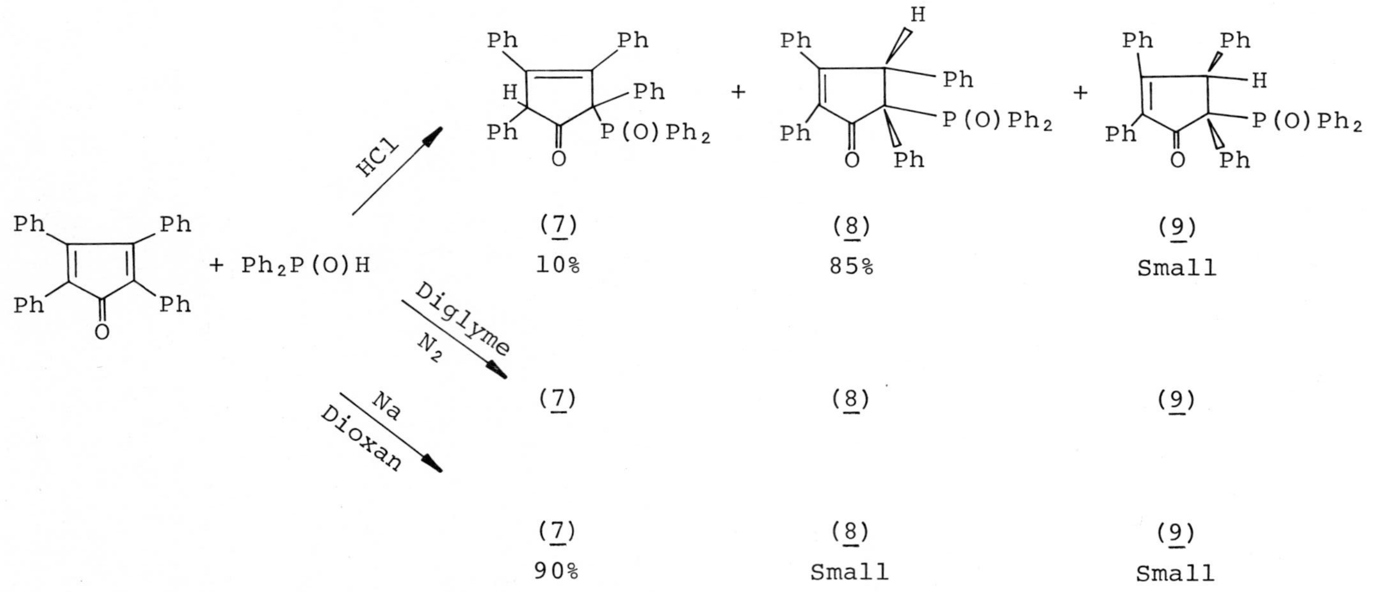
Ph
Ph
Ph
Ph
O
+ $Ph_2P(O)H$
HCl
Diglyme
N_2
Na
Dioxan
(7)
10%
(8)
85%
(9)
Small
(7)
(8)
(9)
(7)
90%
(8)
Small
(9)
Small
H
P(O)Ph_2
+

$$\text{Et}_2\text{P(O)H} + \text{PhCHO} \xrightarrow{\text{NaOEt}} \text{Et}_2\text{P(O)CH(OH)Ph}$$

react aldehydes and ketones with the Grignard complex of dialkylphosphine oxides.[76]

In the absence of catalysts, secondary phosphine oxides add across the carbon-nitrogen double bond of isocyanates to form amides.[94]

$$\begin{array}{c}\text{t-Bu} \\ \diagdown \\ \diagup \\ \text{Ph}\end{array}\text{P(O)H} + \text{FSO}_2\text{N=C=O} \longrightarrow \begin{array}{c}\text{t-Bu} \\ \diagdown \\ \diagup \\ \text{Ph}\end{array}\text{P(O)C(O)NHSO}_2\text{F}$$

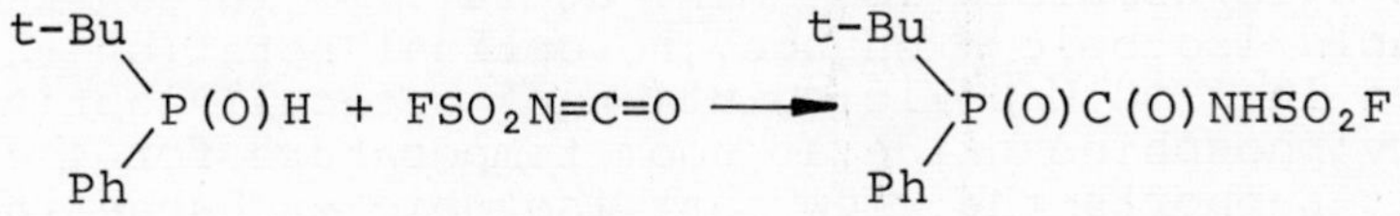

Esters of α-ketocarboxylic or ketophosphinic acid react exothermally with dialkylphosphine oxides in the absence of catalysts.[176] The products are easily isomerized

$$\text{R}_2\text{P(O)H} + \underset{\underset{O}{\|}}{\text{R'CCOOR''}} \longrightarrow \text{R'}-\underset{\underset{\text{P(O)R}_2}{|}}{\overset{\overset{\text{OH}}{|}}{\text{C}}}-\text{COOR''} \xrightarrow{\Delta} \text{R'}-\underset{\underset{\text{OP(O)R}_2}{|}}{\text{CHCOOR''}}$$

$$\text{R}_2\text{P(O)H} + \underset{\underset{O}{\|}}{\text{R'CP(O)(OR'')}_2} \longrightarrow \text{R'}\underset{\underset{\text{P(O)R}_2}{|}}{\overset{\overset{\text{OH}}{|}}{\text{C}}}-\text{P(O)(OR'')}_2 \xrightarrow{\Delta}$$

$$\text{R'CHP(O)(OR'')}_2 \atop \underset{\text{OP(O)R}_2}{|}$$

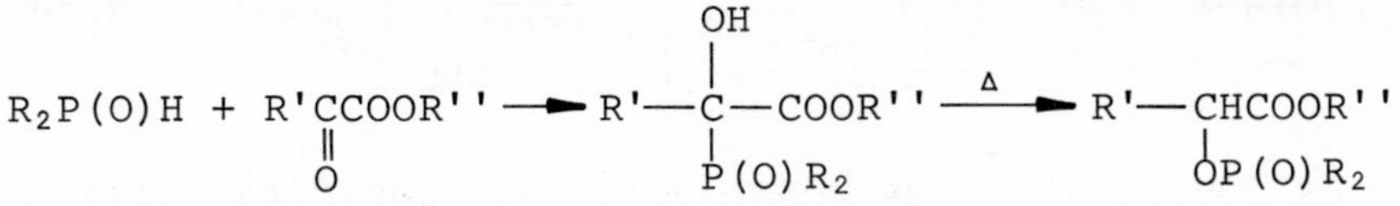

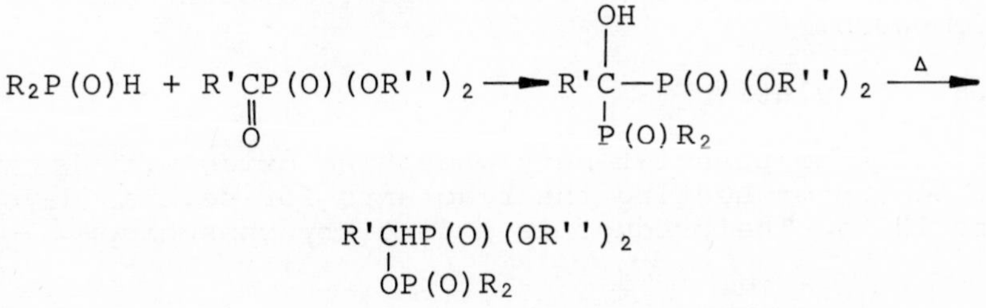

by heat to esters of dialkylphosphinic acid, an easily characterized product.

B.2.3. Conversion to Phosphinyl Halides

Chlorine,[211] thionyl chloride,[99] and N-chlorosuccinimide[99] have all been used to convert secondary phosphine oxides directly to phosphinyl chlorides. This reaction is often used for proof of structure of secondary phosphine

$$\text{Ph}_2\text{P(O)H} + \text{Cl}_2 \longrightarrow \text{Ph}_2\text{P(O)Cl} + \text{HCl}$$

oxides, since hydrolysis of the phosphinyl halides produces phosphinic acids.

Excess phosphorus trichloride converts diarylphosphine oxides to diarylphosphinous chlorides in yields of 60 to 80%.[144] The reaction is of interest as an example of an unusual conversion of tetra- to tricovalent phosphorus.

B.2.4. Hydrogen-Deuterium Exchange and Racemization

Benzylphenylphosphine oxide undergoes rapid hydrogen-deuterium exchange at phosphorus when treated with methanol-d. Exchange is complete in 4 min. at 25°.[61] In contrast to this rapid isotopic exchange, no optical rotation change is observed in a methanolic solution of optically active benzylphenylphosphine oxide at room temperature for 4 days. This result supports the view that isotopic exchange in methanol occurs with retention of configuration, possibly through an intermediate phosphinous acid.

$$\begin{array}{c}Ph\\ \diagdown\\ \diagup\\ PhCH_2\end{array}P(O)H + CH_3OH \rightleftharpoons \begin{array}{c}Ph\quad O\text{---}H\\ \diagdown\diagup\\ P\\ \diagup\diagdown\\ PhCH_2\quad H\text{---}OCH_3\end{array} \rightleftharpoons \begin{array}{c}Ph\\ \diagdown\\ \diagup\\ PhCH_2\end{array}P\text{-OH} + CH_3OH$$

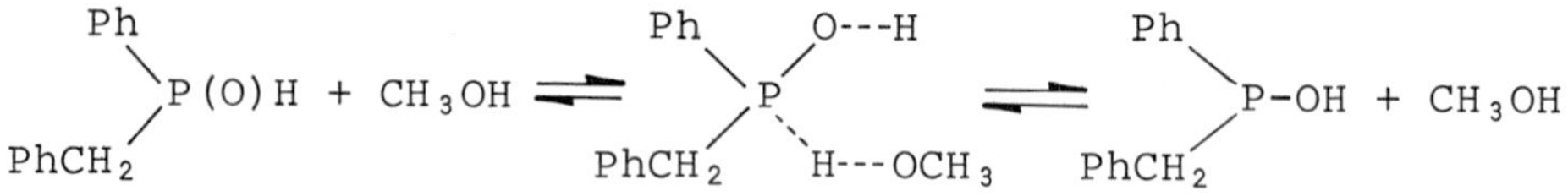

In strongly acidic or basic solution, rapid racemization of benzylphenylphosphine oxide occurs, probably through intermediate formation of a pentacovalent phosphorus compound.[61]

B.2.5. Alkylation

Alkylation of phenyl-t-butylphosphine oxide with benzyl chloride occurs on heating the reactants for several days at 110 to 120°. The product is a tertiary phosphine oxide.[94]

$$\begin{array}{c}Ph\\ \diagdown\\ \diagup\\ t\text{-Bu}\end{array}P(O)H + PhCH_2Cl \longrightarrow \begin{array}{c}Ph\quad O\\ \diagdown\,\diagup\diagup\\ P\\ \diagup\diagdown\\ t\text{-Bu}\quad CH_2Ph\end{array} + HCl$$

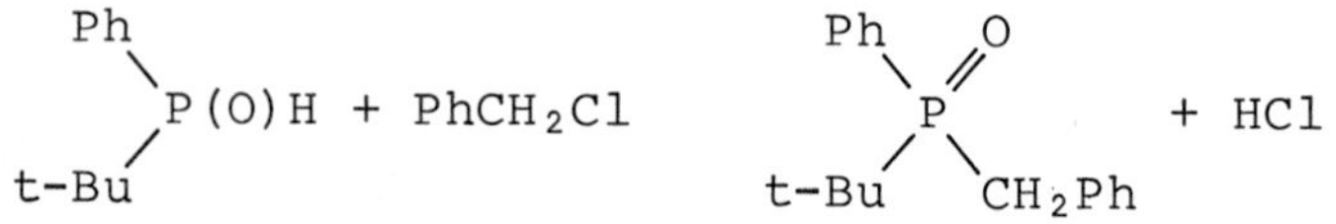

B.2.6. Salt Formation

Secondary phosphine oxides have been shown to act as nucleophiles in the presence of base. It has recently been demonstrated, however, that these oxides can act as electrophiles toward bases. Thus excess sodium hydroxide or sodium ethoxide reacts with dialkyl- or diarylphosphine oxides with rapid evolution of hydrogen, as well as formation of the sodium salt of the phosphinic acid. This reaction is explained by addition of hydroxyl ion to phosphorus followed by expulsion of a hydride ion.[94]

$$Ar_2P(O)H + \quad \overset{\ominus}{OH} \longrightarrow Ar_2P\overset{\displaystyle \overset{O^{\ominus}}{\diagup}}{\underset{\displaystyle H}{\diagdown}}OH \xrightarrow{\quad H_2O \quad} Ar_2P(O)OH + H_2 + \quad \overset{\ominus}{OH}$$

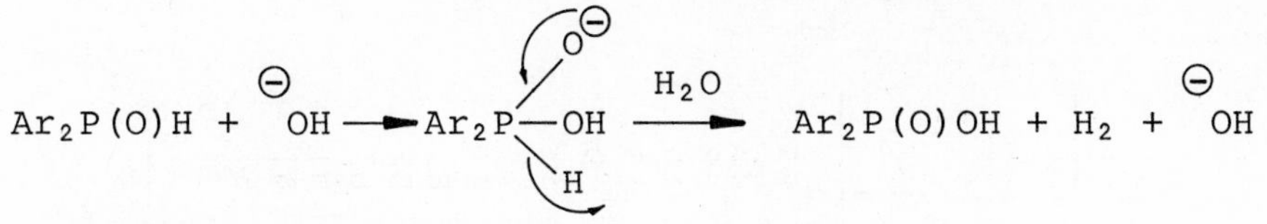

B.2.7.   Absorption Spectra and NMR Data

Several detailed studies of the IR spectra of secondary phosphine oxides have been reported.[69,138,190,210,212] Dialkylphosphine oxides (and dialkylphosphine sulfides) in solution are reported to have strong IR absorption in the 2280-2300 cm^{-1} region (phosphorus-hydrogen) and in the 1190 cm^{-1} region (P→O). These absorptions are found at 1150 to 1155 cm^{-1} and near 2335 cm^{-1} when the spectra are determined as potassium bromide pellets. Intramolecular association is postulated to account for the frequency shift for the phosphoryl group.

$$R_2P\overset{\displaystyle O \text{-----} H}{\underset{\displaystyle H \text{------} O}{\diagdown\diagup}}PR_2$$

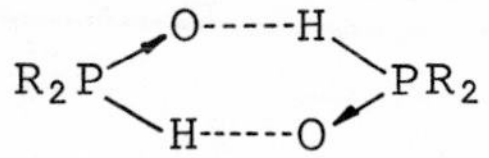

With aromatic secondary phosphine oxides, the phosphoryl group exhibits a doublet in the 1160 to 1190 cm^{-1} region.[69,210] The phosphorus-hydrogen in these compounds absorbs at 2300 to 2400 cm^{-1}.

The pK_a values of a series of diarylphosphine oxides demonstrate that they are weak acids in t-butanol and in 95% ethanol. The pK_a values range from 9.5 to 12.5, depending on the aromatic substituent.[81]

There have been no systematic studies of ^1H or ^{31}P NMR spectra of secondary phosphine oxides. A collection of available proton NMR data appears in Table 2. Generally, the phosphorus-hydrogen resonance is found in the 1.9 to 3.6τ range, with dialkyl derivatives observed at higher fields than diaryl or mixed alkyl aryl derivatives. Phosphorus-hydrogen coupling constants of 450 to 550 Hz are observed. Insufficient data is available on ^{31}P spectra to provide guidance in predicting approximate chemical shifts and coupling constants.[91]

B.3.   Phosphinites

B.3.1.   Rearrangement of Phosphinites to Phosphine Oxides

Perhaps the most thoroughly investigated reaction of phosphinites is their rearrangement to tertiary phosphine oxides. Such rearrangements are well documented and are commonly classified within the scope of the Michaelis-Arbuzov reaction.

$$R_2POR' \longrightarrow R_2P(O)R'$$

Table 2. NMR Data on Secondary Phosphine Oxides

RR'PH(O)		Chemical Shift, τ		
R	R'	(TMS Ref.)	J_{H-P_H} (Hz)	Ref.
Me	Me	2.5	490	68
Me	CF_3	3.1	516	32
Et	Et	2.8	468	68
Me	t-Bu	3.4	444	19
Et	t-Bu	3.6	435	19
n-Bu	t-Bu	3.6	433	19
Ph	t-Bu	3.0	453	19
Ph	t-C_5H_{11}	2.9	543	19
Ph	n-Bu	2.8	461	61
4-ClC_6H_4	n-Bu	3.2	460	61
Ph	Ph	2.0	481	61
Ph	4-PhC_6H_4	1.9	482	61
Ph	4-BrC_6H_4	2.5	476	61
Ph	$PhCH_2$	2.5	474	61
$PhCH_2$	n-Bu	3.2	455	61
$PhCH_2$	$PhCH_2$	3.0	465	61

In the uncatalyzed reaction, the rearrangement occurs with the phosphorus acquiring a fourth substituent in an intramolecular fashion, concomitant with the development of the phosphonyl group. When the hydrocarbon portion of the migrating molecule is prone to rearrange, it too will undergo skeletal changes.

Thermograms, using differential thermal analysis, demonstrate that the ease of rearrangement for various ester substituents[134,173] is

2-alkynyl>2,3-alkadienyl>2-alkenyl>alkyl

The reaction in the latter case has been demonstrated to be intermolecular, and it is catalyzed by alkyl halide (see also Chapter 6).

B.3.1.1. Alkynyl Esters. The most extensively studied phosphinite rearrangement involves 2-alkynyl esters.[134] The thermal transformation of such esters if visualized as taking place by a concerted intramolecular pathway initiated by nucleophilic attack of phosphorus on the terminal acetylenic carbon and concluded by rupture of the carbon-oxygen bond.

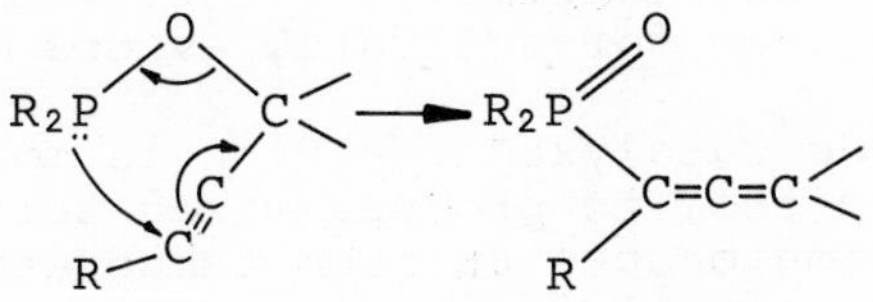

Important parameters in the reaction include:

1. Alkyl substitution on the terminal acetylenic carbon atom, which decreases the rate of reaction (probably by rendering this atom less susceptible to nucleophilic attack by the electron pair on the phosphorus atom).
2. Alkyl substitution on the carbon atom alpha to the oxygen, which increases the rate of rearrangement. This

$$
\begin{array}{ccc}
CH_3 & CH_3 & \\
| & | & \\
HC{\equiv}C\text{-}\,O\text{-} >H\text{-} & C{\equiv}C\text{-}\,O\text{-} >H\text{-}C{\equiv}CH_2\text{-}O\text{-} \\
| & | & \\
CH_3 & H &
\end{array}
$$

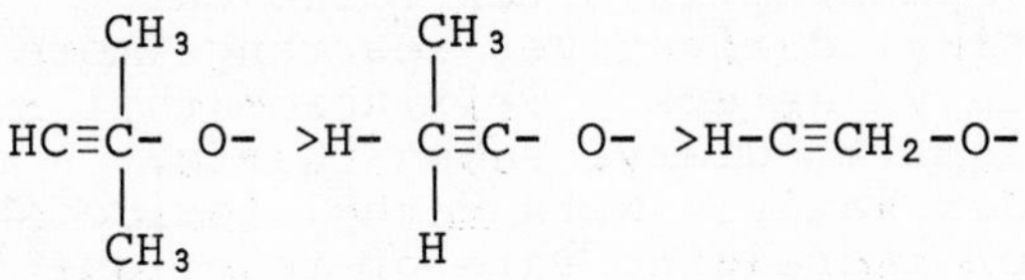

sequence parallels the ease of carbonium ion formation.
3. Dilution with solvent, which has no effect on the rearrangement, no crossed products are observed when mixtures are subjected to rearrangement.[171]
4. Optically active derivatives yield optically active

allenic compounds, and epimeric pairs of acetylenic derivatives yield the stereoisomeric allenylphosphine oxides.[197]

5. When the allenic phosphine oxide products are allowed to age, the acetylenic bond is formed.

$$R_2P(O)CR{=}C{=}CH_2 \longrightarrow R_2P(O)CRH\text{-}C{\equiv}CH$$

B.3.1.2. Allyl Esters. Although allyl derivatives can undergo rearrangement analogous to propargyl derivatives, they are stable under conditions that would allow the propargyl compounds to rearrange readily. Thermolysis at 200° produces phosphine oxide in a reaction[133] that follows first-order kinetics.[92] Thermodynamic data suggest a concerted, allylic shift.

With allyl diphenyl- and diethylphosphinites, reaction occurs more readily than with phosphonites or phosphites.[169,193,194] There is evidence for complete inversion of the allyl group.

B.3.1.3. Alkyl Esters. The isomerization of alkyl dialkylphosphinites to trialkylphosphine oxides has been reviewed in considerable detail[90,119] (see also Chapter 6). The reaction does occur thermally, but elevated temperatures are required even for activated esters such as benzyl derivatives.

The alkyl-halide-catalyzed reaction is more frequently used as a synthetic tool to produce mixed trialkylphosphine oxides. The reaction occurs at room temperature and proceeds in two stages, involving an ionic intermediate whose existence has been demonstrated by dipole moment studies.

$$R'X + Et_2POEt \xrightarrow{0^\circ} [R'Et_2\overset{\oplus}{P}OEt]\overset{\ominus}{X} \xrightarrow{30^\circ} R'Et_2PO + EtX$$

Formation of the intermediate phosphonium compound usually occurs by way of a bimolecular, nucleophilic substitution. However, with reagents capable of facile carbonium ion formation, an SN_1 pathway is often followed.

Esters of dialkyl derivatives rearrange more readily than analogous diaryl esters. Trifluoromethyl substituents are about equivalent to diaryl substituents. A study of trifluoromethyl derivatives with methyl iodide shows the following order of increasing rate of reaction:[29,32,85]

$$\begin{matrix}(CF_3)_2POMe\\ \\(CF_3)_2PSMe\\ \\(CF_3)_2PS\text{-}t\text{-}Bu\end{matrix} < CF_3MePSMe < \begin{matrix}CF_3MePOMe\\ \\CF_3MePS\text{-}t\text{-}Bu\end{matrix} < CF_3MePO\text{-}t\text{-}Bu < Me_2PO\text{-}t\text{-}Bu$$

Generally t-alkyl groups seem to rearrange more readily than methyl or n-alkyl groups.[83,84,90] Temperatures above about 80° must be avoided in the case of t-alkyl dialkylphosphinites in order to prevent a β-elimination reaction to produce olefin.

B.3.1.4. Perkow Reaction. The Perkow reaction is a modification of the Michaelis-Arbuzov reaction wherein an α-haloaldehyde or an activated halogen-containing compound reacts with a phosphinite (or an other trivalent phosphorus species).

Reaction of a phosphinite with dichlorodiphenylmethane and related halides, which has been examined in some detail, proceeds in two steps.[70] After rapid formation of a phosphonium salt, the decomposition to tertiary phosphine oxide only becomes important when the halide is substituted on the same carbon atom or in the alpha position by an electron-withdrawing group. Steric hindrance on the carbon attacked by phosphorus also facilitates salt decomposition. Phosphinites are more reactive toward dichlorodiphenylmethane than are aryl or alkylphosphines, but they are less reactive than phosphites or phosphonates.

The reaction of phosphinites with α-haloaldehydes has been used to prepare dialkylvinylphosphonates.[125] Similar compounds are obtained by the use of dihaloethanes as nucleophiles for the Arbuzov rearrangement.[141] Trichloroacetaldehyde and carbon tetrachloride undergo similar reaction with alkyl diarylphosphinites.[115,117]

$$(4\text{-ClC}_6\text{H}_4)_2\text{POEt} + \text{CCl}_4 \longrightarrow (4\text{-ClC}_6\text{H}_4)_2\text{P(O)CCl}_3 + \text{EtCl}$$

$$(4\text{-ClC}_6\text{H}_4)_2\text{POEt} + \text{Cl}_3\text{CCHO} \longrightarrow (4\text{-ClC}_6\text{H}_4)_2\text{P(O)OCH=CCl}_2 + \text{EtCl}$$

Other active-halogen-containing reagents that have been used for the same type of reaction and produce the corresponding phosphine oxide include acetyl bromide,[8] chloroformic acid,[8] ethyl chloroformate,[8] bromoacetophenone,[10] and chloroacetone.[10] In the latter case, as often happens in these rearrangements, by-products are observed from thermal reaction of the ester itself and from attack by phosphorus on the oxygen atom with concerted elimination of alkyl halide.

$$\text{Ph}_2\text{POEt} + \text{ClCH}_2\text{COCH}_3 \longrightarrow \begin{cases} \text{Ph}_2\text{P(O)CH}_2\text{COCH}_3 + \text{EtCl} \\ \text{Ph}_2\text{EtPO} \\ \text{Ph}_2\text{P(O)OCMe=CH}_2 \end{cases}$$

B.3.2. Oxidation

The phosphinites are relatively unstable in air and undergo easy oxidation to phosphinic acid esters.[7,111] Oxygen is more frequently used in this reaction, which serves to characterize phosphinites.

$$Ph_2POEt + \tfrac{1}{2}O_2 \longrightarrow Ph_2P(O)OEt$$

A smoother oxidation, uncomplicated by side products, results from addition of bromine to the phosphinite followed by treatment with water or alcohol.[136]

Active manganese dioxide is a convenient reagent for oxidizing phosphinites to phosphinic acid esters. The reaction is frequently carried out in acetone and is reported to proceed without side reaction products common to most other oxidation techniques.[107,163]

Yields of 40 to 80% of phosphinic acid esters are reported from oxidation of alkyl dialkylphosphinites with hydrogen peroxide.[205]

Elemental sulfur is frequently used to convert phosphinites to phosphinothioates.[205] When the reaction is carried out in oxygen-free benzene solution using a stoichi-

$$Ph_2POEt + S \longrightarrow Ph_2P(S)OEt$$

ometric amount of sulfur, yields of 50 to 85% are obtained.

B.3.3. Radical Reactions

Irradiation with UV light of a benzene solution of methyldiphenylphosphinite in the presence of tetrazen generates a mixture of N,N- dimethyldiphenylphosphinic amide and methyldiphenylphosphine oxide.[54] Apparently the methyl radical initially produced by attack of the dimethylamine radical on the phosphinite reacts with a second molecule of ester to give another methyl radical in a chain reaction. Yields of amide are generally 35 to 40%, although they can be increased by the use of a large excess of tetrazen.

$$Me_2N\text{-}NMe_2 \longrightarrow 2Me_2N\cdot$$

$$Me_2N\cdot + Ph_2POMe \longrightarrow Ph_2P(O)NMe_2 + Me\cdot$$

$$Me\cdot + Ph_2POMe \longrightarrow Ph_2P(O)Me + Me\cdot$$

In methanolic solution, thiyl radicals derived from irradiation of aryl disulfides oxidize alkyl diphenylphosphinites to nonsulfur-containing phosphorus derivatives.[53]

It is suggested that methanol solvolyzes the intermediate phosphonium salt in an ionic reaction leading to formation of a tertiary phosphine oxide, a mercaptan, and a sulfide.

B.3.4. Hydrolysis

Controlled hydrolysis of phosphinites using stoichiometric amounts of water in neutral or weakly acidic medium produces secondary phosphine oxides in surprisingly high yields.[8,107,111]

B.3.5. Reaction with Organometallic Reagents

It is possible to prepare phosphinites by the reaction of organolithium or Grignard reagents with phosphites, but excess organometallic reagent will displace the alkoxide group from the phosphinite with formation of a trisubstituted phosphine.[14,77] Vinyl metallic reagents are particularly reactive with alkyl dialkylphosphinites, producing high yields of dialkylvinylphosphines.

Triethyl tin iodide[175] and triethylaluminum[192] react exothermally with phosphinites to give intermediate metal-containing complexes in which the trivalent phosphorus has undergone Arbuzov rearrangement. The adducts in a series of trialkyl tin iodides are unstable at 300° with separation of metallic tin.

B.3.6. Transesterification

Transesterification of alkyl dialkylphosphinites proceeds under milder conditions than those characterizing analogous reactions using phosphites or phosphonites.[164] This is in agreement with the formation of an intermediate phosphonium compound. Because the alkyl group is a poorer

$$Me_2POEt + ROH \longrightarrow [Me_2\overset{\oplus}{P}(H)\text{-}OEt]\overset{\ominus}{O}R \xrightarrow{-EtOH} Me_2POR$$

electrophile than the hydroxyl or alkoxyl group, the electron density at phosphorus in these esters is greater than in phosphites or phosphonites, thus facilitating the rate-controlling formation of the phosphonium salt.

Alkyl diarylphosphinites undergo similar transesterification with aliphatic alcohols, but the reaction is slower and requires forcing conditions.

Both primary and secondary alcohols have been used as transesterifying agents, but the reaction has not been effected with tertiary alcohols. Glycerine derivatives have been used to produce mono-, di-, and tri-phosphinites.[163,164]

B.3.7. Miscellaneous Reactions

With α,β-unsaturated ketones or with γ-lactones, the entering nucleophile remains attached to the phosphorus of a phosphinite. The result is the translocation of the negative charge within the intermediate.

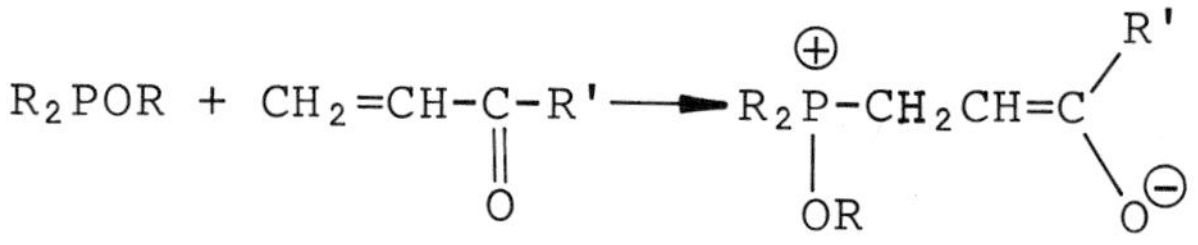

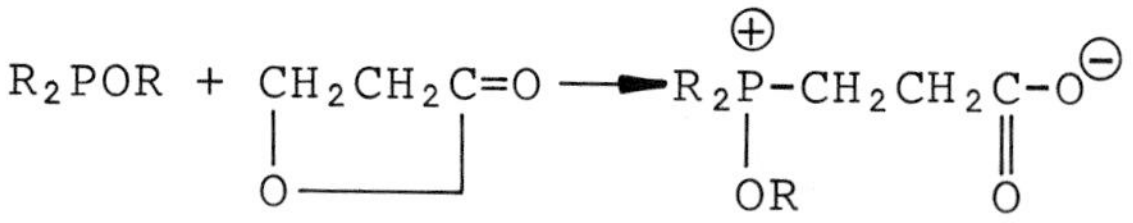

3-Benzylidene-2,4-pentanedione reacts with alkyl diarylphosphinites to produce five-membered ring oxyphosphoranes [e.g., (10)] which are derivatives of the 2,2-dihydro-1,2-oxyphosphol-4-ene ring system. At elevated temperatures, the phosphorus-oxygen bond is broken to produce an open dipolar structure (11).

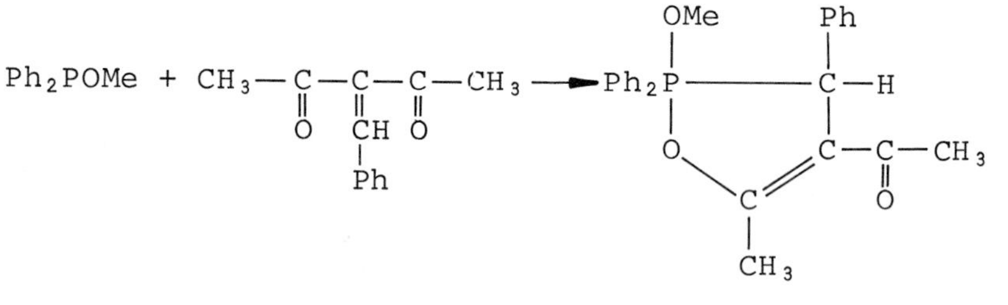

(10)

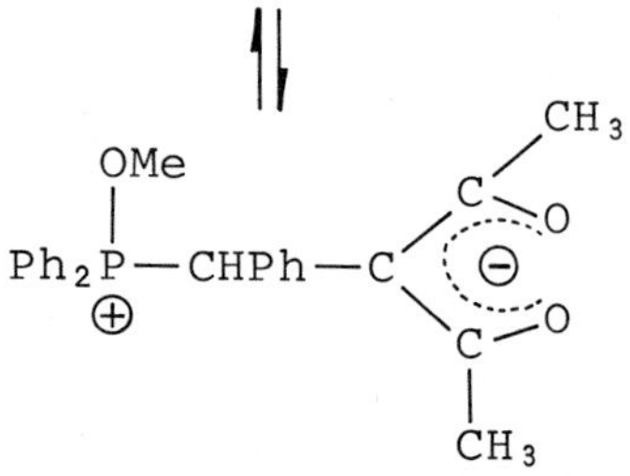

(11)

Both dialkyl- and diarylphosphinites react with O,O-dialkylphosphorodithioates at low temperatures to give high yields of a secondary phosphine oxide and an S-alkyl ester of the phosphorodithioate.[174] Spectral evidence points to a zwitterionic intermediate $[Ph_2P(OR)H]^{\oplus}$ $[(EtO)_2P(S)S]^{\ominus}$.

$$Ph_2POR + (EtO)_2P(S)SH \longrightarrow Ph_2P(O)H + (EtO)_2P(S)SR$$

B.4. Phosphinothioites R_2PSR

B.4.1. Rearrangement to Phosphine Sulfides

Phosphinothioites undergo many of the reactions of their oxygen analogs, the phosphinites. The most commonly studied reaction is the Michaelis-Arbuzov rearrangement, which leads to trisubstituted phosphine sulfides.[8] The reaction has not been studied in such great detail, as have the phosphinite reactions, but mechanistic details of both the thermal and the alkyl halide catalyzed reactions are believed to be similar (see also Chapter 7).

$$R_2PSR' \longrightarrow R_2P(S)R'$$

$$R_2PSR + R'X \longrightarrow R_2P(S)R' + RX$$

The rearrangement of alkyl dialkylphosphinothioites is more rapid than with analogous phosphinites.[32,205] This is perhaps related to the strength of the carbon-sulfur bond compared with the carbon-oxygen bond. When the reaction with alkyl halides is carried out at low temperatures, intermediate phosphonium salts can be isolated.[3]

In a series of alkyl bis(trifluoromethylphosphinothioites),[29,32,83] the following order of increasing rate of transfer of the methyl group from methyl iodide to phosphorus has been observed:

$$\begin{matrix} (CF_3)_2PSMe \\ (CF_3)_2PS\text{-}t\text{-}Bu \end{matrix} < CF_3MePSMe < CF_3MePS\text{-}t\text{-}Bu < Me_2PSMe$$

B.4.2. Oxidation

Oxidation of dialkyl- or diarylphosphinothioites with powerful oxidizing agents such as nitric acid, sodium hypobromite, or N_2O_4 produces secondary phosphinic acids $R_2P(O)OH$, with elimination of sulfur.[119] Under carefully controlled conditions it is possible to stop nitrogen dioxide oxidations of alkyl dialkylphosphinothioites at the intermediate alkyl dialkylthiophosphinate $R_2P(O)SR$. Such oxidations must be carried out in dilute methylene chloride solution at 0 to 5°.[1]

Air[166] and hydrogen peroxide[205] have also been used to oxidize dialkylphosphinothioites to the thiophosphinate. Sulfur in benzene solution adds rapidly and exothermally to alkyl diarylphosphinothioites, forming diphenyl alkylthiophosphinothionates.[2,3,7,205]

B.4.3. Miscellaneous Reactions

The stability of phosphinothioites to acids and bases has not been systematically studied. In trifluoromethyl derivatives, no evidence for intermediate secondary phosphine sulfides has been obtained (in contrast to the facile formation of secondary phosphine oxides from hydrolysis of phosphinites).

In hydrochloric acid, the equilibrium strongly favors the reactants; thus both long reaction times and the removal of mercaptan are required to isolate the reaction

$$(CF_3)_2PSMe + HCl \rightleftharpoons (CF_3)_2PCl + MeSH$$

products.[32] Bulky groups attached to the sulfur are particularly resistant to acid-catalyzed carbon-sulfur bond cleavage.

Ethyl diethylphosphinothioite reacts with ethyl chloroformate to form ethyl (diethylphosphinothioyl) formate and ethyl chloride, along with the expected products of the Arbuzov reaction.[122] An excess of thioite must be avoided

$$Et_2PSEt + Cl{-}C({=}O){-}OEt \longrightarrow Et_2P({=}S){-}C({=}O){-}OEt + EtCl$$

to prevent further reaction of the product to form ethyl diethylphosphinodithioate and ethyl(diethylphosphino)formate.

$$Et_2PSEt + Et_2P({=}S){-}C({=}O){-}OEt \longrightarrow Et_2P({=}S){-}SEt + Et_2P{-}C({=}O)OEt$$

B.5. Secondary Phosphine Sulfides $R_2P(S)H$, Phosphinothious Acids R_2PSH, and Secondary Phosphine Selenides $R_2P(Se)H$

With the exception of the trifluoromethyl derivative, secondary phosphine sulfides exist in the thiono form rather than the thiol form. Thus they exhibit a weak-intensity phosphorus-hydrogen absorption band at 2320 cm^{-1} and a

$$R_2PSH \rightleftharpoons R_2P(S)H$$

P→S band at 600 to 640 cm^{-1}.[157,182] The ^{31}P NMR spectrum shows a doublet usually centered between -5 to -31 ppm (relative to 85% H_3PO_4), consistent with the thiono structure.[132]

Like secondary phosphine sulfides, secondary phosphine selenides also exist in the seleno form rather than the selenol form: $R_2P(Se)H$ and not R_2PSeH. Thus they exhibit in the IR spectrum a phosphorus-hydrogen absorption band at 2340±20 cm^{-1} and they show in the ^{31}P NMR spectrum a doublet centered between -5 to +5 ppm (J_{PH} about 440±10 Hz).[131a]

B.5.1. Bis(trifluoromethylphosphinothious) Acid and its Reactions

Bis(trifluoromethylphosphinothious) acid is the only reported derivative that exists in the thiol form. It is stable at its boiling point, 59°. Infrared spectral data show the presence of sulfur-hydrogen stretching and bending modes that disappear on deuteration.[41] The NMR spectrum also supports the formulation as a tervalent phosphorus compound rather than the quinquevalent structure, since the coupling constant is in close agreement with other P-X-H couplings and quite different from other $R_2P(S)H$ coupling constants. Phosphorus-fluorine coupling constants also support the thiol structure.

Only a few reactions of $(CF_3)_2PSH$ have been reported.[41]

1. Reaction with trimethylamine forms an equimolar complex $Me_3\overset{+}{N}H(CF_3)_2P\overset{-}{S}$.
2. Hydrolysis with caustic at 70° yields fluoroform.
3. Mercuric chloride displaces the sulfhydryl group, forming bistrifluoromethylphosphinous chloride. This reaction is complicated by the reaction of the product with excess thiol at room temperature, forming di(bistrifluoromethylphosphino) sulfide.

$$(CF_3)_2PSH + HgCl_2 \longrightarrow (CF_3)_2PCl + HCl + HgS$$

$$(CF_3)_2PCl + (CF_3)_2PSH \longrightarrow (CF_3)_2PSP(CF_3)_2 + HCl$$

B.5.2. Reactions of Secondary Phosphine Sulfides

B.5.2.1. Oxidation. Direct reaction of secondary phosphine sulfides with sulfur provides high yields of phosphinodithioic acids.[157,158,182] The reaction is

$$Et_2P(S)H + S \longrightarrow Et_2P(S)SH$$

applicable to both aliphatic and aromatic derivatives. The major limitations of the reaction include the following: an excess of sulfur must be avoided, since further oxidation of the acid to disulfide is frequently encountered;[158] high temperatures lead to complex product mixtures, particularly in the case of aliphatic derivatives; direct isolation of the free acid is preferred to isolation through formation of salts.

Oxidation of secondary dialkylphosphine sulfides with hydrogen peroxide or iodine is reported to produce dialkylphosphinic and dialkylphosphinodithioic acids.[158]

$$Et_2P(S)H \xrightarrow[\text{or } I_2]{H_2O_2} Et_2P(O)OH + Et_2P(S)SH$$

B.5.2.2. Halogenation. Chlorination of secondary phosphine sulfides using chlorine or thionyl chloride is not a practical route to dialkylphosphinothioic chlorides. Low yields are observed, and a complex mixture of oxidation products as well as sulfur replacement products is obtained.[158]

In the presence of base, carbon tetrachloride can be used to chlorinate dialkylphosphine sulfides to dialkylphosphinothioic chlorides.[127] The scope of the reaction

$$Et_2P(S)H + CCl_4 \xrightarrow{Et_3N} Et_2P(S)Cl + CHCl_3$$

is limited by competing side reactions, particularly in the case of higher molecular weight analogs. Even in these cases, however, the chloride has been demonstrated to be an active intermediate by trapping experiments using isopropanol. Thus in the reaction of di-n-butylphosphine sulfide with carbon tetrachloride in isopropanol, the major product is O-isopropyl dibutylphosphinothioate.[158]

B.5.2.3. Addition to Carbonyl and Unsaturated Compounds. Dialkylphosphine sulfides undergo base-catalyzed nucleophilic addition to carbonyl compounds forming α-hydroxyalkyl tertiary phosphine sulfides.[158] Since the

$$Et_2P(S)H + Me_2CO \rightleftharpoons Et_2P(S)C(OH)Me_2$$

reaction is reversible and since high temperatures favor the reactants, only solid products lend themselves to easy purification and complete characterization. Usually a trace of base is required to catalyze the reaction, but this is not required with either reactive aldehydes or ketones or with phosphine sulfides containing electronegative groups such as phenyl or cyano.

Isocyanates, olefins, and activated olefins, such as acrylonitrile or triethyoxyvinylsilane, have also been used to provide adducts of secondary phosphine sulfides.

$$Et_2P(S)H + PhNCO \longrightarrow Et_2P(S)CONHPh$$

$$Et_2P(S)H + CH_2{=}CHCN \longrightarrow Et_2P(S)CH_2CH_2CN$$

In addition to double bonds, a free radical mechanism has been postulated because the reaction is catalyzed by UV light or by azobisisobutyronitrile.[132] Terminal addition to the double bond is typical of such additions.

B.5.2.4. Miscellaneous Reactions. Activated halogen compounds react with dialkylphosphine sulfides, eliminating hydrogen halide and forming the tertiary phosphine sulfide. Benzyl halides,[158] α-halocarbonyl compounds,[128] arylsulfenyl chlorides,[48] and phosphinothioyl chlorides[148] have all been used as reactants, usually in the presence of methanol and a catalytic amount of base (see Chapters 2 and 7).

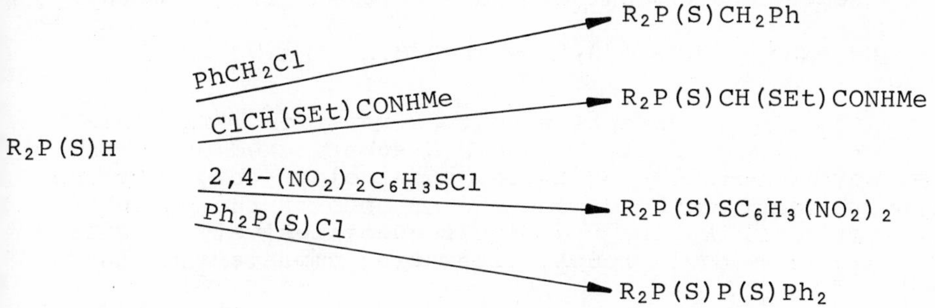

B.6. Phosphinous Amides R_2PNR_2

The phosphorus-nitrogen bond in N-N-disubstituted dialkyl- and diarylphosphinous amides is thermally stable and does not readily undergo the usual tautomeric transformations of trivalent phosphinous compounds to pentavalent phosphorus derivatives.[78] However, phosphinous amides are easily cleaved by hydrolysis, alcoholysis, and ammonolysis, and they undergo a variety of other reactions. The reactions can be grouped into three classes: (1) reactions involving displacement of the $-NR_2$ group (usually involving a stronger nucleophile); (2) oxidative reactions at the phosphorus atom, including addition reactions in which phosphorus undergoes valence changes; and (3) insertion reactions at the phosphorus-nitrogen bond.

There is considerable current interest in the stereochemistry of phosphinous amides. Infrared and NMR spectra indicate the existence of two rotational isomers in thermal equilibrium for the compound $(CF_3)_2PNHMe$, whereas the N-t-butyl and N-phenyl derivatives show only one conformer.[82,100] It has subsequently been suggested that the spectral evidence for isomers in the N-methyl compound is not a conclusive argument, since methyl-nitrogen-hydrogen coupling rather than restricted phosphorus-nitrogen bond rotation could account for the spectral details.[50]

B.6.1. Displacement of the $-NR_2$ Group

Anhydrous hydrogen chloride cleaves the phosphorus-nitrogen bond of N,N-dialkyl dialkylphosphinous amides, forming the dialkylphosphinous chloride and the amine hydrochloride.[35,38]

$$Ph_2PNEt_2 + 2HCl \longrightarrow Ph_2PCl + Et_2NH{\cdot}HCl$$

Aqueous hydrochloric acid hydrolyzes phosphinous amides to form secondary phosphine oxides. Some trifluoromethyl

$$Ph_2PNMe_2 + HCl + H_2O \longrightarrow Ph_2P(O)H + Me_2NH{\cdot}HCl$$

derivatives are an exception to this reaction. The compound $(CF_3)_2PNH_2$ and its N- and N,N-substituted products are all hydrolytically unstable, undergoing rupture of the phosphorus-nitrogen bonds and one or both of the phosphorus-carbon bonds. Alkaline hydrolysis quantitatively removes both trifluoromethyl groups, liberating ammonia and fluoroform.[87]

$$(CF_3)_2PNH_2 \xrightarrow{2H_2O} CF_3PH(O)ONH_4 + HCF_3$$

$$(CF_3)_2PNH_2 \xrightarrow[2H_2O]{HCl} CF_3PH(O)OH + NH_4Cl + HCF_3$$

$$(CF_3)_2PNH_2 + 2NaOH + H_2O \longrightarrow Na_2HPO_3 + NH_3 + 2HCF_3$$

The secondary amide $[(CF_3)_2P]_2NH$ is stable to hydrogen chloride even at 100°, an observation serving to emphasize the N→P π bonding that utilizes the pair of electrons on nitrogen. In the tertiary amide $[(CF_3)_2P]_3N$, the P-N-P bonding must be nearly planar, with residual σ acceptor character on the phosphorus atom so that the molecule is quantitatively cleaved to $[(CF_3)_2P]_2NH$ by hydrogen chloride.[127]

The dialkylamino group can be substituted by alkoxy groups by heating with excess alcohol at 70 to 150°.[39,129,130] Yields of 95 to 98% are reported.

$$Me_2PNEt_2 + C_{12}H_{25}OH \longrightarrow Me_2POC_{12}H_{25} + Et_2NH$$

A halogen atom from phosphorus trichloride similarly displaces an amino group, producing a dialkylphosphinous chloride and a dichlorophosphinous amide in high yields.[129]

$$Et_2PNEt_2 + PCl_3 \longrightarrow Et_2PCl + Cl_2PNEt_2$$

The reaction is analogous to the reaction of PCl_3 with trialkylaluminum compounds.

Anhydrous ammonia displaces an alkylamino group from N,N-dialkyl dialkylphosphinous amides. The resulting unstable primary phosphinous amide undergoes subsequent condensation to a bis(dialkylphosphinous amide).[35]

$$Me_2PNMe_2 \xrightarrow[-Me_2NH]{+NH_3} [Me_2PNH_2] \longrightarrow Me_2PNHPMe_2 + \text{other products}$$

The reactivity of $(CH_3)_2P$-, CF_3CH_3P- and $(CF_3)_2P$- amide derivatives to ammonia offers some contrasts that seem to be related to the electron density on phosphorus. The compounds $(CF_3)_2PNH_2$ and $[(CF_3)_2P]_2NH$ are stable to ammonia either in the liquid or gas phase at 200°.[26] The methyl analog $(CH_3)_2PNH_2$ has never been isolated, probably because of its easy equilibration in excess ammonia to $[(CH_3)_2P]_2NH$ and $[(CH_3)_2P]_3N$. The chemistry of $CF_3CH_3PNH_2$ is inter-

mediate to the dimethyl and bistrifluoromethyl compounds
and is easily interconverted to the bis- and tris-deriva-
tives by ammonia.[31]

Anions derived from Grignard reagents,[103] organolithium
reagents,[63,153,207a] and alkylaluminum compounds[129] will
displace the dialkylamino group from most dialkylphosphinous
amides forming phosphines. It is because of this reaction
that excess Grignard reagent is to be avoided in the
preparation of dialkylphosphinous amides from dichlorophosphinous amides.

Acid anhydrides react with N,N-dialkyl dialkylphosphinous amides, displacing the amino group and forming acyl
dialkylphosphinites.[108] The products are generally charac-

$$Et_2PNEt_2 + (CH_3CO)_2O \longrightarrow Et_2POCOCH_3 + Et_2NCOCH_3$$

terized by reaction with elemental sulfur in benzene,
forming solid acetyl dialkylthiophosphinates $R_2P(S)OCOCH_3$.

Dialkylphosphines undergo an equilibrium reaction with
N,N-dimethyl dimethylphosphinous amide with formation of
a tetraalkylbiphosphine. At 100° this equilibrium favors

$$Me_2PNMe_2 + Me_2PH \rightleftharpoons Me_2P-PMe_2 + Me_2NH$$

the biphosphine to the extent of 85%. Addition of hydrogen
chloride allows separation of the biphosphine in 96% yield
(see Chapter 2).

B.6.2. Oxidation and Addition Reactions

Like most trivalent phosphinous acid derivatives, phosphinous amides are unstable in air and can be converted in
high yields to phosphinic amides by passing air through
dilute benzene solutions of the amide.[149] Active oxidizing
agents, including manganese dioxide and hydrogen peroxide,
convert the phosphinous amides to phosphinic amides.[202]
Secondary phosphine sulfides are produced by the action of
hydrogen sulfide.[132] Addition of sulfur yields thiophosphinic amides.[130,200-202]

Mono-N-substituted diarylphosphinous amides with a
mobile hydrogen atom on the nitrogen add rapidly to α,β-
unsaturated nitriles. This exothermic, uncatalyzed reaction
yields 70 to 97% of the phosphinimidylpropionitrile.[172]

$$Ph_2PNHPh + CH_2=CHCN \longrightarrow \overset{\oplus}{Ph_2PCH_2}-CH=C=\overset{\ominus}{N} \longrightarrow Ph_2\overset{\|}{P}-CH_2CH_2CN$$
$$\qquad\qquad\qquad\qquad\qquad | \qquad\qquad\qquad\qquad\qquad \|$$
$$\qquad\qquad\qquad\qquad\quad NHPh \qquad\qquad\qquad\qquad NPh$$

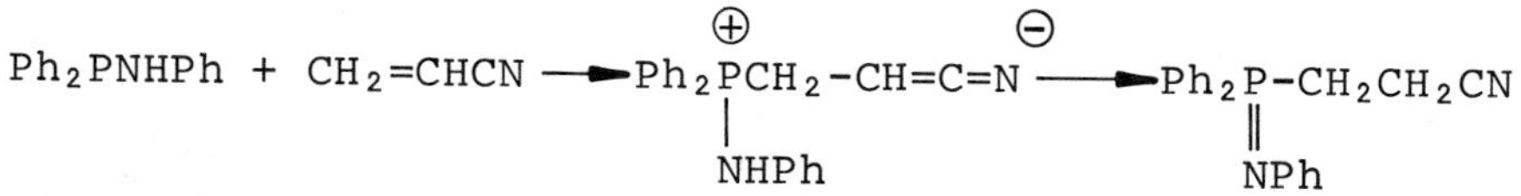

In this reaction acrylonitrile is more reactive than

methacrylonitrile, suggesting the importance of steric
effects in the migration of hydrogen from nitrogen to carbon. The aromatic derivatives are crystalline substances
with sharp melting points, soluble in benzene and acetone.
The products are characterized by reaction with carbon
disulfide, which produces a mixture of an arylisothiocyanate
and a β-cyanoethyldiarylphosphine sulfide.

$$Ph_2PCH_2CH_2CN + CS_2 \longrightarrow Ph_2P(S)CH_2CH_2CN + PhNCS$$
$$\overset{\|}{N}Ph$$

Reactions of phosphinous amides with isocyanates are
complex, leading to a variety of products. A most striking
difference in behavior between mono- and dinitrogen-sub-
stituted derivatives is observed. With derivatives of
secondary amines R_2P-NR_2, phenyl isocyanates react to give
dimer or trimer $(PhNCO)_2$ or $(PhNCO)_3$, with no adduct being
formed.[145] Primary amine derivatives containing an acidic
hydrogen add to phenyl isocyanate to form a zwitterionic
intermediate, which undergoes hydrogen migration to form
the urea derivative.[146]

$$Ph_2PNHPh + PhNCO \rightleftharpoons Ph_2P\overset{\oplus}{-}N(Ph)H \overset{H^{\oplus}}{\longrightarrow} Ph_2P-N-Ph$$
$$\underset{\underset{\ominus}{O=C-NPh}}{|} \qquad \underset{O=C-NHPh}{|}$$

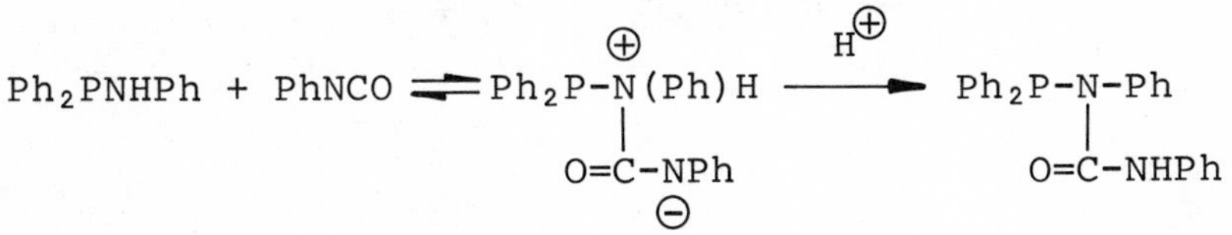

N,N-Dimethyl diphenylphosphinous amide reacts with
phenyl cyanate by way of a phosphonium salt intermediate,
producing phenyl diphenylphosphinite in 75 to 83% yield.[118]

$$Ph_2PNMe_2 + PhCNO \longrightarrow Ph_2POPh + Me_2NCN$$

N-Alkyl and N-aryl diarylphosphinous amides react
rapidly with carbon tetrachloride or bromotrichloromethane
forming phosphinimines and chloroform. The reaction proceeds by initial attack of phosphorus (assisted by electron
release from the nitrogen atom) onto positive halogen
followed by deprotonation of the halophosphinimine.[97]

$$Ph_2PNHPh + BrCCl_3 \rightleftharpoons Ph_2\overset{\oplus}{P}(Br)NHPh \; \overset{\ominus}{C}Cl_3$$
$$\downarrow -HCCl_3$$
$$Ph_2P(Br)=NPh$$

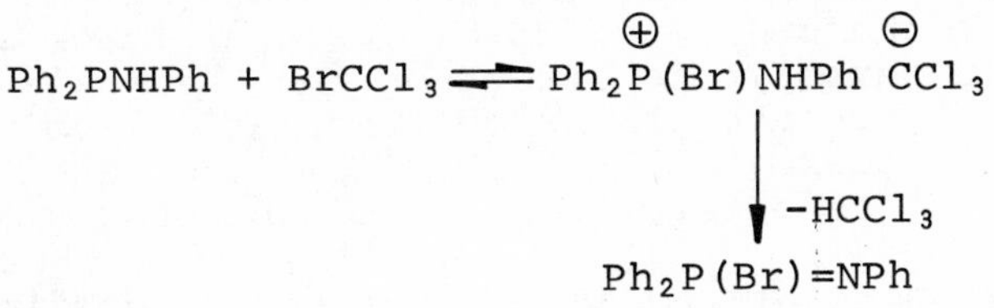

Although there are extremes in the basic character of the nitrogen atom in phosphinous amides, many of these compounds undergo complex formation. One or two BH_3 groups can be combined with N,N-dimethyl dimethylphosphinous amide,[36] whereas the higher molecular weight derivatives (including larger alkyl groups on phosphorus or nitrogen) apparently yield only mono BH_3 complexes.[36,37] The borane complexes are colorless liquids which can be distilled without decomposition. Above 250° they decompose by elimination of the amino group. The bonding in these complexes is believed to involve phosphorus-boron bonds based on the observation that $Me_2PNMe_2 \cdot BH_3$ and $Me_2PNMe_2 \cdot 2BH_3$ produce the phosphinoborane $(Me_2PBH_2)_3$ on pyrolysis.[36]

The reaction of chloramine with N-substituted diaryl- and dialkylphosphinous amides[44,89] produces the corresponding aminophosphonium chlorides. Chemical and physical data indicate that chloramination occurs on the phosphorus atom rather than the nitrogen atom. This observation re-emphasizes the reduced basicity of the nitrogen atom in phosphinous amides brought about by partial donation of

$$Ph_2PNR_2 + NH_2Cl \longrightarrow [Ph_2P(NH_2)NR_2]^{\oplus} Cl^{\ominus}$$

$$(Ph_2P)_2NR + 2NH_2Cl + NH_3 \longrightarrow \begin{bmatrix} \begin{matrix} NH_2 \\ | \\ Ph_2P \\ \\ \\ Ph_2P \\ \| \\ NH \end{matrix} \Big\rangle NR \end{bmatrix}^{\oplus} Cl^{\ominus} + NH_4Cl$$

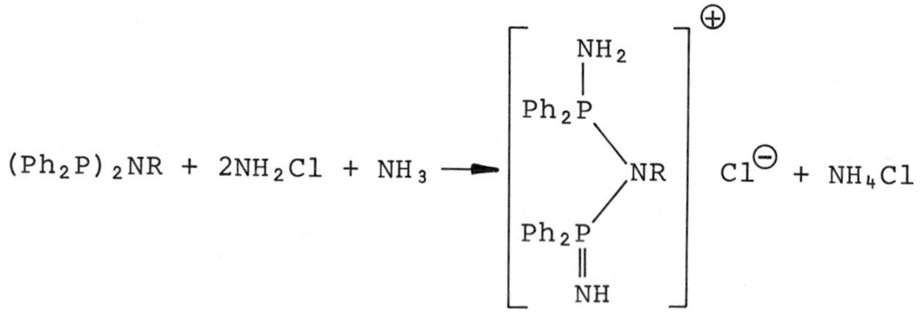

the nitrogen electron pair into vacant orbitals of the phosphorus atom.

A comparison of the ^{31}P NMR chemical shift values[44] of the phosphinous amides and their chloramination products shows that the phosphorus resonances have been shifted downfield for the salts relative to the starting amide. This change is evidence that the environment of the phosphorus atom has changed considerably, since attack at the nitrogen should have little effect on the chemical shift. Also, all the IR spectra of the chloramination products exhibit a strong peak at 1110 to 1120 cm^{-1}, a band that has been assigned to a tetracoordinated phosphorus atom.[89,198]

The halide salts are easily converted to hexafluorophosphate salts by mixing an aqueous solution of the halide with a saturated solution of potassium hexafluorophosphate.[89]

B.6.3. Insertion Reactions

With carbon dioxide or carbon disulfide, an insertion reaction at the phosphorus-nitrogen bond of phosphinous amides is observed. The products are isocyanates[207] or isothiocyanates.[154]

$$Ph_2PNPh_2 \xrightarrow{CO_2} Ph_2POC(O)NPh_2$$

$$Ph_2PNPh_2 \xrightarrow{CS_2} Ph_2PSC(S)NPh_2$$

Carbonyl compounds undergo a similar insertion reaction with phosphinous amides, forming intermediates that are decomposed on heating to form amino-substituted trialkyl- or triarylphosphine oxides.[96]

$$Me_2PNMe_2 + PhCHO \longrightarrow Me_2POCHPhNMe_2 \xrightarrow{\Delta} Me_2P(O)CHPhNMe_2$$

Primary amides are more reactive to the carbonyl reagent than secondary amides, which in turn are more reactive than tertiary amides. It has been demonstrated that the reaction with benzaldehyde proceeds by cleavage of the zwitterionic intermediate to a secondary phosphine oxide and an aldimine, which rapidly recombine to form the amino-substituted tertiary phosphine oxide.[98]

B.7. Compounds Containing P-N-As, P-O-Si, P-O-B, P-N-Si, and P-N-Ge Linkages

Compounds containing the P-N-As linkage offer a direct comparison of the π acceptor ability of an As_{4d} versus a P_{3d} orbital. The compounds $(CF_3)_2PNHAs(CF_3)_2$, $(CF_3)_2PNMeAs(CF_3)_2$, and $[(CF_3)_2P]_2NAs(CF_3)_2$ are easily cleaved by hydrogen chloride or ammonia in contrast to the analogous P-N-P compounds, which are resistant to such reactions.[199] The π acceptor strength of the As_{4d} orbital is weak enough to allow the electron pair on nitrogen to function as an effective base. The Lewis acid boron trichloride is less effective than hydrogen chloride for As-N-P cleavage in these compounds.

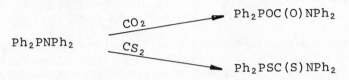

The boron phosphinites $(CF_3)_2POBMe_2$, $[(CF_3)_2PO]_2BMe$, and $[(CF_3)_2PO]_3B$ are hydrolytically and thermally unstable, decomposing on heating to $(CF_3)_2POP(CF_3)_2$ and B-O-B compounds. The order of stability decreases with decreasing alkyl substitution on boron.[28] None of the compounds undergoes Arbuzov rearrangement to form boron-phosphorus bonds.

The P-O-Si bond in a series of bis(trifluoromethyl-phosphinoxysilanes) $(CF_3)_2POSiR_3$ is quite stable against disproportionation to P-O-P and Si-O-Si compounds. Also, the Arbuzov rearrangement has not been observed (at least at temperatures below 70°). Hydrolysis with caustic yields fluoroform.

In contrast to the behavior of the trifluoromethyl derivatives, dialkyl- and diarylphosphinoxysilanes are reported to undergo ready Arbuzov reaction with methyl iodide to form tertiary phosphine oxides and halosilanes.[104]

$$Me_2P\text{-}O\text{-}SiMe_3 + MeI \longrightarrow Me_3PO + Me_3SiI$$

Hindered primary phosphinous amides such as di-t-butyl-phosphinous amide are metallated by n-butyl lithium at 0° in hexane. Further reaction with chlorotrimethylsilane or chlorotrimethylgermane yields the dialkylphosphinyl-aminosilanes or -germanes.[195] Further metallation at room

$$t\text{-}Bu_2PNH_2 + n\text{-}BuLi \longrightarrow t\text{-}Bu_2NHLi \xrightarrow{Me_3SiCl} t\text{-}Bu_2NHSiMe_3$$

$$t\text{-}Bu_2NHLi \xrightarrow{Me_3GeCl} t\text{-}Bu_2PNHGeMe_3$$

temperature with butyl lithium and treatment of the product with chlorotrimethylsilane or -germane yields an imino compound in which the second silyl group is attached to phosphorus as a result of an Arbuzov rearrangement of trimethylsilane from nitrogen to phosphorus.

$$t\text{-}Bu_2PNHSiMe_3 \xrightarrow{BuLi} t\text{-}Bu_2PN(Li)SiMe_3 \xrightarrow{ClSiMe_3} t\text{-}Bu_2P(SiMe_3)\text{=}N\text{-}SiMe_3$$

C. LIST OF COMPOUNDS

Phosphinous Acids, Anhydrides: R_2POH and R_2POPR_2

$(CF_3)_2POH$. I.[83-85] Liq., b. 61.4°,[83] m. -20.8°,[83] Trouton const. 21.4 esu,[83] ΔH_f 1910 cal/mol,[83] $\log P_{mm} = 9.6968 - 0.01099\,T - 2528.3/T$,[83] IR.[83]

$(CF_3)_2POD$. I.[83,84] Liq., b. 62.2°,[83] m. -19.2°,[83] IR.[83]

$(CF_3)_2POP(CF_3)_2$. I.[83,84] Liq., b. 78.4°,[83] dec. 250°,[83]

m. -53.1 to -52.6°,[83] Trouton cons. 22.3 esu,[83] log P_{mm} = 6.2498 + 1.75 log T - 0.00529T - 2097/T,[83] IR,[83] UV,[83] adduct with Me_3N log P_{mm} = 8.752 - 1979/T.[83]
Ph_2POAg. I.[99,179] White crystals, insol. in H_2O,[99] IR.[191]
$(4\text{-}Me_2NC_6H_4)_2POH$. I.[18,137,181] Crystals, m. 169° (from benzene),[18] dihydrate m. 165° (from H_2O).[181] Structure is questionable.

Secondary Phosphine Oxides: $R_2P(O)H$

$Me_2P(O)H$. II.[91] Crystals, m. 39-41°, b_6 65-67°, IR, ^{31}P -25.8 (J_{PH} 456 Hz), ^1H.
$MeCF_3P(O)H$. VI.[32] Liq., b. 169°, m. -9.0 to -9.3°, Trouton cons. 22.6 esu, log P = 7.7278 + 1.75 log T - 0.006T - 3016/T, ^1H, ^{19}F, IR, mass spect.
$MeCF_3P(O)D$. VI.[32] Liq., m. -10 to -9.7°, IR.
$Et_2P(O)H$. VI.[110] Liq., $b_{1.5}$ 52-53°,[110] n_D^{20} 1.4549,[110] d_4^{20} 0.9698,[110] ^{31}P -47.7 (in H_2O), -41 (in $CHCl_3$), -37 (neat),[91] (J_{PH} 463 Hz),[91] ^1H,[91] IR.[91]
Me-t-BuP(O)H. II.[19] Liq., b_4 68°, n_D^{25} 1.4530, d_4^{25} 0.9431, IR, ^1H.
Et-t-BuP(O)H. II.[19] III.[120] Liq., $b_{0.1}$ 26°,[19] b_{26} 116-116.5°,[120] n_D^{25} 1.4494,[19] d_4^{25} 0.9227,[19] IR,[19] ^1H.[19]
n-$Pr_2P(O)H$. VI.[110] Crystals, m. 48-50°, $b_{1.5}$ 71-72°.
i-$Pr_2P(O)H$. VI.[110] Liq., $b_{1.5}$ 54-55°, n_D^{20} 1.4538, d_4^{20} 0.9359.
i-Pr(2-OH-i-C_3H_6)P(O)H. VII.[21] Crystals, m. 71-72° (from hexane), ^{31}P-59 (J_{PH} 422).[134a]
$(NCCH_2CH_2)_2P(O)H$. V.[81,184,185] Crystals, m. 98-99°,[185] 95-96.5°[81] (from acetone); ^{31}P -25 (J_{PH} 356).[134a]
n-$Bu_2P(O)H$. V.[81,814,185] White crystals, m. 66° (from hexane).[185] m. 53-56° (from benzene),[81] chloral hydrate derio. m. 132-133°,[185] ^{31}P -41.8 (J_{PH} 462).[68]
n-$Bu_2P(O)D$. II.[68] Crystals, ^{31}P -40.7 (J_{PD} 72 Hz).
i-$Bu_2P(O)H$. VI.[110] V.[184] Crystals, m. 31-33°, b_2 74-75° chloral hydrate derio. m. 175° dec.
t-$Bu_2P(O)H$. III.[51] Hygroscopic crystals, m. 55-57°, b_9 112°, IR, mass spect. ^{31}P + 3.8 (J_{PH} 678).[134a]
n-Bu t-BuP(O)H. II.[19] Liq., $b_{0.025}$ 73°, ^1H, IR.
$(n\text{-}C_5H_{11})_2P(O)H$. II.[210] White crystals, m. 65-66° (from hexane).
t-BuPhP(O)H. III.[94,120] II.[19] Solid, m. 53-55°,[120] b_6 147-148°,[120] $b_{0.75}$ 103°,[19] $b_{0.1}$ 104-109°,[94] ^1H,[19] IR,[19] hygroscopic, oxidizes in air.
cyclo-C_5H_9(1-OH-cyclo-C_5H_8)P(O)H. VII.[21] Crystals, m. 147-148°; ^{31}P -51 (J_{PH} 464).[134a]
n-BuPhP(O)H. II.[61] Liq., $b_{2.9}$ 136-138°, ^1H.
n-Bu(4-Cl-C_6H_4)P(O)H. II.[61] Crystals, m. 83.5-84.5° (from benzene-hexane), ^1H.

$Ph(C_2H_5CMe_2)P(O)H$. II.[19] Liq., $b_{0.05}$ 87°, ^1H, IR.
n-$BuPhCH_2P(O)H$. II.[61] Crystals, m. 61.5-62° (from benzene-hexane), ^1H.
n-$C_8H_{17}(H_2NC(O)CH_2CH_2)P(O)H$. ^{31}P -39 (J_{PH} 454).[134a]
n-$C_8H_{17}(NCCH_2CH\ )P(O)H$. III.[184] Solid, m. 68-69° (from benzene-petrol. ether).
(n-$C_6H_{13})_2P(O)H$. II.[210,211] White crystals, m. 76.5° (from hexane),[210] m. 75-76°,[190] IR.[190]
(cyclo-$C_6H_{11})PhP(O)H$. ^{31}P -34 (J_{PH} 423).[134a]
(cyclo-$C_6H_{11})_2P(O)H$. V.[81] VII.[62] Crystals, m. 72.5-74.5° (from benzene),[81] m. 76°,[190] IR,[81] ^{31}P -46.4 (J_{PH} 427).[134a]
cyclo-C_6H_{11}(1-OH-cyclo-$C_6H_{10})P(O)H$. VII.[21] Crystals, m. 151-152° (from benzene).
$Ph_2P(O)H$. V.[81] [99] Crystals, m. 53-55°,[81,99] 51-53°,[178] 57°,[190] ^{31}P -25.9 (J_{PH} 513 Hz),[68,190] IR.[190]
$Ph_2P(O)D$. II.[68] Crystals, IR, ^{31}P -25 (J_{PD} 78 Hz).
$Ph(4\text{-}ClC_6H_4)P(O)H$. IV.[178] Crystals, m. 48-50° (from pentane), IR.
$Ph(3\text{-}ClC_6H_4)P(O)H$. IV.[178] Crystals, m. 35-36.5° (from pentane), IR.
$Ph(4\text{-}BrC_6H_4)P(O)H$. IV.[178] Crystals, m. 53-55° (from pentane), IR.
$(4\text{-}ClC_6H_4)_2P(O)H$. V.[81] Crystals, m. 131-133° (from benzene).
$(4\text{-}BrC_6H_4)_2P(O)H$. V.[81] Crystals, m. 139-141° (from benzene).
$(3\text{-}ClC_6H_4)(4\text{-}ClC_6H_4)P(O)H$. IV.[178] Liq., IR.
$PhPhCH_2(O)H$. II.[61] Crystals, m. 83.5-84.5° (from benzene-hexane), ^1H.
$Ph(4\text{-}BrC_6H_4CH_2)P(O)H$. II.[61] Crystals, m. 160.5-161.5° (from benzene-hexane), ^1H.
$Ph(4\text{-}NCC_6H_4)P(O)H$. IV.[178] Liq., IR.
$Ph(3\text{-}CF_3C_6H_4)P(O)H$. IV.[178] Liq., IR.
n-$C_8H_{17}(EtO_2CCH_2CH_2)P(O)H$. IV.[184] Solid, m. 50-52° (from petrol. ether).
(n-$C_7H_{15})_2P(O)H$. II.[210,211] White crystals, m. 80.6-81.6° (from hexane).[210]
$(PhCH_2)_2P(O)H$. II.[143] Crystals, m. 109.3-110.1° (from benzene-hexane),[143] m. 109°,[190] IR,[190,212] ^{31}P -41 (J_{PH} 341 Hz).[134a]
$(4\text{-}MeOC_6H_4)_2P(O)H$. V.[81] Crystals, m. 125-126° (from benzene).
$(2\text{-}MeOC_6H_4)_2P(O)H$. II.[209] Crystals, m. 135-136°.
$(4\text{-}MeC_6H_4)_2P(O)H$. V.[81] Crystals, m. 102-103° (from benzene).
$(4\text{-}CF_3C_6H_4)_2P(O)H$. V.[81] Crystals, m. 65-67° (from ether).
$(3\text{-}CF_3C_6H_4)_2P(O)H$. V.[81] Crystals, m. 53.5-54.5° (from ether).
$Ph(PhCHMe)P(O)H$. VII.[62] Crystals, m. 114-115° (from hexane).

$n\text{-}C_8H_{17}$(1-OH-cyclo-C_6H_{10})P(O)H. VII.[21] Crystals, m. 101-103° (from EtOH); ^{31}P -52 (J_{PH} 454 Hz).[134a]
$n\text{-}C_8H_{17}$(4-$NO_2C_6H_4$NHCO)P(O)H. ^{31}P -11 (J_{PH} 340 Hz).[134a]
($n\text{-}C_8H_{17}$)$_2$P(O)H. II.[210 211] VII.[184 185] Crystals, m. 85-86° (dec.) (from hexane);[210] ^{31}P -28 (J 308 Hz).[134a]
($PhCH_2CH_2$)$_2$P(O)H. V.[184] Solid, m. 67-71° (from heptane).
(PhCHMe)(PhCOHMe)P(O)H. VII. Crystals, m. 124-126° (from H_2O).
[3,5-(CF_3)$_2C_6H_3$]$_2$P(O)H. V.[81] Crystals, m. 122-125° (from ether).
($n\text{-}C_9H_{19}$)$_2$P(O)H. II.[210] White crystals, m. 88-89° (dec.) (from hexane).
Ph(4-PhC_6H_4)P(O)H. II.[61] Crystals, m. 95.5-97° (from benzene-hexane), ^{1}H.
(2,4,6-$Me_3C_6H_2$)$_2$P(O)H. III.[69] Liq., insol. H_2O, NaOH, sol. EtOH, benzene.
($n\text{-}C_{10}H_{21}$)$_2$P(O)H. II.[210] White crystals, m. 91.5-92.5° (from hexane).
(cyclo-C_5H_5Fe-cyclo-C_5H_5)$_2$P(O)H. II.[203] Orange needles, m. 194-195° (from H_2O), IR.
(2,3,5,6-Me_4C_6H)$_2$P(O)H. III.[69] Crystals, m. 150° (dec.) (from benzene).
($n\text{-}C_{11}H_{23}$)$_2$P(O)H. II.[210] White crystals, m. 96-97° (from benzene-hexane).
(2,3,4,5,6-Me_5C_6)$_2$P(O)H. III.[69] Crystals, m. 240° (dec.) (from benzene).
($n\text{-}C_{12}H_{25}$)$_2$P(O)H. II.[210] V.[184] White crystals, m. 98-100°,[184] 97.2-97.8° (from benzene).[210]
($n\text{-}C_{14}H_{29}$)$_2$P(O)H. II.[210] White crystals, m. 102-103° (from benzene).
($n\text{-}C_{16}H_{33}$)$_2$P(O)H. II.[210] White crystals, m. 106-107° (from benzene).
($n\text{-}C_{18}H_{37}$)$_2$P(O)H. II.[210] White crystals, m. 109-109.5° (from benzene).

Phosphinites: R_2POR

$MeCF_3$POMe. VIII.[32] Liq., b. 60.6°, Trouton const. 20.7 esu. $\log P_{mm} = 5.8501 + 1.75 \log T - 0.005T - 1908/T$, IR, ^{1}H, ^{31}P -91.8, ^{19}F 94.4 (ref. C_6F_6).
(CF_3)$_2$POMe. VIII.[83,84] Liq., b. 55.4°,[83] m. -78.5°,[83] Trouton const. 21.1 esu,[83] $\log P_{mm} = 7.5975 + 1.75 \log T - 0.008T - 2133/T$,[83] IR.[83]
MeEtPOMe. IX.[163] Liq., b_7 87.5-88.5°, n_D^{20} 1.4485, d_4^{20} 1.0043.
(CF_3)$_2$POEt. VIII.[83 84] IX.[60] Liq., b. 74°,[60] Trouton const. 22.9 esu,[60] IR,[83] ^{31}P -92.3,[156] ^{19}F.[156]
MeEtPOEt. VIII.[164] Liq., b_{15} 67-70°, n_D^{20} 1.4275, d_4^{20} 0.8755.
Et_2POEt. VIII.[11 167 188] Liq., b_{15} 80-85°,[188] d_4^{20} 0.8496,[11] n_D^{20} 1.4328.[11]

$MeCF_3PO$-t-Bu. VIII.[32] Liq., b. 113.2°, Trouton const. 20.7 esu, log P_{mm} = 5.8501 + 1.75 log T - 0.005T - 1908/T, IR, ^1H, ^{19}F 92.8 (ref. $CFCl_3$), mass spect.

Me_2POBu. XV. Liq., b_{48} 56-58°.[132a]

Me_2PO-t-Bu. ^{31}P -90.9.[134a]

$(CF_3)_2PO$-t-Bu. VIII.[83,84] Liq., b. 110.3°,[83] m. -26.1°,[83] Trouton const. 20.3 esu,[83] log P_{mm} = 7.3142 + 1.75 log T - 0.00688T - 2422/T,[83] IR.[83]

$Et_2POCH_2CH_2Cl$. VIII.[109,111] Liq., b_2 29-30°,[109] n_D^{20} 1.4670,[109] d_4^{20} 1.0144.[109]

Et_2PO-n-Pr. VII.[11,188] Liq., b_{80} 87-89°,[11] d_4^{20} 0.8523, n_D^{20} 1.4365.[11]

Et_2PO-i-Pr. VIII.[11,188] Liq., b_{96} 78-79°,[11] d_4^{20} 0.8395, n_D^{20} 1.4320.[11]

$Et_2POCH_2CH{=}CH_2$. VII.[169] Liq., b_{17} 46-47°, n_D^{20} 1.4530, d_4^{20} 0.8701.

Pr_2POEt. VIII.[167] Liq., b_{15} 97-103°.

Et_2PO-n-Bu. VIII.[11,111,188] Liq., b_{10} 54-55°,[11] b_{19} 70-70.5°,[111] d_4^{20} 0.8516,[11] d_4^{20} 0.8479,[111] n_D^{20} 1.4410,[11] n_D^{20} 1.4425.[111]

Et_2PO-i-Bu. VIII.[11,188] Liq., b_{42} 80-82°,[11] d_4^{20} 0.8430,[11] n_D^{20} 1.4381.[11]

$MeEtPOCH_2CH_2OPEtMe$. VIII.[164] Liq., b_9 97-99°, n_D^{20} 1.4632, d_4^{20} 0.9665.

$Pr_2POCH_2CH_2Cl$. VIV.[109,111] Liq., b_2 49-50°,[109] n_D^{20} 1.4675,[109] d_4^{20} 0.9825.[109]

$MeEtPO$-n-C_6H_{13}. VIII.[164] Liq., b_7 72-73°,[164] $b_{0.3}$ 138-141°,[205] n_D^{20} 1.4430,[164] d_4^{20} 0.8545.[164]

$MeEtPOCMe_2CH_2CH_2CH_3$. VIII.[164] Liq., b_9 72-73°, n_D^{20} 1.4430, d_4^{20} 0.8450.

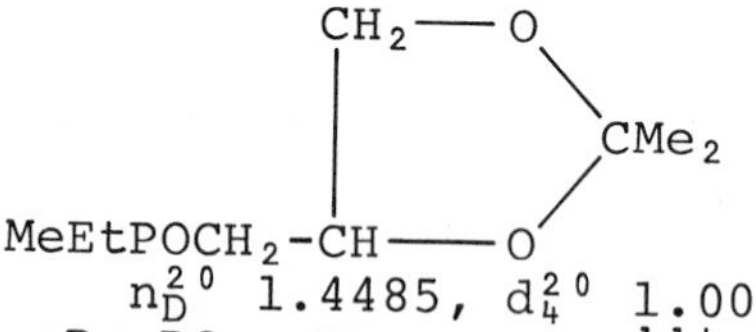

IX.[163] Liq., b_7 87.5-88.5°, n_D^{20} 1.4485, d_4^{20} 1.0043.

n-Pr_2PO-n-Pr. VIII.[114] Liq., dipole moment 3.29 D.

$EtPhPOMe$. VIII.[114] Liq., $b_{3.5}$ 60.5-61.5°, n_D^{20} 1.5357, d_4^{20} 1.0038.

$MePhPOEt$. VIII.[167] Liq., b_{15} 125-130°.

$Et(4\text{-}ClC_6H_4)POMe$. VIII.[116] Liq., $b_{4.5}$ 89.5-90.5°, n_D^{20} 1.5515, d_4^{20} 1.1240.

n-Pr_2PO-n-Bu. VIII.[111] Liq., b_1 47-48°, n_D^{20} 1.4470, d_4^{20} 0.8471.

i-Pr_2PO-n-Bu. VIII.[111] Liq., b_1 36.5-37°, n_D^{20} 1.4452, d_4^{20} 0.8510.

Et_2PO-n-C_6H_{13}. VIII.[11,188] Liq., b_9 88-89°, n_D^{20} 1.4448, d_4^{20} 0.8503.

n-Bu_2POEt. VIII.[167] Liq., b_{15} 112-116°.

$EtPhPOEt$. VIII.[114,167] Liq., b_{15} 137-142°,[167] b_{10} 94.5-95.5°,[115] $b_{2.5}$ 67-68°,[114] n_D^{20} 1.5250,[167] d_4^{20} 0.9781.[114]

$EtPhPOCH_2CH_2Cl$. VIII.[109,111,114] Liq., b_1 81-82°,[109] n_D^{20} 1.5430,[109] d_4^{20} 1.1111.[109]

$Et(4\text{-}MeC_6H_4)POMe$. VIII.[116] Liq., b_{12} 100.5-101°, n_D^{20} 1.5344, d_4^{20} 0.9839.

$Et(4\text{-}ClC_6H_4)POEt$. VIII.[116] Liq., b_7 114.5-115.5°, n_D^{20} 1.5402, d_4^{20} 1.0900.

$Et_2PO\text{-}n\text{-}C_7H_{15}$. VIII.[11,111,188] Liq., b_1 62-63°,[111] b_2 70-71°,[11] d_4^{20} 0.8547,[188] d_4^{20} 0.8486,[11] n_D^{20} 1.4495,[188] n_D^{20} 1.4490.[11]

$MeEtPO\text{-}n\text{-}C_8H_{17}$. VIII.[164] Liq., b_7 94-95°, n_D^{20} 1.4500, d_4^{20} 0.8584.

$MeEtPO\text{-}sec\text{-}C_8H_{17}$. VIII.[164] Liq., b_{11} 105-106°, n_D^{20} 1.4465, d_4^{20} 0.8517.

$EtPhPO\text{-}n\text{-}Pr$. VIII.[114] Liq., b_4 81.5-82.5°, n_D^{20} 1.5190, d_4^{20} 0.9663.

$Et(4\text{-}MeC_6H_4)POEt$. VIII.[116] Liq., b_3 92.5-93°, n_D^{20} 1.5245, d_4^{20} 0.9704.

$Et(4\text{-}ClC_6H_4)PO\text{-}n\text{-}Pr$. VIII.[116] Liq., b_2 98.5-99.5°, n^{20} 1.5333, d_4^{20} 1.0696.

$n\text{-}Bu_2PO\text{-}i\text{-}Bu$. VIII.[111] Liq., b_1 64-65°, n_D^{20} 1.4469, d_4^{20} 0.8428.

$i\text{-}Bu_2PO\text{-}n\text{-}Bu$. VIII.[111] Liq., b_1 57-58°, n_D^{20} 1.4458, d_4^{20} 0.8407.

$Et_2PO\text{-}n\text{-}C_8H_{17}$. VIII.[111,188] Liq., b_3 84.5-85°,[111] b_1 82-83°,[11] n_D^{20} 1.4520,[188] n_D^{20} 1.4509,[11] d_4^{20} 0.8579,[188] d_4^{20} 0.8488.[11]

$n\text{-}Bu_2PO\text{-}n\text{-}Bu$. VIII.[20,111] Liq., $b_{0.5}$ 78-83°,[20] $b_{1.5}$ 68-69°,[111] n_D^{20} 1.4453,[20] n_D^{20} 1.4520,[111] d_4^{20} 0.8472,[20] dipole moment 3.29 D, ^{31}P -126.[134a]

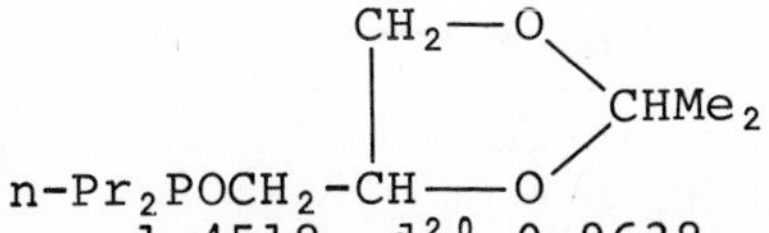

IX.[163] Liq., b_7 125-126°, n_D^{20} 1.4518, d_4^{20} 0.9638.

$(CH_2{=}CH)PhPO\text{-}n\text{-}Bu$. VIII.[106,107,114] Liq., b_2 76-77°,[107] n_D^{20} 1.5310,[114] d_0^{20} 0.9762.[114]

$(EtOCH{=}CH)PhPOEt$. VIII.[4,114] Liq., n_D^{20} 1.5372, d_4^{20} 1.0383.

$Et(4\text{-}MeC_6H_4)PO\text{-}n\text{-}Pr$. VIII.[116] Liq., b_4 102.5-103.5°, n_D^{20} 1.5198, d_4^{20} 0.9556.

$Et(4\text{-}ClC_6H_4)PO\text{-}n\text{-}Bu$. VIII.[116] Liq., b_3 114.5-115°, n_D^{20} 1.5290, d_4^{20} 1.0540.

$EtPhPO\text{-}n\text{-}Bu$. VIII.[114] Liq., b_4 94.5-95°, n_D^{20} 1.5162, d_4^{20} 0.9563.

$Et_2PO\text{-}n\text{-}C_9H_{19}$. VIII.[188] Liq., d_4^{20} 0.8584, n_D^{20} 1.4525.

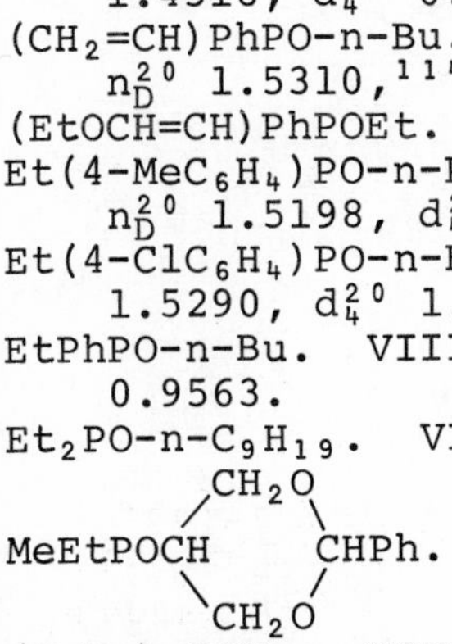

. IX.[163] Solid, m. 49-50.5°.

$(C_6F_5)_2POMe$. VIII.[66] Liq., $b_{0.1}$ 88°, IR.

Ph_2POMe. VIII.[8] Liq., b_{10} 151-152°, d_0^{15} 1.1040, n_D^{20} 1.6038[8]; ^{31}P -115.6;[134a] CuCl salt, m. 135-136°.[8]

$(4\text{-}ClC_6H_4)_2POMe$. VII.[117] Liq., b_8 176°, n_D^{20} 1.6178, d_4^{20} 1.2790, IR.

$EtPhPO\text{-}n\text{-}C_5H_{11}$. VIII.[114] Liq., b_4 108.5-109.5°, n_D^{20} 1.5115, d_4^{20} 0.9478.

$Et(4\text{-}MeC_6H_4)PO\text{-}n\text{-}Bu$. VIII.[116] Liq., b_5 110.5°, n_D^{20} 1.5164, d_4^{20} 0.9482.

$Et(4\text{-}ClC_6H_4)PO\text{-}n\text{-}C_5H_{11}$. VIII.[116] Liq., $b_{3.5}$ 126.5-127°, n_D^{20} 1.5249, d_4^{20} 1.0397.

$Et_2PO\text{-}n\text{-}C_{10}H_{21}$. VIII.[188] Liq., d_4^{20} 0.8531, n_D^{20} 1.4531.

$(n\text{-}C_5H_{11})_2PO\text{-}n\text{-}Bu$. VIII.[111] Liq., b_1 84-85°, n_D^{20} 1.4520, d_4^{20} 0.8475.

Ph_2POEt. VIII.[5] IX.[115] Liq., b_{14} 179°,[5] b_1 127-128°,[165] b_4 130-132°,[115] n_D^{20} 1.5910,[115,165] d_4^{20} 1.0750,[115] d_0^0 1.0896;[5] ^{31}P -109.8,[134a] CuCl salt, m. 190-191°.[5]

$(4\text{-}ClC_6H_4)_2POEt$. VIII.[117,167] Needles, m. 53-60°, b_{15} 169-172°,[167] b_7 183°,[117] IR.[117]

$(2\text{-}ClC_6H_4)_2POEt$. VIII.[167] Crystals, m. 26-29°, b_{15} 132-137°.

$Me_2POC_{12}H_{25}$. XV. Liq., b_2 115-120°, ^{31}P -117.9.[129]

$EtPhPOPh$. VIII.[114] Liq., b_{11} 173.5-174.5°, n_D^{20} 1.5880, d_4^{20} 1.0758.

$(EtOCH=CH)PhPO\text{-}i\text{-}Bu$. VIII.[4,114] Liq., n_D^{20} 1.5240, d_4^{20} 0.9976.

$Et(4\text{-}MeC_6H_4)PO\text{-}n\text{-}C_5H_{11}$. VIII.[116] Liq., b_{26} 166-167°, n_D^{20} 1.5131, d_4^{20} 0.9420.

$(n\text{-}C_5H_{11})_2PO\text{-}n\text{-}C_5H_{11}$. VIII.[12] Liq., d_4^{20} 0.8489, dipole moment 3.19 D.

$Ph_2PO\text{-}n\text{-}Pr$. XI.[115] Liq., b_3 137.5-139°, b_{13} 175.5-176.5°, n_D^{20} 1.5795, d_4^{20} 1.0530.

$Ph_2PO\text{-}i\text{-}Pr$. VIII.[5] Liq., b_{17} 185-189°,[5] b_8 160°,[6] d_0^0 1.0925;[5] CuI salt, m. 114-115°.

$Ph_2POCH_2CH=CH_2$. VIII.[8,169] Liq., $b_{0.02}$ 93-94°,[169] n_D^{20} 1.5940,[169] d_4^{20} 1.0769,[169] ^{31}P -113.4.[134a]

$(4\text{-}ClC_6H_4)_2PO\text{-}n\text{-}Pr$. VIII.[117] Liq., b_4 177.5°, n_D^{20} 1.5938, d_4^{20} 1.2160, IR.

$Ph(4\text{-}MeC_6H_4)POEt$. VIII.[167] Liq., b_{15} 180-190°.

$EtPhPO\text{-}n\text{-}C_7H_{15}$. VIII.[114] Liq., b_4 127.5-128.5°, n_D^{20} 1.5072, d_4^{20} 0.9356.

$Et(4\text{-}MeC_6H_4)POPh$. VIII.[116] Liq., b_4 138.5-139.5°, n_D^{20} 1.5825, d_4^{20} 1.0579.

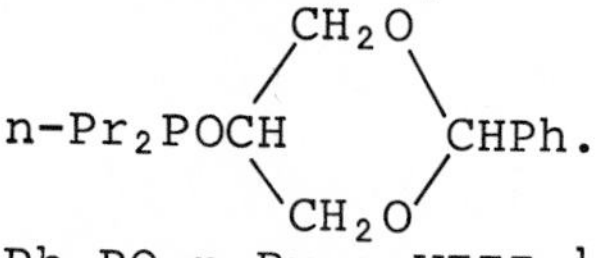

$$n\text{-}Pr_2POCH \overset{\displaystyle CH_2O}{\underset{\displaystyle CH_2O}{\diagup\diagdown}} CHPh.$$ IX.[163] Solid, m. 73-74°.

$Ph_2PO\text{-}n\text{-}Bu$. VIII.[111] IX.[115] XV.[132a] Liq., b_2 127-128°,[111] $b_{6.5}$ 157.5-158.5°,[115] $b_{0.1}$ 109-111°,[132a] n_D^{20} 1.5733,[111] n_D^{20} 1.5728,[115] d_4^{20} 1.0387,[111] d_4^{20} 1.0399,[115] ^{31}P -111.1, ^1H.[132a]

$Ph_2PO\text{-}i\text{-}Bu$. VIII.[5,9] Liq., b_{11} 202-203°,[5] d_0^{17} 1.0311.[5]

$(4\text{-}ClC_6H_4)_2PO\text{-}n\text{-}Bu$. VIII.[117] Liq., b_5 181-182°, n_D^{20}
   1.5881, d_4^{20} 1.1920, IR.
$(2\text{-}MeC_6H_4)_2POEt$. VIII.[167] Liq., b_{15} 176-185°.
$(PhCH_2)_2POEt$. VIII.[111] Liq., b_1 120-121°, n_D^{20} 1.5737,
   d_4^{20} 1.0415.
$Et_2POC_{12}H_{25}$. XV. Liq., $b_{0.008}$ 100-106°, n_D^{20} 1.4546; IR,
   ^{31}P -133.9.[129]
$Et(4\text{-}MeC_6H_4)PO\text{-}n\text{-}C_7H_{14}$. VIII.[116] Liq., b_4 141.5-142.5°,
   n_D^{20} 1.5070, d_4^{20} 0.9325.
$Ph_2PO\text{-}n\text{-}C_5H_{11}$. IX.[115] Liq., b_3 154-155°, n_D^{20} 1.5671,
   d_4^{20} 1.0269.
$(4\text{-}ClC_6H_4)_2PO\text{-}n\text{-}C_5H_{11}$. VIII.[117] Liq., b_5 202°, n_D^{20} 1.5789,
   d_4^{20} 1.1760, IR.
Ph_2POPh. VIII.[135] IX.[118] Liq., b_{62} 265-270°,[135] d_4^{24}
   1.1400,[135] $b_{0.4}$ 165-172.[118]
$Ph_2PO\text{-}n\text{-}C_6H_{13}$. VIII.[164,205] Liq., $b_{0.3}$ 138-141°,[205]
   $b_{0.5}$ 127-128°,[164] n_D^{20} 1.5608,[205] n_D^{20} 1.5712,[164] d_4^{20}
   1.0140,[205] d_4^{20} 1.0475.[164]
$Ph_2PO\text{-}cyclo\text{-}C_6H_{11}$. VIII.[164] Liq., b_1 161-168°, n_D^{20} 1.5905,
   d_4^{20} 1.0935.
$Ph_2POCH_2CHClCH_2O_2CCH=CH_2$. ^{31}P - 115.7.[134a]

$$Ph_2POCH_2CH\overset{\displaystyle CH_2-O}{\underset{\displaystyle\qquad\quad O}{\big|}}CMe_2$$

   IX.[163] Liq., $b_{0.001}$ 106.5-107°,
   n_D^{20} 1.5687, d_4^{20} 1.1508.
$(C_6F_5)_2POPh$. VIII.[66] Liq., $b_{0.5}$ 151°, IR.
Ph_2POCH_2Ph. VIII.[8] Undistillable without isomerization;
   CuCl salt, m. 125-126°.
$(PhCH_2)_2POPh$. X.[191] Liq., $b_{2.5}$ 200-203°.
$(PhCH_2)_2PO\text{-}n\text{-}Bu$. VIII.[111] Liq., b_2 148-149°, n_D^{20} 1.5590,
   d_4^{20} 1.0180.
$Ph_2PO\text{-}n\text{-}C_8H_{17}$. VIII.[164,205] Liq., $b_{0.2}$ 156-162°,[205] $b_{0.4}$
   160-161°,[164] n_D^{20} 1.5508,[205] n_D^{20} 1.5550,[164] d_4^{20} 1.0066,[205]
   d_4^{20} 1.0180.[164]

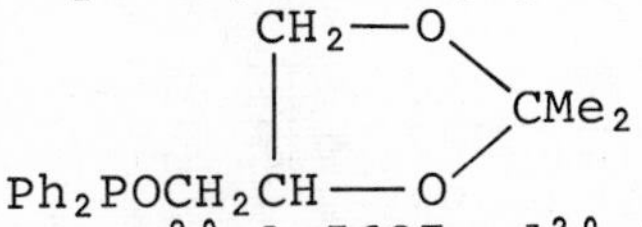

$$n\text{-}Pr_2POCH_2CH\overset{\displaystyle CH_2-O}{\underset{\displaystyle\qquad\quad O}{\big|}}CPh_2$$

   IX.[163] Solid, m. 39-40°, $b_{0.01}$
   104-105°.

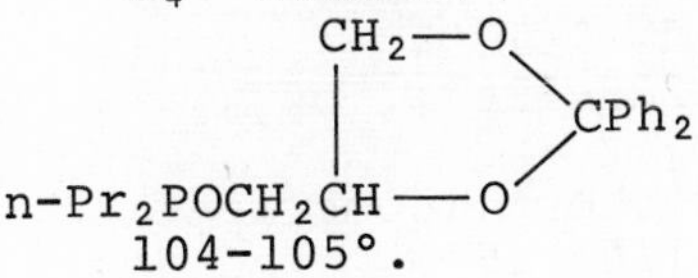

Phosphinothious Acids, Thioanhydrides: R_2PSH and R_2PSPR_2

H_2PSH. XV.[47] Colorless, dec. -130°, IR, mass spect.
D_2PSH. XV.[47] Colorless, dec. -130°.
$(CF_3)_2PSH$. XV.[29,41] Liq., b. 55°,[29] m. -100°,[29] IR,[41]
   ^1H.[41]
$(CF_3)_2PSP(CF_3)_2$. XV.[29,41] Liq., b. 112°,[29] m. -33°,[29]
   IR.[41]

Phosphinothioites: R_2PSR

H_2PSMe. XVI.[47] Colorless, dec. -78°, IR.
$(CF_3)_2PSMe$. XVI.[29,41] Liq., b. 92°,[29] m. -58°,[29] ^{31}P -37.1,[41] IR,[44] ^1H.[41,130] ^{19}F.[41]
$(CF_3)_2PSCF_3$. XVI.[59,156] Colorless liq., b. 50.3°, m. -107°, IR,[59] ^{31}P -12.8,[156] ^{19}F.[156]
$MeCF_3PSMe$. XVI.[32] Liq., b. 107.3°, Trouton const. 21.4 esu, log P_{mm} = 6.0865 + 1.75 log T - 0.005T - 2214/T, ^1H, IR, ^{31}P -27, ^{19}F 62 (ref. Cl_3CF).
Et_2PSEt. XVI.[3,123] Liq., b_9 53-55°,[123] n_D^{20} 1.5050,[123] d_4^{20} 0.9180,[123] b_7 50-52°,[3,115] EtI adduct m. 99-100°,[3] MeI adduct m. 105-106°,[3] $PhCH_2Br$ adduct m. 107-108°.[3]
$(CF_3)_2PS$-t-Bu. XVI.[29] Liq., b. 144°, m. -18°, IR.
$MeCF_3PS$-t-Bu. XVI.[32] Liq., b. 152°, m. -30°, Trouton const. 21.8 esu, log P_{mm} = 5.6743 + 1.75 log T - 0.004T - 2420/T, IR, ^1H, ^{19}F 21.1 (ref. Cl_3CF).
Et_2PS-n-Pr. XVI.[1,3] Liq., b_9 71-71.5°,[1] n_D^{20} 1.5022,[1] d_4^{20} 0.9129, b_7 65-67°.[3]
Et_2PS-i-Pr. XVI.[3] Liq., b_{20} 77-79°, n_D^{20} 1.4963, d_4^{20} 0.9032.
Et_2PS-n-Bu. XVI.[1,3] Liq., b_9 86-87°,[1] b_{20} 104-106°,[3] n_D^{20} 1.4985,[1] d_4^{20} 0.9062.[1]
Et_2PS-i-Bu. XVI.[3] Liq., b_{20} 97-99°, n_D^{25} 1.4958, d_4^{25} 0.8970.
EtPhPSEt. XVI.[2] Liq., $b_{0.05}$ 73-75°, n_D^{20} 1.5832, d_4^{20} 1.0362.
EtPhPS-n-Pr. XVI.[1,115] Liq., $b_{0.02}$ 69-70°,[1] $b_{0.04}$ 78-79°,[115] n_D^{20} 1.5740,[1] d_4^{20} 1.0206.[1]
EtPhPS-i-Pr. XVI.[2] Liq., $b_{0.07}$ 71-73°, n_D^{20} 1.5680, d_4^{20} 1.0131.
EtPhPS-n-Bu. XVI.[2] Liq., $b_{0.05}$ 86-88°, n_D^{20} 1.5665, d_4^{20} 1.0094.
EtPhPS-i-Bu. XVI.[2] Liq., $b_{0.05}$ 79-81°, n_D^{20} 1.5624, d_4^{20} 1.0042.
i-Bu_2PS-n-Bu. XVII.[80] Liq., $b_{0.17}$ 156-166°.
i-$Bu_2PSCH_2CH_2COOEt$. XVII.[80] Liq., $b_{0.18}$ 123-128°.
i-Bu_2PSPh. XVII.[80] Liq., $b_{0.12}$ 94-98°.
Ph_2PSEt. XVI.[5] Liq., b_{13} 196.5-197°, d_0^0 1.133.
Ph_2PS-n-Pr. XVI.[5] XVII.[58] Liq., b_{28} 229-230°,[5] $b_{0.01}$ 154-160°,[58] n_D^{20} 1.6189.[58]
$Ph_2PSCH_2CH{=}CH_2$. XVI.[8] Undistillable without isomerization.
i-Bu_2PS-4-MeC_6H_4. XVII.[80] Liq., $b_{0.17}$ 110-115°.
Ph_2PS-n-Bu. XVII.[58] XVI.[166] Liq., $b_{0.01}$ 154-160°,[58] b_1 183-186°,[166] n_D^{20} 1.6370,[58] n_D^{20} 1.6122,[166] d_4^{20} 1.1210.[166]
Ph_2PS-i-Bu. XVI.[5] Liq., b_8 200.5-201°, d_0^0 1.0892.
(cyclo-$C_6H_{11})_2PS$-n-Bu. XVII.[80] Liq., $b_{0.15}$ 126-133°.
Ph_2PS-i-C_5H_{11}. XVI.[5] Liq., b_{12} 219-220°, d_0^{17} 1.0645.
Ph_2PS-n-C_6H_{13}. XVI.[205] Liq., $b_{0.3}$ 163-165°, n_D^{20} 1.5988, d_4^{20} 1.0776.
Ph_2PSPh. XVI.[128a] Solid, m. 52°.
i-BuPhPS-n-C_8H_{17}. XVII.[58] Liq., $b_{0.08}$ 130-135°, n_D^{20} 1.5400.

$Ph_2PSCH_2CON(CH_2)_5$. XVII.[58] Liq., dec. on distillation at 0.2 mm, n_D^{20} 1.6351.
Ph_2PSCH_2Ph. XVI.[8] Liq., isomerizes on distillation.
$Ph_2PS\text{-}n\text{-}C_8H_{17}$. XVI.[205] XVII.[58] Liq., $b_{0.3}$ 185-188°,[205] n_D^{20} 1.5828,[205] d_4^{20} 1.0269,[205] $b_{0.02}$ 220-225°,[58] n_D^{20} 1.5865.[58]
$Ph_2PSCH_2CH_2SPPh_2$. XVII.[58] Solid, m. 77-80° (from benzene-hexane).

Phosphinoselenites: R_2PSeR

$(CF_3)_2PSeCF_3$. XVI.[59,156] Colorless liq., b. 62°,[59] m. -98°,[59] IR,[59] ^{31}P -14.1.[156] ^{19}F.[156]
Ph_2PSePh. XVI.[128a] Yellow crystals, m. 54°.

Secondary Phosphine Sulfides: $R_2P(S)H$

$Me_2P(S)H$. XIX.[47] XVIII.[131] XXI.[132] Liq., $b_{0.5}$ 50°,[47] $b_{0.5}$ 48-50°,[132] IR,[47] n_D^{20} 1.5570,[132] ^{31}P -5 (J_{PH} 456 Hz).[132]
$Et_2P(S)H$. XIX.[47] XVIII.[131] XXI.[132] Liq., $b_{0.5}$ 69°,[47] $b_{0.3}$ 51-55°,[132] IR,[47,131] n_D^{20} 1.5350,[132] ^{31}P -31 (J_{PH} 437 Hz).[132]
$n\text{-}Bu_2P(S)H$. XIX.[157] XXI.[132] Liq., $b_{1.5}$ 122-125°,[157] IR.[132]
$i\text{-}Bu_2P(S)H$. XIX.[157] XXI.[132] Crystals, m. 61-62°,[157] 62.5-63.5° (from petrol. ether),[132] ^{31}P -13.4 (J_{PH} 427).[132]
$cyclo\text{-}C_6H_{11}(NCCH_2CH_2)P(S)H$. XIX.[157] Liq.
$(EtO_2CCH_2CH_2)_2P(S)H$. XIX.[157] Liq.
$(cyclo\text{-}C_6H_{11})_2P(S)H$. XIX.[182] Crystals, m. 107-108°, ^{31}P -47 (J_{PH} 453.6).
$Ph_2P(S)H$. XVIII.[157] XXI.[132] Solid, m. 95-97°,[157] 98-99.5° (from MeCN),[132] IR,[139,157] ^{31}P -19.55 (J_{PH} 444).[132]

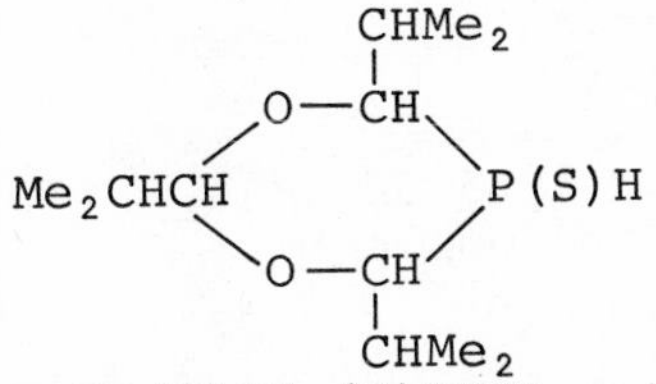
XIX.[157] Solid, m. 37°, $b_{1.5}$ 107-112°.

$EtHP(S)CH_2(S)PHEt$. XIX.[101] Solid, m. 150-153°.
$EtHP(S)CH_2CH_2(S)PHEt$. XIX.[102] Solid, m. 109-110°.
$cyclo\text{-}C_6H_{11}HP(S)CH_2CH_2(S)PH\text{-}cyclo\text{-}C_6H_{11}$. XIX.[102] Form A, m. 125-130°; form B, m. 155-158°.
$EtHP(S)CH_2CH_2CH_2(S)PHEt$. XIX.[101] Solid, m. 86-87°.
$EtHP(S)CH_2CH_2CH_2CH_2(S)PHEt$. XIX.[101] Solid, m. 91-92°.
$PhHP(S)CH_2CH_2CH_2CH_2(S)PHPh$. XIX.[93] Solid, m. 127-129°.

Secondary Phosphine Selenides: $R_2P(Se)H$

i-$Bu_2P(Se)H$. XIX. Crystals, m. 77-78° (from ligroin), IR, ^{31}P +5.35 ppm (J_{PH} 420 Hz).[131a]
$Ph_2P(Se)H$. XIX. Crystals, m. 111-112° (from CH_3CN), IR, ^{31}P -5.8 ppm (J_{PH} 450 Hz).[131a]

Phosphinous Amides: R_2PNR_2

$MeCF_3PNH_2$. XXII.[31] Liq., b. 79.1°, Trouton const. 21.1 esu, IR.
$(CF_3)_2PNH_2$. XXII.[86,87] Liq., b. 67.1°, m. -87.6°.
$CF_3(CHF_2)PNH_2$. XXII.[79] Liq.
$(CF_3)_2PNHMe$. XXII.[87,88] Liq., b. 73.2°, m. -46.5°.
$MeCF_3PNHMe$. XXII.[31] Liq., b. 92.1°, Trouton const. 21.1 esu, IR.
$(CF_3)_2PNHP(CF_3)_2$. XXII.[30] Liq., b. 93°, m. -54°.
$MeCF_3PNMe_2$. XXII.[31] Liq., b. 98.9°, Trouton const. 21.0 esu, IR.
$(MeCF_3P)_2NH$. XXII.[31] Liq., b. 142.8°, Trouton const. 21.9 esu, IR.
$(CF_3)_2PNMe_2$. XXII.[86-88] Liq., b. 83.2°, m. -81.5°,[86-88] ^{31}P -46.3, ^{19}F.[156]
$Me_2PNHPMe_2$. XXII.[35] Colorless solid, m. 39.5°.
Me_2PNMe_2. XXII.[34,124] XXIII.[129] Liq., b. 99.4°,[34] m. -97°,[34] ^{31}P -39,[34] -38.2,[129] ^1H;[34] CH_3I complex, m. 315-320°;[124] mercuric iodide, yellow solid, m. >300°.[124]
$Me_2PNMe_2 \cdot BH_3$. XXII.[124] Liq., b_{15} 92°, d_4^{20} 0.8344, n_D^{20} 1.4660, ^1H, ^{31}P +51, ^{11}B +37.
$(MeCF_3P)_2NMe$. XXII.[31] Liq., b. 164.9°, Trouton const. 21.0 esu, IR.
Et_2PNMe_2. XXIII.[129] Liq., b_{715} 141-143°, d_4^{20} 0.8277, n_D^{20} 1.4550, ^{31}P -63.4.
$(n\text{-}C_3F_7)_2PNH_2$. XXII.[60,156] Liq., b. 143°,[156] m. -23°,[156] Trouton const. 22.2 esu,[156] ^{31}P -46.3.[60]
$(HC{\equiv}C)_2PNMe_2$. XXII.[42] Liq., b_{15} 48°.
$[(CF_3)_2P]_3N$. XXII.[30] Colorless solid, m. 36.5-36.8°.
$(CH_2)_4PNMe_2$. XXII.[36] Liq., b. 170.4°, Trouton const. 21.5 esu, $\log P_{mm} = 5.4766 - 0.0037\ T + 1.75 \log T - 2479/T$, IR.
$Me_2PNMe_2 \cdot BEtH_2$. XXII.[124] Liq., b_1 64°, d_4^{20} 0.8373, n_D^{20} 1.4650.
$(CF_3)_2PNHPh$. XXII.[87,88] Liq., b. 182.1°, m. 1.7°.[88]
Et_2PNEt_2. XXIII.[103,129] Liq., b. 181°,[103] b_{723} 178-183°,[129] n_D^{20} 1.4678,[129] ^{31}P -61.6.[129]
t-Bu_2PNH_2. XXII.[195] Liq., b_2 33-34°, m. -1 to 1°, sol. in benzene, ^1H.
$(CH_3C{\equiv}C)_2PNMe_2$. XXII.[42] Liq., b_{12} 100°.
$MePhPNMe_2$. XXIII.[129] Solid, m. 70-74°, $b_{0.2}$ 102-110°.
$Me_2PNMePh$. XXIII.[129] Liq., b_2 75-78°, ^{31}P -29.8.

n-Pr_2PNEt_2. XXIII. Liq., b_{10} 85°, n_D^{20} 1.4630.[207a]
n-Bu_2PNHEt. XXII.[44] Liq., b_2 75°, IR, ^{31}P -37.
i-Bu_2PNMe_2. XXII.[129] Liq., b_{10} 79-83°, n_D^{20} 1.4559, ^{31}P -49.9.
$Et_2PNMePh$. XXIII.[129] Liq., $b_{0.01}$ 67-73°, n_D^{20} 1.5542, ^{31}P -52.1.
$MePhPNEt_2$. XXII.[129] Solid, m. 70-74°, $b_{0.2}$ 102-110°.[129]
$(HC{\equiv}C)_2PNMePh$. XXII.[42] Liq., b_1 80°.
$(CH_3C{\equiv}C)PhPNMe_2$. XXII.[42] Liq., $b_{0.07}$ 70°.
n-Bu_2PNEt_2. XXIII.[103,207a] Liq., b_{16} 121°,[103] b_{11} 112-114°, n_D^{20} 1.4642.[207a]
Ph_2PNHMe. XXII.[64] Liq., $b_{0.1}$ 114-118°; methiodide, m. 108-110°.
$(CH_3C{\equiv}C)_2PNMePh$. XXII.[42] Solid, m. 40°.
$(n$-$C_5H_{11})_2PNEt_2$. XXIII. Liq., b_{11} 137-141°, $b_{0.25}$ 77-79°, n_D^{20} 1.4662.[207a]
Ph_2PNHEt. XXII.[64] Liq., $b_{0.1}$ 195°.
Ph_2PNMe_2. XXII.[118,130] Solid, m. 31.5-33.5°,[130] $b_{0.1}$ 123-124°,[130] $b_{0.2}$ 105.5-106.5,[118] n_D^{20} 1.6086,[118] d_4^{20} 1.0398,[118] ^{31}P -63.9.[130]
Ph_2PNH-i-Pr. XXII.[88] Liq., $b_{0.09}$ 116-119°, IR, ^1H.
Ph_2PNEt_2. XXII.[65,88,118] Liq., $b_{0.1}$ 126°,[65] $b_{0.01}$ 115.5-116.5°,[118] n_D^{20} 1.5878,[118] d_4^{20} 1.0522,[118] IR,[88] ^1H;[88] ^{31}P -60.8.[132a] ^1H;[132a] methiodide, m. 112-115°;[65] mercuric iodide complex, m. 188-190°;[65] CuI complex, m. 225-227°.[65]
$Ph(C_6F_5)PNEt_2$. XXII. Liq., $b_{0.5}$ 118-20°, ^{31}P -47.[66c]
$(C_6F_5)_2PNEt_2$. XXII. Liq., $b_{0.4}$ 108°, ^{31}P -21.9.[66c]
Ph_2PNH-t-Bu. XXII.[88,201] Solid, m. 38-40°;[201] methiodide, m. 198.5-200°,[201] IR,[201] ^{31}P -22.4.[88]
$(cyclo$-$C_6H_{11})_2PNEt_2$. XXIII.[103] Liq., b_{22} 188°.
$Ph_2PN(CH_2)_5$. XXII.[200] Liq., $b_{0.5}$ 160-164°.
$Ph_2PNEt(CH_2CH{\equiv}CH_2)$. XXII.[170] Liq., b_1 132-133°, d_4^{20} 1.0390, n_D^{20} 1.5950.
Ph_2PN-n-Pr_2. XXI.[88,118] Liq., $b_{0.09}$ 132.8-133.5°,[118] $b_{0.2}$ 143°,[88] d_4^{20} 1.0210,[118] n_D^{20} 1.5814,[118] IR,[88] ^1H,[88] ^{31}P -62.
$Ph_2PN(CH_2CH{\equiv}CH_2)_2$. XXII.[200] Liq., $b_{0.5}$ 141-143°.
Ph_2PN-n-Bu_2. XXII.[88] Liq., $b_{0.2}$ 143°, IR, ^1H.
Ph_2PN-i-Bu_2. XXII.[88] Liq., $b_{0.18}$ 148°, IR, ^1H, ^{31}P -22.4.
$Ph_2PNHC_6H_3$-2,6-Et_2. XXII.[201] Solid, m. 88-89°.
$(cyclo$-C_5H_5Fe-$cyclo$-$C_5H_5)_2PNMe_2$. XXII.[203] Golden needles, m. 120-121° (from petrol. ether), IR.
Ph_2PNPh_2. XXII.[200] Solid, m. 130-132° (from ethanol), IR.
$Ph_2PNHPPh_2$. XXII.[44] Solid, m. 145-146°, IR, ^{31}P -61.
$(Ph_2P)_2NH$. XXII.[67] Solid, m. 142-144°.
$Ph_2PNMePPh_2$. XXII.[44,64] Solid, m. 116-118° (from ethanol);[64] methiodide, m. 170°(d),[64] IR,[44] ^{31}P -72.[44]
$Ph_2PNEtPPh_2$. XXII.[44,64] Solid, m. 99° (from aq. ethanol);[64] methiodide, m. 145°;[64] $NiNO_3$ complex, yellow, m. 160-163°;[64] $PdCl_2$ complex, m. 276-278°;[64] IR,[44] ^{31}P -61.[44]

$Ph_2P-P(Ph_2)=NPPh_2$. XXII.[67] White solid, m. 124-127°.
$(Ph_2P)_3N\cdot HCl$. XXII.[67] Hygroscopic solid, m. 105-110°.
$(Ph_2P)_4NCl$. XXII.[67] Solid, m. 115° (dec.).

Phosphinous Hydrazides: R_2PNRNR_2

$(CF_3)_2PNHNHMe$. XXIV.[162] White solid, m. 35.5°, ^1H, IR, mass spect.
$(CF_3)_2PNMeNH_2$. XXIV.[162] Liq., b_{13} 23.2°, ^1H, IR, mass spect.
$(CF_3)_2PNHNMe_2$. XXIV.[161] Solid, m. 40.4°, Trouton const. 22.2 esu, $\log P_{mm} = 12.51 - 3388/T$.
$(CF_3)_2PNMeNMe_2$. XXIV.[160] Liq., b. 130.5°, m. -47.0°, IR, ^1H,[162] $\log P_{mm} = 6.7566 - 0.005896T + 1.75 \log T - 2245/T$.
$Ph_2PNHNMe_2$. XXIV.[149] White crystals, m. 68.5-69.5° (from hexane), subl. 60° (0.02 mm), IR, ^1H, ^{31}P -37.6.
$Ph_2PNEtNMe_2$. XXIV.[149] Liq., IR, ^1H.
$Ph_2PNHNHPPh_2$. XXIV.[149] White crystals, m. 129-129.5° (from acetone), ^1H, IR.
$Ph_2PN(PPh_2)NMe_2$. XXIV.[149] White solid, m. 125.5-132.5° (from heptane), ^1H, IR.
$(Ph_2P)_2NNMePPh_2$. XXIV.[149] White crystals, m. 152.3-153° (from acetone), ^1H, IR.

Phosphinous Azides, Isocyanates, Isothiocyanates: R_2PN_3, R_2PNCO, R_2PNCS.

$(CF_3)_2PN_3$. XXII.[206] Dec. slowly at 0°.
Ph_2PN_3. XXII.[155] Solid, m. 13.6° (dec.).
$(CF_3)_2PNCO$. XXII.[156] Liq., b. 53.8°, $\log P_{mm} = -1775/T + 8.3$, IR, ^{31}P -34.5, ^{19}F.
$(CF_3)_2PNCS$. XXII.[156] Liq., b. 84°, $\log P_{mm} = -1715/T + 7.68$, IR, ^{19}F.
$Ph(C_6F_5)PNCO$. XXII. Liq., $b_{1.8}$ 138°,[66a] IR.
$Ph(C_6F_5)PNCS$. XXII. Liq., $b_{1.5}$ 152-155°, IR.[66a]
$(C_6F_5)_2PNCO$. XXII. Solid, m. 69°, IR,[66a] ^{31}P -17.[66b]
$(C_6F_5)_2PNCS$. XXII. Liq., $b_{2.6}$ 152-153°, IR,[66a] ^{31}P -12.4.[66b]

Compounds with Phosphorus (III)-N-Metal Bond

$(CF_3)_2PNHAs(CF_3)_2$. XXVI.[199] Liq., b. 110.5°, m. -44.8 to -44.4°, Trouton const. 21.5 esu, IR, Me_3N adduct.
$(CF_3)_2PNMeAs(CF_3)_2$. XXVI.[199] Liq., b. 127.2°, m. -30.7 to -30.0°, Trouton const. 21.4 esu, IR.
$[(CF_3)_2P]_2NAs(CF_3)_2$. XXVI.[199] Liq., b. 161.2°, m. 25.2-25.9°, Trouton const. 21.2 esu, IR.
$t\text{-}Bu_2PNHSiMe_3$. XXVI.[195] Liq., b_1 39-40°, m. -4 to -2°, ^1H.

t-$Bu_2PNHGeMe_3$. XXVI.[195] Liq., b_1 49-50°, m. -6 to -4°, ^1H.

$Ph_2PN(SiMe_3)_2$. XXVI.[67] White solid, m. 49-50°.

Miscellaneous Anhydrides of Phosphinous Acid

$(CF_3)_2POBMe_2$. XXVI.[28] Liq., b. 69.6°, Trouton const. 21.1 esu, IR, ^{31}P -31.2, ^{11}B -42, ^1H -0.17, ^{19}F.

$[(CF_3)_2PO]_2BMe$. XXVI.[28] Liq., b. 131.1°, Trouton const. 21.7 esu, IR, ^1H -0.31, ^{31}P -81.3, ^{11}B -15.3, ^{19}F.

$[(CF_3)_2PO]_3B$. XXVI.[28] Liq., b. 168°, Trouton const. 21.3 esu, IR, ^{31}P -36.3, ^{11}B 1.1, ^{19}F.

$(CF_3)_2POSiH_3$. XXVI.[27] Liq., b. 39.6°, m. -116°, Trouton const. 21.8 esu, IR.

$(CF_3)_2POSiMe_3$. XXVI.[27] Liq., b. 101°, m. -69.3°, Trouton const. 21.4 esu, IR.

$[(CF_3)_2PO]_3SiH$. XXVI.[27] Liq., b. 161.1°, Trouton const. 23.7 esu, IR.

$[(CF_3)_2PO]_4Si$. XXVI.[27] Liq., b. 193.1°, m. -53.8°, Trouton const. 21.5 esu, IR.

$Ph_2POSnPh_3$. XXVI.[196] Solid, m. 120°, IR, ^1H, Raman spect.

REFERENCES

1. Akamsin, V. D., and N. I. Rizpolozhenskii, Izv. Akad. Nauk SSSR, 9, 1987 (1967).
2. Akamsin, V. D., and N. I. Rizpolozhenskii, Izv. Akad. Nauk SSSR, 8, 493 (1966).
3. Akamsin, V. D., and N. I. Rizpolozhenskii, Dokl. Akad. Nauk SSSR, 168, 807 (1966).
4. Anisimov, K. N., and N. E. Kolobova, Izv. Akad. Nauk SSSR, Ser. Khim., 446 (1962).
5. Arbuzov, A., J. Russ. Phys.-Chem. Soc., 42, 549 (1910).
6. Arbuzov, A., and J. Arbusov, J. Russ. Phys.-Chem. Soc., 61, 217 (1929).
7. Arbuzov, A., and G. Kamai, J. Russ. Phys.-Chem. Soc., 61, 619 (1929).
8. Arbuzov, A., and K. V. Nikonorov, Zh. Obshch. Khim., 18, 2008 (1948).
9. Arbuzov, A., and V. Razumov, Zh. Obshch. Khim., 4, 834 (1934).
10. Arbuzov, B. A., N. A. Polezhaeva, V. S. Vinogradova, and A. K. Shamsutdinova, Izv. Akad. Nauk, SSSR, Ser. Khim., 1965, 669.
11. Arbuzov, B. A., and N. I. Rizpolozhenskii, Dokl. Akad. Nauk, SSSR, 89, 291 (1953).
12. Baeteman, N., and J. Baudet, Compt. Rend., 265, 288 (1967).
13. Bennett, F., H. J. Emeleus, and R. N. Haszeldine, J. Chem. Soc., 1953, 1565.

14. Berlin, K., T. Austin, and K. Stone, Abstr. ACS Meeting, New York, September 1963, p. 69Q.
15. Berlin, K. D., and G. B. Butler, Chem. Rev., 60, 243 (1960).
16. Birum, G. H., U.S. Patent 3,014,944 (1961).
17. Boisselle, A. P., and N. A. Meinhardt, J. Org. Chem., 27, 1828 (1962).
18. Bourneuf, M., Bull. Soc. Chim. (France), 33, 1808 (1923).
19. Brown, A. D., Jr., and G. M. Kosolapoff, J. Chem. Soc., 1968, 839.
20. Buckler, S. A., J. Amer. Chem. Soc., 84, 3093 (1962).
21. Buckler, S. A., and M. Epstein, U.S. Patent 3,005,029 (1961).
22. Buckler, S. A., and M. Epstein, Tetrahedron, 18, 1231 (1962).
23. Buckler, S. A., and M. Epstein, U.S. Patent 3,116,334 (1963).
24. Buckler, S. A., and M. Epstein, Tetrahedron, 18, 1211, 1221 (1962).
25. Buckner, B., and L. B. Lockhard, Jr., J. Amer. Chem. Soc., 73, 755 (1951).
26. Burg, A. B., Acct. Chem. Res., 2, 353 (1969).
27. Burg, A. B., and J. S. Basi, J. Amer. Chem. Soc., 90, 3361 (1968).
28. Burg, A. B., and J. S. Basi, J. Amer. Chem. Soc., 91, 1937 (1969).
29. Burg, A. B., and K. Gosling, J. Amer. Chem. Soc., 87, 2113 (1965).
30. Burg, A. B., and J. Heners, J. Amer. Chem. Soc., 87, 3092 (1965).
31. Burg, A. B., K. K. Joshi, and J. F. Nixon, J. Amer. Chem. Soc., 88, 31 (1966).
32. Burg, A. B., and D.-K. Kang, J. Amer. Chem. Soc., 92, 1901 (1970).
33. Burg, A. B., and K. Modritzer, J. Inorg. Nucl. Chem., 13, 318 (1960).
34. Burg, A. B., and P. J. Slota, J. Amer. Chem. Soc., 80, 1107 (1958).
35. Burg, A. B., and P. J. Slota, J. Amer. Chem. Soc., 82, 2145 (1960).
36. Burg, A. B., and P. J. Slota, J. Amer. Chem. Soc., 82, 2148 (1960).
37. Burg, A. B., and R. I. Wagner, J. Amer. Chem. Soc., 75, 3872 (1953).
38. Burg, A. B., and R. I. Wagner, U.S. Patent 2,934,564 (1957).
39. Burgada, R., G. Martin, and G. Mavel, Bull. Soc. Chim. (France), 1963, 2154.
40. Campbell, I. G. M., and I. D. R. Stevens, Chem. Commun., 15, 505 (1966).

41. Cavell, R. G., and H. J. Emeleus, J. Chem. Soc., 1964, 5825.
42. Charrier, C., and M. P. Simonnin, Compt. Rend., 264-995 (1967).
43. Chernyshev, E. A., and E. F. Bugerenko, Organomet. Chem. Rev., 3, 469 (1968).
44. Clemens, D. F., and H. H. Sisler, Inorg. Chem., 4, 1222 (1965).
45. Coates, G. E., and J. G. Livingstone, J. Chem. Soc., 1961, 1000.
46. Coates, G. E., and J. G. Livingstone, J. Chem. Soc., 1961, 5053.
47. Colln, R., and G. Schrader, German Patent 1,138,771 (1962).
48. Colln, R., and G. Schrader, German Patent 1,141,990 (1963).
49. Colthup, N. B., L. H. Daly, and S. E. Wiberly, Introduction to Infrared and Raman Spectroscopy, New York: Academic Press, 1964, pp. 299-402.
50. Cowley, A. H., M. J. S. Dewar, W. R. Jackson, and W. B. Jennings, J. Amer. Chem. Soc., 92, 1085 (1970).
51. Crofts, P. C., and D. M. Parker, J. Chem. Soc., 1970, 332.
52. Crosbie, K. D., C. Glidewell, and G. M. Sheldrick, J. Chem. Soc., 1969, 1861.
53. Davidson, R. S., J. Chem. Soc., C, 1967, 2131.
54. Davidson, R. S., Tetrahedron Lett., 1968, 3029.
55. Davidson, N., and H. C. Brown, J. Amer. Chem. Soc., 64, 316 (1942).
56. Derkach, G. I., and A. V. Kirsanov, Zh. Obshch. Khim., 29, 1815 (1959).
57. Dewar, M. J. S., and V. P. Kubba, J. Amer. Chem. Soc., 82, 5685 (1960).
58. Dietsche, W. H., Tetrahedron, 23, 3049 (1967).
59. Eméleus, H. J., K. J. Packer, and N. Welcmam, J. Chem. Soc., 1962, 2529.
60. Eméleus, H. J., and J. D. Smith, J. Chem. Soc., 1959, 375.
61. Emmick, T. L., and R. L. Letsinger, J. Amer. Chem. Soc., 90, 3459 (1968).
62. Epstein, M., and S. A. Buckler, Tetrahedron, 18, 1231 (1962).
63. Evlath, E. M., L. D. Freedman, and R. I. Wagner, J. Org. Chem., 27, 2192 (1962).
64. Ewart, G., A. P. Lane, J. McKechnie, and D. S. Payne, J. Chem. Soc., 1964, 1543.
65. Ewart, G., D. S. Payne, A. L. Porte, and A. P. Lane, J. Chem. Soc., 1962, 3984.
66. Fild, M., I. Hollenberg, and O. Glemser, Naturwiss., 22, 248 (1961).
66a. Fild, M., O. Glemser, and I. Hollenberg, Naturwiss.,

53, 130 (1966).
66b. Fild, M., Z. Naturforsch., 23b, 604 (1968).
66c. Fild, M., O. Glemser, and I. Hollenberg, Z. Naturforsch., 21b, 920 (1966).
67. Fluck, E., Phosphorus-Nitrogen Chemistry. Vol. 5, Topics in Phosphorus Chemistry, Grayson and Griffiths, Eds., New York: Wiley, 1967, p. 320.
68. Fluck, E., and H. Binder, Z. Naturforsch., 22b, 805 (1967).
69. Frank, A. W., J. Org. Chem., 24, 966 (1959).
70. Freeman, K. L., and M. J. Gallagher, Austral. J. Chem., 21, 145 (1968).
71. Fritz, G., Angew. Chem., 78, 80 (1966).
72. Fritz, G., Z. Naturforsch., 86, 776 (1953).
73. Fritz, G., and G. Becker, Angew. Chem., 79, 1068 (1967).
74. Fritz, G., and G. Pappenburg, Angew. Chem., 72, 208 (1960).
75. Fritz, G., and G. Trenezek, Z. Anorg. Allg. Chem., 313, 236 (1961).
76. Gawron, O., C. Grelecki, W. Reilly, and J. Sands, J. Amer. Chem. Soc., 73, 4101 (1951).
77. Gilman, H., and J. Robinson, Rec. Trav. Chim., 48, 328 (1929).
78. Gilyarov, V. A., "Chemistry and Applications of Organophosphorus Compounds," Proc. 2nd Conf., Moscow, 1962, p. 80.
79. Goldwhite, H., R. N. Haszeldine, and D. G. Rowsell, Chem. Commun., 1965, 83.
80. Grayson, M., and C. E. Farley, J. Org. Chem., 32, 237 (1967).
81. Grayson, M., C. E. Farley, and C. A. Streuli, Tetrahedron, 23, 1065 (1967).
82. Greenwood, N. N., B. H. Robinson, and B. P. Straughan, J. Chem. Soc., A, 1968, 230.
83. Griffiths, J. E., and A. B. Burg, J. Amer. Chem. Soc., 84, 3442 (1962).
84. Griffiths, J. E., and A. B. Burg, J. Amer. Chem. Soc., 82, 1507 (1960).
85. Griffiths, J. E., and A. B. Burg, Proc. Chem. Soc., 1961, 12.
86. Harris, G. S., Proc. Chem. Soc., 1957, 118.
87. Harris, G. S., J. Chem. Soc., 1958, 512.
88. Hart, W. A., and H. H. Sisler, Inorg. Chem., 3, 617 (1964).
89. Hart, W. A., Dissertation, University of Florida, 1963.
90. Harvey, R. G., and E. R. DeSombre, The Michaelis-Arbusov and Related Reactions, Vol. 1, Topics in Phosphorus Chemistry, Grayson and Griffiths, Eds., New York: Interscience, 1964.

91. Hays, H. R., J. Org. Chem., 33, 3690 (1968).
92. Herriott, A. W., and K. Mislow, Tetrahedron Letters, 1968, 3013.
93. Hoffmann, E., British Patent 921,463 (1963).
94. Hoffmann, H., and P. Schellenbeck, Ber., 99, 1134 (1966).
95. Horner, L., P. Beck, and V. G. Toscano, Chem. Ber., 94, 1317 (1961).
96. Hudson, R. F., and R. Searle, Chimia, 20, 117 (1966).
97. Hudson, R. F., R. J. G. Searle, and F. H. Devitt, J. Chem. Soc., C, 1966, 1001.
98. Hudson, R. F., R. J. G. Searle, and F. H. Devitt, J. Chem. Soc., B, 1966, 789.
99. Hunt, B. B., and B. C. Saunders, J. Chem. Soc., 1957, 2413.
100. Imberry, D., and H. Friebolin, Z. Naturforsch., 23b, 759 (1968).
101. Issleib, K., and G. Doll, Z. Anorg. Allg. Chem., 324, 259 (1963).
102. Issleib, K., and G. Doll, Ber., 96, 1544 (1963).
103. Issleib, K., and W. Seidel, Ber., 92, 2681 (1959).
104. Issleib, K., and B. Walther, Angew. Chem. Int. Ed. Engl., 6, 88 (1967).
105. Ivin, S. Z., and K. V. Karavanov, Zh. Obshch. Khim., 29, 3456 (1959).
106. Kabachnik, M. I., C. Jung-yu, and E. N. Tsvetkov, Dokl. Akad. Nauk SSSR, 131, 1334 (1960).
107. Kabachnik, M. I., C. Jung-yu, and E. N. Tsvetkov, Zh. Obshch. Khim., 32, 3351 (1962).
108. Kabachnik, M. I., T. A. Mastryukova, and A. E. Shilov, Zh. Obshch. Khim., 33, 320 (1963).
109. Kabachnik, M. I., T. Medved, and Y. M. Polikarpov, Dokl. Akad. Nauk SSSR, 135, 849 (1960).
110. Kabachnik, M. I., and E. N. Tsvetkov, Izv. Akad. Nauk, SSSR, Ser. Khim., 7, 1227 (1963).
111. Kabachnik, M. I., and E. N. Tsvetkov, Dokl. Akad. Nauk SSSR, 135, 323 (1960).
112. Kamai, G., F. M. Kharrasova, and N. A. Chadaeva, "Chemistry and Applications of Organophosphorus Compounds," Proc. 2nd Conf., Moscow, 1962, p. 220.
113. Kamai, G., and V. S. Tsivunin, Dokl. Akad. Nauk SSSR, 1946, 295.
114. Kharrasova, F. M., and G. Kamai, Zh. Obshch. Khim., 38, 359 (1968).
115. Kharrasova, F. M., and G. Kamai, Zh. Obshch. Khim., 34, 2195 (1964).
116. Kharrasova, F. M., and G. Kamai, Zh. Obshch. Khim., 38, 617 (1968).
117. Kharrasova, F. M., G. Kamai, R. B. Sultanova, and R. R. Shagidullin, Zh. Obshch. Khim., 37, 687 (1967).
118. Koketsu, J., S. Sakai, and Y. Ishii, Kogyo Kagaku

Zasshi, 72, 2503 (1969).
119. Kosolapoff, G. M., Organophosphorus Compounds, New York: Wiley, 1950.
120. Kosolapoff, G. M., and A. D. Brown, Jr., J. Chem. Soc., 1967, 1789.
121. Kosolapoff, G. M., and R. H. Watson, J. Amer. Chem. Soc., 73, 4101 (1951).
122. Krasilnikova, E. A., N. A. Moskva, and A. I. Razumov, Zh. Obshch. Khim., 39, 216 (1969).
123. Krasilnikova, E. A., A. M. Potapov, and A. I. Razumov, Zh. Obshch. Khim., 37, 2365 (1967).
124. Laurent, J. P., G. Jugie, and G. Commenges, J. Inorg. Nucl. Chem., 31, 1353 (1969).
125. Lichtenthaler, F., Chem. Rev., 61, 607 (1961).
126. Linton, H. R., and E. R. Nixon, Spectrochim. Acta, 1959, 146.
127. Lorenz, W., and G. Schrader, German Patent 1,067,017 (1959).
128. Lorenz, W., H. G. Schicke, and G. Schrader, Belgian Patent 616,096 (1962).
128a. McLean, R. A. N., Inorg. Nucl. Chem. Lett., 5, 745 (1969).
129. Maier, L., Helv. Chim. Acta, 47, 2129 (1964).
130. Maier, L., Helv. Chim. Acta, 46, 2667 (1963).
131. Maier, L., Topics in Phosphorus Chemistry, Vol. 2, Grayson and Griffiths, Eds., New York: Interscience, 1965, p. 61.
131a. Maier, L., Helv. Chim. Acta, 49, 1000 (1966).
132. Maier, L., Helv. Chim. Acta, 49, 1249 (1966).
132a. Maier, L., Helv. Chim. Acta, 52, 858 (1969).
133. Mark, V., Abstr. 147th Meeting ACS, Philadelphia, April, 1964, p. 29L.
134. Mark, V., "Mechanisms of Molecular Migrations," B. S. Thyagarajan, Ed., New York: Interscience, 1969.
134a. Mark, V., C. H. Dungan, M. M. Crutchfield, and J. R. VanWazer, in Topics in Phosphorus Chemistry Vol. 5, 227.
135. Michaelis, A., and W. LaCoste, Ber., 18, 2109 (1885).
136. Michaelis, A., and A. Link, Ann., 207, 193 (1881).
137. Michaelis, A., and A. Schenk, Ann., 260, 20 (1890).
138. Miller, C. D., R. C. Miller, and W. Rogers, Jr., J. Amer. Chem. Soc., 80, 1562 (1958).
139. Miller, J. A., Tetrahedron Lett. 1969, 4335.
140. Miller, R. C., J. Org. Chem., 24, 2013 (1959).
141. Miller, R. C., Abstr. 140th Meeting ACS, Chicago, September, 1961, p. 43Q.
142. Miller, R. C., C. D. Miller, W. Rogers, Jr., and L. A. Hamilton, J. Amer. Chem. Soc., 79, 424 (1957).
143. Miller, R. C., J. S. Bradley, and L. A. Hamilton, J. Amer. Chem. Soc., 78, 5299 (1956).
144. Montgomery, R. E., and L. D. Quin, J. Org. Chem., 30,

2393 (1965).
145. Mukaiyama, T., and Y. Kodaira, Bull. Chem. Soc. Japan, 39, 1297 (1966).
146. Mukaiyama, T., and Y. Yokota, Bull. Chem. Soc. Japan, 38, 858 (1965).
147. Niebergall, H., Makromol. Chem., 52, 218 (1962).
148. Niebergall, H., and B. Langenfeld, Ber., 95, 64 (1962).
149. Nielsen, R. P., and H. H. Sisler, Inorg. Chem., 2, 753 (1963).
150. Nöth, H., and W. Schragle, Chem. Ber., 97, 2218 (1964).
151. Nöth, H., and W. Schragle, Chem. Ber., 97, 2374 (1964).
152. Nöth, H., and W. Schragle, Z. Naturforsch., 16b, 473 (1961).
153. Nöth, H., and H. J. Vetter, Chem. Ber., 96, 1109 (1963).
154. Oertel, G., H. Maly, and H. Holtschmidt, Chem. Ber., 97, 891 (1964).
155. Paciorek, K., and R. Kratzer, Inorg. Chem., 3, 594 (1964).
156. Packer, K. J., J. Chem. Soc., 1963, 960.
157. Peters, G., J. Amer. Chem. Soc., 82, 4751 (1960).
158. Peters, G., J. Org. Chem., 27, 2198 (1962).
159. Peterson, L. K., and A. B. Burg, J. Amer. Chem. Soc., 83, 4833 (1961).
160. Peterson, L. K., G. L. Wilson, and K. I. The, Can. J. Chem., 47, 1025 (1969).
161. Peterson, L. K., and G. L. Wilson, Can. J. Chem., 46, 685 (1968).
162. Peterson, L. K., and G. L. Wilson, Can. J. Chem., 47, 4281 (1969).
163. Petrov, K. A., E. E. Nifantev, and L. V. Khorkhayanu, Zh. Obshch. Khim., 31, 2889 (1961).
164. Petrov, K. A., E. E. Nifantev, L. V. Khorkhayanu, and A. I. Trushkov, Zh. Obshch. Khim., 31, 3085 (1961).
165. Petrov, K. A., E. E. Nifantev, T. N. Lysenko, and V. P. Evdakov, Zh. Obshch. Khim., 31, 2337 (1961).
166. Petrov, K. A., V. A. Parshina, B. A. Orlov, and G. M. Tsypina, Zh. Obshch. Khim., 32, 4017 (1962).
167. Plets, V. M., Dissertation, Kazan, 1938.
168. Pudovik, A. N., and B. A. Arbuzov, Dokl. Akad. Nauk SSSR, 73, 327 (1950).
169. Pudovik, A. N., I. M. Aladzheva, and L. V. Spirina, Zh. Obshch. Khim., 37, 700 (1967).
170. Pudovik, A. N., and I. M. Aladzheva, Zh. Obshch. Khim., 37, 2715 (1967).
171. Pudovik, A. N., I. M. Aladzheva, and L. N. Yakovenko, J. Gen. Chem., USSR, 35, 1214 (1965).
172. Pudovik, A. N., and E. S. Batyeva, Zh. Obshch. Khim.,

39, 334 (1969).
173. Pudovik, A. N., and V. K. Krupnov, Zh. Obshch. Khim., 38, 1287 (1968).
174. Pudovik, A. N., and V. K. Krupnov, Zh. Obshch. Khim., 38, 304 (1968).
175. Pudovik, A. N., A. A. Muratova, and E. P. Semkina, Zh. Obshch. Khim., 33, 3350 (1962).
176. Pudovik, A. N., I. V. Guryanova, M. G. Zimin, and A. N. Durneva, Zh. Obshch. Khim., 39, 1018 (1969).
177. Quin, L. D., and H. G. Anderson, J. Org. Chem., 31, 1206 (1966).
178. Quin, L. D., and R. E. Montgomery, J. Org. Chem., 28, 3315 (1963).
179. Quin, L. D., and R. E. Montgomery, J. Inorg. Nucl. Chem., 28, 1750 (1966).
180. Ramirez, F., J. F. Pilot, O. P. Madan, and C. P. Smith, J. Amer. Chem. Soc., 90, 1275 (1968).
181. Raudnitz, H., Ber., 60, 743 (1927).
182. Rauhut, M. M., H. A. Currier, and V. P. Wystrach, J. Org. Chem., 26, 5133 (1961).
183. Rauhut, M. M., and H. A. Currier, J. Org. Chem., 26, 4626 (1961).
184. Rauhut, M. M., I. Hechenbleikner, and H. A. Currier, U.S. Patent 2,953,596 (1960).
185. Rauhut, M. M., I. Hechenbleikner, H. A. Currier, and V. P. Wystrach, J. Amer. Chem. Soc., 80, 6690 (1958).
186. Razumov, A., and A. Bankovskaja, Dokl. Akad. Nauk SSSR, 116, 241 (1957).
187. Razumov, A., J. Gen. Chem., USSR, 29, 1609 (1959).
188. Razumov, A., and O. Mukhacheva, Zh. Obshch. Khim., 26, 1436 (1956).
189. Samuel, D., Pure Appl. Chem., 9, 449 (1964).
190. Sanchez, M., R. Wolf, and F. Mathis, Spectrochim. Acta, 23A, 2617 (1967).
191. Sander, M., Ber., 93, 1220 (1960).
192. Sander, M., Z. Angew. Chem., 79, 67 (1961).
193. Savage, M. P., and S. Trippett, J. Chem. Soc., 1966, 1842.
194. Savage, M. P., and S. Trippett, J. Chem. Soc., 1967, 1998.
195. Scherer, O. J., and F. Schieder, Angew. Chem., Int. Ed., 7, 75 (1968).
196. Schumann, H., Angew. Chem., 23, 970 (1969).
197. Sevin, A., and W. Chodkiewicz, Tetrahedron Lett., 1967, 2975.
198. Sheldon, J. C., and S. Y. Tyree, Jr., J. Amer. Chem. Soc., 81, 6177 (1959).
199. Singh, J., and A. B. Burg, J. Amer. Chem. Soc., 88, 718 (1966).
200. Sisler, H. H., and N. L. Smith, J. Org. Chem., 26, 611 (1961).

201. Sisler, H. H., and N. L. Smith, J. Org. Chem., 26, 4733 (1961).
202. Smith, N. L., and H. H. Sisler, J. Org. Chem., 26, 5145 (1961).
203. Sollott, G. P., and W. R. Peterson, Jr., J. Organomet. Chem., 19, 143 (1969).
204. Stevens, D. R., and R. S. Spindt, U.S. Patent 2,542,370 (1948).
205. Stuebe, C., W. M. LeSuer, and G. R. Norman, J. Amer. Chem. Soc., 77, 3526 (1955).
206. Testi, G., C. P. Haber, and C. M. Douglas, Proc. Chem. Soc., 1960, 219.
207. Vetter, H. J., and H. Noth, Chem. Ber., 96, 1308 (1963).
207a. Voskuil, W., and J. F. Arens, Rec. Trav. Chim., 81, 993 (1962).
208. Wagner, H. I., and A. B. Burg, J. Amer. Chem. Soc., 75, 3869 (1953).
209. Willans, J. L., Chem. Ind. (London), 8, 235 (1957).
210. Williams, R. H., and L. A. Hamilton, J. Amer. Chem. Soc., 77, 3411 (1955).
211. Williams, R. H., and L. A. Hamilton, J. Amer. Chem. Soc., 74, 5418 (1952).
212. Wolf, R., D. Houalla, and F. Mathis, Spectrochim. Acta, 23A, 1641 (1967).